AUGUSTINE

Confessions

II

AUGUSTINE

Confessions

II

Commentary on Books 1–7

JAMES J. O'DONNELL

CLARENDON PRESS · OXFORD
1992

Oxford University Press, Walton Street, Oxford OX2 6DP
Oxford New York Toronto
Delhi Bombay Calcutta Madras Karachi
Petaling Jaya Singapore Hong Kong Tokyo
Nairobi Dar es Salaam Cape Town
Melbourne Auckland
and associated companies in
Berlin Ibadan

Oxford is a trade mark of Oxford University Press

Published in the United States
by Oxford University Press, New York

British Library Cataloguing in Publication Data
Data available

Library of Congress Cataloging in Publication Data
Augustine, Saint, Bishop of Hippo. [*Confessiones*]
Confessions/Augustine; [*commentary by*] *James J. O'Donnell.*
Latin text with English commentary.
Includes bibliographical references and indexes.
Contents: 1. Introduction and text—2. Commentary on Books 1–7—3. Commentary on Books 8–13; Indexes.
1. Augustine, Saint, Bishop of Hippo. 2. Christian saints—Algeria—Hippo—Biography. 3. Catholic Church—Algeria—Hippo—Bishops—Biography. 4. Hippo (Algeria)—Biography. I. O'Donnell, James Joseph, 1950– II. Title.
BR65.A6 1992 242—dc20 92—12361
ISBN 0—19—814074—6 (v. 2)

Typeset by Joshua Associates Ltd., Oxford
Printed in Great Britain by
Bookcraft (Bath) Ltd., Midsomer Norton

Contents

VOLUME I

VOLUME II

VOLUME III

Abbreviations and References

[1] [2] [3] indicate elements in triads of terms or names representing persons of the trinity; see vol. ii, 48 n 17.

THE WORKS OF AUGUSTINE

See Introduction, pp. lxv–lxix.

MANUSCRIPTS

S Rome, Biblioteca Nazionale Centrale, Sessorianus 55. Late 6th cent.
O Paris, Bibliothèque Nationale, lat. 1911. Early 9th cent.
P Paris, Bibliothèque Nationale, lat. 1912. Early 9th cent.
C Paris, Bibliothèque Nationale, lat. 1913. Mid-9th cent.
D Paris, Bibliothèque Nationale, lat. 1913A. Mid-9th cent.
E Paris, Bibliothèque Nationale, lat. 12191. Late 9th–early 10th cent.
G Paris, Bibliothèque Nationale, lat. 12193. 9th/10th cent.
A Stuttgart, Württembergische Landesbibliothek, HB. vii. 15. End 9th cent.
H Paris, Bibliothèque Nationale, lat. 122224. Mid-9th cent.
V Vatican City, Biblioteca Apostolica Vaticana, Vat. lat. 5756. Late 9th cent.

See also Introduction, pp. lvi–lvii.

EDITIONS AND TRANSLATIONS

Editions (*cited by editor's last name except where abbr. is indicated*)

J. Amerbach (Basle, 1506)
Erasmus (Basle, 1528)
T. Gozaeus and J. Molanus (Louvain, 1576) [Louvain]
Benedictines of St Maur (Paris, 1679), repr. with minor alterations in *PL* 32 (1845) [Maur.]
E. B. Pusey (Oxford, 1838)
Raumer (Gütersloh, 1855)
J. Gibb and W. Montgomery (Cambridge, 2nd edn., 1927; repr. New York, 1979) [G–M]
P. Knöll (*CSEL* 33; Vienna, 1896); I occasionally quote (and specify) his ed. min. (Leipzig, 1898)

F. Ramorino (Rome, 1909)

P. de Labriolle (Paris, 1925–6)

M. Skutella (Leipzig, 1934; from the revision by Juergens–Schaub, Stuttgart, 1969) [Skut.]

A. C. Vega (*Biblioteca de autores cristianos*, 11; Madrid, fifth edn., 1968)

Bibliothèque Augustinienne (*BA* 13–14; Paris, 1962: reprinting Skutella, with annotation by A. Solignac) [*BA*]

M. Pellegrino (*Nuova Biblioteca Agostiniana*; Rome, 1975): this edition is mainly a reprinting of Skutella, noted only for a few divergences; the translation is cited as 'Carena'—see below under Translations [Pell.]

L. Verheijen (*CCSL* 27; Turnhout, 1981) [Ver.]

Translations (*cited by translator's last name*)

J. K. Ryan (Image Bks., New York, 1960)

R. S. Pine-Coffin (Penguin, Harmondsworth and New York, 1961)

E. B. Pusey (repr. Everyman Library, London, 1962)

J. Bernhart (Munich, 1955; repr. Insel, Frankfurt-am-Main, 1987)

Bibliothèque Augustinienne (Paris, 1962; see above under Editions—the translation is by E. Tréhorel and G. Boissou)

R. Warner (New York, 1963)

A. C. Vega (Madrid, 1968; see above under Editions)

C. Carena (Rome, 1975; see above under Editions)

Professor Henry Chadwick's version (Oxford, 1991) reached me only after the completed manuscript had been sent to the Press.

BOOKS AND PERIODICALS

For bibliographical guidance, the reader should consult the volumes of the *Fichier augustinien* (Boston, 1970–), which incorporate and systematize, and are in turn supplemented by, the annual bibliographical bulletins that appear in the *Revue des études augustiniennes*. As the *Augustinus-Lexikon* fascicles appear, they too will have valuable bibliography; a computerized bibliography prepared in Würzburg is also promised. The works listed below are by and large those most important and generally useful for the student going further. In general, titles of articles are omitted. The latest edition noted is the one actually consulted by me.

AHDLMA	*Archives d'histoire doctrinale et littéraire du moyen âge* (Paris, 1926–)
AJP	*American Journal of Philology*
Alfaric	P. Alfaric, *L'Évolution intellectuelle de saint Augustin* (Paris, 1918)

ALMA	*Archivium Latinitatis Medii Aevi*
Arts	M. R. Arts, *The Syntax of the Confessions of Saint Augustine* (Washington, DC, 1927)
Atti 1986	*Congresso internazionale su s. Agostino nel XVI centenario della conversione (Roma, 15–20 settembre 1986), Atti* (Rome, 1987)
Aug.-Lex.	*Augustinus-Lexikon* (Basel, 1985–)
Aug. Mag.	*Augustinus Magister* (Paris, 1954)
BA	Bibliothèque augustinienne
Brown	P. Brown, *Augustine of Hippo* (London and Berkeley, 1967)
Brown, *Body and Society*	P. Brown, *The Body and Society* (New York, 1988)
Burnaby	J. Burnaby, *Amor Dei* (London, 1938)
CCSL	*Corpus Christianorum, Series Latina*
CL	Classical Latin
CLA	E. A. Lowe, *Codices Latini Antiquiores* (Oxford, 1934–72)
Courcelle, *LLW*	P. Courcelle, *Late Latin Writers and Their Greek Sources* (trans. H. Wedeck; Cambridge, Mass., 1969)
Courcelle, *Recherches*	P. Courcelle, *Recherches sur les Confessions de saint Augustin* (Paris, 1950; 2nd edn. Paris, 1968)
Courcelle, *Les Confessions*	P. Courcelle, *Les Confessions de saint Augustin dans la tradition littéraire* (Paris, 1963)
CQ	*Classical Quarterly*
CRAI	*Comptes rendus de l'Académie des Inscriptions et Belles-Lettres*
CSEL	*Corpus Scriptorum Ecclesiasticorum Latinorum*
DACL	*Dictionnaire d'archéologie chrétienne et de liturgie* (1907–53)
Decret, *Aspects*	F. Decret, *Aspects du manichéisme dans l'Afrique romaine* (Paris, 1970)
Decret, *L'Afrique*	F. Decret, *L'Afrique manichéene (IVᵉ–Vᵉ siècles)* (Paris, 1978)
De Marchi	V. De Marchi, 'De nonnullis Augustini Confessionum locis', *Rendiconti dell'Istituto Lombardo, Classe di Lettere, Scienze morali et storiche* 96 (1962), 310–16
Dulaey	M. Dulaey, *Le Rêve dans la vie et la pensée de saint Augustin* (Paris, 1973)
du Roy	O. du Roy, *L'Intelligence de la foi en la trinité selon*

	saint Augustin: Genèse de sa théologie trinitaire jusqu'en 391 (Paris, 1966)
Frend, *Donatist Church*	W. H. C. Frend, *The Donatist Church* (Oxford, 1951)
GRBS	*Greek, Roman and Byzantine Studies*
Guardini	R. Guardini, *The Conversion of Augustine* (London, 1960)
Hagendahl	H. Hagendahl, *Augustine and the Latin Classics* (Göteborg, 1967)
Hrdlicka	C. L. Hrdlicka, *A Study of the Late Latin Vocabulary and of the Prepositions and Demonstrative Pronouns in the Confessions of St. Augustine* (Washington, DC, 1931)
HSCP	*Harvard Studies in Classical Philology*
HThR	*Harvard Theological Review*
Isnenghi	A. Isnenghi, 'Textkritisches zu Augustins "Bekenntnissen"', *Augustiniana* 15 (1965), 5–31
Jones, *LRE*	A. H. M. Jones, *The Later Roman Empire 284–602: A Social, Economic, and Administrative Survey* (Oxford, 1964)
JRS	*Journal of Roman Studies*
JThS	*Journal of Theological Studies*
Keil	H. Keil, *Grammatici Latini* (Leipzig, 1857–80; repr. Hildesheim, 1961)
Knauer	G. N. Knauer, *Psalmenzitate in Augustins Konfessionen* (Göttingen, 1955; repr. in his *Three Studies*, New York, 1987)
Kunzelmann	A. Kunzelmann, 'Die Chronologie der Sermones des Hl. Augustinus', *MA* 2. 417–520
Kusch	H. Kusch, 'Studien über Augustinus', *Festschrift Franz Dornseiff* (Leipzig, 1953), 124–200
La Bonnardière, *Recherches*	A.-M. La Bonnardière, *Recherches de la chronologie augustinienne* (Paris, 1965)
La Bonnardière, *Biblia Augustiniana*	A.-M. La Bonnardière, *Biblia Augustiniana* (Paris, 1960–) 1960 Livres historiques 1964 Épître aux Thessaloniciens, à Tite et à Philémon 1964 Douze petits prophètes 1967 Deutéronome 1970 Livre de la Sagesse

	1972 Livre de Jérémie 1975 Livre des Proverbes
Lawless, *Rule*	G. Lawless, *Augustine of Hippo and his Monastic Rule* (Oxford, 1987)
Lectio I–II, III–V, VI–IX, X–XIII	*'Le Confessioni' di Agostino d'Ippona: Lectio Augustini: Settimana Agostiniana Pavese* (Palermo, 1984–7)
LHS	M. Leumann, J. B. Hofmann, and A. Szantyr, *Lateinische Syntax und Stilistik* (Munich, 1972)
Lieu, *Manichaeism*	S. N. C. Lieu, *Manichaeism* (Manchester, 1985)
MA	*Miscellanea Agostiniana* (Rome, 1930)
Madec, *Saint Ambrose*	G. Madec, *Saint Ambroise et la philosophie* (Paris, 1974)
Mandouze	A. Mandouze, *Saint Augustin: L'Aventure de la raison et de la grâce* (Paris, 1968)
Mandouze, *Pros. chr.*	A. Mandouze, *Prosopographie chrétienne du Bas-Empire*, 1. *Afrique (303–533)* (Paris, 1982)
Marrou	H.-I. Marrou, *Saint Augustin et la fin de la culture antique* (4th edn., Paris, 1958)
Mayer, *Zeichen*, 1	C. P. Mayer, *Die Zeichen in der geistigen Entwicklung und in der Theologie des jungen Augustinus* (Würzburg, 1969)
Mayer, *Zeichen*, 2	C. P. Mayer, *Die Zeichen in der geistigen Entwicklung und in der Theologie Augustins*, 2. *Die antimanichäische Epoche* (Würzburg, 1974)
MEFR	*Mélanges d'archéologie et d'histoire de L'École Française de Rome*
Meijering	E. P. Meijering, *Augustin über Schöpfung, Ewigkeit und Zeit: Das elfte Buch der Bekenntnisse* (Leiden, 1979)
Milne	C. H. Milne, *A Reconstruction of the Old Latin Text or Texts of the Gospels used by Saint Augustine* (Cambridge, 1926)
O'Daly	G. J. P. O'Daly, *Augustine's Philosophy of Mind* (London and Berkeley, 1987)
OLD	*Oxford Latin Dictionary*
O'Meara	J. J. O'Meara, *The Young Augustine* (London, 1954; corr. repr. 1980)
Otto, *Sprichwörter*	A. Otto, *Die Sprichwörter und sprichwörtlichen Redensarten der Römer* (Leipzig, 1890; repr. Hildesheim, 1962)

Pellegrino, *Les Confessions*	M. Pellegrino, *Les Confessions de saint Augustin* (Paris, 1960)
Perler	O. Perler (with J.-L. Maier), *Les voyages de saint Augustin* (Paris, 1969)
Pincherle, *Formazione teologica*	A. Pincherle, *La formazione teologica di Sant' Agostino* (Rome, n.d. [1947])
PLRE	A. H. M. Jones *et al.* (eds.), *Prosopography of the Later Roman Empire*, i–ii (Cambridge, 1971–80)
Poque, *Le langage symbolique*	S. Poque, *Le langage symbolique dans la prédication d'Augustin d'Hippone: Images héroïques* (Paris, 1984)
RA	*Recherches augustiniennes*
REA	*Revue d'études anciennes*
REAug	*Revue des études augustiniennes*
REL	*Revue d'études latines*
RHR	*Revue de l'histoire des religions*
RMAL	*Revue du moyen âge latin*
Rousselle, *Porneia*	A. Rousselle, *Porneia* (Oxford, 1988)
RTAM	*Recherches de théologie ancienne et médiévale*
SDHI	*Studia et Documenta Historiae et Iuris*
Signum Pietatis	A. Zumkeller (ed.), *Signum Pietatis: Festgabe . . . C. P. Mayer* (Würzburg, 1989)
SLA	W. Hensellek *et al.* (eds.), *Specimina eines Lexicon Augustinianum* (Vienna, 1987–)
Sorabji, *Time*	R. Sorabji, *Time, Creation and the Continuum: Theories in Antiquity and the Early Middle Ages* (Ithaca, 1983)
Souter	A. Souter, *A Glossary of Later Latin* (Oxford, 1949)
TAPA	*Transactions of the American Philological Association*
TeSelle	E. TeSelle, *Augustine the Theologian* (New York, 1970)
Testard	M. Testard, *Saint Augustin et Cicéron* (Paris, 1958)
Theiler, *P.u.A.*	W. Theiler, *Porphyrios und Augustin* (Halle, 1933; repr. in his *Forschungen zum Neuplatonismus* (Berlin, 1966), 160–248)
TLL	*Thesaurus Linguae Latinae*
TU	Texte und Untersuchungen
van Bavel	T. J. van Bavel, *Recherches sur la christologie de saint Augustin: L'Humain et le divin dans le Christ*

	d'après saint Augustin (Fribourg, Switzerland, 1954)
van der Meer	F. van der Meer, *Augustine the Bishop* (London, 1961)
Verbraken	P. Verbraken, *Études critiques sur les sermons authentiques de saint Augustin* (Steenbrugge and the Hague, 1976) [Where no further reference is given it may be assumed that 'ad loc.' is implied, referring to V.'s catalogue of A.'s sermons at his pp. 53–196]
Verheijen, *Eloquentia Pedisequa*	[L.] M. J. Verheijen, *Eloquentia Pedisequa* (Nijmegen, 1949)
Vg.	R. Weber, *Biblia Sacra iuxta vulgatam versionem* (3rd edn., Stuttgart, 1983)
VL (Beuron)	*Vetus Latina: Die Reste der lateinischen Bibel* (Freiburg, 1951–) ['VL' alone indicates a reading attributed to pre-Vulgate Latin scripture for a book of scripture not yet treated by Beuron]
Warns	G.-D. Warns. I thus refer to several unpublished papers preliminary to a Berlin dissertation that Herr Warns has been kind enough to allow me to see
Weber, *Psautier Romain*	R. Weber, *Le Psautier Romain et les autres anciens Psautiers latins* (Rome, 1953)
Zarb	S. Zarb, *Chronologia operum s. Augustini secundum ordinem Retractationum digesta* (Rome, 1934)
ZPE	*Zeitschrift für Papyrologie und Epigraphik*

COMMENTARY

egoque ipse multa quae nesciebam
scribendo me didicisse confitear.

(*trin.* 3. pro. 1)

ex multiloquio non effugies peccatum.

(Prov. 10: 19)

Title

Aureli Augustini: The *praenomen* is thinly attested. Neither A. nor any of his correspondents or polemical opponents use it, and it has been thought to be a misreading of something in Orosius (A. M. La Bonnardière, *Rev. Bén.* 91 (1981), 231–7, citing Oros. *liber apologeticus* 1. 4); but its frequent occurrence in early manuscript colophons (cf. M. M. Gorman, *JThS* 35 (1984), 475–80) cannot be explained as a vulgar error. See also Mandouze, 71–4, on A.'s name and attempts (mainly drawing on the names of Monnica and Adeodatus) to give A. a non-Roman genealogical and cultural background (e.g. W. H. C. Frend, *JThS* 43 (1942), 188–91, reprised in his *The Donatist Church* (Oxford, 1952), 230). More objective is C. Lepelley, *Atti 1986*, 1. 104, showing the extreme rarity in Africa of the names *Augustinus* (3 others attested in Mandouze, *Pros. chr.*, and *CIL* 8), *Patricius* (2 others), and *Monnica* (1 other), with a certain (perhaps remote) aristocratic pretension in the names of father and son. Whatever the facts, and whatever 'African' character may be discerned in the luxuriance of A.'s style, *this* text is thoroughly Latin, Roman, and Christian.

A. never mentions his own name in *conf.*, and does so rarely elsewhere, usually self-consciously (*en. Ps.* 36. s. 3. 20, saying how he would have his congregation defend him against critics: 'Augustinus episcopus est in ecclesia catholica, sarcinam suam portat, rationem redditurus est deo'; *c. Sec.* 1, 'senti de Augustino quidquid libet, sola me in oculis dei conscientia non accuset').

confessionum libri tredecim: Title directly attested at *retr.* 2. 6. 1, 'confessionum mearum libri tredecim'; cf. also *persev.* 20. 53, *ep.* 231. 6, and Possidius, *indiculum* (ed. Wilmart, *MA* 2. 161–208) 103. 6.

confessionum: It is perhaps a rhetorical gesture when A. says at *en. Ps.* 30. en. 1. 11, 'vita mea est confiteri te', but it is also a warning that 'confession' runs beyond the pages of this text and a suggestion that the relationship of this text to A.'s life is not that of signifier to signified.

On the term, see M. Verheijen, *Eloquentia Pedisequa* (Nijmegen, 1949), 11–81, and J. Ratzinger, *REAug* 3 (1957), 375–92. There are biblical roots; note esp. Ps. 9: 2, 'confitebor tibi, domine, in toto corde meo; narrabo omnia mirabilia tua'; Ps. 31: 18, 'confitebor adversum me iniustitiam meam domino, et tu remisisti impietatem peccati mei'; Ps. 34: 18, 'confitebor tibi in ecclesia magna.' The connotations of praise and thanksgiving are innate in the word's biblical usage (Verheijen, 69), but not in classical usage (Ratzinger, 376 f., cites ps.-Quint. *decl.* 314, 'ego enim confessionem existimo qualemcumque contra se pronuntiationem. . . . immo ea natura est omnis confessionis, ut possit videri demens qui de se confitetur. furore impulsus est: alius ebrietate, alius errore, alius dolore, quidam quaestione. nemo contra se dicit, nisi aliquo cogente'), and other Christian writers at least acknowledge them. Origen, *in Ps.* 135. 2 (PG 12. 1653–5), *ἡ ἐξομολόγησις τὴν εὐχαριστίαν καὶ δοξολογίαν σημαίνει· κεῖται δὲ καὶ ἐπὶ τῆς ἐξομολογήσεως τῶν ἁμαρτιῶν*. We encounter the theme in Hilary of Poitiers (*in Ps*. 66. 6, 'invenimus confessionem duplici ratione tractandam: esse unam confessionem peccatorum ... aliam laudationis dei'—he was probably following Origen or Eusebius (so Jerome at A. *ep.* 75. 6. 20)), and in Jerome as well, e.g. *in Is.* 2. 38, 'confessioque in hoc loco non pro paenitentia sed pro gloria et laude accipitur.' For the influence of Christian martyrdom on the one hand and of penitence on the other in shaping the attitude that A. embodies, see Ratzinger, 380–2.

Confiteri is a verb of speaking, and *confessio* is speech that is made possible, and hence authorized, by God: 1. 5. 5, 'miserere ut loquar' (and cf. *adn. Iob* on 6: 10, 'non alia dixit quam audivit a deo, id est de homine generaliter prophetantis, quia auxilio indiget in confessione'). God is ordinarily the addressee of this speech, but not exclusively (e.g. 10. 37. 62, 'confitear oraturis pro me fratribus meis'). Depending on the subject, the effect may be that of praise (*confessio laudis*), self-blame (*confessio peccatorum*), or (least common in A. and in *conf.*) determined avowal (*confessio fidei*). That there is one *confessio* taking different forms in different circumstances is clear from 10. 1. 1, 'cum enim malus sum, nihil est aliud confiteri tibi quam displicere mihi; cum vero pius, nihil est aliud confiteri tibi quam hoc non tribuere mihi' (sim. at *en. Ps.* 105. 2).

A. knew he was insisting on a breadth of meaning that was not always felt by his listeners, who had our specialized sense clearly in mind, as at *en. Ps.* 141. 19, 'duobus autem modis confessio intellegitur, et in peccatis nostris, et in laude dei. in peccatis nostris nota est confessio, et ita nota omni populo, ut quando auditum fuerit nomen confessionis in lectione, sive in laude dicatur, sive de peccatis dicatur, currant pugni ad pectus' (sim. at *en. Ps.* 78. 17, 137. 2, *ss.* 29. 2. 2, 67. 1. 1.)

The verb *confiteor* and noun *confessio* occur 111 times in *conf.* (Verheijen, 11–21). The origin of *confessio* lies in divine action; sometimes it is the divine act that is treated as the agent (5. 10. 20, 'cui confitentur ex me miserationes tuae'; cf. 6. 7. 12, 8. 1. 1). It is the opposite of *praesumptio* (7. 20. 26), and thus denotes not only the speech itself but the predisposition that makes the privileged speech possible. Hence though it is speech, it is not always spoken (10. 2. 2, 'confessio itaque mea, deus meus, in conspectu tuo tibi tacite fit et non tacite'); and conversely it is treated as speech even when written down (10. 3. 3, 'ut audiant confessiones meas'; 10. 3. 4, 'confessiones ... cum leguntur et audiuntur'; 12. 26. 36; 13. 14. 15, 12. 6. 6; cf. 9. 12. 33, 'confiteor tibi in litteris'). It is associated from the outset with praise (e.g. 1. 6. 9, 'laudari a me iubes et confiteri me tibi?'; 1. 7. 12); it is itself a form of sacrifice (5. 1. 1, 'sacrificium confessionum mearum'; 12. 24. 33); and it is roughly equated to *gratiarum actio* (9. 8. 17).

By presenting his words as a 'confession', A. disclaims authority for his own text (and by so doing does exactly what he claims not to do), but refers to a higher authority those who question his authority for speaking. He turns away sceptics by telling them (at 10. 3. 3) that his text is for those who are joined to him in *caritas* (again invoking the higher authority), and that it is not surprising if others refuse to believe what they read in it. This strategy aims to freeze out the hostile or sceptical reader. A. admits that the text has no authority with such a reader, but implies that this failure is the fault of the reader, not of the writer. His business is with his God, for the edification of those who are chosen by his God to benefit from the text; other readers are left to shift for themselves.

A.'s comments on his own work are few and all date from the last years of his life, when he often found himself going back over old ground, especially in his quarrel with Julian of Eclanum:

Confessionum mearum libri tredecim et de malis et de bonis meis deum laudant iustum et bonum, atque in eum excitant humanum intellectum et affectum. interim quod ad me attinet, hoc in me egerunt cum scriberentur et agunt cum leguntur. quid de illis alii sentiant, ipsi viderint; multis tamen fratribus eos multum placuisse et placere scio. a primo usque ad decimum de me scripti sunt, in tribus ceteris de scripturis sanctis, ab eo quod scriptum est: 'in principio fecit deus caelum et terram,' usque ad sabbati requiem. *retr.* 2. 6. 1 (426/7)

Quid autem meorum opusculorum frequentius et delectabilius innotescere potuit quam libri confessionum mearum? cum et ipsos ediderim antequam pelagiana haeresis extitisset, in eis certe dixi deo nostro et saepe dixi: 'da quod iubes et iube quod vis' [10. 29. 40]. quae mea verba Pelagius Romae, cum a quodam fratre coepiscopo meo fuissent eo praesente commemorata, ferre non potuit et contradicens aliquanto commotius paene cum eo qui illa commemoraverat litigavit. . . . et in eisdem etiam libris quod de mea conversione narravi, deo me convertente ad eam fidem quam miserrima et furiossima loquacitate vastabam, nonne ita narratum esse meministis, ut ostenderem me fidelibus et cotidianis matris meae lacrimis [see on 3. 11. 19–12. 21] ne perirem fuisse concessum? *persev.* 20. 53 (428/9)

Sume itaque, mi fili, sume, vir bone et non in superficie sed christiana caritate christiane, sume, inquam, etiam libros quos desiderasti confessionum mearum; ibi me inspice, ne me laudes ultra quam sum, ibi non aliis de me crede sed mihi, ibi me attende et vide quid fuerim in me ipso per me ipsum [cf. 10. 3. 3–4. 6]. et si quid in me tibi placuerit, lauda ibi mecum quem laudari volui de me, neque enim me, quoniam 'ipse fecit nos et non ipsi nos' [Ps. 99: 3; see on 9. 10. 25, 10. 6. 9]; nos autem perdideramus nos, sed qui fecit refecit. cum autem ibi me inveneris, ora pro me, ne deficiam sed perficiar. *ep.* 231. 6 (to Darius: 428/9)

The last words of *doctr. chr.* (4. 31. 64; written about the same time as *retr.*) reflect a similar spirit: 'ego tamen deo nostro ago gratias, quod in his quattuor libris non qualis ego essem, cui multa desunt, sed qualis esse debeat qui in doctrina sana, id est christiana, non solum sibi sed aliis etiam laborare studet, quantulacumque potui facultate disserui.'

The 13th chapter of Tobias (a book not often quoted by A., but accepted as canonical at *doctr. chr.* 2. 8. 13, and cf. 10. 34. 52, 'o lux quam videbat Tobis'), the canticle of Tobias, resembles *conf.* both for content and structure (Vg. text).[1] Italics here indicate the most notable parallels to *conf.*:

1 Aperiens autem Tobias senior os suum benedixit dominum et dixit:
'magnus es domine in aeternum [1. 1. 1]
et in omnia saecula regnum tuum [Bk. 11 generally].
2 *quoniam tu flagellas et salvas* [2. 2. 4, 3. 3. 5].
deducis ad infernum et reducis [1. 2. 2].
et non est qui effugiat manum tuam.
3 *confitemini* domino filii Israhel
et *in conspectu gentium laudate eum* [1. 16. 26 *et saep.*].
4 quoniam ideo dispersit vos inter gentes quae ignorant eum,
ut vos narretis mirabilia eius et faciatis scire eos [4. 15. 24, 8. 6. 14, etc.],
quia non est alius deus omnipotens praeter eum.
5 ipse castigavit nos propter iniquitates nostras
et ipse salvabit nos *propter misericordiam suam*.
6 *aspicite ergo quae fecit vobiscum*
et cum timore et tremore confitemini illi [see on 2. 3. 6]
regemque saeculorum exaltate in operibus vestris.
7 ego autem *in terra captivatitis* meae *confitebor* illi [see on 2. 9. 18],
quoniam ostendit maiestatem suam in gentem peccatricem.
8 *convertimini itaque peccatores et facite iustitiam coram deo*,
credentes quod faciat vobiscum misericordiam suam.
9 ego autem et anima mea in eo laetabimur.
10 benedicite dominum omnes electi eius,
agite dies laetitiae, *confitemini illi*.
11 *Hierusalem civitas dei*, castigavit te dominus [see esp. 12. 16. 23]:
in operibus manuum tuarum
12 *confitere domino in bonis* et benedic deum saeculorum,
ut reaedificet in te tabernaculum suum
et revocet ad te omnes *captivos* [2. 6. 14, 7. 21. 27, 8. 5. 12]
et *gaudeas in omnia saecula saeculorum* [13. 38. 53].
13 luce splendida *fulgebis* et omnes fines terrae adorabunt te. [10. 2. 2, 10. 6. 8]

. . . .

19 anima mea, benedic dominum,
quoniam liberavit *Hierusalem civitatem suam*.
20 beatus ero si fuerint reliquiae seminis mei
ad videndam claritatem Hierusalem.

. . . . . .

[1] The appositeness is corroborated by Possidius, *v. Aug.*, pr. 7 (quoting Tob. 12: 7 to characterize *conf.*), '"sacramentum igitur regis", ut angelica auctoritate prolatum est, "bonum est abscondere: opera autem domini revelare et confiteri honorificum est."'

23 benedictus dominus qui excitavit eam,
ut sit regnum eius in saecula saeculorum super eam. amen.

A similar set of parallels obtains between *conf.* and the text of Ps. 4; see on 9. 4. 8 for text and see 9. 4. 8–11 for exegesis. Other, earlier Christian analogues to A.'s self-analysis are few, sketchy, and not really comparable to this *confessio*: see J.-M. Le Blond, *Les conversions de saint Augustin* (Paris, 1950), 5 n. 1, and cf. Hil. Pict. *trin.* 1. 1–15. P. Frassinetti, *Gior. Ital. Filol.* 2 (1949), 50–9, adduces Arnob. *nat.* 1. 31, 'o maxime, o summe rerum visibilium et invisibilium procreator, o ipse invisibilis et nullis umquam comprehense naturis, dignus, dignus es vere, si modo te dignum mortali dicendum est ore, cui spirans omnis intellegensque natura et habere et agere numquam desinat gratias.' Liturgical usage and the developing vocabulary of prayer undoubtedly lie behind such expressions. A.'s genius lay in making a book of such discourse, and making a good book.

The claim that A.'s *confessio* is related to the annual Manichean confession of sins at the Bema festival (A. Adam, *Zschr. für Kirchengesch.* 69 (1958), 7: 'Was Augustin als manichäischer Katechumen Jahr um Jahr hat wiederholen müssen, das hat er in den Konfessionen auf seine christliche Existenz übertragen. Er hat sogar ein anderes kennzeichnendes Stilmittel des manichäischen Denkens dabei übernommen: die überquellende Bildhaftigkeit, die in keiner seiner übrigen Schriften in gleicher Weise festzustellen ist.') has not been sustained (see Brown 179 n. 1 and J. Ries, *REAug* 22(1976), 229–30). It was uncircumspect to use as the framework of his discourse a form of expression that could so easily be paralleled in Manichean (but not, at that time, in catholic) practice.

BOOK ONE

1. 1. 1–5. 6	*Proem*
1. 6. 7–7. 12	*Infantia*
1. 8. 13–19. 30	*Pueritia*
1. 20. 31	*Gratiarum actio*

The narrative of childhood is not specially Christian. Cic. *Tusc.* 3. 1. 2–2. 3 displays the same pattern and could stand as a summary of what A. says: 'simul atque editi in lucem et suscepti sumus, in omni continuo pravitate et in summa opinionum perversitate versamur, ut paene cum lacte nutricis errorem suxisse videamur. cum vero parentibus redditi, dein magistris traditi sumus, tum ita variis imbuimur erroribus, ut vanitati veritas et opinioni confirmatae natura ipsa cedat. (3. 2. 3) accedunt etiam poetae, qui cum magnam speciem doctrinae sapientiaeque prae se tulerunt, audiuntur leguntur ediscuntur et inhaerescunt penitus in mentibus.'

1. 1. 1

There have been various attempts to find precedents for this form of opening,[2] but in the history of Latin literature, its originality and oddity are clear. Most Latin prose works begin with a dedicatory epistle or a formal proem: this work has neither.[3] It begins abruptly, with speech directed to a silent God—but speech chosen from the words of that God himself. The first sentence is followed by a reflective pause for inquiry, which should not blur the main purpose: invocation, an opening more appropriate to poetry than prose. The work is not verse, but it is not conventional prose either: study of its rhythms has shown that the pattern of the work's sentence endings is neither quantitative nor accentual, and *sui generis* in A.'s own works (see Prolegomena, n. 126). A. invented a form and style unique in his own *œuvre* and in the traditions he inherited. In retrospect we can see him working towards this style through earlier, far more conventional works (e.g. *sol.*, where the dialogue form prevails, and the prayer of invocation is carefully set off within the dialogue at *sol.* 1. 1. 2–6), but *conf.* represents a breakthrough for A.

This opening can give rise to the disconcerting feeling of coming into a room

[2] Most recently, for instance, Pizzolato, *Lectio I–II*, 9, citing a Greek grammarian's precept to quote authority at the outset of a work; more pertinently, attempts to see Manichean or pagan precedents in hymns and prayers of praise: a collection of such textual analogues is M. Zepf, *Augustins Confessiones* (Tübingen, 1926), 63–96.

[3] A.'s own practice varies: some works have formal separate prefaces, some dedicatory epistles, some prefaces incorporating a dedication, and some simply begin baldly (e.g. *Gn. litt.*): no other work of his begins with direct address to God.

and chancing upon a man speaking to someone who isn't there. He gestures in our direction and mentions us from time to time, but he *never addresses his readers*. As literary text, *conf.* resembles a one-sided, non-fiction epistolary novel, enacted in the presence of the silence (and darkness) of God.[4] What A. attempts is a radical turn away from common sense—seen as tragically flawed by mad self-love—towards the wholly other, and *thus* towards the true self—for to him, we are not who we think we are (see on 10. 8. 15).[5]

Magnus es, domine: The work begins with *confessio laudis*.[6] Strictly speaking, these two lines contain a complete confession (see *en. Ps.* 143. 12 (quoted on 8. 2. 4), which shows the proud of this world giving way and confessing thus: 'dicturi sunt: tu magnus, domine') that renders the remaining 78,000 or so words of the text superfluous. This exclamation is self-sufficient; nothing more need be said, ever. But the fall from eternity into time brought with it the fall from timeless immutability of discourse into a restless and open-ended search for God in the inspired texts. Cf. *en. Ps.* 145. 6, 'quando implet laudator excellentiam laudati?'

Ps. 95: 4, 'quoniam magnus dominus et laudabilis valde'; Ps. 144: 3, 'magnus dominus et laudabilis valde, et magnitudinis eius non est finis'; Ps. 47: 2, 'magnus dominus et laudabilis valde, in civitate dei nostri, in monte sancto eius'; cf. Tob. 13: 1, 'magnus es domine in aeternum' (see above, pp. 6–7, on Tob. 13). *en. Ps.* 95. 4, 'quis "dominus", nisi Iesus Christus, "magnus et laudabilis nimis"?' (The text at *en. Ps.* 95. 4 has *nimis* for *valde* with the Roman Psalter; but the Verona Psalter has *valde*, and the adverb is the same in the LXX at all three texts cited here: σφόδρα.) Psalm citation is not (see Knauer, *passim*) mere ostentation of erudition: *en. Ps.* 26. en. 2. 1, 'voces istae psalmi quas audivimus et ex parte cantavimus, si dicamus quod nostrae sint, verendum est quemadmodum verum dicamus; sunt enim voces

[4] Eliot, 'East Coker': 'So the darkness shall be the light, and the stillness the dancing'; Lawrence Durrell, *Justine*: 'Does not everything depend on our interpretation of the silence around us?'

[5] For further and more ambitious reflections in this vein, see J.-J. Goux, *Tel Quel* 21 (1965), 67–75, e.g. at 68: 'Tendu vers l'Auditeur muet, Augustin est pris par le désir de se dire tout entier'; cf. also F. E. Consolino, *Materiali e discussioni per l'analisi dei testi classici* 6 (1982), 119–46 (esp. on the way *sol.* and *util. cred.* approach in places what A. does in *conf.*); and R. Herzog, in K. Stierle and R. Warning (eds.), *Das Gespräch* (Munich, 1984), 213–50; in a more traditional vein, see R. Guardini, *Anfang* (Munich, 3rd edn., 1953). W. Theiler's display of Hermetic parallels (*Die Vorbereitung des Neuplatonismus* (Berlin, 1930), 128–31) is less trenchant; see G. Fowden, *The Egyptian Hermes* (Cambridge, 1986), placing the Hermetic texts in a context broad enough to show how their ideas and terms entered a wider realm of discourse on which A. and others could draw.

[6] The *words* for 'confession' are lacking in the opening paragraphs; *fateor* not until 1. 5. 6, *confiteor* not until 1. 6. 9. T. Deman, *La vie spirituelle* 39 (1957), 261, noted this to ask whether the notion *confessio* was an original part of the work's conception, suggesting that it only really becomes prominent in Book 10. The reading of the first page proposed here demonstrates that if the exact *word* was not in mind, the thing for which it stood it A.'s mind is amply present.

magis spiritus dei quam nostrae. rursum si dicamus nostras non esse, profecto mentimur.'

domine: A. speaks to God alone. At 11. 1. 1, after canvassing his own past (Bks. 1–9) and present (Bk. 10), he expects his readers to be able to join him in the present: 'sed affectum meum excito in te et eorum qui haec legunt, ut dicamus omnes: "magnus dominus et laudabilis valde"' (see notes there). The second-person direct address to God pervades *conf.*, occurring in 381 of the work's 453 paragraphs (measured by the presence of vocatives, second-person singular pronouns, and verb forms: but the second-person verb form does not seem to occur addressed to God without either a vocative or a pronoun to specify the reference). Paragraphs without such address include some notably desolate and God-less passages (e.g. 6. 15. 25, on the banishment of his mistress), but not all are of that sort; see in particular on 10. 6. 9 for the longest stretch without such address. On the forms of address to God in *conf.*, see Knauer 31–74 and J. Morán, *Aug. Stud.* 4 (1973), 141–57, with a catalogue of 165 passages at 152–7.

magna virtus tua et sapientiae tuae non est numerus: Ps. 146: 5, 'magnus dominus noster et magna virtus eius et sapientiae eius non est numerus'; *en. Ps.* 146. 11, 'conticescant humanae voces, requiescant humanae cogitationes; ad incomprehensibilia non se extendant quasi comprehensuri, sed tamquam participaturi; participes enim erimus.' 1 Cor. 1: 24, 'Christum dei virtutem, et dei sapientiam'; this is the first scriptural text A. cites in his surviving works, at *c. acad.* 2. 1. 1 and *beata v.* 4. 34; from *c. Faust.* 20. 2 (where Faustus cites the text to claim that *virtus* resides in the sun and *sapientia* in the moon) it is clear that to call Christ *virtus* and/or *sapientia* would particularly irritate Manichees—the text is also cited against them at *c. Fort.* 9. Cf. 5. 3. 5, 'sed non noverunt viam, verbum tuum . . . et sapientiae tuae non est numerus.'

virtus . . . sapientiae: God in action . . . God at rest.

non est numerus: *Io. ev. tr.* 39. 4, 'pater et filius et spiritus sanctus trinitas. si tres, quid tres? deficit numerus. ita deus nec recedit a numero, nec capitur numero. quia tres sunt, tamquam est numerus; si quaeris quid tres, non est numerus. unde dictum est, "magnus dominus noster et magna virtus eius, et sapientiae eius non est numerus." ubi cogitare coeperis, incipis numerare: ubi numeraveris, quid numeraveris non potes respondere.' The second person of the trinity is Word *par excellence*, but for A. he is also Number: see on 5. 4. 7.

et laudare te vult homo: Cf. Verheijen, 90–5, on *et* and its frequency of use in *conf.*: abundant in some passages, rare in others, often occurring where a more restrictive conjunction (*sed*, *cum*, *si*) would be expected. V. estimates that in Bk. 8 *et* occurs with more than 'la simple valeur copulative' more than 100 times, in Bk. 1 more than 75 (cautioning that such calculations are subjective). A. on the force of the conjunction, hinting that its part in the struggle to restore integrity to language is not insignificant: *en. Ps.* 4. 4 (on Ps.

4: 4—see on 9. 4. 8–11), 'sed si quem movet coniunctio superaddita, quod ait, "*et* scitote", facile est ut in scripturis animadvertat huius locutionis genus familiare esse illi linguae qua prophetae locuti sunt. nam saepe invenis ita coeptum: "et dixit dominus ad illum; et factum est verbum domini ad illum." quae iunctura coniunctionis, cum sententia non praecesserit cui sequens adnectatur, mirabiliter fortassis insinuat prolationem veritatis in voce cum ea visione quae fit in corde esse coniunctam.'

For comparison, in sample works of Cicero, *et* appears with the following relative frequencies: *Brut.* 3. 6%, *de amicitia* 2. 4%, *pro Caelio* 1. 8%; from A., in *civ.*, approx. 3. 3%, in *en. Ps.* approx. 4. 5%, and in *trin.* approx. 5. 2%; in *conf.* 5. 8% (4585 out of 78,858 words).

laudare . . . vult: Not *laudat*, for the praise in the first two lines is imperfect, the expression in words more wish than deed (cf. 2. 1. 1, 'recordari volo'). A. seeks to praise. In the implicit question whether he succeeds is encompassed the tension of the whole text. At the moment of giving praise, his words fall back into self-reflection and doubt. His mortality might well seem a good reason for not praising the creator who made him mortal: but already in that objection, the explanation begins to obtrude.

homo: A. means not *homo quidam*, but speaks directly of himself, his own act of attempted praise in the first lines. An exact parallel at 1. 7. 11, 'exaudi, deus: vae peccatis hominum! et homo dicit haec, et misereris eius, quoniam tu fecisti eum [cf. here 'fecisti nos ad te'] et peccatum non fecisti in eo.' In both cases the author's voice is heard in spontaneous exclamation, followed immediately by a less spontaneous, more detached reflection on what that first exclamation means in view of the creature/creator relation of humankind to God. Cf. also 7. 1. 1, 'homo et talis homo'. (To be sure, the form of expression emphasizes A.'s representativeness, hence inviting readers to share in the inquiry and praise that follow; a similar representativeness underlies the whole of Bks. 11–13.)

mortalitatem: A fact that speaks of A. ('testimonium peccati'), and of God ('testimonium quia superbis resistis'), evidence of alienation, a pretext perhaps for angry avoidance of praise. But instead, praise prevails. Cf. 10. 4. 6 (of Christian believers), 'sociorum gaudii mei et consortium mortalitatis meae': mortality is the counterweight of the joy from which praise might rightly be expected to rise. Death is punishment already at *mus.* 6. 33, 6. 46, and *Gn. c. man.* 1. 16. 26, 1. 18. 29, 2. 26. 38 and elsewhere, then at *vera rel.* 26. 48 ('erga eos qui peccato mortalitatem meruerunt'), and consistently afterwards. For the expression here, cf. *s. dom. m.* 2. 9. 35, 'istam mortalitatem circumferimus' (sim. at *s.* 49. 3. 3).

quia superbis resistis: Prov. 3: 34 (VL), 'deus superbis resistit, humilibus autem dat gratiam'; quoted at 1 Pet. 5: 5 and Jas. 4: 6 in the same form (following the LXX). The second half of the text is here unspoken (as at 4. 15. 26; spoken at 3. 5. 9, 4. 3. 5, 7. 9. 13, 10. 36. 59). Cf. *civ.* 1. pr., 'rex enim et conditor civitatis huius de qua loqui instituimus in scriptura populi sui sententiam divinae legis aperuit, qua dictum est: "deus superbis resistit, humilibus autem dat gratiam." hoc vero, quod dei est, superbae quoque animae spiritus inflatus adfectat amatque sibi in laudibus dici: "parcere subiectis et debellare superbos".' The Virgilian juxtaposition suggests a partial explanation for the appeal of the scriptural text to A. Cf. La Bonnardière, *Biblia Augustiniana: Proverbes*, 202–4, for a list of 42 other citations of the text in A.'s *œuvre*. The verse may have been a recent discovery: before *doctr. chr.* it appears *only* at *en. Ps.* 18. en. 1. 8, 18. en. 1. 15, and *s. dom. m.* 1. 11. 32. Note *doctr. chr.* 3. 23. 33, 'nulla enim fere pagina est sanctorum librorum in qua non sonet quod "deus superbis resistit, humilibus autem dat gratiam".'[7] On *superbia* in *conf.*, see the catalogue and discussion by M. Testard in *Homo Spiritalis* (Festschrift Verheijen: Würzburg, 1987), 136–70.

et tamen: One of A.'s commonest ways (59 × in *conf.*, 12 × in Bk. 1) of asseverating in the face of apparent contradiction and paradox. See Pizzolato, *Lectio I–II*, 10 (but he undercounts the frequency).

excitas: 16 × in *conf.*; cf. 5. 1. 1, 'accipe sacrificium confessionum mearum de manu linguae meae (quam formasti et excitasti, ut confiteatur nomini tuo)'; 11. 1. 1, 'affectum meum excito in te'; cf. 10. 3. 4, 'confessiones praeteritorum malorum meorum ... excitant cor ne dormiat in desperatione.' Cf. *retr.* 2. 6. 1 (quoted more fully above), 'confessionum ... libri ... excitant humanum intellectum et affectum'; *sol.* 1. 1. 3, 'deus, cui nos *fides* excitat, spes erigit, caritas iungit' (and *fides* is basis of *confessio*, as the remainder of this paragraph demonstrates). See E. Feldmann, *Augustinus* 31 (1986), 113–22.

laudare te delectet: Brown 155, 'Augustine came to view "delight" as the mainspring of human action', citing *div. qu. Simp.* 1. 2. 21; cf. also *doctr. chr.* 1. 33. 37, 'vicinissime dicitur *frui* "cum delectatione uti".' In phrasing the matter that way, Brown approaches the position of Cornelius Jansen, who distinguished *delectatio* from *voluntas*, and made the former a determining force on the latter; for an orthodox critique of that view (taking what is indeed the more defensible position that for A. 'delight' and 'will' are two names for the same thing), see E. Gilson, *The Christian Philosophy of Saint Augustine* (New York, 1960), 321 n. 81–23 n. 84.

The role of *delectatio* is strongly foreshadowed in the rhetorical tradition. Cf. Cic. *or.* 21. 69, 'erit igitur eloquens ... qui in foro causisque civilibus ita

[7] But the privilege conferred by that text is not absolute: *ep.* 249 (before 411? Mandouze, *Pros. chr.* s.v. Restitutus 7), 'ita prope nulla est [sacrorum librorum pagina] quae nos non admoneat intus in ipsa societate sacramentorum quibus imbuimur ad vitam aeternam, cum his qui oderunt pacem esse debere pacificos, donec ingemiscendo nostra longinqua peregrinatio transeat [Ps. 119: 5–7] atque in virtute Hierusalem matris aeternae securissima pace perfruamur.' Or cf. *s.* 46. 14. 33, 'prope omnis pagina [scripturae] nihil aliud sonat quam Christum, et ecclesiam toto orbe diffusam.'

dicet, ut probet, ut delectet, ut flectat' (quoted with slight modifications at *doctr. chr.* 4. 12. 27), with parallels at *Brut.* 49. 185, *de or.* 2. 28. 121, *opt. gen. or.* 1. 3. Between *probare* (the establishment of a fact) and *flectere* (stirring the audience to act), *delectare* is the crucial moment of motivation. See on 1. 6. 7.

inquietum . . . requiescat: This initial disquiet is answered by the adumbration of eternal rest at 13. 35. 50–38. 53, the last lines of the text; cf. also 1. 5. 5, 'quis mihi dabit adquiescere in te?' This restlessness arises from disorder: 13. 9. 10, 'minus ordinata inquieta sunt; ordinantur et quiescunt.' For *ordo*, see on 1. 7. 12 and see further on 13. 9. 10. Cf. *en. Ps.* 38. 5, 'coepit esse inquietum cor meum. . . . et suspirans in finem quendam, ubi ista non erat passurus, in illum, inquam, finem quo dicturus est bono erogatori dominus, "intra in gaudium domini tui".' (Mt. 25: 21: cf. 9. 10. 25, where the same scriptural quotation is the culmination of the Ostia vision); cf. also *en. Ps.* 91. 2, 48. s. 2. 6. This phrase has evoked an abundant literature: A. Di Giovanni, *L'inquietudine dell'anima* (Rome, 1964), esp. 87 n. 8; A. Pincherle, *Augustinus* 13 (1968), 353–68 (on *requies* and the link to the last pages of *conf.*); E. Maccagnolo, *Riv. di Filos. Neo-scolastica* 71 (1979), 314–25; G. Lawless, *REAug* 26 (1980), 45–61 (on 'interior peace'); and generally de la Peza (see next note).

cor: 10. 3. 4, 'ad cor meum, ubi ego sum quicumque sum'. E. de la Peza, *El significado de 'cor' en San Agustin* (Paris, 1962); in summary form in *REAug* 7 (1961), 339–68; and cf. A. Maxsein, *Philosophia cordis: Das Wesen der Personalität bei Augustin* (Salzburg, 1966), 46–52. *Cor* in A. is a word whose use is demonstrably influenced by contact with its scriptural employment; it is in A. an expression for the indivisible, authentic centre of human life, where the tensions of a sinful world are most clearly felt. The term is, as others have observed, unphilosophical, even untheological, but it is eminently scriptural, and Augustinian. Cf. Ps. 37: 11, 'cor meum conturbatum est et deseruit me fortitudo mea'; *en. Ps.* 37. 15, 'haec patitur homo intus, ibi secum, in seipso, et seipsum, de nemine ad neminem praeter se.'

donec: Pizzolato, *Lectio I–II* 13: '*Donec* pare il termine-chiave dell'espressione.' The temporal dimension of disquiet is both an important manifestation of the problem itself (as Bk. 11 will make clear) and at the same time a suggestion here of hope.

da mihi: 11. 3. 5, 'da et mihi haec intellegere'. (See above on 'magnus es domine' for ref. to 11. 1. 1: Bk. 11 marks a renewed beginning in several ways.) Cf. Ps. 118: 34, 'da mihi intellectum, et scrutabor legem tuam, et custodiam illam in toto corde meo'; *en. Ps.* 118. s. 11. 4, 'quid habeat utilitatis etiam hoc ipsum quod "lex subintravit, ut abundaret delictum" [Rom. 5: 20] nemo comprehendit, nisi a domino acceperit intellectum; unde iste adiungit et dicit: "da mihi intellectum, et scrutabor legem tuam, et custodiam illam in toto corde meo." cum enim quisque legem scrutatus fuerit, et ad eius alta pervenerit in quibus tota pendet, profecto debet deum diligere ex toto corde, ex tota anima, ex tota mente, et proximum suum tamquam seipsum [cf. Mt. 22: 37].' On the quest of faith for understanding (*Io. ev. tr.* 29. 6, 'noli quaerere intellegere ut credas, sed crede ut intellegas'), see R. Holte, *Béatitude et Sagesse* (Paris, 1962), and O. du Roy, *L'Intelligence de la foi en la trinité selon saint Augustin* (Paris, 1966); when expressed in its fullest form as in the passage just cited, the understanding is (*a*) scriptural, (*b*) allegorical, and (*c*) directed beyond intellection to action. The inquiry introduced here by these words is itself about the proper form of inquiry, and shares implicitly all three of those qualities. Placed here, it reflects the intrinsic importance to A. of the content of the inquiry, but also sets a paradigmatic model of the search for understanding that ends in praise here on the first page. (For the form of expression, cf. Tiberianus (4th c. AD), *versus Platonis*, ed. Baehrens, *PLM* 3. 268, lines 26 f.: 'da nosse volenti', and further M. Zepf, *Augustins Confessiones* (Tübingen, 1926), 64 n. 4.)

Only God can give this gift: cf. the last words of *conf.* (with an echo of Mt. 7: 7, also echoed below in 1. 1. 1): 13. 38. 53, 'et hoc intellegere quis hominum dabit homini? quis angelus angelo? quis angelus homini? a te petatur, in te quaeratur, ad te pulsetur: sic, sic accipietur, sic invenietur, sic aperitur.'

scire: The sense here is probably not distinguished sharply from *credere*, hence the passage is congruent with A.'s treatment of faith and understanding. Cf. *sol.* 1. 3. 8, 'omne autem quod scimus recte fortasse etiam credere dicimur'; *retr.* 1. 14. 3 (demurring at a sharp distinction between *scire* and *credere* at *util. cred.* 11. 25), 'proprie quippe cum loquimur, id solum scire dicimus quod mentis firma ratione comprehendimus. cum vero loquimur verbis consuetudini aptioribus, sicut loquitur etiam scriptura divina, non dubitemus dicere scire nos et quod percipimus nostri corporis sensibus et quod fide dignis credimus testibus, dum tamen inter haec et illud quid distet intellegamus.'

An alternative view is possible, according to which *scire* and *intellegere* here would be hendiadys: *mag.* 11. 37 (of exegetical difficulties raised by the three boys in Daniel), 'haec autem omnia quae in illa leguntur historia ita illo tempore facta esse ut conscripta sunt, credere me potius quam scire confiteor: neque istam differentiam idem ipsi quibus credimus nescierunt. ait enim propheta, "nisi credideritis, non intellegetis" [Is. 7: 9 (VL)].'

The former view is to be preferred. It imparts to the text a tension between present and future that is at home on this page: *scire* (= *credere*) would apply to the inquiry in the present paragraph, while *intellegere* would anticipate the fuller comprehension to be worked out through the remainder of *conf.* and through the rest of A.'s life. Against that, the hendiadys is flat and far from obvious; such redundancy of expression is out of place with the concise and carefully chosen style of this opening page.

invocare ... laudare ... scire: Ps. 104: 1, 'confitemini domino, et invocate nomen eius'; *en. Ps.* 104. 1, 'praemissa enim laude, invocatio sequi solet... exaudit quippe invocantem, quem laudantem videt: laudantem videt; quem probat amantem.'

aliud ... pro alio: G–M: 'A. probably has in mind the objects of his own worship in his Manichaean days,' when he invoked God without faith. Cf. *sol.* 1. 1. 6, 'fac et me, pater, vindica me ab errore: quaerenti te mihi nihil aliud pro te occurrat'; sim. at *doctr. chr.* 2. 6. 7, *en. Ps.* 78. 9; *qu. hept.* 1. pr., 'nonnulla enim pars inventionis est nosse quid quaeras.' Cf. 5. 5. 9, 'aliud pro alio sentientem'.

quomodo autem: Rom. 10: 13–14, '"qui invocaverit nomen domini salvus erit" [Joel 2: 32]. quomodo autem invocabunt, in quem non crediderunt? (14) aut quomodo credent, quem non audierunt? quomodo autem audient, sine praedicante? aut quomodo praedicabunt, si non mittantur.' Text from *s.* 57. 1. 1, where he explains: 'missi sunt ergo praedicatores, praedicaverunt Christum. illis praedicantibus populi audierunt, audiendo crediderunt, credendo invocaverunt' (the verse is quoted similarly at the outset of each of A.'s sermons on the Lord's Prayer for *competentes* (*ss.* 56. 1. 1, 57. 1. 1, 58. 1. 1, and 59. 1. 1), and cf. *s.* 115. 1. 1, where it is taken to show that *fides* is the *fons orationis*); cf. *div. qu. Simp.* 1. 2. 7 and 1. 2. 10. The *principle* invoked is one that A. much later marked as one that distinguished his mature thought from that which preceded the watershed *div. qu. Simp.*: *praed. sanct.* 3. 7, 'credere non possemus, si non praecederet praeconium veritatis; ut autem praedicato nobis evangelio consentiremus, nostrum esse proprium, et nobis ex nobis esse arbitrabar. quem meum errorem nonnulla opuscula mea satis indicant, ante episcopatum meum scripta'; in the next paragraph there A. retracts *exp. prop. Rom.* and endorses *div. qu. Simp.*

credent CDGO² Maur. Knöll (*ss.* 56. 1. 1, 57. 1. 1; *c. Faust.* 12. 45; *div. qu. Simp.* 1. 2. 10) : *credunt* O¹S Eug. Skut. Ver. (*s. Guelf.* 1. 1; *div. qu. Simp.* 1. 2. 7). The Greek NT has the aorist subjunctive (πιστεύσωσιν), as for the other verbs in the passage. A. regularly omits the *ei* that occurs after *credent* in the Vg. The passage was apparently condensed for rhetorical effectiveness.

et laudabunt ... requirunt eum: Ps. 21: 27, 'edent pauperes, et saturabuntur. et laudabunt dominum qui requirunt eum. vivent corda eorum in saeculum saeculi'; *en. Ps.* 21. en. 2. 27, 'quare sunt pauperes? quia dominum laudant, et dominum quaerunt. dominus est divitiae pauperum; ideo inanis est domus, ut cor plenum divitiis sit. ... pauperes quaerunt unde cor impleant; et cum impleverint, laudant dominum qui requirunt eum.' The verse recurs in the last words of Bk. 10 (10. 43. 70—see there for important eucharistic overtones), completing what is begun here, preparatory to beginning again in 11. 1. 1.

quaerentes enim inveniunt eum: Mt. 7: 7, 'petite et dabitur vobis; quaerite et invenietis; pulsate et aperietur vobis.' Echoes and citations in *conf.*: 6. 4. 2, 6. 11. 18, 6. 11. 20, 11. 2. 3, 11. 22. 28, 12. 1. 1, 12. 12. 15, 12. 15. 22, 12. 24. 33,

13. 38. 53 (last words of *conf.*). At *s.* 51. 5. 6, he describes the errors of his youth as a failure to knock and seek.

The verse was significant from before his baptism: *c. acad.* 2. 3. 9 (one of very few scriptural echoes at Cassiciacum), 'nam mihi credite, vel potius illi credite qui ait, "quaerite et invenietis", nec cognitionem desperandam esse, et manifestiorem futuram quam sunt illi numeri'; cf. *sol.* 1. 1. 3, 1. 1. 6. A. throws the verse up to the Manichees as one they were fond of: *mor.* 1. 17. 31, 'nam si sapientia et veritas non totis animi viribus concupiscatur, inveniri nullo pacto potest. at si ita quaeratur ut dignum est, subtrahere sese atque abscondere a suis dilectoribus non potest. hinc est illud quod in ore habere etiam vos soletis, quod ait, "petite et accipietis, quaerite et invenietis, pulsate et aperietur vobis. nihil est occultum quod non revelabitur".' It recurs in anti-Manichean polemic, e.g. *Gn. c. man.* 2. 2. 3 (of exegesis).

Linked to the search of faith for understanding, it also appears at *lib. arb.* 2. 2. 6, 'nisi enim et aliud esset credere aliud intellegere et primo credendum esset quod magnum et divinum intellegere cuperemus, frustra propheta dixisset: "nisi credideritis, non intellegetis."[8] ipse quoque dominus noster ... iam credentibus dicit, "quaerite et invenietis", nam neque inventum dici potest quod incognitum creditur, neque quisquam inveniendo deo fit idoneus, nisi ante crediderit quod est postea cogniturus.' Cf. *lib. arb.* 3. 20. 58. See also A.'s plea to Valerius for time for strengthening his grasp of scripture (*ep.* 21. 4), and elsewhere in the years before *conf.* it occurs with ease and frequency, e.g. *ep.* 23. 7, *util. cred.* 14. 30, with formal exposition at *s. dom. m.* 2. 21. 71,[9] 'petite et dabitur ... pulsanti aperietur: petitio pertinet ad impetrandam sanitatem firmitatemque animi, ut ea quae praecipiuntur implere possimus, inquisitio autem ad inveniendam veritatem. cum enim beata vita actione et cognitione compleatur, actio facultatem virium, contemplatio manifestationem rerum desiderat. horum ergo primum petendum, secundum quaerendum est, ut illud detur, hoc inveniatur. sed cognitio in hac vita viae prius quam ipsius possessionis est. sed cum quisque veram viam invenerit, perveniet ad ipsam possessionem, quae tamen pulsanti aperietur.' Cf. also Jer. 29: 12–13, 'et invocabitis me, et ibitis; et orabitis me, et ego exaudiam vos. (13) quaeretis me, et invenieritis, cum quaesieritis me in toto corde vestro.'

The questions posed here by A. (from 'da mihi, domine' to 'ut sciaris') are thus met by a sequence of authoritative scriptural dicta, collected from three different sources (Rom. 10: 13–14, Ps. 21: 27, Mt. 7: 7), in part themselves questions—but only rhetorical questions (on that device see below on 1. 3. 3).

[8] Is. 7: 9 (VL)—that thematic verse first in A. at *lib. arb.* 1. 2. 4, frequent thereafter: du Roy, 213 n. 3.

[9] The distinction between *petitio*, *inquisitio* and *pulsatio* (which does not otherwise play a part in A.'s use of the verse) is renounced by *retr.* 1. 19. 9.

These citations together provide the data required to answer the questions that precede. (The phrase 'invenientes laudabunt eum' does not come with clear scriptural warrant: it appears as a conclusion drawn by A. himself.)

The scriptural data may be schematized thus:

credere (= *scire*?) → *invocare*
praedicare → *credere*
requirere (= *quaerere*) → *laudare*
quaerere → *invenire*.

These reduce without difficulty to:

praedicare → *credere* → *invocare*
requirere/quaerere → *invenire/laudare*.

The inquiry arose from a moment of unfulfilled intention ('et laudare te *vult* homo') that is now situated in its context. Such a quest leads to discovery and praise, but arises from invocation, which arises from belief, which in turn finally arises from the antecedent act of preaching (the only act in the sequence performed by someone other than the speaker). Invocation, therefore, is the possible and necessary place to begin for a search that will end, for purposes of this book, at 13. 38. 53, and that will end for A., in the fullness of divine *requies*. But invocation itself (as the next paragraph will develop) is problematical.

Listening precedes speech elsewhere in *conf.*: 12. 15. 22, 'quid igitur ex his quae clamavit cor meum ad deum meum, cum audiret interius vocem laudis eius, quid tandem falsum esse contenditis?'; 11. 8. 10, 'sic in evangelio per carnem ait, et hoc insonuit foris auribus hominum ut *crederetur* et intus *quaereretur* et *inveniretur* in aeterna veritate, ubi omnes discipulos bonos et solus magister docet.' So also, grace precedes invocation: 1. 10. 16, 'libera etiam eos qui nondum te invocant, ut invocent te et liberes eos.' Cf. *s.* 17. 4. 4, 'qui vult audiri a deo, carissimi, prius audiat deum.'

inveniunt GO[1]S Knöll Skut. Ver. : *invenient* CDO[2] Maur.

per humanitatem filii tui: The incarnation of the Word is the fundamental act of revelation.

per ministerium praedicatoris tui: Who is the *praedicator*? Ambrose? (So Pusey ad loc. in his translation.) Christ? (So Courcelle, *Recherches*, 43 n. 5.) Paul? (*div. qu. Simp.* 1. 2. 22, 'repente ex evangelii mirabili persecutore mirabilior praedicator effectus est', with the same transformation of Paul from *persecutor* to *praedicator* at least 10 × elsewhere in A., and cf. 10. 23. 34, where Paul is 'homo tuus verum *praedicans*'; suggested to me by Professor Paula Fredriksen.) Perhaps the generic should be allowed to stand unspecified.

1. 2. 2

Invocation is postponed by consideration of the problems it raises. Every half-assertion to be inferred from a rhetorical question seems cancelled by later questions. This paragraph and the next pursue the inquiry preliminary to the attempt at invocation itself beginning in 1. 4. 4. The organization of the pattern is similar to a series of concentric circles: *homo* within *caelum et terra* within *quidquid est*. The paradox is that of calling down into a man a God who is already there. This concern with the 'place' of God, is tied up with A.'s pre-conversion notion of God permeating all matter (see on 7. 1. 2, and cf. 7. 5. 7, 'spongiam quamlibet magnam'). See also 13. 1. 1 for renewed invocation (and confirmation that the speaker's invocation is preceded by divine speech). Of many parallel passages (see others below), note *en. Ps.* 74. 2, 'noli antequam confitearis invocare; confitere, et invoca' (reading Ps. 74: 2, 'confitebimur tibi, deus, confitebimur tibi, et invocabimus nomen tuum').

A. acts up to his training as grammarian when he chooses to explore difficulty through etymology; cf. Marrou, 127–8, but even Marrou underestimates the practicality of the device for A. as a way of opening discussion of a difficult subject. For an exact parallel, cf. *en. Ps.* 30. en. 2. s. 3. 4, 'invocas deum, quando in te vocas deum. hoc est enim illum invocare, illum in te vocare, quodam modo eum in domum cordis tui invitare.' Sim. at *en. Ps.* 30. en. 2. s. 3. 4, 52. 8, 85. 8, *s.* 47. 7. 8. Two passages offer further nuances: *en. Ps.* 79. 14, '"et nomen tuum invocabimus" [Ps. 74: 19]; id est, te diligemus'; *en. Ps.* 144. 22, 'ergo qui deum ipsum, a quo accepit de quibus gaudet, praeponit his omnibus rebus quas accepit, ipse invocat deum in veritate. . . . tunc eris rectus, cum in omnibus bonis quae facit, deus tibi placet; in omnibus malis quae pateris, deus tibi non displicet. hoc est invocare deum in veritate.'

vocabo CDO Maur. Ver. : *invocabo* GS Eug. Knöll Skut.

qui fecit caelum et terram: Gn. 1: 1, 'in principio fecit deus caelum et terram.' The *fecit* is characteristic of the version in which Augustine quotes Genesis: for text, see on 13. 1. 1. Because Bks. 11–13 comment at length on the first chapter of Genesis, this verse offers the most frequently repeated verbal pattern in *conf.*

caelum et terram: See on 12. 2. 2 ff. for full discussion (and cf. also *Gn. litt.* 1). The phrase may mean either the visible heaven and earth as representative of all of visible creation, or it may be taken to represent both invisible and visible creation—on that reading, *terra* represents all that is material and visible, while *caelum* becomes the *caelum caeli* (12. 2. 2). A.'s interpretation ensures that the phrase remains all-embracing, even for readers who do not share the cosmology of the author of Genesis.

inferi CDGO[1]S Knöll Skut. : *inferis sum* O[2] : *in infernis* Maur. : *in ⟨profundis⟩ inferi* Verheijen (cf. 3. 6. 11, 'in profunda inferi'). Interpretation is controversial:

—*inferi* as possessive genitive: Knauer, 131, citing *TLL* 7. 1. 1390. 66; Pizzolato, *Lectio I–II*, 20 n. 65, comes to the same conclusion independently.

—as locative: *BA* trans., 'je ne suis pas encore aux enfers.'

—nominative plural: G–M, 'A. in fact does not mean merely "I am not *in* hell", but more boldly and pointedly "I am not, after all, hell; and even if I were, Thou wouldst still be in me—witness the Psalmist".' Pizzolato, loc. cit., cites parallels for that construction (*civ.* 22. 30, 'dies enim septimus etiam nos ipsi erimus'; *conf.* 13. 2. 3, 'fuimus . . . tenebrae' (Eph. 5: 8)); and for the sense: *inferi* as place of the dead: *vera rel.* 52. 101, 'cavendi sunt ergo inferiores inferi, id est post hanc vitam poenae graviores'; *qu. hept.* 4. 29 (on Num. 16: 32: 'et descenderunt ipsi . . . viventes ad inferos'), 'notandum secundum locum terrenum dictos esse inferos, hoc est in inferioribus terrae partibus. varie quippe in scripturis et sub intellectu multiplici, sicut rerum de quibus agitur sensus exigit, nomen ponitur inferorum et maxime in mortuis hoc accipi solet.' Cf. also *Gn. litt.* 12. 33. 62, *ep.* 187. 2. 6.

nam etsi descendero: Ps. 138: 8, 'si ascendero in caelum, tu ibi es; si descendero in infernum, ades'; *en. Ps.* 138. 11, 'etiam illic ades, ut vindices. quo ergo iturus sum, ut a facie tua fugiam, id est, iratum te non sentiam?' This descent is not ethically neutral, but implies sin, flight, and avoidance. On that, cf. Cassian *conl.* 1. 14. 2 quoting Ps. 113: 17–18 and adding a parenthetical gloss: '"non mortui laudabunt te, domine: neque omnes qui descendunt in infernum" (sine dubio: peccati).'

As in 1. 1. 1, the method is to proceed by questioning (*quaerere*); adducing scriptural authority, this is the first declarative sentence in the paragraph. The second declarative sentence that follows makes explicit the conclusion authorized by scripture. A. then ('an potius . . . omnia') revises the conclusion just reached through further inquiry culminating in renewed citation of scriptural authority. The revision is affirmed in the elliptical sentence to follow ('etiam sic . . .').

ex quo omnia: Not an echo only of Rom. 11: 36, 'quoniam ex ipso et per ipsum et in ipso sunt omnia: ipsi gloria in saecula saeculorum. amen.' Du Roy, 479–85 shows that this is a liturgical doxology based on the Romans verse but fused with 1 Cor. 8: 6 ('ex quo omnia . . . per quem omnia'). Both the 'doxology' and a more accurate 'citation' of the Romans verse occur at all periods of A.'s career, but the 'doxology' form appears as early as *quant. an.* 34. 77 and *mor.* 1. 14. 24, hence (on du Roy's argument) attests a liturgical influence at an early period that scholars have tended to see as intellectualist and Platonic rather than churchly. (Only here so fully in *conf.*; echoes with a light touch at 1. 6. 10, 4. 12. 18, 4. 15. 24, 7. 20. 26, 12. 7. 7, 12. 19. 28, some so indirect as to leave us unable to tell whether they reflect the 'doxology' or the 'citation'; where it is possible to judge, these echoes favour the 'doxology', but they are not on du Roy's list).

qui dixit: Jer. 23: 24, 'numquid non caelum et terram ego impleo? dicit dominus.' This passage is regularly used as an affirmative statement, not even a rhetorical question, by Augustine: *cons. ev.* 1. 23. 31, *civ.* 12. 26, 22. 29, *trin.* 2. 5. 7, 'ubique enim erat qui ait "caelum et terram ego impleo".'

The collocation of Ps. 138: 8 + Rom. 11: 36 (both as 'citation' and as 'doxology') + Jer. 23: 24 is frequent in A. as evidence for the ubiquity of God-as-trinity: Knauer, 132. So *trin.* 2. 5. 7, *en. Ps.* 49. 18, *ep.* 187. 4. 14, and (only Ps. + Jer.) *Io. ev. tr.* 36. 8. (See also 4. 9. 14, Jer. with Ps. 138: 7.) Ps. + Rom. at *trin.* 5. 8. 9, 'nam et . . . dicitur . . . et "si ascendero in caelum tu ibi es", quod ad locum [dicitur]. . . . itaque omnipotens pater, omnipotens filius, omnipotens spiritus sanctus, nec tamen tres omnipotentes, sed unus omnipotens, "ex quo omnia, per quem omnia, in quo omnia; ipsi gloria." quidquid ergo ad se ipsum dicitur deus et de singulis personis ter dicitur patre et filio et spiritu sancto, et simul de ipsa trinitate non pluraliter sed singulariter dicitur.'

1. 3. 3

Argument by rhetorical question: There are almost 700 question marks in the text of *conf.*: many are explicitly 'rhetorical' ('put not to elicit information, but as a more striking substitute for a statement of contrary effect': Fowler, *Modern English Usage*), and most of the rest are like those in the last two sentences here, open-ended when asked, but swiftly resolved by A. This is a high frequency of interrogation, but ancient practice was more abundant in this regard than modern: in 31 OCT pages of Cic. *har. resp.*, for instance, there are 38 question marks (about half the *conf.* rate), while in a comparable piece of modern expository prose, there are only 15 (less than a quarter the *conf.* rate). Where it is difficult to speak of God (cf. the end of 1. 4. 4), the rhetorical question has the effect of sharing the burden of direct assertion with the audience; and where *quaerere* is a theme, the audience can share that as well.

ergone: Enclitic interrogative *-ne* occurs 33 times in *conf.*, 17 in the form of *nonne* (then always in first position in phrase or clause); in every case save this one, the word to which *-ne* is attached stands first in the *caesum* or *membrum* (twice the word is *ergone*: 4. 14. 22, 10. 37. 61).

refundis OCD Maur. Skut. Ver. : *refundes* GS Knöll. Present time is required by 'restat', 'imples', and the context.

quoquam CDO^1S^1 edd. : *a quoquam* GO^2S^2.

effunderis: Acts 2: 17, 18, 'effundam de spiritu meo': Peter's first speech, quoting Joel 2: 28–9 (Vg.), 'effundam spiritum meum super omnem carnem; . . . (29) in diebus illis effundam spiritum meum'; in the divine voice also elsewhere in OT, e.g. Is. 44: 3, 'effundam spiritum meum'.

erigis: Ps. 145: 8, 'dominus erigit elisos'; *en. Ps.* 145. 17, 'quare elisi sunt? quia erecti erant. quare eriguntur? quia humilati sunt. cecidit atque elisus est

Adam: ille cecidit, Christus descendit. quare descendit qui non cecidit, nisi ut levaretur qui cecidit?' Cf. 9. 10. 24 (Ostia), 'erigentes nos ardentiore affectu in idipsum'.

conligis nos: Is. 11: 12, 'et levabit signum in nationes et congregabit profugos Israhel et dispersos Israhel conliget.' (The context is messianic: cf. 11: 1 ff., 'radix Iesse'.) Ps. 146: 2, 'aedificans Hierusalem dominus, et dispersiones Israhel conligens'; *en. Ps.* 146. 4, 'induit enim se ille mortalitatem carnis, et ibi erat sanguis quo fuso redimeremur. illo sanguine conlegit dispersiones Israel'; *s.* 96. 6. 6, 'a multis curre ad unum, dispersa conlige in unum.' The notion recurs in *conf.*: cf. 1. 4. 4, 2. 1. 1, 'conligens me a dispersione' (sim. at 10. 11. 18, 12. 16. 23), 10. 29. 40, 'per continentiam quippe conligimur et redigimur in unum, a quo in multa defluximus', 10. 40. 65. The scriptural language has a neo-Platonic correlate (e.g. Plot. 4. 8. 4. 14–15, [ψύχη] *μέρος γενομένη μονοῦταί τε καὶ ἀσθενεῖ καὶ πολυπραγμονεῖ καὶ πρὸς μέρος βλέπει*; Plot. 6. 9. 1. 1–14, *πάντα τὰ ὄντα τῷ ἑνί ἐστιν ὄντα. . . . καὶ δὴ καὶ τὰ τῶν φυτῶν καὶ ζῴων σώματα ἓν ὄντα ἓκαστα εἰ φεύγοι τὸ ἓν εἰς πλῆθος θρυπτόμενα, τὴν οὐσίαν αὑτῶν, ἣν εἶχεν, ἀπώλεσεν οὐκέτι ὄντα ἃ ἦν*), but neither scripture nor Plotinus is ever a wholly sufficient explanation for such expressions.

(1) Though it may seem self-evident that A. found such notions in his neo-Platonic sources, then sought out the scriptural warrants that would support them, the process is likely to have been more complicated. What he found in the neo-Platonists that appealed to him had some correspondence to what he knew and thought before ever he read the *platonicorum libri*, for he did not come to those texts in any specially naïve or untaught way. What he selected of the neo-Platonists to retain by the time of writing *conf.* was further influenced by what of their doctrine he thought compatible with (better: thought to be a reflection of) Christian doctrine. Moderns find his mature doctrine (that Christian teaching is the antecedent and lucid whole, neo-Platonism the derivative and imperfect reflection) to be the reverse of what we expect, and we now portray an A. who manufactured Christian doctrines from neo-Platonic cloth: the argument may perhaps be sustained, but we should never forget that to A. it seemed otherwise.

(2) Here, moreover, a doctrine of dispersion of divine being into material fragments, and the hope of restoration of wholeness, was central to the Manichees: *ep.* 236. 2, 'animas non solum hominum sed etiam pecorum de dei esse substantia et omnino partes dei esse arbitrantur. deum denique bonum et verum dicunt cum tenebrarum gente pugnasse et partem suam tenebrarum principibus miscuisse eamque toto mundo inquinatam et ligatam per cibos electorum suorum et per solem ac lunam purgari asseverant et, quod purgari de ipsa dei parte non potuerit, in fine saeculi aeterno ac poenali vinculo conligari.' The coincidence points to a late antique habit of thought that perceived the world-as-experienced as a place

of shards and fragments, and supplemented that perception with a yearning for wholeness. A. knew Manichean, Platonic, and Christian forms of that perception and that yearning and chose to use Christian ones here, without having to abjure echoes of other forms.

ubique totus: Again at 3. 7. 12, 6. 3. 4 ('ubique totus es'), 12. 2. 2. For the phrase in A., see du Roy, 469–70, at 470: 'Chez Augustin en tout cas, ce thème est certainement d'origine plotinienne et provient sans doute de *Enn.* 6. 4–5. ... Ce thème n'apparaît donc qu'en 388, au moment du séjour d'Augustin à Rome.' For a more ambitious attempt to find roots in Plotinian thought, see R. J. O'Connell, *REAug* 9 (1963), 1–39. Note also Mart. Cap. 7. 731, predicating 'ubique totum' of the Monad (see on 4. 15. 24), and Macrob. *somn. Scip.* 1. 17. 11, applying 'ubique tota, ubique perfecta' to the 'world soul' in a paragraph expressly citing Plotinus (prob. Plot. 2. 2. 1. 39–40, ἔτι πανταχοῦ οὖσα ἡ ψυχὴ ὅλη; the same idea, less verbally apposite, at Porph. *sent.* 2, ἀσώματα ... πανταχῇ ἐστιν ... ἀμερῶς).

Of the numerous parallels in fourth-century Christian writers cited by du Roy, the most important is Amb., *de fide* 1. 16. 106 (in form not unlike *conf.* 1. 4. 4), 'complens omnia, nusquam ipse confusus, penetrans omnia, nusquam ipse penetrandus, ubique totus eodemque tempore vel in caelo, vel in terris, vel in novissimo maris praesens'. Since A. attributes his doctrine of divine being (and the confutation of the notion that Christianity required anthropomorphism) to Ambrose, the bishop's formulation may have had a more proximate influence than the philosopher's. (Other Christian texts cited by du Roy include: the Latin Origen, Basil, Chrysostom, Arnobius *nat.* 6. 4, 'ubique esse totos', Hil. Pict. *trin.* 2. 6, 'totus ubicumque est' (and 3 × in his Psalm comms.), Hier., *in Eph.* 1. 2, 'totus ubique'; the thought without the *ipsissima verba* is in Paulinus of Nola, *ep.* 24. 1, 'qui unus atque idem ubique terrarum'; *ep.* 30. 2, 'unus atque idem ... ubique'.)

Appearances in A. begin in 388 and continue throughout his career (this list substantially expands du Roy, 469–70, but is probably not complete): *mor.* 1. 11. 19 ('postremo quis me locus ab eius caritate divellet, qui non ubique totus esset si ullo contineretur loco'); *lib. arb.* 2. 14. 38, *c. ep. fund.* 15. 20 ('ubique integram, ubique praesentem'), 16. 16, *div. qu. Simp.* 2. 6, *trin.* 5. 1. 2, 13. 19. 24, and 14. 15. 21, *Gn. litt.* 11. 33. 43, *Io. ev. tr.* 1. 8, 31. 9, 34. 6, 78. 1, *civ.* 1. 29, 7. 30, 11. 5, 11. 10, 16. 5, and 22. 29 (quoting Jer. 23: 24: see on 1. 2. 2 above), *qu. hept.* 2. 151, 5. 10, *Dulc. qu.* 8. 1, *haer.* 86, *epp.* 92. 3, 118. 4. 23 ('qui enim didicerit deum non distendi aut diffundi per locos neque finitos neque infinitos, quasi in aliqua parte maior sit in aliqua minor, sed *totum ubique* esse praesentem'), 120. 3. 14–17 (3 ×), 137. 2. 4–7 (2 ×), 137. 3. 12, 140. 3. 6, 147. 12. 29, 148. 1. 1–3 (4 ×), 155. 4. 13 ('qui ubique praesens est et ubique totus, non pedibus licet ire licet sed moribus': on 'non pedibus', see on 1. 18. 28), 162. 9, 166. 2. 4, 187. 4. 14, 187. 5. 17, 187. 6. 18, 187. 8. 29, 187. 12. 35, 187. 13. 38, and 187. 13. 41, *ss.* 12. 12. 12, 52. 5. 15, 53. 8. 7, 120. 2–3

(3 ×: here it is the *Verbum* that is *ubique totum*), 264. 4, 277. 13. 13 (2 ×), 277. 19. 18, 341. 2. 2, and 377. 1, *s. Mai* 158. 4, *en. Ps.* 18. en. 2. 3, 67. 7, 99. 5.

1. 4. 4

Invocation at last: A. calling on his God to give power to his speech. Once again inquiry finds its answer in a scriptural text—again a rhetorical question.

The paragraph is a tissue of paradoxes, with a submerged polemical purpose. The commonest antitheses are between the apparent mutability and this-worldly action of God and the immutability and otherwordliness that orthodoxy proclaims. One effect of asserting those pairs of opposites is to rule out Manichean criticism of the God of the Old Testament. For the criticism as A. saw it, cf. *mor.* 1. 10. 16, 'desinite errare, non colimus paenitentem deum, non invidum, non indigum, non crudelem, non quaerentem de hominum vel pecorum sanguine voluptatem, non cui flagitia et scelera placeant, non possessionem suam terrae quadam particula terminantem. in has enim atque huiusmodi nugas graviter copioseque invehi soletis.'

For the form of the paragraph, cf. Amb., *de fide* 1. 16. 106 (quoted above on 1. 3. 3, 'ubique totus'). Detailed discussion of this paragraph by W. Simon, *Wissenschaft und Weisheit* 45 (1982), 130–57; he collects many more biblical parallels for the expressions here, confining himself less strictly to specific verbal echoes.

es CDGO Maur. Ver. : *est* S Knöll Skut. For the comma after 'ergo', see B. Löfstedt, *Symb. Osl.* 56 (1981), 105.

quis enim dominus: Ps. 17: 32, 'quoniam quis deus, praeter dominum? et quis deus praeter deum nostrum?' The MSS are unanimous in favour of 'dominus' against the Ps.-text.

summe, optime . . .: Cf. Wisd. 7: 22, 'sanctus, unicus, multiplex, subtilis, disertus, mobilis, incoinquinatus, certus, suavis, amans bonum, acutus'. See below on 'innovans omnia'.

summe: 'furthest above', viewed from beneath, not from afar (contrast 'altissimus': Ps. 7: 18 and often elsewhere in OT); 6. 3. 4, 'altissime et proxime'.

omnipotentissime: Language here is pressed beyond its own extremes; *omnipotens* strictly should have no comparative or superlative, but it is used, e.g. by A. at *civ.* 21. 9, 'per miraculum omnipotentissimi creatoris', and by Macrob., *somn. Scip.* 1. 17. 12, 'deus ille omnipotentissimus'.

misericordissime et iustissime: Paradox is another way beyond the limits of language. Here, as often, it has scriptural warrant: Ps. 114: 5, 'misericors dominus et iustus, et deus noster miseretur'; *en. Ps.* 114. 5, 'misericors primum, quia inclinavit aurem suam mihi; et nesciebam propinquasse aurem dei mei ori meo, nisi per illos speciosos pedes excitarer, ut invocarem; quis enim eum invocavit, nisi quem ipse prior vocavit? hinc ergo primum misericors. iustus autem, quia flagellat.' Cf. also *c. adv. leg.* 1. 20. 40, quoted below on 'paenitet te'.

Though the comparative and superlative of *misericors* are not unexampled elsewhere, A. is noticeably fond of the superlative in particular. *TLL* 8. 1128 reports the comparative once each in Plautus and Cicero as well as ps.-Quintilian and 3 × in *civ.*; the superlative is only attested from A. 'saepe' (56 ×; in *conf.* at 6. 5. 7, 6. 8. 13, 9. 2. 4), ps.-Amb. *s.* 17. 2, and Leo Magnus, *s.* 2. 2.

secretissime et praesentissime: 'Most hidden away and most at hand'. Not merely facts about place, but attributions of mystery and benevolence; cf. 6. 3. 4, 'secretissime et praesentissime', *quant. an.* 34. 77, 'quo nihil sit *secretius*, nihil *praesentius*', and *civ.* 1. 29, 'deus meus ubique praesens, ubique totus [see on 1. 3. 3], nusquam inclusus, qui possit adesse secretus, abesse non motus'. Both the remoteness and the presence of God are evoked later, in approximately equal proportions. *secretus*: 1. 18. 29 ('quam tu secretus es'), 5. 6. 11, 9. 7. 16 ('in thesauro secreti tui'), 10. 42. 67, 10. 43. 68, 11. 2. 3, 11. 31. 41. *praesens*: 5. 2. 2, 5. 8. 14 ('et praesentissima in nos misericordia tua'), 5. 9. 16 ('tu autem ubique praesens'), 7. 1. 2, 9. 8. 18, 9. 9. 22, 12. 11. 12.

pulcherrime: See on 10. 27. 38 ('pulchritudo').

incomprehensibilis: Jn. 1: 5, 'et tenebrae eum non comprehenderunt'; Jer. 32: 19, 'magnus consilio, et incomprehensibilis cogitatu'; Rom. 11: 33, 'quam incomprehensibilia sunt iudicia eius!'

G–M: 'A's rhetorical ingenuity finds play in the series of antitheses which follow.' The matter is more serious than that. Here it is the elusiveness—the incomprehensibility—of God that reduces A. to the rhetoric of paradox (see *en. Ps.* 146. 11, quoted on 1. 1. 1, 'magna virtus tua').

immutabilis: The immutability of God is a principle firmly held, and often stated, by A. at all periods—even, to hear him tell it, in his Manichean phase; cf. on 1. 6. 10, 'non mutaris', and on 7. 1. 1, 'incommutabilem'. *en. Ps.* 43. 5, '"tu es ipse rex meus et deus meus." tu es ipse, non enim mutatus es. tempora mutata video, creator temporum non mutatur.'

innovans omnia: Wisd. 7: 27, 'in se ipsa manens innovat omnia' (cf. *trin.* 2. 8. 14, 'eadem quippe incommutabilitas eius commemorata est ut diceretur, "in se ipsa manens innovat omnia"'). Also at 7. 11. 17, 9. 10. 24, and often echoed and cited, e.g. *lib. arb.* 2. 17. 45, 'conficitur itaque ut corpus et animus forma quadam incommutabili et semper manente formentur.'

in vetustatem perducens: Job 9: 5, 'qui in vetustatem perducit montes et nesciunt'; *adn. Iob* on 9: 5, 'qui in vetustatem perducit montes et nesciunt: id est ad infirmitatem, ut est, "inveteravi in omnibus inimicis meis" [Ps. 6: 8].'

Cf. Gal. 6: 15, 'nova creatura', et sim. For the substitution of *superbos* for *montes*, *en. Ps.* 143. 12, 'montes superbos' (see on 8. 2. 4).

quietus: 13. 38. 53, 'tu autem bonum *nullo indigens* [cf. 'et non egens' here] bono semper quietus es, quoniam tua quies tu ipse es'; cf. 13. 36. 51.

conligens: See on 1. 3. 3, 'conligis nos'.

non egens: Acts 17: 25 (Paul on the Areopagus), 'nec manibus colitur indigens aliquo, cum ipse det omnibus vitam, et inspirationem, et omnia.' Divine self-sufficiency is classical from Plato *Tim.* 29 D on (cf. Plot. 5. 4. 1); cf. 13. 4. 5, 'non ex indigentia', and see on 13. 8. 9.

implens: Cf. 1. 3. 3.

zelas . . . paenitet . . . irasceris: *c. adv. leg.* 1. 20. 40, 'non enim sicut hominem *paenitet* deum; sed sicut deum, quemadmodum non sicut homo *irascitur* nec sicut homo miseretur nec sicut homo *zelat*, sed omnia sicut deus. paenitentia dei non est post errorem; ira dei non habet perturbati animi ardorem; misericordia dei non habet compatientis miserum cor, unde in latina lingua nomen accepit; zelus dei non habet mentis livorem. . . . [see quotation below on 'diximus'] . . . sic ergo quando eum paenitet, non mutatur et mutat; sicut quando irascitur, non movetur et vindicat; et quando miseretur, non dolet et liberat; et quando zelat, non cruciatur et cruciat.' The parallel between the present passage of *conf.* and the quotation (from 420) suggests that the emotions predicated of God, which do not seem to occur together in any scriptural passage, were grouped this way in controversy known to A. before he wrote *conf.*, probably from Manichee days. Cf. the following scriptural occurrences (samples only):

zelas: Least common: in Vg. only at Joel 2: 18, 'zelatus est dominus,' Zach. 1: 14, 8: 2, 'zelatus sum,' Exod. 20: 5, 34: 14, 'zelotes' (cf. *c. Adim.* 7. 1 on Ex. 20: 5: the metaphor A. uses is that of the husband jealous of his wife's chastity (cf. also *c. Adim.* 11. 1 and 13. 1–2); hence *securus* as the antithesis).

paenitet: Gn. 6: 6–7, 'paenituit eum . . . paenitet enim me fecisse eos' (at *loc. hept.* 1. 14 (quoted on 2. 1. 1, 'recogitationis'), A. thinks that the Gk. of that passage [διενοήθη] had better been translated 'recogitavit'); 1 Kgs. 15: 35, dominum paenitebat quod constituisset eum regem super Israhel.' On the appropriateness of *paenitentia* of God, see also *div. qu. Simp.* 2. 2. 1–5, arguing (esp. at 2. 2. 1) from ineffability.

irasceris: Exod. 4: 14, 'iratus dominus in Moysen'; on which, *qu. hept.* 2. 10, 'quemadmodum possit intellegi irascens deus, quia non sicut homo per inrationabilem perturbationem, per omnia tenendum est, ubi tale aliquid scriptura dicit, ne de hoc eadem saepe dicenda sint.' Deut. 9: 20, 'adversum Aaron quoque vehementer iratus'.

usuras exigis: Mt. 25: 14–30, the parable of the talents; for the conventional expression, cf. Amb. *off.* 1. 32. 168, 'si ipse non exigas avarus'.

supererogatur: Cf. Lk. 10: 35, 'quodcumque supererogaveris, ego cum rediero reddam tibi.'

et quis habet: 1 Cor. 4: 7, 'quid autem habes quod non accepisti?' Cf. *gr. et lib. arb.* 6. 15, 'dona sua coronat deus, non merita tua; si tibi a te ipso, non ab illo sunt merita tua.'

reddis CGO Maur. Skut. Ver. : *reddens* S Knöll. Cf. 9. 13. 36, 'dimissa debita sua ab eo cui nemo reddet quod pro nobis non debens reddit'; Mt. 6: 12, 'dimitte nobis debita nostra.'

donas CDGOS[1] Maur. Skut. Ver. : *donans* S[2] Knöll.

diximus: i.e. in this very paragraph. Invocation is inadequate. *Io. ev. tr.* 13. 5, 'omnia possunt dici de deo et *nihil digne dicitur de deo*'; *c. adv. leg.* 1. 20. 40 (ellipsis from passage quoted above: the 'adversary of the Law' shared with the Manichees a disdain for the Old Testament), 'unde iste qui tam *loquaciter* exagitavit dei paenitentiam, discat primo *vix inveniri aliquid quod digne de deo dici possit*, sed plurima et paene omnia nos de illo dicere loquendi necessitate, quae magis homines ex hominibus metiuntur. . . . propter quod providentissime scriptura divina de illo ineffabili loquens ad quaedam etiam verba descendit, quae iam hominibus et ipsis carnalibus, cum de deo sermo est, videntur absurda et indigna: ut, cum timentur ista sic accipi quomodo in hominibus solet et discutiuntur quemadmodum de deo bene accipi possint, ibi discatur etiam illa quae humanis sensibus in eisdem scripturis deo digna videbantur, non secundum hominum mores intellegi vel credi oportere.' Sim. at *doctr. chr.* 1. 6. 6 ('diximusne aliquid et sonuimus aliquid dignum deo?'), *Gn. c. man.* 1. 8. 14 ('nihil enim de deo digne dici potest'), *s.* 117. 10. 15 ('illud verbum, de quo tanta et tam multa diximus, et nihil dignum diximus'), *s.* 341. 7. 9 ('quid ergo de deo digne dicitur? . . . qui autem et ista transcenderit et de deo, quantum homini conceditur, digne cogitare coeperit, inveniet silentium ineffabili cordis voce laudandum'). For concise summary of an important subject, R. Flores, *Aug. Stud.* 6 (1975), 1–13; sim. in M. Colish, *The Mirror of Language* (rev. edn. Lincoln, Nebr., 1983), 17–48.

vita mea: Jn. 11: 25, 'ego sum resurrectio et vita'; Jn. 14: 6, 'ego sum via veritas et vita'. As vocative to God again at 1. 13. 20, 1. 17. 27, 3. 6. 10, 7. 1. 2, 9. 13. 35, 10. 17. 26, 12. 25. 34.

dulcedo mea sancta: See on 1. 6. 9. 'dulcedo mea'.

tacentibus: Cf. Ps. 31: 3, 'quoniam tacui, inveteraverunt ossa mea'; *en. Ps.* 31. en. 1. 3 paraphrases: 'quoniam non protuli ore confessionem ad salutem'; *en. Ps.* 31. en. 2. 13, 'tacuit confessionem, clamavit praesumptionem. "tacui," enim dixit, "non sum confessus".'

quoniam loquaces muti sunt: 'For though they say much [about other things] they are mute [in all that matters]';[10] *BA*: 'Et malheur à ceux qui se

[10] The view taken here was also that expressed, forcefully, by Wittgenstein (quoted by M. O'C. Darcy, in *Ludwig Wittgenstein: Personal Recollections*, ed. R. Rhees (Totowa, NJ, 1981), 105). The sentence 'et vae . . . muti sunt' was used as a dedicatory inscription in a copy of his own *Stunden-Buch* by Rilke in 1912/13, at a time when he had been translating this part of Book 1 (J. Ferreiro, *Rilke y San Agustin* (Madrid, 1966), 7).

taisent sur toi puisque, bavards, ils sont muets.' The oxymoron (*mutus* offers the natural antonym for *loquax* at *civ.* 4. 19, 'Fortuna loquax et muta Felicitas') has puzzled. The preceding sentence ('et quid diximus . . . cum de te dicit') treats those who, like A., do speak of God; this sentence refers to those who say nothing at all *of God*—or better, who utter many words but succeed in saying nothing that genuinely speaks of the God who is. The phrase recurs at 7. 2. 3, 'sat erat mihi . . . adversus . . . loquaces mutos [sc. manichaeos] . . . illud quod . . . proponi solebat'. Its use there does not suggest that the persons referred to in the present passage are well-intentioned. The Manichees are models of *curiositas* (see on 3. 6. 10); at 7. 6. 8, there is a parallel case of *curiosi* (astrologers) who rattle on, their language out of control, so much so that they even foretell the future correctly by pure chance sometimes, and cf. *civ.* 18. 24, 'philosophorum subtilis et acuta loquacitas'; *en. Ps.* 144. 7, '*eloquentes muti*, laudantes creaturam, obliviscentes creatorem'. Cf. 5. 7. 12, where 'loquaces . . . dicentes nihil' also refers to the Manichees; cf. also 8. 10. 22, 'vaniloqui et mentis seductores'—of Manichees (sim. association at *civ.* 5. 26, 'quid est loquacius vanitate?': *vanitas* linked with *loquacitas* at least 17 × in A.'s works); they are the frequent, but not exclusive, targets of the same reproach often elsewhere (e.g. *Gn. c. man.* 1. 16. 26–17. 27, 2. 25. 38 ('nulli enim loquacius'); of himself as a Manichee at *persev.* 2. 55, 'quam [fidem catholicam] miserrima et furiosissima *loquacitate* vastabam').

An alternative interpretation is not uncommon: Pusey: 'Yet woe to him that speaketh not, since mute are even the most eloquent.' J. M. Campbell and M. R. P. McGuire, *The Confessions of Saint Augustine* (New York, 1931, ad loc.: 'even the most eloquent in his praise are as if mute.' Others equivocate (e.g. G–M: 'since those who say most are no better than dumb'). W. Simon, *Wissenschaft und Weisheit* 45 (1982), 156–7, turns the expression on its head by seeing in 'loquaces muti sunt' allusion to Gospel passages where the mute are made to speak (Pellegrino/Carena suggest the same link), but he does not discuss the force of *loquax* in A.

Pizzolato, *Lectio I–II*, 26 and 26 n. 95, wavers between two translations: 'dal momento che anche chi è muto ne parla' (silence speaks volumes—this the version he mainly prefers: but if our silence praises God anyway, then why worry whether we speak or not?), and 'dal momento che anche chi ne parla è muto' (speaking much says nothing). Neither reading takes into account the pejorative quality of *loquaces*. Pizzolato also cites several parallel passages, corroborating A.'s insistence on the ineffability of God, but not pertinent here: *doctr. chr.* 4. 19. 38, *en. Ps.* 32. en. 2. s. 1. 8, 134. 11, 148. 3, ('ipsum voluisse laudasse est. non enim verba a te quaerit deus, sed cor'). To the same end, see also *en. Ps.* 102. 8, 'ergo si non possumus dicere, et prae gaudio non permittitur tacere, nec loquamur, nec taceamus. quid ergo faciamus, non loquentes et non tacentes?'

If *loquacitas* is most at this period a fault charged to the Manichees, it becomes for A. the pre-eminent flaw of Julian's character: from the first paragraph of *c. Iul. imp.*, the accusation occurs no fewer than 78 times in that work. Thus faults of speech were at all times ones that A. felt strongly about, and feared to fall into (note also late in life, *retr.* pr. 2 quoting Prov. 10: 19, 'ex multiloquio non effugies peccatum' as a caution prior to reviewing his own abundant literary output).[11]

1. 5. 5

adquiescere: Cf. 1. 1. 1, 'inquietum'; 13. 38. 53.

inebries: *en. Ps.* 22. 5, 'inebrians . . . oblivionem praestans priorum vanarum delectationum'; see further on 5. 13. 23, 'sobriam vini ebrietatem'.

amplectar: See on 10. 6. 8.

quid mihi es? . . . quid tibi sum ipse: *sol.* 1. 1. 7, '(A.) ecce oravi deum. (Ratio) quid ergo scire vis? (A.) haec ipsa omnia quae oravi. (Ratio) breviter ea conlige. (A.) *deum et animam scire cupio*. (Ratio) *nihilne plus?* (A.) *nihil omnino*'; *sol*. 2. 1. 1, 'deus semper idem, *noverim me, noverim te*.'

miserere ut loquar: *cont.* 1. 1, 'nam qui eam [continentiam] donat continentibus fidelibus suis, ipse dat sermonem de illa loquentibus ministris suis'; *s. Caill.* 2. 5. 1, 'inde loquamur quod ipse donabit.' (See on 6. 3. 3 for other expressions of the way a bishop preaches not his own words but those given him by God for the occasion.) A.'s regular recourse to scriptural language is a sign of a deliberate attempt to accept—or to appropriate—the divine language for his own discourse.

miserere . . . irascaris: *en. Ps.* 59. 3, 'merito "iratus es, et misertus es nostri." non misertus nisi irascereris. destruxisti nos in ira tua; sed ira tua in vetustatem nostram fuit, ut destrueretur vetustas. sed misertus es nostri propter novitatem.'

per miserationes tuas: Here, it is not A. that confesses, but the merciful acts of God in A. speaking for him: Ps. 106: 8, 'confiteantur domino miserationes eius.' The verse occurs 4 × in the Psalm as A. reads it (verses 8, 15, 21, 31); he makes this the pretext for an extended discussion of four kinds of temptation (liberation from which leads to this repeated prayer: Solignac, *Lectio X–XIII*, 13, describes parallels to *conf.*): *en. Ps.* 106. 8, 'prima temptatio erroris et famis verbi [e.g. of one completely untouched by the word of God—A. when young]; secunda difficultatis vincendarum concupiscentiarum [e.g. A. in Bk. 8]; tertia taedii atque fastidii [cf. *en. Ps.* 106. 6, 'iam non te illicit adulterium, nec tamen delectat dei verbum'—an anticipation of monastic *accidie*]; quarta tempestatis et periculorum in gubernandis ecclesiis [cf. 10. 36. 59 ff.]; et in

[11] Eugippius implicitly reads the passage in yet another way by presenting: 'quoniam loquaces muti *facti* sunt'.

his omnibus exclamationes, et liberationes, et miserationum dei confessiones.' Elsewhere, with variations, at 1. 15. 24, 5. 10. 20, 7. 6. 8.

ecce: 115 × in *conf.*, evenly distributed; less common in his other works of sustained exposition (over a comparable amount of text in *civ.*, 17 ×, in *trin.*, 24 ×), but common in *en. Ps.* (85 × in one sample of comparable length). A spoken punctuation mark, adding emphasis.

aures cordis: Cf. 1. 20. 31, 'interiore sensu' (*lib. arb.* 2. 4. 10, 'arbitror etiam illud esse manifestum, sensum illum interiorem non ea tantum sentire quae accepit a quinque sensibus corporis, sed etiam ipsos ab eo sentiri'); see on 10. 6. 9–10 for the *interior homo* and *interior sensus*. The connection, giving reality to what began as metaphor, is already implied in 'amplectar'.

dic animae meae: Ps. 34: 3, 'dic animae meae, salus tua ego sum'; *en. Ps.* 34. s. 1. 5, 'cum dixeris animae meae, "salus tua ego sum", iuste vivet [anima], ut neminem in adiutorium praeter te quaeram.' A. seeks divine help to speak; to signify his intention of accepting that help, he cites scripture; the scripture he cites contains the word of another man asking God to speak to him. The verse recurs significantly at 10. 4. 6, where it marks another transition from hearing to independent speech. On this verse in *conf.*, see M. Pellegrino, *Augustinianum* 7 (1967), 221–57.

noli abscondere: Ex. 33: 20 (on Sinai), 'non poteris videre faciem meam: non enim videbit me homo et vivet'; *en. Ps.* 12. 1, 'avertere autem deus faciem dicitur, dum non dat animae notitiam sui, quae adhuc purum mentis oculum non habet'; *en. Ps.* 9. 5, '"a facie" dei, id est a cognitione dei'. See on 1. 18. 28, 'quaesivi vultum tuum' (Ps. 26: 9). On *facies*, cf. on 1. 18. 28.

moriar, ne moriar, ut eam videam: *s.* 231. 3. 3, 'qui autem nondum mortuus est, nec resurrexit, male adhuc vivit: et si male vivit, non vivit: moriatur, ne moriatur. quid est, moriatur, ne moriatur? mutetur, ne damnetur. . . . "mortui enim estis, et vita vestra abscondita est cum Christo in deo" [Col. 3: 3].' That passage is decisive against misreadings, e.g. G–M: 'Lit. I would die to see it, lest I die (by not seeing it).' Pusey's note is closer to the truth ('i.e., Let me see the face of God, though I die, since if I see it not, but it be turned away, I must needs die, and that "the second death"'); the alternative punctuation of G. Wijdeveld, *Vig. Chr.* 10 (1956), 229–31 (as already in an old note by J. le Clercq (1657–1736) reprinted at *PL* 47. 204), is likewise excluded.

1. 5. 6

domus animae meae: *en. Ps.* 100. 4, 'domus enim nostra interior, cor nostrum est'; cf. Is. 49: 20, 'angustus est mihi locus, fac spatium mihi ut habitem'; Mt. 8: 8, 'non sum dignus ut intres sub tectum meum.' In a different sense, but not long before *conf.*, cf. *ep.* 27. 1 (to Paulinus of Nola), 'quo modo ergo non doleam, quod nondum faciem tuam novi, hoc est domum animae tuae, quam sicut meam novi?'

For metaphor introduced as epexegetic genitive, see Verheijen, *Eloquentia Pedisequa*, 135–6, and cf. 1. 7. 12, 'oblivionis tenebrae'; 1. 16. 26, 'vinum erroris'; also common in biblical Latin, e.g. Ps. 59: 5, 'vinum compunctionis', Job 3: 5, 'tenebrae mortis'.

quo: Wijdeveld (*REAug* 5 (1959), 33), proposed repunctuation: 'angusta est domus animae meae. Quo venias ad eam, dilatetur abs te.' Against that punctuation is the arrangement of the following two sentences ('refice eam' parallels 'dilatetur abs te', while 'quae offendant oculos tuos' offers a parallel for the suordinate clause) and the construction itself: in *conf.*, quo introduces a purpose clause without comparative (LHS 679–80) at least three times (6. 13. 23, 8. 3. 7, 9. 12. 32), but in those cases the *quo*-clause comes second in the sentence. It is not easy to find parallels for the construction and arrangement Wijdeveld proposes.

ruinosa: Ezech. 36: 10, 'et habitabuntur civitates, et ruinosa instaurabuntur'; Ezech. 36: 33, 'et instauravero ruinosa.' At *s.* 86. 1. 1, 'domus terrena ruinosa est: domus caelestis aeterna est' offers a metaphor for the contrast between the mortal body and the resurrected body. The image in different context: *s.* 311. 17. 14, 'sed mala, inquis, fiunt in mundo, aspera immunda, odiosa. foedus est, non ametur. ecce talis est, et sic amatur. ruinosa est domus, et piget migrare.'

fateor: first occurrence of *fateor*; *confiteor* not until 1. 6. 9.

ab occultis: Ps. 18: 13–14, 'delicta quis intellegit? ab occultis meis munda me, domine, et ab alienis parce servo tuo'; *lib. arb.* 3. 10. 29, 'cum enim duae sint origines peccatorum, una spontanea cogitatione, alia persuasione alterius, quo pertinere arbitror quod propheta dicit, "ab occultis meis munda me, domine, et ab alienis parce servo tuo."' Identical interpretation with noteworthy continuity of opinion at *en. Ps.* 18. en. 1. 14 (392) and *en. Ps.* 18. en. 2. 13 (411/12). See also on 2. 9. 17, 'delicta quis intellegit?'

propter quod: speech the earnest of faith: Ps. 115: 10, 'credidi propter quod locutus sum, ego autem humilatus sum nimis'; 2 Cor. 4: 13, 'habentes autem eundem spiritum fidei, sicut scriptum est, "credidi propter quod locutus sum", et nos credimus, propter quod et loquimur'; *en. Ps.* 115. 2, 'hoc est, perfecte credidi. non enim perfecte credunt qui quod credunt loqui nolunt.' Cf. 11. 22. 28. The punctuation is from B. Löfstedt, *Symb. Osl.* 56 (1981), 105, with evidence on position of vocatives.

tu scis: Knauer, 76–7, lists the nineteen occurrences of 'tu scis', an expression with strong biblical and Psalmish overtones. The phrase emphasizes the personal tone, and is markedly 'confessional'. God could tell this story better than A., if only he would.

prolocutus sum: The rare verb means 'reveal, expound, expose'; cf. 5. 3. 3, 'proloquar in conspectu dei mei annum illum undetricensimum aetatis meae.' The word is a curio here. It is not infrequent in comedy, but it is entirely unattested in classical prose except in archaic quotations (e.g. Cic.

or. 43. 147, quoting 'pervulgatissimus ille versus, qui vetat "artem pudere proloqui quam factites".' (Sandys ad loc.: '*proloqui* = *profiteri, non dissimulare*'); cf. Cic. *Tusc.* 3. 26. 63 for an instance from Ennius). Varro, *ling. lat.* 6. 56 (Müll.) defines: 'prolocutum [dicimus], cum animo quod habuit, extulit loquendo'; so one would read a line of Ennius cited by Nonius 232. 24 (Lindsay).

But we should not assume that the word in A. reflects the conclusions of either ancient or modern lexicographers. Note a related word in a work nearly contemporaneous with *conf.*: Amm. Marc.29. 1. 38, 'imperator cognitorum consultationi respondens, sub uno proloquio cunctos iubet occidi'; see also 28. 1. 11, '[imperator] efferatus . . . uno proloquio . . . omnes . . . statuit tormentis affligi.' Better to take the verb as to 'speak judgement against'—here, his own past sins, in 5. 3. 3, a dismal year of spiritual disarray.

In support of that, note that Ps. 31: 5 is clearly cited 5 × in A. with *proloquor* for *pronuntiabo*: *s. Caill.* 2. 11. 5, *s. Denis* 9. 4, *ss.* 323. 3. 4, 352. 1. 2, and *s. Lambot* 10 (reading in every case: 'dixi, *proloquar* adversum me delictum meum domino deo meo, et tu dimisisti impietatem cordis mei'); at *en. Ps.* 31. en. 1. 5 and *en. Ps.* 31. en. 2. 15–16, A. reads with the Ps. Rom. *pronuntiabo*—the closest reading in any of the versions reported by Weber, *Psautier Romaine*, is *eloquar*, in a sixth-century Paris MS and in the Psalt. Mediol.

The word itself occurs in A. 18 ×, sometimes neutrally, sometimes with an overtone of 'accusation' filtered in from the Psalm reading. Neutral readings include *dial.* 4, *cat. rud.* 2. 4, *c. litt. Pet.* 3. 27. 32, *adult. coniug.* 1. 4. 4, and *s.* 37. 1. 1. More interesting are those that seem to parallel the present use: *adn. Iob* on 19: 7, '"ecce rideo opprobrio et non loquor": utilitatem dicit *confessionis*, quia, si vellet ridere peccatum suum et non illud *proloqui*, clamaret et non exaudiretur'; also *c. Cresc.* 4. 36. 43, *c. Iul. imp.* 1. 21. See also on 5. 3. 3.

non iudicio contendo tecum: Jer. 2: 29, 'utquid vultis mecum iudicio contendere?' *en. Ps.* 42. 7, 'noli ergo iudicio contendere; da operam esse iustus; et quantumcumque fueris, confitere te peccatorem.' Cf. Job 9: 3 (VL), 'si enim velit iudicio contendere cum eo, non respondebit ei unum de mille.'

qui veritas es: Jn. 14: 6, 'ego sum via, et veritas, et vita.' The nouns of that sentence, esp. *via* and *veritas*, are extremely common in *conf.*. Are we to assume that at each occurrence A. means us to think of *via* or *veritas* as another name of the second person of the trinity? Had he so imbibed the teaching of scripture that there is no question of what he 'means us to think' but simply of what he himself habitually thought? On the available evidence, it is hard to rule out the Christological overtones of those words; see on *via* at 7. 7. 11; on *veritas* at 10. 23. 33. But questions of author's intention are difficult (as A. himself says at length: 12. 23. 32) and lead to no verifiable conclusions. Better not to prescribe to these words what they *must* mean, but to allow them at every turn to mean what, on the purely textual evidence of

A.'s writings and without speculation as to his intentions, they certainly *may* have meant. If in this way we arrive at a reading of *conf.* that is more expressly Christ-centred than some would think A. intended, but a reading demonstrably in the spirit of his texts, something of value has been gained.

The line of interpretation suggested here is one A. himself practised, e.g. *en. Ps.* 39. 18, '"veritatem tuam et salutare tuum dixi" [Ps. 39: 11]. Christum tuum dixi, hoc est, "veritatem tuam et salutare tuum dixi." unde veritas Christus? "ego sum veritas." unde salutare ipsius Christus? Simeon agnovit infantem in manibus matris in templo, et dixit: "quoniam viderunt oculi mei salutare tuum".'

et ego nolo: The same sentiment generalized at 10. 23. 33 ('multos expertus sum . . .'); cf. Jas. 1: 22, 'estote autem factores verbi, et non auditores tantum, fallentes vosmetipsos.'

ne mentiatur: Ps. 26: 12, 'et mentita est iniquitas sibi.'

si iniquitates: Ps. 129: 3, 'si iniquitates observaveris, domine, domine, quis sustinebit? ne intres in iudicium cum servo tuo, quia non iustificabitur in conspectu tuo omnis vivens'; *en. Ps.* 129. 2, 'ecce aperuit de quo profundo clamaret. . . . vidit enim prope totam vitam humanam circumlatrari peccatis suis, accusari omnes conscientias cogitationibus suis, non inveniri cor castum praesumens de sua iustitia.'

The sequence of verses here is typical of A.'s habits of scriptural association: in some cases, the connection is substantive (from 'veritas' to 'fallere' to 'mentiatur'), in others verbal (from 'iniquitas' to 'iniquitates'). But notice also that Ps. 129: 3 occurs elsewhere in connection (suggested by the second half of the verse) with Jer. 2: 29 (cited again here in the words immediately preceding), as at *en. Ps.* 42. 7 ('"si iniquitates observaveris domine, . . . omnis vivens." . . . nam ex alio propheta arrogantes et superbos identidem sic obiurgat, "utquid vultis mecum iudicio contendere?"') and at *en. Ps.* 128. 9 ('sed iactas te de meritis tuis, "et vis mecum", inquit deus, "iudicio contendere, intrare mecum ad iudicum"; cum debeas in reatu tuo satisfacere deo, et clamare ad illum, quod clamatur in alio psalmo, "si iniquitates observaveris, domine, domine, quis sustinebit?"').

1. 6. 7

The most sensitive and successful attempt to read A.'s treatment of infancy in the light of modern psychological interpretation is M. Miles, *Jour. Amer. Acad. Rel.* 50 (1982), 349–64. See also B. Shaw, *Past and Present*, 115 (May, 1987), 3–51.

The emphasis in *conf.* is on the inclination to irrational conduct already in infants; later A. chose to emphasize their suffering as a sign of the burden of original sin. Non-polemically, at *trin.* 14. 5. 7 he inquires whether the mind of an infant can be said to know itself—unlikely, for they are so eager to gobble up the sensible world that if you leave them with a night-light where they can see it, they will develop a squint from staring at it all night long. Polemically, a raft of texts from the anti-Pelagian period and especially against Julian can be instanced, e.g. *c. Iul. imp.* 2. 22, 'catholica illa est, quae ostendit iustum deum in tot ac tantis poenis et cruciatibus parvulorum'. See also *c. Iul.* 4. 16. 83, *c. Iul. imp.* 3. 48, 3. 154 (reminiscent of *conf.*: 'natosque ipsos omnes flentes, sero ridentes, serius loquentes et hoc balbutientes, in scholas postea duci ut litteras discant, sub loris ferulis virgisque plorantes, pro varietatibus ingeniorum distributa varietate poenarum'), and 3. 187. At *c. Iul.* 4. 12. 60, A. is careful to show that his melancholy view finds precedent in Cicero's *de republica*; on the classical view and its Christian permutations, see A. Goulon, *REAug* 18 (1972), 3–26.

sine me loqui apud misericordiam tuam: Cf. 1. 5. 5, 'miserere ut loquar.' For divine control of recalcitrant speech, cf. Mk. 1: 34, 'non sinebat [Jesus] ea [daemonia] loqui, quoniam sciebant eum' (in the parallel at Lk. 4: 41: 'quia sciebant eum esse Christum').

me terram et cinerem: Gn. 18: 27, 'loquar ad dominum meum, cum sim terra et cinis'; Job 42: 6 (VL), 'ideo despexi memetipsum et distabui et aestimavi me terram et cinerem' (Job's last spoken words: at *civ.* 22. 29, A. wonders whether Job was at this moment seeing God as he really is); *en. Ps.* 147. 24, 'dixit enim quaedam scriptura "despexi meipsum et distabui, et aestimavi me terram et cinerem." haec est humilitas paenitentis. quando Abraham loquitur ad deum suum et quando vult sibi aperiri incendium Sodomorum: "ego", inquit, "terra et cinis sum".' Also Sirach 10: 9, 'quid superbit terra et cinis?', Sirach 17: 31, and Ps. 102: 14, 'memento, quia pulvis sumus.' Also at 7. 8. 12 and 10. 5. 7.

inrisor: At 10. 36. 59 ff., A. acknowledges lingering vulnerability to the temptations of vanity; he presents himself indeed, like 'Jean-Baptiste Clamence' in Camus's *La Chute*, as a man who could hear the laughter behind him even when it was not there. Some confirmation may be found in the frequency and context in *conf.* of real and imagined mockery directed at A., particularly as a spur to *confessio* (as at 4. 1. 1): *First*, A.'s own sensitivity to mockery (even that from God) in the narrative books: 1. 6. 9 ('an inrides me ista quaerentem teque de hoc, quod novi, laudari a me iubes et confiteri me tibi?'), 1. 7. 11, 1. 9. 15, 3. 3. 6, 3. 10. 18, 4. 1. 1, 4. 3. 6, 5. 10. 20, 5. 14. 24, 6. 5. 7, 6. 6. 9, 6. 14. 24, 8. 11. 27, 9. 12. 33, 10. 12. 19, 10. 12. 19. *Second*, his own mockeries past: 3. 10. 18 ('inridebam illos sanctos servos et prophetas tuos. et quid agebam, cum inridebam eos, nisi ut inriderer abs te?'), 4. 4. 8, 6. 7. 12, 7. 6. 10. *Finally*, a revealing passage in which 'mockers' are a ready example of evil undertaken for the gratuitous pleasure of doing evil: 3. 8. 16, 'libido est nocendi . . . sola voluptate alieni mali, sicut spectatores gladiatorum aut inrisores aut inlusores quorumlibet.'

conversus misereberis mei: Ps. 70: 20, 'et conversus vivificasti me. . . . et conversus consolatus es me'; Jer. 12: 15, 'et cum evulsero, convertar et miserebor eorum, et reducam eos.'

vitam mortalem an mortem vitalem: 'in this life that dies, this death that lives.' Cf. Lucr. 3. 869, 'mortalem vitam mors cum immortalis ademit'.

nescio: At every stage of A.'s life, he refused to take a definite position on the origin of the soul, an ambivalence he later compared to the canniness of the sailor avoiding Scylla and Charybdis: *nat. et or. an.* 2. 13. 18. The fundamental doubt is expressed early: *lib. arb.* 3. 21. 59, 'harum autem quattuor de anima sententiarum, [1] utrum de propagine veniant, [2] an in singulis quibusque nascentibus novae fiant, an in corpora nascentium iam alicubi existentes [3] vel mittantur divinitus, [4] vel sua sponte labantur, nullam temere adfirmare oportebit. aut enim nondum ista quaestio a divinorum librorum catholicis tractatoribus pro merito suae obscuritatis et perplexitatis evoluta atque inlustrata est, aut si iam factum est, nondum in manus nostras huiuscemodi litterae pervenerunt.' For his late insistence on his ignorance, *retr.* 1. 1. 3, 'nam quod attinet ad eius originem, qua fit ut sit in corpore, utrum de illo uno sit qui primum creatus est, quando "factus est homo in animam vivam", an similiter fiant singulis singuli, nec tunc sciebam nec adhuc scio.' For the best statement of what he was willing to affirm, dating from not too long after *conf.*, see *Gn. litt.* 7. 28. 43; see generally *Gn. litt.* bks. 7 and 10 (with note at *BA* 49. 537–9 on the significance of the problem for infant baptism) and the later (anti-Pelagian) *nat. et or. an.* For the most authoritative presentation and interpretation of the evidence (including the nuances that emerged in A.'s later works), see O'Daly, 15–20 and 199–203. For the minority position of R. J. O'Connell, who holds that at Cassiciacum, and for some time after, A. inclined to the third or fourth of these views (which on that reading he owed mainly to Plotinus), see his *St. Augustine's Early Theory of Man,* A.D. *386–391* (Cambridge, Mass., 1968), 148–52; the thesis is revised, expanded, and defended with vigour if not circumspection in his *The Origin of the Soul in St. Augustine's Later Works* (New York, 1987). O'Connell's position refuses to acknowledge A.'s willingness to leave such a significant question radically open and *not* feel a philosopher's qualm that his *system* is thus undermined. A. had no system, despite what we try to make of him, but found himself following the teachings of a God who did not always make everything perfectly clear. A. managed, with great difficulty, to make his peace with the uncertainties of that condition: not all his students since have been so successful.

A.'s agnosticism gave him both advantage and disadvantage in polemical situations. To be noncommittal on an essential issue left him open to one kind of criticism, but to have asserted any form of transmission of souls from Adam would have given Pelagians more grounds for attacking A. as a crypto-Manichee (cf. *c. Iul. imp.* 2. 178), and he was also eager himself to avoid taking a position the Manichees could attack (*c. Fort.* 11, 'si quaeris utrum a deo descenderit anima, magna quidem quaestio est; sed sive a deo descendit sive non, illud de anima respondeo non esse deum'). Cassiodorus, *de anima* 9. 22 (*CCSL*), speaks of A.'s refusal to declare himself and describes him as 'religiosissima dubitatione laudandus'.

consolationes: Cf. 11. 2. 2, 'hortamenta tua et omnes terrores tuos et consolationes et gubernationes'; *en. Ps.* 93. 22, 'multi dolores, sed multae consolationes: amara vulnera, sed suavia medicamenta.'

ex quo et in qua: *ex Patricio et in Monnica.*

formasti: At *civ.* 13. 24, A. prefers to render Gn. 2: 7, ἔπλασεν (of the creation of Adam and Eve), by *formavit* rather than *finxit*. See on 12. 3. 3 for *species/forma* and Bk. 12 generally for 'formatio' in creation.

non enim ego memini: 'non . . . memini' again at 1. 6. 8, 1. 6. 10, 1. 7. 11, and 1. 7. 12—all dealing with infancy; see on 6. 5. 7 for the importance of such examples in persuading the adult A. that faith has a place in our scheme of knowledge. At *util. cred.* 12. 26 he argues that *fides* has a place in human affairs because you have to believe your mother for who your father is, and nurses and midwives and servants for who your mother is; sim. at *ep.* 147. 5. On the numerous inadequacies of memory in *conf.*, see the express discussion of forgetfulness at 10. 16. 24 and Courcelle, *Recherches*, 32–5.

lactis humani: For the nursing practices, see A. Rousselle, *Porneia*, 51–8.

mater mea vel nutrices meae: For a less direct way of making a mother's nurture stand for divine generosity, cf. *en. Ps.* 30. en. 2. s. 1. 9, 'sicut enim mater lactans eandem escam, cui sumendae idoneus infans non est, per carnem traicit, et lac infundit . . ., sic dominus sapientiam suam ut lac nobis faceret, carne indutus venit ad nos.'

A.'s mother plays, as many know who never read this work, a large part in his narrative and in his life. We must forgo that knowledge to see the slow development of her character in the narrative. This first allusion to her is noteworthy for its ambivalence: in the mist of infant oblivion, not surely distinguished from the wet nurses she employed. (With younger siblings—at least one sister and one brother: cf. Mandouze, *Pros. chr.* s.v. Monnica—A. could recall clearly other nurses like those who had tended him.) For implicit criticism, cf. *en. Ps.* 49. 27, 'sunt enim matres quae cum pepererint, dant nutricibus; illae quae pepererunt, non fovent filios suos, quia nutriendos dederunt.'

alimentum: L. C. Ferrari, *Aug. Stud.* 9 (1978), 1–14, counts 'no less than one hundred and fifty different words connected with the acts of eating and drinking' in *conf.*

ordinatum affectum: On the trinitarian view of humanity, *affectus* represents the will in action, and as such the part of the human being that corresponds to the third person of the trinity (see on 13. 11. 12); *ordo*, in another of A.'s triads (*modus/species/ordo*: see on 1. 7. 12), represents the action of the same third person of the trinity. *Ordinatus affectus*, then, is carefully phrased to identify the place of the impulse to offer nourishment both in human affairs (whence it arises) and in divine affairs (whence it is governed)—the love of a mother that offers this nurture is human love in harmony with divine love.

salus: Cf. 1. 5. 5 (= Ps. 34: 3), 'salus tua ego sum.'

clamante: In OT common, always of humans (90%+: addressing God); only exception, Mic. 6: 9, 'vox domini ad civitatem clamat.' In Mt./Mk./Lk. of Jesus only on the cross; but frequent in John, as 7: 28, 'clamabat ergo Jesus in templo docens'; cf. Gal. 4: 6, 'misit deus spiritum filii sui in corda vestra clamantem, abba pater.' For the underlying idea, cf. Rom. 1: 20 (see on 7. 9. 14): God gives all things (Gn. 33: 11, 'deus tribuens omnia') and speaks through them.

intus et foris: 34 × and 19 × respectively in *conf.*, signally at 10. 27. 38 (q.v.). It would be possible to ascribe the prominence of this disjunction in A. and his successors to a philosophical influence (comparanda can be found in neo-Platonism) or to a late antique psychic alienation; but both hypotheses are in the end unverifiable. There is no comprehensive treatment of the theme in A.; best is a collection of texts with some comments and a few parallels at Courcelle, *Recherches*, edn. 2, 393–404; for *conf.*, see W. Schmidt-Dengler, *REAug* 14 (1968), 69–89; and there is much in P. Cambronne, 'Recherches sur la structure de l'imaginaire dans les Confessions de saint Augustin' (microfiche thèse, Paris, 1982). See also 'Foris-intus', *Aug.-Lex.* (forthcoming).

tunc: The tension between past and present is marked by the adverbs *tunc* and *nunc*, for whose intricate interplay see J. Morán, *Augustinianum*, 8 (1968), 147–54.

delectationibus: sc. *carnis meae*. The infant is almost an Epicurean: cf. Cic. *fin.* 1. 16. 54, for the appeal of such a view by a proponent as A. would have known it: 'quodsi . . . voluptas autem est sola quae nos vocet ad se et alliciat suapte natura, non potest esse dubium quin id sit summum atque extremum bonorum omnium, beateque vivere nihil aliud sit nisi cum voluptate vivere.' For the licit role of *delectatio*, see on 1. 1. 1; in Bk. 1, *delectare* and *delectatio* are mainly negative and self-indulgent: 1. 9. 15, 1. 15. 24, 1. 16. 26, 1. 19. 30. Anticipation of better things returns near the end of the book at 1. 18. 28 ('sitientem delectationes tuas') and 1. 20. 31 ('veritate delectabar').

1. 6. 8

First consciousness is accompanied by the attempted expression of will, frustrated by the discovery of the distance separating inner from outer worlds. This first attempt to participate in human society (by asserting the autonomous authority of the self) ends in failure and tears. Cic. *fin.* 5. 20. 55, 'omnes veteres philosophi, maxime nostri, ad incunabula accedunt, quod in pueritia facillime se arbitrantur naturae voluntatem posse cognoscere. videmus igitur ut conquiescere ne infantes quidem possint.' Cf. Cic. *Tusc.* 3. 1. 2–2. 3, quoted on 1. 1. 1 above.

ridere: Modern medicine ascribes the apparent smile of a sleeping newborn to flatulence; see on 1. 7. 11. *sol.* 2. 20. 34, ' (Ratio) at si tibi quispiam dicat te post paucos dies risisse quam natus es, non audes dicere falsum esse et si auctor sit cui fides habenda est, non recordaturus, sed crediturus es; totum enim tempus illud validissima tibi oblivione sepultum est.' Sim. at *civ.* 21. 14, buttressed by an anecdote about Zoroaster in the cradle also known from Pliny the elder (cf. *BA* 37. 803–4).

indicatum est: A favoured verb (28 ×) in *conf.*

nec . . . valebant introire: The dilemma recurs at 10. 2. 2–3. 3, fortified there by an echo of 1 Cor. 2: 11, 'nemo scit hominum quid agatur in homine, nisi spiritus hominis, qui in ipso est.'

iactabam membra et voces: See on 8. 12. 28, 'iactabam voces'.

signa: *doctr. chr.* 1. 2. 2, 'omnis doctrina vel rerum est vel signorum, sed res per signa discuntur.' All knowing is interpretation of signs, a risky business: *doctr. chr.* 3. 5. 9, 'ea demum est miserabilis animi servitus, signa pro rebus accipere; et supra creaturam corpoream oculum mentis ad hauriendum aeternum lumen levare non posse' (so cf. Rom. 1: 20: see on 7. 9. 14). In *conf.*, cf. esp. 1. 8. 13, 13. 23. 34–25. 38. The classic study is R. A. Markus, *Phronesis* 2 (1957), 60–83, with further discussion, particularly of Stoic antecedents, by B. D. Jackson, *REAug* 15 (1969), 9–49 (both repr. in Markus, *Augustine* (Garden City, NY, 1972)). Cf. U. Duchrow, *Sprachverständnis und biblisches Hören bei Augustin* (Tübingen, 1965), with a thoroughgoing attempt to show that an inherently suspicious attitude towards language and its powers underlies even A.'s most logocentric doctrines; and Mayer, *Zeichen* (two volumes published, three more projected). Despite the contemporary critical interest, there is no satisfactory study to integrate the Augustinian data with contemporary theory and discussion. The present passage has had an influence arising from its appearance on the first page of Wittgenstein's *Philosophische Untersuchungen*, on which see F. Kerr, *Theology after Wittgenstein* (Oxford, 1986), 38–42, with M. Burnyeat, *Proceedings of the Aristotelian Society*, Supplementary Vol. 1987, 1–24.

vere similia scripsi : *veresimilia* CDGO[1]S Skut. Ver. : *verisimilia* O[2] Maur. Knöll Pell. The question here is not the relation of the *signa* to the truth (verisimilitude was a central concern of the Academics, to whom A. turned later: see on 5. 10. 19), but their resemblance to (and hence effectiveness at communicating) Augustine's velleities. He claims that his first discovery in the realm of human communication was the arbitrariness of signs and the inability of signs in and of themselves to communicate *voluntates* (without some prior convention of agreement). For similar wordplay, cf. *c. acad.* 3. 10. 23, 'istam sententiam Carneades *falsae* esse *similem* doceat.'

non subditis maioribus et liberis non servientibus: The oxymorons underscore the misdirection of the infant's first expressions of *voluntas*.

1. 6. 9

The juxtaposition of present and past time is a leitmotif, recapitulated in full in Bk. 10. The evocation of the divine present (from 'tu autem domine') gives those juxtapositions a context. There is, nevertheless, a perceptible difference in texture between what we may call the narrative juxtaposition (see, for instance, on 1. 6. 7, 'non enim ego memini'), where the tension is between the two versions of the past, as lived and as recalled, and this more substantive contrast, between the past as lived and the present state of the author. One effect of the frequent recourse to this substantive contrast is to keep the reader's awareness rooted in the author's present, not in the narrative past. Some narratives are framed by a narrator's present, where an author sets a scene, presents one or more characters, and creates a mood within which to place the narrative put into the mouth of one of his own characters (even, or especially, when that narrator is identified with the author). The effect of that device may be to make the narrative itself seem yet more authentic: one thinks of such disparate works as the Sherlock Holmes stories, *Moby-Dick*, and *Heart of Darkness*. This book, however, is different, and the analogues in our literature fewer—Dante's *Vita Nuova* comes to mind.

ante omne: Cf. 11. 12. 14, answering those who ask 'quid faciebat deus, antequam faceret caelum et terram?'

instabilium . . .: Cf. 1. 4. 4, 'stabilis . . . immutabilis . . . numquam novus, numquam vetus'; note below, 'misericors misero'. A.'s zeal for distinguishing and defining leads him to risk that God will seem to be no more than the antithesis of creation (or creation only the antithesis of God); these juxtapositions, however, depend on the verbal sense of paradox they evoke to suggest the relationship between God and creature—it is in that relationship (and only in that relationship, A. hints) that both are known.

inrationalium OS Knöll Skut. Ver. : *inrationabilium* CDG Maur. 13. 32. 47 ('inrationabilibus') and 13. 34. 49 ('rationabilem') both imply creatures (or the actions of creatures) possessing the faculty of reason (the same is true of the three occurrences in *en. Ps.*: 42. 7, 77. 3, 82. 3); *rationalis*, by contrast, means more nearly 'formed according to *ratio*,' which suits the context here. (The line between the words is, however, not hard and clear in A.'s use over time.) The words, indeed the notion of *ratio* itself (see on 1. 7. 11), are notoriously un-Ciceronian: reason and reasonableness arise from other philosophical traditions, however indirectly.

sempiternae . . . rationes: On the 'rationes causales', see the lucid and extensive note at *BA* 48. 653–68, with bibliography. The connections to

Plotinian thought are controversial, and investigation has been derailed repeatedly by attempts to juxtapose A.'s teaching to that of evolutionary biology. In A., the idea (as a *tertium quid*) reconciles the immutability and eternity of the creating God with the mutable, temporal (that is to say, sequential) multifariousness of creation. This need is raised most forcefully by the particular form of the Genesis narratives of creation, featuring both simultaneous creation of all things and sequential creation through six days. Cf. *Gn. litt.* 4. 33. 51, 'quapropter quam facilis ei efficacissimus motus est, tam facile deus condidit omnia, quoniam per illam sunt condita, ut hoc, quod nunc videmus temporalibus intervallis ea moveri ad peragenda quae suo cuique generi competunt, ex illis rationibus insitis veniat quas tamquam seminaliter sparsit deus in ictu condendi, cum "dixit, et facta sunt; mandavit et creata sunt" [Ps. 32: 9].'

vidi: The first person perfect is strong and vivid, used 4 × in *conf.* of eyewitness testimony (1. 7. 11 (2 ×), 6. 3. 3); see on 7. 10. 16 for its adaptation to mystical experience.

dulcedo: Not used to address God in Vg., but cf. Ps. 30: 20, 67: 11 ('in dulcedine tua'); used both in address (1. 4. 4, 1. 20. 31, 'dulcedo mea et honor meus et fiducia mea' (see notes there), 2. 1. 1, 'dulcedo non fallax, dulcedo felix et secura') and of abstract nouns and metaphors closely associated with God (8. 1. 2, 9. 6. 14, 10. 3. 4 (sim. at 13. 23. 33), 10. 40. 65, 'introrsus ad nescio quam dulcedinem', 12. 9. 9, 13. 30. 45); identified with divine *iustitia* at *civ.* 21. 24; much earlier, cf. *ep.* 11. 4 (388/91, to Nebridius: *suavitas* associated with the third person of the trinity). Such metaphor is slippery when applied to God: it seems that we know of sweetness in ordinary life, and predicate a greater, more perfect sweetness of our conception of God. A. would hold that this procedure is a necessary consequence of the fall of language, but that in so far as the expression is true and useful it is rather that the sweetness that we know in ordinary life is a pale reflection of the authentic and original sweetness of God. See J. Ziegler, *Dulcedo dei* (= *Alttestamentliche Abhandlungen* 13. 2, Münster 1937); on A., pp. 88–98 (with abundant references).

inrides: See on 1. 6. 7. Cf. Ps. 36: 13, 'dominus autem inridebit eum'; *en. Ps.* 36. s. 2. 2, '"dominus autem inridebit eum." quem? utique peccatorem stridentem super iustum dentibus suis.'

confiteri: First in *conf.* here, where A. faces the possibility that he may have strayed from his purpose.

1. 6. 10

This 'confessional' paragraph is interposed between the more or less objective description of infancy and the (more famous) moral assessment of 1. 7. 11. The meditation on time anticipates Bk. 11 (Knauer, 108 n. 2).

mulierculanum: The diminutive is pejorative; elsewhere in *conf.* only at 6. 14. 24, derisively of the wives and wives-to-be whose objections derailed the plan for a community in philosophical retirement; elsewhere in A. rare and always pejorative; e.g. *en. Ps.* 103. s. 4. 7.

eram ... vivebam ... et signa ... quaerebam: Cf. 1. 20. 31, 'eram ... vivebam atque sentiebam' (note 'sensa' immediately here); as here A. assesses his *infantia*, so at 1. 20. 31, he judges his *pueritia*.

sensa: Not only 'perceptions' but 'ideas, opinions' as matter for speech: *doctr. chr.* 2. 3. 4, 'signorum igitur quibus inter se homines sua sensa communicant'; sim. at *nat. et or. an.* 2. 4. 8. Cic. *de or.* 3. 14. 55, 'sensa mentis et consilia verbis sic explicat'; cf. *de or.* 1. 8. 32, Quint. 8. 5. 1. Cf. 1. 8. 13, 9. 6. 14 ('sensa omnia quae inseruntur ibi [in libro de magistro]').

iam in fine infantiae: A. is mindful of the etymology of *infantia*, implying that speech properly belongs to the next stage of his life.

artifex: Wisd. 7: 21, 'omnium ... artifex docuit me sapientia', 14: 2, 'artifex sapientia fabricavit sua'; cf. Heb. 11: 10, 'cuius artifex et conditor deus'.

summe esse: Mayer, *Zeichen*, 2. 118 n. 36: 'Die adverbialen Bestimmungen des Seins als summe esse, vere esse, optime esse etc. entstammen dem Vulgärlatein der Jahrhunderte um die Zeitwende.' See on 1. 4. 4, 'summe, optime'—the beginning of the list there; cf. *mor.* 2. 1. 1, 'hoc enim *maxime esse* dicendum est, quod semper eodem modo sese habet, quod omnimodo sui simile est, quod nulla ex parte corrumpi ac mutari potest, quod non subiacet tempori, quod aliter nunc se habere quam habebat antea non potest. id enim est quod *esse verissime* dicitur'; *lib. arb.* 2. 15. 39, 'est enim deus et vere summeque est'; *c. ep. fund.* 40. 46 (quoted on 1. 7. 12).

idipsum: Ps. 4: 9, 'in pace in idipsum obdormiam et somnum capiam'; tied, through Ex. 3: 14, 'ego sum qui sum' (see on 7. 10. 16), to A.'s doctrine of the being of God; see on 9. 4. 11 and 9. 10. 24.

es CDO² Maur. : *est* GO¹S Knöll Skut. Ver.

et non mutaris: See on 1. 4. 4 and 7. 1. 1.

peragitur: See on 4. 10. 15.

quia in te sunt: Rom. 11: 35, 'et in ipso sunt omnia', hence 'et ista' here.

anni tui non deficiunt: Ps. 101: 28, 'tu autem idem ipse es, et anni tui non deficient' (variously echoed also at 11. 13. 16, 12. 11. 13, 13. 18. 22); *en. Ps.* 101. s. 2. 12–14, 'illi anni tui non deficient, illi anni tui qui erunt in generatione generationum, non deficient. haec ergo sciens, exiguitatem dierum meorum quaererem a te, nisi scirem omnes dies saeculi ab initio usque in finem exiguos esse in comparatione aeternitatis tuae. ... (14) speremus ergo etiam immutationem corporum nostrorum, sed tamen ab illo qui erat et ante nos, et manet post nos; a quo sumus quod sumus, ad quem veniemus cum fuerimus mutati; mutantem, non mutatum; facientem, non factum; et moventem, sed manentem; et quomodo intellegi a carne et sanguine potest: "ego sum qui sum" [Ex. 3: 14]; "tu vero idem ipse es, et anni tui non deficient." sed nos ad illos annos cum his pannosis annis quid sumus? . . . ergo speremus venturos nos ad hos annos stantes, in quibus non circuitu solis *peraguntur* dies, sed manet quod est sicuti est, *quia hoc solum vere est.*'

ex illo acceperunt modos: See on 1. 7. 12, 'a quo est omnis modus'.

tu autem idem ipse es: Ps. 101: 28 again; cf. 7. 21. 27, 'te, qui es semper idem', and passages cited above.

omniaque D²O² (ut vid.; O¹ unclear) Maur. Knöll Skut. Ver. : *omniaquae* CD¹GS.

quid ad me: *c. acad.* 1. 3. 7, 'placuit enim Ciceroni nostro [in Hortensio], beatum esse qui veritatem investigat, etiamsi ad eius inventionem non valeat pervenire'; *c. acad.* 1. 2. 6, 'quoniam igitur alteri vestrum videtur beatam vitam sola investigatione veritatis, alteri non nisi inventione posse contingere . . .'; but the flavour here is distinctly Augustinian rather than Academic; cf. *ord.* 2. 18. 47, 'ipsum parentem universitatis, cuius nulla scientia est in anima nisi scire quomodo eum nesciat'; sim. e.g. at *lib. arb.* 3. 2. 5, *Io. ev. tr.* 2. 3. Cf. 11. 31. 41, 'qui intellegit, confiteatur tibi, et qui non intellegit, confiteatur tibi.'

quid est hoc: Ex. 13: 14, 16: 15, Sirach 39: 26; see on 7. 6. 10.

1. 7. 11

Exaudi, deus: Ps. 54: 2, 'exaudi, deus, deprecationem meam, et ne despexeris precem meam: intende mihi, et exaudi me'; *en. Ps.* 54. 4, 'satagentis, solliciti, in tribulatione positi, verba sunt ista. orat multa patiens, de malo liberari desiderans.' Cf. Ps. 60: 2, Ps. 63: 2. In *conf.* 17 × where God is the subject (approx. 183 × in Vg. OT, 7 × in NT, almost always with God as subject); in CL, only poetic in this sense and not common. Knauer, 71 n. 4: 'wohl antwortend auf 1. 6. 10 "confiteor . . . terrae".' See on 1. 15. 24.

vae peccatis hominum!: Is. 1: 4, 'vae genti peccatrici, populo gravi iniquitate, semini nequam, filiis sceleratis'; Mt. 23: 13, 'vae vobis scribae et pharisaei, hypocritae.' *en. Ps.* 130. 2, '"vae eis qui trahunt peccata velut restem longam" [Is. 5: 18]. illi autem trahunt peccata sicut restem longam, qui addunt peccata peccatis'; sim. at *en. Ps.* 139. 9. Though in Latin *vae* may be a general imprecation without a specific forboding implication (as Gk. οὐαί in Mt. 23: 13, just cited), for A. the word seems always to have the force of 'alas for [usu. dative object], *because* evil [perhaps specified] will befall you.' The effect is thus more compassionate than angry. Hence in *conf.*, 1. 4. 4, 1. 16. 25, 2. 9. 17, 3. 6. 11, 6. 16. 26, 9. 13. 34, and esp. 10. 28. 39 ('vae prosperitatibus saeculi semel et iterum. . . . vae adversitatibus saeculi semel et iterum et tertio').

et homo: See on 1. 1. 1, 'et laudare te vult homo' for parallel.

commemorat: With the personal accusative: 'unusual' (G–M, who instance *mag.* 1. 1, 'ut commemoremus vel alios vel nos ipsos'); perhaps influenced by the use of the passive in scriptural texts (Gn. 19: 29 apud Tyconius, *reg.* 7, and Gn. 42: 9 apud A. *qu. hept.* 1. 138, and Sirach 49: 11 (= LXX 49: 9)—in each case translating Gk. ἐμνήσθη).

nemo mundus: Job 14: 4–5 (VL), 'quis enim erit mundus absque sorde? nec unus quidem, etiamsi unius diei fuerit vita eius super terram.'

A.'s mature doctrine of original sin was far from inevitable in 397. From the vast literature, the most pertinent and interesting here is A. Sage's study at *REAug* 13 (1967), 212–48, though many would differ with some details of his reconstruction; the case for greater continuity and less development is made by P. Rigby, *Original Sin in Augustine's 'Confessions'* (Ottawa, 1987). More venturesome is the attempt of P. F. Beatrice, *Tradux peccati* (Milan, 1978), to affiliate A.'s doctrine in second-century Encratite teachings; such a context helps explain the intensity of the reaction A. stirred up. The shifts over time are perhaps more clearly visible when looked at in a less theological way: see W. Eborowicz, *Studia Patristica*, 14 (1976), 410–16. A. himself speaks in defence of the consistency of his position, but he is unlikely to convince the most careful readers of his own works: *c. Iul.* 6. 13. 39, 'nam ego per unum hominem in mundum intrasse peccatum, et per peccatum mortem, et ita in omnes homines pertransisse, in quo peccaverunt omnes, ab *initio conversionis meae* sic tenui semper ut teneo. exstant libri quos adhuc laicus recentissima mea conversione conscripsi, etsi nondum sicut postea sacris litteris eruditus, tamen nihil de hac re iam tunc sentiens, et ubi disputandi ratio poposcerat dicens, nisi quod antiquitus discit et docet omnis ecclesia.' The *term* 'original sin' enters A.'s writings in *div. qu. Simp.* 1. 1. 10–11, expounding Rom. 7: 18–20 (2 ×; at 1. 1. 11, 'quod perficere bonum non est in potestate, ad meritum pertinet originalis peccati'), recurring next at *conf.* 5. 9. 16, and infrequently thereafter until the Pelagian controversy.[12] A useful comparandum for *div. qu. Simp.* is *lib. arb.* 3. 20. 55 (prob. 395), where everything *but* the term is already present.

On infant baptism, see J.-C. Didier, *REAug* 2 (1956), 109–29; there is no suggestion of the practice in *conf.* (A. was himself as infant only registered as a catechumen: see 1. 11. 17). In *quant. an.* 36. 80, there is credulous incomprehension: 'iamvero etiam puerorum infantium consecrationes quantum prosint, obscurissima quaestio est, nonnihil tamen prodesse credendum est. inveniet hoc ratio, cum quaeri oportuerit'; at *lib. arb.* 3. 23. 67 (written as priest), 'etiam illud perscrutari homines solent, sacramentum baptismi Christi quid parvulis prosit, cum eo accepto plerumque moriuntur

[12] Either the authenticity or the commonly accepted date (391) or the integrity from later contamination of *s.* 351. 2. 2 (which says that the faith of the sponsors suffices 'ad consecrationem remissionemque originalis peccati' in infant baptism) must be contested; the authenticity has drawn various opinions (see Verbraken).

priusquam ex eo quicquam cognoscere potuerint. qua in re satis pie recteque creditur prodesse parvulo eorum fidem a quibus consecrandus offertur. et hoc ecclesiae commendat saluberrima auctoritas.' By 411/12, he has discovered and accepted the encouragement of Cyprian (*pecc. mer.* 3. 5. 10 quotes Cypr. *ep.* 64), even as he doubts infants guilty of actual sin; cf. *pecc. mer.* 1. 17. 22 and 1. 35. 65, concluding (in rhetorical question): 'nihil mali commiserint infantes, qui propter hoc vocantur ab omnibus innocentes? nonne tanta infirmitas animi et corporis, tanta rerum ignorantia, tam nulla omnino praecepti capacitas, nullus vel naturalis vel conscriptae legis sensus aut motus, nullus in alterutram partem rationis usus, hoc multo testatiore silentio quam sermo noster proclamat atque indicat?' He shows himself more tolerant of infant iniquity than one might expect from the present passage: *pecc. mer.* 1. 35. 66, 'vellem tamen quisquis hoc sapit diceret quod peccatum viderit vel putarit infantis recentis ab utero, cui redimendo fatetur iam baptismum necessarium, quid mali in hac propria sua vita per animum proprium corpusve commiserit. si forte quod plorat taedioque est maioribus, mirum si hoc iniquitati non infelicitati potius deputandum est. an quod ab ipso fletu nulla sua ratione, nulla cuiusquam prohibitione compescitur? at hoc ignorantiae est, in qua profundissima iacet, qua etiam matrem, cum post exiguum tempus valuerit, percutiet iratus et saepe ipsas eius mammas quas, dum esurit, exigit. haec non modo feruntur, verum etiam diliguntur in parvulis, et hoc quo affectu nisi carnali.' He seems there to have in mind opponents who would claim that infant baptism is needed only because of actual sins, and not for original sin, and *hence* he forces the distinction and recovers his tolerance.

But the Job text is prominent in 411: *pecc. mer.* 1. 24. 34, 'porro si a salute ac vita aeterna hominem nisi peccata non separant, per haec sacramenta non nisi peccati reatus in parvulis solvitur, de quo reatu scriptum est neminem esse mundum "nec si unius diei fuerit vita eius". unde est et illud in psalmis: "ego enim in iniquitatibus conceptus sum et in peccatis mater mea me in utero aluit" [Ps. 50: 7].' Noteworthy here are the liturgical frame of reference and the link—here mediated by the text from Ps. 50 (the juxtaposition recurs at *c. ep. pel.* 4. 4. 4 and 4. 10. 27)—between the culpability of the newborn infant and the culpability of the procreative process. The same link has been forged already by Jerome, as A. himself quotes at *pecc. mer.* 3. 7. 13, where from *adv. Iov.* 2. 1, Job 14: 5 appears with Ps. 50: 7, in a nest of proof-texts (for the interpretation in this context of Ps. 50: 7, see below on 1. 7. 12); cf. *pecc. mer.* 2. 10. 14–15, 3. 6. 12. By 415, A. can quote Pelagius' disciple Caelestius (at *perf. iust.* 11. 23) as thinking that Job 14: 4–5 is a common (but, Caelestius thinks, ill-chosen) proof-text of the enemies of free will; Caelestius seems to think Job 12: 4 an adequate response: 'iustus enim vir et sine querela factus sum in derisu.' A. counters that Caelestius misunderstands a figure of speech, 'non intellegens posse dici hominem iustum, qui perfectioni iustitiae plurimum accessit, ita ut ei proximus fieret: quod multos etiam in hac vita potuisse, in qua ex fide vivitur, non negamus.' (Job 14: 4–5 quoted with a frequency to confirm Caelestius: *gr. et pecc. or.* 2. 32. 37, *nupt. et conc.* 2. 29. 50, *civ.* 20. 26, and *c. Iul.* 1. 3. 10, 5. 13. 49, 6. 26. 83, etc.)

uberibus inhiabam plorans: The infant's avidity at the breast can also be positive: *c. acad.* 1. 1. 4, 'philosophia est enim, a cuius uberibus se nulla aetas queretur excludi. ad quam avidius retinendam et hauriendam quo te incitarem, quamvis tuam sitim bene noverim, gustum tamen mittere volui'; sim. at *util. cred.* 1. 2.

reprehendi $CD^2G^2O^2$ edd. : *reprehendendi* $D^1G^1O^1S$.

ratio: Can be both the faculty (10. 6. 10 ('iudex ratio'), 10. 33. 49 ('dum rationi sensus non ita comitatur'), and 13. 24. 37 ('et dominatur ei ratio')) and (more classical and more frequent, as here) the substance of its function (5. 3. 6 ('et occurrebat mihi ratio . . . et non mihi occurrebat ratio'); 9. 9. 19 ('rationem facti sui reddebat')). The evidence of the senses might inspire anger and reprimand, but by an odd collusion, ordinary human weakness and a superior human faculty conspire to hold it in check.

nam extirpamus: Cf. Jn. 15: 2, 'omnem [palmitem] qui fert fructum, purgabit [pater meus] eum, ut fructum plus afferat.'

indignari: Recalls 1. 6. 8, 'indignabar'; the verb specifies anger directed at another apart from any action taken to give the anger force—cf. 9. 4. 8. As with the first smiles (see on 1. 6. 8), A. oversteps himself by failing to distinguish observation from interpretation. For him an infant is a small adult, lacking various powers but experiencing the world just as an adult would. He cannot speak or make his *indignatio* efficacious, but he is capable of all the emotions and the velleities that arise from them. Such a view of the infant is eminently compatible with a doctrine of infant baptism, but is philosophically problematic.

dicunt: In the absence of exact parallels, it is unclear whether some particular superstition is implied; the use of *expiare* leads in that direction. Of 91 × in A., only here is expiation in any sense trivial; the only passage even remotely comparable in weakness of offence is *ep.* 167. 6. 20, 'peccata cotidiana, sine quibus hic non vivitur, cotidianis remediis *expientur*'). The ordinary mode of 'expiation' is sacrifice, often in OT contexts, and sometimes contrasted to the regeneration of baptism: e.g. *nat. et or. an.* 2. 15. 21, 'nec abluendas baptismo nec *expiandas* Christi corporis et sanguinis sacrificio et in aeternum damnandas'; though baptism can have 'expiatory' effect as well when the contrast is not being emphasized: e.g. *ep.* 157. 3. 13, 'a quo delicto parvuli per baptismum *expiantur*'. Thus it is likely that the nurses found the infant behaviour disturbing and used means to 'expiate' it that were some form of 'pagan' ritual. (The one 'pagan' practice to which A. applies the word elsewhere is at *en. Ps.* 57. 4, 'respondet forte aliqua dura et mala mathematicus; curritur ad aruspicem ut *expietur*; respondet haruspex non se posse *expiare*;

maleficus quaeritur'. There the sacrificial implication is probably also present; cf. *conf.* 4. 2. 3, where consulting a *haruspex* implies consenting to sacrifice.) Cf. *en. Ps.* 30. en. 2. s. 2. 12, 'forte utiliter et in malum incidisti quem bonum putasti, ut in ubere quasi materno amaritudinem invenires, et ea offensione repellereris, et ad cibum validiorem invitareris. faciunt enim hoc nutrices mammothreptis, ut aliqua amara ponant in papillis suis, quibus offensi parvuli ab ubere resiliant et ad mensam *inhient*.' (The same lore at *s.* 311. 17. 14.)

nisi vero et ista innocentia est: In cataloguing the errancy of boyhood, he comes to the same ironic phrase: 1. 19. 30, 'istane est innocentia puerilis?'

licet probes: The generic second person is rare in *conf.* (here a momentary lapse in the nominal address of the whole text to God), and here with the only occurrence of *licet* governing subjunctive in a main clause (but *licet* only 5 × in *conf.*).

1. 7. 12

This apostrophe marks the conclusion of the treatment of *infantia*, the speechless time before memory, to be followed (1. 8. 13) by *pueritia* that brought speech and initiated memory. There are parallels here to 1. 20. 31, the concluding paragraph on *pueritia*.

quod: 'for the reason that', explaining 'iubes me laudare' below: note 'in istis', resuming the content of the *quod*-clause. Arts, 95–6, counts 38 *quod*-causal with indicative, 32 with subjunctive; she claims that his use of the moods follows regular classical practice, but here the subjunctive could have been employed, to ascribe the reason to the will of God—'ut videmus', however, emphasizes the factuality of the reason. *Quia* occurs 444 × (15 × already in Bk. 1, where this is the first *quod*-causal).

instruxisti . . . : Theological preoccupation and rhetorical habit join forces: three actions of God regarding individual aspects of the infant's bodily life ('instruxisti sensibus, compegisti membris, figura decorasti') are followed by a single action affecting the whole ('proque eius universitate': cf. 1. 20. 31, 'meamque incolumitatem, vestigium secretissimae *unitatis*'—and see on 1. 20. 31 for the trinitarian sense there). To ascribe a conscious trinitarian intent and to deny a trinitarian resonance would be equally rash.

proque: 'and on behalf of', i.e. 'in order to assure' its 'completeness in oneness' and 'safety'—all the efforts of the 'living being' (*animantis*: a careful word here, used elsewhere in the singular only at 1. 20. 31, 'in tali animante'; in plural elsewhere of 'living things', e.g. 4. 2. 3, 'necaturus . . . in sacrificiis suis animantia').

psallere: Only here in *conf.*, but *laudare*, *confiteri*, and *psallere* are closely related. Ps. 91: 1–2, 'psalmus cantici in diem sabbati. bonum est confiteri domino et psallere nomini tuo, altissime'; *en. Ps.* 91. 3, 'psallere autem quid est, fratres? psalterium organi genus est; chordas habet. opus nostrum, psalterium nostrum est; quicumque manibus operatur opera bona, psallit deo; quicumque ore confitetur, cantat deo. canta ore, psalle operibus.'

une: Appropriate of the Father, as *unicus* of the Son: see *BA* 72. 234 n. 34 with parallels.

a quo est omnis modus, formosissime, qui formas omnia et lege tua ordinas omnia: The choice of words is careful and precise, constituting an invocation of all three persons of the trinity. The triad *modus/species/ordo* is significant in A.'s thought and expression in the late 390s and recurs, more and less obtrusively, throughout *conf.* See also on 5. 4. 7, where the parallel triad from Wisd. 11: 21 (*mensura/numerus/pondus*) is discussed (there are frequent conflations, e.g. *Gn. c. man.*, 1. 21. 32, 'mensuras et numeros et ordines', as at *lib. arb.* 2. 20. 54 and *c. ep. fund.* 29. 32; cf. *Gn. litt.* 4. 3. 7, *civ.* 5. 11, 11. 15); see du Roy, 421–4, and TeSelle, 118–19; W. J. Roche, *New Scholasticism*, 15 (1941), 350–76, treats the two triads as one, with useful discussion. There are numerous permutations of the basic triad:[13] *ep.* 11. 3, *causa/species/manentia*; *mus.* 6. 17. 56, *unum/species/bonitas*; *vera rel.* 7. 13, *esse/species/ordo* and *unum/species/ordo*; *vera rel.* 55. 113, *principium/forma/gratia*; *nat. b.* 3, *modus/species/ordo*; *c. Faust.* 21. 6 (quoted on 2. 5. 10), *modus/pulchritudo/ordo*, *Gn. litt.* 4. 3. 7, *modus/species/quies*; *trin.* 3. 2. 8, *moles/species/ordo*; *trin.* 6. 10. 12, *unitas/species/ordo*; *civ.* 5. 11, *modus/species/ordo*; *civ.* 11. 28, *modo esse/specie contineri/ordinem appetere*.

That God's threeness is reflected in the *modus*, *species*, and *ordo* of created things is adumbrated in *c. ep. fund.* (just cited, and also 30. 33, 31. 34, 33. 36, 41. 47) and becomes a central idea of the anti-Manichean *nat. b.* of 398: *nat. b.* 3, 'nos enim catholici christiani deum colimus, a quo omnia bona sunt seu magna seu parva; a quo est omnis *modus*, sive magnus sive parvus; a quo omnis *species*, sive magna sive parva; a quo omnis *ordo*, sive magnus sive parvus. . . . deus itaque supra omnem creaturae *modum* est, supra omnem *speciem*, supra omnem *ordinem*; . . . a quo omnis *modus*, omnis *species*, omnis *ordo*. haec tria . . . ubi nulla sunt, nulla natura est. omnis ergo natura bona est.' Hence *nat. b.* 4 defines *malum*: 'corruptio vel modi vel speciei vel ordinis naturalis'. The triad recurs in that work in chs. 9, 10, 13, 18, 23, 30, 36, 37, 38; at ch. 21 the main triad is linked to the alternate triad of *mensura/numerus/pondus*; and at ch. 41 it is linked to a counter-triad, 'immoderationem, deformitatem, perversitatem'.[14] See also *vera rel.* 43. 81, quoted below.

[13] This list is based on and supplements Mayer, *Zeichen* 2. 131–3.

[14] *nat. b.* is probably contemporary with *conf.* (*retr.* 2. 9 places it after *c. Fel.* of 404, but that work is out of order; see Mutzenbecher in *CCSL* 57, pp. xviii-xix). Neither the *retr.* notice nor the work itself give any hint of circumstances, addressee, etc. Its *Tendenz* is anti-Manichean (it quotes their *ep. fund.* and *thesaurus* (including some strikingly 'explicit' (*ut nostrates aiunt*) texts that discreet translators pass over)), and its emphasis on the ontology of *modus/species/ordo* makes explicit

By *modus* A. understands 'manner of being' and thinks of the initial creation of unformed matter and hence sees a parallel to the creative act of the first person of the trinity. The concept arises in Ciceronian texts and is present in A. as a unique synthesis of Platonic and Ciceronian ideas from the first works at Cassiciacum: see du Roy, 152–8, and B. J. Cooke, *Modern Schoolman*, 23 (1946), 180–2, and 24 (1947), 44–5.

Species was interchangeable with *forma* in Latin to represent Plato's ἰδέα as early as Cicero; in A. at *div. qu.* 46. 2, 'ideas igitur latine possumus vel formas vel species dicere', and cf. *Tusc.* 1. 24. 58 and *or.* 3. 10 and 29. 101. In particular, *forma* occurs for the plural (where *species* is rare: cf. Cic. *top.* 7. 30) and *formare* supplies the lack of any verb derived from *species*. By *species* A. means (from *sol.* 2. 18. 32 on) 'principle of individuation or differentiation', both the outward appearance and the intrinsic structure (Mayer, *Zeichen*, 2. 148), by which the unformed matter is made into a unique created thing. Hence he sees a parallel to the action of the creative Word; *species* is the *sine qua non* of existence (*imm. an.* 8. 13–15). The dialectical antithesis of *species* is therefore *corruptio* (*c. ep. fund.* 40. 46, 'species aucta cogit esse et deum fatemur summe esse; corruptio vero aucta cogit non esse, et constat quod non est nihil esse'). Cf. also du Roy, 288.[15]

Modus and *species*, products of the first two stages of creation, present a world of isolated things, which is given pattern and dynamic (anything but the docile stasis we might expect in medieval ideas of order) by the addition of *ordo*, the animating and governing force of the third person of the trinity. The *ordo* of *ord.* is cosmic and irresistible—even evil is subsumed in it (*ord.* 2.7.23; cf. *mor.* 2. 6. 8), but in the watershed *div. qu. Simp.* 1. 2. 18 (see R. A. Markus, *Conversion and Disenchantment* (Villanova, 1989), 25) *inordinatio* is explicitly identified with *peccatum*; from then on in A.'s thought, the old high notion of cosmic order exists side by side with a dynamic view that allows for momentary disorder arising from sin, to be put right quickly enough by God. So *civ.* 15. 22, 'unde mihi videtur quod definitio brevis et vera virtutis ordo est amoris; propter quod in sancto cantico canticorum cantat sponsa Christi, civitas dei: "ordinate in me caritatem" [Cant. 2: 4].' See also the close ties between *ordo* and *pax* expounded at length at *civ.* 19. 13–20; for the

something often implicit in *c. ep. fund.* The last paragraph (*nat. b.* 48) is an *oratio Augustini*, heavily scriptural, that would not be out of place in *conf.* Many of the scriptural proof-texts occur frequently in *conf.*, and it would not be surprising if *nat. b.* were at least in part a sketch of ideas (most of the work, chs. 1–39, is repetitive, as if A. were going over ideas and getting them straight) to be used elsewhere.

[15] The notion animates a passage like *s.* 27. 6. 6, 'deformitas Christi te format. ille enim si deformis esse noluisset, tu formam quam perdidisti non recepisses. pendebat ergo in cruce deformis: sed deformitas illius pulchritudo nostra erat. in hac ergo vita deformem Christum teneamus. . . . huius deformitatis signum in fronte portamus: de ista deformitate Christi non erubescamus. hanc viam teneamus, et ad speciem perveniemus.'

association of *ordo* and *pondus*, cf. *s.* 242. 3. 5–7. 9, on the *ordo ponderum* (see also on 13. 9. 10 below), and *s. Casin.* 1. 133. 1, '*inordinate* amans deficit sub *pondere* passionis'. The standard discussion of the *ordo amoris* is Burnaby, 113–37. So *vera rel.* 41. 77, 'nihil enim est ordinatum quod non sit pulchrum. et, sicut ait apostolus, omnis *ordo* a deo est' (where the latter phrase, representing Rom. 13: 1, is a misquotation under the influence of the idea of 'order'; at *retr.* 1. 13. 8 A. owns to the error, for Paul actually wrote 'quae autem sunt a deo ordinata sunt').

For the triad read in light of the neo-Platonic tradition (Gk.: *οὐσία, εἶδος, τάξις*), see W. Theiler, *P. u. A.*, 11–12, and esp. 32–4, who makes of Wisd. 11: 21 a mere pretext, or scriptural corroboration. Even if we accept Theiler's (questionable) willingness to see in disparate elements attested in Porphyry a source for a clear and repeated pattern of expression in A., surely it was as much the coincidence of scriptural and Platonic doctrine that impressed A. in a matter such as this as it was either element independently.[16] (The most careful refutation of Theiler's assertion of Porphyry as the source of A.'s triad theology is du Roy, 402–8.) The triad has evident connections with a view ascribed to Aristotle on the first page of Amb.'s *exameron* (1. 1. 1): 'alii quoque, ut Aristoteles cum suis disputandum putavit, duo principia ponerent, materiam et speciem, et tertium cum his, quod operatorium dicitur, cui subpeteret competenter efficere quod adoriendum putasset.' The origins of that doctrine and its transmission are studied at length in J. Pépin, *Théologie cosmique et théologie chrétienne* (Paris, 1964), but he does not explore the connection to A.; for *operatorium*, note especially *mor.* 1. 16. 27, 'cumque ibi dictum sit, dei filium, dei virtutem esse atque sapientiam: cumque virtus ad operationem [3], sapientia vero ad disciplinam [2] pertinere intellegatur . . .';[17] also *mor.* 2. 6. 8, *Gn. litt. imp.* 4. 12 quoted below, and *Gn. litt. imp.* 15. 51. The *exameron* consists of sermons delivered by Ambrose, probably during Holy Week, perhaps in 386.[18] Another possible

[16] For a similar point of dispute, see on 1 Jn. 2: 16 discussed at 1. 10. 16 and 10. 29. 41 below; see on 10. 29. 41 for the relevance of Cic. *off.* 1. 4. 14 (quoted in next paragraph here), where *modus*, *pulchritudo*, and *ordo* appear in a passage alongside foreshadowings of the triad of temptations presented by 1 Jn. 2: 16.

[17] Throughout this commentary, the figures [1], [2], [3] mark words or phrases that form part of trinitarian triads. Sometimes they represent ideas that are predicated directly of the three persons of the trinity, and sometimes they represent parts of the world of creature (esp. of human nature) that reflect (or distort) the trinitarian structure of the creating God. The point in marking these triads is not so much theological or philosophical as rhetorical: to show how phraseology of this sort runs through A.'s texts. For the deeper issues, see du Roy *passim*.

[18] For 386, see Courcelle, *Recherches*, 101, and *Les Confessions*, 31, with concurrence of J.-R. Palanque in his review of Courcelle's *Recherches* at *Revue d'histoire de l'Église de France* 28 (1952), 134 (Palanque, *Saint Ambroise et l'empire romain* (Paris, 1933), 519 had argued for 387); other readers have held out for 387 (e.g. Madec, *Saint Ambroise* (Paris, 1974), 72, arguing that the holy week of 386 was too turbulent a period (see on 9. 7. 15) for such sermons—not a compelling point.

partial precedent is Hilary of Poitiers, whose *de trinitate* 2. 1 is quoted and discussed in *trin.* 6. 10. 11 (twenty years after *conf.*) for the triad 'aeternitas in patre [1], species in imagine [2], usus in munere [3].' A.'s discussion there refines that triad to *unitas/species/ordo* (*trin.* 6. 10. 12).

The whole triad is anticipated in two Ciceronian texts: *off.* 1. 4. 14, 'nec vero illa parva vis naturae est rationisque, quod unum hoc animal sentit quid sit *ordo* [3], quid sit quod *deceat* [2], in factis dictisque qui *modus* [1]'; *fin.* 2. 14. 47, 'atque his tribus generibus honestorum notatis quartum sequitur et in eadem *pulchritudine* [2] et aptum ex illis tribus, in quo inest *ordo* [3] et *moderatio* [1]. cuius similitudine perspecta in formarum specie ac dignitate transitum est ad honestatem dictorum atque factorum'; cf. Cic. *leg.* 3. 24, Cic. *Tim.* 3. 9.

Adumbrations of this triad in the earliest works can be identified, but not all are equally compelling: *beata v.* 4. 34–5, 'sed quid putatis esse sapientiam, nisi veritatem? etiam hoc enim dictum est: "ego sum veritas" [Jn. 14: 6]. *veritas autem ut sit, fit per aliquem summum modum, a quo procedit et in quem se perfecta convertit*. ipsi autem summo modo nullus alius modus imponitur; . . . sed etiam summus modus necesse est ut verus modus sit. ut igitur veritas modo gignitur, ita modus veritate cognoscitur. neque igitur veritas sine modo, neque modus sine veritate umquam fuit. quis est dei filius? dictum est, "veritas". *quis est qui non habet patrem? quis alius quam summus modus?* quisquis igitur ad summum modum per veritatem venerit, beatus est. . . . (35) hoc est beata vita, pie perfecteque cognoscere a quo inducaris in *veritatem* [3], qua *veritate* perfruaris [2], per quid conectaris *summo modo* [1]. *quae tria unum deum intellegentibus unamque substantiam* exclusis vanitatibus variae superstitionis *ostendunt*';[19] *c. acad.* 2. 2. 4, 'quod quaero intentissimus *veritatem* [2], quod invenire iam ingredior, quod me *ad summum ipsum modum* [1] perventurum esse confido' (and behind that passage lurks the shadow of Jn. 14: 6, 'ego sum via et veritas et vita; *nemo venit ad patrem, nisi per me*'). *ord.* 2. 19. 51, 'cum autem se composuerit [1] et ordinarit [3] atque concinnam pulchramque [2] reddiderit, audebit iam deum videre.' Cf. also *imm. an.* 8. 15, 'mutabilitas non adimit corpori corpus esse [1], sed de specie in speciem [2] transire facit motu ordinatissimo [3].' Clearer is *Gn. litt. imp.* 4. 12, 'chaos graece dicitur. . . . haec materies [1] quae consequenti operatione [3] dei in rerum formas [2] ordinata [3] distinguitur.' Sim. at *ep.* 11. 3; undatable and perhaps early at *div. qu.* 18, 'omne quod est aliud est quo constat, aliud quo discernitur, aliud quo congruit'; cf. *div. qu.* 6, 'ubi autem aliqua *species*, necessario est aliquis *modus*; et modus aliquid boni est.' Then in 392, *f. et symb.* 2. 2, 'nullo modo credendum est illam ipsam materiam . . . quamvis informem . . . per se ipsam esse potuisse tamquam coaeternam et

[19] On controversies over the reading of the triad in that last sentence, see J. Doignon, *BA* 4. 1 (rev. edn. of 1986), note at 151–2.

coaevam deo; sed quemlibet *modum* suum quem habebat ut quoquo modo esset et distinctarum rerum *formas* posset accipere, non habebat nisi ab omnipotente deo. . . . si de aliqua materia factus est mundus, eadem ipsa materia de nihilo facta est, ut *ordinatissimo* dei munere prima capacitas *formarum* fieret, ac deinde *formarentur* quaecumque *formata* sunt.' Sim. at *lib. arb.* 3. 12. 35, 3. 21. 60 ('domino deo auctore et formatore et ordinatore rerum omnium').

An instructive passage is *vera rel.* 43. 81, 'ipse autem *ordinis* [3] *modus* [1] vivit in *veritate* [2] perpetua nec mole vastus nec protractione volubilis, sed potentia supra omnes locos magnus, aeternitate supra omnia tempora immobilis, sine quo tamen nec ullius molis vastitas in unum redigi nec ullius temporis productio potest ab errore cohiberi et aliquid esse vel corpus, ut corpus sit, vel motus, ut motus sit. ipsum est unum principale nec per finitum nec per infinitum crassum.' This is the best statement of what *modus* means; as it appears here in conjunction with two words evoking the other two persons of the trinity, the allusion seems clear and definite. Cf. *vera rel.* 7. 13, where the substitution is Being and Oneness for *modus*: 'id est unum ipsum deum patrem et filium et spiritum sanctum, qua trinitate quantum in hac vita datum est cognita, omnis intellectualis et animalis et corporalis creatura ab eadem trinitate creatrice [1] esse in quantum est et *speciem suam habere*[20] [2] et ordinatissime administrari [3] sine ulla dubitatione perspicitur. . . . omnis enim res vel substantia vel essentia vel natura vel si quo alio verbo melius enuntiatur simul habet haec tria: ut et [1] *unum* aliquid sit, et [2] *specie* propria discernatur a ceteris, et rerum [3] *ordinem* non excedat.' Cf. also *agon.* 14. 16, of those who do not know the trinity: 'nesciunt enim principium ex quo *sunt* omnia, et imaginem eius per quam *formantur* omnia, et sanctitatem eius in qua *ordinantur* omnia.'[21]

The pattern of this triad also matches another triad imprinted on A.'s way of thinking: the rhetorical distinction of three kinds of *quaestiones* from at least Cicero on in Latin: Cic. *or.* 14. 45, 'quidquid est quod in controversia aut in contentione versetur, in eo aut *sitne* aut *quid sit* aut *quale sit* quaeritur' (see on 10. 10. 17, 'tria genera esse quaestionum, *an sit*, *quid sit*, *quale sit*,' for other references to Cicero, Quintilian, and Martianus Capella). A. himself mentions the three kinds of questions in works (*div. qu.* 18 and *ep.* 11. 4, both passages cited above) almost contemporaneous with the extensive use of the

[20] The italicized words are missing in all known MSS of *vera rel.*, but appear in all modern editions beginning with Amerbach (1506) and are defended by the most recent editor (K.-D. Daur, *Sac. Erud.* 12 (1961), 338); only Green in *CSEL* 77 rejects them; as du Roy 384 argues, they are necessary, logical, and probable, not least because Amerbach did not indulge in conjecture and must therefore have had a manuscript containing them.

[21] The triad even invades characterization of someone else's doctrine: *c. prisc. et orig.* 8. 9, 'unde illud quod aiunt [priscillianistae], in eius sapientia iam fuisse facta omnia, antequam in istas *formas* et *modos* proprios proferentur, atque in suis *ordinibus* apparerent, non sobrie dicitur.'

triad in *vera rel.* (du Roy 385–6: du Roy does not want to see here a 'source', *pace* A. Schindler, *Wort und Analogie in Augustins Trinitätslehre* (Tübingen, 1965), 56–60; it is best to take this as an example of the practice du Roy (386) describes: 'Augustin charge souvent des expressions de sa culture latine et surtout cicéronienne d'un sens néo-platonicien qui en est une réinterprétation').

A.'s penchant for 'triad theology' is one aspect of his thought that rarely found a sympathetic response in other writers, either late antique or modern. Hence a formula such as *modus/species/ordo* tends to be overlooked more often than not; for *conf.* see esp. discussions below on 1. 10. 16, 5. 4. 7, and 13. 9. 10. There are numerous more or less close parallels in *civ.*; cf. *civ.* 12. 6 ('res ipsa quaecumque . . . procul dubio bona est, habens modum et speciem suam in genere atque ordine suo'), *civ.* 8. 4, ('[deus], *causa* [1] subsistendi, *ratio* [2] intellegendi, et *ordo* [3] vivendi'), *civ.* 11. 24 ('origo [1] et informatio [2] et beatitudo [3]'). For one sign of afterlife, cf. Aquinas, *s. t.* 1. q. 39. a. 8, 'ad pulchritudinem tria requiruntur. primo quidem, integritas sive perfectio [1]; . . . et debita proportio sive consonantia [3]; et iterum claritas [2]'—paraphrased by Joyce in *Portrait of the Artist* (New York, 1964), 212.

The three dialogues of Cassiciacum are not without interest in light of this triad. *c. acad.* is devoted to the nature of truth; *beata v.* ends (4. 34–5: quoted above) with an extensive and important discussion of *modus* that links that concept to the first person of the trinity; and the third dialogue is expressly about *ordo*. Note further that *beata v.* and *ord.* each record dialogues that extended over three days, while *c. acad.* is broken in two sections of three days each. There are twelve days of discussion in all, and seven days elapse between the two halves of *c. acad.* See on 9. 4. 7 for the probability that *beata v.* and *c. acad.* are constructed to end, with expressly Christian remarks, on Sundays.

quam (*me vixisse*) GO Maur. Ver. Pell. : *qua* SC Knöll Skut. : *de qua* D. At 2. 5. 10, 'vita quam hic vivimus', OS have *quam*, CDG have *qua*. *en. Ps.* 104. 3, 'tota vita ista qua hic vivitur' is appropriate for the passive.

non memini: *trin.* 14. 5. 8, 'sed hanc aetatem omittamus quae nec interrogari potest quid in se agatur et nos ipsi eius valde obliti sumus.'

quam (*vivo*) GOS Maur. Knöll Skut. Ver. : *qua* CD.

quam vivo in hoc saeculo: Tit. 2: 12–14, 'erudiens nos, ut abnegantes impietatem et saecularia desideria sobrie et iuste et pie *vivamus in hoc saeculo*, (13) expectantes beatam spem et adventum gloriae magni dei et salvatoris nostri Iesu Christi, (14) qui dedit semet ipsum pro nobis.'

in hoc saeculo: For comparison of *mundus* (more spatial in original meaning) and *saeculum* (more temporal) in *conf.*, see L. Verheijen, *Studi in onore di Alberto Pincherle* (= *Studi e materiali di storia delle religioni* . . .), 38 (1967), 665–82, though on Verheijen's evidence one might better use objective/subjective as the distinction (or world-as-constructed-by-reason against world-as-lived-by-fallen-men); cf. R. A. Markus, *Saeculum* (Cambridge, 1969); in *conf.* found with *hoc* only at 1. 9. 14, 8. 6. 13, and 4 times in Bk. 13 (13. 13. 14, 13. 21. 30, 13. 21. 31, 13. 22. 32) quoting Rom. 12: 2, 'nolite conformari huic saeculo.'

oblivionis: *civ.* 16. 43, 'quam profecto aetatem primam demergit oblivio, sicut aetas prima generis humani est deleta diluvio. quotus enim quisque est, qui suam recordetur infantiam?' Sim. many years earlier at *Gn. c. man.* 1. 23. 35.

si et in iniquitate: Ps. 50: 7, 'ecce enim in iniquitatibus conceptus sum, et in peccatis mater mea me in utero aluit'; *en. Ps.* 50. 10, 'quid est quod se dicit in iniquitate conceptum, nisi quia trahitur iniquitas ex Adam? . . . [quotes Job 14: 4: see on 1. 7. 11] . . . non ergo ideo in iniquitate concipiuntur homines, et in peccatis in utero a matribus aluntur, quia peccatum est misceri coniugibus; sed quia illud quod fit, utique fit de carne poenali. poena enim carnis mors est, et utique inest ipsa mortalitas. . . . opus hoc castum in coniuge non habet culpam, sed origo peccati trahit secum debitam poenam.'

ego, servus tuus: Ps. 115: 16 (7), 'o domine, ego servus tuus; ego servus tuus et filius ancillae tuae.' See on 9. 1. 1, and cf. 12. 24. 33.

omitto: Infancy in the end eludes him, unrecoverable and shadowed, and so he lets it slip from his hands here to go on to surer subjects.

1. 8. 13

pueritiam: Already marked as the age of consent: *nat. et or. an.* 1. 10. 12 (with sim. at *nat. et or. an.* 3. 9. 12), 'nam illius aetatis [aet. 7] pueri et mentiri et verum loqui et confiteri et negare iam possunt. et ideo cum baptizantur, iam et symbolum reddunt et ipsi pro se ad interrogata respondent.' Easy and trivial memories of this time are elided, appearing infrequently elsewhere: *en. Ps.* 76. 20, 'solebamus pueri suspicari, cum audiremus tonitrua de caelo, quasi vehicula de stabulo processisse'; *quant. an.* 21. 36, 'cur ergo puer multo amplius itineris conficiebam sine defectu, cum aucupandi studio in ambulando exercerer, quam adolescens, cum me ad alia studia, quibus sedere magis cogebar, contulissem, si accedenti aetati et per hanc animae crescenti vires ampliores tribuendae sunt?'

The six days of creation, the six ages of human history, and the six ages of a man: each of these schemes throws light on the others for A., and each is finally referred to the sabbath they share, the day of rest without beginning or end that is reached in anticipation on the last page of *conf.* A.'s narrative of his early life in Bks. 1–7 follows this pattern, which is then abandoned (a reason for this is suggested below). L. Pizzolato, *Le 'Confessioni' di sant'Agostino* (Milan, 1968), saw in the six-age pattern the underlying structure of *conf.* as a whole: in so doing, he (*a*) minimizes A.'s abandonment of the pattern after Bk. 7, (*b*) fails to do justice to other important structural elements in the text, and (*c*) finally succumbs to the temptation to think that a work of the breadth and complexity of *conf.* can be interpreted authoritatively in terms of a single key pattern. With those reservations, however, it must be said that the pattern he detects *is* present, one among several.

The other explicit texts from this scheme are 2. 1. 1 ('exarsi enim aliquando satiari inferis in adulescentia') and 7. 1. 1 ('iam mortua erat adulescentia mea mala et nefanda et ibam in iuventutem'). No further mention is made of the pattern, but it should be borne in mind that A. at the time of writing *c.*397 was approaching the next juncture of ages on that scheme, the one separating *iuventus* from the fifth age (variously named: see below) at age 45;[22] the last boundary, crossed at age 60, found A. at another decisive moment in his literary career, finishing several major works while beginning *civ.* and the anti-Pelagian campaign. The pattern employed here is entirely literary and textual, perhaps anachronistic (P. Veyne, *A History of Private Life* (Cambridge, Mass., 1987), 1. 23–5): A. interpreted his life by a textual tradition.

For concise treatment with refs., see B. Kötting and W. Geerlings, *Aug.-Lex.* 1. 150–8. For backgrounds, see K. H. Schwarte, *Die Vorgeschichte der augustinischen Weltalterlehre* (Bonn, 1966), A. Luneau, *L'Histoire du salut chez les Pères de l'Église: La Doctrine des âges du monde* (Paris, 1964), and G. Ladner, *The Idea of Reform* (Cambridge, Mass., 1959), 222–38. Ladner traces (224) the correlation between the six days and the six historical ages to Ps.-Barnabas, Justin, and Irenaeus; its origins are related to millenarism—six historical ages of 1000 years each, ending in the final millennium; A. himself has millenarian tendencies in his earlier writings (e.g. *s.* 259. 2 (393? certainly not after 400), 'octavus ergo iste dies in fine saeculi novam vitam significat: septimus quietem futuram sanctorum in hac terra'; he qualifies his earlier opinions at *civ.* 20. 7, 'etiam nos hoc opinati fuimus aliquando'; cf. G. Folliet, *REAug* 2 (1956), 371–90 and see on 13. 35. 50, 'pacem sabbati'). More recently J. A. Burrow, *The Ages of Man* (Oxford, 1986), 61, underscores the significance of the parable of the workers in the vineyard (Mt. 20: 1–8), treated since Origen as a five-age scheme. (A. once adapts this by skipping *infantia*: *s.* 49. 2. 2; at another, he forces the pattern by omitting *adulescentia*: *s.* 87. 5. 7.) Burrow is good on the medieval literary afterlife, while E. Sears, *The Ages of Man* (Princeton, 1986), emphasizes visual materials in the tradition.

The most explicit and decisive texts on the subject in A. come in *vera rel.* For the basic six-day pattern, *vera rel.* 26. 48: 'prima huius [aetas] *infantia* in nutrimentis corporalibus agitur, penitus obliviscenda crescenti. eam *pueritia* sequitur, unde incipimus aliquid meminisse. huic succedit *adolescentia*, cui iam propagationem prolis natura permittit, et patrem facit.

[22] The fifth age was appropriate for involvement in the business of the secular world (*Gn. c. man.* 1. 25. 43, 'incipiat quinto die in actionibus turbulentissimi saeculi, tamquam in aquis maris operari'), business whose burden A. acutely felt in the late 390s.

porro adolescentiam *iuventus* excipit, iam exercenda muneribus publicis et domanda sub legibus. ... post labores autem iuventutis, *seniori* pax nonnulla conceditur. inde usque ad mortem *deterior aetas ac decolor*, et morbis subiectior debilisque perducit. haec est vita hominis viventis ex corpore, et cupiditatibus rerum temporalium conligati. hic dicitur vetus homo, et exterior, et terrenus.' This version of the scheme comes without a seventh day.

The scheme is clearly not carried beyond the turn from *adulescentia* to *iuventus* at 7. 1. 1. Then conversion supervenes and the pattern is abandoned. One relevance of this is that the conversion occurs when A. is about the same age as that at which Christ was crucified, as A. noted at *s.* 88. 10. 9, 'natus est de virgine Maria. ... per aetates cucurrit usque ad iuventutem. ... infantiae pueritia, pueritiae adolescentia, adolescentiae iuventus transeunti cedentique successit.' Since conversion entails death, burial, and rebirth in Christ (i.e. baptism: see on 9. 1. 1), the pattern is complete and parallel—howbeit only implicitly so—in Christ and in A.

There is, moreover, an important variant of the basic pattern, a seven-day pattern for the New Man, discussed only at *vera rel.* 26. 49: 'iste dicitur novus homo, et interior, et caelestis, habens et ipse proportione non annis sed provectibus distinctas quasdam spiritales aetates suas: *primam* in uberibus utilis historiae quae nutrit exemplis;[23] *secundam* iam obliviscentem humana et ad divina tendentem, in qua non auctoritatis humanae continetur sinu sed ad summam et incommutabilem legem passibus rationis innititur;[24] *tertiam* iam fidentiorem et carnalem appetitum rationis robore maritantem gaudentemque intrinsecus in quadam dulcedine coniugali, cum anima menti copulatur et velamento pudoris obnubitur, ut iam recte vivere non cogatur sed, etiamsi omnes concedant, peccare non libeat;[25] *quartam* iam id ipsum multo firmius ordinatiusque facientem et emicantem in virum perfectum, atque aptam et idoneam omnibus et persecutionibus et mundi huius tempestatibus ac fluctibus sustinendis atque frangendis;[26] *quintam* pacatam atque omni ex parte tranquillam, viventem in opibus et abundantia incommutabilis regni summae atque ineffabilis sapientiae;[27] *sextam* omnimodae mutationis in aeternam vitam et usque ad totam oblivionem vitae temporalis transeuntem, perfecta forma quae facta est ad imaginem et similitudinem

[23] In Bk. 6, A. learns to read the Old Testament as a series of such *exempla*: cf. 6. 4. 6 f.

[24] Cf. 7. 10. 16 ff.

[25] Cf. 8. 12. 29.

[26] The vision of Ostia (9. 10. 24–25) is perhaps one sign of this stage, but it may be taken as extending beyond the narrative of *conf.* through Cassiciacum and the first years of conversion.

[27] An apt description of the life at Thagaste A. enjoyed at the time of *vera rel.*: cf. *ep.* 10. 2, written to Nebridius from that retirement: 'magna secessione a tumultu rerum labentium, mihi crede, opus est, ut non duritia, non audacia, non cupiditate inanis gloriae, non superstitiosa credulitate fiat in homine, nihil timere. hinc enim fit illud etiam solidum gaudium, nullis omnino laetitiis ulla ex particula conferendum.'

dei;[28] *septima* enim iam quies aeterna est et nullis aetatibus distinguenda beatitudo perpetua. ut enim finis veteris hominis mors est, sic finis novi hominis vita aeterna. ille namque homo peccati est, iste iustitiae.' The seven-day scheme is as importantly parallel to the narrative of *conf.* as is the six-day scheme. A.'s conversion marks a shift from the scheme appropriate to the Old Man to that of the New. In one early text, he characterizes the transition as natural and appropriate to the third age: *div. qu.* 49, 'quorum duorum populorum differentiam etiam in unoquoque nostrum licet advertere, cum quisque ab utero matris veterem hominem necesse est agat, donec veniat ad iuvenilem aetatem, ubi iam non est necesse carnaliter sapere, sed potest ad spiritalia voluntate converti et intrinsecus regenerari. quod ergo in uno homine recte educato ordine naturae disciplinaque contingit, hoc proportione in universo genere humano fieri per divinam providentiam peragique pulcherrimum est.' The repression of the six- (or seven-) age scheme in *conf.* reflects a decline in A.'s willingness to impose an optimistic determinism on the life of the redeemed Christian, and with it his growing aversion to millenarian interpretations of scripture. Pincherle, *Aug. Stud.* 7 (1976), 132–3, links the repression of the six-age scheme to the 'new course his thoughts had taken' in the *div. qu. Simp.*

The fullest elaboration of the common scheme is *Gn. c. man.* 1. 23. 35–41:[29] 1. *infantia* (Adam → Noe): (35) 'quasi vespera huius diei fit diluvium, quia et infantia nostra tamquam oblivionis diluvio deletur.' 2. *pueritia* (Noe → Abraham): (36) 'sed nec ista aetas secunda generavit populum dei, quia nec pueritia apta est ad generandum.' 3. *adulescentia* (Abraham → David): (37) 'haec enim aetas potuit iam generare populum deo, quia et tertia aetas, id est adolescentia, filios habere iam potest.' 4. *iuventus* (David → Babylonian captivity): (38) 'haec aetas similis iuventutis est. et re vera inter omnes aetates regnat iuventus, et ipsa est firmum ornamentum omnium aetatum.' 5. no special name here—see below (Babylonian captivity → Christ): (39) 'declinatio a iuventute ad senectutem, nondum senectus sed iam non iuventus. . . . et re vera sic ista aetas a regni robore inclinata et fracta est in populo Iudaeorum, quemadmodum homo a iuventute fit senior.' 6. *senectus* (from the time of Christ → the end of the age): (40) 'incipit sextus, in quo senectus veteris hominis apparet. . . . tunc fit homo ad imaginem et similitudinem dei, sicut in ista sexta aetate nascitur in carne dominus noster. . . . et quemadmodum in illo die masculus et femina, sic et in ista aetate Christus et Ecclesia. . . . ut ab eo [Christo] domarentur atque mansuescerent homines, vel *carnali concupiscentiae* dediti sicut pecora, vel

[28] This is the next stage anticipated from the tranquillity of Thagaste; it is this transformation that he thought he had lost and for which he shed tears at his ordination in Hippo shortly after writing *vera rel.*

[29] This text and *div. qu.* 58 (cited below) were the most frequently cited in the middle ages by those who took up the pattern from A.: Burrow, 80–5.

tenebrosa *curiositate* obscurati quasi serpentes, vel elati *superbia* quasi aves.'[30] 7. (41) 'huius autem aetatis quasi vespera ...' *div. qu.* 58 has a similar list, specifying *gravitas* to name the fifth age;[31] on the historical six ages, A.'s opinions do not vary significantly through his career (the influence of Mt. 1: 17, the genealogy, is often acknowledged by A.); for the same pattern, see *div. qu.* 64. 2, *s.* 259. 2, *cat. rud.* 22. 39, *c. Faust.* 12. 8, *Io. ev. tr.* 9. 6, 15. 9, *trin.* 4. 4. 7, *ss.* 125. 4, 259. 2, and frequently in *civ.* 15–18, where the pattern provides the frame for A.'s survey of human history, and cf. *civ.* 16. 43, 21. 16, and 22. 30 (last paragraph of *civ.*).

The native Roman tradition (see Kötting and Geerlings, art. cit.) knew at most five ages. So Seneca quoted by Lact. *inst.* 7. 15. 14–17, and Varro according to Censorinus *de die natali* 14. 2, 'Varro quinque gradus aetatis aequabiliter putat esse divisos, unumquemque scilicet praeter extremum in annos quindecim. itaque primo gradu usque annum quintum decimum pueros dictos, quod sint puri, id est impubes. secundo ad tricensimum annum adulescentes, ab alescendo sic nominatos. in tertio gradu qui erant usque quinque et quadraginta annos, iuvenes appellatos eo quod rem publicam in re militari possent iuvare. in quarto autem adusque sexagensimum annum seniore vocitatos, quod tunc primum senescere corpus inciperet. inde usque finem vitae uniuscuiusque quintum gradum factum, in quo qui essent, senes appellatos, quod ea aetate corpus iam senio laboraret.' Cf. Varro according to Servius on *Aen.* 5. 295: 'aetates omnes Varro sic dividit: infantiam, pueritiam, adulescentiam, iuventutem, senectam'; Aul. Gell. 10. 28, 'de aetatum finibus pueritiae iuventae senectae ex Tuberonis historia sumptum'. A similar tradition is reflected in Amb. *Abr.* 2. 9. 65, 'quattuor ... aetates sunt hominibus; pueritia, adulescentia, iuventus, maturitas.'

farer: Imperf. subj. < *fari*: *figura etymologica* with *infans*. *TLL* 6. 1029 lists the form only from the present passage (cf. G–M).

et memini hoc: After the string of things that he does not remember ('non ... memini' at 1. 6. 7, 1. 6. 8, 1. 6. 10, 1. 7. 11, 1. 7. 12), speech and memory enter together. On language learning, cf. *mag.* 1. 2, 'qui enim loquitur, suae voluntatis signum foras dat per articulatum sonum,' and *mag.* 10. 33, 'etenim cum primum istae duae syllabae, cum dicimus "caput", aures meas impulerunt, tam nescivi quid significarent quam cum primo audirem legeremve "sarabaras".[32] sed cum saepe diceretur "caput", notans atque

[30] For the last three clauses, cf. 1 Jn. 2: 16 (see on 1. 10. 16).

[31] *en. Ps.* 70. s. 2. 4, 'nam *presbytes* dicitur gravis, et *geron* senex. quia vero in latina lingua duorum istorum nominum distinctio deficit, de senectute ambo sunt posita, senecta et senium'; the latter distinction A. tries to force at *qu. hept.* 1. 35, 'seniorum aetas minor est quam senum, quamvis et senes appellentur seniores,' and *qu. hept.* 1. 70. See also *en. Ps.* 127. 15 (no distinct fifth age), *s.* 216. 8. 8 (fifth age called *senium*, the sixth *senectus*; with an interesting correlation to stages of spiritual growth not otherwise attested) and *ep.* 213. 1 (fifth age called *gravitas*).

[32] Cf. G. N. Knauer, *Glotta* 38 (1954), 100–18.

animadvertens quando diceretur, repperi vocabulum esse rei quae mihi iam erat videndo notissima.' See below.

didiceram OS Knöll Skut. Ver. : *didicerim* CDG Maur. Indicative in indirect question is well-attested in *conf.* (35 ×, against *c.* 120 × subjunctive), but it is a laxity he does not allow himself so frequently in *civ.* (indicative only 6 × in the whole work: figures from Arts, 94).

non . . . docebant me: 1. 14. 23, 'non a docentibus'.

cum gemitibus et vocibus variis: Cf. 1. 6. 8 ('itaque iactabam et membra et voces') and 1. 6. 10 ('signa quibus sensa mea nota aliis facerem').

prensabam O Z Ver. Solignac Pell. Pizzolato (*Lectio I–II*, 45 n. 170) : *pensabam* CDS AFH[2] Knöll Skut. : *pensabo* H[1] : *praesonabam* EG Maur. Pusey (that verb occurs nowhere else in A. and is rare besides) : *personabam*, M : *praesentabam* V : *praestabam* BP.

Prensabam is preferable for the palaeographical reason that all the other readings may be explained as deriving from it and for the substantive reason that it well depicts the infant who would speak grasping wilfully at the words, as well as the things, of his world—and A.'s memory is a storehouse of such acquisitions (10. 8. 12–14). The verb is not otherwise attested in A., but occurs 4 × in the *Aeneid*, a reputable authority for A.'s use. Note 'tenebam' just below, common in A. in the sense of 'hold in memory' (see on 1. 13. 20), and continuing the implied metaphor. (Note also that *penso* is regularly used without a specifying instrumental ablative in the sense 'to weigh in the mind, ponder, consider' (*OLD*); to use it here with *mente* and *memoria* would be doubly redundant. In A., moreover, the word is used 40+ times, most often in contexts where something is to be evaluated; the only cases with an instrumental ablative: *ep.* 258. 2, '[res humanas] rerum divinarum cognitione pensamus'; *civ.* 9. 4, 'pari tamen aestimatione pensentur').

The punctuation is difficult. The Maurists read *praesonabam* (Pusey translates: 'practise the sounds in my memory') and placed the full stop after 'memoria'. Knöll read *pensabam* with his favourite manuscript, and put the period after 'omnibus' (G–M accepted his reading, but restored the original punctuation). Skutella retained the reading and the punctuation. Verheijen and Pellegrino abandon the reading—but all editors since 1935 retain the period before 'p(r)ensabam'. (In its defence, see M. Burnyeat, *Proceedings of the Aristotelian Society*, Supplemental Volume 1987, 2 n. 3; but he did not realize that 'prensabam' was an alternative, nor what it entailed.)

In avoiding a difficulty of syntax, however, these editors introduce a difficulty of sense. If their punctuation is followed, then 'ego' has no expressed verb and one must be supplied. Sc. *me docebam*? 'My elders taught me not, but rather I [taught myself], when the circumstances of infancy left me frustrated at being unable to express my will.' To which G–M: 'it would perhaps have been more in accordance with A.'s style to write "sed ego me ipsum", but the sentence ends awkwardly with the long subordinate clause, and the next begins abruptly.' Is not the logical connection between the frustration expressed by the *cum*-clauses and the grasping reaction expressed by 'prensabam'? 'My elders taught me not, but rather I, when the circumstances of infancy left me frustrated at being unable to express my will, grasped with my memory: when they named something, I . . .' It is true (and indeed notable) that *mens* and *memoria* overlap significantly for A. (*mens* does not appear at all in the discussion of memory in Bk. 10: when A. needs a word there with similar meaning, he uses *animus*); if the two appear here guiltily yoked in syntax, the best explanation seems to be a slight anacoluthon caused by the length of the intervening clauses. (The reading *memoriam* (in MSS AHVG) is a vulgar error independently arrived at; note that it makes nonsense of G's own reading of the verb, *praesonabam*.)

cum ipsi . . . enuntiabam: These lines are quoted at the outset of Wittgenstein's *Philosophische Untersuchungen* (New York, 1953), 2. He goes on: 'In diesen Worten erhalten wir, so scheint es mir, ein bestimmtes Bild von dem Wesen der menschlichen Sprache. Nämlich dieses: Die Wörter der Sprache benennen Gegenstände—Sätze sind Verbindungen von solchen Benennungen. —In diesem Bild von der Sprache finden wir die Wurzeln der Idee: Jedes Wort hat eine Bedeutung. Diese Bedeutung ist dem Wort zugeordnet. Sie ist der Gegenstand, für welchen das Wort steht. Von einem Unterschied der Wortarten spricht Augustinus nicht.' This 'Vorstellung einer primitiven Sprache', as W. characterizes it, then forms the basis for extended reflections on the philosophical defects of such a system. The weakness of W.'s meditation is that it does not do justice to the ethical context in which A. is careful to situate his own recollections: the aim of language-learning for the infant A. was not any disinterested attempt to utter true propositions about the world, but a much more self-interested attempt to control the world and other people, using whatever rough-and-ready tools were at hand. (A. himself partly forestalls Wittgenstein's criticism many years later in a passage not known to W.: *ep.* 187. 7. 25 (to Dardanus in 417), 'annon videmus, etiam cum articulatae vocis qualiacumque signa edere coeperint atque ad initium fandi transire ab infantia, adhuc eos talia sentire ac dicere, in quibus si remansissent annisque accedentibus tales esse persisterent nullus eos vel fatuus nimium fatuos esse dubitaret?' It remains true, none the less, that the Augustinian view of language learning presumes a coherent 'self' who learns to speak, where Wittgenstein emphasizes the effect of learning to speak in shaping the 'self' that emerges.)

videbam et tenebam: 'I saw and kept in memory' (see on 1. 13. 20).

ceterorumque CDO^2 VBPZEFGM Maur. Knöll Skut. : *ceteroque* O^{1}S AH Ver. Pizzolato, *Lectio I–II*, 45 n. 171: 'A ulteriore sostegno di questa scelta si può invocare il costrutto, che precede di poco questo: "variis membrorum motibus", dove "membrorum" sta ancora racchiuso tra attributo e complemento di cui esso è specificazione.' But *ceterorum* is more precise ('the activity of the other limbs' rather than 'the rest of the activity of the limbs') and *cetero* is the wrong number: the singular occurs nowhere else in *conf.* (55 × attributively or substantively).[33]

voluntatum enuntiandarum: Speech begins in the fallen man as the assertion of *voluntas*; cf. 1. 6. 8 and note that the present paragraph contains: 'vellem', 'voluntati', 'volebam', 'volebam', 'vellent', 'velle', 'voluntates', 'voluntatum', as well as like expressions (e.g. 'affectionem animi in petendis, habendis, reiciendis fugiendisve rebus'). The speech Augustine acquired as a *puer* was thus not the divine gift that he prayed for at 1. 5. 5 ('miserere ut loquar').

communicavi: 'shared', but in the active sense from which arises the sense retained in English 'communicate'.

procellosam societatem: Speech makes us human, introduces us to society, even the dangerous society that arises after the fall. Though in CL *procella* can be used of a torrent of speech, there is little to suggest that A. makes so explicit a connection between speech and the misfortunes of human life. See on 13. 20. 28, 'genus humanum profunde curiosum [2] et procellose tumidum [1] et instabiliter fluuidum [3],' where the trinitarian pattern and juxtaposition of *tumidum* make it likely that the vice particularly in mind is pride. *beata v.* 1. 1, 'cum enim in hunc mundum, sive deus sive natura sive necessitas sive voluntas nostra sive coniuncta horum aliqua sive simul omnia . . . velut in quoddam procellosum salum nos quasi temere passimque proiecerit . . .' Cf. also *en. Ps.* 76. 20 ('vita humana amara est, et procellis ac tempestatibus subdita'); sim. at *en. Ps.* 64. 9, 25. en. 2. 4, 118. s. 19. 3.

auctoritate: 32 × in *conf.*, frequent in *civ.*, less obtrusive in *en. Ps.* The usage here is not identical with the common Christian application to scripture, but that is also common in A. (6. 5. 8, 'auctoritate sanctarum litterarum'). As used here, it represents a characteristic concern of A., one that emerges in his writing from his lifelong dialogue with the classical past. At every stage of the search he narrates in *conf.*, A. is seeking a reason to abandon himself to authoritative doctrine, and seeking a doctrine to which to attribute this authority. *Auctoritas* is notably, as here, a parental quality.

Auctoritas is the crutch that supports the weakened faculty of *ratio*: *vera rel.* 24. 45, 'auctoritas fidem flagitat et rationi praeparat hominem.' See Holte, *Béatitude et Sagesse* (Paris, 1962), esp. 304–10, who measures the extent and limits of A.'s originality here. The yoking of *ratio* and *auctoritas* (Holte 308) 'n'a pas de parfait équivalent chez les théologiens occidentaux antérieurs

[33] I have not examined all 3114 instances of *ceterus, -a, -um* in A., but note that unambiguously singular forms of the word are extremely rare: *ceterus* appears nowhere, *ceteram* only 6 ×, and of the 74 appearances of *cetero*, at least 65 are in the adverbial phrase *de cetero*. By contrast, the unambiguously plural forms are common (e.g. *ceteris* 786 ×, *ceterorum* 85 ×). The lexica confirm the infrequency of the singular as adjective.

comme Tertullien, Lactance ou Ambroise, quoique la terminologie elle-même soit attestée chez ce dernier.'

The juxtaposition of *ratio* and *auctoritas* is familiar in Cicero, e.g. Cic. *Lucullus* 18. 60, 'quae sunt tandem ista mysteria, aut cur celatis quasi turpe aliquid sententiam vestram? "ut qui audient", inquit, "ratione potius quam auctoritate ducantur." quid si utrumque, num peius est? unum tamen illud non celant, nihil esse quod percipi possit. an in eo auctoritas nihil obest? mihi quidem videtur vel plurimum.' If it is doubtful whether A. knew this earlier version of that work (cf. Hagendahl, 1. 60), it is not impossible that similar expressions occurred in the *acad. post.*, which survives only in fragments; and there are other similar passages in C. (see W. Görler, *Untersuchungen zu Ciceros Philosophie* (Heidelberg, 1974), 22). At *c. acad.* 1. 3. 8, Trygetius seems to share Cicero's preference for *ratio* in claiming to have thrown off the *iugum auctoritatis* in the name of that *libertas* which *philosophia* promises. But a characteristically Augustinian view is also present: *c. acad.* 3. 20. 43, 'nulli autem dubium est gemino pondere nos impelli ad discendum, auctoritatis atque rationis. mihi autem certum est nusquam prorsus a Christi auctoritate discedere; non enim reperio valentiorem.'

In most of A.'s early works, *auctoritas* appears in benevolent guise (e.g. *vera rel.* 29. 52, 'auctoritatis beneficentia'), but a particularly clear series of texts revealing ambiguity and marked by the lingering anxiety of the search for authority, appears in *util. cred.*: *util. cred.* 8. 20–11. 25, 'occurrebat igitur *inexplicabilis silva*, cui demum inseri *multum pigebat*; atque inter haec *sine ulla requie* cupiditate reperiendi veri animus *agitabatur*. . . . (9. 21) nam vera religio . . . omnino sine quodam *gravi auctoritatis imperio* inire recte nullo pacto potest. . . . (11. 25) quod intellegimus igitur, debemus rationi, quod credimus, auctoritati, quod opinamur, errori.'

nutuque CDS Maur. Knöll Skut. Ver. : *natuque* GO.

1. 9. 14

A. spent a quarter of a century in the schools of literary culture of late antiquity. The standard work on the implications of this immersion is Marrou, *Saint Augustin et la fin de la culture antique*; that can now be supplemented on some aspects from R. A. Kaster, *Guardians of Language* (Berkeley, 1988).

deus meus: Knauer, 31–2: 'Es sei nur bemerkt, daß Augustin sie [die Wendung *deus meus*] gern an den Anfang neuer Abschnitte setzt, besonders dann, wenn er sein eigenes Handeln, das er schildern will oder geschildert hat, von Gottes Handeln absetzen möchte.'

quas ibi miserias expertus sum: Ter. *adelphoe* 867, quoted to epitomize the ills of social life at *civ.* 19. 5: 'audiant apud comicos suos hominem cum sensu atque consensu omnium hominum dicere, "duxi uxorem; quam ibi miseriam vidi! nati filii, alia cura".' The text is Demea's palinode for a life wasted on ill-tempered selfishness. Here, cf. 1. 8. 13, 'vitae humanae procellosam societatem'.

ludificationes: Apparently 'mockings, triflings'; little if any other attestation in *TLL* for the sense here, but when found elsewhere in A., usu. of demons: *Gn. litt.* 7. 11. 16 ('ludificationibus daemonum'), *civ.* 6. 7, 7. 35, 8. 23, 10. 16, 10. 27, 18. 18; *qu. hept.* 2. 21, *s.* 214. 8.

excellerem: Cf. 3. 3. 6, 'excellerem in eis [studiis]'.

linguosis artibus: *s.* 9. 2. 2, 'ante ultimum iudicium dei tui compone causam tuam. non est unde praesumas; cum ille venerit, nec falsos testes adduces, quibus ille fallatur, nec patronum fraudulentis circumventionibus et *linguosis artibus* adhibebis, nec ambies aliquo modo ut iudicem possis corrumpere.' Cf. Ps. 139: 12, 'vir linguosus non dirigetur super terram'; *en. Ps.* 139. 15, 'vir linguosus amat mendacia. . . . gaudium tibi sit in auditione dei, necessitas tibi sit in locutione tua; et non eris vir linguosus.'

in scholam: On elementary and secondary education, see R. A. Kaster, *TAPA* 113 (1983), 323–46, arguing against a rigid distinction between the two categories. H. Chadwick, *Augustine* (Oxford, 1986), 7: 'He never wrote with admiration or gratitude about any of his teachers.'

litteras: To balance the rhetorical posturing of this memoir, cf. the more conventional and sober *sol.* 2. 11. 19, 'est autem grammatica vocis articulatae custos et moderatrix disciplina; cuius professionis necessitate cogitur humanae linguae omnia etiam figmenta conligere quae memoriae litterisque mandata sunt, non ea falsa faciens, sed de his veram quandam docens adserensque rationem.'

vapulabam: Such punishment was common and disdain for it conventional (Quint. 1. 3. 13–17, 'caedi vero discipulos, quamlibet et receptum sit . . ., minime velim'); with A., cf. *disc. chr.* 11. 12, 'quare vapulasti? quare tanta mala in pueritia pertulisti? ut disceres. quid disceres? litteras. quare? ut haberetur pecunia, aut ut compararetur honor, et teneatur sublimitas dignitatis. . . . novi, dicturus es mihi: quare et vos, episcopi, non litteras legistis; quare non divinas scripturas in ipsa litteratura tractatis. ita vero, sed non ad hoc didicimus litteras. non enim parentes nostri, quando nos in scholam mittebant, hoc nobis dicebant: discite litteras, ut habeatis unde legere possitis codices dominicos.' Sim. at *s.* 70. 2. 2; as proverbial terror, cf. *sol.* 2. 11. 20, *s. Guelf.* 1. 11 ('ferulas et virgas grammaticorum'), *civ.* 21. 14 and 22. 22, and *nat. et gr.* 1. 1 ('paedagogi timor'). Hardness of heart and sympathy in a single formulation: *s.* 56. 13. 17, 'puto enim quod filium tuum diligis et quando caedis. lacrimas vapulantis non curas, quia ei haereditatem servas.' Punishment is probably part of the monastic life for A. as well: cf. *ord. mon.* 10. That text is probably not by A., but may have been known to him: see Verheijen, *La Règle de saint Augustin* (Paris, 1967), 2. 125–74, and Lawless, *Rule*, 167–71. The longer and more authentic *praec.* does not contain any provision for corporal punishment. Cf. also the whole of *correpi.*

L. C. Ferrari, *Aug. Stud.* 5 (1974), 1–14, links these beatings to the image of the scourging God found in A.; elsewhere in *conf.* at 2. 2. 4, 3. 3. 5, 4. 1. 1, 6. 6. 9 ('baculo disciplinae tuae confringebas ossa mea'), 8. 11. 25, 9. 4. 12; earliest, according to Ferrari, in quotations of Heb. 12: 6 in *div. qu.* 73 and *Gn. c. man.*; see Ferrari for other references to God as the scourge of human wickedness in A. Ferrari thinks Patricius was the sort of father likely to beat his children (since he seems to have beaten his wife: 9. 9. 19), but admits there is no evidence. A. makes much of schoolhouse beatings and not of any at home.

et multi ante nos . . . filiis Adam: To be reminded of Adamic descent is not a pleasant thing: 8. 9. 21 ('et tenebrosissimae contritiones filiorum Adam'); cf. 8. 10. 22, 13. 21. 30. Cf. Sirach 40: 1, 'iugum grave super filios Adam, a die exitus de ventre usque in diem sepulturae in matrem omnium'; Jer. 32: 19, 'cuius oculi aperti sunt super omnes vias filiorum Adam'; Gn. 3: 16, 'et mulieri dixit: multiplicans multiplicabo tristitias tuas et suspiria tua et in doloribus paries filios tuos' (text follows *Gn. c. man.* 2. 1. 2.); cf. 6. 6. 9, 'aerumnosis anfractibus et circuitibus ambiebam'.

invenimus: The first narrated encounter with the Christian religion, in the observation of others' prayer and the attempt to imitate it. He ascribes the failure of that first attempt, the first in a long series of false starts to be narrated, to the faulty knowledge of God on which it was based ('sentientes te, ut poteramus, esse magnum aliquem'). A. regularly emphasizes accurate knowledge of God as a prerequisite for spiritual progress, with the most explicit discussion from 7. 1. 1. Note 'invocationem' and cf. with discussion above on 1. 1. 1 ff.—especially for the way *invocatio* unties his tongue and makes speech possible where before there was silence; *confessio* could have begun here, but did not.

exaudire: 17 × in *conf.*; see on 1. 7. 11.

subvenire: Less common than *exaudire*: 6 × in *conf.* (3 × in Vg. OT, not in NT; never combined with *exaudire*); not specially associated with answered prayers.

auxilium et refugium meum: The phrase has a scriptural ring, but no one definite antecedent. Closest: Ps. 93: 22, 'et factus est mihi dominus in refugium et deus meus in auxilium spei meae.' Knauer, 36, catalogues both *auxilium* and *refugium* in Psalm language, but *auxilium* is used there not in address to God but as something predicated of him. *Refugium* occurs in collocations of addresses, as Ps. 17: 3, 'firmamentum meum et refugium meum et liberator meus, deus meus adiutor meus'. 'Dieses Beispiel zeigt, daß Augustin sich nicht an wörtliche Zitate zu binden braucht, sondern dazu neigt, aus verschiedenen Elementen, die ihm aus den Psalmen geläufig sind, neue Gruppierungen zu bilden, die den Eindruck eines wörtlichen Zitates hervorrufen.' *en. Ps.* 17. 3, 'refugium meum . . . quia refugi ad te; refugi autem quia liberasti me. . . . deus meus, qui mihi adiutorium prius vocationis tuae praestitisti, ut sperare in te possim.'

nam puer coepi rogare te: This first mimetic prayer goes astray; scripture provides a context and an explanation. Ps. 21: 2–3, 'verba delictorum meorum: (3) deus meus, clamabo ad te per diem et non exaudies et nocte et non ad insipientiam mihi' (text follows *en. Ps.* 21. en. 2. 3–4); *en. Ps.* 21. en. 1. 3, 'nec exaudies. deus meus, clamabo ad te in rebus prosperis huius vitae, ut non mutentur; nec exaudies, quia verbis delictorum meorum ad te clamabo. et nocte . . . et in adversis utique huius vitae clamabo ut prosperentur, et similiter non exaudies. neque hoc facis ad insipientiam mihi, sed potius ut sapiam quid clamare me velis; non verbis delictorum ex desiderio temporalis vitae, sed verbis conversionis ad te in vitam aeternam.'

non erat ad insipientiam mihi: The phrase is difficult, reflecting the LXX: καὶ οὐκ εἰς ἄνοιαν ἐμοί; the Hebrew has a different sense. A.'s own translators have not achieved either clarity or mutual agreement; best is Pusey, 'not thereby giving me over to foolishness'. The leap of thought for A. seems to have been: 'you heeded not my prayer, so that I would not end in folly (which would have been the result had you granted my wish).' *en. Ps.* 21. en. 2. 4, 'sed non ad insipientiam, sed ad sapientiam'. (The same etymological opposition at e.g. Cic. *fin.* 1. 14. 46 and *Tusc.* 3. 5. 10 ('ita fit ut sapientia sanitas sit animi, insipientia autem quasi insanitas quaedam').) The word is rare in CL, but more common in scripture representing Gk. ἄνοια. Elsewhere in A. in passages echoing this verse: *en. Ps.* 43. 2, 53. 5 ('ut intellegerem quid a te petere deberem'), 59. 7, *civ.* 21. 14, *ep.* 140. 7. 19 (and often in that letter).

mali: A. has no rigorous distinction between *malum* as a term for moral evil ('wrong-doing') and *malum* as a term for natural evil ('suffering': here 'who wanted nothing bad to happen to me'); the former meaning greatly preponderates. Though he repeatedly asks (from 3. 7. 12 to 7. 7. 11) 'unde malum,' in all of *conf.* only the following passages besides the present one certainly refer to natural evil: 3. 8. 16 ('vel evitandi mali . . . vel sola voluptate alieni mali'), 7. 5. 7 ('certe vel timor ipse malum est') and in a medical context 8. 3. 7 ('et vena eius malum renuntiat').

plagae: Cf. 'vapulabam'; 1. 17. 27, 'plagarum metu' (also of school).

1. 9. 15

When he wrote these lines, A. would have recognized with a wry smile the sentiments of a child's prayer at baptism recorded in Eustratius, *vita Eutychii* 8 (*PG* 86. 2283–4; sixth century): κύριε, ἀγαθὸν νοῦν χάρισαί μοι, ἵνα μάθω τὰ γράμματα καὶ νικῶ τοὺς ἑταίρους μου.

Estne . . . affligebamur?: G–M: 'An unusually long, invertebrate sentence'. We are to imagine a devout man, so devout that when faced himself with the tortures of the state, he thinks nothing of them (though others fear them bitterly)—the measurement of how little he thinks of them is the unconcern of parents over the punishment of their children in school. Just as we marvel at the unconcern of the martyr, so the child marvels at the unconcern of his parents.

eculeos et ungulas: For particularly harsh torture, as is made clear at *ep.* 133. 2, to the imperial commissioner Marcellinus: 'noli perdere paternam diligentiam, quam in ipsa inquisitione servasti, quando tantorum scelerum confessionem non extendente eculeo, non sulcantibus ungulis, non urentibus flammis sed virgarum verberibus eruisti.' Cf. also *ep.* 43. 4. 13 and *s.* 13. 6. 7.

diligens: G–M: 'concessive, "though he loves"'. *BA* follows for its translation the correction found in one manuscript (and printed by the Maurists), *deridens*.

minus ea metuebamus aut minus: sc. *quam homines qui tormenta metuunt*.

et peccabamus tamen: The admission of actual sin, well-remembered: not for inability did he not study, but out of a delight in play. It never occurs to him that the child's idleness could be anything but culpable. The right of the teachers to punish this sin, on the other hand, is brought severely into question. The system in which young A. was being brought up was profoundly disordered and ungodly; it was not for his failure to participate in that system as such that he was worthy of punishment, but for other reasons (see next paragraph).

delectabat ludere: Cf. 'amore ludendi' (1. 10. 16, 1. 19. 30); on *delectare*, see on 1. 6. 7.

maiorum nugae negotia vocabantur: Cf. 1. 10. 16 ('in spectacula, ludos maiorum'); *s.* 3, 'omnis lusus puerorum simulacrum est negotii maioris.' Lactant. *inst.* 2. 4. 14, 'merito igitur etiam senum stultitiam Seneca [frg. 121] deridet: "non", inquit, "bis pueri sumus, ut vulgo dicitur, sed semper: verum hoc interest, quod maiora nos ludimus.".'

bonus rerum arbiter: Ironic.

deformius: Adj./adv. in *conf.* used of those things that do not bear the pattern that God imparts (cf. 10. 27. 38, 'et in ista formosa quae fecisti deformis inruebam')—a negative counterpart of the second person of the trinity (see above on 1. 7. 12).

quaestiuncula: The talent for *quaestiunculae* is one that A. exploited as a Manichee (3. 12. 21, 'quod . . . nonnullis quaestiunculis iam multos imperitos exagitassem'). Ironic depreciation of pedantry is as essential to an educational system as pedantry itself; *quaestiuncula* in Latin has no other force (cf. Cic. *de or.* 1. 22. 102, '"quid? mihi vos nunc", inquit Crassus, "tamquam alicui graeculo otioso et loquaci et fortasse docto atque erudito quaestiunculam . . . ponitis?"'). The irony can be pleasant and indulgent, or dismissive: in A., indulgent at *epp.* 13. 2 (to Nebridius, still in the atmosphere of Cassiciacum), 37. 3 (to Simplicianus), and 80. 2 (to Paulinus and Therasia), dismissive at *mag.* 8. 21, *util. cred.* 6. 13, and *c. Faust.* 33. 8 (and directed against A. by Petilian at *c. litt. Pet.* 3. 52. 64 and by Faustus at *c. Faust.* 1. 2).

1. 10. 16

tamen: For all that he can criticize the punishment he received at the time, he still deserved it; cf. 1. 9. 15, 'peccabamus tamen' of the same event.

ordinator [1]: See on 1. 7. 12. Earliest at *Gn. litt. imp.* 5. 25, 'et umbrae in picturis eminentiora quaeque distinguunt, ac non specie sed ordine placent. nam et vitiorum nostrorum non est auctor deus, sed tamen ordinator est. . . . iustos et facit et ordinat; peccatores autem in quantum peccatores sunt non facit sed ordinat tantum. . . . privationes autem specierum defectusque naturarum non facit sed ordinat tantum.' Sim. at *en. Ps.* 7. 19, 9. 20, *c. Faust.* 22. 78, *Gn. litt.* 1. 17. 33, 1. 17. 34, 3. 24. 37, 8. 9. 18, *civ.* 11. 15–18, esp. 11. 17 ('sed deus sicut naturarum bonarum optimus creator est, ita malarum voluntatum iustissimus ordinator'), and *civ.* 14. 26, with discussion at *civ.* 19. 13.

amore ludendi [3] **. . . superbas victorias** [1] **. . . curiositate** [2]: At 10. 30. 41, A. uses 1 Jn. 2: 16 ('quoniam omne quod in mundo est, concupiscentia carnis est, et concupiscentia oculorum, et ambitio saeculi') to arrange the examination of conscience he there undertakes. This commentary will identify the earlier occasions where that triple pattern reflects itself in A.'s analysis of his fall into sin and dereliction and later in his narrative of his reascent to God's favour; see Kusch, and above, introduction. W. Theiler, *P. u. A.* 37, insists that the pattern gains its hold over A. via Porphyry; as with the triad *modus/species/ordo* at 1. 7. 12, he holds that the scriptural text is pretext or corroboration for a philosophical doctrine. But here Theiler cannot show the three temptations *as a triad* in Porphyry; he begs the question by arguing, to prove that the origin is Porphyrian, that the triad occurs without the biblical citation at *lib. arb.* 2. 19. 53; similarly to show (as Theiler tries to do at *P. u. A.* 41–2) that A.'s treatment of the temptations in Bk. 10 reflects various unconnected passages of Porphyry's *de abstinentia* would not show a priority of influence, only a happy confluence (and some originality on A.'s part). R. J. O'Connell, *Traditio* 19 (1963), 24 ff., claims descent from Plotinus. See further on 10. 30. 41.

Here it seems certain that 'amore ludendi' reflects the *concupiscentia carnis* [3] that will be A.'s most acute temptation, 'superbas victorias' seems self-evidently to describe the fruit of the *ambitio saeculi* [1] (in other biblical translations sometimes *superbia vitae*), and 'curiositate' [2] (note 'magis magisque per oculos emicante') represents *concupiscentia oculorum*. The presence of all three temptations *in nuce* here shows that the *puer* Augustine was completely innocent in no essential way: the decline traced through Bks. 2–5 is then not the primeval fall from grace, but a further fall into actual sin, leaving a double redemption for Bks. 6–9 to recount.

scalpi aures meas falsis fabellis: 2 Tim. 4: 3–4, 'erit enim tempus cum sanam doctrinam non sustinebunt, sed ad sua desideria coacervabunt sibi magistros prurientes auribus, (4) et a veritate quidem auditum avertent; ad fabulas autem convertentur'; also echoed at 4. 8. 13. Cf. *adn. Iob* on 31: 40, '"pro tritico germinet mihi urtica": pro spiritalibus doctoribus magistri prurientium auditu per confricationes hominum mente corruptorum et a veritate destitutorum.' The itching ears belong to the Manichees particularly at *c. Sec.* 26.

ludos OS Knöll Skut. Ver.: *ludosque* CDG Maur. (blunting the point). On *ludi* and *spectacula*, see on 3. 2. 2 and 6. 8. 13.

ad talia edenda: Whatever the *niveau* of A.'s family (see on 2. 3. 5), his education and career led him among the higher orders of late Roman society, where his education was naturally supposed to lead to a social eminence entailing public responsibilities. A. never completely joined that class, but never completely quit it either. Every page of *civ.*, for example, vaunts his *savoir faire*, even as he protests his aloofness.

vide ista: Ps. 24: 16–18, 'respice in me, et *miserere* mei, quoniam unicus et pauper sum ego. (17) tribulationes cordis mei multiplicatae sunt; de necessitatibus meis *libera* me. (18) *vide* humilitatem meam et laborem meum.'[34] Ps. 78: 9, 'propter gloriam nominis tui, domine, libera nos'; *en. Ps.* 78. 12, 'sic ergo nos liberas, id est, eruis a malis, cum et adiuvas nos ad faciendam iustitiam, et propitius es peccatis nostris, sine quibus in hac vita non sumus.' Cf. Mt. 6: 13 (the Lord's Prayer), 'sed libera nos a malo'.

The sense of 'vide' here ('to look upon with favour': cf. 'vidisti' in 1. 11. 17) is biblical rather than classical. The same imperative recurs through *conf.*, corresponding to the assertions that confession proceeds *in conspectu dei* (see on 1. 16. 26): 1. 18. 29, 4. 6. 11, 6. 6. 9, 10. 33. 50, 11. 2. 4, 12. 25. 35, 13. 18. 22.

libera . . . eos . . . ut invocent te: See on 1. 1. 1.

1. 11. 17

audieram: The verb occurs 175 × in *conf.*, repeatedly, as here, for the means of transmission of Christian teachings, not always completely understood—

[34] D. De Bruyne, *MA* 2. 553, uses this passage to argue that A. deliberately revised/corrected against the LXX Greek a Psalter close to the Veronensis (whose reading the citation here follows for its parallel to the text). A. in *en. Ps.* 24. 17 reads *educ* (following LXX ἐξάγαγε) for *libera*. Here the Veronensis text seems to persist in memory.

cf. 7. 3. 5, 'quod audiebam, liberum voluntatis arbitrium causam esse'. Here he emphasizes that already in boyhood he had heard the core of the gospel message and undergone sacramental initiation short of baptism. As many have said, A. was never a 'pagan' in the conventional acceptation of that word; his assertion here is designed to make it clear that he was even more 'without excuse' (Rom. 1: 20, 'ut sint inexcusabiles': see on 7. 9. 14) than the traditional 'pagans'.

per humilitatem domini dei nostri: i.e. by the incarnation: cf. Phil. 2: 8, 'humilavit se factus oboediens usque ad mortem' (cited at 7. 9. 14); cf. 7. 9. 13 ('via humilitatis'), 8. 2. 3 ('humilitatem Christi'), 8. 2. 4 ('de sacramentis humilitatis verbi tui'), and 10. 43. 68 ('verax autem mediator, quem ... misisti ut eius exemplo etiam ipsam discerent humilitatem').

signabar: Catechumenate represented sacramental (*bapt.* 4. 21. 28, 'catechumeni sacramentum'; *s. Mai* 94. 1, 'catechumeni, quos iam nonnullo sacramento mater concepit ecclesia') admission to the Christian community, but not full membership. This 'of-but-not-in' status 'was not an anomaly in Greco-Roman antiquity, but a familiar religious posture', even in Judaism (J. Reynolds and R. Tannenbaum, *Jews and Godfearers at Aphrodisias* (Cambridge, 1987), 88). Even infants at the breast were eligible (*s.* 324). The only mention of his infancy after Bk. 1 recalls this event when he has resumed the status acquired thereby: 6. 4. 5 ('mihi nomen Christi infanti est inditum'); the mention here in the section devoted to *pueritia* is out of narrative place. See van der Meer, 365–6, on the ritual (he may understate its force at 366: 'None of this seems very precise, and one cannot help thinking that this particular rite had, by Augustine's day, become little more than a ceremony. ... No doubt the majority of catechumens will have thought of some kind of safeguard against spiritual corruption—a kind of antidemonic specific'). The salt may have been received repeatedly (a eucharist surrogate for catechumens?—it is the least documented part of the rite): *brev. Hipp.* canon 3 (*CCSL* 149. 33), 'ut etiam per sollemnissimos paschales dies sacramentum catechumenis non detur, nisi solitum sal.' The best source is *cat. rud.* 26. 50 (how the rite was modified for *infantes* is unclear): 'his dictis, interrogandus est an haec credat atque observare desideret. quod cum responderit, solemniter utique signandus est et ecclesiae more tractandus.' In that passage, A. expressed a desire for what he had not himself obtained, that the rite be supplemented by doctrine to good effect; he continues, 'deinde monendus est ex hac occasione, ut si quid etiam in scripturis *audiat* quod carnaliter sonet, etiamsi non intellegit, credat tamen spiritale aliquid significari quod ad sanctos more futuramque vitam pertineat. hoc autem ita breviter discit, ut quidquid *audierit* ex libris canonicis quod ad dilectionem aeternitatis et veritatis et sanctitatis et ad dilectionem proximi referre non possit, figurate dictum vel gestum esse credat; atque ita conetur intellegere, ut ad illam geminam referat dilectionem.' At *s.* 215. 5 (authenticity doubtful), he speaks only of the signing with the cross: 'quando primum credidisti, signum Christi in fronte tamquam in domo pudoris accepisti.'

Later, he qualifies the force of the rite in the heat of anti-Pelagian controversy, but there is little anachronism: *pecc. mer.* 2. 26. 42, 'non unius modi est sanctificatio; nam et catechumenos secundum quendam modum suum per signum Christi et orationem manus impositionis puto sanctificari. . . . sanctificatio catechumeni, si non fuerit baptizatus, non ei valet ad intrandum regnum caelorum aut ad peccatorum remissionem.' (That matches closely what Ambrose would have said: Amb. *myst.* 4. 20, 'credit autem etiam catechumenus in crucem domini Iesu qua et ipse signatur, sed nisi baptizatus fuerit . . . remissionem non potest accipere.')

The application of the sign of the cross (not necessarily only at the moment of entering the catechumenate) was often reflected later: *Io. ev. tr.* 118. 5, 'postremo quid est, quod omnes noverunt, signum Christi, nisi crux Christi? quod signum nisi adhibeatur sive frontibus credentium [i.e. catechumenate], sive ipsi aquae ex qua regenerantur [baptism], sive oleo quo chrismate unguntur [anointing], sive sacrificio quo aluntur [the mass], nihil eorum rite perficitur.' Cf. also *cat. rud.* 20. 34 (the cross like the sign of passover on the doorposts), *Io. ev. tr.* 11. 4, 50. 2, *s. Den.* 17. 8, *s. Mai* 94. 1, and *s.* 160. 5. See also Morin's note at *MA* 1. 89 n. 6: 'equidem scio nequaquam id proprie, sed de ductis digito lineis accipiendum: verumtamen saepe miratus sum, cum in Africa septentrionali brevi tempore versarer, plerosque indigenas, etsi multis saeculis Islamitico iugo subditos, ac religioni christianae infensos, crucis signum caeruleis notis fronte compuncta aperte gestare; quo se a parentibus iuxta morem a maioribus receptum protinus a nativitate insignitos esse, haud sine evidenti fastidio mihi confessi sunt.'

quae multum speravit in te: The first hint of Monnica's piety.

custos meus: Skut. and Ver. adduce Job 7: 20, 'quid faciam tibi o custos hominum?' and Gn. 28: 15, 'et ero custos tuus quocumque perrexeris,' but in VL neither text contained the word *custos* (Gn. 28: 15 quoted at *civ.* 16. 38 with 'custodiens te').

dei et domini mei: Jn. 20: 28, 'respondit Thomas et dixit ei, "dominus meus et deus meus".'

flagitavi: The same verb recalls this event at 5. 9. 16. This rush to baptism was unsatisfactory to some, though A. himself *c.*408 was tolerant: *ep.* 98. 5, 'nec illud te moveat, quod quidam non ea fide ad baptismum percipiendum parvulos ferunt ut gratia spiritali ad vitam regenerentur aeternam, sed quod eos putant hoc remedio temporalem retinere vel recipere sanitatem.'

matris omnium nostrum: Gal. 4: 26, 'illa autem quae sursum est Hierusalem libera est, quae est mater nostra.' This is the only *scriptural* warrant for what is a common form of expression for A. (who is notably more favourable toward mothers than fathers): 6. 3. 4 ('matre catholica'), 7. 1. 1, 9. 13. 37, 12. 15. 20, 12. 16. 23; elsewhere cf. e.g. *en. Ps.* 26. en. 2. 18, *ep.* 243. 8. The image is ancient and widespread (J. C. Plumpe, *Mater Ecclesia* (Washington, DC, 1943)) and frequent in A. (P. Rinetti, *Aug. Mag.* 2. 827–34). See Excursus below on mothers and fathers in *conf.*

sacramentis: For *sacramentum* in A., see C. Couturier, in H. Rondet, *et al.* (eds.), *Études augustiniennes* (Paris, 1953), 161–332; see also Mayer, *Zeichen*, 1. 284–331 and 2. 454–7. The word occurs first at *mor.* 1. 7. 12 (of incarnation: 'suscepti hominis sacramento'), recurs importantly in *vera rel.*, a sign of the transformation already occurring in that work; and as a word that helps bridge the gap between things and signs: sacraments are signs that are also things (cf. *doctr. chr.* 1. 2. 2, etc.).

confitens in remissionem peccatorum: Mk. 1: 4–5, 'baptismum paenitentiae in remissionem peccatorum; (5) . . . et baptizabantur ab illo in Iordane flumine, confitentes peccata sua.' Not for the last time, a chance for *confessio* lost.

recreatus essem: He was revived in a lesser but not in a greater way: cf. e.g. *en. Ps.* 45. 14, 142. 17 ('"tu es deus meus", qui creator meus, qui creasti me per verbum tuum, et recreasti me per verbum'), *Io. ev. tr.* 1. 12.

dilata est: See 1. 11. 18 for fuller discussion of the deferral.

mundatio mea: Lev. 16: 30 (prescribing Yom Kippur), 'in hac die expiatio erit vestri atque mundatio ab omnibus peccatis vestris coram domino mundabimini.' *en. Ps.* 94. 11, '"ascendentium de lavacro" [Cant. 6: 5: note 'lavacrum' here]; id est, ascendentium de mundatione' (the Cant. passage has baptismal overtones for A.: *doctr. chr.* 2. 6. 7).

quasi necesse esset . . .: Apoc. 22: 10–12, 'et dicit mihi, ne signaveris verba prophetiae libri huius: tempus enim prope est. (11) qui nocet, noceat adhuc; et qui in sordibus est, sordescat adhuc; et iustus iustitiam faciat adhuc; et sanctus sanctificetur adhuc; (12) ecce venio cito.' NB here, 'quasi'.

post lavacrum illud: Tit. 3: 5–6, 'salvos nos fecit per lavacrum regenerationis et renovationis spiritus sancti, (6) quem effudit in nos abunde per Iesum Christum salvatorem nostrum.'

ita iam credebam: The observation is almost casual—for to A., without baptism, 'faith' was of little value.

nisi pater solus: Patricius' status in a household where women led the way to Christianity was not uncommon (cf. P. Brown, *JRS* 51 (1961), 1–11), nor was his deathbed baptism (9. 9. 22); Possidius, who knew *conf.* but was not overly attentive, could say of A. (*v. Aug.* 1.1) that he was 'parentibus honestis et christianis progenitus'. There is no evidence that there was ever anything 'pagan' (i.e. devoted to cults other than Christian) about Patricius.

ut tu mihi pater esses: Ps. 26: 10, 'pater meus et mater mea dereliquerunt me; dominus autem adsumsit me'; *en. Ps.* 26. en. 2. 18, 'exceptis etiam illis duobus parentibus, de quorum carne nati sumus . . . habemus hic alium patrem et aliam matrem, vel potius habuimus. pater secundum saeculum diabolus est, et fuit nobis pater cum essemus infideles; nam infidelibus dicit dominus, "vos a patre diabolo estis" [Jn. 8: 44]. . . . cognovimus alium patrem, deum; reliquimus diabolum.' We have thus an obscure description of a struggle that may not have been entirely the product of imagination. The tension of the sentence arises from M.'s duty to serve her husband, a duty that in some sense she violated in the struggle over A.'s paternity. A higher authority is invoked to justify her, but if the struggle was more than figurative (quarrels about religion, paternal opposition to distracting the boy with religion, etc.), her part must have seemed at least a daring one. 'Tamen' and 'non evicit' in the preceding sentence both seem to speak of the same struggle.

serviebat: See on 13. 32. 47 for A.'s habit of substituting *servire* for *subdi* when speaking of M. in terms of this scriptural language; as at 9. 9. 19, 9. 13. 37.

EXCURSUS: MOTHERS AND FATHERS IN THE *CONFESSIONS*

The other real-life mothers in *conf.* are noticeably less spiritual than Monnica, but the metaphorical applications are mainly generous if various (for metaphorical application to the church, see above, pp. 68–9): 3. 10. 18 (Manichean doctrine about plants), 3. 12. 21 (ex-Manichee Christian bishop given in youth to the Manichees by his own 'seducta mater'), 5. 5. 9 ('a caritate matre sustinetur'), 6. 1. 1 (the mother of the son raised from dead at Lk. 7: 12–15), 6. 10. 17 (Nebridius' mother not following him to Italy), 7. 6. 8 (mother of Firminus having a dream); 8. 11. 27 ('ipsa *continentia* nequaquam sterilis, sed fecunda mater filiorum gaudiorum'), 9. 7. 15 (mother of Valentinian), 9. 8. 17 (Monnica's mother, who neglected her daughter's welfare), 9. 9. 20 (Monnica's mother-in-law), 13. 6. 7 ('obsecro te per matrem caritatem'), 13. 14. 15 ('nox, mater iniquorum'), 13. 19. 24 ('honorem matris et patris et dilectionem proximi'), 13. 32. 47 ('herbarumque atque arborum mater'). On a specific maternal image applied to God, see on 4. 1. 1, 'sugens lac tuum'.

On the other hand, a non-mortal *pater* is always God—there are no other metaphors: a much more limited role! A complete list of *patres* other than God or Patricius follows: most appear in no good light.

1. 6. 10 (ancestors generically), as also 3. 2. 3 ('deo patrum nostrorum'), 3. 7. 14 ('reprehendebam caecus pios patres [Veteris Testamenti]'), 6. 7. 11 f. (the father of Alypius, with whom A. quarrelled), 7. 6. 8 f. (the father of Firminus, who had taught him astrology), 8. 2. 3 (the only non-physiological father: 'Simplicianum, patrem in accipienda gratia tunc episcopi Ambrosii et quem vere ut patrem diligebat'), 9. 8. 17 f. (Monnica's father (also not in a good light)), 10. 34. 52 (Jacob, son of Isaac and father of Joseph), 13. 18. 23.

Most interesting, Patricius' death is only noted in passing at 3. 4. 7 ('iam defuncto patre ante biennium'), and at 3. 6. 10 we have the first vocative to God the Father ('mi pater summe bone'). The only earlier ref. to God as A.'s father is anticipatory here, 'nam illa satagebat ut tu mihi pater esses'. See on 12. 16. 23 then for the eventual displacement of Monnica by 'Hierusalem mater'.

On fathers and sons in A., see B. Shaw, *Past and Present*, 115 (1987), 19–28, interpreting A.'s 'distant, formal and somewhat fearful' relations with Patricius as not uncommon in the time and place: 'his concomitant attachment to his mother, brother and sister, do[es] not seem so unusual.' Indeed not, if we think of another ancient nuclear family that has had much attention in after times: the household of Jesus, Mary, and Joseph.

For the most responsible psychological interpretation arising out of these data, see C. Kligerman, *Jour. Amer. Psychoan. Assoc.* 5 (1957), 469–84. He lays emphasis on the Dido/Monnica parallel (see on 5. 8. 15), suggesting that A. interprets his life as living out the same story he wept over as a child and sees in Faustus and Ambrose potential father figures, the first of whom disappoints, the second of whom never fully satisfies. (A reader who sets a higher value on baptism than Kligerman does would disagree: Ambrose does become A.'s father, but in an unexpected way, and in another way it is God who becomes A.'s father, in accord with M.'s expressed wish here—and the fulfilment of such a mother's wish is significant.) On the other hand, the suggestion made by Rebecca West in her *Saint Augustine* (London, 1930), and revived by E. TeSelle, *Jour. Sci. Stud. Rel.* 25 (1986), 93–4, that A. came to resemble Patricius more and more as he aged ('increasingly high-handed, officious, authoritarian') is an elegant example of the cloud-ranching to which psychohistory can easily lead.

If of the three temptations (see above on 1. 10. 16), the one that A. least securely vanquishes is *ambitio saeculi* (see on 10. 36. 59 ff.), the implication remains in the air like mist that it is God *the father* with whom A. is never on completely comfortable terms. And of course, the theme of the prodigal son is as much about fathers and estrangement as it is about sons, and there are some surprising ramifications to that observation: see on 1. 18. 28.

1. 11. 18

vellem: The imperfect of an optative subjunctive denotes the wish as unaccomplished in present time, but the following clause, 'si tu etiam velles', creates a present contrary-to-fact condition: A. does not know, because God does not will him to know. *Confessio* forgoes wanting to know out of deference to that will of God, and thus separates itself from *curiositas*. Only the earthly reason for the delay (M.'s prudence) is specified below.

(*non laxata*) **sunt** CDGO[1] Maur. Ver. : *sint* O[2]S Knöll Skut. The subjunctive is appropriate only if punctuated to include this clause in the preceding indirect question. The arrangement printed here (after Verheijen) is correct: 'Was it for my good that the reins of sin were, after a fashion, loosed for me? Or were they really?' (De Marchi, 310–11, had previously suggested deleting the second 'laxata sunt/sint'.)

etiam nunc: Cf. van der Meer, 148–51, on the practice of deferral and A.'s episcopal campaigns against it.

sonat undique: Voices on all sides drove him in directions he should not have gone; the one voice he should have heeded was inaudible; cf. 1. 13. 21, 'et fornicanti sonabat undique'.

recepta salus animae meae: Ps. 34: 3, 'dic animae meae, "salus tua ego sum".' See on 1. 5. 5.

melius vero: A bittersweet reflection on the past. What follows is not reason but rationalization, reason manipulated by worldly prudence.

noverat eos iam illa mater: She foresees (but escapes reproach) what Patricius is criticized for observing (2. 3. 6).

terram: Cf. Gn. 1: 2 as interpreted throughout Bk. 12 (where it = *informis materia*) and employed at 13. 12. 13, 'et terra nostra antequam acciperet formam doctrinae invisibilis erat et incomposita et ignorantiae tenebris tegebamur'; Gn. 2: 7, 'et finxit deus hominem de limo terrae' (regularly but not always in this form in A.; see *VL* (Beuron)). Cf. also 9. 11. 28, 'ut coniuncta terra amborum coniugum terra tegeretur', of the mortal remains of Monnica and Patricius.

per eos: *potius* coni. Knöll, followed by Vega.

effigiem: Infrequent in A., only here in *conf.* Take the sentence thus: 'But how many and what great waves of temptation were to be seen threatening the other side of boyhood: that mother of mine already knew them, and wanted to risk in them the earth from which I would afterwards be given shape, rather than the new image itself.' The parallel at 13. 12. 13 just quoted and the 'iam' are decisive in what has been a *crux interpretum*. But 'unde' need not refer to 'terram': it may perhaps to 'eos' [sc. *fluctus*], making the 'waters' both the floods of temptation and the water from which he would be baptized. Grammatically this solution (suggested to me by Professor Henry Chadwick) is slightly superior to all others.

Cf. Mk. 16: 12 (of the risen Christ), 'post haec autem duobus ex eis ambulantibus ostensus est in alia effigie euntibus in villam.' *Effigies* is therefore the person created by the regeneration of baptism, replacing the old, outworn earth, formless and void, emphasizing the physical form of the baptized body: clearest at *trin.* 11. 1. 1, where the 'vestigium trinitatis' is sought no longer in the inner person but now in the outer: 'in hoc ergo qui corrumpitur quaeramus quemadmodum possumus quandam trinitatis effigiem'; cf. *civ.* 20. 9, where he speaks of sham Christians thus: 'fingunt enim se esse quod non sunt, vocanturque non veraci effigie sed fallaci imagine christiani' (see also *civ.* 17. 5); *Io. ev. tr.* 31. 3, 'secundum carnem et effigiem hominis quam gerebat [Christus]'. The incarnate Christ and carved statues can both be *effigies*. The operative Genesis echoes are 1. 2 and 1. 26; 2. 7 only indicates the way in which the human condition can be read into 1. 2 (see preceding note for quotations). Cf. van Bavel, 45: 'effigies hominis' occurs 2 × in Aug.'s *œuvre* of Christ, and should be seen in a series of rough incarnational synonyms: *filius hominis*, *forma servi*, *forma hominis*, *effigies hominis*, *forma humana*. (This interpretation anticipated by J. le Clercq, *PL* 47. 204.)

committere: Cf. *en. Ps.* 32. en. 1. 16, 'nec quisquis . . . salvus erit, si se suae fortitudini multum commiserit'; *en. Ps.* 36. s. 3. 7, 'hoc est semen tuum; . . . terrae committis, et tanto amplius conligis; Christo committis, et perdis?' And so forth, abundantly with the dative. On close examination of every occurrence of *committo* in *en. Ps.* and *conf.*, the only senses are (1) with the acc. in the sense of 'committing sins' and (2) with acc. + dat. in the sense of entrusting *x* to *y*. From *OLD* and *TLL* 3. 1909, it is clear that it can also occur with acc. + *in* + acc., 'entrusting *x* to *y*.' There seems to be no parallel for the use here, with acc. + *per* + acc., which is the difficulty that has exercised the critics.[35] Knöll's emendation at least goes to the likeliest focus of the problem, but to read *potius* takes away the indirect object; and in the sense of 'entrust/risk' *committo* always takes an explicit or implied indirect object, which is the function that in spite of its oddity *per eos* does indeed fulfil (Arts, 24, counts 170 × in *conf.* where *per* + acc. expresses instrument or means).

1. 12. 19

This paragraph makes one twofold thematic statement about A. and his God: that A. did not like to study ('non amabam litteras'), and that God guided A. through his studies to a good end ('et bene mihi fiebat abs te'). The rest of the paragraph is a meditation on those assertions in a way characteristic of *conf.*: other people appear in it, but as bit players in the main drama between A. and God. Where human actions seem to be determinative, A. seeks to show that they are not. A meditation of this nature is more attractive to us than when he preaches predestinate grace against the Pelagians; but the latter crusade is of a piece with this passage.

tamen: Boyhood has perils enough; cf. 'adulescentia' with 'post pueritiam' just above in 1. 11. 18.

fiebat: The (virtual) passive for contrast with what follows: 'nec faciebam', 'nec . . . faciebant'). With 'tu bene faciebas mihi', the shift from passive to active shifts attention to God's action away from A.'s experience of that action.

non enim discerem: 'enim' introduces the minor premise; 'autem' the major premise; the conclusion is what has already been stated, 'bene mihi fiebat', and will be restated with the addition of 'abs te' presently.

[35] Any such collocation is extremely rare in A. and when found has a different sense, e.g. *c. litt. Pet.* 2. 78. 174, 'quae vestri cotidie committunt per scelera furiosorum', where *per* is clearly 'by means of'; so too the only other example, *s.* 162. 2.

nemo autem invitus . . . quod facit: The sentiment resembles the tenor and development of Amb. *ep.* 37, addressed to Simplicianus (see on 8. 1. 1), itself heavily influenced by Philo. Amb. *ep.* 37. 19–20, 'sapiens ergo liber, quoniam qui ea facit quae vult liber est. . . . sapiens omnia quae facit bene facit. . . . (20) etenim qui neque cogi potest neque prohiberi, is nequaquam servus. sapientis autem est neque cogi neque prohiberi non est igitur servus sapiens. . . . quod si cogitur, manifestum est quod invitus faciat. . . . *unde liquet quia nihil invitus facit sapiens neque cogitur,* qui si servus esset, cogeretur.' Cf. the gloss of Prosper of Aquitaine, who quotes the lemmatized words of *conf.* and adds (*sent.* 173), 'quia nihil prodest spiritus timoris, ubi non est spiritus caritatis.'

autem CDGO Ver. : *enim* S Knöll Skut. This is one of many places where Verheijen has altered the text he received from Skutella (and, in this case, Knöll) without reconsidering the punctuation.

quo referrem: 'to what use I would put'. What is stated in *doctr. chr.* 1. 22. 20 ff. about 'use' and 'enjoyment' appears more frequently in A. as a matter of the goal to which something is 'referred'. Cf. *civ.* 10. 6, 'quanto magis anima ipsa cum se *refert* ad deum, ut igne amoris eius accensa formam concupiscentiae saecularis amittat eique tamquam incommutabili formae subdita reformetur, hinc ei placens, quod ex eius pulchritudine acceperit, fit sacrificium.'

cupiditates: 31 × in *conf.*, never in a good sense; cf. 10. 41. 66, 'in cupiditate triplici'.

copiosae: For the antithesis with *inopia*, 2. 4. 9, 'nec copiosus adactum inopia'; for a similar oxymoron, 12. 1. 1, 'copiosa est egestas humanae intellegentiae'.

capilli GOS Knöll Skut. Ver. : *capilli capitis* CD Maur., by assimilation to Mt. 10: 30, 'capilli capitis vestri omnes numerati sunt.' The text offers an argument against fear of death (see on 6. 16. 26), in favour of resurrection of the flesh: *s.* 65. 2. 3, 'quid timeo ne carnem perdam, qui nec capillum perdo? . . . ipsum corpus quod percuti et occidi potest ad tempus cinis erit, in aeternum immortale erit'; *en. Ps.* 78. 5, 'ex occultis naturae sinibus totum redintegraturo illo, cui et capilli nostri numerati sunt.'

meo: sc. *errore*. These lines (through 'omnis *inordinatus* animus') demonstrate the *ordinatio peccatorum* suggested above at 1. 10. 16, and hence the role of *ordo* (1. 7. 12).

tantillus . . . tantus: Cf. *en. Ps.* 36. s. 2. 16, 'miraberis tantum dari pro tantillo labore.'

de non Maur. Pell. (supported by two minor manuscripts—one offering the reading only in a correction) : *non de* CDGOS Skut. Ver. The text insists on being taken in the way the Maurists' correction implies. The sense of the paragraph is not (as *non de* would make it) that there were people doing good whom God disdained to use, but that there were people doing no good but God used them anyway to do good for A. A. could not have written *non de facientibus* to mean *de non facientibus*. The sequence *non de* is common enough

(14 × in *conf.*; see below) to facilitate the change from *de non*; the latter is rare,[36] but does occur once, for a precisely similar antithetical effect to that which is required here: 10. 39. 64, 'sed sibi placentes multum tibi displicent non tantum de non bonis quasi bonis, verum etiam de bonis tuis quasi suis.' (With the correction, 'non bene facientibus' may perhaps be taken to include both A. and those who pressed him to his studies.) For *non* between preposition and its object, cf. also 3. 8. 16, 'vel in non concessa'; 6. 7. 11, 'quae in non magna aetate satis eminebat'. It is clear on the other hand that the sequence *non de* occurs elsewhere only where the *non* modifies some *other* element than the object of *de*. Occurrences in *conf.*: 3. 6. 10, 8. 4. 9, 10. 23. 33 ('quoniam qui non de te gaudere volunt'), 10. 23. 33, 10. 31. 46, 11. 2. 4, 12. 7. 7 ('fecisti enim caelum et terram non de te'), 12. 17. 25, 12. 28. 38, 13. 23. 33, 13. 30. 45, 13. 33. 48.

retribuebas: Ps. 141: 8, 'me sustinebunt iusti, quoadusque retribuas mihi.' The verb occurs 7 × in *conf.*, recurring importantly to voice the same idea as here, God's influence in forestalling A.'s wrongdoing and bringing good out of evil: 13. 1. 1, 'tu enim, domine, delevisti omnia mala merita mea, ne retribueres manibus meis, in quibus a te defeci, et praevenisti omnia bona merita mea, ut retribueres manibus tuis, quibus me fecisti.'

1. 13. 20

A.'s past remains unintelligible to him except as a history of self-will. His hostility to Greek is momentarily called to mind, then dropped (to return at 1. 14. 23) in favour of reflection on his course of study in Latin. The habits of misreading implanted here obstructed his approach to scripture for a long time.

quid autem: Usually punctuated thus: 'quid autem erat causae, cur graecas litteras oderam, quibus puerulus imbuebar, ne nunc quidem mihi satis exploratum est.'

graecas litteras: On A.'s Greek studies, the consensus (after Marrou, 27–46 and 631–7, Courcelle, *LLW* 149–65, B. Altaner, *Kleine Patristische Schriften* (Berlin, 1967), 129–53, along with Altaner's numerous articles on A.'s specific debts to Greek patristic literature) is that A. had limited but usable knowledge throughout his career (sufficient, for instance, to check readings in the Greek scriptures: see esp. D. De Bruyne, *MA* 2. 521–606), and that he enhanced that skill later in life (in the 410s) to expand his knowledge of Greek patristic literature. His own remarks are strongly marked by rhetorical self-depreciation, e.g. *c. litt. Pet.* 2. 38. 91, 'et ego quidem graecae linguae perparum assecutus sum et prope nihil.' Greek texts never came to

[36] The phrase *de non* in that order approx. 25 × in Aug., with a handful of cases (*ep.* 186. 11. 38, *gr. et pecc. or.* 1. 10. 11, *en. Ps.* 58. s. 1. 16) followed by a present participle.

A. unmediated. He resembles Ausonius, who blamed himself for his lack of progress in Greek:

> obstitit nostrae quid, credo, mentis
> tardior sensus neque disciplinis
> adpulit Graecis puerilis aevi
> noxius error. (*prof. Burd.* 8. 13-16)

(What was *possible* in Africa at this time is another story; Ferrandus' life of Fulgentius of Ruspe (b. *c.*467) makes ambitious claims for his Greek studies under a Monnica-like influence: 'Fulgentium religiosa mater moriente celeriter patre graecis litteris imbuendum primitus tradidit et quamdiu totum simul Homerum memoriter reddidisset, Menandri quoque multa percurreret, nihil de latinis permisit litteris edoceri' (*PL* 65. 119). But, as Courcelle, *LLW* 221–2, shows, his demonstrable competence is modest; perhaps the Ferrandus text is best taken as evidence of the survival of the prestige of the language.)

adamaveram: Similarly of the liking for a learned study at *util. cred.* 7. 16, 'cum igitur ad haec studia nos multitudo invitaverit, quae imperitorum parte copiosa est, ut id quod pauci adipisci possunt adamaremus . . .' Hensellek, *Anzeiger Akad. Wien* 115 (1978), 17, renders 'affectare, appetere'.

primi magistri . . . grammatici: Cf. R. Kaster, art. cit. on 1. 9. 14.

legere et scribere et numerare: The same triad at 1. 15. 24; foreshadowing the colloquial English 'Three R's' (in English they do not antedate a jocular toast offered in 1795 by the Lord Mayor of London, Sir William Curtis, distinguished in the *DNB* as a man the object of more ridicule than any other of his time).

vanitate vitae: = *ambitio saeculi* [1] < 1 Jn. 2: 16 (see on 1. 10. 16). Successes in his schooling and teaching are regular examples of his weakness before this temptation.

qua CDGO^1S Knöll Skut. Ver. : *quia* O^2 Maur. and Ps. Rom. (except the Veronensis, which has *quoniam*); LXX has ὅτι. Here A. has grafted the psalm citation into his own context: 'the life by which I was flesh' is periphrasis for *vita carnalis*. Ps. 77: 39, 'et recordatus est quia caro sunt, spiritus vadens et non revertens' (text follows *en. Ps.* 77. 22; Ps. Rom. has *memoratus* for *recordatus*—which is in the Vulgate).

ambulans: *vadens* in Ps. Rom. and Vg., πορευόμενον in LXX; but in *en. Ps.* cited below he quotes Prov. 2: 19, 'omnes qui ambulant in ea, non revertentur,' as a parallel text. *en. Ps.* 77. 24, 'per seipsos redire non possent. quomodo enim redit caro, spiritus *ambulans* et non revertens, urgente se *in ima atque longinqua* [see on 2. 3. 6, 'inclinatae in ima', and 1. 18. 28, 'in longinqua regione'] malorum pondere meritorum, nisi per electionem gratiae?'

illae quibus: i.e. *litterae grammaticorum*.

tenere: Of memorizing; cf. 'oblitus' in the next line and 1. 13. 22, 'oblivisci errores.' The expression is common in A., esp. in the discussion of memory in Bk. 10, e.g. 10. 13. 20 ('haec omnia memoria teneo'); cf. 10. 13. 20 (2 × more), 10. 15. 23, 10. 16. 25, 10. 16. 25, 10. 17. 26, 10. 18. 27 (2 ×), 10. 19. 28, 10. 20. 29 (2 ×).

Aeneae nescio cuius: This phrase and 'cuiusdam Ciceronis' at 3. 4. 7 (more controversial: see notes there) contain an undeniable air of mild disdain (cf. Cic. *de or.* 1. 20. 91, where Antonius is deprecating the learned tradition among rhetoricians: 'cum repeteret usque a Corace nescio quo et Tisia', where the names were scarcely less familiar than Aeneas' here), but there is no pretence (even a merely rhetorical one) not to know the name; the disdain must not be overemphasized. (Such disdain need not be absolute and his practice could vary: cf. 5. 3. 6, 'cum dictis Manichaei', with 5. 5. 8, 'Manichaeum nescio quem'.) Testard, 1. 14 n. 3, wants this to hint at the fabulous quality of the Aeneas story; in view of the other figures treated in a similar way elsewhere, this is unlikely. Virgil's power to linger in memory is expressly recognized at *civ.* 1. 3, 'apud Vergilium, quem propterea parvuli legunt, ut videlicet poeta magnus omniumque praeclarissimus atque teneris ebibitus animis non facile oblivione possit aboleri.' And it was not only as a text that Aeneas came to A.: *s.* 242. 5. 5, 'pauci nostis in libris, multi in theatris, quia Aeneas descendit ad inferos . . .'.

Some useful parallels: *s.* 105. 7. 10 ('poeta illorum quidam' (= Virgil), where A. goes on to quote 'imperium sine fine dedi' in order to disagree); *Io. ev. tr.* 43. 6 ('ait quidam' (Virgil)), *civ.* 10. 1 ('quidam latini eloquii magnus auctor' (= Virgil)), *doctr. chr.* 4. 10. 24 ('unde ait quidam' (= Cicero)), *doctr. chr.* 4. 12. 27 ('dixit enim quidam' (= Cicero)), *vera rel.* 34. 64 ('Roma . . . quam circa Tiberim nescio quis Romulus dicitur condidisse'), *ep.* 125. 3 ('nescio quis ille Regulus': writing to Alypius), *c. Faust.* 13. 15 ('nescio quis Hermes'), *c. Faust.* 28. 2 and 28. 4 ('nescio cui Persae' (i.e. Mani)). For a *more* disdainful expression, cf. *civ.* 14. 8, 'nam et Alcibiadem ferunt (si me de nomine hominis memoria non fallit) . . .'

errores: 29 × in *conf.*; the antonym is *peregrinatio*. Used equally and without distinction for the sense represented by English 'error' and that here of 'wandering astray'; as here, 4. 1. 1, 'praeteritos circuitus erroris mei'; clearly in the other sense, 4. 7. 12, 'sed vanum phantasma et error meus erat deus meus'; somewhere between, 1. 20. 31, 'atque ita inruebam in dolores, confusiones, errores'. See G. J. P. O'Daly, *Augustinus* 26 (1981), 187*–94* (summarized at O'Daly, 93–4).

plorare: His second tears of this book: 1. 7. 11, 'an quia uberibus inhiabam plorans'. For tears shed again for imaginary characters when there are better reasons to weep, see on 3. 2. 4.

Didonem: H. Jacobson, *HThR* 65 (1972), 296–7, shows similar self-reproach for taking fiction seriously in ps.-Andocides, Plutarch, and Aelian.

vita mea: see on 1. 4. 4.

1. 13. 21

Quid enim: The rhetorical question re-establishes tension between A.'s past and God's authority, but the idea is still conventional: cf. Sen. *ep.* 88. 7, 'quaeris Ulixes ubi erraverit potius quam efficias ne nos semper erremus?' The implicit approval of an ordinate self-love may owe something to Sirach 30: 24, 'miserere animae tuae placens deo' (frequently quoted by A.): cf. O. O'Donovan, *The Problem of Self-Love in St. Augustine* (New Haven, 1980), 46 and 178 n. 22.

lumen cordis mei: Jn. 1: 9, 'erat lumen verum quod inluminat omnem hominem venientem in hunc mundum' (cf. 4. 15. 25); *Io. ev. tr.* 2. 7, 'veniebat enim ad mentes informes, ad corda saucia, ad aciem animae lippientis.' Cf. Jn. 3: 19–21, 'lux venit in mundum, et dilexerunt homines magis tenebras quam lucem . . . (21) qui autem facit veritatem, venit ad lucem, ut manifestentur eius opera, quia in deo sunt facta'; 1 Jn. 1: 5, 'deus lux est.' A partial list of parallel expressions in *conf.*: 3. 4. 8 ('tu, lumen cordis mei'), 4. 16. 30, 5. 3. 4, 7. 9. 13, 7. 9. 13, 8. 10. 22, 9. 4. 11, 10. 17. 26 ('te, dulce lumen'), 12. 10. 10 ('o veritas, lumen cordis mei'), 13. 2. 3, 13. 4. 5, 13. 6. 7, 13. 16. 19, 13. 24. 36.

panis: Jn. 6: 35, 'ego sum panis vitae: qui venit ad me non esuriet.' Like *lumen*, *panis* designates the second person of the trinity. The same image also only at 6. 3. 3 ('de pane tuo').

virtus: Texts such as 1 Cor. 1: 24, 'Christum dei virtutem et dei sapientiam', and 2 Cor. 6: 7, 'in verbo veritatis, in virtute dei', suggest that the focus on the second person of the trinity is maintained. Cf. 10. 1. 1, 11. 2. 3 ('*lux* caecorum *et virtus* infirmorum, statim *lux* videntium *et virtus* fortium').

maritans: *maritare* elsewhere in A. chiefly in the perfect participle, of a married person; except for such occurrences, only 2 × in all the rest of A., *s. Étaix* 2. 1, 'mox viri nesciam sermo dei maritat', and *vera rel.* 26. 49, 'tertiam [aetatem] iam fidentiorem et carnalem appetitum rationis robore maritantem gaudentemque intrinsecus in quadam dulcedine coniugali, cum anima menti copulatur et velamento pudoris obnubitur'.

sinum cogitationis meae: *en. Ps.* 57. 17, '"alligabit quis ignem in sinu suo, et vestimenta sua non comburet?" [Prov. 6: 27–9]. gestas in sinu prunas, perforatur tunica; gestas in cogitatione adulterium, et integra est anima?' Cf. 9. 2. 3 ('sinum cogitationis'), 10. 8. 13 ('sinus memoriae'), 10. 8. 14 ('ipso ingenti sinu animi mei').

fornicabar abs te: Ps. 72: 27–8, 'ecce qui longe se faciunt a te, peribunt; perdidisti omnem qui fornicatur abs te. (28) mihi autem adhaerere deo bonum est' (vs. 28 cited at 7. 11. 17 (where see notes), 12. 15. 22, 13. 2. 3). The image here elsewhere in *conf.* at 2. 6. 14, 4. 2. 3, 5. 12. 22. Since God must be spoken of in metaphor, exegesis has the task of delimiting metaphor: *en. Ps.* 72. 33, 'illi vero longe recesserunt, quia non solum terrena desideraverunt, sed ea a daemonibus et a diabolo petierunt. . . . et quid est, longe a deo fieri? "perdidisti omnem qui fornicatur abs te." huic fornicationi contrarius est amor castus. quis est amor castus? amat iam anima sponsum suum.' Sim. at *en. Ps.* 78. 8; at *s. dom. m.* 1. 16. 43 he debates whether 'fornication' in divorce cases is purely sexual or applies to 'omnem inlicitam corruptionem, sicut est idolatria vel avaritia, et ex eo iam omnis transgressio legis propter inlicitam concupiscentiam.' Though he softens his eventual position at *retr.* 1. 19. 6, the text shows how seriously he took metaphorical *fornicatio*, considering it even where the literal sense is obviously more apt.

'euge! euge!': Exclamations of worldly (and therefore false) praise. Ps. 39: 16, 'ferant confestim confusionem suam, qui dicunt mihi: euge, euge'; *en. Ps.* 39. 26, 'respexit ad aliud genus dolose malevolum et falso benedicum: "ferant . . . euge." laudant falso: magnus vir, bonus vir, litteratus, doctus, sed quare christianus? et tua laudant, quae nolles laudari.' Also at Ps. 34: 21, 34: 24–5, 69: 4, and cf. *en. Ps.* 69. 5, 'quare enim me laudant? laudent deum. quis enim ego sum, ut lauder in me? aut quid ego feci? quid habeo quod non accepi?' For the link to his self-reproaches at the time of *conf.*, see *ep.* 95. 2, quoted on 10. 36. 59, and note there the repetition of 'euge! euge!'.

amicitia enim mundi huius: Jas. 4: 4, 'adulteri, nescitis quia amicitia huius mundi inimica est dei?'

et (*flebam*) CDGS Knöll Skut. Ver. : *sed* O Maur. (cf. 'autem' above).

exstinctam ferroque extrema secutam: *Aen.* 6. 456–7: 'infelix Dido, verus mihi nuntius ergo | venerat exstinctam ferroque extrema secutam?' Why the literal citation? He says thereby not merely 'for the dead Dido', but 'for the dead Dido as incorporated in the text of the *Aeneid*'. It was not just a story or a figure of myth that appealed to him, but a text: the authorized version. He invokes scriptural texts the same way all the time. (Donatus thought highly of this passage—which is not from the narrative of Dido's fate, but the pathetic encounter of Aeneas with her shade in the underworld: *interp. verg.* ad 6. 457, 'dictio ista mira arte concepta est et pro personarum et pro causarum ratione composita.' He emphasizes Aeneas' acknowledgement of his own blame for Dido's death.) J. Bernhart, in a note ad loc. in his German translation, offers a further suggestion: '*Extrema sequi* im Vergil-Zitat wird hier unvermittelt im Sinne neuplatonischer Eschatologie vom moralischen Tod der Hingabe an die Materie verstanden.' This cannot be ruled out if A. knew any full or partial (on the underworld parts of *Aeneid* 6 only?) Neoplatonizing Vergil commentary (see on 8. 2. 3).

sequens ipse extrema: G–M: 'the point of connexion being purely verbal'. Better, given the verbal connection, the result is a transmutation of Virgil's text into something A. finds better and truer. The *primae litterae* thus demonstrate their usefulness apart from the misuses of the schools.

relicto te: Cf. 2. 2. 4, 'sed efferbui miser, *sequens* impetum fluxus mei relicto te.'

terra iens in terram: Gn. 3: 19 (VL), 'terra es, et in terram ibis'; this is the death A. was pursuing to correspond to that of Dido: *civ.* 20. 20, '"in terram" quippe "ibis" est: "in hoc ibis amissa vita, quod eras antequam sumeres vitam"; id est, hoc eris exanimatus, quod eras antequam esses animatus (terrae insufflavit deus in faciem flatum vitae, cum factus est homo in animam vivam); tamquam diceretur, "terra es animata, quod non eras; terra eris exanimis, sicut eras".' That interpretation also appeared at *civ.* 13. 23, but at *civ.* 13. 15 ('mors altera, de qua deus peccatum adhuc puniens homini dixerat, "terra es et in terram ibis"'), he took this phrase to allude to the *secunda mors* of the soul, at least as apt a reading here.

tali CDG : *talis* OS Maur. Knöll Skut. Ver. The want of a demonstrative pronoun (e.g. *haec*) with 'litterae' has led editors to retain with the better manuscripts the improbable reading *talis*. Nothing in A., or in Latin, parallels the construction, difficult first for the awkward change of number, and second, for the odd equation *dementia* = *litterae*, the more so because the verb of the sentence is a verb of mental activity that *dementia* would qualify very well in the ablative. Latin *dementia* denotes activity of mind, which might perhaps with difficulty be extended to apply to sayings uttered in a state of *dementia* (though there is no parallel even for that, e.g. *haec sententia dementia est*). Even examples of *dementia* in the nominative with a copulative verb are rare, and then the only example in A. equates *dementia* to a verb of mental activity: *civ.* 1. 3, 'quae dementia est existimare his tutoribus Romam sapienter fuisse commissam'. Further, *tali (s)* denotes what has immediately gone before, the folly of the young A. sorry when he could not read what would give him sorrow. It is impossible to equate that *dementia* with 'honestiores et uberiores litterae'.

Translate thus: 'by madness such as that evinced by the student A., that literary study is thought more highly of than the study of the elements of reading and writing.'

litterae: The disdain for such 'litterae' is part of A.'s post-conversion habits of thought and expression: *mus.* 6. 1. 1, 'qui litteris saecularibus dediti magnis implicantur erroribus . . . nescientes quid ibi delectet.'

1. 13. 22

in anima mea clamet deus meus, et veritas tua dicat: Anticipates a main topic of Bk. 10; see 10. 24. 35. Cf. Mic 6. 9, 'vox domini ad civitatem clamat.'

prorsus: A very Augustinian adverb; emphatic: Hensellek, *Anzeiger Akad. Wien* 120 (1983), 98–100. In *conf.* 13 × in all.

doctrina: Common in A. in both secular and religious senses. Biblical: in Latin OT, mainly in Proverbs and Sirach; in NT infrequent in Gospels (and once only in the mouth of Jesus: Jn. 7: 16, 'mea doctrina non est mea, sed eius qui misit me'), but common in Paul. The word furnishes A. and those of his cultural background (cf. Marrou 554–8) with an instrument by which Christian teaching was gradually assimilated to the norms and techniques of the prevailing secular culture. At Mt. 16: 12, the *doctrina* of the Pharisees and Sadducees is deprecated, and no one mistook Jesus for a schoolmaster or conventional rabbi. The difference between A. the bishop and A. the schoolmaster is not so pronounced (for the parallel development in the Greek east, see C. Schäublin, *Untersuchungen zur Methode und Herkunft der Antiochenischen Exegese* (Cologne/Bonn, 1974)).

oblivisci errores Aeneae: Cf. on 1. 13. 20, 'tenere'.

at enim: 'To be sure': presenting a possible objection to his view. His response, 'sed non illa magis honorem secreti . . . significant', makes explicit the implication of the objection. He has found these studies disreputable; he imagines the objection, 'but the higher studies are carried on in the dignified privacy provided by the *velum*'; and responds, 'the *velum* is there not to provide privacy but to conceal *error*.' This expression is classical (*OLD* s.v. *at* 4) and is the exclusive use of the collocation in A. The phrase does not occur elsewhere in *conf.*, once only in *en. Ps.* (31. en. 2. 16), but is common in *civ.* (13 ×), particularly the earlier books of rhetorical parry and thrust. In every case, *at enim* introduces a single sentence or clause containing a possible objection to A.'s position, and is followed immediately by A.'s rejoinder, often marked by an adversative conjunction: e.g. *civ.* 1. 10, '*at enim* quidam boni etiam christiani tormentis excruciati sunt, ut bona sua hostibus proderent. illi *vero* nec prodere nec perdere potuerunt bonum quo ipsi boni erant.'

vela: The adaptation here parallels that of *doctrina*. The under-master watched by the curtain at the entrance: *s.* 178. 7. 8, 'pauperrimus homo . . . nobis apud Mediolanum constitutis . . . tam pauper ut proscholus esset grammatici: sed plane christianus, quamvis ille esset paganus grammaticus; melior ad velum, quam in cathedra.' Cf. *s.* 51. 4. 5, 'quanto enim quisque honoratior est, tanto plura vela pendent in domo eius. *vela faciunt honorem secreti*: sed honorantibus levantur vela.'

quos: i.e. 'venditores grammaticae vel emptores' (below).

malarum viarum mearum: Ps. 118: 101, 'ab omni via maligna prohibui pedes meos, ut custodiam verba tua' (text follows *en. Ps.* 118. s. 22. 5; Ps. Rom. has *mala* for *maligna*, while Veronensis reads *maligni*; LXX πονηρᾶς). Jer. 18: 11, '[Dominus dicit] revertatur unusquisque a via sua mala et dirigite vias vestras et studia vestra'; Jer. 26: 3, 'si forte audiant et convertantur unusquisque a via sua mala.'

respondent CDOS Knöll Skut. Ver.: *respondebunt* G Maur.

quo CDO Maur. Knöll Skut. Ver.: *quod* GS.

poetica illa figmenta: *civ.* 4. 26 (quoting the same passage of Cic. *Tusc.* quoted below: see on 1. 16. 25, 'dicentem'), 'hic exclamet Cicero non contra figmenta poetarum, sed contra instituta maiorum'; sim. at *civ.* 2. 14; see also on 4. 2. 3, 'figmentis'.

et dulcissimum spectaculum [2] **vanitatis** [1].

ipsius umbrae Creusae: *Aen.* 2. 772, 'infelix simulacrum atque ipsius umbra Creusae'. As at 1. 13. 21, the *ipsissima verba* emphasize that the folly is not merely in the story, but in the received textual version. The episodes evoked by these echoes are among the most pathetic in Virgil: the irrevocable loss by Aeneas of the two women who loved him. (A. also abandons two women in *conf.*: Monnica temporarily at 5. 8. 15 (with strong echoes of Dido—see notes there), his concubine permanently at 6. 15. 25 (and she, like Creusa, leaves A. with a son).)

1. 14. 23

There is no scripture in this paragraph and little in the one before. The unscriptural text is at centre stage, infecting the narrative. The implicit assertion, exemplified by this paragraph, is that divine power goes even where the scriptural word is missing.

graecam . . . grammaticam: See on 1. 13. 20.

Homerus: Wherever A. has specific information about Homer's text, it proves on examination to have come to him from the scholiastic tradition on Virgil (e.g. at *civ.* 3. 2, cf. Serv. on *Aen.* 5. 810). None of the few and scattered references to Homer elsewhere in A. gives any sign of direct knowledge or of a more than conventional respect. Outside of *civ.* and *conf.*, H. is named only at *haer.* 7, on the Carpocratians, one of whom, one Marcellina, 'colebat imagines Iesu et Pauli et Homeri et Pythagorae adorando incensumque ponendo'.

peritus: *peritia* for A. is always a secular or worldly skill (particularly apt therefore to *artes* such as medicine: cf. 4. 3. 5), and may even apply to non-Christian religious expertise, but he seems never to speak of it as an excellence associated with the Christian religion. Expertise is often misdirected; hence, though *peritus* may be used in a neutral and even laudatory way (e.g. *civ.* 18. 43, 'temporibus nostris presbyter Hieronymus, homo doctissimus et omnium trium linguarum peritus'), it often carries a slight overtone of criticism (e.g. *en. Ps.* 101. s. 1. 1, 'etiamsi periti senes, quid illorum peritia ad verbum dei? quid illorum peritia ad sapientiam dei?').

mihi CDGO Ver. (preceded by a period) : *et mihi* S Maur. Knöll Skut. (with no punctuation preceding).

difficultas[2] CDS Knöll Skut. Ver. : om. GO.

cum me urgeret: Wilfulness again at the origins of human speech; cf. 1. 6. 8, 1. 6. 10, 1. 8. 13. 'parienda' → 'parturiebam' (below).

†et qua† Skut. : *et qua* CDS EFM[1] PZ AH Vega Pell. Ver. : *et quae* BG : *et quia* M[2]OV : *quae* Maur. : *id quod* Knöll.

esset CDGS EFM PZ AHV Knöll Skut. Ver. : *essent* BO : *possem* Maur.

A. Sizoo, *Mnemosyne* ser. 3, 13 (1947), 141–2, emended to *atque ea nata non essent*, instancing *nascor* in *trin*. 9. 7. 12–9. 14, and 15. 10. 19.

Verheijen edn., p. xxxvii: '*et qua non esset*: et il n'y aurait pas de *via qua*, c'est-a-dire cela serait impossible.' Carena: 'perché il mio cuore stesso mi sollecitava a dare alla luce i suoi pensieri. Ma non c'era la via, se non avessi imparato qualche vocabolo, piú che a scuola da chi insegnava, dalla voce di chi parlava, nelle cui orecchie a mia volta deponevo i miei sentimenti.' Verheijen and Carena are probably right as to the sense, but wrong as to the language. The ellipsis is too harsh; on their view, the traditional emendation *quae non possem*, is far simpler and clearer, if palaeographically improbable. Skutella is correct to obelize the passage; one or more words have fallen out.

in quorum et ego auribus: The hyperbaton of an extraneous element (here 'et ego') in mid-phrase ('in quorum . . . auribus') is fairly common with *quorum/quarum* and especially where the interjected element is a first person pronoun. Cf. 4. 16. 28, 6. 11. 20 ('quarum mihi non eram conscius'), 6. 12. 21 ('a quorum ego quidem granditate animi longe aberam'), 8. 2. 4, 8. 6. 15 ('quorum nos nihil sciebamus'), 10. 3. 3 ('quorum mihi aures caritas aperit'), 10. 11. 18, 12. 14. 17 ('quorum ecce ante nos superficies blandiens parvulis'), 12. 27. 37, 13. 20. 26.

curiositatem . . . illius fluxum: On *curiositas*, see on 1. 10. 16 and on 10. 35. 54. *Fluxus* always represents for A. the instability of the material world; cf. particularly 9. 8. 18, 'fluxum saeculorum *ordinate* turbulentum' (see on 1. 10. 16, 'ordinator peccatorum'); *defluere* provides the corresponding verb and is used in similar ways: 2. 10. 18 ('defluxi abs te ego et erravi'), 10. 29. 40, 12. 10. 10, 13. 7. 8.

meticulosam: Here = *metuendus*, for which sense the only other attestation in *TLL* is Plaut. *most.* 1101, 'nescis quam meticulosa res sit venire ad iudicem'; the ordinary active sense, 'timorous, anxious', is not apt. Occurs only 3 × in all A.'s works; *ep.* 28. 4. 6, 'meticulosam potius . . . quam iustam tulisse sententiam' (sense as here); at *c. Iul.* 3. 1. 5, Julian contrasts the adj. to *prudens* (the conventional sense). (Perhaps better spelled *metuculosus*: see *OLD* and *TLL* and cf. the reading of S here: *metulosam*.)

legibus tuis . . . legibus tuis . . . legibus tuis: The pleonasm compels translators either to omit one occurrence of the phrase or to supply a verb with 'a magistrorum . . . martyrum'.

usque ad temptationes martyrum: For similar melodramatic link between the schoolboy's and the martyr's torments, see 1. 9. 15.

recessimus a te: A recurrent theme (on *reditus*, see on 1. 18. 28): cf. 1. 2. 2, 2. 6. 14, 3. 3. 5 ('ad longe recedendum a te'), 4. 16. 31, 5. 3. 4, 6. 16. 26, 8. 10. 22, 10. 26. 37 ('et nusquam locus, et recedimus et accedimus, et nusquam locus'), 10. 40. 65.

1. 15. 24

The memoirs of infancy and boyhood are at an end; A. returns to the confessional mode, to comment rather than remember, to judge rather than to describe. How artificial is the high-minded austerity that values in school studies nothing but a kind of education that could have been got less damagingly elsewhere? A similar doubt intrudes for us at *civ.* 1. 9, when he suggests that the sufferings of the just arise from no active fault but from their failure to reproach the wickedness of others. His weakest arguments seem to have been excogitated but not felt.

Exaudi: Ps. 60: 2, 'exaudi, deus, deprecationem meam, intende orationi meae'; cf. Ps. 54: 2; see on 1. 9. 14; for the imperative, cf. 1. 7. 11, 9. 4. 8, 9. 13. 35, 10. 33. 50, 11. 2. 4, and see Knauer, 64–74.

deficiat: Ps. 83: 3, 'desiderat et deficit anima mea in atria domini.'

disciplina: Pizzolato, *Le fondazioni dello stile delle 'Confessioni'* (Milan, 1972), 31 n. 3, aptly cites *Gn. c. man.* 1. 25. 43, 'secundo die tamquam firmamentum disciplinae, quo discernit inter carnalia et spiritalia'; if *pueritia* is the second age, then *disciplina* is the appropriate gift here (but not received in the right way); see on 2. 1. 1 for a similar correlation with the third day in *adulescentia*. Pizzolato, *Lectio I–II*, 63–4: 'La disciplina è quindi correttivo della *curiositas* esasperata, che tutto ricerca "experiendi noscendique libidine" (10. 35. 55), finendo col rovinare lo stesso piacere; ed è ugualmente un correttivo della *meticulosa necessitas*, che dà ordine al sapere, ma negandone la piacevolezza e facendo leva sulla paura (*metus*).' Cf. below, 'tu disciplinam dabas mihi'; elsewhere often in a secular sense (esp. *disciplinae liberales*; see excursus on 4. 16. 30) and applied to bringing up a child in the Christian religion (9. 6. 14, of Adeodatus); for something closer to the sense here, see in *conf.* only 6. 6. 9, 'baculo disciplinae tuae'. The *disciplina* of God is, however, important in A., with particular connection to the second person of the trinity (cf. *div. qu.* 38, 'natura [1], disciplina [2], usus [3]').

deficiam in: Hensellek, *Anzeiger Akad. Wien* 114 (1977), 150: the construction is unusual, but cf. *util. cred.* 6. 13, 'in quaestiuncula magistro deficienti . . . suscensemus'.

in confitendo tibi: 'Immer aber, wenn Augustin das Wort "confiteri" benutzt, bedeutet es ein erneutes Hinwenden zu Gott' (Knauer, 78).

omnibus viis meis pessimis: 4 Kgs. 17: 13, 'dominus . . . dicens revertimini a viis vestris pessimis et custodite praecepta mea et caerimonias'; sim. at 2 Chron. 7: 14, Jer. 36: 3 and 36: 7, 'revertatur unusquisque a via sua pessima'. N.B. 'via tuta est' below.

dulcescas: See on 1. 6. 9, 'dulcedo'.

totis praecordiis meis: Cf. Sirach 33: 5, 'praecordia fatui sicut rota carri.'

et eruas me ab omni temptatione: Ps. 17: 30, 'quoniam a te eruar a temptatione' (text from *en. Ps.* 17 and Psalt. Veronense); cf. Mt. 6. 13, 'ne nos inferas in temptationem'; 10. 31. 46, 'eripe me ab omni temptatione'.

eruas: Almost always in *conf.* of God's rescuing hand: 1. 18. 28 ('et nunc eruis de hoc immanissimo profundo'), 3. 11. 19, 6. 8. 13, 9. 4. 7.

usque in finem: See Knauer, 75 n. 1. Ps. 15: 10, 'adimplebis me laetitia ... usque in finem'; cf. Ps. 37: 7, 67: 17, 1 Cor. 1: 8. In *conf.*, at 8. 3. 8, 13. 15. 18, but with the addition of a genitive, which has a limiting (or literalizing) force absent in the present phrase and the scriptural texts. The common phrase 'in finem' in the *tituli* of the Psalms is interpreted by A. at its first appearance thus: *en. Ps.* 4. 1, 'in finem ... "finis legis Christus ad iustitiam omni credenti" [Rom. 10: 4]. hic enim finis perfectionem significat, non consumptionem.'

rex meus et deus meus: Ps. 5: 3, 'intende voci obsecrationis meae, rex meus et deus meus'; Ps. 43: 5, 'tu es ipse rex meus et deus meus.' The first of these texts is linked to the second person of the trinity: *en. Ps.* 5. 3, 'quamvis et filius deus, et pater deus, et simul pater et filius unus deus; et si interrogemur de spiritu sancto, nihil aliud respondendum est, nisi quod deus sit; ... tamen regem filium solent appellare scripturae.' But the interpretation given of the second twenty years later (*en. Ps.* 43. 5, '"tu es ipse rex meus et deus meus." tu es ipse, non enim mutatus es. tempora mutata video, creator temporum non mutatur') leans slightly towards directing the reference to the first person. *Rex* of the first person at 1. 19. 30, 8. 4. 9; for the second person see on 10. 35. 56.

quod loquor: 1. 8. 13, 'iam puer loquens eram'.

et scribo et lego et numero: See on 1. 13. 20.

in eis vanis: G–M connect with 'delectationum' by hyperbaton; unlikely, because that reading would make the demonstrative 'eis' superfluous. The more obvious reading, 'the sins of my delights', should be preferred.

via: Jn. 14: 6, 'ego sum via et vita et veritas.' See on 7. 7. 11.

ambularent: Earlier editors, seeing here a connection with 'dimisisti' in the last sentence, hark back to the parable of the paralytic (cf. Mk. 2: 5–10, Lk. 5: 23): Mt. 9: 5, 'quid enim est facilius, dicere: "dimittuntur peccata tua", aut dicere: "surge et ambula".' If there is a hint of that, it would suggest again the second person of the trinity.

1. 16. 25

flumen: Cf. 1. 16. 26, 'o flumen tartareum'; *en. Ps.* 65. 11, 'quid est flumen? flumen est omnis mortalitas saeculi' (sim. at *en. Ps.* 143. 2).

quis resistet tibi?: Ps. 75: 8, 'quis resistet tibi [domine] tunc ab ira tua?'

Evae filios: *en. Ps.* 126. 8, 'in cuius figura etiam Eva "mater vivorum" appellata est.' Cf. Gn. 3: 20 (apud *Gn. litt.* 11. 1. 1), 'et vocavit Adam nomen mulieris suae "Vita", quoniam haec est mater omnium viventium.' 'Vita' is an occasional idiosyncrasy in VL texts, but the more familiar name prevailed.

transeunt: *ep.* 55. 1. 2, 'transitur, ut dixi, de morte ad vitam ... in quo [hebraeo] eloquio pascha transitus dicitur.'

lignum: Gn. 3: 1 (apud *Gn. litt.* 11. 1. 1), 'de fructu *ligni* quod est in medio paradiso'; to which A. connects (at *civ.* 13. 20) Prov. 3: 18, 'lignum vitae est amplectentibus eam'; cf. Wisd. 14: 7, 'benedictum est enim lignum per quod fit iustitia.' That *lignum* in turn suggests the ark, the cross, and the church: *civ.* 15. 26, 'procul dubio [arca] figura est peregrinantis in hoc saeculo civitatis dei, hoc est ecclesiae, quae fit salva per lignum, in quo pependit mediator dei et hominum, homo Christus Iesus.' In A., the image that predominates is the cross: *Io. ev. tr.* 2. 2, 'sic ad illam stabilitatem nostram ubi quod est est ... volumus pervenire, interiacet mare huius saeculi qua imus, etsi iam videmus quo imus: nam multi nec quo eant vident. ... nemo enim potest transire mare huius saeculi, nisi cruce Christi portatus.' Sim. at *cat. rud.* 19. 32, *en. Ps.* 103. s. 4. 4, 136. 4.

in te legi: L. Herrmann, *Aug. Mag.* 1. 136–7, thinks *te* an abbreviation for 'Terentio'; this is unnecessarily literal-minded.

tonantem: *civ.* 7. 9, 'unde satius esset eis alium aliquem Iovis nomine nuncupare, dignum turpibus et flagitiosis honoribus, supposito vano figmento quod potius blasphemarent ... quam istum deum dicere *et tonantem et adulterantem*.' *Tonans* is a standard epithet of Jupiter from the cult tradition, common in inscriptions and formal prose; *adulterans* offers a mocking parallel from the mythic tradition (cf. *ep.* 138. 4. 18, 'tot stuprorum auctor et perpetrator, quem Iovem nominant'). Christ steals Jove's thunder at 4. 12. 19, 'et descendit huc ipsa vita nostra ... et tonuit clamans, ut redeamus hinc ad illum.'

ad imitandum GO Maur. Pell.: *imitandum* CDS Knöll Skut. Ver. But 'verum adulterium' is probably the subject of 'haberet'; cf. W. Schmidt-Dengler, 'Stilistische Studien zum Aufbau der Konfessionen Augustins' (Diss., Vienna, 1965), 69. Cf. 8. 5. 10, 'exarsi ad imitandum'.

paenulatorum: The *paenula* was originally a close-fitting hood to protect against wet weather; we find it put to more formal use at Tac. *dial.* 39. 1, 'quantum humilitatis putamus eloquentiae attulisse paenulas istas, quibus adstricti et velut inclusi cum iudicibus fabulamur?' Cf. Ambstr. *in 2 Tim.* 4. 13, of Paul's appeal as Roman citizen: 'idcirco necesse est eos et curiam habere, in qua more Romanorum paenulati conveniant.' Of *paenulatos discipulos* (less well-turned-out than *togatos*, but presentable): Gell. 13. 22. 1. At Mart. Cap. 3. 223, Grammatica enters: 'ritu Romuleo ... senatum deum ingressa est paenulata' (cf. 3. 325).

ex eodem pulvere: Not the dust of which Adam is created in Genesis, but the dust of the competitive arena—the other *magistri* will be aroused to contentious denial by such a statement. The metaphor is implicit: it was commonplace (e.g. Hor. *carm.* 1. 1. 3 and 1. 8. 4) to speak of the *pulvis* of the real arenas.

dicentem: Cic., *Tusc.* 1. 26. 65, 'fingebat haec Homerus et humana ad deos transferebat: divina mallem ad nos.' Here the authority of the line is implicit, to be recognized or not by one reader or another; contrast *civ.* 4. 26, 'sed "fingebat haec Homerus", ait Tullius, "et humana ad deos transferebat: divina mallem ad nos." merito displicuit viro gravi divinorum criminum poeta confictor.' See on 1. 13. 22, 'poetica illa figmenta'.

flagitia: See on 3. 8. 16.

imitatus: Mimesis is a recognized path to behaviour acquisition (22 × in *conf.*), but how models are treated determines whether the behaviour is good or bad. It is wrong to imitate God in the wrong way (2. 6. 13, 'nam et superbia celsitudinem imitatur'; 2. 6. 14, 'perverse te imitantur omnes'), but very right to imitate Christ or even those who imitate him (13. 21. 31, 'imitando imitatores Christi tui'), and careful presentation of models can inspire praiseworthy zeal (8. 5. 10, 'exarsi ad imitandum [Victorinum]').

1. 16. 26

o: The interjection occurs frequently, preponderantly in addresses to the nominal second-person audience of the text—God, but with a wide and curious miscellany of other persons and things as well, particularly in the treatment of the pear-theft (2. 6. 12 ff.). Here A. presents himself as uncomfortable with the personification implied, and in a few more lines turns emphatically to formal address directed again to God.

o addressed to God: 1. 17. 27, 2. 2. 2, 3. 3. 5, 3. 6. 10, 3. 11. 19, 4. 16. 31, 7. 10. 16, 9. 4. 11 (3 ×), 10. 29. 40, 11. 2. 3, 11. 11. 13, 11. 31. 41, 12. 15. 21, 12. 25. 34, 13. 6. 7, 13. 24. 35, 13. 29. 44. Addressed to other than God: here and 2. 6. 12 ('o furtum meum, o facinus illud meum nocturnum'), 2. 6. 14 ('o putredo, o monstrum vitae et mortis profunditas'), 2. 9. 17 ('o nimis inimica amicitia'), 4. 7. 12 ('o stultum hominem immoderate humana patientem'), 6. 11. 18 ('o magni viri academici'), 6. 16. 26 ('o tortuosas vias'), 12. 11. 12 ('o beata [sc. creatura]), 13. 13. 14 ('o stulti Galatae'), 13. 26. 40 ('o Paule magne'); but several of those cases are more exclamatory that apostrophic, like a few other cases where the interjection expresses a fervent wish, with no literal expression of any particular auditor: 9. 4. 10 ('et o si fatigentur inedia et dicant'), 9. 4. 10 ('o si viderent internum aeternum'), 9. 13. 34 ('o si cognoscant se homines homines'), 12. 14. 17 ('o si occidas eos de gladio bis acuto').

tartareum: Cf. 3. 1. 1, 'de tartaro libidinis', and 8. 4. 9, 'tartaro caecitatis'. *en. Ps.* 140. 2, 'amor iste tartareus est: viscum habet quo deiciat in profundum, non pennas quibus levet in caelum.' For the rivers of hell in metaphor in this life, cf. Macrob. *somn. Scip.* 1. 10. 11, 'pari interpretatione Phlegethontem ardores irarum et cupiditatum putarunt, Acherontem quidquid fecisse dixisseve usque ad tristitiam humanae varietatis more nos paenitet, Cocytum quidquid homines in luctum lacrimasque compellit, Stygem quidquid inter se humanos animos in gurgitem mergit odiorum.'

filii C^2GO Maur. Pell.: *fili* C^1DS Knöll Skut. Ver. The shorter form, if indeed it represents a lingering trace of A.'s practice, is a nicety too poorly attested to be insisted upon; *fili* for A. is a regular form of the vocative singular (9. 10. 26, 13. 12. 13 (2 ×)). It is printed once by Skutella and Verheijen for the vocative plural (4. 12. 19) inconsistently, since they also print the exact same citation of Ps. 4: 3 at 9. 4. 9 with *filii*; the text at 4. 12. 19 is attested as here by C^1DS for *fili* and C^2GO for *filii*; the other ninth-century manuscripts are similarly consistent in their treatment of the two passages with one minor exception. But *filii hominum* occurs elsewhere and is uniformly attested in the MSS in the longer form and so printed at 1. 18. 29 and 12. 8. 8. Where the nominative plural is required elsewhere, *filii* is unanimously attested: 8. 4. 9, 13. 14. 15 (4 ×).

cum mercedibus: 'with the fees' they will pay.

cum hoc agitur publice in foro: Cf. 1. 13. 22, 'at enim vela pendent liminibus grammaticarum scholarum', where the emphasis was on the comparative privacy sought in the public place. Marrou, *History of Education* (New York, 1964 (part 3, chapter 5)), 370 summarizes their status; cf. Kaster (art. cit. on 1. 9. 14) for qualifications.

supra mercedem salaria: *merces* was what the student paid privately to a teacher (5. 12. 22, 'ne mercedem magistro reddant'; sim. at *civ.* 2. 19), the *salarium* what the state provided over and above: *civ.* 1. 3, 'nec suscensent auctoribus suis, quos ut ediscerent *mercedem* dederunt doctoresque ipsos insuper et *salario* publico et honoribus dignissimos habuerunt'; these passages supplement the evidence collected by Kaster, *Guardians of Language*, 118 n. 96.

saxa: *en. Ps.* 136. 3, 'multum sonat, strepitus aquae saxa percutit.'

imbrem aureum CDGO^2 Maur. Skut. : *imbrem et aureum* O^1S Knöll Ver.

Terentius: Ter. *eun.* 583–91:[37]

dum apparatur, virgo in conclavi sedet
su*spectans tabulam quandam pictam: ubi inerat pictura haec, Iovem*
quo pacto Danaae misisse aiunt quondam in gremium imbrem aureum.
egomet quoque id spectare coepi, et quia consimilem luserat
iam olim ille ludum, impendio magis animu' gaudebat mihi
deum sese in hominem convortisse atque in alienas tegulas
venisse clanculum per impluvium *fucum factum mulieri.*
at quem deum! 'qui templa caeli summa sonitu concutit'.
ego homuncio hoc non facerem? ego vero illud ita feci ac lubens.

[37] Text follows OCT but is assimilated to that of A. where the OCT apparatus indicates that readings are in the MSS of Terence; italics mark words cited here by A.

The quotation marks in 590 enclose words that Donatus describes as 'parodia de Ennio'. The character admiring this picture attacks a young woman ten lines later. Note particularly this image of Jupiter *adulterans* taking human form: wickedly like and unlike Christian incarnation. A.'s moralizing reading of the passage is matched by that of Donatus (on *eun.* 3. 5. 36–8: cf. C. Bennett, *REAug* 34 (1988), 58 n. 21). So the same passage provokes similar reflections at *ep.* 91. 4–5 (91. 5: 'tot locis pingitur, funditur, tunditur, sculpitur, scribitur, legitur, agitur, cantatur, saltatus Iupiter, adulteria tanta committens') and at *civ.* 2. 7–8. A. also knew of Jove's example simultaneously mocked and presented for emulation in the *ludi Romani* (*civ.* 4. 26, 'in illis ludis corruptorem pudicitiae Iovem turpissimi histriones cantabant, agebant, placebant'). The myth is given a moral reading at *civ.* 18. 13, 'quicumque finxerunt . . . Danaes per imbrem aureum appetisse concubitum, *ubi intellegitur pudicitia mulieris auro fuisse corrupta*'. For wall painting as a standard art-form (of which A. has a low opinion), see *vera rel.* 51. 100 and *c. ep. fund.* 34. 39.

All four standard school authors of A.'s time—Cicero (1. 16. 25), Virgil (1. 13. 21–2), Terence (here), and Sallust (2. 5. 11)—are expressly cited in the early pages of *conf.* (They were canonized in a celebrated work of Arusianus Messius (a senator and *comes primi ordinis*; the work was dedicated to the consuls Olybrius and Probinus in 395—hence contemporary[38]), *exempla elocutionum ex Virgilio, Sallustio, Terentio, Cicerone digesta per litteras* (Keil, 7. 449 ff.); cf. Cassiod. *inst.* 1. 15. 7, 'regulas . . . elocutionum latinorum, id est quadrigam Messii,' treating it as a standard textbook.)

libidinem: 27 × in *conf.*, never in a good sense; see on 2. 2. 2.

vero illud feci GO²S Maur. Knöll Skut (the MSS of *civ.* 2. 7 quoting this passage are unanimous for this reading) : *illud vero feci* CDO¹ Ver.

electa CDGO Ver. Pell. (comparing Acts 9: 15, 'vas electionis', and 1 Pet. 2: 6, 'lapidem electum, pretiosum') : *lecta* S Knöll Skut. Contrast *Io. ev. tr.* 1. 8, 'cotidie dicendo verba viluerunt nobis, quia sonando verba et transeundo viluerunt et nihil aliud videntur quam verba.'

pretiosa . . . erroris . . . propinabatur . . . biberemus: *s.* 312. 2. 2, 'quid enim ei pagano proderat eloquentia, qua tamquam poculo *pretioso* et *bibebat* mortiferos, et *propinebat errores*?' Pizzolato, *Lectio I–II*, 21, rightly sees an analogy, if unconscious, to 1. 3. 3, 'non enim vasa quae te plena sunt stabilem te faciunt, quia etsi frangantur non effunderis'; words 'contain' God as inadequately as a human being does. The apparent container is really the thing permeated, underpinned, and itself contained by the divine.

[38] And perhaps closer to A. than usually thought. If A. Cameron (in *Christianisme et formes littéraires de l'antiquité tardive en occident* (Entretiens Hardt, 23: Geneva, 1976), 7 n. 2) is correct in seeing an allusion to A.M. in Amb. *fuga saec.* 3. 16 (of approx. 387), then some earlier version of the work may have been in circulation in Christian circles by the time A. was in Milan.

ebriis: See on 5. 13. 23.

in cuius conspectu: This work itself exists before the silent, seeing presence of God, which is often evoked in these pages: 3. 11. 20, 5. 3. 3 ('proloquar in conspectu dei mei'), 6. 2. 2, 7. 8. 12, 9. 2. 2, 9. 7. 16, 9. 12. 33, 10. 2. 2 ('confessio itaque mea, deus meus, in conspectu tuo tibi tacite fit et non tacite'; cf. Ps. 95: 6, 'confessio et pulchritudo in conspectu eius'), 10. 4. 5, 10. 5. 7, 10. 35. 57, 11. 2. 4, 12. 11. 11 (2 ×), 12. 11. 13, 13. 21. 29. (Here there is the additional contrast with 'in conspectu legum' above.) Knauer, 75 n. 3, discusses, but his list derives from Skutella's citations and is not reliable.

delectabar: See on 1. 1. 1 and 1. 6. 7.

et ob hoc bonae spei puer appellabar: *doctr. chr.* 3. 18. 26, 'spei bonae homines'; the phrase is applied to a minor child in a convent at *s.* 355. 2. 3, 'filia . . . bonae spei est', and cf. *persev.* 21. 55; it occurs in scripture (Wisd. 12: 19) but may be commonplace (cf. Cic. *fin.* 2. 35. 117, 'ergo in iis adolescentibus bonam spem esse dicemus et magnam indolem'; Tac. *Agr.* 9. 7, 'egregiae . . . spei filiam'; Jordanes, *Getica*, 52. 269, 'bonae spei . . . puerolus').

1. 17. 27

C. Bennett, *REAug* 34 (1988), 48: 'The Bible should have replaced Vergil (1. 17. 27). This position is not surprising; what is, is that there was also a safe way to read Vergil—a way the *Confessions* itself illustrates.'

Sine me . . . dicere: Prayer for the gift of speech marked the beginning of the main exposition of Bk. 1 (1. 6. 7, 'sine me loqui apud misericordiam tuam'), and now recurs to introduce the concluding section.

in quibus: The asyndeton of the relative clause is a little harsh, because the construction is not quite parallel: 'Let me say something even of my *ingenium*, your gift, [and let me say] by what acts of madness it was squandered.'

deliramentis: Of specific instances (thoughts, words, or deeds) of a disordered intellect, where *dementia* (see on 1. 13. 21) is used of a pattern of such activity indicating, and virtually identified with, an underlying disorder.

inquietum: In an active sense, 'disquieting' (G–M).

plagarum metu: See on 1. 9. 14.

irascentis et dolentis: Cf. below, 'irae ac doloris'; *Aen.* 1. 8–11:

> Musa, mihi causas memora, quo numine laeso
> quidve *dolens* regina deum tot volvere casus
> insignem pietate virum, tot adire labores
> impulerit. tantaene animis caelestibus *irae*?

Juno dwells in the regions of the air (*cons. ev.* 1. 29. 45: Jupiter is in the ether or, variously the heaven, above) and was to be identified with Astarte (*qu. hept.* 7. 16).

quod non posset: *Aen.* 1. 37–8: 'haec secum: "mene incepto desistere victam | *nec posse Italia Teucrorum avertere regem!*"' A.'s meditation on the fall of Rome in *s.* 81 (taking a rather harder anti-Roman line than in *civ.*) ends with the recollection of Juno's speech to Aeolus, to argue that the Romans themselves were, as Trojans, once fugitives from a sacked city, and that a mighty goddess even then hounded them and opposed the founding of their new city: *s.* 81. 9, 'inducitur a poeta ipsorum Iuno irascens Aeneae et Troianis fugientibus, et dicit, "gens inimica mihi Tyrrhenum navigat aequor, | illum in Italiam portans victosque penates" [*Aen.* 1. 67–8].' The same lines also at *civ.* 1. 3.

The exercise was traditional: Quint. 1. 9. 2, '[condiscant] versus primo solvere, mox mutatis verbis interpretari, tum paraphrasi audacius vertere, qua et breviare quaedam et exornare salvo modo poetae sensu permittitur.' Another memory of a similar exercise occurs at *civ.* 1. 19: discussing the history of Lucretia, A. asks rhetorically, 'quid dicemus? adultera haec an casta iudicanda est? quis in hac *controversia* laborandum putaverit? egregie quidam ex hoc veraciterque *declamans* ait, "mirabile dictu, duo fuerunt et adulterium unus admisit".' For surviving examples on this theme, cf. the undistinguished African pastiche, the *epistula Didonis ad Aeneam* (E. Baehrens, *Poetae Latini Minores* (Leipzig, 1879-83), 4. 271–7), or Ennodius, *dictio* 28 ('nec tibi diva parens').

quae numquam . . . audieram: De Marchi, 311–12, is right that the phrase is meant to specify that the exercise was creative, not merely recitative.

errantes: The *errores* of Aeneas (1. 13. 20, 1. 13. 22) imitated by the student.

solutis verbis . . . versibus: 'prose . . . verse.' As in 'verbis sententias' (cf. e.g. Cic. *or.* 13. 42, 'dulce igitur orationis genus et solutum et fluens, *sententiis* argutum, *verbis* sonans'), A. employs the vocabulary of the schools.

vera vita: Jn. 14: 6 etc. (see on 1. 4. 4); the last similar expression, at 1. 13. 20 ('deus, vita mea'), also occurred in a meditation on the way he read the *Aeneid*.

quod CDGO Maur. Vega Ver. Pell. : *quid* S Knöll Skut.

conlectoribus: Apparently first here in the surviving literature (cf. only Evod. ad Aug. in A. *ep.* 158. 10, 'condiscipulus et conlector') and rare (next attestations two hundred years later in Greg. Mag. and Ildef. Tolet.). The same word in the more obvious cognate sense 'gatherer' also appears, approximately as rare and as late: cf. *TLL* s.v.

fumus et ventus: Cf. *s.* 22. 8. 8, 'ergo superbus sic se erigit contra deum, quomodo *fumus* contra caelum; restat ut ita deficiat, et tamquam in *ventos* suae vanitatis elatione dissipatus intereat'; *Io. ev. tr.* 10. 6, 'quid enim prodest adquirere in hoc saeculo quodlibet temporale et transitorium, sive sit pecunia, sive sit voluptas ventris et gutturis, sive sit honor in laude humana? *nonne omnia fumus et ventus?*'

laudes tuae: i.e. 'your praises, had I exercised my talent and tongue in uttering them.'

palmitem: Jn. 15: 4, 'sicut palmes non potest ferre fructum a semetipso, nisi manserit in vite; sic nec vos, nisi in me manseritis. ego sum vitis, vos palmites.' The scriptural language usually evokes from A. discussions (e.g. *Io. ev. tr.* 80–81) of pruning fruitless branches; here the modification is slight but optimistic. The passage has proved difficult for translators to render with much literal fidelity. Pine-Coffin rarely strives for literal fidelity, and so perhaps expresses the metaphor most effectively here: 'I might have used them [wits and tongue] O Lord, to praise you in the words of your Scriptures, which could have been a prop to support my heart, as if it were a young vine, so that it would not have produced the crop of worthless fruit, fit only for the birds to peck at.'

praeda volatilibus: *volatilia* is the regular word in *conf.* for birds (two exceptions only, the *volucrum* of Rom. 1: 23 at 5. 3. 5 and 7. 9. 15).

non . . . uno modo: The conventional acceptation of 'paganism' leaves no room for the young A., brought up in a Christian milieu, always flirting with the church (Brown, 41: 'Paganism meant nothing to Augustine'); but his view was that anything but full church membership was rank heathenism, and he stretches to assimilate his behaviour to his own preconception. Cf. his use of Jer. 2: 27 at 2. 3. 6.

sacrificatur: 4. 2. 3, 'daemonibus, quibus me illa superstitione ipse sacrificabam'.

angelis: for the fall of the angels, cf. 7. 3. 5, 'ex bono angelo diabolus factus est'; see *civ.* 12. 1–9, and for the Christian identification of the fallen angels as *daemones*, *civ.* 9. 19; for the further inference that the gods of 'paganism' are those same *transgressores angeli*, *civ.* 2. 24, 'nempe intelleguntur daemones, sicut saepe dixi notumque nobis est in litteris sacris resque ipsae satis indicant, negotium suum agere, ut pro diis habeantur et colantur.'

1. 18. 28

foras: 'outwards', as against *foris*, 'outside': see on 1. 6. 7. See *s.* 96. 2. 2, 'quid is foras'—quoted further below on 'filius ille tuus minor'.

barbarismo aut soloecismo: Suetonius frg. 176, 'soloecismus in sensu fit, barbarismus in voce'; cf. *doctr. chr.* 2. 13. 19, quoted below on 1. 18. 29. In *cat. rud.* 9. 13, A. advises the catechist to take care that well-educated proselytes do not take offence at what they encounter in church: 'ita enim non inridebunt, si aliquos antistites et ministros ecclesiae forte animadverterint vel cum barbarismis et soloecismis deum invocare, vel eadem verba quae pronuntiant non intellegere perturbateque distinguere.' But as bishop he loses his fastidiousness: *en. Ps.* 36. s. 3. 6, 'quid ad nos quid grammatici velint? melius in barbarismo nostro vos intellegitis quam in nostra disertitudine vos deserti eritis'; sim. at *en. Ps.* 138. 20. He was himself when young very fastidious, but the transformation can be plotted with a curious passage, stepping outside the tradition while drawing on and paying homage to it: *ord.* 2. 17. 45, 'soloecismos autem quos dicimus fortasse quisque doctus diligenter attendens in oratione mea reperiet; non enim defuit qui mihi nonnulla huius modi vitia ipsum Ciceronem fecisse peritissime persuasisset. barbarismorum autem genus nostris temporibus tale compertum est ut et ipsa eius oratio barbara videatur qua Roma servata est.' (Poetic licence turned what might otherwise be *barbarismi et soloecismi* into figures of speech—*schemata et metaplasmi*: *ord.* 2. 4. 13 and *c. Faust.* 22. 25.) For collection and discussion of texts, see G. Bellissima, *Aug. Mag.* 1. 39–40.

copiose ornateque: *c. acad.* 3. 18. 41, 'copiosissime atque ornatissime'; *doctr. chr.* 4. 5. 7, 'acute, ornate, vehementer'; cf. *Gn. c. man.* 1. 1. 1, 'ornato politoque sermone'; *c. litt. Pet.* 1. 1. 1, 'cultum ornatumque'. The categories are Ciceronian: Cic. *Tusc.* 1. 4. 7, 'hanc enim perfectam philosophiam semper iudicavi, quae de maximis quaestionibus copiose posset ornateque dicere'; Cic. *or.* 28. 97, 'tertius est ille amplus copiosus gravis ornatus, in quo profecto vis maxima est. hic est enim, cuius ornatum dicendi et copiam admiratae gentes eloquentiam in civitatibus plurimum valere passae sunt.' Cf. Cic. *Sull.* 4. 12, *Brut.* 5. 21.

longanimis . . . et verax: Ps. 85: 15, 'et tu domine deus meus miserator et misericors, longanimis et multum misericors et verax'; *en. Ps.* 85. 20, 'quare "longanimis et multum misericors et miserator"? quia in cruce pendens ait, "pater, ignosce illis, quia nesciunt quid faciunt".'

et verax: *Io. ev. tr.* 39. 8, 'verax est deus, non participando sed generando veritatem.'

de immanissimo profundo: Ps. 85: 13, 'et eruisti animam meam ex inferno inferiore.'

immanissimo: Superlative in *conf.*: 1. 18. 29, 6. 8. 13, 7. 9. 13.

sitientem: Ps. 41: 3–4, 'sitivit anima mea ad deum vivum: (4) quando veniam et apparebo ante faciem dei?'; *en. Ps.* 41. 5, 'sitio in peregrinatione, sitio in cursu; satiabor in adventu.' Ps. 62: 2, 'sitivit tibi anima mea'; *en. Ps.* 62. 5, 'omnis qui sibi vult aliquid praestari, in ardore est desiderii: ipsum desiderium sitis est animae.'

quaesivi vultum tuum: Ps. 26: 8–9, 'tibi dixit cor meum, "quaesivi vultum tuum, vultum tuum domine requiram, (9) ne avertas faciem tuam a me, ne declines in ira a servo tuo".' (Verse 8 is echoed at 9. 3. 6.) *en. Ps.* 26. en. 2. 16, '"tibi dixit cor meum, quaesivi vultum tuum, vultum tuum domine, requiram. ne avertas faciem tuam a me; ne declines in ira a servo tuo." magnifice, nihil dici divinius potest. sentiunt hoc qui vere amant.' Cf. *en. Ps.* 26. en. 2. 15, 'cor nostrum cui dicit, "quaesivi vultum tuum", nisi ei qui pertinet ad oculum cordis? . . . vis illam lucem quae videtur oculis cordis videre, quia ipsa lux deus est.'

As in CL, *vultus* is properly 'expression' (cf. 1. 8. 13, 'quae fiunt vultu et nutu oculorum ceterorumque membrorum actu') and *facies* is properly 'face' (i.e. the front of the head—cf. 2. 3. 6, 'qui ponunt tergum ad te et non faciem'); but already in CL the two were loosely interchangeable in many contexts, and that holds true even more for A., who is under the influence of scriptural texts where the Latin words are inconsistently used to translate Greek translations of the Hebrew; here, at Ps. 26: 8, the Greek is πρόσωπον; at Ps. 104: 4, 'quaerite faciem eius semper,' the Greek is still πρόσωπον. Whether A. noticed that inconsistency or not, the substantive context is so similar that his exegesis is relevant here. Similar longing for the face of God (in a work that transpires *in conspectu dei*—see on 1. 16. 26) is expressed (with various scriptural warrants noted ad loc.) frequently—*facies* is the most common word: 1. 5. 5, 4. 10. 15, 8. 10. 22, 9. 4. 10, 10. 5. 7, 10. 41. 66, 12. 15. 21, 12. 17. 24, 13. 8. 9, 13. 14. 15, 13. 15. 18, 13. 21. 30.

For A.'s exegesis concerning the 'face of God', see *en. Ps.* 104. 3, '"quaerite", inquit, "faciem eius semper." quae est facies domini, nisi praesentia dei? . . . scio quidem quia mihi adhaerere deo bonum est [Ps. 72: 28: see on 7. 11. 17]; sed si semper quaeritur, quando invenitur? an "semper" dixit, in tota vita ista qua hic vivitur, ex quo id nos facere debere cognovimus, quando et inventus quaerendus est? iam quippe illum invenit fides, sed adhuc eum quaerit spes. caritas autem et invenit eum per fidem, et eum quaerit habere per speciem, ubi tunc sic invenietur ut sufficiat nobis et ulterius non quaeratur. . . . et nimirum hoc est, "quaerite faciem eius semper", ut non huic inquisitioni, qua significatur amor, finem praestet inventio, sed amore crescente inquisitio crescat inventi.'

nam longe: '*eram* must be understood' (G–M).

longe: Over and over in the first 10 books, the distance separating God from his human creature, variously to be bridged or minimized or denied, is denoted with this poignant word. The last three books are, in this way and others (see on 3. 2. 4, 'lacrimae') more optimistic: 2. 2. 2, 2. 2. 3, 2. 2. 4 ('ubi eram? et quam longe exulabam a deliciis domus tuae'), 2. 6. 14, 3. 3. 5 (2 ×), 3. 6. 10, 3. 6. 11, 3. 6. 11 ('et longe peregrinabar abs te'), 4. 5. 10, 4. 12. 18, 4. 16. 31, 5. 2. 2, 5. 3. 3, 5. 13. 23, 7. 10. 16, 9. 11. 28 ('"nihil", inquit [Monnica] "longe est deo"'), 10. 27. 38, 10. 35. 54, 10. 35. 56, 10. 42. 67, 10. 42. 67, 11. 31. 41, 12. 11. 13. See *Io. ev. tr.* 2. 4: Platonists understand the doctrine of the Logos: 'illud potuerunt videre quod est, sed viderunt *de longe*: noluerunt tenere humilitatem Christi . . . et sorduit eis crux Christi.'

non enim pedibus: This paragraph ends with a vivid synthesis of two texts that epitomize A.'s principal spiritual authorities. For an early expression of the image, with no clear verbal echo of either Plot. or scripture, see *sol.* 1. 1. 5, 'recipe, oro, fugitivum tuum, domine, clementissime pater: iamiam satis poenas dederim, satis inimicis tuis, quos sub pedibus habes, servierim, satis fuerim fallaciarum ludibrium. accipe me ab istis fugientem famulum tuum, quia et isti me quando a te fugiebam acceperunt alienum ad te mihi redeundum esse sentio: pateat mihi pulsanti ianua tua.'

Plot. 1. 6. 8. 16–27, *φεύγωμεν δὴ φίλην ἐς πατρίδα* [*Iliad* 2. 140], *ἀληθέστερον ἄν τις παρακελεύοιτο. τίς οὖν ἡ φυγὴ καὶ πῶς; . . . πατρὶς δὴ ἡμῖν, ὅθεν παρήλθομεν, καὶ πατὴρ ἐκεῖ. τίς οὖν ὁ στόλος καὶ ἡ φυγή; οὐ ποσὶ δεῖ διανύσαι· πανταχοῦ γὰρ φέρουσι πόδες ἐπὶ γῆν ἄλλην ἀπ' ἄλλης· οὐδέ σε δεῖ ἵππων ὄχημα ἤ τι θαλάττιον παρασκευάσαι, ἀλλὰ ταῦτα πάντα ἀφεῖναι δεῖ καὶ μὴ βλέπειν, ἀλλ' οἷον μύσαντα ὄψιν ἄλλην ἀλλάξασθαι καὶ ἀνεγεῖραι, ἣν ἔχει μὲν πᾶς, χρῶνται δὲ ὀλίγοι.*

We are missing the Latin version in which A. encountered the text, but there is a clear echo at *civ.* 9. 17, 'ubi est illud Plotini, ubi ait, "fugiendum est igitur ad carissimam patriam, et ibi pater, et ibi omnia. quae igitur, inquit, classis aut fuga? similem deo fieri" [see on 7. 10. 16].' In spite of that echo, *a priori* argument has been employed to trace allegiance here to Porphyry (W. Theiler *P. u. A.* 60, 'unserer Auffassung nach Zitat des Porphyrios': see also ibid. 43–53), but no consensus arises from such methods; cf. Courcelle, *Recherches*, 128–9. Ambrose independently follows Plotinus: Amb. *Isaac* 8. 78–9, 'fugiamus ergo in patriam . . . (79) non utique pedum, qui sunt corporis . . . nec navibus fugiamus, aut curribus aut equis.'

Other Augustinian passages corroborate the echo, but do not further clarify how it came to him: 8. 8. 19, 'et non illuc ibatur navibus aut quadrigis aut pedibus'; *mor.* 1. 12. 21, 'quanto ergo magis longe discedit a deo, non loco sed affectione atque cupiditate ad inferiora quam est ipse, tanto magis stultitia miseriaque completur. dilectione igitur reditur in deum'; *doctr. chr.* 1. 10. 10, 'purgandus est animus, ut et perspicere illam lucem valeat et inhaerere perspectae. quam purgationem quasi ambulationem quandam et quasi navigationem ad patriam esse arbitremur. non enim ad eum qui ubique praesens est locis movemur, sed bono studio bonisque moribus.' Cf. *ss.* 141. 4. 4, 344. 1, *en. Ps.* 94. 2, 144. 3, 149. 5, *epp.* 145. 4, 155. 8. 13, and see du Roy, 98 n. 1.

itur abs te aut reditur ad te: A phrase echoed constantly through the prodigal Augustine's story (*redeo* in the ordinary sense of motion from place to place occurs only half as often as this metaphorical sense in *conf.*): 2. 6. 14 ('nisi cum redit ad te'), 3. 4. 7, 3. 8. 16, 4. 11. 16, 4. 12. 18, 4. 12. 19 (3 ×), 4. 16. 31, 5. 12. 22, 7. 7. 11 (2 ×), 8. 3. 8, 8. 4. 9, 10. 42. 67, 11. 8. 10, 12. 10. 10.

filius ille tuus minor: The image of the prodigal recurs in many ways throughout *conf.*, as throughout A.'s writings; the echoes almost always recall the first part of the story, the going astray.[39] Lk. 15: 11–18: 'homo quidam

[39] The story itself seems to invite reflection on the prodigality of the father as much as of the son, but this reading seems alien to the interpretation of A. and his time (or is at least ruled out if the

habuit duos filios; (12) et dixit adulescentior ex illis patri, "pater, da mihi portionem substantiae quae me contingit", et divisit illis substantiam, (13) et non post multos dies congregatis omnibus adulescentior filius peregre profectus in regionem longinquam et ibi dissipavit substantiam suam vivendo luxuriose. (14) et postquam omnia consummasset facta est fames valida in regione illa et ipse coepit egere, (15) et abiit et adhesit uni civium regionis illius et misit illum in villam suam ut pasceret porcos, (16) et cupiebat implere ventrem suum de siliquis quas porci manducabant et nemo illi dabat. (17) in se autem reversus dixit, "quanti mercennarii patris mei abundant panibus, ego autem hic fame pereo? (18) surgam et ibo ad patrem meum".'

The best treatment is G. N. Knauer, *Hermes* 85 (1957), 216–48; see also L. Ferrari, *RA* 12 (1977), 105–18 (who offers the interesting suggestion that A.'s own brother Navigius (who appears at 9. 11. 27) bears some resemblances to the elder brother of the parable). There is a suggestive note by A.-M. La Bonnardière in *Annuaire de l'École pratique des Hautes Études, v^e^ section, Sc. Relig.* 73 (1965–6), 154–5. For a provocative reading that may not be far wrong, bear in mind Rilke, *Malte Laurids Brigge*: 'Man wird mich schwer davon überzeugen, daß die Geschichte des verlorenen Sohnes nicht die Legende dessen ist, der nicht geliebt werden wollte.'[40] But note the essential difference. The acquisition of the Bible by cultured Greeks and Latins imported a huge collection of narratives against which to measure their own experience. These stories, for all their subtleties of construction, were at bottom unsubtle, black-and-white without shades of grey, cast in the desert sunlight of divine judgement. The breathing room for a re-reading of the story that Rilke finds was not there for A. and is indeed utterly alien to a *scriptural* reading of the story.

The 'literal sense' of the parable is the story told by Jesus about a man and his two sons, without special explanation. The two sons can then represent those who have been faithful and those who stray—in some contexts, this can imply the contrast between Jew and Gentile, but A. also takes the story back to the garden and asserts that the departure from home is equated with original sin (as at *en. Ps.* 24. 5, 'neque enim dimissus a te de paradiso et in longinquam regionem peregrinatus, per meipsum redire possum, nisi

suspicion arises: cf. Amb. *in Luc.* 7. 213, 'vides quod divinum patrimonium petentibus datur, nec putes culpam patris quod adulescentiori dedit').

[40] The prodigal was long familiar to Rilke's pen, from his *Geschichten vom lieben Gott* to his 'Der Auszug des verlorenen Sohnes' of 1906 (in *Neue Gedichte*), his translation of Gide's treatment of the story, and elsewhere. *Malte* was published in 1910. A year later, Rilke began translating *conf.* in Paris, breaking off to go up to Duino for the winter of 1911–12, where he began the *Duineser Elegien*. The translation made little progress, and the papers were among those left in Paris during the war. He never returned to the task, but there are surviving fragments that may eventually be published. See J. Ferreiro, *Rilke y San Agustín* (Madrid, 1966).

occurras erranti'). By the flexibility of that allegorical reading, the story then becomes the story of A. himself, a man whose relationship with his own father was difficult and strained, apparently unmarked by any final reconciliation. When A. is baptized (which on the theological level represents a return of the prodigal to the paternal God), he is at the same time being reconciled with Patricius—or at least with Patricius' own final disposition to accept baptism. Hence the possibility of prayer invoked for both Monnica and Patricius at 9. 13. 37.

A.'s fullest exposition of this parable, almost contemporary with *conf.*, deserves to be quoted somewhat fully as well, and parts of it will be cited again in other contexts:[41] *qu. ev.* 2. 33, 'homo habens duos filios, deus habens duos populos intellegitur, tamquam stirpes duas generis humani, unam eorum qui permanserunt in unius dei cultu, alteram eorum qui usque ad colenda idola deseruerunt deum. ... minorem in regionem longinquam profectum esse dicit. petiit enim ut sibi pars substantiae quae eum tangeret daretur a patre, tamquam anima potestate sua delectata, id quod illi est vivere, intellegere, meminisse, ingenio alacri excellere. omnia ista divina sunt munera, quae in potestatem accipiens per liberum arbitrium ... minor filius in regionem longinquam profectus est ... relicto ipso creatore. quod non post multos dies dixit factum, ut congregatis omnibus peregre proficisceretur in regionem longinquam, quia non multo post institutionem humani generis placuit animae per liberum arbitrium ... deserere eum a quo condita est. ... regio itaque longinqua oblivio dei est. fames autem in illa regione est indigentia verbi veritatis. unus civium regionis illius, aliquis aerius princeps ad militiam diaboli pertinens. ... siliquae quibus porcos pascebat[42] saeculares doctrinae sterili vanitate resonantes, de quibus laudes idolorum fabularumque ad deos gentium pertinentium vario sermone atque carminibus percrepant, quibus daemonia delectantur. ... "in se autem reversus", iam scilicet ab eis quae forinsecus frustra inliciunt et seducunt in conscientiae interiora reducem faciens intentionem suam. ... meretrices autem cum quibus dissipasse substantiam suam filius iunior accusatus est recte intelleguntur superstitiones relicto uno connubio legitimo verbi dei cum turba daemoniorum cupiditate turpissima fornicari.'

For other treatments of the parable of varying length, see esp. *s. Caill.* 2. 11 (*c.*400), then *Gn. litt.* 6. 20. 31, 6. 27. 38, 8. 4. 8, *Io. ev. tr.* 107. 2, *civ.* 11. 28, *en. Ps.* 18. en. 2. 3, 95. 5, 118. s. 5. 2, 138. 5–7, *ss.* 96. 2. 2, 330. 3, *adn. Iob* on 9: 25. For a hint of the elements of the story's interpretation that survived A.'s whole career, cf. *retr.* 1. 8. 3, '"vellem", inquam, "hinc plura dicere ac me ipsum constringere, dum quasi tibi praecipio ut nihil aliud agerem quam redderer mihi, cui me maxime debeo" [*quant. an.* 28. 55], ubi videor dicere potius

[41] His discussion of this pericope is by far the longest such essay in *qu. ev.*

[42] See on 3. 6. 11, 'a siliquis porcorum'.

debuisse, "redderer deo, cui me maxime debeo." sed quoniam prius sibi ipse homo reddendus est, ut illic quasi gradu facto inde surgat atque attollatur ad deum, sicut filius ille minor prius reversus est ad semet ipsum et tunc ait, "surgam et ibo ad patrem meum", ideo sic sum locutus.'

pinna visibili: Cf. 8. 7. 18 ('umerisque liberioribus pinnas recipiunt') and Amb., *fuga saec.* 5. 30, 'qui vult igitur manu Christi levari ante ipse evolet, habeat pennas suas (qui fugit saeculum pennas habet) et si suas non habet (ne forte ille solus habeat suas qui potestatem habet volandi), si suas inquam non habet, ab eo qui habet accipiat.'

regione: See on 4. 16. 31.

1. 18. 29

Two patterns of order are superimposed on each other here, with their similarities drawn out to contrast the values they represent. Order and law exist in human language as it is used here, but it is all shown to be a fraud, and a killing one at that, by the hidden order and law of God that works in and through even the deeds of his most perverse opponents.

Vide: See on 1. 10. 16.

pacta . . . pacta . . . placita: Cf. 1. 13. 22, 'secundum id pactum et placitum quo inter se homines ista signa firmarunt'. Caution remains in spite of this principled disdain for such fastidiousness; at *doctr. chr.* 4. 10. 24 (from the section of the work completed *c.*427), he advises that it is better to say *ossum* than *os*, because African ears are not good at catching longs and shorts; he goes on: 'qui ergo docet, vitabit verba omnia quae non docent; et si pro eis alia quae intellegantur integra potest dicere, id magis eliget; si autem non potest, sive quia non sunt sive quia in praesentia non occurrunt, utetur etiam verbis minus integris, dum tamen res ipsa doceatur atque discatur integre'; cf. *doctr. chr.* 3. 3. 7. Cf. also *Io. ev. tr.* 2. 14, 'non timeamus ferulas grammaticorum, dum tamen ad veritatem solidam et certiorem perveniamus.'

hominem (*dixerit*) CDGO[1]S Knöll Skut. Ver. : *ominem* O[2] Maur. Pell.

inimicando: Deponent (Souter renders 'have hostile feelings'); originally biblical Latin (Sirach 28: 6, 'desine inimicari' (LXX παῦσαι ἐχθραίνων)); in A. it occurs at *en. Ps.* 58. s. 1. 21, *ep.* 93. 12. 50, and cf. *s.* 90. 9, 'ecce homo est inimicus tuus: responde mihi, quid in illo inimicetur tibi? numquid hoc quod homo est, inimicatur tibi? non. sed quid? quod malus est. quod homo est, quod ego feci, non tibi inimicatur. . . . quod ipse fecit, hoc tibi inimicatur: unde malus, tibi inimicus est; non unde homo est.'

scripta conscientia: Rom. 2: 15, 'qui ostendunt opus legis scriptum in cordibus suis, testimonium reddente illis conscientia ipsorum et inter se invicem cogitationum accusantium aut etiam defendentium' (for 'reddente' *en. Ps.* 5. 13 and 9. 9 suggest 'perhibente'). Cf. 2. 5. 11, 4. 9. 14 ('ut rea sibi sit humana conscientia'), 5. 6. 11, 8. 7. 18 ('et increparet in me conscientia mea'), 10. 2. 2, 10. 3. 4, 10. 6. 8, 10. 30. 41, 12. 18. 27. *util. cred.* 16. 34, '*interior* nescio quae conscientia deum quaerendum deoque serviendum meliores quosque animos quasi publice privatimque hortatur'; *Io. ev. tr.* 32. 4, 'venter interioris hominis conscientia cordis est.'

se alteri facere CDGOS Knöll Skut. Ver. Pell. : *se alteri non facere* Z[2] Ramorino : *ne alteri facere* Vega. The emendations are pointless: though the allusion to the scriptural texts is clear, 'conscience' reminds us when we *have* done unto others what we would rather not suffer. The OT text is closest here: Tob. 4: 16, 'quod ab alio odis fieri tibi, vide ne alteri tu aliquando facias'; Mt. 7: 12, 'omnia ergo quaecumque vultis ut faciant vobis homines, et vos facite eis'; Lk. 6: 31, 'et prout vultis ut faciant vobis homines, et vos facite illis similiter'. Cf. Otto, *Sprichwörter*, 16: the sentiment is not securely attested in Latin outside scripture before the 390s ('Lampridius', *Alex. Sev.* 51—and there attributed to 'Iudaeis sive christianis'). But cf. also an odd inscription from Salona, usually dated fourth-century, emended to make it perhaps more ecumenical in spirit than it originally was: 'quitquit [pag]ani sive iude/i vel crissi[ani colunt] / colent e[t deo]s manis. unus / quisque quot sibi fi/[e]ri non vu[lt] facere non debet.' (*AE* 1959, no. 251, with much discussion; see L. Cracco Ruggini, *Augustinianum* 14 (1974), 427 n. 59). A more likely supplement for the second word is *[Rom]ani*, but the golden rule is undoubtedly present in the second sentence.

quam tu secretus es, habitans in excelsis in silentio: The remoteness of God again (cf. 'taces' in 1. 18. 28 above).

secretus: See on 1. 4. 4, 'secretissime et praesentissime'.

in excelsis: Is. 33: 5 (VL), 'sanctus deus qui habitat in excelsis'.

in silentio: *en. Ps.* 7. 1, 'silentio . . . id est altissimo secreto . . . ad hoc profundum secretum altumque silentium cum venisset apostolus, exclamavit quasi quodam ipsius altitudinis horrore perculsus: "o altitudo . . . quis consiliarius illius fuit?" [Rom. 11: 33–6] ita magnum illud silentium non expositione magis aperit, quam admiratione commendat.'

cupiditates: 1 Jn. 2: 16 (see on 1. 10. 16); but no threefold distinction is suggested here.

quaeritans GOS Knöll Skut. Ver. : *quaerit stans* CD : *quaerit astans* Maur. (supported by two minor MSS).

(*inter*) **hominibus** CDO Maur. Knöll Ver. : *homines* SG : *omines* Knöll (ed. min.) Skut. Pell. G–M think *omines* a better example of 'barbarism' (see below, 1. 19. 30, 'ubi magis timebam barbarismum facere') than *hominibus*, which would be a 'solecism', and they have A.'s own authority, almost contemporary with *conf.*, to support them: *doctr. chr.* 2. 13. 19, 'nam soloecismus qui dicitur, nihil est aliud quam cum verba non ea lege sibi coaptantur qua coaptaverunt qui priores nobis non sine auctoritate aliqua locuti sunt. utrum enim "inter homines" an "inter hominibus" dicatur, ad rerum non pertinet cognitorem. item barbarismus quid aliud est nisi verbum non eis litteris vel sono enuntiatum quo ab eis qui ante nos latine locuti sunt enuntiari solet? utrum autem "ignoscere" producta an correpta tertia syllaba dicatur ...' (At *mus.* 2. 2. 2, he confines *barbarismus* to an error in pronunciation of vowel quantity.)

But (1) it is hard to imagine a scribe finding an unexceptionable text (*homines*) and, having been paying attention closely to what he was copying, conjecturally emending to insert a grammatical fault—likelier that *hominibus* was there, and that it was regularized to *homines* by a solicitous but inattentive scribe; and (2) he has given an example of barbarism in this paragraph already ('sine adspiratione primae syllabae hominem dixerit') and is here complementing it with an example of solecism, still using the word *homo* as his touchstone, to give moral weight to his comparison above ('quam si contra tua praecepta hominem oderit, cum sit homo').

1. 19. 30

ego: 224 × in *conf.*; with *miser* at 2. 6. 12, 2. 8. 16, 3. 2. 4, 6. 16. 26, 8. 7. 17 ('at ego adulescens miser valde'), and cf. 10. 34. 53 ('nam ego capior miserabiliter'). On the hyperbaton, see on 1. 14. 23.

palaestra: The metaphor is common and Ciceronian (e.g. *or.* 56. 186, 68. 228–9).

proiectus: Ps. 30: 23, 'ego dixi in ecstasi mea, "proiectus sum a facie oculorum tuorum." exaudisti domine vocem orationis meae'; *en. Ps.* 30. en. 2. s. 3. 10, 'quia confessus sum, quia dixi: "proiectus sum ... tuorum", quia non superbus exstiti, sed cor meum accusavi, et in tribulatione mea titubans ad te exclamavi, exaudisti orationem meam.' Cf. 10. 41. 66.

paedagogum: Presumably the family servant who accompanied the young boy to school.

amore ludendi, studio spectandi nugatoria et imitandi ludicra inquietudine: There are three ways to take the last three words. (1) 'ludicra inquietudine' as ablative, parallel to 'amore' and 'studio' and governing the genitive gerund 'imitandi' (the reading of the *BA* translation); (2) 'inquietudine' as ablative, parallel to 'amore' and 'studio' and governing the genitive gerund 'imitandi', with 'ludicra' as neuter plural object of 'imitandi' (the reading of Arts, 118); (3—assumed for the punctuation here) 'imitandi' as genitive governed by 'studio' (and parallel with 'spectandi') and 'ludicra inquietudine' as ablative of the manner in which this gawping and aping takes place.

furta ... faciebam: Anticipation of the pear-theft of 2. 4. 9. The sort of sin that A. thought possible even among children: *Gn. litt.* 10. 13. 23, 'non enim de pueris grandiusculis agimus, quibus quidem peccatum proprium nolunt adtribuere quidam nisi ab anni quarti decimi articulo, cum pubescere coeperint. quod merito crederemus, si nulla essent peccata nisi quae membris genitalibus admittuntur. quis vero audeat adfirmare *furta*, mendacia, periuria non esse peccata, nisi qui talia vult inpune committere? at his plena est puerilis aetas, quamvis in eis non ita ut in maioribus punienda videantur, quod sperentur annis accedentibus, quibus ratio convalescat, posse praecepta salutaria melius intellegere eisque libentius oboedire.' The errors of boyhood include *curiositas* ('studio spectandi nugatoria'), *concupiscentia carnis* ('gula imperitante'), and *superbia* ('vana excellentiae cupiditate victus'): 1 Jn. 2: 16 (see on 1. 10. 16).

excellentiae: See on 6. 3. 3.

deprehenderem CDO²S Maur. Knöll Skut. Ver. Pell. : *deprehenderer* GO¹.

id quod aliis faciebam: The golden rule, as above, 1. 18. 29.

innocentia puerilis: What he said of infancy (1. 7. 11, 'ita imbecillitas membrorum infantilium innocens est, non animus infantium'), he says again of boyhood. P. De Labriolle, *Rev. de Philologie* 54 (1928), 47–9 (see Courcelle, *Recherches*, 50) cites parallels to Seneca, *const. sap.* 12. 1 (not acknowledged by Hagendahl): 'non ideo quicquam inter illos puerosque interesse quis dixerit, quod illis talorum nucumve et aeris minuti avaritia est, his auri argentique et urbium, quod illi inter ipsos magistratus gerunt et praetextam fascesque imitantur, hi eadem in campo foroque et in curia serio ludunt. . . . ergo par pueris longiusque progressis, sed in alia maioraque error est.' (For the rarity of A.'s allusions to Seneca, see on 5. 6. 11.)

a nucibus et pilulis et passeribus: 'nuts [= marbles?], balls, and sparrows': only the first regularly attested as a plaything for small boys (as at *Io. ev. tr.* 26. 5).

talium est: Mt. 19: 14, 'sinite pueros venire ad me: talium est enim regnum caelorum': note *pueros*, not *infantes*, hence the applicability here. A. regularly insists that the gospel text uses children as symbols, by their small stature, of humility, not of childish innocence: *qu. ev.* 1. 24, '"qui autem scandalizaverit unum ex pusillis istis" [Mt. 18: 6], id est, ex humilibus, quales vult esse discipulos suos.' Sim. at *pecc. mer.* 1. 19. 24, 'cum verax sit ille intellectus quod humilitatis similitudinem in parva aetate posuerit'; cf. *en. Ps.* 112. 1.

1. 20. 31

The first book closes with a tentative closure of the whole work: 'had I died in childhood, then that childhood would have been matter enough for confession such as this.' The paragraph begins and ends with thanks to God. Just as Bk. 13 ends with anticipation of the afterlife, so see here 'ero ipse tecum'.

rectori universitatis: Phrase also at 3. 8. 16; cf. 'unitatis' (below).

eram . . . vivebam . . . sentiebam: Another parallel to his infancy narrative: 1. 6. 10, '*eram* enim et *vivebam* etiam tunc et signa, quibus *sensa* mea nota aliis facerem, iam in fine infantiae quaerebam.' The triple statement implies another image of the trinity (cf. du Roy 172–7), *esse/vivere/intellegere* (cf. *sol.* 2. 1. 1, 'esse vis, vivere et intellegere, sed esse ut vivas, vivere ut intellegas'). Cf. *adn. Iob* on 38: 20, 'quis autem meminit fuisse se aut, cum esset, scire potuit esse se in lumbis patris sui? quandoquidem nec illud tempus quisquam recolit, quo non in parentibus, sed in se ipso natus est; quo tempore certe nemo dubitat quod iam erat et vivebat et sentiebat.'

incolumitatem: 1. 7. 12, 'proque eius [infantis] universitate atque incolumitate omnes conatus animantis insinuasti.'

vestigium secretissimae unitatis: This emphasis on *unitas* [sc. *dei*] immediately after a triadic image confirms the trinitarian intent of the passage. On 'secretissimae' see on 1. 4. 4. Plot. 6. 9. 1. 1, *πάντα τὰ ὄντα τῷ ἑνί ἐστιν ὄντα*. *mor.* 2. 6. 8, 'itaque in quantum quidque unitatem adipiscitur, in tantum est. unitatis est enim operatio, convenientia, et concordia, qua sunt in quantum sunt ea quae composita sunt: nam simplicia per se sunt, quia una sunt: quae autem non sunt simplicia, concordia partium imitantur unitatem, et in tantum sunt in quantum assequuntur. quare *ordinatio* [3] esse cogit, inordinatio vero non esse, quae perversio etiam nominatur atque corruptio. quidquid igitur corrumpitur, eo tendit ut non sit.'

interiore sensu: See on 10. 6. 8 (esp. *lib. arb.* 2. 3. 8–9 quoted there).

delectabar: See on 1. 6. 7.

falli nolebam: 10. 23. 33, 'multos expertus sum, qui vellent fallere, qui autem falli, neminem.'

fugiebam dolorem, abiectionem, ignorantiam: Cf. 'voluptates, sublimitates, veritates' below, and note; cf. 2. 2. 2, '*superba* deiectione'.

illi exulto: Ps. 2: 11, 'et exsultate illi cum tremore.'

peccabam: Plot. 1. 6. 8. 3–6 (before the passage quoted on 1. 18. 28 above), *ἴτω δὴ καὶ συνεπέσθω εἰς τὸ εἴσω ὁ δυνάμενος ἔξω καταλιπὼν ὄψιν ὀμμάτων μηδ' ἐπιστρέφων αὐτὸν εἰς τὰς προτέρας ἀγλαίας σωμάτων*. Cf. 10. 27. 38.

ceteris CDGO Maur. Skut. Ver. : *ceteras* S Knöll

voluptates [3], **sublimitates** [1], **veritates** [2]: The desire for these things manifests itself as, respectively, *concupiscentia carnis*, *ambitio saeculi*, *concupiscentia oculorum* (1 Jn. 2: 16: see on 1. 10. 16).

dolores [3], **confusiones** [1], **errores** [2]: The search for pleasure (*concupiscentia carnis*) brings pain, the search for loftiness (*ambitio saeculi*) brings confusion, the search for knowledge (*concupiscentia oculorum*) brings error. See on 3. 2. 4, 'dolere amabam'.

dulcedo [3] . . . **honor** [1] . . . **fiducia** [2]: The authentic goals of those three ambitions should be the sweetness that comes from God, the honour (in a client relationship: God honours me, and so I honour him—hence an expression of humility), and the faithfulness (active sense mutually: from A., the opposite of *curiositas*).

dulcedo: See on 1. 6. 9, and cf. 1. 4. 4 (invocation) and 2. 1. 1 ('ut tu dulcescas mihi, dulcedo non fallax, dulcedo felix et secura': beginning the book dominated by concupiscence). Cf. 2. 6. 13, where *luxuria* is the vice that imitates God: 'luxuria satietatem atque abundantiam se cupit vocari: tu es autem plenitudo et indeficiens copia incorruptibilis suavitatis.' For explicit link to the third person of the trinity, see *ep.* 11. 4 (quoted on 1. 6. 9).

fiducia: The usage (of the person in whom one reposes confidence) is not unparalleled: see L. G. Engels, *Graecitas et Latinitas Christianorum primaeva, Supplementa* III (Nijmegen, 1970), 73–4 (not biblical as far as A. knew, but Jerome used it twice in his version of the Psalter from the Hebrew, at Ps. 21: 10 and 70: 5 (in both cases rendering Gk. ἐλπίς)).

gratias tibi de donis tuis: 2 Cor. 9: 15, 'gratias deo super inenarrabili dono eius.'

BOOK TWO

Bk. 2 depicts the falling away from God that comes from adherence to the flesh. The book falls naturally into two parts:

2. 1. 1

If the six ages of man correlate to the six days of creation (see on 1. 8. 13), then *adulescentia* corresponds to the third day (Gn. 1: 9), the gathering of the waters. The imagery of this book, in contrast, vividly depicts waters of various kinds flowing unchecked and out of control (cf. Mandouze 75 n. 1). In the 'fall' of the young Augustine, the particular failure of *adulescentia* is thus the failure to acquire the divine order characteristic of the third day.

There is little here that is so unusual: adolescent sexual profligacy of a sort that seemed inconceivable to some of our pious ancestors, but that is less so today and was certainly far less so in late antiquity. It would be easy for Patricius to be an indulgent, self-satisfied spectator: *he* is the unremarkable member of his family. Imagery thus becomes an instrument by which A. records a judgement that derives from his profound conviction that his past is to be rewritten, that he is the one to give it an authorized interpretation. What gives power to these books that narrate the personal fall of A. is just the tension they create between the past as lived and the past as rewritten.

Every modern generation has found its own formulas for keeping A.'s description of his sexual conduct at arm's length. He may be taken as utterly profligate (an older view); or as a hyper-conscious religious fanatic exaggerating a few peccadilloes (a refreshingly modern view: e.g. Brown, 39); or as a helpless victim of a sexual addiction—and hence as no model for others less afflicted (a brave and orthodox post-modernist view: M. Miles, *Jour. Amer. Acad. Rel.* 50 (1982), 358). To those who knew him superficially in the early days, he seemed rather high-minded and sedate, as in the description of the Rogatist Vincentius (*ep.* 93. 13. 51, 'studiis olim deditum litterarum quietis et honestatis fuisse cultorem') or of the Manichee Secundinus (*ep. Sec.* 3, 'novi ego haec te semper odio habuisse, novi ego te semper magna amavisse, quae terras desererent, quae caelos peterent, quae corpora mortificarent, quae animas vivificarent'). The Pelagians, on the other hand, thought A. in his catholic days was more libertine than prude, against modern prejudgements (*c. Iul. imp*. 1. 98 (Julian accuses), 'ad Ioviniani consortium confugisti' and 2. 15).[1]

A conventionally dismissive view of this book (e.g. W. E. Mann, *Apeiron*, 12 (1978), 51: 'One is inclined to say that if this is the foulest sin that Augustine could muster from his repertoire, then he led an exemplary childhood') hints at something more important, that the synecdoche by which the pear theft stands for all A.'s other sins is a device that allows him, in the midst of his ardent profession of *confessio*, to maintain an extraordinary reticence. The *confessio peccati* contained in this work is incomplete, inexplicit, and indirect. (See on 4. 2. 2 for the possibility that A.'s liaison with his concubine began as early as his sixteenth year at Thagaste (cf. 2. 3. 5).)

Recordari volo: Bk. 8, connected to Bk. 2 by several threads, begins (8. 1. 1), 'deus meus, *recorder* in gratiarum actione tibi et *confitear* misericordias tuas super me.'

corruptiones: First here: God is incorruptible, man corruptible—the difference furnishes a line of contrast at Rom. 1: 23, echoed in 5. 3. 5 (and see on 7. 9. 14 for the whole Rom. passage in which it occurs). The cure for *corruptio* is *renovatio* (e.g. 13. 34. 49, 'renovasti ad imaginem et similitudinem tuam').

amore amoris tui facio istuc: The 'love of love', an unclassical phrase, is powerful but risks going astray if wrongly focused (cf. 3. 1. 1, 'nondum amabam et amare amabam . . . amans amare'), but at 11. 1. 1, its power is sublimated to the present task: 'iam dixi et dicam, amore amoris tui facio istuc.' J. Balogh, *Didaskaleion* NS 4 (1926), 5–8, rightly insists that 'tui' is objective genitive (cf. 'amem te').

istuc S Ver. Knöll Pell. : *istud* CDGO Maur. Skut. Decisive here is Knauer, 154 n. 1, adducing 10. 40. 65, 4. 5. 10, 8. 9. 21, 11. 22. 28; though CDGO often read *istud*, it is clear that *istuc* is A's own preference.

recolens: The verb 23 × in *conf.* (9 × in Bk. 10 on memory).

vias: Cf. 1. 13. 22, 'malarum viarum mearum'.

nequissimas: Infrequent in *conf.* and always in a context of cupidity: at 2. 4. 9 the pear-thieves are 'nequissimi adulescentuli', and cf. 4. 16. 30, 'nequissimus malarum cupiditatum servus'. Elsewhere in A. the association is not exclusive, but the tendency persists, e.g. *en. Ps.* 149. 15, 'progressus libidinum suarum in quaeque nequissima'.

amaritudine: *amarus* and derivatives 23 × in *conf.*, of the condition of man apart from the sweetness (cf. on *dulcescas* below) of God. One of his worst moments comes on the death of his friend at 4. 6. 11, 'requiescebam in amaritudine'. (Both there and here, the phrase is probably a veiled way of saying that A. savoured such bitterness in weeping.) Cf. Ex. 15: 22–6, where

[1] Cf. J. McLaughlin, *Jefferson and Monticello* (New York, 1988), 150–1, 'the question of whether Jefferson was a sexual adventurer [is] an historical Rorschach test: It tells more about the one making judgments than it does about the historical figure.'

the bitter waters of Mara are turned to sweetness by casting in a *lignum* (a word A. regularly uses of the redeeming cross). See on 6. 10. 17.

recogitationis: 'thinking over', usu. with remorse; only here in *conf.*, as noun elsewhere only of God's seeming penitence (*civ.* 15. 25) and the prodigal's incipient remorse (*qu. ev.* 2. 33); not classical, but attested from Tertullian on, perhaps influenced by Gn. 6: 6 (Vg.), 'et paenituit eum quod hominem fecisset in terra,' on which see *loc. hept.* 1. 14, 'quod scriptum est in quibusdam latinis codicibus: "et paenituit . . .", in graeco invenitur διενοήθη, quod magis "recogitavit" quam "paenituit" significare perhibetur. quod verbum etiam nonnulli latini codices habent' (so A.'s citation of the verse at *civ.* 15. 24 has 'recogitavit'—the verb is also rare in A.). The only other citations showing *recogitavit* for this infrequently-cited verse in *VL* (Beuron) are from Rufinus.

dulcescas . . . dulcedo non fallax, dulcedo: To furnish contrast with *amaritudo*: see on 1. 6. 9. At 1. 20. 31, *voluptas*, *dolor*, and *dulcedo* constitute a triad of vain aspiration, bitter discovery, and sweet alternative. There the divine person corresponding to 'sweetness' is the Spirit, personification of authentic love; Bk. 2 is the book of love gone wrong. Cf. *div. qu. Simp.* 1. 1. 5, '*fallax* enim dulcedo est quam plures atque maiores poenarum *amaritudines* consequuntur. quia ergo ab hominibus nondum spiritalem gratiam percipientibus suavius admittitur quod vetatur, fallit peccatum falsa dulcedine.'

felix: *felix* and its derivatives occur 26 × in *conf.*; of those, the overwhelming majority are in passages that associate *felicitas* with worldly satisfactions whose authenticity A. denies (e.g. 3. 2. 3, 'et amissione miserae felicitatis', 5. 8. 14, 'falsam felicitatem'), and most of the few that seem to associate *felicitas* with a higher happiness are open to explanation on other grounds. The exceptions: 2. 1. 1, 'dulcedo felix et secura' (predicated of God, not man; probably evoked by assonance with the preceding phrase, 'dulcedo non *fallax*'); 2. 2. 3, 'felicior expectarem amplexus tuos' (the word facilitates comparison to the state of worldly *felicitas* he sought at the time); 9. 3. 6, 'sapientiam pro aviditate sua sine fine felix' (where again there is an implied comparison, describing eternal happiness in terms of earthly); 12. 9. 9, 'valde mutabilitatem suam prae dulcedine felicissimae contemplationis tuae cohibet' (even here, *contemplatio* is *felicissima* by comparison to the *mutabilitas* it rivals).

frustatim discissus . . . ab uno te . . . in multa evanui: Is. 11: 12, 'et dispersos Israhel conliget' (see on 1. 3. 3), and cf. *adn. Iob* on 38: 2, 'significat ad hoc pati dura et amara dei servos in hoc saeculo, ut omnes affectiones suas a terrenarum delectationum *fluxu* [see on 2. 2. 2] conligant atque constringant.' From the praise of God for lingering goodness (as in 1. 20. 31, 'vestigium secretissimae unitatis') to the fragmentation that threatens that unity: 10. 29. 40, 'per continentiam quippe conligimur et redigimur in unum, a quo in multa defluximus'; *trin.* 4. 7. 11, 'quia enim ab uno deo summo et vero per impietatis iniquitatem resilientes et dissonantes defluxeramus et evanueramus in multa, *discissi per multa* et inhaerentes in multis.' The neo-Platonic echo is loud and clear: Plot. 6. 9. 1. 11–12, *εἰ φεύγοι τὸ ἓν εἰς πλῆθος θρυπτόμενα*; Plot. 4. 8. 4. 14–15, *μέρος γενομένη μονοῦταί τε καὶ ἀσθενεῖ καὶ πολυπραγμονεῖ καὶ πρὸς μέρος βλέπει* (and cf. Plot. 4. 2. 1. 34–5); Porphyry, *Marc.* 10, *συλλέγουσα ἀπὸ τοῦ σώματος πάντα τὰ διασκεδασθέντα σου μέλη καὶ εἰς πλῆθος κατακερματισθέντα ἀπὸ τῆς τέως ἐν μεγέθει δυνάμεως ἰσχυούσης ἑνώσεως*; cf. also Porph. *sent.* 5 and 11.

exarsi: A potent word, usually (as at 8. 5. 10, 10. 27. 38 ('tetigisti me, et exarsi in pacem tuam'), 11. 22. 28) of intense desire for that which is God or is associated with God. Rare occurrences in a negative sense at *civ.* 1. 1, 'multi vero in eam [civitatem dei] tantis exardescunt ignibus odiorum . . .'; sim. at *civ.* 2. 25, 3. 26; it is even used once of the fires of hell: *civ.* 21. 10. Most apposite here, from 392, with Plotinian resonance: *en. Ps.* 4. 9, 'cum dedita temporalibus voluptatibus anima semper exardescit cupiditate nec satiari potest, et multiplici atque aerumnosa cogitatione distenta simplex bonum videre non sinitur.'

satiari inferis: Cf. 1. 2. 2, 'non enim ego iam inferi.' The metaphor of satiety is predominantly used of carnal cupidity, secondarily of the alternative satisfactions to be known from God. Contrast 1. 12. 19, 'ad satiandas insatiabiles cupiditates', 2. 10. 18, 'et insatiabili satietate', and 6. 12. 22, 'consuetudo satiandae insatiabilis concupiscentiae', with 9. 6. 14, 'nec satiabar illis diebus dulcedine mirabili', and 11. 9. 11, 'satiabis in bonis desiderium meum'.

adulescentia: The third age now begins, to endure through Bks. 2–6; *infantia* began implicitly at 1. 6. 7, *pueritia* explicitly at 1. 8. 13 (see note there), and *iuventus* follows at 7. 1. 1.

silvescere: The word is rare and in origin botanical: Columella, *re rust.* 4. 11, Cic. *sen.* 15. 52; metaphorical (for hair) at Arnob. *nat.* 3. 15. In A., rare and always metaphorical (only exception for pure botany at *Gn. litt.* 8. 1. 4); at *gest. Pel.* 35. 65, *epp.* 36. 8. 18, 159. 2, 194. 10. 44, *civ.* 2. 18, 21. 8, *s.* 203. 3. 3, *c. Iul. imp.* 5. 15 (by Julian, mocking A.), *c. Faust.* 22. 70, *dial.* 10, *qu. ev.* 2. 6. Similar foliage metaphors at e.g. *c. acad.* 2. 2. 6, *Io. ep. tr.* 2. 9. The underlying neo-Platonic sense probably derives from the use of *silva* to = Gk. *ὕλη* ('matter'); cf. Macrob. *somn. Scip.* 1. 6. 9, 'anima aliena a silvestris contagione materiae tantum se auctori suo ac sibi debens'. But it is probably also worth keeping in mind *en. Ps.* 37. 15, 'quem [deum] cum [Adam] offendisset, fugit ad umbram, et abscondit se inter ligna paradisi' (cf. Gn. 3: 8). A forest of this sort can be 'bitter' (cf. *amaritudine* above) as well: *Io. ev. tr.* 16. 6, 'amara silva mundus hic fuit.'

contabuit species [2] **mea:** A. is saying nearly the opposite of the literal truth to make his case. He means (as 'placere cupiens oculis hominum' makes clear) that his outward appearance was something over which he took particular care as he entered the lists of love; but his real *species* (that which reflects the second person of the trinity: see on 1. 7. 12) was deteriorating as his outward show was enhanced—Jekyll and Hyde. Dan. 10: 8, 'sed et species mea immuta est et emarcui.' See on 3. 1. 1, 'elegans et urbanus'.

coram oculis tuis: Cf. Ps. 78: 10, 'coram oculis nostris', Lk. 15: 18 (the prodigal), 'pater, peccavi in caelum, et coram te'; see also on 10. 1. 1.

placens . . . hominum: Ps. 52: 6, 'deus dissipavit ossa hominibus placentium'; *en. Ps.* 52. 9, 'Christus . . . maluit displicere hominibus talibus, quales illi erant; filiis hominum, non filiis dei maluit displicere.'

2. 2. 2

delectabat: See on 1. 6. 7. In Bk. 2, *delectatio* incites both virtue and vice: 2. 5. 10 ('habent enim et haec ima delectationes'), 2. 5. 11, 2. 6. 12.

amare et amari: Cf. Cic. *Catil.* 2. 10. 23, 'hi pueri tam lepidi et delicati non solum *amare et amari* neque saltare et cantare, sed etiam sicas vibrare et spargere venena didicerunt'; Catiline occurs below (see on 2. 4. 9 and 2. 5. 11).

modus [1]: The flesh's lure brings clouds and darkness into the bright world of the mind, blurring boundaries and intermingling things better kept separate—not out of any abstract or arbitrary ethical imperative, but in respect of the innate quality of the beings themselves, the *modus* with which they were created (see on 1. 7. 12). For similar boundaries, cf. 2. 2. 3, 'ut *usque ad* coniugale litus exaestuarent fluctus aetatis meae . . . *fine* procreandorum *contentam*', 2. 2. 4, 'excessi omnia legitima tua', 2. 3. 8, 'cohercere terminis coniugalis affectus'; then see 13. 17. 20, 'tu enim coherces etiam malas cupiditates animarum et figis limites, quousque progredi sinantur aquae ut in se comminuantur fluctus earum, atque ita facis mare ordine imperii tui super omnia.' That passage depends on Job 38: 8–11, which A. expounds thus (*adn. Iob* on 38: 11): 'sicut ipse diabolus *modum* accepit *quo usque* affligeret Iob, ita illud mare *quo usque* persequeretur ecclesiam.' Note at 4. 10. 15, 'usque' in represented discourse is again a way of stating *modus*.

luminosus limes amicitiae: As G–M observe, a boundary so bright and clear should also have been readily visible and unmistakeable. Cf. 3. 1. 1 (which recapitulates in several ways the developments of Bk. 2), 'venam igitur amicitiae coinquinabam sordibus concupiscentiae candoremque eius obnubilabam de tartaro libidinis.' Cf. *ep.* 109. 2 (Severus of Milevis writing *to* A.: the passage is conceivably an undetected allusion to the present passage of *conf.*), 'vides quid facias, quod sic bonus es, quam nos rapias in amorem proximi, qui nobis primus ad dilectionem dei et ultimus gradus est et quasi limes quo sibi uterque adnectitur dei et proximi, in quo nos, ut dixi, quasi limite stantes amborum calore tangimur et amborum flagramus amore.' Cf. Cic. *amic.* 16. 56, 'constituendi sunt qui sint in amicitia fines et quasi termini diligendi.'

The importance to A. of friendship is self-evident from his writings (in *conf.* notably the loss of his unnamed friend in Bk. 4, the gathering of comrades on the verge of discovery in Bk. 6, the community at Cassiciacum in Bk. 9—always marked by the words *amicus/amicitia* (62 × in *conf.*, 59 in Bks. 1–9; first here except for the metaphorical 'amicitia huius mundi' at 1. 13. 21)), but here he also discreetly awards friendship an important place in his review of his moral past. In the background is the traditional Ciceronian definition, which A. quotes at *ep.* 258. 1, 'amicitia est rerum humanarum et divinarum cum benevolentia et caritate consensio.'[2]

Friendship for A. is undoubtedly a higher form of association than marriage, as suggested by the rhetorical climax at 8. 3. 8 ('hoc [est verum] in turpi et exsecranda laetitia, hoc in ea quae concessa et licita est [i.e. marriage], hoc in ipsa sincerissima honestate amicitiae'). Not long after *conf.*, A. connects the two again with ironic undertones: it was God who, to ensure the good will of human beings for each other, reinforced the natural possibility of friendship with the kinship that arises from sexual love and procreation. The reinforcement is in practice so strong that it tends to blur the original purpose. The model of continent friendship between man and woman he then proposes was not an obvious one to his contemporaries: *b. coniug.* 1. 1, 'quoniam unusquisque homo humani generis pars est et sociale quiddam est humana natura magnumque habet et naturale bonum, vim quoque *amicitiae*, ob hoc ex uno deus voluit omnes homines condere, ut in sua societate non sola similitudine generis, sed etiam cognationis vinculo tenerentur. prima itaque naturalis humanae societatis copula vir et uxor est. ... poterat enim esse in utroque sexu etiam sine tali commixtione [carnis] alterius regentis, alterius obsequentis amicalis quaedam et germana coniunctio.' (Cf. *b. coniug.* 9. 9, 'quaedam [necessaria sunt] propter amicitiam, sicut nuptiae vel concubitus; hinc enim subsistit propagatio generis humani, in quo societas amicalis magnum bonum est.') Sim. at *s. Den.* 16. 1 and 16. 7; at *Gn. litt.* 11. 42. 59, A. says that Adam ate the forbidden fruit not out of *concupiscentia*, 'quam nondum senserat', but 'amicali quadam benevolentia': this keeps the antithesis between friendship and concupiscence, and further suggests the presence of friendship in a husband–wife relationship.

Several writers, most noticeably in English Rebecca West, in her popular (and largely hostile) biography, *St. Augustine* (London, 1933), take this phrase as an admission of homosexual involvement, preceded by W. Achelis, *Die*

[2] On A. and C.'s view of friendship, see now T. J. van Bavel, *Augustiniana Traiectina* (Paris, 1987), 59–72.

Deutung Augustins (Prien-am-Chiemsee, 1921), who cited 3. 1. 1, and G. Papini, *S. Agostino* (Florence, 1930), 40 ff., and followed by J. Dittes, *Jour. Sci. Stud. Rel.* 4 (1965/66), 134 (citing also 4. 4. 7); J. Boswell, *Christianity, Social Tolerance, and Homosexuality* (Chicago, 1980), 135, is uncharacteristically reticent, but takes the parallel phrase at 3. 1. 1 ('venam igitur amicitiae coinquinabam sordibus concupiscentiae candoremque eius obnubilabam de tartaro libidinis') as hinting at homosexual acts. Others (see Mandouze, 79 n. 5, for references) find the same admission in 2. 1. 1, 'variis et umbrosis amoribus'. The topic elicits displays of high nervousness from all sides. If A. can be claimed for homosexuality, something apparently is assumed to change in contemporary debates over Christian teaching on the subject. But the nervousness is entirely absent in A., and that is why his narrative so artlessly leaves itself open to the interpretation.

limosa: *en. Ps.* 39. 3, 'limus luti, concupiscentiae carnales, tenebrae iniquitatum'.

concupiscentia carnis: First here in *conf.*; the phrase is biblical in 1 Jn. 2: 16 (see on 1. 10. 16 and esp. on 10. 30. 41).

scatebra: The noun only here in A.; in CL a 'gushing spring'; but cf. Oribasius *syn.* 5. 6 (quoted by Souter), 'cutaneous eruption'. Cf. also Ex. 16: 20, '[manna] et scatere coepit vermibus atque conputruit' (cf. 'conputrui' above here), and Hab. 3: 16, 'ingrediatur putredo in ossibus meis et subter me scateat.'

pubertatis: Of the age capable of begetting children; a medical term for A., hence infrequent.

obnubilabant: Also at 3. 1. 1, 4. 14. 23, 7. 1. 1—after that, the clouds begin to lift.

serenitas: Chaste (8. 11. 27, 'casta dignitas *continentiae*, serena et non dissolute hilaris') and divine (2. 3. 8, '*caligo* intercludens mihi, deus meus, serenitatem veritatis tuae,' 3. 2. 3, 'de caelesti serenitate').

caligine: See on 2. 3. 8.

libidinis: 27 × in *conf.*, never in a good sense. Some characteristic images occur, almost all in Bk. 2 and the first pages of 3: 2. 2. 4 ('vesania libidinis'), 2. 3. 6 ('vepres libidinum'), 3. 1. 1 ('de tartaro libidinis'), 3. 2. 3 ('aestus immanes taetrarum libidinum'), 9. 1. 1 ('scabiem libidinum').

gurgite flagitiorum: 6. 7. 11, 'gurges tamen morum Carthaginiensium'. Think also of the drowning of Pharaoh's men in the Red Sea in Ex. 14.

ira tua: At 13. 14. 15, God's *ira* is associated with the night of wickedness and sin, and the two pass away together. The vice of fleshly concupiscence, here adumbrated, is healed at the end of Bk. 8 in a scene whose master-text from scripture is Rom. 13: 12–14, 'nox praecessit, dies autem appropiavit. abiciamus ergo opera tenebrarum et induamur arma lucis' (see on 8. 12. 30; cf. 8. 12. 28, 'usquequo, domine, *irasceris* in finem?').

obsurdueram: 10. 27. 38, 'et rupisti surditatem meam'—in a famous passage of sensory images of restored life, and 13. 29. 44, 'perrumpens meam surditatem et clamans'.

stridore catenae mortalitatis meae, poena superbiae animae meae: The parallels between the two phrases are perfect, and explain themselves by comparison to 1. 1. 1, 'homo circumferens mortalitatem suam, circumferens testimonium peccati sui et testimonium quia superbis resistis'. *Mortalitas* is not (as G–M would have it) 'bodily senses . . . mortal body, abstract for concrete', but the presence in the person of the power of God resisting sin; once again, the young Augustine did not see the pattern that was so obvious to the confessor ('obsurdueram'). Cf. e.g. *Io. ep. tr.* 2. 10, 'sic in ista mortalitate, nos reatu tenebamur.'

longius a te: See on 1. 18. 28 ('non enim pedibus') and on 9. 11. 28 ('"nihil", inquit, [Monnica] "longe est deo"'): the separation from God is entirely in the will of the sinner. Cf. 2. 2. 3, 'non enim longe est a nobis omnipotentia tua, etiam cum longe sumus a te', and 2. 2. 4, 'quam longe exulabam', and note esp. 2. 6. 14, 'qui longe se a te faciunt'.

effundebar: 9. 4. 10, 'volentes enim gaudere forinsecus facile vanescunt et effunduntur in ea quae videntur et temporalia sunt.'

diffluebam et ebulliebam: These specially vivid metaphors of loss of control arising from self-will and concupiscence, leading to loss of self, prepare the paradox of restoration of self and of control through surrender to a power outside the self. 'Flux' is a characteristic of what is bad for A., though the valuation can be reversed by metaphor (as 2. 6. 13, 'sed bonorum omnium largitor affluentissimus tu es'); cf. also 1. 14. 23, 2. 2. 3, 2. 2. 4 (see on 'sequens impetum fluxus mei'), 2. 10. 18, 4. 11. 16, 8. 7. 18, 9. 8. 18, 10. 29. 40, 6. 16. 26 ('in quantalibet affluentia carnalium voluptatum'), 8. 7. 17 ('et ad nutum circumfluentibus corporis voluptatibus'), 9. 8. 18, 11. 2. 3 ('vox tua super affluentiam voluptatum'), 12. 10. 10, 12. 17. 25, 13. 7. 8, 13. 8. 9, 13. 10. 11. Elsewhere, see Hensellek, *Sitzungsber. Akad. Wien* 376 (1981), 5–6, on *vera rel.* 4. 6, 'a cupiditate bonorum temporalium atque affluentium'; *ord.* 2. 20. 52, *c. acad.* 2. 2. 4, 'a superfluarum cupiditatium vinculis evolavi.'

tacebas: See on 1. 18. 28, 'taces'.

tardum gaudium meum: 10. 27. 38, 'sero te amavi.'

sterilia semina: L. C. Ferrari, *Aug. Stud.* 8 (1977), 55–70, connects this phrase to the *regio egestatis* with which this book ends (2. 10. 18). That connection further links this passage to the theme of the prodigal (see on 1. 18. 28), and to other metaphors of sterility (and the farmer's struggles) at 2. 2. 3, 'ad temperamentum spinarum', 2. 3. 5, 'desertus potius a cultura tua', 2. 3. 6, 'vepres libidinum', 2. 3. 8, 'resecari ad vivum'; cf. also 4. 16. 29, 'terra spinas et tribulos pareret', 10. 16. 25, 'terra difficultatis', and 13. 19. 24, 'contristata est terra sterilis, et spinae offocaverunt verbum.' Ferrari shows (64) that 'thorns and briars are concerned not merely with sexual lust, but with the larger domain of the concupiscence of the senses in general.' See also Job

4: 8, 'quin potius vidi eos qui operantur iniquitatem et seminant dolores.' To 'sterilia semina dolorum', an important counter-image is that of 8. 11. 27, 'continentia . . . fecunda mater filiorum gaudiorum', where each element of the phrase is contradicted: sterility answered by fecundity, male seed (seed that does beget something, viz. *dolores*, but to no avail) by motherhood, *dolor* by *gaudium*.

superba deiectione et inquieta lassitudine: Two more oxymorons, where the adjective again suggests the cause and the noun depicts the paradoxical effect to which the cause gives rise (or fall). Cf. 1. 20. 31, 'fugiebam dolorem [3], *abiectionem* [1] (note the explicit *superba* here), ignorantiam' [2]. Cf. 2. 3. 6, 'inquieta adulescentia'.

lassitudine: 5. 1. 1, 'ut exsurgat in te a lassitudine anima nostra'; and see on 7. 18. 24 for a link between *lassitudo* and *mortalitas* (with a further link back to 1. 1. 1 for both *mortalitas* (cf. on 'stridore catenae mortalitatis meae' above) and *inquietum*); cf. 10. 34. 53, 'nec eam spargant in deliciosas *lassitudines*.'

2. 2. 3

What might have been is sketched, to pass judgement on what was. Marriage? Celibacy? Both honourable possibilities. God could have led A. to either; the presentation of the possible past is put in terms of a *hearing* ('sonitum nubium tuarum'; 'voces exaudirem') of scripture that might have been but was not. The clarity of this paragraph has been obscured by mispunctuation and misreading of the scriptural echo from 1 Cor. 7: 28 (see below). Either his parents could have arranged a marriage, or he himself could have heeded God's call. (The punctuation has been clarified. The mistake of earlier punctuation is to link the *si*-clause with what goes before, thus leading A. to the improbability of saying effectively, 'if it were impossible for me to be happy with employing my sexuality to the end of begetting children, someone might have encouraged me to marry.') O. Tescari (reviewing Skutella in *Riv. filol. istr. class.* NS 13 (1935), 525–9) offers unnecessary emendation and another punctuation.

Quis mihi modularetur [1] **aerumnam meam et novissimarum rerum fugaces pulchritudines** [2] **in usum** [3] **verteret:** A self-answering deliberative question (by no means uncommon in *conf.*)—the triune God, or a creature in his image (say, A.'s mother), could or would have acted, but did not. For the imperfect subjunctive of past time, see LHS 338.

novissimarum: Cf. 5. 2. 2, 12. 2. 2, for the same sense.

fluctus: Job 38: 11 (VL), Yahweh from the whirlwind: 'et dixi, "hucusque venies, et non transibis, sed in temetipso comminuentur fluctus tui".' Cf. on 2. 2. 4, 'sequens impetum fluxus mei'.

aetatis meae: i.e. *adulescentiae*.

sicut praescribit: Gn. 1: 28, 'crescite et multiplicamini et replete terram.'

lex tua: See on 2. 4. 9.

spinarum a paradiso tuo seclusarum: Gn. 3: 17–18, 'Adae autem dixit, . . . "maledicta terra erit in omnibus operibus tuis . . . (18) spinas et tribulos pariet tibi."' (This Genesis passage is quoted in a slightly different form at Heb. 6: 8; cf. 4. 16. 29.) Of importance is also Hosea 2: 5–6, esp. 6, 'propter hoc ecce ego saepiam viam tuam spinis' (in a context of metaphorical adultery and fornication).

longe: See on 2. 2. 2 above.

aut certe: 'at all events, this at any rate is certain': cf. *OLD* s.v. *certe* 2c, *TLL* 2. 1572–3 and 3. 938–9. Caesar, *b. civ.* 1. 85. 10, 'aut cum honore aliquo aut certe sine ignominia domum revertantur'; Livy 9. 37. 4, 'aut nocte aut certe luce prima'. Closest parallel in *conf.*, 9. 2. 4, 'aut, si curari et convalescere potuissem, certe intermittere'. *Others* might have guided him, or at least *he* could have heeded God's word himself. The phrase would stand better before the protasis, clarifying the disjunction with the other possibility (outlined in the first sentence of the paragraph); the length of the intrusive parenthesis probably justifies the anacoluthon. On this reading, A. does not realize at this moment in the text that by regarding what goes before as parenthetical, he has already set up a perfectly apt protasis that would make of the following clause a satisfactory apodosis without the *aut*.

nubium tuarum: *en. Ps.* 35. 8, 'quae sunt nubes? praedicatores verbi dei. . . . quando minatur per praedicatores deus, tonat per nubes. quando miracula facit per praedicatores deus, coruscat per nubes, terret per nubes, et inrigat per pluviam.' (The same equation at *en. Ps.* 56. 17, 76. 19, 88. s. 1. 7, 103. s. 1. 11, and *Gn. c. man.* 2. 3. 5.) Cf. 13. 15. 18, 'transeunt *nubes*, caelum autem manet. transeunt *praedicatores verbi* tui.'

There follows a spate of scriptural voices advocating continence, unheard at the time. A further scriptural voice with a similar exhortation is finally heard and heeded at 8. 12. 29, in the garden at Milan. A.'s life from here to there is deliberately pictured as an unnecessary detour—against which there is the mostly unspoken picture of how his life could have been or should have been, and at the same time doubts (as in the comparable recollection of a lost opportunity at 1. 11. 17, the deferred baptism). (The three Pauline texts recur as a group in the central chapters of A.'s *virg.*: see below for some details.)

tribulationem autem: 1 Cor. 7: 28, 'si autem acceperis uxorem non peccasti, et si nupserit virgo non peccavit; tribulationem tamen carnis habebunt huius modi, ego autem vobis parco' ('huius modi' specifies the subject of 'habebunt' (θλίψιν . . . ἕξουσιν οἱ τοιοῦτοι), 'those who marry'). Modern readers, who read Paul eagerly seeking permission to love and do as they will, may take this text mainly as allowing marriage; A. could read it otherwise not long after *conf.*—his reading of 'ego autem vobis parco' is particularly noteworthy in this regard: *virg.* 16. 16, 'addidit tamen, "tribulationem autem carnis habebunt huius modi, ego autem vobis parco", *hoc modo exhortans ad virginitatem continentiamque perpetuam*, ut aliquantulum a nuptiis etiam deterreret modeste sane, non tamquam a re mala et inlicita, sed tamquam ab onerosa ac molesta. . . . quod autem se dicit eis parcere, quos ait tribulationem carnis habituros, nihil mihi interim sanius occurrit quam eum noluisse aperire et explicare verbis eandem ipsam carnis tribulationem, quam praenuntiavit eis qui eligunt nuptias in suspicionibus zeli coniugalis, in procreandis filiis atque nutriendis, in timoribus et maeroribus orbitatis.' Indeed, at *virg.* 20. 20–21. 21, A. feels he has to go further and defend 'ego autem vobis parco' from a charge that it masks a *condemnation* of marriage—that what goes unsaid is a catalogue of punishments visited *in the afterlife* upon those who marry.

bonum est: 1 Cor. 7: 1–2, 'bonum est homini mulierem non tangere, (2) propter fornicationes autem unusquisque suam uxorem habeat.'

qui sine uxore est: 1 Cor. 7: 32–3, 'volo autem vos sine sollicitudine esse; qui sine uxore est cogitat ea quae sunt dei [domini], quomodo placeat deo [domino]; (33) qui autem matrimonio iunctus est, cogitat ea quae sunt mundi, quomodo placeat uxori.' Text from *civ.* 21. 26, corroborated by *virg.* 22. 22 (which reads *domini* and *domino*).

abscisus propter regnum caelorum: Mt. 19: 12, 'sunt enim spadones qui ita nati sunt; sunt autem alii qui ab hominibus facti sunt, et sunt spadones qui se ipsos castraverunt propter regnum caelorum: qui potest capere capiat'; *virg.* 23. 23 (giving text as here), 'quid veracius, quid lucidius dici potuit? Christus dicit, veritas dicit, virtus et sapientia dei dicit eos, qui pio proposito ab uxore ducenda se continuerint, castrare se ipsos propter regnum caelorum'.

felicior expectarem amplexus tuos: Cf. 13. 8. 9, 'ut currat vita mea in amplexus tuos nec avertatur'; on erotic metaphors for divine union in *conf.* see on 12. 16. 23, 'amatoria [cantica]', and the less explicit *deliciae* (e.g. 2. 2. 4, 2. 5. 10).

2. 2. 4

The juxtaposition of two patterns: he lived one way, and the presence of divine *flagella* in his life gave silent testimony to the other pattern he was violating, a pattern he could have found (cf. on 'legitima') from listening to what God had said through Moses. Implicit in such passages is an Old Testament/New Testament reading of his own life: the Law was known to him and could/should have changed his life; but it required the coming of Christ (see 7. 18. 24–7. 19. 25) to set him free.

sequens impetum fluxus mei: *en. Ps.* 74. 6, '"defluxit terra" [Ps. 74: 4]. si defluxit terra, unde defluxit nisi peccatis? ideo et delicta dicuntur. delinquere est tamquam de liquido quodam defluere a stabilitate firmamenti virtutis atque iustitiae. cupiditate enim inferiorum quisque peccat; sicut roboratur caritate superiorum, sic deficit, et quasi liquescit cupiditate inferiorum.'

legitima: A common expression in OT, but never as a substantive in NT; hence a way of saying that it was not only the evangelical or apostolical advice (quoted in 2. 2. 3) that spoke against his ways at this time, but even the Mosaic law. Cf. e.g. Lev. 10: 11, 'et instruere filios Israhel omnia legitima quae locutus est dominus ad eos per manum Moysi' (text from *qu. hept.* 3. 33); Lev. 18: 26, 'custodite legitima mea atque iudicia'; Num. 18: 18, 'tradidi tibi et filiis tuis . . . legitima sempiterna.'

misericorditer saeviens: Oxymoron again; paradox a sign of truth.

aspergens O Ver. : *aspargens* CGS Knöll Skut. Pell. A.'s practice, both elsewhere in *conf.* and in other works, is consistent in favour of *aspergo*. See on 9. 10. 23 for the verb's mystic sense.

sine offensione iucundari: As often, a form of *delectatio* (see on 1. 6. 7 and above on 2. 2. 2) is the legitimate mainspring of action; the way God manipulates life to exploit this motive foreshadows A.'s theories about the mechanics of predestination.

qui fingis dolorem in praecepto: Ps. 93: 20, 'numquid adhaeret tibi sedes iniquitatis, qui fingis dolorem in praecepto.' (The Psalm text's difficulty stems from an underlying mistranslation of the LXX; the precedent of the relative should be 'sedes', not 'tibi'; cf. Knauer, 58–60, showing a Greek antecedent for the reading that led to the incorrect translation.) *en. Ps.* 93. 23, 'formas, inquit, dolorem in praecepto, id est de dolore praeceptum nobis facis, ut ipse dolor praeceptum sit nobis.' The general topic of *en. Ps.* 93. 23–5 is the misery of the human condition. A. asserts God present in the world through the suffering he inflicts: *en. Ps.* 93. 24, 'qui fingis dolorem in praecepto, qui et istos filios sic voluisti exercere et erudire, sic voluisti eis praecepta dare, ut non essent sine timore, ne amarent aliquid aliud, et obliviscerentur te verum bonum suum? bonus est deus; si cessaret deus, et non misceret *amaritudines* felicitatibus saeculi, oblivisceremur eum.' Sim. at *en. Ps.* 38. 17.

occidis nos: Cf. Deut. 32: 39, 'videte quod ego sim solus et non sit alius deus praeter me. ego occidam et ego vivificabo, ego percutiam et ego sanabo, et non est qui de manu mea possit eruere.'

a deliciis domus tuae: Mic. 2: 9, 'mulieres . . . eiecistis de domo deliciarum suarum.' There is through these lines a general analogy to the condition of the prodigal son (see on 1. 18. 28): far from his father's house, he surrenders to the authority of another; but there is no verbal echo.

anno illo sexto decimo: A. was born 13 November 354, hence this sixteenth year ran from November 369 to November 370.

licentiosae . . . inlicitae: Cf. 'legitima' (above).

non fuit cura meorum: A. sees himself surrounded by misdirection. The people who cared for him did not see the divine pattern (or at least the course of least resistance: marriage (cf. 2. 2. 3); he does not expect them to have recommended continence), but had another plan for him, concentrating not on hearing what has been spoken by God but on speaking powerful words (speaking: 'sermonem facere'; powerful words: 'persuadere dictione' (cf. 1. 16. 26, 'hinc adquiritur eloquentia, rebus *persuadendis* . . . necessaria': the task that had been set for A. in boyhood)). There is implicit criticism of his parents for not seeing that the *adulescens* was different from the *puer*—criticism made explicit in remarking the callousness of his father in 2. 3. 6.

2. 3. 5

The education A. received set him apart even from his family: Brown 21, 'His cousins were less fortunate: they remained without a proper education (*beata v.* 1. 6, 'nec Lartidianum et Rusticum consobrinos meos, quamvis nullum vel grammaticum passi sint, desse volui'), and would have to face the poverty and boredom of a narrow world of unlettered squireens.' See on 6. 1. 1 for the way A.'s career prospects were expected to make the fortunes of the whole family.

Madauris: At *ep.* 16. 4, the elderly, 'pagan' teacher Maximus of Madauros addresses A. as 'vir eximie, qui a mea secta deviasti'; he may indeed have been A.'s teacher (as argued by Frend, *The Donatist Church* (Oxford, 1952), 235). We may flesh out his memories and add some local colour here (but A. deliberately omitted such things, and we run the risk of distorting *this* text by throwing additional light upon it) from *epp.* 16–17 (390?), from and to Maximus. Contrast Maximus' version of the city centre at *ep.* 16. 1 ('at vero nostrae urbis forum salutarium numinum frequentia possessum nos cernimus et probamus') with A.'s own memory at *ep.* 17. 1 ('in isto foro recordarer esse in duobus simulacris unum Martem nudum, alterum armatum, quorum daemonium infestissimum civibus porrectis tribus digitis contra conlocata statua humana comprimeret'). See P. Mastandrea, *Massimo di Madauros: Agostino, Epistulae 16 e 17* (Padua, 1985); C. Lepelley, *Les Cités de l'Afrique romaine au Bas-Empire*, 2 (Paris, 1981), 127–39, esp. 135–7 on Maximus (esp. 135 n. 27, where he quotes a funerary inscription (*ILAlg.* 1. 2209) dedicated to a grammarian named Marius by his daughter Maxima; if that Marius is not identical with A.'s correspondent, he might as well be for the long service and love of native city that the inscription attests) and on A.'s *ep.* 232, 'ad Madaurenses' (no date). *ep.* 232. 1, 'miror quod "patri" et "in domino salutem" scripsistis, quorum mihi superstitiosus cultus idolorum, contra quae idola facilius templa vestra quam corda clauduntur vel potius quae idola non magis in templis quam in vestris cordibus includuntur, cum magno est dolore notissimus.' It is conventional to surmise some line of intellectual contact between A. and Madauros' most famous citizen,

Apuleius; against such inclinations is the lack of mention of Apuleius and his works in A. until at least 408/9 (< *ep.* 102. 6. 32), and perhaps not until 412 (see my argument at *RA* 15 (1980), 149–50).

animositate: Not CL, biblical (rendering Gk. θυμός), but not exclusively so, not common (first in third century). Cf. Gal. 5: 19–21, 'opera carnis, quae sunt fornicationes, . . . (20) idolorum servitus, veneficia, inimicitiae, contentiones, aemulationes, animositates, dissensiones, haereses, (21) . . . et his similia' (cited at *civ.* 14. 2, 21. 25; a sim. list at 2 Cor. 12: 20); cf. Sirach 1: 28 and 31. 40 and Heb. 11: 27. A.'s use of the word does not have positive ring: 5. 10. 19, 'pristina animositate', may be neutral (though in a bad cause), but 6. 4. 5, 'puerili errore et animositate', is anything but positive. Cf. *civ.* 6. pr. ('levissimae temeritatis et perniciosissimae animositatis'; sim. at *ep.* 93. 5. 17, quoted in full on 'Thagastensis' below); cf. *retr.* 2. 33 ('iam haereticus [Pelagius] pertinacissima animositate defendit'; sim. at *haer.* 81), *c. ep. fund.* 27. 29, *ss.* 46. 16. 40, 164. 10. 14, 292. 2. 2 ('arrogantia tumidae animositatis'), *adult. coniug.* 2. 19. 20, *vera rel.* 53. 103, *cat. rud.* 8. 12 ('animositate acerrima et pervicaci arrogantia defensitantes'), *duab. an.* 9. 11, *c. Iul.* 6. 26. 83. The most positive use is of the adjective and is scarcely more than neutral: *sol.* 1. 1. 1, 'hoc ipsum litteris manda, ut prole tua fias animosior' ('livelier'?). Lexical definitions suggested include 'vehemence, impetuosity, ardor' (Lewis and Short), 'wrathful spirit, anger' (Souter); *TLL* offers no definition but quotes a gloss that equates it with *audacia*.

The translators here offer a wide variety of readings. Of the ones examined, only Carena ('ambizioni'), Hefele ('der stolze Ehrgeiz'), and Warner's paraphrase ('This was rather because my father had big ideas than because he was rich') carry any possibility of implied criticism; the others, all perhaps moved by a respect for fathers who make efforts for their sons, find generous versions: De Capitani, *Lectio I–II*, 107, discusses and prefers 'coraggio e tenacia' to 'ambizione'; Ryan and Pine-Coffin, 'determination'; Pusey, 'resolution'; Vega, 'animosa resolución'; *BA*, 'mon père y apportait plus de cœur que de moyens.' Perhaps in this context the nearest equivalent might be 'chutzpah'.

municipis: P.'s social status as a 'town-councillor' means that A.'s career is to be seen against the backdrop of the endless legislative attempts to keep the sons of *curiales* in their fathers' places. A. remarkably escaped twice, once to Carthage and Rome, then a second time, having come back to Thagaste, when he was ordained at Hippo. On the legal situation, see Jones, *LRE* 737–57.

P.'s financial status is difficult to assess; see C. Lepelley, *Bull. litt. eccl.* 88 (1987), 231–3, and *Atti 1986*, 1. 101–5, and B. Shaw, *Past and Present* 115 (1987), 8–10; among the wealthy, he was impecunious, but he was of the *honestiores* (cf. Possidius *v. Aug.* 1.1, 'parentibus honestis'), and A. (no less than Monnica) would have grown up with servants (1. 19. 30, and the gossips of

9. 9. 20) and a sense of place in the community. Shaw even suggests that P.'s inability to finance A.'s education represents a 'periodic liquidity problem' arising from a lean year on the land; but *s.* 356. 13 represents some recalled constraints: 'Augustinum, id est, hominem pauperem, de pauperibus natum. modo dicturi sunt homines quia inveni pretiosas vestes, quas non potuissem habere vel in domo patris mei vel in illa saeculari professione mea' (Lepelley, *Atti 1986*, 101, neglects the force of that last clause). Cf. *ep.* 126. 7, '... paucis agellulis paternis contemptis ... ego quippe secundum multorum sensum comparantium semet ipsos sibimet ipsis non divitias dimisisse sed ad divitias videor venisse. vix enim vigesima particula res mea paterna existimari potest in comparatione praediorum ecclesiae quae nunc ut dominus existimor possidere.'

Thagastensis: For concise description of history and position, see C. Lepelley, art. cit. 230. *ep.* 93. 5. 17 (to Vincent the Rogatist), 'mihi opponebatur civitas mea, quae cum tota esset in parte Donati, ad unitatem catholicam legum imperialium timore conversa est, quam nunc videmus ita huius vestrae *animositatis* perniciem detestari, ut in ea numquam fuisse credatur. ... quam multi enim, quod certo scimus, iam volebant esse catholici manifestissima veritate commoti et offensionem suorum reverendo cotidie differebant! quam multos non veritas, qua numquam praesumpsistis, sed obduratae consuetudinis grave vinculum conligabat.' This passage, taken together with *ep.* 52 (which is addressed to a Donatist named Severinus, who is some sort of blood relative of A. (*ep.* 52. 1, 'qui secundum carnem fratres sumus'; 52. 4, 'consanguinitas temporalis')) compels us to consider the sectarian history of Monnica herself. Frend, *The Donatist Church*, 184 (followed by R. Crespin, *Ministère et sainteté* (Paris, 1965), 140), dates the conversion of Thagaste to the anti-Donatist party to the period 348–61 ('the most probable time ... having regard to Augustine's own upbringing in a Catholic household'). If M. herself was a native of Thagaste (this we do not know for certain), she would have been between 16 and 29 (Mandouze, *Pros. chr.* s.v. Monnica), hence in all probability a young bride and mother, when the change came. There is no slightest whisper anywhere in A.'s writings that she may ever have been, willing or unwilling, a Donatist communicant, nor is there any insinuation raised by any of his Donatist opponents, nor is the question of her past discussed by scholars. At any rate, she is unlikely to have been completely untouched by such controversies, and A.'s childhood ('in a Catholic household') was surely not completely insulated either. (Frend now (*Atti 1986*, 1. 141) takes for granted that M. had been brought up Donatist, and thinks her only superficially converted.)

cui narro haec?: The authorial voice breaks off the 'confession' proper and speaks as it were in a void, to no one. This excursion begins from the preposterousness of 'narration' directed to an all-knowing God. The fact of narrative compels him to admit that he has a human audience, with which he shares at least the frustration implied in the paradox of the silent God to whom men wish to speak, who is immeasurably far above and at the same time intimately accessible. The accessibility of God is conditioned not, however, on finding the right pattern of words, but on finding the right pattern of life ('si cor confitens et vita ex fide est').

de quam profundo: Ps. 129: 1, 'de profundis clamavi ad te, domine, domine, exaudi vocem meam'; *en. Ps.* 129. 1, 'profundum enim nobis est vita ista mortalis. quisquis se in profundo intellexerit, clamat, gemit, suspirat, donec de profundo eruatur, et veniat ad eum qui super omnes abyssos sedet et super Cherubim.'

quam OS edd. : *quo* CDG.

vita ex fide: Hab. 2: 4, 'iustus autem in fide sua vivet'; close echoes/quotations at Rom. 1: 17, Gal. 3: 11, and Heb. 10: 38.

quis enim: The thought is connected both to the beginning of the present paragraph (the *cura* taken by his father for his education) and to the thrust of the last paragraph (the inadequacy of the 'cura meorum').

cum interea: Concessive: adducing again the contrast from the last paragraph, 'even though'.

satageret: Used once of the *cura* of each parent: 1. 11. 17, 'nam illa [Monnica] satagebat, ut tu mihi pater esses.'

dummodo . . . a cultura tua: The sequence of metaphors is instructive: 'disertus' represents the goal his father sought; a purely accidental verbal echo suggests 'desertus', which can be taken not only in the personal sense of abandonment but in an agricultural sense, hence 'a cultura tua' (implying that 'disertus' also had something to do with a different kind of *cultura*—that pair of metaphors forms a circle), but thence on a direct line it is possible to speak of God as the farmer: cf. 1 Cor. 3: 9 ('dei agricultura estis') and Mt. 13: 24–30 (the parable of the sower). There is want of strict logic here, but each step of the sequence of verbal association makes sense as it is taken, and the effect of the whole is strong. (The same word-play at *en. Ps.* 36. s. 3. 6, 'melius in barbarismo nostro vos intellegitis, quam in nostra disertitudine vos deserti eritis.')

2. 3. 6

This is narrative without action: A. himself does nothing in this paragraph. We see him through his parents' eyes, and A. contrives to make only the mother's view (fearing what may come) trustworthy—she is the only one of the three at the time who already has the faith, and the scriptural citation describing A. is imposed on her thoughts. The young A. did not go to a temple to worship rock and wood, but he put himself in the camp of those who do: this is Augustine's 'paganism' in his own words.

sexto illo et decimo anno: Cf. 2. 2. 4.

vepres libidinum: Cf. on 2. 2. 4, 'vesania libidinis'. The liaison with the mother of Adeodatus probably began during 370/1, the first year at Carthage (see on 4. 2. 2).

nulla . . . eradicans manus: The image of absent (and agricultural) authority recurs (cf. 2. 2. 3, 'quis mihi modularetur . . .'), and now criticism of A.'s parents is more explicit.

ille: i.e. the same one who was collecting funds for A.'s education.

in balneis vidit: If A. did not know Amb. *off.* 1. 18. 79, 'mos vetus et in urbe Roma et in plerisque civitatibus fuit ut filii puberes cum parentibus vel generi cum soceris non lavarent, ne paternae reverentiae auctoritas minueretur,' he had surely read Amb.'s source, Cic. *off.* 1. 35. 129, 'nostro quidem more cum parentibus puberes filii, cum soceris generi non lavantur. retinenda igitur est huius generis verecundia.' There is a similar assertion in Plut. *Cato maior* 20 (*τοῦ παιδὸς παρόντος συλλούσασθαι μηδέποτε· καὶ τοῦτο κοινὸν ἔοικε Ῥωμαίων ἔθος εἶναι*), but he suggests that the custom had faded.

inquieta indutum adulescentia: The epithet may be accurate, as far as it goes, but its choice integrates this episode into the pattern of restlessness and repose within which A. sets *conf.* The disquieting revelation of nakedness (even if disquieting only in recollection) at the origin of sin echoes that of Adam and Eve (Gn. 3: 7), whose fig-leaves give meaning to the fig-tree under which A.'s crisis is finally resolved (8. 12. 28), and who were eventually given *tunicae pelliceae* by God (Gn. 3: 21), which represent the mortality that comes of sin (see on 7. 18. 24). Whether it was more than simple nakedness that Patricius saw is a question most leave veiled in discretion; for indiscretion, see C. Kligerman, *Jour. Amer. Psychoan. Assoc.* 5 (1957), 473, giving a separate force to *inquieta*; and R. Brändle and W. Neidhart, *Theol. Zschr.* 40 (1984), 173, who go even further. Patricius' enthusiasm was, at any rate, perfectly typical of his class and station in life: Rousselle, *Porneia*, 59: 'The appearance of pubic hair and his first ejaculations were a cause for celebration for the whole household, particularly the father. . . . When he came to take a wife, his father would have to give assurances as to his son's virility, a kind of certificate that he was not impotent. At these first manifestations of sexual maturity the young man would be the object of renewed attentions.'

in nepotes gestiret: Hrdlicka, 198, lists five other verbs of 'eager desire, longing, etc.', with which A. uses *in* + acc. as here: *deficio* (3. 6. 10), *flagro* (3. 8. 16, 5. 7. 13, 7. 6. 8), *anhelo* (5. 11. 21), *flammo* (11. 22. 28); *exardesco* (10. 27. 38); only the last shows the construction in classical Latin.

gaudens matri indicavit: The same words at 8. 12. 30 describe reporting exactly the opposite news (the end of A.'s active sexual life) to the same person: 'inde ad matrem ingredimur, indicamus: gaudet.'

vinulentia: Only here metaphorical. Used in *conf.* of Patricius here, twice of Monnica (6. 2. 2, and see on 9. 8. 18), and once of the beggar seen from his carriage in Milan (6. 6. 10). Elsewhere, *vinulentia* is commonly listed by A. among serious sins: *en. Ps.* 58. s. 2. 5, 75. 16.

oblitus est . . . amavit: Rom. 1: 25 (see on 7. 9. 14), 'qui transmutaverunt veritatem dei in mendacium et coluerunt et servierunt creaturae potius quam creatori.' The implication here is that Patricius is in the camp of the idolators; see on 'qui ponunt ad te tergum' below.

inclinatae in ima DG edd. : *inclinata in ima* C : *inclinatae** in ima* O : *inlicitae in anima* S. See on 2. 5. 10, 'ima', and cf. 7. 16. 22, 'a summa substantia, te deo, detortae in infima voluntatis perversitatem'.

inchoaveras: Nothing here indicates great or long-standing piety on M.'s part; quite the contrary. In view of the contrast with Patricius in the next line, this may mean only that she had already been baptized. Her actions so far are those of a worldly mother.

templum tuum: 1 Cor. 3: 16–17, 'nescitis quia templum dei estis et spiritus dei habitat in vobis? (17) si quis autem templum dei violaverit, disperdet illum deus. templum enim dei sanctum est quod estis vos.' A. does not see the individual as a separate temple, but as part of a community: *en. Ps.* 126. 3, 'domus enim dei, populus dei; quia domus dei, templum dei. et quid dicit apostolus? "templum enim dei sanctum est, quod estis vos." omnes autem fideles, quae est domus dei.'

exordium sanctae habitationis tuae: Sirach 24: 14 (Wisdom speaks), 'ab initio ante saeculum creata sum, et usque ad futurum saeculum non desinam, et in habitatione sancta coram ipso ministravi.'

catechumenus: P.'s baptism is mentioned at 9. 9. 22; his death is dated at 3. 4. 7 to 370/1, the year following the episode here (esp. if A. were at Carthage when his father died, the episode recounted here may be among A.'s last strong memories of him).

exilivit CDGO Maur. Ver. ('started up'—often with verbs of fearing) : *exsiluit* S Knöll Skut.

pia trepidatione ac tremore: 2 Cor. 7: 15, 'quomodo cum timore et tremore excepistis eum.' The echo answers the citation of Rom. 1: 25, by way of saying that M. is on the side of the faithful; but her attitude is that of fear ('trepidatione . . . timore . . . timuit': Ps. 110: 10, 'initium sapientiae timor'), i.e. imperfect (1 Jn. 4: 18, 'perfecta caritas foras mittit timorem'). Contrast the serenity with which she meets her end at 9. 10. 26, 'quid hic facio?'

quamvis . . . tamen: If he is not a Christian, then his moral failings are irrelevant; if he had been baptized, then they would have been critical.

fideli: A.'s use of this word through the narrative books (1–9) of *conf.* strongly suggests that he means it particularly to mean 'full, sacramental member of the Christian community'. The complete register of other occurrences in *conf.* where the adjective/substantive is used in a religious context: 2. 3. 7 (of

Monnica); 3. 11. 19 (again of M.); 4. 15. 26, 'et dicebam parvulis fidelibus tuis, civibus meis'; 5. 5. 8, 'consolatorem et ditatorem fidelium tuorum'; 5. 9. 17, 'quae illa [sc. Monnicae] fideli pectore tenebat'; 6. 1. 1, 'me visura esset fidelem catholicum'; 8. 2. 5, 'in conspectu populi fidelis Romae' (where Marius Victorinus, not yet *fidelis* in this sense, professes his faith in the presence of the baptized); 8. 6. 14, 'christianus quippe et fidelis erat' (not, on this interpretation, merely an additional fact about the sincerity of Ponticianus, but a way of specifying that he had taken the step, baptism, that A. was deferring); 9. 3. 5, 'nondum christianus coniuge fideli'; 9. 3. 5, 'et in ea christianus et fidelis factus'; 9. 3. 6, 'tuum ipsum etiam fidelem catholicum'; 9. 8. 17, 'in domo fideli, bono membro ecclesiae tuae'; 9. 9. 22, 'nec in eo iam fideli planxit, quod in nondum fideli toleraverat' (*eo* = *Patricio*, now baptized on his deathbed); 9. 11. 28, 'quae immittis in corda fidelium tuorum'. A less clearly marked group of texts occurs in Bk. 13, where A.'s allegorical reading of Genesis correlates the days of creation with the stages of the life of the visible church. On the reading suggested here, those passages also take on new implications, e.g. 13. 21. 29, 'sed et fideles exhortantur et benedicuntur eis', 13. 23. 34, 'in operibus moribusque fidelium'; see also 13. 21. 29, 13. 21. 30, 13. 34. 49 (4 ×). The only apparent exception to this pattern in *conf.* may be taken without difficulty as corroboration: Moses was, by what would later be remarked as 'baptism of desire', a full member of the heavenly city *sub lege*: 12. 26. 36, 'de Moyse fidelissimo famulo tuo'. In other works of A., there are passages that do not insist on being taken this way (see e.g. *praed. sanct.* 5. 10, 'fides discernit ab infideli fidelem'), but numerous texts offer some corroboration (and in support of the view advanced here, see B. Quinot at *BA* 30. 751–2): *ep.* 98. 10, 'ipsius fidei sacramentum fidelem facit'; *s.* 181. 3. 3, 'christiani, inquiunt, sumus. fideles ergo estis? baptizati estis? baptizati, inquiunt, sumus'; *s.* 151. 2. 2, 'fidelibus loquens, baptizatis loquens, quibus utique in sancto lavacro omnia fuerant dimissa peccata'; *mor.* 1. 35. 80, 'nolite iam dicere, catechumenis licere uti coniugibus, fidelibus autem non licere'; *f. et op.* 7. 11, 'utrumque catechumenis, utrumque fidelibus, utrumque baptizandis, utrumque baptizatis'; sim. at *epp.* 47. 2 ('fideles vocantur qui baptizantur in Christo'), 2*. 4, 151. 14, *en. Ps.* 90. s. 2. 6, 132. 4, 134. 21, 137. 4, *Io. ev. tr.* 44. 2, *ss.* 49. 8. 8 ('post sermonem fit missa catechumenis: manebunt fideles'), 181. 5. 7, 232. 7. 7, 294. 13. 14, 392. 2. 2, 392. 3. 3, 'fideles, id est baptizati', *s. Den.* 17. 8 (where catechumens are clearly included in the category *christiani*), *Rom. inch. exp.* 19, *pecc. mer.* 3. 12. 21 (and at *pecc. mer.* 1. 20. 28 and 1. 33. 62, baptism puts infants among the *fideles*); note also *praed. sanct.* 14. 27, suggesting three categories: 'fideles, paenitentes, catechumenos'. An explicit statement contemporary with A. is found in the fragment of Hier. *tract. in ps.* 91, printed in *PL* 39. 2173–4 as ps.-Aug. *s.* 232. 2 app. (*PL* 39. 2173: cf. *Clavis Patrum Latinorum*, 593): 'denique accepto baptismo hoc dicimus: fidelis factus est, sive factus sum'; and cf.

Amb. *sacr.* 1. 1. 1, 'in christiano enim viro prima est fides. ideo et Romae [v.l.: ideo recto nomine] fideles dicuntur, qui baptizati sunt.' *TLL* s.v. *fidelis* shows the word opposed to catechumen at Tert. *praescr.* 41, *itin. Eger.* 24. 6, Hier. *in Is.* 19. 18 (perhaps), and Nicetas Remes. *instr.* frg. 2, but gives no indication that the usage is general.

qui ponunt ad te tergum et non faciem: Jer. 2: 27, 'dicentes ligno, "pater meus es tu", et lapidi, "tu me genuisti", verterunt ad me tergum et non faciem.' *en. Ps.* 113. s. 2. 1, 'nam dictum est alio loco scripturarum de *simulacrorum cultoribus*, "dicentes ligno, 'pater meus es tu', et lapidi, 'tu me genuisti'."' *en. Ps.* 65. 21, 'qui dicebam "ligno, 'pater meus es tu,' et lapidi, 'tu me genuisti'," modo dico, "pater noster, qui es in caelis".' *lib. arb.* 2. 16. 43, 'vae qui se avertunt a lumine tuo et obscuritati suae dulciter inhaerent! tamquam enim dorsum ad te ponentes [cf. 4. 16. 30, 'dorsum enim habebam ad lumen'] in carnali opera velut in umbra sua defiguntur.' Father, mother, and son now each have their own characterizing scriptural citation in this paragraph: father and son are on the side of the idolators, mother has begun to follow a different path.

2. 3. 7

et audeo dicere tacuisse te: This calls into question the silence attributed to God at 2. 2. 2 and evokes again the distance from God that yawned repeatedly through 2. 2. 2–2. 2. 4. The two motifs are intertwined: what seems to be the silence of God is the heedlessness of the one who turns his back on God (2. 3. 6, 'qui ponunt ad te tergum'). Cf. Is. 42: 14 (VL), as quoted in *s.* 47. 3. 4, 'audio ergo te loquentem in tot praeceptis, in tot sacramentis, in tot paginis, in tot libris: audio denique in hoc ipso quod dicis, "tacui: numquid semper tacebo?" quomodo ergo tacuisti?'

ne adulterarem: M.'s attitude is reminiscent of the conventional wisdom of Cic. *Cael.* 18. 42, 'parcat iuventus pudicitiae suae, ne spoliet alienam, ne effundat patrimonium, . . .'; sim. at *Cael.* 12. 28.

monitus muliebres: The same disdain in another context at *util. cred.* 1. 2, 'spernentem scilicet quasi aniles fabulas'.

a me, a me: OS Knöll Skut. Pell. Ver. : *a me* CDG Maur.

filio . . . servo tuo: Cf. 5. 10. 18 and see on 9. 1. 1. Ps. 115: 16 (7), 'O domine, ego servus tuus; ego servus tuus, et filius ancillae tuae.' The sequence here is to be noted: 'her son, son of your handmaid, your servant.' The middle term of the three is ambiguous; the *ancilla* is both Monnica and the church; the first meaning looks back to 'filio eius', the second forward to 'servo tuo'.

sed nesciebam: These lines anticipate and comment on the pear-tree incident, giving it a context clearly congruent with that of the sexual transgressions implicit through the first half of the book. Pellegrino, *Les Confessions*,

75 n. 5, adduces Amb. *Noe* 22. 81, 'ex illa enim aetate crescit malitia, licet alibi legimus quod non sit sine peccato nec unius diei infans [Job 14: 6 (VL): echoed at *conf.* 1. 7. 11, where also, 'ita imbecillitas membrorum infantilium innocens est, non animus infantium'], sed et infantia sine peccato non sit propter corporis infirmitatem, diligentia autem et studium peccandi incipiat a iuventute, ut puer quasi infirmus peccet, iuvenis tamquam inprobus, qui studiose cupiat peccata committere et in criminibus glorietur [cf. here 'gloriantes']; apud plerosque enim innocentia pro ignavia et culpa pro laude [cf. here 'laudis'] habetur. . . . crescit ergo cum aetatibus culpa.'

praeceps . . . caecitate: For the juxtaposition (both words implying failure of vision and attention), cf. 6. 7. 12, 'caeco et praecipiti studio'; cf. also 4. 1. 1, 'dux in praeceps'.

caecitate: Blindness to beauty and loss of light (Jn. 1: 9, 'lumen verum') is the metaphor for A.'s separation from the second person of the trinity: cf. 6. 16. 26, 'ita demersus et caecus cogitare non possem lumen honestatis et gratis amplectendae *pulchritudinis* [2]'; cf. 7. 8. 12 (where see on 'collyrio'). Blindness elsewhere in *conf.* as a sign of ignorance and separation from God: 2. 3. 7, 3. 3. 6, 5. 3. 5, 5. 8. 14, 6. 4. 5, 6. 7. 12, 8. 4. 9, 10. 27. 38 ('et fugasti caecitatem meam'), 10. 34. 52, 11. 2. 3, 13. 23. 34.

libebat . . . libidine: Word-play as argument (and cf. 'vituperatione . . . vitium' just below). The words themselves contain a truth to which the adolescent was obtuse. The appeal of etymology is that it furnishes argument that is little more than irrefutable tautology.

2. 3. 8

Ecce . . .: The willing captive in Babylon. The elements of this sentence all have scriptural, specifically OT (i.e. pre-redemption) analogues, but no 'citation' is to be imputed here. Rather, the scriptural language gives a texture of authority and a range of reference to what A. wishes to say.

iter agebam: *agere* with *iter* not especially CL (cf. perhaps *Aen.* 6. 384, 'iter inceptum peragunt'); it occurs several times in Latin OT: 1 Kgs. 28: 22 ('ut possis iter agere'), Tobias 8: 21, 2 Macc. 9: 4.

platearum: The word is common in OT, representing Gk. πλατεῖα; in CL, not so common. See Lam. 4: 18, 'lubricaverunt vestigia nostra in itinere platearum nostrarum,' and cf. Sirach 9: 7–8, in a passage of warnings against dangerous associations with women, 'noli circumspicere in vicis civitatis, nec oberraveris in plateis illius. (8) averte faciem tuam a muliere compta, et ne circumspicias speciem alienam.'

Babyloniae: Apoc. 17: 5, 'Babylon magna, mater *fornicationum* Babylon,' for A. the biblical type of the earthly city. It is anachronistic to insist on the full development of that idea at the time of *conf.*, but the foundations of what would emerge are already laid down: *en. Ps.* 136. 1 of 412, 'duas civitates permixtas . . . unam cui finis est pax aeterna, et vocatur Hierusalem; alteram cui gaudium est pax temporalis, et vocatur Babylonia. interpretationes etiam nominum, si non fallor, tenetis: Hierusalem interpretari "visionem pacis"; Babyloniam "confusionem."' (No text earlier than *s. Den.* 20. 9 (no earlier than 405, probably 411) contains the interpretation *Babylon* = *confusio*.)

caeno: Jer. 38: 22, 'demerserunt in coeno et in lubrico pedes tuos.'

in cinnamis et unguentis pretiosis: Cant. 4: 14, 'nardus et crocus, fistula et cinnamomum cum universis lignis Libani, murra et aloe cum omnibus primis unguentis'; not elsewhere in A. according to La Bonnardière, *REAug* 1 (1955), 225–37.

umbilico: G–M suggest the word is used here to avoid repetition with 'medio' below; cf. Ezech. 38: 12, 'habitator umbilici terrae'.

calcabat: Ps. 55: 3, 'conculcaverunt me inimici mei tota die.'

seductilis: First in Latin in A.? So G–M and Souter; clear that it = *seductibilis*: so at *en. Ps.* 106. 14, '"vaccas" dicit *seductiles* animas, quae facile consentiunt seductoribus'; but at *en. Ps.* 54. 22 and *en. Ps.* 67. 39, the identical interpretation of 'vaccas' is given with the adjective *seductibiles*; but such an argument, given our inadequate edition of *en. Ps.* is not certain—the less familiar form may have occurred more often.

non enim: To be taken with *ita curavit*.

iam de medio Babylonis fugerat: Jer. 51: 6, 'fugite de medio Babylonis, et salvet unusquisque animam suam.' *civ.* 18. 18, 'quod praeceptum propheticum ita spiritaliter intellegitur, ut de huius saeculi civitate, quae profecto et angelorum et hominum societas impiorum est, fidei passibus, quae per dilectionem operatur, in deum vivum proficiendo fugiamus.'

in ceteris: i.e. 'in partibus Babyloniae non mediis' (G–M): *en. Ps.* 136. 6, 'multum enim interest inter medium Babylonis, et exteriora Babyloniae. sunt qui non sunt in medio eius, id est, non tanta concupiscentia saeculi et delectationibus obruuntur. qui vero, ut aperte dixerim et breviter, multum mali sunt, in medio Babylonis sunt.' There remains a touch of criticism for M.'s attitude.

pestilentiosum: Implies a moral judgement analogous to that which the mature A. might impose on the event, hence is perhaps just slightly anachronistic of M. here; see on 4. 10. 15, 'pestilentiosis'.

cohercere: Governed by 'curavit' and in turn taking as its object the compound relative clause 'quod . . . sentiebat'.

ad vivum: Otto, *Sprichwörter*, s.v. *vivus* 4, p. 377–8: Columella, *re rustica*, 6. 12. 3, 'extrema pars ipsius unguis ad vivum resecatur.'

non curavit hoc: Finally the explicit answer to the question posed at the beginning of 2. 2. 3, 'quis mihi modularetur aerumnam meam': no human agent in the image of God would so act, not even his mother, who still cherished a vain hope.

nonnullo adiumento . . . studia doctrinae: This view, though censured by A. at the time of *conf.*, is not entirely incompatible with his own view *c.*386/7, when he devoted time to the *libri disciplinarum* (see excursus on 4. 16. 30). Presumably M.'s view of the help liberal studies could offer was less subtle than A.'s later one, but the coincidence suggests an underlying common attitude.

caligo: The same smog elsewhere in *conf.*: 2. 2. 2 ('a caligine libidinis'), 3. 11. 19, 3. 11. 20, 11. 9. 11, 12. 28. 39 (the only case where A. himself is not the source and victim of the phenomenon: 'terram caliginosam').

serenitatem veritatis tuae: Cf. *c. litt. Pet.* 3. 21. 24, 'sed ecce angustatus [Petilianus] in causa rursum in me impetum facit nebulosum atque ventosum, ut veritatis serenitas obscuretur.'

ex adipe iniquitas mea: Ps. 72: 7, 'prodiet quasi ex adipe iniquitas eorum'; *en. Ps.* 72. 12, 'aliud est de necessitate peccare, aliud in abundantia' (i.e. the sins of the rich are less excusable than those of the poor). Knauer, 117 n. 1, 'In diesem wiederum sehr kunstvollen Gebilde wird in Ps. 72. 7 . . . das Tempus . . . auf Augustins Situation bezogen, aus dem Futur wird ein Imperfektum und die allgemeine "iniquitas" als die eigene bezeichnet ("eorum" > "mea").'

The psalm citation marks the structural middle of the book and summarizes the first half: Knauer, 135, 'Das Zitat leitet deutlich durch die doppelte Wiederholung von "iniquitas" in den neuen Abschnitt über, beide werden dadurch eng verbunden.' So far there has been (implied) sexual profligacy, an absence of guidance from his parents, and a hovering divine presence unfelt at the time.

2. 4. 9

Why so much space for the theft of the pears? Is it a metaphor for sexual sin?

That he did not need what he stole is important to the problematic of sin in the second part of the paragraph and the rest of the book; the case is logically simpler than any fornication could have been. Of no sexual transgression could he have said unequivocally (or had any hope of convincing us) that what appealed to him was not the thing itself but the wrongness itself.

The parallel to the theft of the fruit of the tree of good and evil in the Garden dominates. The aim of Bk. 2 is to present sins of concupiscence of the flesh, the prevalence of the lower will over the higher. A tree and its fruit (stolen though unneeded) is too perfect a parallel to be avoided: an unequivocal divine command, followed by a snatching and eating of the fruit of a forbidden tree. The result for Adam and Eve was that their eyes were opened 'ad invicem concupiscendum, ad peccati poenam carnis ipsius morte conceptam' (*Gn. litt.* 11. 31. 40).

Therefore, the pear-theft is transgression pure and simple, of the fundamental kind that explains and re-enacts the fall into sexual concupiscence. It is not that they stole the pears and then, in chronological succession, fell prey to their lusts; the sequence of the book is against that. Rather, the concupiscence was assuredly present even without the theft. The theft occurs at a moment in Augustine's life where it represents the fall of the will whose effects were visible elsewhere in his life at the same moment.

Ferrari (*REAug* 16 (1970), 236) is right to say (following F. C. Burkitt, *Religion of the Manichees* (Cambridge, 1925)) that a Manichee reader would find this episode particularly shocking. The plucking of the fruit was wrong in itself, but the casting to swine was worse. But Ferrari seems to think that the Manichee shock explains A.'s own exaggerated feelings of guilt; better to imagine that he dramatized the episode in part to shock his old co-religionists.

It is conventional to read the pear-theft narrative with naïve bemusement. (See on 4. 7. 11 for Nietzsche's reaction.) It may help to bear in mind a late antique parallel, perhaps almost contemporary with A., though the possibility of indebtedness to A. cannot be ruled out: Ps.-Macarius, *apophthegma* 37 (*PG* 65. 277–80 = *PG* 34. 258—cf. *Clavis Patrum Graecorum* 2. 2417): *εἶπεν ὁ ἀββᾶς Παφνούτιος ὁ μαθητής τοῦ ἀββᾶ Μακαρίου, ὅτι ἔλεγεν ὁ γέρων, ὅτι ὅτε ἤμην παιδίον, μετὰ τῶν ἄλλων παιδίων ἔβοσκον βοΐδια· καὶ ἀπῆλθον κλέψαι συκίδια· καὶ ὡς τρέχουσιν, ἔπεσεν ἕν ἐξ αὐτῶν, καὶ λαβὼν ἔφαγον αὐτό· καὶ ὅτε μνημονεύω αὐτοῦ, κάθημαι κλαίων.* For a late antique judgement, cf. Ennodius, *ep.* 1. 4. 6: 'doctorem Libycum adseritis sublata a se piri poma flevisse. merito lamentis expiandum est quod cum pudoris dispendio venter adquirit. vilia fuerint forte quae sustulit aut neglegentia aut usu aut tempestate peritura, non fuit culpa vacuus tamen iuxta apostolum raptor: carnem quam animam plus amavit.'

The most important treatment of this episode is that of Courcelle, *REA* 73 (1971), 141–50 (repr. in his *Opuscula Selecta* (Paris, 1984), 319–28), concentrating on the way Catiline's wrongdoing is invoked and echoed through these paragraphs (details in notes below): at 150, 'Augustin agit même en rhéteur lorsqu'il met en parallèle, volontairement, une peccadille quelconque de son adolescence avec le crime le plus illustre de l'histoire romaine.' Catiline had been a model of pre-eminent evil as early as *c. acad.* 3. 16. 36 (see Courcelle, art. cit. 143, for parallels there with Sallust and Cicero), and his name came up as well at *ord.* 2. 7. 22 and recurs conventionally elsewhere (most notably in *civ.*). Courcelle's insistence, however, that the classical model dominates the episode to the exclusion of any scriptural prefiguration (which had been suggested by L. Ferrari, *REAug* 16 (1970), 233–42) is excessive. That 'le crime le plus illustre de l'histoire romaine' should stand side by side with Adam's fall in the background to this episode is unsurprising. Courcelle's claim that the link to the fall of man would make sense only if the tree in paradise had been a pear tree is flatly absurd, and overlooks the clear indications of scriptural allusion (see on 2. 6. 14). See further on the explicit mention of Catiline at 2. 5. 11. (Ferrari defended his view adroitly, with some additional interesting material, at *REAug* 25 (1979), 35–46, especially on the symmetry of the trees of Eden and Golgotha in A. (and the later middle ages would cheerfully believe that the cross itself was cut from the wood of the Tree of Knowledge).) Long years later, A. invokes the image of a tree for adolescent sexual sin against Julian: *c. Iul. imp.* 2. 105, 'cum ad illud delictum quod generatio trahit usus voluntatis accesserit, quem non habent parvuli, in multas et varias cupiditates arbor multorum pullulat peccatorum.'

B. Hopkins, *American Imago*, 38 (1981), 97–104, quotes Freud on this type of behaviour to the effect that the vandal is secretly guilty on account of his oedipal fantasies and performs the deed to have something to which to attach his feelings. Hence (she quotes Freud) 'the sense of guilt was present before the misdeed, . . . it did not arise from it, but conversely—the misdeed arose from the sense of guilt.' 'On account of his oedipal fantasies' is doctrinaire, but the question whether the guilt or the theft came first is intriguing.

Furtum: Ex. 20: 15, 'non furtum facies'; Deut. 5: 19, 'furtumque non facies'; hence in his typical lists of serious sins, e.g. *en. Ps.* 75. 16.

lex tua: In *conf.* often with direct reference to OT, at least as often with indirect OT reference (as here), only occasionally resisting such focus (29 × referring to divine law in all); most common in the books where he is conscious of his transgressions. In the background through this book: cf. 2. 2. 3, 2. 2. 4 (see on 'legitima'), and 2. 5. 10 (see on 'lex tua'). *s.* 153. 8. 10, 'o vita dulcis! dulcis est quidem voluptas concupiscentiae: verum est, nec eam homines sequerentur nisi dulcis esset. theatrum, spectaculum, meretrix lasciva, turpissima cantilena, dulcia sunt ista concupiscentiae, dulcia plane, suavia, delectabilia; sed . . . non sicut lex tua, domine.'

scripta CDO^{2}S edd. : *inscripta* GO^1. Rom. 2: 14–15, 'cum enim gentes quae legem non habent naturaliter quae legis sunt faciunt, eiusmodi legem non habentes ipsi sibi sunt lex, (15) qui ostendunt opus *legis scriptum in cordibus suis*, testimonium reddente illis conscientia ipsorum.' The notion is characteristically Pauline and New Testament, but it occurs also at Jer. 31: 33, 'sed hoc erit pactum quod feriam cum domo Israhel; . . . dans leges meas in cordibus eorum; in cordibus eorum superscribam eas. et ero eis in deum, et ipsi erunt mihi in populum' (it may be apposite that the Prophets stood in relation to the Law in somewhat the relation that the New Testament did to the Old). On the 'natural law' against the Manichees, *c. Faust.* 22. 27, 'lex vero aeterna est ratio divina vel voluntas dei ordinem naturalem conservari iubens, perturbari vetans.' From the earliest period, *ord.* 2. 8. 25, 'haec autem disciplina ipsa dei lex est, quae apud eum fixa et inconcussa semper manens in sapientes animas quasi *transcribitur*. . . . adulescentibus ergo studiosis eius ita vivendum est ut a venereis rebus, ab inlecebris ventris et gutturis . . . se abstineant'; cf. also *duab. an.* 12. 16. A passage from the early anti-Pelagian period, on the other hand, contradicts the text of *conf.* here in an illuminating way: *spir. et litt.* 27. 47, 'proinde naturaliter homines quae legis sunt faciunt; qui enim hoc non faciunt, vitio suo non faciunt. quo vitio lex dei *est deleta de cordibus* ac per hoc vitio sanato, cum illic scribitur, fiunt quae legis sunt naturaliter.' (The interplay of written and oral verbal authority throughout *conf.* is a delicate balance, perhaps resisting any systematic inquiry and description; it is worth noting that *scribere* occurs 88 × in *conf.* (but only this once in Bk. 2), *legere* 83 ×, while *audire* occurs 179 × and *dicere* approx. 520 ×.)

iniquitas: Cf. below, and 2. 3. 8, 'prodiebat tamquam ex adipe iniquitas mea.'

inopia: Sall. *Catil.* 18. 4 (cf. also 17. 5), 'inopia atque mali mores stimulabant.'

furtum facere: This passage has been anticipated at 1. 19. 30, 'furta etiam faciebam de cellario parentum.'

nulla compulsus egestate: But Bk. 2 ends (2. 11. 18), 'et factus sum mihi regio egestatis.' Concupiscence is the origin of want, not its potential cure. Cf. Sall. *Catil.* 20. 15, 'egestas'.

penuria et fastidio: G–M: 'are perhaps to be taken as a Hendiadys—the loathing that springs from emptiness', citing 3. 1. 1, 'sine desiderio . . . non quia plenus eis eram, sed quo inanior, fastidiosior'.

sagina: The contrast is designedly coarse: see *en. Ps.* 102. 8, *en. Ps.* 103. s. 4. 19, 'ructate saginam vestram'. Cf. 'adipe' at the end of 2. 3. 8 above.

nec forma nec sapore inlecebrosis: Gn. 3: 6, quoted at *Gn. litt.* 11. 30. 38: 'denique verbis non contenta serpentis consideravit [mulier] lignum, "viditque bonum ad escam, et decorum ad aspectum".' Gn. 3. 6 quoted at *Gn. c. man.* 2. 15. 23: 'et vidit mulier quia bonum est lingum ad escam; et quia bonum est oculis ad videndum et cognoscendum.' A. denies himself even the excuse that Adam and Eve had.

adulescentuli: The same word repeatedly of Catiline's conspirators: Sall. *Catil.* 14. 5, 18. 4; Cic. *Catil.* 1. 6. 13.

nocte intempesta: *en. Ps.* 118. s. 29. 3 (coinciding with a secular tradition: Macrob. *sat.* 1. 3. 15), 'nox quoque intempesta, id est media, quando quiescendum est, hinc procul dubio nuncupata est, quia inopportuna est actionibus vigilantium'—a time when 'boys will be boys,' perhaps: cf. here 'quousque . . . produxeramus'. At 2. 6. 12, the crime is characterized as 'facinus illud meum nocturnum'. Cf. Sall. *Catil.* 27. 3, 'intempesta nocte coniurationis principes convocat (Catilina)'; 32. 1, 'nocte intempesta cum paucis in Manliana castra profectus est (Catilina).' Courcelle, *REA* 73 (1971), 145 n. 2: '*intempesta nocte* désigne, en principe, le moment de la nuit où les honnêtes gens dorment; mais par suite, dans les textes littéraires, c'est aussi l'heure classique des ruses de guerre et des crimes, ainsi que de l'extase pour les saints.' So a later text suggests the transformation the phrase would undergo among the Christians, into a time of spiritual loneliness, marked by both vulnerability and possibility: Greg. Mag. *dial.* 3. 4. 2, 'itaque intempestae noctis silentio cum vir dei quiescerit, antiquus hostis inmensis vocibus magnisque clamoribus coepit imitari rugitus leonum, balatus pecorum, ruditus asinorum, sibilos serpentium, porcorum stridores et soricum.'

de . . . more: Virg. *Aen.* 3. 65, 'Iliades crinem de more solutae'; prosaic in A., cf. 9. 8. 18, 9. 12. 31; as G–M remark, the genitive here is 'practically equivalent to an adjective'.

areis: The word *can* mean 'playing fields' (*OLD* s.v. 1 d), which would be apt here, but nothing suggests A. would have recognized the usage.

proicienda porcis: The obtrusion of this line (*vel*) even though they did eat some of the pears invites us to listen for the echoes. Mt. 7: 6, 'neque mittatis margaritas vestras ante porcos.' The prodigal son fed pigs as well: Lk. 15: 15, 'et misit illum in villam suam ut pasceret porcos (16) et cupiebat implere ventrem suum de siliquis quas porci manducabant et nemo illi dabat'; when A. depicts himself in the prodigal role, he feeds pigs again: 3. 6. 11, 'et longe peregrinabar abs te, exclusus et a siliquis porcorum quos de siliquis pascebam'—see there for the way feeding pigs is equated with catering to demons.

ecce cor meum: See on 1. 5. 5, 'ecce'.

abyssi: First in Latin in Gn. 1: 2? *TLL* 1. 243: 'non legitur ante Christianos'; outside scripture, there is one lone text in Tert. *adv. Marc.* 4. 20, and then next is Hil. Pict. *in Ps.* 2. 32; also at Commod. *inst.* 1. 27. 19, with the wrong gender ('abyssus noster').

dicat . . . cor meum: Praying again for the gift of speech (see on 1. 5. 5, 'miserere ut loquar'): 'give to my heart the words it needs to say what was going on in that theft.' This speech from the heart, relying on divine guidance, is what fills the rest of Bk. 2.

amavi defectum meum: This is the view of evil that A. learned from his contact with Christian neo-Platonism at Milan (7. 11. 17–16. 22). Cf. 10. 38. 63, 'donec reficiatur defectus meus'.

anima: Best taken (against most translators, who make it nominative) as ablative with *turpis*, as Ryan: 'base in soul was I, and I leaped down.'

dissiliens: 11. 29. 38, 'at ego in tempora dissilui'; in the sense 'leap down' as at 1 Macc. 9: 48 (according to some Vg. manuscripts).

firmamento: Ps. 70: 3, 'quoniam firmamentum meum et refugium meum es tu'; *en. Ps.* 70. s. 1. 5, 'ut fiam firmus ex te, sicubi fuero infirmatus in me, refugiam ad te. . . . ut firmus sim in hoc saeculo adversus omnes temptationes. . . . ad hoc enim confitebor infirmitatem meam.'

exterminium: 'destruction'; earliest in Tertullian, but commonest as VL translationese (= LXX ὄλεθρος, ἀφανισμός, etc.); else in A. only in his text of Deut. 7: 2 ('et percuties eos, exterminio exterminabis eos') at *qu. hept.* 6. 21 and of Jos. 17: 13 ('et fecerunt Chananaeos obaudientes, exterminio autem eos non exterminaverunt') at *qu. hept.* 6. 30; in Vg. at Sirach 39: 36, Jdt. 4: 10, and 5 × in Wisd., 5 × in 1 and 2 Macc., and at 4 Ezra 10: 10. (Could there be a VL text in the background here that has so far escaped detection?)

2. 5. 10

Etenim species: These lines speak first to the sense of sight ('species'), then to the sense of touch and the others ('contactu carnis ... ceterisque sensibus'), then to the lust for worldly glory ('honor ... potentia'): together ('cuncta haec') they embrace the categories of 1 Jn. 2: 16 (see on 10. 30. 41): *concupiscentia oculorum, concupiscentia carnis, ambitio saeculi*. The hallmarks of the three persons of the trinity are also present here. 'Species ... pulchris' [2] here (and 'decus' and 'decoris', and cf. 2. 6. 12, 'et ecce species nulla est'), 'modum' [1] (and 'adcommodata modificatio' and 'immoderata'), and 'convenientia' [3] (on *convenientia*, cf. du Roy 354 ('la volupté s'oppose pour lui à la recherche de la vraie "harmonie (*convenientia*)", fruit de l'Esprit') and 354 n. 3 (citing *mor.* 2. 6. 8, 'ordo enim ad convenientiam quandam quod ordinat redigit'; *vera rel.* 39. 72, 'recognosce igitur quae sit summa convenientia')). The three hallmarks occur together here, predicated of 'life' ('vita'): 'modum [1] decoris [2] ... convenientiam [3]'.

in contactu carnis: By a process of elimination it is clear even at 10. 30. 41–2 that the sense of touch predominates for A. in the imagery of sexual transgression (elimination: 10. 31. 45 ff. on taste, 10. 32. 48 on smell, 10. 33. 49 f. on hearing, 10. 34. 51, '*restat* voluptas oculorum istorum carnis meae'). That coincidence, in the midst of this synoptic glance at all temptation, reinforces the idea that a congruence with sexual sin is evoked for the pear-theft, but no crude identification. Cf. 3. 1. 1, 'anima ... miserabiliter scalpi avida *contactu* sensibilium.'

congruentia: The notion of 'congruence' is at the heart of A.'s notions of beauty and the attractions that arise therefrom; cf., e.g, *mus.* 1. 13. 27 ('congruentia delectari'), *mus.* 6. 13. 38 ('quod nostris oculis congruit appetimus'), *c. Faust.* 21. 13 ('pro cuiusque corporis congruentia vel delectet esca vel offendat'), and cf. the concept of the *aptum* (4. 13. 20 below). Cf. also 'convenientiam' just below, and cf. Cic. *fin.* 3. 6. 21, on Stoic ὁμολογία, which Cicero renders *convenientia*.

non est egrediendum abs te, domine: The metaphor drawn from the prodigal is so pervasive that a phrase such as this, with no verbal echo of any authoritative text, continues the metaphor without even (probably) consciously invoking it for most readers. *Egredi* is common in OT and NT, and perhaps adds a flavour of authoritative language; *deviare* is rare in scripture.

quam OS edd. : *qua* CDG. See on 1. 7. 12.

inlecebram: See on 2. 4. 9, 'inlecebrosis'.

amicitia: See on 2. 2. 2, 'luminosus limes amicitiae'. Here, the introduction of the idea suggests indirectly the explanation of the pear-theft that will eventually win A.'s assent.

immoderata in ipsa inclinatione: 2. 3. 6, 'inclinatae in ima' (see below on 'ima'); 'ipsa' is acc.

extrema . . . summa: On the three levels of being (*summum* (God), *medium* (e.g. *voluntas*), *extrema* (e.g. material things)), see the note at *BA* 13. 664. The ethical implication of this ontology is that the will turns upwards or downwards (downwards: it ceases to be *voluntas*: *s. dom. m.* 2. 22. 74, 'voluntas namque non est nisi in bonis, nam in malis flagitiosisque factis cupiditas proprie dicitur, non voluntas'); this opens the way for metaphor from gravity as well (see on 13. 9. 10): at 5. 4. 7 and 13. 9. 10 (see on both passages), A. connects *pondus* to his trinitarian structure, through Wisd. 11: 21, as at 5. 4. 7, 'qui omnia in mensura in numero et pondere disposuisti'; cf. 13. 9. 10, 'pondus meum amor meus'. Cf. here *lib. arb.* 2. 19. 50, 3. 1. 2.

domine deus noster, [1] **et veritas tua** [2] **et lex tua** [3]. On *lex* here, see on 1. 7. 12, and cf. *c. Faust.* 21. 6, 'cuius unitate [1] omnis modus sistitur, cuius sapientia [2] omnis pulchritudo formatur, cuius lege [3] omnis ordo disponitur'; *c. Faust.* 20. 7 has 'illud autem incommutabili voluntate [1], veritate [2], aeternitate [3] persistit, et inde nobis est initium existendi [1], ratio cognoscendi [2], lex amandi [3].'

There is, to be sure, evidence that A. could insist on identifying Truth and Law: Ps. 118: 142, 'iustitia tua iustitia in aeternum, et lex tua veritas' (also echoed at 4. 9. 14); *en. Ps.* 118. s. 28. 5, 'quomodo enim non veritas lex, per quam cognitio peccati, et quae testimonium perhibet iustitiae dei?' The Law, i.e. scripture, is certainly meant in A.'s reading of the Psalm, and the overtone is here as well (Knauer, 127 n. 2).

But the two readings (distinguishing Truth [2] and Law [3] vs. insisting on their identity) contain within themselves the possibility of reconciliation. For on A.'s own trinitarian doctrine, the distinction of the persons in the trinity does not entail a distinction of essence—Three in One. To press towards that reconciliation, however, the reader of *conf.* is beginning not merely to observe but to participate in the working of this text: but *that* was A.'s intention (*retr.* 2. 6. 1).

ima: Cf. 2. 3. 6 and 3. 3. 5, 4. 15. 27, 8. 3. 8, 9. 2. 3.

deus meus, qui fecit omnia: God the creator (as often in *conf.* from 1. 2. 2, 'deus, qui fecit caelum et terram'); the half-echo of Gn. 1: 1 implies not merely 'God the creator' but 'God the creator of whom we hear in scripture'. On any assessment of A.'s use of language, it is clear that it is God the immutable creator who looms largest in his mind, ahead of God the incarnate redeemer, or God the spirit of love. (Note, for instance, that in the transports of mystical ascent, it is God the maker, not God the redeemer or

God the loving spirit, to whom attention is directed: 9. 10. 25, 'non ipsa nos fecimus, sed fecit nos qui manet in aeternum'—more fully developed at 10. 6. 9.) His defence of predestination against the Pelagians is surely as much a testimony to his reverence for the original creative act of God as for individual acts of redemptive grace. That emphasis in turn may help define the difference between A. and Paul.

rectorum corde: Ps. 63: 11, 'et laudabuntur omnes qui recti sunt corde.'

2. 5. 11

Why Catiline? And why the explicit quotation repeatedly in space of a few lines? He offered an entirely 'pagan' exemplum of evil, wicked among the wicked. His is a story in which no impiety against the divine is part of the tradition, hence there is no complicating question of love or hatred for a divine law-giver (note citation below from *mor.* 2. 13. 28, where A. makes C. *sacrilegus*; on the other hand, at *civ.* 3. 2, A. could emphasize the corruption of the city that allowed C.'s conspiracy to flourish: in different contexts the story could be used with different emphases). Only the natural law applies to Catiline, and his action is explicable in rational terms. So much the more likely that Augustine's action is explicable in similar terms.

facinore: 'criminal act', only occurring in Bks. 2, 3, and 4. See on 3. 8. 16 (with comparison to *flagitium*).

causa: For an author reputed to have sacrificed free will on the altar of predestination, A. is singularly insistent that evil is an entirely uncaused and free act of the will (see *civ.* 12. 7, quoted below on 2. 6. 12). Here such evil occurs as a simple fact of his own youth—for all that it seems unlikely that anyone would be so enamoured of evil for its own sake.

appetitus: The noun and the verb *appetere* are outwardly less morally charged than *concupiscentia*, but, though less frequent than *concupiscentia*, almost always in *conf.* in a negative sense : 6 × in all in the pear-theft episode (here and 2. 4. 9 (2 ×), 2. 5. 11, 2. 6. 13, 2. 9. 17), also at 5. 8. 14, 8. 10. 24, 9. 2. 3, 9. 8. 18, 10. 20. 29, 10. 35. 54, 13. 21. 30. The only partial exceptions occur where a moral neutrality is superficially maintained, e.g. 10. 40. 65, 'docens quid caveam et quid appetam', and 13. 32. 47. *div. qu.* 35. 1, 'nihil aliud est amare quam propter seipsam rem aliquam appetere.'

beatificis: See on 7. 20. 26.

homicidium fecit: The examples given, both hypothetical and historical, illustrate the practice just suggested: a murderer must have either wanted or feared something.

homicidium: *mend.* 9. 14, 'quamvis enim gravius sit homicidium quam furtum, peius est tamen facere furtum quam pati homicidium.'

de quo dictum est: The postponement of the name clearly expects the reader to identify the familiar villain from the few words of allusive quotation presented. Sall. *Catil.* 16. 3, 'si *causa* peccandi in praesens minus suppetebat, nihilo minus insontis sicuti sontis circumvenire iugulare: scilicet *ne per otium torpescerent manus aut animus, gratuito potius malus atque crudelis erat*.'[3] Cf. *en. Ps.* 108. 3, 'sicut autem pii gratis amant Christum, sic impii gratis oderunt; ... unde et apud auctores saecularium litterarum dictum est de quodam pessimo, "gratuito potius malus atque crudelis erat".' Cf. also Sall. *Catil.* 16. 1, 'sed iuventutem, quam, ut supra diximus, inlexerat, multis modis mala facinora edocebat'; A. gives us himself at this stage as in training under Catiline, one of the young men going astray (see above on 2. 4. 9, 'adulescentuli', and below on 'exercitatione').

With this citation, A. has now explicitly quoted all four of the authors on which the secondary education of his time was founded: see on 1. 16. 26. A. had a high opinion of Sallust as stylist (*beata v.* 4. 31, 'Sallustius, lectissimus pensator verborum'); Sallust's jaundiced view of Roman history furnishes much welcome ammunition for A. in the first books of *civ.*; Julian of Eclanum quotes and cites S., and A. gladly caps quotations with him: see Hagendahl, 43 ff. and 226 ff. for details. Catiline is a paragon of crime at *c. acad.* 3. 16. 36, 'legant orationem Catilinae [Sall. *Cat.* 20. 2–17] qua patriae parricidium, quo uno continentur omnia scelera, persuasit.' As a different kind of paragon: *mor.* 2. 13. 28, 'de Catilina memoriae proditum est, quod "frigus, sitim, famem ferre poterat" [Cic. *Cat.* 3. 7. 16—also quoted at *pat.* 5. 4]. haec erant illi spurco sacrilegoque etiam cum apostolis nostris communia.'

quaere CD^{1}GO Ver. : *quare* D^{2}S Maur. Knöll Skut. Punctuation is based on Verheijen's; the idea is that Augustine will infer another *causa* beyond the one made explicit in Sallust's text. Vega reads *quare* but suggests: 'fortasse legendum sit, "quaeres id quoque, cur ita?" ut consonet verbis init. cap.: "cum de facinore quaeritur".'

exercitatione: Referring to the training of young men in Catiline's band (as G–M saw: Cic. *Catil.* 2. 5. 9, 'stuprorum et scelerum exercitatione adsuefactus'), with a parallel to A.'s situation.

inopiam: Sall. *Catil.* 5. 7: 'agitabatur magis magisque in dies animus ferox *inopia rei familiaris et conscientia scelerum*.'

2. 6. 12

o furtum meum: Apostrophe; see on 1. 16. 26.

pulcherrime omnium [2]: 1. 4. 4, 'pulcherrime et fortissime'.

[3] The edd. of Sallust do not offer variants for 'torpescerent', while the MSS of A. do not offer variants for 'torpesceret': A. may, as often, be quoting from memory.

creator omnium [1]: See on 9. 12. 32, quoting Amb. *hymn.* 1. 2. 1–2, 'deus, creator omnium polique rector'.

deus bone [3?]: 8. 3. 6, 'deus bone,' 10. 17. 26, 'te ... vere bone,' 13. 38. 53, 'deus une bone'; 3. 6. 10, 'mi pater summe bone'.

summum bonum: Introduces explicitly the idea suggested at 2. 5. 10, 'cum extrema bona sint, meliora et summa deseruntur'; God as *summum bonum* in *conf.* (and cf. 1. 4. 4, 'summe, optime'): 3. 6. 10, 4. 15. 24, 7. 4. 6, 7. 5. 7, 12. 15. 19, 12. 16. 23, 13. 2. 2. The phrase makes its appearance in our surviving Latin literature once in Lucretius (6. 26, 'bonum summum quo tendimus omnes'), then often in Cicero. In Cic. with a specifically ethical reference: *acad. post.* 1. 5. 19, 'ac primum illam partem bene vivendi a natura petebant [sc. the immediate successors of Plato in the Academy] eique parendum esse dicebant, neque ulla alia in re nisi in natura quaerendum esse illud summum bonum quo omnia referrentur.'

nam et si quid ... intravit in os meum: Cf. 2. 4. 9, 'etiamsi aliquid inde comedimus'. The meditation here returns to the details of the narrative from that paragraph at several points, followed by the specific problems posed by the theft.

species: *BA*: 'séduisante beauté'; see on 2. 5. 10 for role of *species* in connection with this theft, and for *species* generally, see on 1. 7. 12 and 12. 3. 3.

non dico: sc. 'speciem nullam esse'.

aequitate ... mente ... memoria ... sensibus ... vita: G–M, citing *quant. an.* 33. 70–6 (on which see O'Daly, 13–15), see here traces of the seven 'degrees' of ascent to contemplative vision set out there. The sequence, in ascending order, in the passage cited is: organic ('corpus ... vivificat'), sensory ('quid possit anima in sensibus'), mental ('memoria'—the first stage in this sequence that human beings enjoy apart from the animals), moral in intention ('ex quo bonitas incipit atque omnis vera laudatio'), moral in act ('in seipsa laetissime tenet ... aliud est enim efficere, aliud tenere puritatem'), contemplative in intention ('appetitio intellegendi ea quae vere summeque sunt'), and contemplative in act ('in ipsa visione atque contemplatione veritatis ... neque iam gradus sed quaedam mansio, quo illis gradibus pervenitur'). That passage in turn should be compared to other Augustinian schemes, notably the seven ages of the redeemed person at *vera rel.* 26. 49 (see on 1. 8. 13); it also reflects other ancient philosophical and religious schemes, with various degrees of precision. Compare, for instance, the four faculties that go back to Aristotle, *de anima* 3. 3. 304, and recur in Boethius (*cons.* 5. P4. 27, 'ipsum quoque hominem aliter sensus, aliter imaginatio, aliter ratio, aliter intellegentia contuetur').

Here, the scheme occurs silently, in descending order, beginning from the middle, the stage of life that is moral in intention ('sicut in aequitate atque prudentia'), descending through the mental ('sicut in mente hominis atque memoria'), through the sensory ('et sensibus') and the corporeal ('et vegetante vita'), to the subhuman level of inert matter ('neque sicut speciosa sunt sidera et decora locis suis et terra et mare plena fetibus, qui succedunt nascendo decedentibus'), and beyond to the void of non-existence—at least where there is some lingering semblance of existence. When that seven-stage scheme is placed against the later moral and spiritual history of A. as recounted in *conf.*, it is perhaps most suggestive that the first book of *conf.* to show positive improvement (Bk. 6) ends with a passage that matches the fourth of these seven stages (*quant. an.* 33. 73, 'in ipso enim purgationis negotio subest metus mortis saepe non magnus, saepe vero vehementissimus': cf. *conf.* 6. 16. 26, 'metus mortis et futuri iudicii tui'); from there, it is easy enough to match the fifth stage to the moral renovation marked by the garden scene at 8. 12. 29 and the sixth to the contemplative aspiration of Ostia (9. 10. 24 f.)—and the seventh to the seventh day of creation, limned on the last pages of *conf.*

Such speculative correlations are useful to this extent: The first four stages of ascent in this scheme are ones which A. apparently believes he had attained even in adolescence—and on that reading, the only doubt is that he would be crediting himself with the desire for goodness, even if he believed that its attainment was remote. The first three stages of ascent (corporeal, sensory, and mental) are reflected in various ways in what has already gone before (cf. perhaps esp. 1. 20. 31, finding the best that could be said of his infancy and boyhood).

saltem: *OLD*: '(in negative . . . sentences) Even, so much as'.

defectiva species et umbratica: *civ.* 12. 7, 'nemo igitur quaerat efficientem causam malae voluntatis; non enim est efficiens sed deficiens, quia nec illa effectio sed defectio. deficere namque ab eo quod summe est ad id quod minus est, hoc est incipere habere voluntatem malam. causas porro defectionum istarum, cum efficientes non sint, ut dixi, sed deficientes, velle invenire tale est ac si quisquam velit videre tenebras vel audire silentium, quod tamen utrumque nobis notum est, neque illud nisi per oculos, neque hoc nisi per aures, non sane in specie, sed in speciei privatione.' A.'s sin was worse than merely evil, it was evil for evil's sake. This paragraph sharpens the issue, and makes the dilemma of the one who would explain the problem worse.

2. 6. 13

The vices appear here in three clusters, which match the three temptations of 1 Jn. 2: 16 (see on 1. 10. 16 and 10. 30. 41). The second of the three in John is *curiositas*, the third *superbia*: here the order is reversed. The principal vice in each category is accompanied by others that represent defects of the same sort, even pairs incompatible with each other (e.g. *effusio liberalitatis* and *avaritia*).

The sin of the tree of Genesis is not here, for these are the temptations to sin that arise from that original fall.

Each vice is defined by its aspiration to possess outside of God something that is authentically possessed only by (and thus in) God. The absence of scriptural language marks the passage as philosophically reflective, A's own observations rather than a direct message from the scriptural text.

superbia . . . imitatur: *vera rel.* 45. 84, 'quid enim aliud in ea homo appetit, nisi solus esse si fieri possit, cui cuncta subiecta sint, perversa scilicet imitatione omnipotentis dei?' *Gn. litt.* 8. 14. 31, 'id est superbiae et contumaciae, perversae imitationis dei et noxiae libertatis.' Sim. at *civ.* 19. 12.

celsitudinem: 11. 31. 41, 'quorum celsitudo tu es'; 12. 26. 36, 'deus meus, celsitudo humilitatis meae et requies laboris meae'. *civ.* 14. 13, 'quid est autem superbia, nisi perversae celsitudinis appetitus?'

Resemblance to the creator is both innate in man (Gn. 1: 26: see on 13. 22. 32) and potential (through redemptive restoration). Here the young A. is seen aspiring to resemblance, but of a forbidden sort. Cf. *trin.* 11. 5. 8, quoted on 2. 6. 14 below.

cum tu (3 × in the paragraph): Circumstantial, but almost concessive. Translate, e.g. the first two lines: 'For even pride mimics loftiness, when (in spite of the fact that) you are one God, highest above all things.'

blanditiae: Not necessarily in a bad sense, but admitting, and perhaps half-requiring, the specification *honeste* when in a good sense (8. 11. 27, 'honeste blandiens'); cf. 5. 10. 20 ('blande et amanter'), 9. 12. 30 ('obsequiis meis interblandiens'), 12. 14. 17 ('superficies blandiens parvulis'), 13. 15. 18 ('et blanditus est et inflammavit').

quam illa prae cunctis formosa et luminosa veritas tua [2].

te simplicius quicquam non reperitur: Cf. *civ.* 11. 10, 'est itaque bonum solum simplex *et ob hoc* solum incommutabile, quod est deus.' The connection is often made: cf. 4. 16. 29, 'te, deus meus, mirabiliter simplicem atque incommutabilem', 13. 3. 4, 'quia solus simpliciter es'; hence the juxtaposition at 9. 4. 10, 'in aeterna simplicitate'.

opera sua malis inimica sunt: *en. Ps.* 7. 16, 'ipsa peccata sic ordinare ut quae fuerunt delectamenta homini peccanti, sint instrumenta domino punienti.' The effect of this argument is to absolve God of responsibility for the punishment itself, which is made to be self-inflicted; taken to an extreme, the idea verges on *apocatastasis*, as in Eriugena, *periphyseon* 5, where all men return to their divine origin, but those who are wrapped up in *phantasiae* derive no pleasure from what they find in heaven.

ignavia . . . quietem: *ignavia* not elsewhere in *conf.*; for the purposes of the present passage, it seems to be taken as idleness, 'incuriosity' (though elsewhere, *quies* is the goal of concupiscence of the flesh: *vera rel.* 52. 101, 'appetit voluptas corporis . . . quietem'). The link to *curiositas* is perhaps suggested by *div. qu.* 58. 2, 'nemo autem ignorat hominum vitam iam aliquid administrantem cognitione et actione fulciri. nam et actio temeraria est sine cognitione, et sine actione *ignava cognitio*'—'idle curiosity' leading to 'idle knowledge.'

luxuria: Elsewhere in *conf.* only at 5. 3. 4, 8. 10. 24, 'utrum emat voluptatem luxuria', 13. 21. 30, 'ab inerti voluptate luxuriae'.

invidentia de excellentia litigat: quid te excellentius?: The vices interpenetrate. The direct desire for *excellentia* would be an example of worldly ambition, but the desire for *another's* rank is cupidity.

vindictam: Cf. Rom. 12: 19, 'mihi vindicta: ego retribuam, dicit dominus'; also at 2. 5. 10.

quis a te separat: Cf. Rom. 8: 35, 'quis ergo nos separabit a caritate Christi?'

2. 6. 14

Seeing the vices as perverse imitation is important because imitation acknowledges the authority of the model. (Hence the serpent's offer of divinity to Eve; see below.) From 'quid ergo' a return to active consideration of the pear-theft.

Ita fornicatur anima: As described in 2. 6. 13. Ps. 72: 27, 'perdidisti omnem qui fornicatur abs te; *en. Ps.* 78. 8, 'ne anima legem domini sui contemnat, et a deo suo fornicando dispereat.' See on 1. 13. 21, 'fornicabar abs te' (cf. 4. 2. 3, and 5. 12. 22).

redit ad te: See on 1. 18. 28.

perverse te imitantur omnes: *trin.* 11. 5. 8, 'male itaque vivitur et deformiter secundum trinitatem hominis exterioris. . . . nam et animae in ipsis peccatis suis non nisi quandam similitudinem dei, superba et praepostera et, ut ita dicam, servili libertate [cf. 'mancam libertatem' below] sectantur. ita nec primis parentibus nostris persuaderi peccatum posset, nisi diceretur, "eritis sicut dii" [Gn. 3: 5].' Cf. 10. 36. 59, 'ut te perversa et distorta via imitanti tenebrosi frigidique servirent,' and contrast good imitation: 13. 21. 31, 'imitando imitatores Christi tui'.

longe . . . a te: See on 2. 2. 2, 'longe a te'.

quid ergo: His own vice is imagined to be vicious at second remove: not merely violation of God's law and thus perverse imitation of God, but a secondary perverse imitation of perverse imitation, out of acknowledged weakness.

vel vitiose atque perverse: *vel* here is concessive: 'quamvis, licet'. Cf. Hensellek, *Anzeiger Akad. Wien* 120 (1983), 106, with texts.

potentatu: Stronger than *potestas* ('potentiality, capacity')—closer to 'strength, power'. The word allows A. to omit the infinitive (e.g. *facere*, *efficere*) that even now would not be entirely superfluous. Cf. *en. Ps.* 31. en. 2. 1, 'si se in audaciam quandam praesumptionis erexerit, et de suis viribus iustitiaque praesumserit, et proposuerit animo implere iustitiam, et omnia quae praecipiuntur in lege ita facere ut in nullo offendat, atque in *potestate* sua se habere vitam suam, ut omnino nusquam labatur, nusquam deficiat, nusquam titubet, nusquam caliget, sibique hoc tribuat, et *potentatui* voluntatis suae . . .' Not elsewhere in *conf.*, rare in A. except in biblical (esp. Ps.) citations.

mancam libertatem: *libertas* four other times in *conf.*: 3. 3. 5 ('fugitivam libertatem'), 3. 8. 16 ('falsae libertatis'), 4. 4. 8 ('mirabili et repentina libertate': describes the dying friend's rebuke), 9. 3. 6, ('libertatis otiosae'); *en. Ps.* 101. s. 2. 2, 'perniciosa libertas'. It is not a word to which A. naturally applies much positive value (easily neglected in studies of his views on *liberum arbitrium voluntatis*); it is important in a good many scriptural passages, but he always takes the *libertas* he finds there as something special, contrary to the appearances of *libertas* that are arrogated by pride here.

tenebrosa: See on 2. 6. 12.

ille servus: A.'s sin involving a tree directly parallels the sin of Adam: A. becomes Adam. Job 7: 2 (VL), 'tamquam servus metuens dominum suum et consecutus umbram'; *adn. Iob* on 7: 2, 'quod significat absconditio Adae a facie domini et tectio foliorum de quibus umbra fit, quam relicto deo consecutus est homo.' Sim. at *en. Ps.* 142. 16 and *s.* 351. 3. 3. Kusch, 152, quotes the above texts and adds: 'Die Schilderung des Birnendiebstahls ist also keine selbstquälerische moralische Deklamation, sondern exakte Analyse, die den Griff nach der verbotenen Frucht als eine Widerholung des historischen Abfalles von Gott begreift.'

Strong words follow:

putredo: Post-classical, cf. Prov. 12: 4. (*OLD*: 2 × in Apuleius). At 4. 3. 4, astrologers seek to exculpate man, 'ut homo sine culpa sit, caro et sanguis et superba putredo'; sim. at 2. 1. 1, 4. 11. 16, 6. 15. 25.

monstrum: Again, imprecisely, of a disorder of the will in the face of concupiscence of the flesh at 8. 9. 21: 'unde hoc monstrum? et quare istuc? . . . non igitur monstrum partim velle, partim nolle, sed aegritudo animi est.'

profunditas: Frightening depth, preponderantly in a bad and dangerous sense in *conf.*, with a few examples applied to God in a way that transfers the sense of awe. Bad and dangerous: 1. 18. 28, 2. 3. 5, 3. 6. 11, 3. 11. 19, 3. 11. 20, 4. 13. 20, 4. 14. 22, 6. 1. 1, 6. 16. 26, 7. 3. 5, 8. 3. 8, 8. 4. 9, 9. 1. 1, 9. 4. 12, 9. 8. 18, 11. 2. 3 (2 ×), 12. 3. 3, 12. 8. 8, 13. 8. 9, 13. 20. 28, 13. 21. 29 (3 ×), 13. 23. 34, 13. 24. 37, 13. 34. 49. More or less of God: 6. 5. 8 ('et secreti sui dignitatem in intellectu profundiore'), 10. 17. 26 ('profunda et infinita multiplicitas'), 12. 14. 17 ('mira profunditas eloquiorum tuorum, . . . sed mira profunditas, deus meus, mira profunditas').

2. 7. 15

The inquiry—what sense to make of the pear-theft—is left aside for another abstracted reflection on the difference between the past as lived and the past as now interpreted. In this text, interpretation of the past becomes the life of the present, but even in this life there is the alternation between narrative of and meditation on the past and, on the other hand, meditation on the act of meditation (as here). Such interruptions occur where the ordinary flow of the text encounters an incoherence, at a moment of unresolved tension. At such a node, A.'s interpretative tactic is to invoke the divine authority, ever capable of introducing order and pattern where there seems to be none. The attainment of such order makes it possible to return to the original, less abstractly reflective, level of the text's movement.

The tension arises here, as often, between A.'s own text and the scepticism with which he professes to expect it to be read. While the formal address remains directed to God, there begins here a veiled dialogue with an imaginary audience, in which A. attempts to define the audience with which he will find favour (or, read another way, to prescribe to his audience the interpretative principles according to which they will be able to receive his text favourably)—cf. 10. 3. 3. His sensitivity to mockery (see on 1. 6. 7) is near the surface.

The thanksgiving here is not ironic, and is to be compared to that which concludes Bk. 1 (1. 20. 31).

Quid retribuam: Ps. 115: 12 (3), 'quid retribuam domino pro omnibus quae retribuit mihi?' *en. Ps.* 115. 4 explains '*re*-tribuit': 'quae igitur praecesserant hominis, ut omnium donorum dei non attributio, sed retributio vocari possit? quae praecesserant hominis nisi peccata? retribuit ergo deus bona pro malis.' The *quid*-clause is answered by 'diligam . . . tuo', while the *quod*-clause is paralleled and explained by the *quoniam* clause. Memory without fear is the fruit of forgiveness; love, thanksgiving, and confession are repayment to God for the forgiveness that has preceded all confession.

confitear nomini tuo: Cf. Ps. 53: 8, 'confitebor nomini tuo, domine, quoniam bonum est.'

tamquam glaciem solvisti: Sirach 3: 17, 'sicut in sereno glacies ita solventur peccata tua.' *en. Ps.* 125. 10 (after quoting Sirach), 'ergo peccata ligabant nos. quomodo? quomodo frigus ligat aquam ne currat. et illigati frigore peccatorum gelavimus. . . . flavit auster spiritus sanctus; dimissa sunt nobis peccata, soluti sumus a frigore iniquitatis; tamquam glacies in sereno, solvuntur peccata nostra.'

gratiae tuae: An omnipotent God is one who has forgiven A. all sins, even the ones that he would have committed but did not, forestalled in his love of gratuitous evil only by divine act. The rhetorical effect is to portray the adolescent A. in the worst of lights; A.'s biographers labour to verify the description.

conversis ad te: Ps. 50: 15, 'et impii ad te convertentur.'
languoribus: 14 × in *conf.*; see on 10. 3. 3.

2. 8. 16

Only now, after accusing himself dramatically of willingness to accomplish all manner of evil for its own sake, does A. allow a thought to surface that shows him taking pleasure in *something* in the pear-theft. His principle, after all, is that nothing is nothing save evil, and that there is no thing-ness to evil that could attract even the wickedest of souls. To be sure, he settles for the slightest of attractions, the camaraderie of thieves—and amateur thieves at that. The principle is saved, at the price of his dignity.

Quem fructum habui: Rom. 6: 21, 'quem ergo fructum habuistis tunc in illis in quibus nunc erubescitis? nam finis illorum mors est.' (Often cited by A. with *gloriam* for *fructum*.) *en. Ps.* 85. 23 (citing this verse), 'videtis esse et modo confusionem salubrem in loco paenitentiae; tunc autem seram, inutilem, infructuosam'; cf. *en. Ps.* 125. 3. Death is the threat in the scriptural text; here it is unspoken, but recall 2. 6. 14, 'o putredo, o monstrum vitae et *mortis* profunditas!'.

in his . . . maxime in illo furto: The pear-theft is only one of many sins, standing for others, and in all of them it was the false society (*civitas terrena*, to anticipate a later development of A.'s doctrine) that appealed to him, the *confricatio consciorum animorum*.

et tamen: See on 1. 1. 1, 'et tamen laudare te vult homo'. The meditation on the pear-theft is the longest sustained passage on any topic so far in *conf.*; now, after painting himself entirely into a corner since 2. 4. 9, he finally opens the way out.

consortium: As the end of this paragraph insists, the pleasure was in the *facinus*, enhanced by the companionship. A. avoids here the word *amicitia*, postponing it until the apostrophe in 2. 9. 17, 'o nimis inimica amicitia'; but right and wrong friendship and use of friendship thus bracket Bk. 2 and measure his fall: see on 2. 2. 2, 'luminosus limes amicitiae'.

non ergo . . . amavi?: The question mark is necessary; cf. Ryan trans.: 'Then it was not only the theft that I loved? No, truly, nothing else.'

illud: sc. *consortium*.

quid est re vera?: sc. *illud consortium*. This reading of these lines, contradicting most existing editions (see below) is necessary to make sense of the continuation through 2. 9. 17, beginning 'quid erat ille affectus animi?'.

quis est . . . umbras eius?: The punctuation (following Vega against Knöll, Skut., Ver.) calls attention to this brief moment of meditation on meditation and to clarify the structure of the last half of the paragraph, divided between asking 'what is that *consortium*?' and meditation on the act of asking that question—which is not answered until 2. 9. 17–18.

qui inluminat cor meum: Sirach 2: 10, 'diligite illum et inluminabuntur corda vestra'; Eph. 1: 18, 'inluminatos oculos cordis vestri'; cf. Jn. 1: 9, 'quod inluminat omnem hominem venientem in hunc mundum' (4. 15. 25, 7. 9. 13, 9. 4. 10). On divine illumination, see V. Warnach, *Aug. Mag.* 1. 429–50; in addition to echoes of Jn. 1: 9, cf. esp. 10. 23. 33, 'deus inluminatio mea,' and 11. 2. 2, 'primordia inluminationis tuae'.

discernit umbras eius: Cf. on 13. 14. 15, 'discernit', for the way this is an echo of Gn. 1: 4, 'et divisit deus inter lucem et tenebras.'

quid est?: All twentieth-century editors have printed what follows thus: 'quid est, quod mihi venit . . .' with a question mark after 'pruritum cupiditatis meae': but then the *quia*-clause has already answered the question, forcing many translators (Pusey, Ryan, *BA*, Carena) to put the question mark implicitly after 'considerare' (where the Maurists put it in Latin); but then the question asked is not one answered by the *quia*-clause/sentence. The logic is that A. asks the question and then explains the dilemma that gives the question force: 'if it were just the pears . . . but on the other hand if it were just the thrill . . ., in either case I could have acted alone.' The last sentence of the paragraph clarifies the issue, and the next paragraph begins by asking further just what *was* the appeal.

confricatione: Hrdlicka, 11, (following *TLL*) reports doubtful instances at Chiron 491 and Veg. *mulom.* 2. 86. 2; otherwise in *adn. Iob* on 31: 40 and here; essentially a neologism; probably some sexual connotation, esp. in view of the epithet 4. 8. 13, 'adulterina confricatione'.

facinore: *doctr. chr.* 3. 10. 16, 'ut alteri noceat' (see on 3. 8. 16).

2. 9. 17

vae mihi: Cf. Job 10: 15 (VL), 'quod si impie fecero, vae mihi.'

delicta quis intellegit: Ps. 18: 13, 'delicta quis intellegit?'; *en. Ps.* 18. en. 2. 13, 'pater, ignosce illis, quia nesciunt quid faciunt. . . . si videntur tenebrae, intelleguntur delicta. denique quando nos delicti paenitet, in luce sumus. nam quando quisque ipso delicto involutus est, quasi obtenebratis opertisque oculis non videt delictum; quia si tibi tegatur et carnis oculus, nec aliud vides, nec hoc unde tegitur vides.' See on 2. 6. 12, 'defectiva species', esp. *civ.* 12. 7, quoted there. The echo here signals that the answer to the question at hand is that no fully satisfactory answer to the question will be found.

non facerem, non facerem: The repetition is solemn and conclusive: what began in 2. 8. 16 in the subjunctive ('non . . . *fecissem*') is now indicative and certain.

viva recordatio animae meae: Num. 10: 9, 'et erit recordatio vestri coram domino deo vestro, ut eruamini de manibus inimicorum vestrorum.' There the genitive *vestri* seems to be objective; in the text here, it seems to be subjective, but that proves nothing. The acts of memory are salvific, by whomever A. intends us to see them performed. The echo marks a boundary, the completion of the investigation that has led to its necessarily unsatisfactory result.

solus . . . nec facerem: Summary of the results of the investigation into the nature of the sin.

o nimis . . . impudentem: Declamation, representation of the attitude that brought about the sin.

o nimis inimica amicitia: Here is the heart of Bk. 2: friendship gone awry, from 2. 2. 2 to here.

investigabilis: 'incomprehensible' (G–M). Cf. 'delicta quis intellegit?'. G–M further: 'always used in this negative sense though, so far as its form goes, it might mean the opposite'. Tertullian (*adv. Hermog.* 45, cited by Souter) uses the word in the positive form (Souter: 'traceable, comprehensible'), but scriptural texts, esp. Rom. 11: 33 (rendering ἀνεξιχνίαστοι), but also Prov. 5: 6, 'vagi sunt gressus eius, et investigabiles' (οὐκ εὔγνωστοι), and Eph. 3: 8, 'evangelizare investigabiles divitias [τὸ ἀνεξιχνίαστον πλοῦτος] Christi,' gave the word its form and meaning in Christian Latin. See also 4. 4. 8.

et (*pudet*): = *etiam*.

2. 10. 18

A. turns away from the repulsive sight of his unintelligible sin, to look upon God (with turns of vocabulary that anticipate the mystical features of Bks. 9 ff.). The last sentence of the book is for him the last word on the wanderings of the prodigal in adolescence.

exaperit: Christian Latin (Souter cites Ex. 34: 19 in one VL manuscript, and cf. Iren. 2. 19. 8 and Tertull. *apol.* 18. 8 (= 'translate')): 'to disclose, disentangle, explain'.

nodositatem: 'knottiness, intricacy' (Hrdlicka, 22: also in Rufinus).

foeda est: Repeats the judgement from 2. 4. 9.

honestis luminibus: May be either dative (with what goes before) or ablative (as punctuated here, with what goes after). Qualities attractive to vision are predicated of God, and A. then wills to look upon them with righteous vision.

quies: *vera rel.* 52. 101, summarizing the three temptations: 'quid enim appetit curiositas nisi cognitionem [2] . . . quid appetit superbia nisi potentiam [1] . . . quid appetit voluptas corporis nisi quietem [3], quae non est nisi ubi nulla est *indigentia* et nulla corruptio?' Cf. 13. 38. 53.

intrat in gaudium: Mt. 25: 21, 'intra in gaudium domini tui' (cf. 9. 10. 25); *en. Ps.* 139. 15, 'si autem hoc te delectat quod foris agis, vide ne tumescas foris, et non possis redire per angustam, et non possit tibi dicere deus tuus: "intra in gaudium domini tui".'

defluxi: No biblical resonance (though perhaps a Plotinian one), but cf. 10. 29. 40, 'a quo in multa defluximus', 12. 10. 10, 'defluxi ad ista et obscuratus sum', 12. 15. 19, 'ab illo se resolvat et defluat', 13. 7. 8, 'defluens inferius amore curarum', 13. 8. 9, 'defluxit anima hominis.' See on 2. 2. 2, 'diffluebam'.

erravi: Ps. 118: 176, 'erravi, sicut ovis perdita.'

regio egestatis: Rewrites 2. 4. 9, 'nulla compulsus egestate'; Lk. 15: 14 (the prodigal), 'et postquam omnis consummasset, facta est fames valida in *regione* illa, et ipse coepit *egere*'; cf. Lk. 15: 13, 'in regionem longinquam'. *s. Caill.* 2. 11. 3, '"erat autem egestas in illa regione": non egestas visibilis panis, sed egestas invisibilis veritatis'; *s. Caill.* 2. 11. 5, 'iam ergo iste obtriverat cor in regione egestatis: redierat enim ad cor, ut obtereret cor; reliquerat cor superbus, redierat ad cor iratus.' (See on 1. 18. 28, esp. for quotation from *qu. ev.* 2. 33.) Cf. 13. 8. 9, 'omnis mihi copia, quae deus meus non est, egestas est'; *ord.* 1. 2. 3, 'idcircoque illam [pulchritudinem universitatis] videre non licet animae quae in multa procedit [cf. 'defluxi'] sectaturque aviditate pauperiem. . . . nec mirere quod eo *egestatem* patitur magis, qui magis appetit plura complecti'; also early at *beata v.* 4. 28, 'est ergo animi egestas . . . nihil aliud quam stultitia. . . . ergo ut omnis egestas miseria, ita omnis miseria egestas esse convincitur' (in the following discussion of the word, A. had occasion to quote the practice of Sallust, noted from *Catil.* 52. 22); *ep.* 3. 2, 'merito philosophi in rebus intellegibilibus divitias ponunt, in sensibilibus egestatem'; cf. *vera rel.* 52. 101 on the three temptations, quoted above on 'quies'. For the approach to *egestas* through sin, cf. *doctr. chr.* 3. 10. 16 (after defining *flagitia* and *facinora*: see on 3. 9. 17), 'cum [flagitia] exinaniverint animum et ad quandam egestatem perduxerint, in facinora prosilitur'. Cf. *vera rel.* 12. 23, 'trahitur ergo ad poenas, quia diligendo inferiora in *egestate voluptatum suarum* et in doloribus apud inferos ordinatur.' Cf. *en. Ps.* 52. 8, 'interior . . . egestas'; *s.* 105. 2. 2, 'in omni cupiditate et egestate saeculi'. In the Augustinian geography of salvation, the *regio sine egestate* is at *civ.* 14. 26: 'in paradiso . . . vivebat sine ulla egestate'.

There are roots in Porphyry: *sententiae* 37, *ἐπεὶ δὲ πρὸς μὲν ὕλην ῥεπούσῃ ἀπορία πάντων καὶ τῆς οἰκείας δυνάμεως κένωσις, εἰς δὲ τὸν νοῦν ἀναγομένῃ τὸ πλῆρες αὐτῆς κατὰ ⟨τὸ⟩ τὴν δύναμιν ἔχειν τῆς πάσης εὑρίσκετο, τὴν μὲν εἰκότως Πενίαν, τὴν δὲ Πόρον οἱ τοῦτο πρῶτον γνόντες τῆς ψυχῆς τὸ πάθος ᾐνίξαντο.* See A. Solignac, *BA* 13. 664–5, and du Roy, 150.

The end of the book echoes the beginning: 2. 1. 1, 'conligens me a dispersione, in qua frustatim discissus sum, dum ab uno te aversus in multa evanui'.

BOOK THREE

Book 3 is tripartite:

3. 1. 1–3. 6	*Student life at Carthage: from* flagitiosi amores *to the* spectacula
3. 4. 7–10. 18	*Quest for 'Wisdom'*
3. 4. 7–5. 9	*In Cicero's* Hortensius
3. 6. 10–10. 18	*Among the Manichees*
3. 11. 19–12. 21	*Monnica's attitude*

Curiositas predominates, but all three temptations are present; note *concupiscentia carnis* here in 3. 1. 1 and *superbia* in 3. 3. 6.

3. 1. 1

This dense and vivid passage reprises the opening of Bk. 2. (The beginnings of books of *conf.* are more often closely related to the conclusions of the books just finished, as at the junctures of 3/4, 5/6, 7/8, 8/9, 10/11; cf. *civ.* 2. 2, recapitulating the first book.). Here the opening paragraph recapitulates to facilitate forward motion, and so only seems retrospective rather than prospective.

BA 13. 665–7 gives a detailed note on this passage, from which extracts are given under the lemmata below; they react against G. Wijdeveld, *Vig. Chr.* 10 (1956), 231–5. *BA* 665: 'Augustin en effet ne décrit pas l'expérience telle qu'il la vécut en son temps, mais bien telle qu'il la réfléchit et la juge en évêque et en philosophe.' This paragraph is so vivid and memorable that it is often treated as if it were narrative in content, when it is in reality a tissue of meditative abstractions.

Carthaginem: Carthage, not Rome, remained for him the familiar epitome of urban life: *ep.* 43. 3. 7, 'erat Carthago civitas ampla et inlustris, unde se per totum Africae corpus malum quod ibi esset exortum tamquam a vertice effunderet. erat etiam transmarinis vicina regionibus et fama celeberrima nobilis'; *ep.* 118. 2. 9, 'cum duae tantae urbes latinarum litterarum artifices, Roma atque Carthago, nec taedio tibi sint . . . nec taedia tua curent.' The reputation of the city changed with time and with the outlook of the observer. Apul., *flor.* 20, had the old veneration for the place: 'Carthago provinciae nostrae magistra venerabilis, Carthago Africae Musa caelestis, Carthago Camena togatorum'; but Salvian, two generations younger than A., was notoriously critical of the mores of the city (though it is far from clear how reliable his information was): *gub.* 7. 16–17, 'scaturrientem vitiis civitatem . . . quae enim fuit pars civitatis non plena sordibus, quae intra urbem platea aut semita non lupanar?'

sartago: Probably still 'frying pan' (Lev. 6: 21 (quoted at *qu. hept.* 3. 15), 'in sartagine, in oleo fiet, consparsam offeret eam'), but perhaps of the contents rather than the instrument at *s.* 273. 8. 8, 'oderunt martyres lagenas vestras, oderunt martyres sartagines vestras, oderunt martyres ebrietates vestras.' Souter s.v. takes it as 'a heated instrument of torture,' citing Cyprian *Fort.* 11, but that is clearly not what it is in A.'s sermon.

amare GO[1] Knöll S Skut. Ver.: *amari* CD[1]O[2]. See on 2. 1. 1; for the paradoxes, cf. *div. qu.* 35. 1, 'amandum est sine metu vivere. . . . an amor quoque ipse amandus est? ita vero, quando sine hoc illa non amantur. sed si propter alia quae amanda sunt amor amatur, non recte amari dicitur. nihil enim aliud est amare quam propter se ipsam rem aliquam appetere. . . . est enim et turpis amor, quo animus se ipso inferiora sectatur, quae magis proprie cupiditas dicitur, omnium scilicet malorum radix.'

secretiore indigentia (= 'fames . . . ab interiore cibo'): *BA*: 'c'est l'indigence de l'homme qui, faute de se connaître soi-même et de discerner la dignité de sa véritable essence, se disperse dans le sensible au lieu de se concentrer en soi pour y découvrir Dieu'; cf. 2. 10. 18, 'regio egestatis', and texts cited there.

oderam me: As often in Latin (as with English 'hated'), *oderam* is not much stronger than 'disliked' (E. Fraenkel, *Horace* (Oxford, 1957), 263). For the implications here (when is self-love 'really' a form of self-hatred?), see O. O'Donovan, *The Problem of Self-Love in Augustine* (New Haven, 1980), esp. 37–59; but he does not discuss the present passage.

minus indigentem: Take the clause thus: 'In the midst of the *regio egestatis* I suffered an *indigentia* that affected me deep within (thus *secretiore*), and so I hated the very thought of what I might be like if I did not suffer that *indigentia* ('me minus indigentem'), hence I was entirely averse to anything that might have had a good effect on me. I "hated" the one thing that was good, "in love" with "Love", foolishly.' Hence the desperate need to love wrongly below. *en. Ps.* 118. s. 1. 1, 'nam quisquis libidinibus deditus luxuria stuprisque corrumpitur, in hoc malo beatitudinem quaerit et se miserum putat, cum ad suae concupiscentiae voluptatem laetitiamque non pervenit, beatum vero non dubitat iactare cum pervenit.'

securitatem et viam sine muscipulis: *BA*: 'la *securitas* serait la consistance, la certitude, la sérénité de la vie chrétienne, ou du moins celle qui résulterait d'une légitime conception de l'amour humain; ce serait la *vena* et le *candor* d'une *amicitia* qu'aucun sentiment trouble ne viendrait ternir. La *via sine muscipulis* est celle où l'âme est à l'abri des corruptions (ou mieux des corrupteurs) de l'amour.' Wijdeveld, *Vig. Chr.* 10 (1956), 233, cites *en. Ps.* 36. s. 3. 13 ('ergo est securitas, sed si intus sit deus') and *s.* 40. 5. 7 ('quae est autem securitas, fratres, vel mea vel vestra, nisi ut domini iussa intente et diligenter audiamus et promissa fideliter expectemus?').

viam sine muscipulis: The editors adduce Wisd. 14: 11, 'propter hoc et idolis nationum non erit respectus, quoniam creaturae dei in odium factae sunt et in temptationem animis hominum et in muscipulum pedibus insipientium.' But the 'path without pitfalls' is Christ (Ps. 141: 4, 'in via hac qua ingrediebar absconderunt mihi muscipulam'; *en. Ps.* 141. 9, 'via haec qua ingrediebatur, Christus est; ibi illi absconderunt muscipulam, qui persequuntur in Christo, propter nomen Christi'), who either did or did not encounter *muscipulae* himself. One consistent line of interpretation says he did: *s.* 130. 2 ('ad pretium nostrum tetendit muscipulam crucem suam'), *s. Guelf.* 21. 2 (= *s.* 263. 1: 'muscipula diaboli, crux domini'), *s. Mor.* 17. 5 ('crux Christi muscipula fuit'). On the other hand there is *en. Ps.* 90. s. 1. 4: 'et longe ab ipsis muscipulis ambulant homines qui in Christo ambulant; non audet enim in Christo tendere muscipulam; circa viam ponit, in via non ponit. via autem tua Christus sit, et tu non cades in muscipulam diaboli. aberranti a via, iam ibi est muscipula.' Sim. at *ss.* 134. 5. 6 and 216. 6. 6.

Muscipula is common in A.'s Psalter, though rare in the Roman and Gallican versions. The flavour is important to catch: Ps. 90: 3, 'sperabo in eum, ipse eruet me de muscipula venantium et a verbo aspero' (cf. *en. Ps.* 90. s. 1. 4 just quoted); Ps. 9: 16, 'in muscipula ista quam occultaverunt comprehensus est pes eorum' (*en. Ps.* 9. 15, 'muscipula occulta est dolosa cogitatio. pes animae recte intellegitur amor; qui cum pravus est, vocatur cupiditas aut libido; cum autem rectus, dilectio vel caritas.'); Ps. 30: 5, 'educes me de muscipula ista, quam occultaverunt'; Ps. 123: 7, 'anima nostra sicut passer eruta est de muscipula venantium; muscipula contrita est, et nos eruti sumus' (*en. Ps.* 123. 12, 'muscipula erat dulcedo vita huius'). Cf. also Ps. 9: 31, 34: 7–8, 63: 6, 139: 6, 140: 9; and 1 Tim. 6: 9 (VL). For treatments of the image without direct discussion of the present passage, see S. Poque, *Le langage symbolique dans la prédication d'Augustin* (Paris, 1984), 1. 22–8, and J. Rivière, *RTAM* 1 (1929), 484–96; the latter is an excellent survey, but makes no reference to *conf.* or this passage.

Other snares: 4. 6. 11 ('evellens de laqueo pedes meos'), 6. 12. 21 ('dulces laqueos in via eius'), 10. 34. 52 ('ut tu evellas de laqueo pedes meos').

ab (*interiore cibo*): 'for want of.' *BA*: 'ce n'est pas une faim *de* la nourriture intérieure, mais une faim *par privation de* cette nourriture.' He hungers for God, but does not know his hunger ('ea fame non esuriebam') and hence has no longing for incorruptible things ('sed eram sine desiderio alimentorum incorruptibilium'), not out of satiety ('non quia plenus eram'), but vacuity ('inanior').

fastidiosior C^1DO^1S Knöll Skut. Ver. Pell. : *o fastidiosior* C^2O^2 Maur.

ulcerosa: Cf. Job 2: 7–8 (VL), 'egressus diabolus a facie domini percussit Iob vulnere pessimo, a pedibus usque ad caput. (8) et tulit sibi testam, ut raderet saniem. et ipse sedebat in stercore.' The scene immediately precedes the temptation Job hears in the words of his wife: *en. Ps.* 133. 2, 'tunc Eva [!] ausa est eum temptare, "dic aliquid in deum tuum, et morere."' But *ulcerosus* for A. regularly refers to the story of Dives and Lazarus (Lk. 16: 20 ff.: *en. Ps.*

145. 7, 'duo erant: pauper ulcerosus ad ianuam iacens divitis, et dives indutus purpura et bysso, in epulis quotidie splendidis'); cf. 'elegans et urbanus' below.

foras: See on 1. 18. 28, 'ibam foras'; cf. also 4. 15. 27 and 7. 16. 22.

contactu: Touch is the sense to which sexual temptation appeals: see on 2. 5. 10.

si non haberent animam: G–M: 'we do not speak of "loving" inanimate objects. A's meaning is that, even in this association, stained as it was with lust, he was seeking, however mistakenly, the satisfaction of a real soul hunger.' With the sentence that follows, it seems clear that he means that at this time, avid for contact with the sensible, he wanted to Love, and could not love what was without *anima*; hence, lust as the particular focus of his evil, the perverse parody of a central virtue. He could not stop loving, but he could only love in an evil way.

amare et amari: See on 2. 2. 2.

magis: The preceding comma was introduced by Vega; *magis* occurs elsewhere at the beginning of a clause or extended phrase (1. 6. 8, 'et me talem fuisse magis mihi ipsi indicaverunt nescientes quam scientes nutritores mei'; other examples at 1. 9. 15, 1. 18. 29, 3. 1. 1, 3. 10. 18, 4. 15. 26, 5. 12. 22, 5. 14. 25, 6. 10. 16, 9. 4. 7). Other editors (Knöll, Skut., Pell., Ver.) have put the comma after *magis*, (1) making 'dulce . . . magis' a quasi-comparative (not impossible in later Latin, but not attested in *conf.*; for A.'s variations on traditional comparative forms, see Arts 43–5) and (2) reducing the affirmation to a single statement ('Love was sweeter if . . .'), breaking up the sequence ('Love was sweet, indeed even sweeter if . . .').

venam igitur amicitiae . . . obnubilabam: The wording brings us back to the opening of Bk. 2, reinforcing the prevalence of concupiscence in what has gone between. Cf. 2. 2. 2, 'luminosus limes amicitiae' and 'obnubilabant atque obfuscabant cor meum,' and 3. 2. 3, 'vena amicitiae' (with context that makes clear that a channel for liquid is envisioned by the metaphor). The *vena amicitiae* is close in sense to the *vena caritatis* in similar context at *b. coniug.* 16. 18 and *Io. ep. tr.* 6. 2.

de tartaro libidinis: 8. 4. 9, 'de tartaro caecitatis'.

foedus atque inhonestus: *imm. an.* 8. 13, 'quod si non id quod est in mole corporis, sed id quod in specie facit corpus esse, quae sententia invictiore ratione approbatur—tanto enim magis est corpus, quanto speciosius est atque pulchrius, tantoque minus est, quanto foedius ac deformius, quae defectio non praecisione molis . . . sed speciei privatione contingit.'

elegans et urbanus: J. V. Fleming, *Reason and the Lover* (Princeton, 1984), 95–6, renders 'fine and courtly', with comments. At *c. acad.* 2. 2. 6, the true beauty of philosophy, hidden from one who is mired down in worldly delights ('baias et amoena pomeria et delicata nitidaque convivia et domesticos histriones'), though he struggles towards true beauty: 'inde est illa hospitalitas, inde in conviviis multa humanitatis condimenta, inde ipsa *elegantia*, nitor, mundissima facies rerum omnium, et undique cuncta perfundens adumbratae venustatis *urbanitas*.' See the rest of that passage quoted on 2. 1. 1. On *deformiter*, cf. 10. 27. 38, 'et in ista formosa quae fecisti deformis inruebam.'

misericordia mea: *mea* corresponds to an objective genitive. Ps. 58: 18, 'adiutor meus, tibi psallam, quia deus susceptor meus es, deus meus misercordia mea'; *en. Ps.* 58. s. 2. 11, 'totum quidquid sum, de misericordia tua est.' Cf. Ps. 143: 1–2; cf. also 3. 3. 5, bracketing this section: 'et circumvolabat super me fidelis a longe *misericordia tua*.' Knauer, 37, 'Beide Stellen sagen das gleiche aus und sind wohlverständlich—Gott als "refugium" oder "auxilium". Es kommt aber noch hinzu, daß sie den Abschnitt einrahmen, in dem Augustin die Theaterleidenschaft und ihre Motive ... behandelt (3. 2. 2–4). Im Theater, wo seine eigenen "miseriae" dargestellt werden, wird doch nur eine "mirabilis insania" erregt (3. 2. 2). Je stärker man selber von dem vorgestellten Affekten betroffen ist, um so mehr verfällt man dieser Leidenschaft, die, erleidet man sie selber, "miseria", leidet man aber für andere mit, "misericordia" genannt zu werden pflegt.'

felle ... aspersisti (see on 9. 10. 23, 'aspersi'): 1. 14. 23, 'difficultas ... quasi felle aspergebat omnes suavitates graecas fabulosarum narrationum.'

occulte: Vega: 'Distinguíanse entre los romanos dos clases de concubinato: uno legal, público, en el que la mujer no tenía derechos jurídicos de esposa, y otro privado, o simple concubinato. San Agustín quiere dar a entender—más adelante lo dice claramente—que su unión fue de este segundo modo.' *Per contra*, Peter Brown, 'Augustine and Sexuality', *Colloquies of the Center for Hermeneutical Study*, Protocol 46 (Berkeley, 1983), 2: 'it was a relationship beyond moral reproach, even among the Catholic clergy' (which leaves *occulte* unexplained).

vinculum ... nexibus: The attachment to sexual desire is often expressed as a bondage; see on 8. 11. 25. For 'conligabar', cf. 8. 1. 2, 'conligabar ex femina'; 8. 8. 20, 'conligata vinculis' (again of sexual attachment, but more metaphorical).

ut caederer: To what is A. referring? He is at least deliberately taking upon himself the conventional ideas of the quarrels and the troubles of obsessed young lovers. That he uses hints of biblical language tacitly passes judgement on the relationship.

virgis ferreis: Ps. 2: 9, 'reges eos in virga ferrea'; *en. Ps.* 2. 8, 'in inflexibili iustitia'.

suspicionum ... rixarum: Gal. 5: 20, 'manifesta autem sunt opera carnis quae sunt ... contentiones, aemulationes, irae, rixae, dissensiones, sectae.'

3. 2. 2

Curiositas is always a vice for A. For a definition using the word itself, *vera rel.* 52. 101, 'quid enim appetit *curiositas* nisi cognitionem quae certa esse non potest nisi rerum aeternarum et eodem modo se semper habentium?'; though the word is absent, more revealing perhaps is *vera rel.* 33. 62, 'ille [animus] autem vult mentem convertere ad corpora, oculos ad deum. quaerit enim intellegere carnalia et videre spiritalia, quod fieri non potest.'

For bibliography, see on 10. 35. 54. There is much originality to A.'s development, but the possible influence of Ambrose has not been sufficiently considered. Not that Ambrose has a concept of *curiositas*, but some of his suspicions of the excesses of the philosophers are at least apposite to A.'s own thoughts, though A. develops them further. See Amb. *off.* 1. 26. 122 (written 386), 'itaque tractant in veri investigatione tenendum illud decorum, ut summo studio requiramus quid verum sit, non falsa pro veris ducere, non obscuris vera involvere, *non superfluis vel inplexis atque ambiguis occupare animum*. quid tam indecorum quam venerari ligna, quod ipsi faciunt? quid tam obscurum quam de astronomia et geometria tractare, quod probant, et profunda aeris spatia metiri, caelum quoque et mare numeris includere; relinquere causam salutis, erroris quaerere?' Such ill-guided search for knowledge Amb. then contrasts to Moses' superior wisdom.

The subject emerges early and pervades the years before *conf.*; at *ord.* 1. 8. 26 he recalls watching a cock-fight: 'cur . . . nos ipsa pugnae facies aliquantum et praeter altiorem istam considerationem duceret in voluptatem spectaculi?' Later in the same work, he cautions measure in inquiry: *ord.* 2. 5. 17, 'si quis temere ac sine ordine disciplinarum in harum rerum cognitionem audet inruere, pro studioso illum *curiosum*, pro docto credulum, pro cauto incredulum fieri' (cf. *ord.* 1. 11. 31, 'curiosi vel nimium studiosi'); specific examples are offered: *ord.* 2. 12. 37 (the trivial pursuits of grammarians), 2. 15. 42 (on astronomy: 'astrologiam genuit, magnum religiosis argumentum tormentumque curiosis'); cf. *quant. an.* 19. 33. The noun *curiositas* itself is introduced at *mus.* 6. 13. 39, 'avertit [a contemplatione aeternorum] denique amor vanissimae cognitionis talium rerum. . . . curiositas nascitur ipso curae nomine inimica securitati, et vanitate impos veritatis'. Connection with 1 Jn. 2: 16 (see on 1. 10. 16 and on 10. 30. 41) is slower coming, though it may underlie *mor.* 1. 21. 38: 'quamobrem recte etiam curiosi esse prohibemur, quod magnum *temperantiae* munus est. . . . reprimat igitur se anima ab huiusmodi vanae cognitionis cupiditate, si se castam deo servare disposuit.' Similar discussions occur later: cf., e.g. *trin.* 10. 1. 3, 'aut si tam curiosus est ut non propter aliquam notam causam sed solo amore rapiatur incognita sciendi, discernendus quidem est ab studiosi nomine iste curiosus; sed nec ipse amat incognita, immo congruentius dicitur, odit incognita, quae nulla esse vult dum vult omnia cognita.'

A clear echo of 1 Jn. 2: 16 occurs at *lib. arb.* 2. 19. 53 (nothing in the work can be surely dated before it was put in final form, which may have been as late as 395), 'ad proprium convertitur [1], cum suae potestatis vult esse; ad exterius [2], cum aliorum propria vel quaecumque ad se non pertinent cognoscere studet, ad inferius [3] cum voluptatem corporis diligit. atque ita *homo superbus* [1] *et curiosus* [2] *et lascivus* [3] effectus excipitur ab alia vita quae in comparatione superioris vitae mors est.' But comparable echoes may be found at *Gn. c. man.* 1. 23. 40 (quoted on 10. 30. 41) and 2. 18. 27 ('genus tertium temptationis his verbis figurare, quod est *curiositas*').

The last thing A. wrote before his ordination in 391 was *vera rel.*, whose structure and contents are heavily influenced by 1 Jn. 2: 16; there are clear attacks on *curiositas* at *vera rel.* 3. 4, 4. 7 ('nam tertio vitio curiositatis in percontandis daemonibus'), 29. 52 ('in quorum consideratione non vana et peritura *curiositas* exercenda est, sed *gradus* ad immortalia et semper manentia faciendus'), 38. 70, 38. 71 (quoted on 10. 30. 41 for the link between the temptations of 1 Jn. and the three temptations of Christ in the desert), 49. 94 ('iam vero cuncta *spectacula* et omnis illa quae appellatur *curiositas*, quid aliud quaerit quam de rerum cognitione laetitiam?'), and cf. also *vera rel.* 52. 101–54. 105, and see *div. qu.* 68. 1, quoted on 5. 3. 5 below.

When we bring our modern incomprehension to A.'s disdain for what is now an unquestioned virtue, we forget that for him *curiositas* led directly to demons: *cat. rud.* 25. 48, 'qui christianum nomen oderunt . . . et adhuc simulacris et daemoniorum curiositatibus servire desiderant,' and *Io. ep. tr.* 2. 13 (on 1 Jn. 2: 16), 'iam quam late patet curiositas? ipsa in spectaculis, in theatris, in sacramentis diaboli, in magicis artibus, in maleficiis ipsa est curiositas.' See also *trin.* 4. 11. 14–4. 12. 15, *civ.* 10. 26, 10. 28 (where he presents Porphyry playing to an audience of the curious: 'ut talium quoque rerum quasi peritus appareas et placeas inlicitarum artium curiosis, vel ad eas facias ipse curiosos'), and 10. 29. (A familiar villain for fourth-century Christians was similarly led astray by *curiositas*, acting in concert this time with *ambitio saeculi*: *civ.* 5. 21, 'apostatae Iuliano, cuius egregiam indolem decepit amore dominandi sacrilega et detestanda curiositas'.) For the word and the thing, see further on 10. 35. 54.

Curiositas emerges here in narrative as A. comes to the great city and finds himself surrounded by all manner of marvels. One innocent recollection may date to this awed time: *civ.* 16. 8, 'quosdam [homines] sine cervice oculos habentes in umeris, et cetera hominum vel quasi hominum genera, quae in maritima platea Carthaginis musivo picta sunt, ex libris deprompta velut curiosioris historiae.'

spectacula . . . imaginibus: Students were discouraged by the local authorities from too much spectacle-going: *cod. theod.* 14. 9. 1 (12 March 370), 'neve spectacula frequentius adeant'. The same law declared that indiscipline could be punished (at least at Rome and Constantinople) by whippings and forced rustication. A similar moralizing restriction was enjoined upon the young Julian by his tutor Mardonius (Julian, *misopogon* 351 C-D), and Libanius (*ep.* 976) thought the theatre a distraction for students. A.'s remarks here make it clear that it was the enacted stories that appealed to him most, as later the circus (6. 7. 11–12) and the gladiatorial combats (6. 8. 13) would appeal to Alypius; these seem to have been the three main classes of entertainment available to A. (and classed by him as *spectacula*): *s.* 198. 3, 'delectantur nugatorio spectaculo et turpitudinibus variis theatrorum, insania circi, crudelitate amphitheatri, certaminibus animosis eorum qui pro pestilentibus hominibus lites et contentiones usque ad inimicitias suscipiunt, pro mimo, pro histrione, pro pantomimo [these three are from the *theatra*], pro auriga [from the circus], pro venatore [from the gladiatorial amphitheatre].'[1]

Spectaculum in A. is almost always accompanied by verbs of seeing, frequently with word-play on *spectare*; in this paragraph note 'spectacula', 'spectat', 'spectator', 'spectat', and add 3. 2. 3, 'spectaculi', 3. 2. 4, 'spectare', 3. 8. 16, 'spectatores', 3. 8. 16, 'principandi [1] et spectandi [2] et sentiendi [3] libidine'; the recapitulation at 4. 1. 1 (paralleling the recapitulation of Bk. 2 at 3. 1. 1) speaks of 'spectaculorum nugas'.

The appeal is to the *concupiscentia oculorum* (*s. Den.* 14. 3, 'quae mala facit turpis curiositas, concupiscentia vana oculorum, aviditas nugacium spectaculorum, insania stadiorum, nullo praemio conflictus certaminum!'; cf. *Io. ep. tr.* 2. 13, quoted above). Compare Alypius covering his eyes at the gladiatorial *spectaculum*, but yielding, 'curiositate victus' (6. 8. 13); cf. 1. 10. 16 ('eadem *curiositate* magis magisque per oculos emicante in spectacula'), 1. 13. 22, 10. 35. 54 ('ex hoc morbo [curiositatis] in spectaculis exhibentur quaeque miracula'); *theatra* also appeal to *curiositas* (10. 35. 56). Cf. *vera rel.* 22. 43, 'nec ob aliud a talibus prohibemur spectaculis, nisi ne umbris rerum decepti ab ipsis rebus quarum illae umbrae sunt aberremus'; sim. at *vera rel.* 49. 94, 54. 105; and again at *trin.* 4. 11. 14. *civ.* 2. 4, 'veniebamus etiam nos aliquando adulescentes ad spectacula ludibriaque sacrilegiorum, spectabamus arrepticios, audiebamus symphoniacos, ludis turbissimis qui diis deabusque exhibebantur oblectabamur, Caelesti virgini et Berecynthiae matri omnium, ante cuius lecticam die sollemni lavationis eius talia per publicum cantitabantur a nequissimis scaenicis qualia non ... matrem ipsorum scaenicorum deceret audire. ... quae si inlecta *curiositate* adesse potuit circumfusa, saltem offensa castitate debuit abire confusa'; *civ.* 2. 26, 'ante ipsum tamen delubrum, ubi simulacrum illud [Caelestis] locatum conspicebamus ... intentissime *spectabamus*, intuentes alternante conspectu

[1] The same classification at *en. Ps.* 39. 9. See H.A. Kelly, *Traditio* 35 (1979), 21–44, and (broader in scope) H. Jürgens, *Pompa diaboli* (Stuttgart, 1972); see also W. Weismann, *Kirche und Schauspiele* (Würzburg, 1972), esp. 33–68 and 134–48.

hinc meretriciam pompam, illinc virginem deam.' See also *civ.* 1. 32–3, 1. 35, 2. 8, 7. 26.

G–M, Theiler, *P. u. A.* 60, and *BA* all attempt to situate this text in the tradition of ancient discussions of the emotional impact of the theatre. There is nothing here in *conf.* to connect A.'s views with any of the surviving discussions, but he is surely their heir at some distance (at *civ.* 8. 13 he even invokes Plato's suspicion of poets in support of his views). Dominant is surely his own notion of the connection to *curiositas*. His works nevertheless (including the passages just cited from *civ.*) offer some glimpses of what the life of the *spectacula* entailed in the Carthage of his day. See also *en. Ps.* 103. s. 1. 13 ('videtis quid faciat civitas ubi abundant spectacula: in agro securius loquerer'), 146. 4, 147. 8, *s.* 241. 5 (quoted on 1. 13. 20, 'Aeneae nescio cuius'). Alfaric, 32–3, offers additional texts.

imaginibus: Cf. 'imaginum' below, and 'imaginarie' at 3. 2. 3 immediately following. The counterfoils of reality, *imagines* loom in *conf.* in two ways. (1) Not used until the present passage, the word and its corresponding verb *imaginare* are common from here on (but especially here in Bk. 3—the book of the temptations to vision—and in Bk. 10 (41 × from 10. 8. 12–10. 25. 36) in the discussion of memory, which is the faculty of the soul replete with *imagines* (10. 8. 12, 'campos et lata praetoria memoriae, ubi sunt thesauri innumerabilium imaginum')), as the appearances that both suggest a reality beyond appearance and at the same time veil it from direct sight. (See further on 4. 10. 15.) (2) In a few contexts, the word marks the special relationship with God that is authorized by the text of Gn. 1: 26, 'faciamus hominem ad imaginem et similitudinem nostram'. This topic is first introduced briefly at 3. 7. 12, recurs strongly at 6. 3. 4–6. 4. 5, appears at 7. 7. 11, and then returns for full development from 13. 22. 32 to the end of the work. In the present context, note that *imagines* recur as powerful incitements to *concupiscentia carnis* when that temptation is reviewed at 10. 30. 41 f. In that light, 'fomitibus' here is probably further explanation of 'imaginibus': 'images that called to mind my misery and further kindled the fire of my libido'.

et dolor ipse est voluptas eius: A. was not always so insightful and usually settled for a facile opposition of *voluptas* and *dolor*; cf. *div. qu.* 36. 1, 'nemo est qui non magis dolorem fugiat quam appetat voluptatem'; *qu. ev.* 1. 47, 'cupiditati voluptatis opponitur timor doloris.' Even at 1. 20. 31, the two are opposites: 'voluptates . . . quaerebam, atque ita inruebam in dolores' (see notes there). A variation on the synergy of pleasure and pain is proposed at 10. 31. 43, 'nunc autem suavis est mihi necessitas . . .'.

mirabilis CDOS Skut. Ver. : *miserabilis* G Maur. Knöll Pel. *Miserabilis* is thinly attested and is also the *facilior lectio*. Elsewhere in *conf.* madness itself can be marvellous (4. 15. 26, 'mira dementia'), and a variety of other abstract substantives (notably both *continentia* at 6. 10. 16 and *abstinentia* at 10. 31. 46) are *mirabilis*: see also 4. 4. 8, 4. 14. 21, 9. 4. 12, 9. 6. 14.

miseria . . . misericordia: Word-play concentrates the effects of one form of *curiositas*, extended through 3. 2. 4. The real fall through *curiositas* of this book will be the lure of Manicheism, but here at the outset we get a self-contained display of another dimension of that same fault, not unlike—with different rhetorical proportions—the pear-theft from Bk. 2. He interpreted *misericordia* etymologically: *c. Adim.* 11, 'ex eo appellatam misericordiam dicunt, quod miserum cor faciat dolentis aliena miseria'; sim. at *mor.* 1. 27. 53, *civ.* 9. 5.

actori O Maur. Isnenghi Pell. Ver.: *auctori* CDGS Knöll Skut. But *auctor* here would have to apply to the author of a text being enacted (*civ.* 2. 14, 'qua ratione rectum est, ut poeticorum figmentorum et ignominiosorum deorum infamentur actores, honorentur auctores?'; cf. 'agantur' here, 3. 2. 4, '*actio* histrionis'); *actor* with appropriate genitives is common in A.'s discussion of theatrical representations in *civ.* (*civ.* 2. 11, 'scaenicos actores earundem fabularum'; 2. 13, 'actores talium fabularum'; 2. 14, 'actores poeticarum fabularum'). In *conf.*, *auctor* appears 3 × of God (3. 8. 15, 5. 5. 9, 11. 13. 15), once hypothetically of the devil (7. 3. 5), and once of the writers of the books of scripture (13. 24. 37).

imaginum: At *civ.* 6. 9, *imago* is the word for the relation between the theatrical portrayal of the gods and the purportedly serious civil theology on which it is based: 'illam theatricam et fabulosam theologian ab ista civili pendere noverunt et ei de carminibus poetarum tamquam de speculo resultare, et ideo ista exposita, quam damnare non audent, illam eius imaginem liberius arguunt et reprehendunt.'

gaudens lacrimat. ergo CDEFGV Ver. : *gaudens lacrimat * ergo* O[1] (Ver., p. xlii: 'Avant la correction O a peut-être donné *lacrimatur*. Mais *lacrimat*, appuyé malgré tout par la leçon de O, a l'avantage de mieux correspondre au *doleat* de la même phrase') : *gaudens lacrimat. lacrimae ergo* AH Pell. : *gaudens. lacrimae ergo* O[2]S BPZ Knöll Skut. Vega B. Löfstedt, *Symb. Osl.* 56 (1981), 106 (with a question mark after the following *dolores*.) : *gaudens lacrimatur. ergo* Maur. : *gaudens lacrimator. ergo* M.

The context offers 'ergo amentur dolores aliquando' a few lines below in 3. 2. 3, bracketing (with 'ergo amantur et dolores' beginning that paragraph) the demonstration (with no mention of *lacrimae*). On palaeographical probability, the reading could be that of AH: *gaudens lacrimat lacrimae*, but the want of a subject for *amantur* could inspire haplography, and the presence of *lacrimae* robs the 'et' before 'dolores' of its emphatic force (see *en. Ps.* 4 cited in notes on 1. 1. 1). On weight of MSS, one would incline to the reading of O[2]SBPZ: *gaudens. lacrimae*. (*Ergo* in first position in its clause is less frequent in *conf.* (26 × against 167 × post-positioned), but well attested (as esp. here with iteration later).) But on content, the authentic reading was that of CDEFGVO[1], *gaudens lacrimat ergo*; a simple error, repeated later and encouraged by the context (the oxymoron of 'gaudens lacrimat' would discourage reading the two words together, taken with the availability of 'amantur et' to attract a subject), created 'lacrimae'. *lacrimat lacrimae* is haplography reinventing the same error. (On tears in *conf.*, see on 3. 2. 4.)

3. 2. 3

The derangement of *misericordia* by *concupiscentia carnis* and *curiositas* is the pretext for another measurement of the distance between present and past (presented as identical with the tension between right and wrong). The paragraph is circular in composition: First, 'ergo amantur dolores ... ergo amentur dolores'; then, 'cave immunditiam ... cave immunditiam'; then, 'neque enim nunc non misereor ... nunc vero magis misereor', leading to the conclusion, 'haec certe verior misericordia, sed non in ea delectat dolor'.

ergo amantur: On text and punctuation, see on 3. 2. 2.

quod quia ... hac una causa: some redundancy; 'is it for this reason, that *misericordia* is not without *dolor*, for this reason alone that *dolores* are loved?' Translators generally dispense with one of the conjunctions (e.g. *BA*, '... puisqu'on prend plaisir pourtant à être miséricordieux et que cela ne va pas sans souffrance, ne serait-ce pas pour cette unique raison que l'on aime les souffrances?').

et hoc: emphatic, 'and this too.'

vena amicitiae: Cf. 3. 1. 1, 'venam igitur amicitiae coinquinabam'. What follows ('ut quid decurrit ...') makes it clear that 'vena' is taken as a channel for liquid.

torrentem picis: Isaiah 34: 9 (of the *dies ultionis*), 'et convertentur torrentes eius in picem, et humus eius in sulphur, et erit terra eius in picem ardentem.'

taetrarum libidinum: For other epithets with *libido*, see on 2. 2. 2.

ergo amentur dolores aliquando: Wijdeveld, *Vig. Chr.* 10 (1956), 235, punctuates (paralleling the immediately previous question and answer): 'ergo amentur dolores? aliquando.'

immunditiam: First occurrences in *conf.* here; strongest contrast at 13. 7. 8, 'affectus sunt, amores sunt, immunditia spiritus nostri defluens inferius amore curarum et sanctitas tui attollens nos superius amore securitatis'; cf. also 10. 31. 46, 'immunditiam cupiditatis'; elsewhere, 4. 6. 11 of his grief for the death of his friend ('talium affectionum immunditia'); 7. 1. 1 of his Manichean-influenced ideas of God ('circumvolantem turbam immunditiae'), and cf. 8. 11. 27 where Continentia advises, 'obsurdesce adversus immunda illa membra tua super terram, ut mortificentur.'

anima mea: This brief apostrophe to his own soul is repeated in a much longer and more complex passage: 4. 11. 16–19 (see on 4. 11. 16), and see also 10. 6. 10.

tutore: God as *tutor* also at 10. 4. 6, 12. 16. 23. Taken with 'deo patrum nostrorum' here, the appeal is to the first person of the trinity. The word is otherwise rare in both A. and scripture, and there are no useful parallels. Note, however, that the word occurs a number of times in *civ.* (1. 3, 3. 9, 3. 13, 3. 20, 18. 41) applied to the 'pagan' divinities to whom Rome look in vain for protection (infrequent but classical: *CIL* 14. 25, 'Iovi tutori').

superexaltato in omnia saecula: Cf. Dan. 3: 52, 'benedictus es domine, deus patrum nostrorum, et laudabile et superexaltatum in omnibus saeculis'—the first words of the Song of the Three Children, i.e. the first words spoken by young men of virtue trapped in a pit of fire from which, through divine assistance, they will escape; cf. the 'torrentem picis' above.

congaudebam amantibus: G–M: 'The plays to which A. refers were evidently not of a quality to suggest the Aristotelian point of view.' (!)

imaginarie . . . ludo spectaculi: 3. 2. 2, 'spectacula theatrica plena *imaginibus*'. A. is the first writer attested for *imaginarie*: *c. ep. fund.* 43. 49, 'phantasmata, quae de carnali sensu tracta imaginarie cogitatio nostra versat et continet'.

gererent O Maur. Skut. Ver. Pell. : *gererentur* CDG : *gerent* S : *agerent* coni. Vega, adducing 3. 2. 2, 'si calamitates illae hominum vel antiquae vel falsae sic agantur' (and cf. also 3. 2. 2, 'actori'); *civ.* 2. 26, 'ludos qui agebantur intentissime spectabamus' (both refs. given inaccurately by Vega).

sese amittebant: not reflexive elsewhere in *conf.*; cf. here, 'amissione verae felicitatis'.

misereor CDGO Maur. Ver. Pell. : *miseror* S Knöll Skut. Vega. The question here is morphology rather than semantics. On available evidence, the following may be said of A.'s practice. In *conf.* passages where the reading is not in dispute, *misereor* or its derivatives appear 22 ×, *miseror et al.* 8 ×; but of those *miseror* appears 6 × in perfect and imperative forms, where it has euphony on its side. In imperative, imperfect, and present forms (both ind. and sub.), the prevalence of *misereor* stands at 19 × to 2 ×.

caritatis: Second occurrence in *conf.* here, first where *caritas* is a quality of human actions (earlier: 2. 6. 13, 'neque blandius est aliquid tua caritate'); first with scriptural echo not until 4. 4. 7, 'caritate diffusa in cordibus nostris'. 34 × in all.

germanitus: *ep.* 140. 34. 79, 'neque enim germanitus dixerunt: apud te laus mea . . . cum ignorantes dei iustitiam iustitiam suam constituerent'; *nat. et or. an.* 3. 14. 20, 'intus et germanitus in corde sensisti'. Rare and archaic, a grammarian's keepsake.

(*misericors*) **est:** where *esset* would be normal in view of the next line; but A. wishes to grant the preposterous hypothesis, to emphasize the distastefulness of the conclusion to which it leads. Vega puts a question mark at the end of the next sentence (after 'misereatur'), unnecessarily.

hoc: i.e. 'quod nullo dolore sauciaris'.

et ad haec quis idoneus: 2 Cor. 2: 16–17, 'et ad haec quis tam idoneus? (17)

non enim sumus sicut plurimi adulterantes verbum dei, sed sicut ex sinceritate, sed sicut ex deo coram deo in Christo loquimur.' At *en. Ps.* 77. 30, A. uses the same exclamation to hint at the ineffability of divine ways; if we would follow his thoughts through the next verse of Paul, we would hear him claiming authority for what he says, despite its inadequacies. Paul's text, of course, compares awkwardly with Lk. 18: 11, 'pharisaeus stans haec apud se orabat, "deus, gratias ago tibi quia non sum sicut ceteri hominum".'

3. 2. 4

miser . . . amabam: Cf. 3. 1. 1, '*amare* amabam . . . quaerebam *quid amarem*'. The intervening exposition of his reaction to the *spectacula* makes it clear that the quest for love leads to *dolor*. Cf. 1. 20. 31, where the quest for *voluptates* ends in *dolores*.

lacrimae: A.'s tears over the *spectacula* are his first adult tears (as *infans* and *puer*, he wept regularly: 1. 6. 7, 'flere autem offensiones carnis meae'; 1. 13. 21, 'et flebam Didonem extinctam'), but they are not his last. Episodes of weeping in *conf.*: Monnica's tears for A., 3. 11. 19–3. 12. 21 (see on 3. 11. 19, 'fleret'); the death of A.'s friend, 4. 4. 9–4. 7. 12; Monnica abandoned at Carthage, 5. 8. 15; the garden scene at Milan, 8. 12. 28–30; A.'s emotional reaction to the hymns and psalms of the church not long thereafter, 9. 6. 14, (cf. the tears shed in the Cassiciacum dialogues: *ord.* 1. 8. 22 and 1. 10. 30, *c. acad.* 2. 7. 18, *sol.* 1. 14. 26 and 2. 1. 1); the death of Monnica, 9. 11. 27–9. 12. 33. The 'confessional' parts of Bk. 10 speak of tears and weeping (10. 1. 1, 10. 28. 39 ('laetitiae meae flendae'), 10. 40. 65 ('et resorbeor solitis et teneor et multum fleo')), but Bks. 11–13 are free of tears except for two mentions of ways in which tears will pass away (12. 11. 13, 'si iam factae sunt ei lacrimae suae panis'; 13. 13. 14, 'transierint lacrimae'). The most important tears that he had not yet shed were the 'lacrimas confessionis' of 7. 21. 27. Weeping and prayer explicitly connected: 5. 8. 15 (of Monnica), 8. 12. 28–29 (the garden scene), 9. 6. 14, 9. 7. 16, 10. 4. 5. On tears in Christian prayer in antiquity, see J. Balogh, *Didaskaleion*, NS 4 (1926), 10–21, on the way licit tears for A. draw us nearer to God, not to earthly things, and with numerous good texts from Christian antiquity.

infelix pecus: Virg. *ecl.* 3. 3, 'infelix o semper, oves, pecus!' The Vergilian echo both evokes the lost sheep and includes the most powerful Dido epithet, *infelix* (see C. Bennett, *REAug* 34 (1988), 59–60). Here again the secular and Christian motifs intertwine inextricably. The lost-sheep parable occurs in the same chapter of Luke as the prodigal son: Lk. 15: 4, 'quis ex vobis homo qui habet centum oves, et si perdiderit unam ex illis, nonne dimittit nonaginta novem in deserto et vadit ad illam quae perierat donec inveniat illam?' The parallel synoptic version is at Mt. 18: 12; cf. also Ps. 118: 176, 'erravi sicut ovis quae periit; quaere servum tuum, quia mandata tua non sum oblitus'; 1 Pet. 2: 25, 'eratis enim sicut oves errantes, sed conversi estis nunc ad pastorem.'

turpi scabie ... sanies horrida: The deliberately graphic imagery of a festering wound, the foothold that death can take in a sound body. Was this life?

foedarer: 2. 1. 1, 'transactas foeditates meae'; 3. 1. 1, 'foedus atque inhonestus'.

spectare: the detachment of *curiositas* enables the soul, already fallen through *concupiscentia carnis*, to participate vicariously in deformities yet more perverse than those in which it really participates.

3. 3. 5

Et circumvolabat . . . misericordia tua: The summary here turns from the encapsulated exemplum of *curiositas* in the theatre to the wider issues of his life at this time, and so underlines the persistence of *concupiscentia carnis* (cf. 'concupiscere'). The judgemental present intrudes only in the apostrophe 'o tu praegrandis . . .'.

longe: Cf. 'ad longe' below, with reversal of point of view; on *longe* elsewhere, see on 1. 18. 28.

sacrilega curiositate CDGO^{2}S Knöll Skut. Vega : *sacrilegam curiositatem* O^1 Maur. Ver. Pell. Ablative of manner; cf. 4. 16. 31, 'cum deformiter et sacrilega turpitudine in doctrina pietatis errarem'. Cf. *c. Cresc.* 4. 10. 12, 'sacrilegium vero tanto est gravius peccatum, quantum committi non potest nisi in deum.' The epithet elsewhere in *conf.:* 5. 9. 16, 'corde sacrilego', 8. 2. 4, 'non erubescendo de sacris sacrilegis', 8. 7. 17, 'superstitione sacrilega', 10. 35. 56, 'omnia sacrilega sacramenta detestor'. *c. Faust.* 13. 15, 'philosophi gentium de filio dei aut de patre deo vera praedixisse seu dixisse perhibentur. . . . quamobrem quantum distat de Christi adventu inter praedicationem angelorum et confessionem daemoniorum, tantum inter auctoritatem prophetarum et curiositatem sacrilegorum.'

The word *sacrilegus* (rare in scripture, only 2 × in NT (Act. 19: 37, Rom. 2: 22), in both cases deriving from Gk. ἱεροσυλέω, 'to rob a temple'; already in Tertullian) is another example of the surprisingly large vocabulary of 'pagan' religion taken over with little modification by Christianity.

ima: See on 2. 5. 10.

circumventoria: First (only?) here in Latin (*TLL*).

daemoniorum: Cf. 1. 17. 27, 'non enim uno modo sacrificatur transgressoribus angelis'; *c. Faust.* 20. 22, 'illi quippe superbi et impii spiritus non nidore ac fumo, sicut nonnulli vani opinantur, sed hominum pascuntur erroribus.' For the 'demonic' in his own conduct at the time, see on 3. 3. 6. Deut. 32: 17, 'immolaverunt daemonibus et non deo', echoed at 1 Cor. 10: 20.

The Gk. original of the word came over to Latin by two different paths at about the same time: as a technical term of Platonic philosophy with Apuleius, and as a pejorative in Latin translations of scripture (and in Christian writers from Tertullian on). It is difficult to see any distinction for A. between *daemon* (in *conf.* only at 4. 2. 3, 9. 7. 16, in the form *daemonibus*) and *daemonium* (in *conf.* here and at 3. 3. 6 and 8. 2. 4), not least because the two words share a nom./acc. plural (*daemonia*, as at 4. 2. 3).

We shall never know what A. really thought of demons, even as we catalogue his pronouncements. One document takes us a little closer to actuality, his *divin. daem.* (406/11?). This pamphlet begins as transcript of a discussion that took place in A.'s episcopal *secretarium* before divine service during the octave of Easter. The *laici* with whom he was speaking (*divin. daem.* 1. 1) probably included some newly baptized at the Easter vigil. The discussion (and the more connected exposition that follows it) assumes in explicit material terms that demons exist, that they perform actions in the material world, and that they have various advantages (including a light, ethereal body that has keener senses and swifter motion than anything human or animal, and a wealth of ancient experience besides that enables them to predict future events more accurately than we (*divin. daem.* 3. 7–4. 7)) over mortals in the attempt to know the future. They can also foretell events that they will themselves cause. The least that must be admitted is that the bishop catered to a belief that he had no particular polemical or apologetical reason for accepting. His limit is reached only by the naïve suggestion from one of his interlocutors that the ancient religious rites dictated by the *libri pontificales* were in some way licit and approved by God (*divin. daem.* 2. 5), and that it is only secret and illicit sacrifice by night to demons that merits censure. A.'s theoretical discussion of demons is more familiar (esp. *civ.* 8. 14 f.) but brings us perhaps less close to the actuality.

flagellabas: Rare in script. in any positive sense, but cf. Ps. 72: 14, 'et fui flagellatus tota die' (Knauer, 76 n. 1), Tob. 13: 2 (from the *canticum Tobiae*: see above, pp. 6–7), 'quoniam tu flagellas et salvas'.

in celebritate sollemnitatum tuarum: Becoming a specific term for 'the celebration of Mass'; *celebrare* is usual for important services, and *sollemnitas* for 'feast, holy day, service'. *un. eccl.* 25. 74, 'quid si ergo fictus accedit atque adversus veritatem et ecclesiam cor inimicissimum gerit; quamvis peragatur in eo *solemnitas*, numquid reconciliatur, numquid inseritur? absit'; *en. Ps.* 54. 19, 'festa martyrum *celebrabamus*, erant tibi mecum; paschae *sollemnitatem* frequentabamus, erant ibi mecum'; *sollemnitas* often elsewhere, e.g. *en. Ps.* 61. 10, 69. 2, 73. 7, *s.* 267. 1. 1 ('ideo enim *sollemnitas celebratur*, ne quod semel factum est, de memoria deleatur'). See also on 8. 3. 6.

This incident is regularly extracted by biographers and commentators to capture the flavour of A.'s disorderly life. In context, two features become noteworthy. (1) He was a churchgoer in these pre-Manichean days at Carthage. (2) The lines before and after make it an example of the way in which God's punishing flail was never far away; the specifics of the incident and the concrete form that punishment took are left to our imagination.

intra parietes: The *parietes ecclesiae* recur emphatically at 8. 2. 4.

agere negotium procurandi fructus mortis: A. shies away from this act with recourse not merely to euphemism, but to the construction of this elaborately ambiguous but unambiguously judgemental phrase, which has the ring of scriptural authority. But no plausible parallel can be found other than Rom. 7: 5, 'cum essemus in carne, passiones peccatorum quae per legem erant operabantur in membris nostris ut fructificarent morti.'

misericordia mea, deus meus: Ps. 58: 18, 'deus meus misericordia mea'; Ps. 143: 2, 'misericordia mea'. Knauer, 118: '3. 1. 1 und 3. 3. 5—sie rahmte die Verurteilung der falschen "misericordia" ein.' Virtually the same phrase ('deus meus, misericordia mea') occurred at 3. 1. 1 as a guide to the passions of his early days at Carthage; now that they have been recounted and understood, the citation is repeated with this asseveration and the emphatic 'praegrandis'.

refugium: See on 1. 9. 14, 'puer coepi rogare te, auxilium et refugium meum'.

longe: See on 1. 18. 28.

amans vias meas: i.e. not the 'viam sine muscipulis' of 3. 1. 1?

fugitivam libertatem: The flight of the prodigal. For *libertas*, see on 2. 6. 14.

3. 3. 6

For the first time, the third temptation of 1 Jn. 2: 16, *ambitio saeculi*, exercises an influence directly on A. (in Bk. 1, it was his parents whose ambition directed his education). These opening paragraphs set a stage for the intellectual drama presented in the rest of the book. Their central focus is A.'s reaction to the *spectacula*, and hence his *curiositas*, but both of the other two Johannine temptations have been given cameo roles. A. paints himself both grasping and detached, and not a little vulnerable to outside influence.

honesta: 1. 13. 21, 'honestiores et uberiores litterae', 5. 3. 3 (of Faustus' reputation), 'honestarum omnium doctrinarum peritissimus', 7. 6. 9, 'honestam educationem liberalesque doctrinas'.

fora litigiosa: Ovid, *fasti* 4. 187–8: 'spectate, Quirites | et *fora* Marte suo *litigiosa* vacent'—from the description, s. d. 4 March, of the *ludi Matris Magnae*; though A. never mentions Ovid and barely cites him (see Hagendahl, 213–14: he may have known the story of Pyramus from O.'s *metamorphoses*, and there is one probable echo of *met.* 1 in *civ.* 22. 24), it was at about this time in A.'s life that he attended the rites of the Magna Mater at Carthage. The phrase here may hint that he looked into Ovid to learn more of those rites. The context reveals that the memory was associated for A. with his adolescent passion for the *spectacula*, and suggests again the connection between the stage and non-Christian religion that explains some of A.'s own hostility to these memories: see *civ.* 2. 4, quoted above on 3. 2. 2, 'spectacula ... imaginibus'.

caecitas: Cf. 2. 3. 7, where *caecitas* is also the metaphor for a heedlessness inspired by camaraderie.

typho: A leitmotif for ambitious pride; see on 7. 9. 13 ('typho turgidum'); often associated with 'swollenness'; with *tumidus* or *tumesco* at *civ.* 11. 33, *c. Iul.* 4. 3. 28, *en. Ps.* 149. 10, *s.* 4. 30. 33.

remotus omnino ab eversionibus: In *ep.* 93, addressed to his old school-friend Vincentius, now (408) a bishop of the small Rogatist sect of Donatists, the young A. appears similarly restrained (see on 2. 1. 1).

eversionibus: At 5. 12. 22, the *eversiones* of the next generation of students (perhaps 10 years later) so disturb A. the teacher that he explains his departure for Rome in part as an attempt to escape to a more peaceful academic atmosphere. Cf. *vera rel.* 40. 75, 'his [angelis iracundiae] similes sunt homines qui gaudent miseriis alienis et risus sibi ac *ludicra spectacula* exhibent vel exhiberi volunt *eversionibus* et *erroribus* aliorum.' The theme of Bk. 3 is *curiositas*, which feeds on the *spectacula* and leads to *error:* there is thus an odd appropriateness to the ways of the *eversores*. The words *eversio/eversor* have a legal use of squandering paternal property (Tac. *ann.* 6. 17, 'eversio rei familiaris', *cod. theod.* 12. 6. 1), the sort of thing that a prodigal son might do, or that one who saw himself nervously as a possible prodigal might frown upon; in one case, the devil is 'fidei eversor' (*s. Guelf.* 31. 1) and in another perhaps inadvertently revealing case, Christ is seen by the Jews as an 'eversor legis' (*Io. ev. tr.* 20. 2), though more common is a use A. follows elsewhere, of famous generals who sacked ancient cities (e.g. *civ.* 1. 6, 'Fabius, Tarentinae urbis eversor'; cf. *civ.* 1. 34, 3. 15); evidently a vivid expression, whether it is merely A.'s term of abuse or an authentic piece of local slang. On student life in this period generally (mixing evidence from east and west), see A. Müller, *Philologus*, 69 (1910), 292–317.

urbanitatis: See *c. acad.* 2. 2. 6, quoted on 3. 1. 1.

pudore impudenti: Oxymoron heightens the perversity of his values, as before. But here the love of comradeship that led him into sin in the pear-theft seems to have faded a little. He still enjoys the comrades, but does not share their wrong-doing (which has taken on a specific malice directed against other people). The passage is almost exculpatory, but now his sinfulness separates him from his fellows.

ignotorum: 'Here might almost be rendered *freshmen*' (G–M). Ryan settles for 'new students', *BA* for 'nouveaux'.

daemoniorum: See above on 3. 3. 5. To go by this account, A. was cautious about invoking these powers. He resists a suggestion at 4. 2. 3 and sees them defeated by the power of the relics of Gervasius and Protasius at 9. 7. 16.

Here they offer a threatening simile for the deeds of his boisterous friends. Cf. 'diabolicum'.

3. 4. 7

A. reads philosophy, and turns (3. 5. 9) to scripture, ending in frustration. The same sequence of readings occurs at 7. 9. 13, in changed circumstances, with a different result.

A.'s dialogue with Cicero is a subtext of the next books of *conf.* In 46/5 BC, Cicero wrote the main body of his philosophical *œuvre*, in the order clearly recorded at his *div.* 2. 1. 1, a text A. knew (*civ.* 3. 17): *Hortensius*, *academica*, *de finibus bonorum et malorum*, *Tusculanae disputationes*, *de natura deorum*. Here in Bk. 3, A. reports his encounter with the *Hortensius*; in 5. 10. 19, he turns his attention to the Academics; at 6. 16. 26, he and Alypius discuss between themselves the issues 'de finibus bonorum et malorum', where A. says he would have sided with the Epicureans (cf. Cic. *fin.*, Bks. 1–2) except that he was oppressed by 'metus mortis'—the first subject of the *Tusculans*. Till then there seems a clear pattern, a hint that A.'s philosophical investigations parallel those of Cicero. It is in Bk. 7 that new masters, the Platonists and Paul, are found, but even beyond that traces of the Ciceronian sequence may be descried: the *Tusculans* match in spirit and to some extent in substance (note particularly *Tusc.* 2, 'de tolerando dolore', and cf. 9. 2. 4 and 9. 3. 5 on A.'s ailments—the pain in his chest and a bad toothache) the discussions at Cassiciacum, especially as both lead to a doctrine concerning the *beata vita*, which is also a central concern of *conf.* 10 (see on 9. 4. 7 for more parallels). To urge a parallel between the last three books of *conf.* and Cicero's three books *de natura deorum* is obviously to go beyond the bounds of evidence: but to consider the possibility is to measure not only the similarities, but also the divergences, between Cicero and his most imaginative disciple. (For a similar co-ordination between the progress of the text of *conf.* and a different authoritative text, see the remarks in notes on 8. 1. 1 concerning the Pauline echoes in Bk. 8.)

Modern readers are generally willing to infer from A.'s enthusiasm—we have not much else to go on—that the *Hortensius* was a powerful and important book. A salutary minority view comes from O'Meara, 57–8, who belittles the *Hortensius* and thinks that the effect here is all that of A.'s personality, not the book itself: 'If it was a great book, how explain its comparative obscurity until Augustine read it, and its eventual disappearance? . . . Or was it after all just an ordinary book that happened to set off a flame in Augustine's mind when that mind was prepared to be inflamed?' Even A. was willing to remember that Cicero was an orator who praised philosophy, not a philosopher *per se*: *trin.* 14. 9. 12, 'ita ille tantus orator cum philosophiam praedicaret recolens ea quae a philosophis acceperat et praeclare ac suaviter explicans . . .'; cf. *civ.* 2. 27, 'philosophaster Tullius'; on the other hand, his contemporaries certainly knew the text (*ep. Sec.* 3 shows that A.'s Manichee critics had it).

Since the reading of texts is so important for A. and others in *conf.*, it is worth sketching a short catalogue of readings explicitly reported: 1. 13. 20, the *Aeneid*; Hortensius and scripture here; 5. 3. 3, 'multa philosophorum'; 7. 9. 13, 'platonicorum libros'; 7. 21. 27 ff., Paul (esp. 8. 12. 29–30, garden scene); 9. 4. 8, Psalms, esp. Ps. 4; and Gn. 1 in Bks. 11–13 (with most of Bk. 12 (12. 14. 17–32. 43) discussing proper methods of reading). Reading as solution to his problems: 3. 12. 21, 'ipse legendo reperiet quis ille sit error et quanta impietas.' The courtiers at Trier read the life of Antony (8. 6. 15); Ambrose is a model of reading at 6. 3. 3 (silent) and 6. 4. 4 (public exegesis: and already at 5. 14. 24); and the role of the reader in relation to *conf.* is frequently evoked and discussed: e.g. 10. 3. 3 f., 12. 24. 36. See R. Flores, *Aug. Stud.* 6 (1975), 1–13.

Bks. 2 and 3 have depicted the adolescent A. as a particular prey to sexual temptation and transgression. It is important to the structure of *conf.* that he not encounter—to our eyes—decisive advice to elect *continentia* until Bk. 8, where *continentia* is the focus of the garden scene in Milan. Signs there (see on 8. 7. 17) hint that the issue, and the possibility of *continentia*, were not new to him. But he makes no mention here that one of the things we know he would have found in the *Hortensius* was just this advice: frg. 81 M (*c. Iul.* 4. 14. 72 and 5. 10. 42, partly corroborated by Nonius 412. 8 Lindsay—given here in form conflated from the two passages in *c. Iul.*): 'an vero voluptates corporis expetendae, quae vere et graviter a Platone dictae sunt inlecebrae esse atque escae malorum? quae enim confectio est valetudinis, quae deformatio coloris et corporis, quod turpe damnum, quod dedecus quod non evocetur atque eliciatur voluptate? cuius motus, ut quisque est maximus, ita est inimicissimus philosophiae. congruere enim cum cogitatione magna voluptas corporis non potest.' Sim. at frg. 74 M (Nonius 33. 7 Lindsay) and frg. 80 M (Nonius 503. 15 Lindsay).

This paragraph and the next are the focus of an unpublished study by E. Feldmann, 'Der Einfluß des Hortensius und des Manichäismus auf das Denken des jungen Augustinus von 373' (Diss., Münster, 1975), with detailed commentary on these paragraphs at 1. 381–513. The work merits publication, perhaps in briefer compass. In the meantime, Feldmann's note at *Atti 1986*, 316–30, usefully presses the question, earlier raised by Theiler (*P. u. A.* 5 n. 1), of how far A. read the *Hortensius* 'mit den Augen des Neuplatonikers'. Better to say that A. read all of Cicero with late antique eyes.[2]

[2] There seems no satisfactory study of the readership of Cicero's philosophical works in antiquity and late antiquity, not even T. Zielinski, *Cicero im Wandel der Jahrhunderte* (4th edn., Leipzig, 1929). A.'s intensity of attention may have been more idiosyncratic than we have assumed. Note e.g. *ep.* 118, where Dioscorus takes his questions to A. after finding the masters at Carthage ill-equipped to handle them; and esp. *ep.* 118. 2. 9, where A. claims not to be able to lay his hand on the relevant texts in Hippo.

eminere cupiebam . . . per gaudia vanitatis: i.e. *per ambitionem saeculi*; see 'omnis vana spes' below.

usitato iam discendi ordine: Cf. *c. acad.* 3. 4. 7, quoted below.

cuiusdam Ciceronis: The phrase has evoked abundant discussion; concise survey at Pellegrino, *Les Confessions*, 84 n. 12; more briefly, *BA* 13. 667; more extensively, Testard, 1. 11–19 (who sees nothing pejorative). Mohrmann, *Vig. Chr.* 13 (1959), 235–40 (reviewing Testard) invokes a Christian habit of citing as *quidam* a 'pagan' author cited anonymously—but as Pellegrino, loc. cit. (on 86), noted, *here* it is anonymity that is ruled out. According to Hagendahl, 579, the word 'clashes with the rest of the account'. Vega ad loc. sees in it a 'manera algún tanto despectiva—no ignorativa—de citar a un escritor tan conocido y estimado de el; pero no es extrano al estilo del Santo'. *Quidam* alone need imply no derogation: 11. 23. 30, 'et cuiusdam [= Joshua, otherwise unnamed] voto cum sol stetisset, ut victoriosum proelium perageret, sol stabat, sed tempus ibat.'

But though his usage is diverse, it is hard to charge inconsistency; at *c. acad.* 1. 3. 7 A. could speak of *Cicero noster* and at *c. acad.* 3. 18. 41 of 'Tullius noster', while at *conf.* 1. 16. 25 he introduces a direct quotation from *Tusc.* with the expression 'ex eodem pulvere hominem clamantem et dicentem'; in 413, 'philosophaster Tullius' (*civ.* 2. 27); from about the same time, *Io. ev. tr.* 58. 3, 'cuiusdam saecularis auctoris verba laudantur . . . ille homo eloquentissimus'; in the fourth book of *doctr. chr.* (not written in this form until many years after *conf.*), Cicero becomes *quidam* again: *doctr. chr.* 4. 10. 24, 'unde ait quidam'; 4. 12. 27, 'dixit enim quidam'; but 4. 17. 34, 'ipse Romani auctor eloquii'. At *civ.* 14. 18, a quotation from *Tusc.* is introduced by a *quidam* juxtaposed with high praise from a Roman source: 'sicut ait etiam quidam "Romani maximus auctor eloquii"' (Lucan 7. 62–3). See also on 1. 13. 20, 'Aeneae nescio cuius'. Of the commentators, *BA* comes closest; there is something slightly arch about the expression, but the derogatory tone is no more than is, surely, Cicero's due; in an address to God, the expression signifies the vanity of a fame like that of Cicero in the presence of God.

pectus: *s.* 108. 7. 7, 'quod est intus in corde tuo, hoc dicatur foris. non aliud pectus tegat, et aliud lingua proferat.' Cf. Testard, 1. 18, on the Ciceronian opposition of *lingua* and *pectus* (not as abundantly attested in C. as Testard seems to think, however). *Pectus* is the organ whence arise the thoughts and sentiments that take shape as words on the *lingua*: Aul. Gell. 1. 10. 4, quoting his teacher Favorinus, 'vive ergo moribus praeteritis, loquere verbis praesentibus atque id, quod a C. Caesare, excellentis ingenii ac prudentiae viro, in primo de analogia libro scriptum est, habe semper in memoria atque in *pectore*, ut "tamquam scopulum, sic fugias inauditum atque insolens verbum".' The *pectus* was in a direct physical sense a deeper source of speech than the *lingua*, so in a metaphorical application it would play a similar, superior, role. Cf. 4. 14. 23 ('aurae linguarum flaverint a pectoribus opinantium') and 6. 3. 4 ('de tam sancto oraculo tuo, pectore illius' [sc. Ambrosii; cf. *ep.* 31. 8, 'pectus tuum tale domini oraculum est']). Secundinus in *ep. Sec.* 5, after much praise of A.'s eloquence at *ep. Sec.* 3, praises A. for his 'aureum pectus'. Also to be read in that context are the reports of his illness in the summer of 386 at 9. 2. 4 and 9. 5. 13 ('dolore pectoris').

To admire someone's *pectus* is then to praise him for sincerity and ingenuousness—for an ability to match words to thoughts truthfully and truly. In this light, other passages of *conf.* take the eye (7. 5. 7, 'talia volvebam pectore misero,' 8. 8. 19, 'illuc me abstulerat tumultus pectoris,' 9. 9. 21, 'docente te magistro intimo in schola pectoris'), as does the frequency with which *pectus* is used in a less explicit metaphorical sense for the seat of feelings close to speech (2. 3. 6, 5. 9. 17, 6. 1. 1, 6. 2. 2, 6. 16. 26, 7. 2. 3, 8. 2. 4, 8. 4. 9, 10. 42. 67).

One moralizing interpretation is congruent and should be kept in mind: *Gn. c. man.* 2. 17. 26, 'nomine enim pectoris significatur superbia, quia ibi dominatur impetus animi: nomine autem ventris significatur carnale desiderium, quia haec pars mollior sentitur in corpore.' If that passage is taken as determinative, then to criticize Cicero's *pectus* is to accuse him of excessive *ambitio saeculi:* not an unverisimilar charge in Cicero's case.

philosophiam: Though the early A. did not scruple to use the word of Christian doctrine (e.g. *beata v.* 1. 1), he uses it in *conf.* only here and in the next paragraph, and only in a sense that is at least ambivalent. There are *philosophi* further on in *conf.* (but only as far as 8. 2. 3), but again only figures ambivalent at best and not identifiably Christian. In Ambrose, the word never applies to Christian doctrine or life, though that sense is not uncommon in earlier Christian writers (see Madec, *Saint Ambroise*, 41, noting the one possible exception).

Hortensius: The majority of references and allusions to the *Hortensius* in A. occur at Cassiciacum; from 387–413, there are only repetitions of previously quoted passages; new and important fragments occur in late books of *trin.* (416 and after) and in *c. Iul.* (421). His reading of the *Hortensius* lingered in his mind for half a century, ever vivid where he gives a long quotation on the mind's ascent to God in *trin.* 14. 19. 26. That fragment clearly shows that the *philosophia* with which the *Hortensius* inspired A. was a philosophy of the mind's ascent, shaping his taste when he came to the Platonists years later: note here 'surgere coeperam ut ad te redirem' and at 3. 4. 8, 'quomodo ardebam, deus meus, quomodo ardebam revolare a terrenis ad te!'

Might A. have modelled the first book of *c. acad.* on Hortensius? At *c. acad.* 1. 1. 4, he calls that work *inductorium*, and it was certainly a dialogue (*trin.* 14. 9. 12, etc.). See *beata v.* 1. 4, quoted in Introduction, and cf. *sol.* 1. 10. 17 ('prorsus unus mihi Ciceronis liber facillime persuasit nullo modo appetendas esse divitias'), *c. acad.* 1. 1. 4 (of Cassiciacum and the young men:

'praesertim cum Hortensius liber Ciceronis iam eos ex magna parte conciliasse philosophiae videretur'), *c. acad.* 3. 4. 7 (to Licentius he says, 'et ad scholam redeas nostram, si tamen aliquid iam de te Hortensius et philosophia meretur, cui dulcissimas primitias iam vestro illo sermone libasti, qui te vehementius quam ista poetica incenderat ad magnarum et vere fructuosarum rerum scientiam'); an allusion is obvious at *util. cred.* 1. 1, 'de invenienda ac retinenda veritate ... cuius, ut scis, ab ineunte adulescentia magno amore flagravimus'.

There is a considerable literature; see first Testard, 1. 19–39, then Hagendahl, 79–94 (fragments/testimonia) and 486–97 (discussion). Of specialist studies on A.'s use of the Hortensius, the most recent is R. Russell in *Aug. Stud.* 7 (1976), 59–68, and cf. Feldmann's dissertation cited in introd. n. above. The Hortensius is quoted by the fragment numbers of Müller's Teubner edition, but M. Ruch, *L'Hortensius de Cicéron: Histoire et reconstitution* (Paris, 1958), with texts (but non-standard numbers of fragments), is also of interest.

mutavit affectum meum: Cf. 10. 3. 4, 'mutans animam meam fide et sacramento tuo'. Brown, 169: 'An intellectual event, such as the reading of a new book, is registered only, as it were, from the inside, in terms of the sheer excitement of the experience, of its impact on Augustine's feelings: of the *Hortensius* of Cicero, for instance, he would never say "it changed my views" but, so characteristically, "it changed my way of feeling"—*mutavit affectum meum*.'

mutavit preces meas: Ryan: 'it turned my prayers to you, Lord'; *BA*: 'm'orientant vers toi, Seigneur, il changea mes prières'; but *mutavit* does not take *ad* + obj. in the sense 'changed *x* [by turning] towards *y*'; G–M rightly take 'ad te ipsum' closely with 'preces meas': 'changed the character of the prayers that I offered to thee.'

concupiscebam: This word hints that something was not right with his attitude towards philosophy.

surgere coeperam: Lk. 15: 18–20 (the prodigal again), 'surgam, et ibo ad patrem meum, et dicam ei: pater, peccavi in caelum, et coram te ... (20) et surgens venit ad patrem suum.'

redirem: See on 1. 18. 28.

maternis: The word strikes sharply and unexpectedly: the last we heard, it was Patricius who was struggling to provide the funds (2. 3. 5, 'animositate magis quam opibus patris'). Now we hear of his death only in an ablative absolute. The natural interpretation of the present passage is that M. succeeded to control of the property.

Readers of *c. acad.* here expect mention of the long and generous support of Romanianus for A.'s ambitions, something congruent to the praise at *c. acad.* 2. 2. 3: 'tu me adulescentulum pauperem ad studia pergentem et domo et sumptu, et quod plus est, animo excepisti. [This is generally and rightly taken as indicating where A.'s father made up the deficiency of his own funds.] tu patre orbatum amicitia consolatus es, hortatione animasti, ope adiuvisti. tu in nostro ipso municipio, favore, familiaritate, communicatione domus tuae paene tecum clarum primatemque fecisti.' See on 6. 14. 24 for possible reasons for this silence.

annum aetatis undevicensimum: A.'s nineteenth year ran from November 372 to November 373; his father thus died as early as late 370, as late as 371; the episode of Bk. 2 (in his sixteenth year) fell in 369/70 (2. 3. 6). Incautious narrators often postdate these events by assigning them to the year following A.'s nineteenth birthday, i.e. his *annum aetatis vicensimum*.

locutio B. Löfstedt, *Symb. Osl.* 56 (1981), 106: *locutionem* MSS edd.

locutio . . . quod loquebatur: See above on *lingua/pectus*; distrust of showy outward form in default of significant content is a leitmotif in the upward progress of A.'s conversion; he puts his disappointment with Faustus in those terms at 5. 6. 10, and finds himself vulnerable to Ambrose when the bishop's eloquence turns out to harbour a more valuable truth (5. 14. 24).

3. 4. 8

This paragraph must be allowed to have its surprises for us. Did A. at the age of 18 *consciously* hanker after the flight from earthly things to God—and did he do so in *those* words (not so different from those of the Platonists that would, he alleges, come as a surprising revelation more than a decade later)?

A.'s preference for the *nomen Christi* is not completely surprising, but must have been a mildly unusual reaction to this particular Ciceronian text. It was not entirely obvious (in spite of his 'itaque' opening 3. 5. 9) that the reader of the *Hortensius* would turn first to scripture to pursue the quest encouraged there.

The spirit of religious enthusiasm is clearly meant to portray the event (how faithfully, we cannot tell) as the forerunner of all the attempts A. will recount at an intellectual ascent to God (first concerted effort: 4. 13. 20 ff.). See du Roy 25–9.

revolare: The metaphor of flight, common among non-Christian speculations of this period for the salvific journey occurs only once otherwise in *conf.*, and then to emphasize its metaphorical quality: 1. 18. 28, 'non . . . filius ille tuus . . . avolavit pinna visibili'.

apud te est enim sapientia: Job 12: 13 (VL), 'apud eum est sapientia et virtus.' The topic of philosophy evokes from A. a scriptural tag, to which is appended, by an *autem*, the first element in a brief synopsis of the contents of the *Hortensius*.

amor autem sapientiae: Boethius *diff. top.* 2 (*PL* 64. 1187–8), 'philosophia amor sapientiae est; huic studendum nemo dubitat; studendum igitur est philosophiae. hic enim non definitio rei, sed nominis interpretatio argumentum dedit, quo Tullius enim in Hortensio in eiusdem philosophiae usus est defensione.' The etymology evidently played a fairly large part in the argument of the *Hortensius*; here it introduces A.'s summary of the contents of the lost dialogue. (The fragment is not accepted by Müller and Ruch, but recognized and promoted by H. Diels and H. Hagendahl (see Hagendahl, § 196).)

sunt qui seducant per philosophiam: The caution was a constant of protreptic, e.g. Boethius, *cons.* 1. P3. 7, on the 'epicureum vulgus et stoicum.'

manifestatur: A favoured word for A. (counting verb, adj., and related forms, 28 × in *conf.*, comparably frequent in other works), as in scripture (e.g. vb. *manifesto* 52 × in Vg.).

A. 'reads' a pre-Christian philosophical text and claims to find therein a message he thinks he can present fairly using the *ipsissima verba* of scripture. This device will recur with notable effect at 7. 9. 13 ff. (see notes there); the difference here is that at age 18 he did not know the scriptural text to juxtapose with the philosophical one (see 'et ego illo tempore' below).

videte . . .: Col. 2: 8–9, 'videte ne quis vos decipiat per philosophiam et inanem seductionem, secundum traditionem hominum, secundum elementa mundi, et non secundum Christum, (9) quia in ipso inhabitat omnis plenitudo divinitatis corporaliter.' The citation is picked up at 3. 6. 10 ('de istis elementis mundi, creatura tua') and recurs in Simplicianus' praise of the Platonists at 8. 2. 3. The verse in A. always speaks to those who fail in the achievement of philosophy, and specifically of those who depend on *sensus* and *imaginatio* rather than *intellegentia* (TeSelle, 74). *c. acad.* 3. 19. 42 is unmistakable: 'non enim est [sc. una verissimae philosophiae disciplina] ista huius mundi philosophia quam sacra nostra meritissime detestantur, sed alterius intellegibilis cui animas multiformibus erroris tenebris caecatas et altissimis a corpore sordibus oblitas numquam ista ratio subtilissima revocaret'; sim. at *ord.* 1. 11. 32, co-ordinating the Platonic realm of the intelligible with Jn. 18: 36, 'regnum meum non est de hoc mundo' (underlining the *hoc*). The passage from *c. acad.* goes on to invoke the incarnation as the divinely provided means for the soul to return to the *patria* 'etiam sine disputationum concertatione'. The Pauline *corporaliter* strengthens the argument: *ep.* 149. 2. 25, 'ideo "corporaliter" dixit quia illi umbraliter seducebant'; cf. *en. Ps.* 67. 23. Cf. Madec, *Saint Ambroise*, 200–7, 'l'inspiration Paulinienne'. This is the only occurrence of the word *philosophia* in scripture; for Ambrose the passage is a caution against excessive reliance on secular wisdom (e.g. *exam.* 2. 1. 2–3, *Abr.* 2. 8. 54). In dealing with the prodigal son, Amb. proposes an interpretation invoking this verse that is congruent with A.'s own exposition of his past, whether A. knew the Ambrosian text or not: Amb. *in Luc.* 7. 218, quoted below on 3. 6. 11.

At *mor.* 1. 21. 38, the Pauline text is a warning against the pre-eminent sin here in Bk. 3, *curiositas*, and contains a warning that parallels the course A. describes for his own career from here on: 'recte etiam curiosi esse prohibemur, quod magnum temperantiae munus est. hinc illud est, "cavete ne quis vos seducat per philosophiam". et quia ipsum nomen philosophiae si consideretur rem magnam totoque animo appetendam significat, siquidem philosophia est amor studiumque sapientiae, cautissime apostolus ne ab amore sapientiae deterrere videretur, subiecit "et elementa huius mundi". . . . tali enim amore plerumque decipitur, ut aut nihil putet esse nisi corpus; aut etiamsi auctoritate commota, fateatur aliquid esse incorporeum, de illo tamen nisi per imagines corporeas cogitare non possit, et tale aliquid esse credere, quale fallax corporis sensus infligit.' (As Manichee, A. had just the difficulty with imagining incorporeal reality that he warns against here: see particularly on 3. 7. 12 and 7. 1. 1.)

lumen cordis mei: Cf. Jn. 1: 9 (VL), 'lumen verum' (see on 1. 13. 21, 'lumen cordis mei').

delectabar: See on 1. 1. 1 for *delectatio*; here, as commonly in *conf.*, its presence is the sign that a moral suasion from outside has struck sparks and will have effect—here, mainly for the good.

quaererem: The word is a reminder to compare this attempted ascent to 'Wisdom' with the pattern prescribed at 1. 1. 1; the obvious defect here is that there has been no antedecent *praedicatio*, hence no accurate knowledge of what A. was seeking. Thus A. falls into the trap foretold at 1. 1. 1: 'aliud enim pro alio potest invocare nesciens'.

amplexarer GO²S Maur. Knöll Skut. : *amplexarem* CDO¹ Ver.

quod nomen Christi non erat ibi: At the age of 18, not yet a Manichee, encountering philosophy in the attenuated and mostly notional form of the Ciceronian protreptic (which he read without context or philosophical guidance and instruction), A. had the Christian expectations of his environment in a pronounced form. What the Manichees had that Cicero did not was just that for which he now pined: the *nomen Christi*. The phrase is common in the first books of *civ.*, and cf. *cons. ev.* 1. 14. 22, where A. thinks of heretics and 'pagans' who are so impressed with the figure of Christ that they use his name on their pseudonymous works: 'ita sentiunt etiam inimici Christi ad suadendum quod proferunt contra doctrinam Christi nullum sibi esse pondus auctoritatis, si non habeat nomen Christi.' For the structure of *conf.* it is important that it is the second person of the trinity whose absence he feels and seeks to remedy (see on 7. 18. 24 and 8. 12. 29). (See also *c. Faust.* 13. 17, quoted below on 3. 6. 10, 'viscum'.)

The depth of A.'s Christian attachments at the time he encountered the Manichees is hard to measure, but there is one revealing text: at *c. ep. fund.* 8. 9, he tells how the Manichees played down the Pascha, because it was only seeming passion, and played up their own feast of the Bema, because it was real: 'hoc enim nobis erat in illa bematis celebritate gratissimum, quod pro pascha frequentabatur, quoniam vehementius desiderabamus illum diem festum subtracto alio qui solebat esse dulcissimus.'[3]

secundum misericordiam tuam: Cf. Ps. 24: 7, 'delicta iuventutis et ignorantiae meae ne memineris, secundum misericordiam tuam . . . domine.'

in ipso adhuc lacte matris: 1. 11. 17, 'audieram enim ego adhuc *puer* (not *infans*) de vita aeterna promissa nobis per humilitatem domini dei nostri descendentis ad superbiam nostram.'

3. 5. 9

The encounter with Cicero leads to A.'s first reported direct encounter with scripture, which disappoints. He could not accept in an appropriate sense what he approached in an inappropriate way. He recounts his pride, and his disdain for the literary quality of the text, and no more (*what* he read he leaves unstated); at least one of his problems with the substance of scripture may be surmised (cf. on 'inclinare cervicem' below), but that is not his concern here. His approach to Manicheism came from a bad reading of scripture, tainted by curiosity and underlying pride. We do not see him approach scripture again until 5. 11. 21, when he begins to get an inkling from Elpidius that things might be other than they seemed. (Bear in mind that access to copies of the scriptural texts was not easy and universal. The reading recounted here may have been his first opportunity to approach the texts seriously.) Another report on his disappointment: *util. cred.* 6. 13, 'nihil me existimare prudentius, castius, religiosius, quam sunt illae scripturae omnes quas testamenti veteris nomine catholica ecclesia retinet. miraris, novi. non enim dissimulare possum *longe aliter nobis fuisse persuasum*. sed nihil est profecto temeritatis plenius, *quae nobis tunc pueris* [see *duab. an.* 15. 24 quoted below] *inerat*, quam quorumque librorum expositores deserere, qui eos se tenere ac discipulis tradere posse profitentur, et eorum sententiam requirere ab his qui conditoribus illorum atque auctoribus acerbissimum nescio qua cogente causa bellum indixerunt.'

The pride here is not *ambitio saeculi/superbia vitae* but the underlying Pride that begets all sin. The range of expressions is remarkable: 'superbis', 'non eram ego talis . . . gressus', 'tumor . . . meus', 'dedignabar esse parvulus', and 'turgidus fastu'.

Itaque: See on 3. 4. 8; he turns to scripture to combine his new zeal for philosophy with his need for the *nomen Christi*.

videre . . . video [cf. 'visa est', 'acies mea']**:** The approach to scripture is made out of *curiositas*, i.e. *concupiscentia oculorum*.

[3] On the Bema, see C. R. C. Allberry, *Zschr. für die neutest. Wiss.* 37 (1938), 2–10, and J. Ries, *REAug* 22 (1976), 218–33. Mani died in prison on 26 February 277; the feast celebrated his suffering and death, and therewith the forgiveness of sins (and as such reflected an annual institution from Mani's own lifetime), and finally offered a renewal of the great revelation of the Manichean gnosis.

ecce video: Courcelle, *Recherches*, 38 n. 6, takes 'non enim sicut modo loquor . . .' here as proof that in *this* line 'ecce video' is only a manner of speech; *hence* he wants to take the 'ecce audio' of the garden scene (8. 12. 29) likewise. Here 'ecce video' is true present (cf. the next sentence), whereas in 8. 12. 29 'ecce audio' is historic present.

compertam CDO Maur. Knöll Skut. Ver. Pell. : *confectam* GS Vega (who incorrectly reads this in *O*).

velatam mysteriis: *s.* 51. 4. 5, 'honora in eo quod nondum intellegis; et tanto magis honora, quanto plura vela cernis. quanto enim quisque honoratior est, tanto plura vela pendent in domo eius. vela faciunt honorem secreti, sed honorantibus levantur vela. inridentes autem vela, et a velorum vicinitate pelluntur. quia ergo transimus ad Christum, aufertur velamen.'

The description of scripture, in terms reminiscent of the theophany of a goddess, is curious and forceful. One must bow the head to enter, but then the space revealed opens out in height and mystery. 'Tumor' below answers to 'humilem' here, and 'interiora' to 'velatam'.

non eram ego talis: *s.* 51. 5. 6, 'loquor vobis, aliquando deceptus, cum primo puer ad divinas scripturas ante vellem affere acumen discutiendi quam pietatem quaerendi: ego ipse contra me perversis moribus claudebam ianuam domini mei: cum pulsare deberem, ut aperiretur [cf. Mt. 7: 7, and see on 1. 1. 1], audebam [so Pellegrino, *Les Confessions*, 93 n. 26, against Maur. *addebam*], ut clauderetur. *superbus* enim audebam quaerere quod nisi humilis non potest invenire. quanto vos beatiores estis modo! quam securi discitis, quam tuti, quicumque adhuc *parvuli estis in nido fidei* [cf. 4. 16. 31, 'parvulis tuis'], et spiritalem escam accipitis! ego autem miser, cum me ad volandum idoneum putarem, reliqui nidum, et prius cecidi quam volarem. sed dominus misericors me, a transeuntibus ne conculcarer et morerer, levavit et in nido reposuit. haec enim me perturbaverunt, quae modo vobis securus in nomine domini et propono et expono.'

On *s.* 51 cf. Courcelle, *Recherches*, 61–2 (with 62 n. 2 on the use here of *puer* to describe A. at age 18). He goes further (63) to say that the 'principale difficulté sur laquelle ait achoppé le jeune Augustin' was the discordance of the nativity stories and the genealogies. The Manichees made much of this (cf. Alfaric, 199–203), for if they could impugn the nativity, they could dispense with the virgin birth and the physical incarnation of God. (Such criticism was regarded later by A. as an attempt to play upon *curiositas* to lead the naïve astray: *cons. ev.* 2. 1 promises to discuss the relations of the gospel narratives 'ne quid ex hoc in fide christiana offendiculi patiantur qui *curiosiores* quam capaciores sunt, quod non utcumque perlectis sed quasi diligentius perscrutatis evangelicis libris inconvenientia quaedam et repugnantia se deprehendisse existimantes magis ea contentiose obiectanda quam prudenter consideranda esse arbitrantur.' Manichean tactics are more explicitly but more briefly described at *agon.* 4. 4.)

But it goes too far to call this the principal difficulty—there were others, and to judge by the analysis he gives in a few paragraphs of the threads that bound him to Manicheism, others were equally pressing. The retrospective selection of questions undermining the historical incarnation is consistent, on the other hand, with a central feature of *conf.*: from 7. 9. 14 through Bk. 8, the central issue is A.'s attempt to find a place—and the right place—for the incarnate Christ in his theology. If we are justified in reading anything of *s.* 51 into the narrative here, it is that the mature A. thought that it was lack of faith in the incarnate Christ that sent him astray in youth. (At 1. 11. 17, his childhood religion did contain the incarnation ('per humilitatem domini dei nostri descendentis'), but was not taken to its logical conclusion in baptism.)

At *doctr. chr.* 2. 8. 12 ff., A. outlines the proper way to approach scripture, which has the effect of offering indirect negative commentary on his approach here. He emphasizes reading all of canonical scripture, avoiding non-canonical books, emphasizing first 'praecepta vivendi vel regulae credendi' (2. 9. 14), and putting off more obscure passages for later consideration.

inclinare cervicem: *en. Ps.* 8. 8, 'inclinavit ergo scripturas deus usque ad infantium et lactentium capacitatem, sicut in alio psalmo canitur, "et inclinavit caelum et descendit". et hoc fecit propter inimicos, qui per superbiam loquacitatis inimici crucis Christi, etiam cum aliqua vera dicunt, *parvulis* tamen et lactentibus prodesse non possunt.'

indigna: cf. Hier. *ep.* 22. 30, to whom it was the language that offended: 'sermo horrebat incultus'. *doctr. chr.* 2. 14. 21, 'tanta est vis consuetudinis etiam ad discendum ut, qui in scripturis sanctis quodammodo nutriti educatique sunt, magis alias locutiones mirentur easque minus latinas putent quam illas quas in scripturis didicerunt, neque in latinae linguae auctoribus reperiuntur.' At *cat. rud.* 8. 12–9. 13, he envisions two types of catechumen, the most learned and the somewhat learned: he himself matches the first category, but it is from the second category that he expects fastidiousness: (9. 13) 'maxime autem isti docendi sunt scripturas audire divinas, ne sordeat eis solidum eloquium, quia non est inflatum, neque arbitrentur carnalibus integumentis involuta atque operta dicta vel facta hominum, quae in illis libris leguntur, non evolvenda atque aperienda ut intellegantur, sed sic accipienda ut litterae sonant; deque ipsa utilitate secreti, unde etiam mysteria vocantur, quid valeant aenigmatum latebrae ad amorem veritatis acuendum discutiendumque fastidii torporem, ipsa experientia probandum est talibus, cum aliquid eis quod in promptu positum non ita movebat, enodatione allegoriae alicuius eruitur. . . . ita enim non inridebunt, si aliquos antistites et ministros ecclesiae forte animadverterint vel cum barbarismis et soloecismis deum invocare, vel eadem verba quae pronuntiant non intellegere perturbateque distinguere.'

autem CDGO Ver. : *tamen* S Knöll Skut. Vega Pell.

parvulis: Cf. *en. Ps.* 8. 8, quoted above.

turgidus fastu: See on 7. 9. 13, 'typho turgidum', of the unnamed individual who procured for him the *platonicorum libri* (that individual, at a time when A. is reading philosophy and turning to scripture, has the qualities that made A. a bad reader here); cf. 3. 3. 6, 'tumebam typho' (→ here, 'tumor meus').

3. 6. 10

An early reader, Secundinus in *ep. Sec.* 3, 'legit enim aliquanta exile meum et qualecumque Romani hominis ingenium reverendae tuae dignationis scripta, in quibus sic irasceris veritati [cf. 'veritas et veritas' here] ut philosophiae Hortensius. . . . in medium solis ac lunae inventus es accusator [cf. 'sol et luna' here].'

The extensive food-metaphor reflects the practice of the Manichean elect, who consumed particles of the divine in their banquets. At 3. 1. 1, a parallel use of metaphor ('famis mihi erat') marked A.'s isolation from authentic nourishment; here now he ingests all manner of false victuals. The emphasis on *phantasmata* is likewise apt: *curiositas* has led him into a world of images—eye-food, images of things that never existed. The paragraph thus moves from imagery drawn from *concupiscentia carnis* (wolfish feeding on food that does not satisfy) to imagery drawn from *concupiscentia oculorum* (greedy gazing at *phantasmata* that are hollow and empty): all against a backdrop of empty words.

In this commentary no attempt will be made to provide systematic exposition of Manichean doctrines or assessment of the accuracy of A.'s reports—others have done that; it will be enough to provide parallels from A.'s writings that illustrate how he understood the particular doctrines that he mentions or to which he alludes here. A brief summary[4] from A.'s hand: *ep.* 236. 2, 'auditores autem qui appellantur apud eos et carnibus vescuntur et agros colunt et, si voluerint, uxores habent, quorum nihil faciunt qui electi vocantur. sed ipsi auditores ante electos genua figunt, ut eis manus supplicibus imponatur non a solis presbyteris vel episcopis aut diaconis eorum sed a quibuslibet electis. solem etiam et lunam cum eis adorant et orant. die quoque dominico cum illis ieiunant et omnes blasphemias cum illis credunt, quibus manichaeorum haeresis detestanda est, negantes scilicet Christum natum esse de virgine nec eius carnem veram confitentes fuisse sed falsam ac per hoc et falsam eius passionem et nullam resurrectionem fuisse contendunt. patriarchas prophetasque blasphemant. legem per famulum dei Moysen datam non a vero deo dicunt sed a principe tenebrarum. animas non solum hominum sed etiam pecorum de dei esse substantia et omnino partes dei esse arbitrantur. deum denique bonum et verum dicunt cum tenebrarum gente pugnasse et partem

[4] A.'s own preference among his anti-Manichean works fell upon the *c. Sec.* (*retr.* 2. 10), which is among the least well-known to moderns.

suam tenebrarum principibus miscuisse eamque toto mundo inquinatam et ligatam per cibos electorum suorum et per solem ac lunam purgari adseverant et, quod purgari de ipsa dei parte non potuerit, in fine saeculi aeterno ac poenali vinculo conligari, ut non solum violabilis et corruptibilis et contaminabilis credatur deus, cuius pars potuit ad mala tanta perduci, sed non possit saltem totus a tanta coinquinatione et immunditia et miseria vel in saeculi fine purgari.'

We may surmise the appeal the Manichees had for a candidate like A. from some of his later writings, e.g. *mor.* 1. 10. 16, 'longe prorsus aliter quam putatis lex et prophetae intelleguntur a nobis. desinite errare, non colimus paenitentem deum, non invidum, non indigum, non crudelem, non quaerentem de hominum vel pecorum sanguine voluptatem, non cui flagitia et scelera placeant, non possessionem suam terrae quadam particula terminantem. in has enim atque huiusmodi nugas graviter copioseque invehi soletis.' Most importantly, Manicheism presented itself to A. as a form of Christianity; recent work (see Koenen's article cited below, esp. p. 163) makes it easier to understand how that could be.

How far A. penetrated into the cult is far from clear; only in Italy does he suggest that he had to do with the elect (5. 10. 18), and even there he never became one. Near the end of his life, he will put together a sketch of Manicheism (at *haer.* 46) that includes the charge that the elect took the eucharist sprinkled with semen (interjecting the intriguing note that the Manichees deny the practice but claim that it is done by other people who *claim* to be Manichees), this on the evidence of an eleven-year-old girl from Carthage (hence, though most of *haer.* is derivative, the charge probably comes from within A.'s own ken, whether before or after his separation from the sect). Similarly at *c. Fort.* 3, A. says that he has heard that the Manichees celebrate eucharist, but as an auditor he never saw it himself. For the possibility that he thought of becoming one of the elect, see on 5. 7. 13, 'ceterum conatus omnis meus'. Note that the Manichees A. knew denied the efficacy of baptism by water (cf. A. *haer.* 46. 17). If there had been any inclination on A.'s part to take that sacrament then (aged 18) or not long after, this Manichee rejection undoubtedly helped postpone it. This had the unexpected effect of avoiding theological scruples later as to the validity of an earlier baptism, and made the sacrament itself the issue of A.'s crisis recounted in Bk. 8.

For fuller treatments, recent works complement each other in various ways. Most accessible in English, with narrative ranging far beyond A.'s period, is S. N. C. Lieu, *Manichaeism* (Manchester, 1985). The most exciting work on the origins and early history of Manicheism has been the rediscovery and edition of the Cologne Mani-Codex (A. Henrichs, L. Koenen, and C. Römer, *Der Kölner Mani-Kodex* (Opladen, 1988), a critical edition of the complete text with references to numerous earlier publications since the codex was discovered). See generally A. Henrichs, *HSCP* 83 (1979), 339–67, and for its import for

Augustinian studies, L. Koenen, *Illinois Class. Stud.* 3 (1978), 154–95. More directly applicable are the works of F. Decret: *Aspects du manichéisme dans l'Afrique romaine* (Paris 1970), dealing mainly with *c. Fort.*, *c. Faust.*, and *c. Fel.*, and his *L'Afrique manichéenne (IVe–Ve siècles)* (Paris, 1978), founded on a study of the other anti-Manichean works of A. See also P. Brown, *JRS* 59 (1969), 92–103 (= Brown, *Religion and Society in the Age of Saint Augustine* (London, 1972), 94–118). A.'s attacks on Manicheism are placed in a wider late Roman (both Christian and non-Christian) context by the articles of S. N. C. Lieu, *Bull. Rylands*, 68 (1986), 434–72; 69 (1986), 235–75. Manicheism in Africa is a fourth-century phenomenon, but its extent is hard to grasp; see Decret, *L'Afrique*; note especially his observation that there is no mention of the Manichees in Africa owning church buildings or slaves (though (Decret, 203) in Turkestan they had veritable monasteries). It is possible (as Decret, 182, suggests) that the scandal of the Donatist/catholic schism turned some, abandoning the literal-minded gospel that appealed to their countrymen, aside to Manicheism. Finally, Alfaric's *L'Évolution intellectuelle*, for all that its central thesis has been exploded, is rich in its integration of materials drawn from the whole range of A.'s works, and especially from his anti-Manichean writings of all periods, into the biographical enterprise, and on those counts it has not yet been surpassed.

incidi: His other reports of this fall[5] offer perspective; see esp. *beata v.* 1. 4 (quoted in Introduction: note '*superstitio* quaedam puerilis me ab ipsa inquisitione terrebat' and '*incidi* in homines'). Also in Introduction are *util. cred.* 1. 2 (note 'in tales homines *incidisse*' and '*superstitione* terreri') and *util. cred.* 8. 20 ('in illa secta, in quam me *incidisse* paenitebat'). See also *duab. an.* 1. 1, 'multa enim erant quae facere debui, ne tam facile ac *diebus paucis* [!] religionis verissimae semina mihi a pueritia salubriter insita errore vel fraude *falsorum fallaciumve* hominum effossa ex animo pellerentur'; *duab. an.* 15. 24 speaks of the Manichees as 'homines cum quibus mihi a *pueritia* in omni convictu fuit summa consensio', adjusting the facts slightly to make himself look a little less responsible for his fall. He suspects the Manichees, as he would later suspect the Academics, of concealing a secret doctrine; he would go to Ambrose privately (see on 6. 3. 3), expecting a secret doctrine, only to be disabused.

The *superstitio* in the passages quoted has given scandal to those who cannot accept that A. would ever apply the term—as Alfaric, 70 n. 7, seems to have been the first to insist he did—to the Christianity of his childhood. Courcelle, *Recherches*, 64–5, even assumes that *util. cred.* represents his later attitude (dislocating the awkward word into represented discourse), but that *beata v.* shows him sharing the attitude of the Manichees in the face of those

[5] It might be fruitful to compare A.'s use of the imagery of 'weight' (see on 13. 9. 10) with his repeated use of this verb for leaving one's appropriate 'place' in the world in favour of a less suitable one.

orthodox Christians who found intellectual inquiry into the truths of faith disquieting. Cf. also *BA* 13. 126–7 n.; *SLA* s.v. takes *beata v.* as innocuous and particular, without reference to any sect. But at all periods A. saw that the boundary between *superstitio* and *vera religio* was not exactly the same as the boundary between Christianity and all that lay beyond. His fastidiousness was, to be sure, greater at Cassiciacum and he became more tolerant (perhaps he would have suggested 'more discriminating') when he was bishop of a socially diverse church. For our immediate purposes, it seems clear that a *Manichean* charge of superstition (see also on 5. 6. 11) levelled against the religion of A.'s boyhood could have found in the eighteen-year-old student a sympathetic ear.

(1) At *ord.* 1. 8. 23, Licentius uses the same word to describe one aspect of Monnica's religion, her distaste at his singing a psalm verse in the outhouse.

(2) A more conventional attitude is attested at *c. acad.* 2. 3. 8 ('si quid *superstitionis* in animum revolutum est, eicietur profecto'), but it shows the hostility that could have fastened on aspects of traditional Christianity.

(3) In *ep.* 10. 2, addressed to Nebridius and famous otherwise for its Platonic vocabulary ('deificari in otio': cf. Folliet, *RA* 2 (1962), 225–36), the same fastidiousness is present: 'magna secessione a tumultu rerum labentium, mihi crede, opus est, ut non duritia, non audacia, non cupiditate inanis gloriae, non *superstitiosa* credulitate fiat in homine nihil timere.'

(4) By way of paying back old debts, in the *mor.* the Manichees were branded as superstitious (1. 29. 59, 1. 34. 74, 2. 15. 36, 2. 16. 52, 2. 19. 68), but even there the possibility of superstition as one of the defects to be found among the orthodox is admitted: 1. 34. 75, 'nolite consectari turbas imperitorum, qui vel in ipsa vera religione *superstitiosi* sunt, ... novi multos esse sepulcrorum et picturarum adoratores.' Cf. *Gn. litt.* 1. 21. 41, 'ut neque falsae philosophiae *loquacitate* [see on 1. 4. 4] seducamur neque falsae religionis superstitione terreamur.' (See *ep.* 54. 2. 3 cited below. For further application of the term to the Manichees, see 4. 1. 1.)

(5) By 391, his definition is ruled by Christian interpretations (esp. of Rom. 1: 25): *vera rel.* 20. 39, 'superstitio malum est qua creaturae potius quam creatori servitur.'

(6) But shortly before *conf.*, it shows more nuance and room for ambiguity: *doctr. chr.* 3. 12. 18, 'quisquis autem rebus praetereuntibus restrictius utitur quam sese habent mores eorum cum quibus vivit, aut temperans aut superstitiosus est.'

(7) At *ep.* 54. 2. 3 (400), A. recounts to Januarius the story of Monnica's difficulties adjusting to Milanese church customs (see on 6. 2. 2) and attributes some of the scruples that arise in such cases to *superstitiosa timiditas*.

(8) At various times, *gentiles* (6. 2. 2, 'superstitioni gentilium essent simillima'), idol-worshippers, Donatists (*brev.* 3. 2. 2), and Jews are also classed as 'superstitious.' A.'s programme as bishop is to leave *superstitio* with

the old ways: *s. Frang.* 8. 5, 'ut vetus superstitio consummetur et nova religio perficietur'. In after years, we also find echoes of Seneca's *de superstitione* at *civ.* 6. 10–11, the only significant appearance of Seneca in all A.'s work—probably a specific work sought out because of the subject (see on 5. 6. 11).

superbe delirantes [1], **carnales nimis** [3] **et loquaces** [2]: That trinitarian counter-pattern anticipates the rejection of the Manichean trinity that follows. A. means to couple *concupiscentia carnis* and *curiositas* here, as the vices of the Manichees to which A. was drawn; they were not without their *ambitio saeculi*, either—they were after all the ones to get A. his interview with Symmachus (5. 13. 23). *civ.* 1. 20, 'num igitur ob hoc, cum audimus "non occides", virgultum vellere nefas ducimus et manichaeorum errori insanissime adquiescimus? his igitur *deliramentis* remotis cum legimus, "non occides", . . . restat ut de homine intellegamus.'

loquaces: The use of *loquax* to depreciate both his own rhetorical achievements and the babblings of the sectarians he now loathed is a sign of how closely those two phases of his life were linked in memory and obloquy: of his own skills, 4. 2. 2, 8. 5. 10 ('loquacem scholam'), 9. 2. 2; of the Manichees, 1. 4. 4 (see on 'quoniam loquaces muti sunt'), 5. 7. 12, 5. 9. 17, 7. 2. 3.

viscum: *util. cred.* 1. 2, 'itaque nobis faciebant, quod insidiosi aucupes solent, qui *viscatos* surculos propter aquam defigunt, ut sitientes aves decipiant.' Sim. at *c. Faust.* 13. 17 ('hoc melle scilicet venenati sui poculi').

laquei diaboli: 1 Tim. 3: 7, 'ut non incidat in opprobrium et laquem diaboli'; 1 Tim. 6: 9, 'nam qui volunt divites fieri, *incidunt* in temptationem et laqueum et *desideria multa* et nociva'; sim. at 2 Tim. 2: 26. These *laquei* are reminiscent of the *muscipulae* that litter false paths at 3. 1. 1 (see on 'viam sine muscipulis', esp. for *en. Ps.* 90. s. 1. 4). Other *laquei* at 4. 6. 11, 5. 7. 13, 6. 12. 21, 10. 31. 44, 10. 34. 52, 10. 36. 59.

nominis tui et domini Iesu Christi et paracleti consolatoris nostri spiritus sancti: The Manichean trinity: *c. Faust.* 20. 2 (quoting Faustus), 'igitur nos patris quidem dei omnipotentis et Christi filii eius et spiritus sancti unum idemque sub triplici appellatione colimus numen. sed patrem quidem ipsum lucem incolere credimus summam ac principalem, quam Paulus alias inaccessibilem vocat [1 Tim. 6: 16], filium vero in hac secunda ac visibili luce consistere. qui quoniam sit et ipse geminus, ut eum apostolus novit Christum dicens esse dei virtutem et dei sapientiam [1 Cor. 1: 24], virtutem quidem eius in sole habitare credimus, sapientiam vero in luna. necnon et spiritus sancti, qui est maiestas tertia, aeris hunc omnem ambitum sedem fatemur ac diversorium; cuius ex viribus ac spiritali profusione terram quoque concipientem gignere patibilem Iesum, qui est vita ac salus hominum, omni suspensus ex ligno. quapropter et nobis circa universa et vobis similiter erga panem et calicem par religio est.'

paracleti: Cf. Jn. 14: 26, 'paracletus autem spiritus sanctus, quem mittet pater in nomine meo'. Decret, *Aspects*, 294: 'Mais rien n'autorise à conclure que

Felix confonde ce Mani-Paraclet avec le *Spiritus sanctus* qui, selon la déclaration de Faustus, siège dans l'"Air ambiant", et moins encore avec le Saint-Esprit qui, dans la doctrine catholique, constitue la troisième personne de la Trinité divine. . . . Il semble que Felix considère Mani comme le dernier apôtre promis et envoyé par le Christ et *en* cet Apôtre parle l'"Esprit-Saint".' Cf. *c. Faust.* 13. 17, 'cum enim Christus promiserit suis missurum se paracletum, id est consolatorem vel advocatum, spiritum veritatis, per hanc promissionis occasionem hunc paracletum dicentes *esse Manichaeum vel in Manichaeo*' (note here 'in se'). So also Lieu, *Bull. Rylands* 68 (1986), 444: 'Augustine does not seem to have fully perceived Mani's identification with the Paraclete through his *syzygos*. Instead he saw Mani's claim purely in terms of the Catholic understanding of the Trinity and the Incarnation. In the same way as in Catholic doctrine the Eternal Son of God had taken on humanity in Jesus Christ, who was therefore called the Son of God, so in Augustine's eyes Mani claimed the title of Paraclete because in his person the Holy Ghost had taken on humanity. . . . On the other hand, Augustine's understanding of Mani's identification with the Paraclete might not have been too distant from the contemporary Manichaean view. In the Manichaean *Psalmbook*, the Father of Light, Jesus the Splendour and Mani the Paraclete were seen as a form of Trinity.' (The only documented surviving Manichean place of worship, near the coast of the China Sea, presents a Mani who has been transfigured into the 'Buddha of Light'; Lieu, *Manichaeism*, 212–13, reports that shrine, but further publication of that (and possibly other surviving sites) should be eagerly awaited; unpublished photographs taken by travellers are of great interest.)

non recedebant de ore eorum: Cf. Jos. 1: 8, 'non recedat volumen legis huius ab ore tuo. sed meditaberis in eo diebus ac noctibus, ut custodias et facias omnia quae scripta sunt in eo.' Empty words (cf. 'commixtione syllabarum nominis' and 'voce sola'), letter without spirit; the remedy to come in Bk. 7, where *curiositas* is cured. Where the Joshua text speaks of the *volumen legis*, the Manis only manage the *nomina*. J. Ries, *Lectio I–II*, 9, has numerous quotations from the *Kephalaion* of the Manichees for the role of Wisdom in their thought. In following pages of that article, he provides similar nests of citations for other names/phrases (God the Father, Jesus) instanced in *conf.* from Manichean sources. Those citations offer confirmation that A. speaks correctly, but show that Manicheism left little permanent trace on the *imagination* of A.

cor inane veri: Cf. 4. 14. 23, 'cor vanum et inane soliditatis tuae'.

veritas et veritas: Jn. 14: 6, 'ego sum via et veritas et vita.' On Truth among the Manichees, Ries, *Lectio I–II*, 14–16: 'La fête annuelle de Bêma est un jugement où le pardon est accordé à chaque élu en conformité avec ses relations avec la vérité.'

huius mundi CDGO Maur. Ver. Pell. : *mundi* S Knöll Skut. Cf. Col. 2: 8, quoted at 3. 5. 9, esp. 'secundum elementa mundi'; but cf. also 8. 2. 3, where Simplicianus praises the Platonists in the same terms as the philosophers A. means to indicate here.

prae amore tuo [3], **mi pater summe bone** [1], **pulchritudo pulchrorum omnium** [2]: Invocation of trinity (countering the Manichean trinity above) precedes extended address to the second person of the trinity. *For pater summe bone* cf. 10. 31. 46, 10. 43. 69, 11. 22. 28, 13. 15. 17.

veritas . . . tibi: The address to the second person of the trinity continues to 'et non pascebar', marked by such expressions as 'sicut nunc mihi *locuta es*'.

suspirabant tibi: With dative in Latin, *suspiro* gives 'soupirer après' (Mohrmann, *Vig. Chr.* 5 (1951), 252).

sol et luna: *c. Faust.* 20. 2 quoted above; at *c. Fort.* 3 A. reports that the Manichees pray 'contra solem'. *util. cred.* 6. 13 reports an anecdote from Manichean days of a woman who was irked by Manichean praise of the sun and so crossed to where the sun shone through a window and stomped on the ground, crying 'ecce solem deumque tuum calco'.

nec ipsa prima: not themselves the first creatures (Gn. 1: 14–18: the fourth day of creation; cf. 13. 19. 24); cf. lines following here.

te veritas CDGO Maur. Ver. : *veritas* S Knöll Skut. Pell.

in qua . . . obumbratio: Jas. 1: 17, 'descendens a patre luminum, apud quem non est commutatio nec moment obumbratio' (see on 4. 15. 25). Sensitivity to the context from which a scriptural echo comes is often essential. It is hard to believe that A. does not fasten on this passage at least in part because of the 'father of lights' it invokes, taken in an anti-Manichean sense.

esuriebam et sitiebam: The combination 'hunger and thirst' is so conventional that biblical parallels are at best dimly illustrative, e.g. Mt. 5: 6, 'beati qui esuriunt et sitiunt iustitiam', or 1 Cor. 4: 11, 'usque in hanc horam et esurimus et sitimus'.

phantasmata splendida: *nat. b.* 42 quotes the authoritative Manichean text A. knew as *epistula quam vocant 'fundamenti'*: 'ita autem fundata sunt eiusdem *splendidissima* regna supra lucidam et beatam terram, ut a nullo unquam aut moveri aut concuti possint. . . . lucis vero beatissimae pater . . .' Ries, *Lectio I–II*, 17, specifies further links to the Manichean 'théologie solaire'; cf. 'inferebatur pro te sol et luna' above.

animo decepto per oculos: By *curiositas/concupiscentia oculorum.*

terrestria . . . volatilibus: Punctuation follows Vega; others punctuate strongly after *terrestria*, then not at all after *volatilibus*.

imaginamur: See on 3. 2. 2, 'imaginibus'.

suspicamur: The verb implies preliminary and incomplete vision, inchoate knowledge (12. 6. 6, 'sed nosse cupiebam, non suspicar'). *ep.* 7. 3. 6 (to Nebridius), 'licet igitur animae imaginanti ex his quae illi sensus invexit demendo, ut dictum est, et addendo ea gignere quae nullo sensu attingit tota.'

amor meus: This vocative, coming after an extended address to the second person of the trinity, personified as Truth, strongly suggests that A. now turns to address the third person, personified as Love.

in quem deficio: Cf. Ps. 118: 81, 'defecit in salutare tuum anima mea, et in verbum tuum speravi'; *en. Ps.* 70. s. 1. 14, 'hoc illis expedit ut peccata sua cognoscant, unde confundantur et *deficiant*, male enim de viribus suis praesumebant; et ipsi *defecti* dicant, "quando infirmor, tunc potens sum" [2 Cor. 12: 10].'

nec ea quae non videmus ibi: G–M: '"corpora" is *not* to be understood. The reference is to the "caelum caeli", the abode of the angels.' We need scarcely insist upon that equation here; the phrase 'spiritalia opera tua' above is sufficient reference; cf. Rom. 1: 20, 'invisibilia dei' (see on 7. 9. 14).

phantasmatis[1] GOS Knöll Skut. Ver. Pell. : *phantasmatibus* CD.

phantasmatis[2] OS Knöll Skut. Ver. Pell. : *a phantasmatis* GV : *phantasmatibus* CD (Skutella's apparatus is incomplete here).

Phantasma and *phantasia* in A.: *sol.* 2. 20. 34 does not yet distinguish (cf. *ep.* 7. 2. 4 (quoted below on 3. 6. 11) and *lib. arb.* 2. 8. 23): 'quid intersit inter veram figuram, quae intellegentia continetur, et eam quam sibi fingit cogitatio, quae graece sive phantasia sive phantasma dicitur'. *mus.* 6. 11. 32 introduces the distinction that is afterwards authoritative, 'aliter enim cogito patrem meum quem saepe vidi, aliter avum quem numquam vidi. horum primum *phantasia* est, alterum *phantasma*'; sim. at *trin.* 8. 6. 9, 9. 6. 10.

See du Roy, 180 n. 7, for the debate whether the distinction is Porphyrian or Plotinian in its influence on A., confuting J. Pépin on the significance of 'Alexandria' in the example A. regularly provides (of a place he has not seen, hence a *phantasma*): Pépin, *REA* 86 (1954), 102–3, takes the name of the city as a sign of Porphyrian origin, but du Roy counters that the city is mentioned elsewhere (and early: *c. acad.* 1. 4. 11–12) under mediated Stoic influence; du Roy also offers a long list of Augustinian texts in which it recurs. du Roy is followed by O'Daly, 106–30, esp. on the origins of the terms at 106–7. He well remarks that the history of *phantasia* is ambiguous, inasmuch as it (O'Daly, 107) 'can refer to mental faculties or processes as well as to their products': Nebridius at *ep.* 6. 2 uses it in the sense of mental power or faculty—'perhaps the less surprising if we recall the technicality of N.'s philosophical interest in such problems as the possible demonic cause of dreams (*ep.* 8) and the question of the soul's vehicle (*ep.* 13. 2). . . . We can only conclude that A. either adopted a scholastic or doxographical distinction unknown to us, or that he himself adapted the Stoic distinction, referred to above, between *phantasia* and *phantasma* to his own technical purposes.' Note that Cicero treated *phantasia* as an un-naturalized Grecism, regarding *visum* as the appropriate translation (*acad. post.* 1. 11. 40, 'phantasian, nos visum appellemus licet'—see on 6. 1. 1).

Phantasmata are of importance in three ways to A.'s discussion of his

Manichean past. (1) The Manichees themselves believed that the crucified Jesus was a docetic illusion, not the real death of the God-man (9. 3. 6, 'ut veritatis filii tui carnem phantasma crederet'; at *Io. ev. tr.* 50. 5, A. interprets the hesitations of the disciples in the presence of the risen Christ as arising from a suspicion that they see only a *phantasma*, and at *s.* 75. 1. 1–8. 9, he discusses and dismisses the possibility that Christ seen walking on the water in Mt. 14: 24–33 was only a *phantasma*). (2) A. himself in retrospect believed that the God he worshipped among the Manichees was a wraith of imagination and no substance (4. 7. 12, 'sed vanum phantasma et error meus erat deus meus'). (3) Particularly in *c. ep. fund.*, he applied the term broadly to Manichean doctrines: *c. ep. fund.* 18. 20, 19. 21, 32. 35, 36. 41, 43. 49 ('detestemur istam haeresim, quae suorum phantasmatum fidem secuta'). Cf. *vera rel.* 55. 108 (*retr.* 1. 13. 1, 'maxime tamen contra duas naturas manichaeorum liber hic loquitur'): 'non sit nobis religio in phantasmatibus nostris. melius est enim qualecumque verum quam omne quidquid pro arbitrio fingi potest.'

vita es animarum: Sim. at 7. 1. 2, 10. 6. 10, 10. 20. 29; and at *en. Ps.* 70. s. 2. 3, *Io. ev. tr.* 19. 12, *ss.* 62. 1. 2, 65. 4. 5, 156. 6. 6, 161. 6. 6, 180. 7. 8 (most of the passages collected by Madec *BA* 6. 352 n.). G–M and Vega suggest Plot. 5. 3. 14. 14–5, ὁ ἔνδον νοῦς, ὁ δοὺς οὐσίαν καὶ τὰ ἄλλα.

vita vitarum: For this apostrophe, see on 1. 4. 4. Read with 'veritas, veritas' (< Jn. 14: 6, 'vita, veritas et via'), earlier in this paragraph; and cf. 1. 4. 4, 'immutabilis'.

et non mutaris: See on 1. 6. 10, 'et non mutaris'.

3. 6. 11

Rather than provide a bald summary of Manichean doctrine, A. sets in motion an impressionistic sketch, juxtaposing representatives of the Manichee, the classical, and the orthodox Christian textual traditions.

longe: See on 1. 18. 28.

peregrinabar: 2 Cor. 5: 6, 'dum sumus in corpore, peregrinamur a domino.' *Peregrinus/peregrinor* in this characteristically Augustinian sense occurs only here in the first nine books of *conf.*, except that the last lines of 9. 13. 37 ('peregrinatio populi tui') anticipate its more frequent use in 10–13: cf. 10. 4. 6, 10. 5. 7, 11. 2. 4, 12. 11. 13, 12. 15. 21, 12. 16. 23, 13. 14. 15.

a siliquis porcorum: Lk. 15: 16, 'et cupiebat implere ventrem suum de siliquis quas porci manducabant.' Deliberate reinvocation of the prodigal just here gives shape to the narrative. The *siliquae* are interpreted with remarkable consistency among the Latin fathers (a reading Origenian in origin? Cf. B. Blumenkranz, *Mélanges . . . Louis Halphen* (Paris, 1951), 11–17): Amb. *in Luc.* 7. 218, 'sunt qui porcos accipiant pro gregibus daemonum, siliquas pro exili virtute inanium hominum sermonumque iactantia qui nihil prodesse possunt, inani quadam philosophiae seductione et quodam sonorum facundiae plausu pompam magis quam utilitatem aliquam demonstrantes'; v. sim. at Hier. *ep.* 21. 13 (21. 4–38 expounds the prodigal to Damasus), and Caes. Arel. *s.* 163 (closely modelled on A.? Cf. B. Blumenkranz, *Vig. Chr.* 2 (1948), 102–5). In A.'s works, *qu. ev.* 2. 33, 'siliquae quibus porcos pascebat, saeculares doctrinae, sterili vanitate resonantes'; sim. at *s. Caill.* 2. 11. 3, *en. Ps.* 16. 13, 18. en. 2. 3 ('daemonum cultor, tamquam porcorum pastor'), 95. 5 ('pascebat porcos, id est, colebat daemonia'), 138. 7.

Medea volans: Typical example of the pernicious falsehood of fiction: *sol.* 2. 15. 29, '[A.] non enim, cum dicitur iunctis alitibus anguibus Medeam volasse, ulla ex parte res ista verum imitatur, quippe quae nulla sit nec imitari aliquid possit ea res quae omnino non sit. . . . itane tandem cum audio, "angues ingentes alites iuncti iugo", non dico falsum? . . . [Ratio] similiter enuntiaretur, etiamsi vere illud Medea fecisset. imitatur ergo ipsa enuntiatione veras sententias falsa sententia.' The link to the present passage is *ep.* 7. 2. 4 (of 389, to Nebridius), 'ego enim mihi, ut libet atque ut occurrit animo, Aeneae faciem fingo, ego Medeae cum suis anguibus alitibus iunctis iugo, ego Chremetis et alicuius Parmenonis. in hoc genere sunt etiam illa quae sive sapientes aliquid veri talibus involventes figuris sive stulti, variarum superstitionum conditores, pro vero attulerunt, ut est tartareus Phlegethon et *quinque antra gentis tenebrarum* et stilus septentrionalis continens caelum et alia poetarum atque haereticorum mille portenta.' A. gets the quotation (a line of Pacuvius) and the treatment from Cic. *inv.* 1. 27. Medea represents particularly heinous evil: *c. Sec.* 26, 'in ipsis nocentibus pauciores sunt homicidae quam fures, pauciores sunt incestatores quam adulteri; denique etiam ipsae antiquorum vel fabulae vel historiae pauciores habent Medeas et Phaedras quam facinorum aliorum flagitiorumque mulieres.'

quinque elementa: *haer.* 46. 7, 'quinque enim elementa quae genuerunt principes proprios genti tribuunt tenebrarum, eaque elementa his nominibus nuncupant: fumum, tenebras, ignem, aquam, ventum. in fumo nata animalia bipedia, unde homines ducere originem censent; in tenebris, serpentia, in igne, quadrupedia, in aquis, natatilia, in vento, volatilia. his quinque elementis malis debellandis alia quinque elementa de regno et substantia dei missa esse, et in illa pugna fuisse permixta fumo aerem, tenebris lucem, igni malo ignem bonum, aquae malae aquam bonam, vento malo ventum bonum.' For the same doctrine, cf. *c. ep. fund.* 28. 31, with excoriation by A. at *c. Faust.* 2. 3–4.

quinque antra tenebrarum: *c. ep. fund.* 18. 20, 'quid iam de illa loquar potentia qua intellegitur veritas, qua istis ipsis quae de corporis sensu haustae figurantur imaginibus sese pro veritate opponentibus magna vivacitate resistitur? . . . ex eadem venire facilitate istam terram lucis per spatia infinita diffusam et *quinque antra gentis tenebrarum* cum habitatoribus suis, in quibus manichaei *phantasmata* veritatis sibi nomen ausa sunt usurpare.'

omnino CDS Knöll Skut. Ver. Pell. : *omnia* GO.

versum et carmen: G–M: 'According to the Benedictine editors it was upon this passage that Petilianus . . . founded his charge against A. of administering love charms ("amatoria maleficia data mulieri"). *c. litt. Pet.* 3. 16. 19. The identification does not appear certain. A.'s meaning seems to be that he could gain his bread by literary pursuits.' (1) The Maurists were certainly wrong on this. (2) The meaning here may be metaphorical: he could get *some* real nutrition from the *siliquae*, but none at all from the Manichee lore.

cantabam: Stories that are told for the sake of the telling, and whose untruth is therefore not penal, and those that are told to be believed, in which untruth therefore is disastrous: at the time of writing *conf.*, neither sort appeals to him, but the former—just then losing its appeal for him—seemed less pernicious.

gradibus: The *gradus* are ordinarily (since at least *quant. an.* 33. 70 ff.) the benchmarks of the ascent towards God (as later in *conf.:* 7. 17. 23, 9. 2. 2, 9. 10. 24, 10. 8. 12 ('transibo ergo et istam [vim] naturae meae, gradibus ascendens ad eum, qui fecit me'), 13. 9. 10). The phraseology here suggests parodic inversion in his descent to Manicheism.

in profunda inferi: Prov. 9: 18, 'et ignoravit quod gigantes ibi sint et in profundis inferni convivae eius.' Cf. on 1. 2. 2, 'inferi'; what was hypothetical there is here represented as in some sense true, for he was in hell, but God was there, 'interior . . . et superior'.

intellectum . . . quaererem: The dissociation of *quaerere* from *intellegere* (see on 1. 1. 1; this is the only significant occurrence of these terms in Bk. 3) is another mark of the disorder A. sees in this particular quest for wisdom.

interior intimo meo et superior summo meo: Esp. of second person of trinity (the antidote to *curiositas*): *vera rel.* 20. 38, 'summa et intima veritas'; *ep.* 101. 3, 'superna intima veritatis'; *trin.* 12. 3. 3, 'supernam et internam . . . veritatem'; cf. *conf.* 9. 1. 1, 'omni secreto interior'; *en. Ps.* 118. s. 22. 6, 'tu interior intimis meis'; *Gn. litt.* 8. 26. 48, 'ipse . . . et interior omni re'.

illam mulierem audacem: Prov. 9: 13–17 (VL), 'mulier insipiens et audax, inops panis effecta, . . . (14) sedit in foribus domus suae super sellam in excelso urbis loco . . . . (16) "qui stultus est divertat ad me . . . (17) panes occultos libenter edite, et aquam furtivam dulcem bibite".' On this passage, see E. Peters, *Augustiniana* 36 (1986), 48–64, and La Bonnardière, *Biblia Augustiniana: Proverbes*, 51–4 (text reconstructed from A.'s citations).

A.'s interpretation is already in place for *Gn. c. man.* 2. 27. 41, 'haereticorum quippe *superba et curiosa cupiditas* in libro proverbiorum clamat sub mulieris imagine, et dicit, "qui stultus est divertat ad me"'; sim. at *exp. prop. Rom.* 32 (39), *doctr. chr.* 3. 25. 36. Most fully developed at *Io. ev. tr.* 97. 2–3, 'cavete timendo et orando, ne inruatis in illud *aenigma Salomonis*, ubi mulier insipiens et audax, inops panis effecta, convocat praetereuntes dicens: "panes occultos libenter attingite et aquae furtivae dulcedinem". haec enim mulier vanitas est impiorum, cum sint insipientissimi, aliquid se scire opinantium. . . . cum sit inops panis, promittit panes; id est, cum sit ignara veritatis, promittit scientiam veritatis. occultos tamen panes promittit, quos dicit libenter attingi, et aquae furtivae dulcedinem, ut ea scilicet libentius et dulcius audiantur et agantur quae palam in ecclesia dici credique prohibentur. ipsa quippe occultatione condiunt quodammodo nefarii doctores sua venena *curiosis*. . . . (3) hinc et nefario ritus suos hominibus sacrilega *curiositate* deceptis vel decipiendis magicarum artium doctrina commendat.'

On *aenigma* generally, A.'s most detailed and programmatic explanation is at *trin.* 15. 9. 15, concluding: 'aenigma est autem ut breviter explicem obscura allegoria sicuti est, "sanguisugae tres erant filiae" [Prov. 30: 15], et quaecumque similia.' Also helpful is *div. qu. Simp.* 2. pr., 'velamen quippe omnimodo intercludit aspectum, aenigma vero tamquam per speculum, sicut idem apostolus ait, "videmus nunc per speculum in aenigmate," [1 Cor. 13: 12] nec evidentissimam detegit speciem, nec prorsus obtegit veritatem.' The word is used 10 × in *conf.*; closest parallel to the sense here is 5. 14. 24, in a description of Ambrose's way of preaching: 'et saepius aenigmate soluto de scriptis veteribus'. See with references C. Mayer, *Aug.-Lex.* 1. 140–1.

inopem prudentiae: Also at 4. 2. 2.

habitantem in oculo carnis meae: i.e. wallowing in *curiositas*. D. Sohlberg, *Mus. Helv.* 28 (1971), 176–9 emends 'in oculo carnis meae' to *in excelso Carthaginis meae*; H. Fuchs responded at *Mus. Helv.* 29 (1972), 62, with the the slightly more plausible suggestion *in habitaculo carnis meae* (but he admitted that this change required the further change below to *per illam* [sc. *per carnem*] *vorassem*). La Bonnardière, *Biblia Augustiniana: Proverbes*, 53, rightly approves the paradosis: 'Augustin a dévoré par l'œil qu'il a fixé sur la femme effrontée—la Folie—une nourriture qu'il a ensuite ruminée en lui-meme.' Sim. by G. Luongo, *Parola del Passato* 27 (1972), 414–15.

3. 7. 12

It was *curiositas*, the *libido sciendi*, that led A. to Manicheism; accordingly, he now (3. 7. 12–3. 10. 18) presents a sustained analysis of the attraction that the cult had for him measured by his ignorances of that time. Had he known certain things, the *phantasmata* of the Manichees would not have held an attraction to his vagrant and concupiscent eyes. The three topics under which he arranges his discussion all recur thematically later in *conf.* as he shows himself finding alternate solutions to those the Manichees proposed.

A. does not say that the three problems outlined here were the ones for which he found the most attractive solutions in Manicheism, but rather that they were three questions for which he could not provide satisfactory answers based on anything else he knew. (When in after years A. would be accused of retaining a Manichean outlook, the justice of the charge would lie in the way he remained in the power, not of the answers the Manichees offered, but of the questions (e.g. *unde malum*) on which they insisted with such effect.) The presentation is thus deliberately partial and schematic. The Manichees also attacked orthodox Christianity for deliberate falsification of scriptures, but A. claims that this argument never made any headway with him (*util. cred.* 3. 7).

One important way in which the limitations of this report interfere with our understanding: it was violently contrary to A.'s intentions at the time of writing *conf.* to do full justice to the appeal the Manichees had for him. He did not concede them the possession even of whatever virtues and strengths had held appeal for him. Hence, although Brown, 50, and G. R. Evans, *Augustine on Evil* (Cambridge, 1982), 13, are quick to emphasize the benefits of the Manichean balm for an aching conscience, in this formal description of his encounter with the Manichees no such ethical question arises: *unde malum* appears only as an intellectual question to which he cannot provide a better answer. Only later (5. 10. 18) will he hint at the ethical side of his Manicheism.

In the narrative of his conversions, the solutions to the three problems are uncovered in reverse order to that posed here:

unde malum: 7. 12. 18;
utrum forma corporea deus finiretur: 7. 1. 1–2. 3—but see notes there;
utrum iusti existimandi essent: 5. 14. 24 and 6. 4. 6.

The three 'non noveram . . . et non noveram . . . et non noveram' statements of this paragraph and the beginning of the next outline the solutions *in nuce*, by way of clarifying the problems.

acutule: Quite rare; as adj. at Cic. *nat. deor.* 3. 7. 18; 3 × else where in A. (*ep.* 205. 1. 4, 'infirmitas . . . humana . . . acutule se garrire arbitratur', and *nat. et gr.* 28. 32, 'ne videantur ei nimis acuta, quae acutule sonant et discussa inveniuntur obtunsa', and *c. Iul.* 6.9.26). Also used by Julian at *c. Iul. imp.* 1.71 and 3.188, in unflattering characterizations of A.'s style of argument.

unde malum: *lib. arb.* 1. 2. 4, 'eam quaestionem moves quae me admodum adulescentem vehementer exercuit et fatigatum in haereticos impulit atque deiecit. quo casu ita sum afflictus et tantis obrutus acervis inanium fabularum [see 5. 3. 3, 'fabulis'] ut nisi mihi amor inveniendi veri opem divinam impetravisset, emergere inde atque in ipsam primam quaerendi libertatem respirare non possem'; *duab. an.* 8. 10, 'hic fortasse quis dicat: "unde ipsa peccata et omnino unde malum? si ab homine, unde homo? si ab angelo, unde angelus?" . . . *hac in quaestione illi regnare se putant, quasi vero interrogare sit* *scire*. utinam id esset; nemo me scientior reperiretur'; *c. Iul. imp.*, 6. 9, 'si enim quaerente illo a nobis unde sit malum, qua quaestione consueverunt [manichaei] corda inerudita turbare'.

The question itself, pressed in this way, was less obvious to ancients than to moderns and bespeaks the concerns of late antique philosophy (e.g. Plot. 1. 8) rather than those of earlier times. It is insistence on the goodness of God that makes the question a pressing one—indeed, makes a hypostasized 'Evil' emerge from the multitude of 'evils' that beleaguer mortal life (viz. Plat. *rep.* 2. 379 B-C, *polit.* 273 B-E, *Tim.* 41 D-42 E). If the question did not originate in gnosticism, it was certainly most at home there.

et utrum forma corporea deus finiretur: *Gn. c. man.* 1. 17. 27, 'istam maxime quaestionem solent manichaei loquaciter agitare et insultare nobis, quod hominem credamus factum ad imaginem et similitudinem dei. attendunt enim figuram corporis nostri, et infeliciter quaerunt utrum habeat deus nares et dentes et barbam, et membra etiam interiora, et cetera quae in nobis sunt necessaria.' It was as a Christian of sorts that A. had thought of God in anthropomorphic terms; his 'conversion' to orthodoxy was helped when he realized that Catholics did not believe this: see on 7. 1. 1 ff.

et utrum iusti existimandi essent: *c. Faust.* begins with the genealogies and contradictions of the Gospels (and returns to the subject: *c. Faust.* 2–3, 7. 1, 23, and 27–8); the two longest books of *c. Faust.* (12 and 22) both deal with the vices of the OT patriarchs. See also *c. Faust.* 6. 2, quoted on 3. 9. 17, 'ad futura praenuntianda', below. That the Manichees rubbed a raw nerve is best seen in *qu. hept.*, written when A. was 65 and long secure in his grasp of the Christian approach to the OT. That work bespeaks the intensity with which the literalist approach made the OT a wearying thing for many late antique readers. How was it possible for angels come to lie with the daughters of men and beget giants? Could the ark have been that big? Could four men have built it? How could Lot have treated his daughters that way? (*qu. hept.* 1. 3–5, 1. 42) The disparity between the genealogies of Matthew and Luke (see on *s.* 51 at 3. 5. 9) rises again, a wound healed but showing scar-tissue, at *qu. ev.* 2. 5, *qu. hept.* 1. 121, and *s.* 83. 4. 5 (*c.*409). For early treatments, cf. *div. qu.* 49, *s.* 1. 4. 4.

de animalibus: For this characteristically Late Latin use of *de* with ablative in place of partitive genitive (see Arts, 17), cf. 9. 5. 13, 'quid mihi potissimum de libris tuis legendum esset'.

recedens: Another expression for the way we are separated from God in *conf.*, e.g. 3. 3. 5, 'ad longe recedendum a te'; cf. also 1. 2. 2, 2. 6. 14, 4. 16. 31, 5. 3. 4, 10. 26. 37.

non noveram: The anaphora of 'non noveram' connects the three clarificatory expositions: '*non noveram* malum non esse nisi privationem boni . . . et *non noveram* deum esse spiritum . . . (3. 7. 13) et *non noveram* iustitiam veram interiorem.' There are three later passages where 'non noveram' is a reminder of this discussion: in all three cases, it is the second ignorance, of the spiritual nature of God, that is recalled: 4. 2. 3, 5. 10. 19, 5. 10. 20.

privationem boni: The doctrine is Platonic in origin (but A. takes the idea further than P.: see on 7. 12. 18), and offers the possibility of an aesthetic theory of evil (e.g. *nat. b.* 16, 'quae tamen etiam privationes rerum sic in universitate naturae ordinantur, ut sapienter considerantibus non indecenter vices suas habeant') which escapes the hard antitheses of dualism at the cost of incurring new difficulties.

viderem . . . videre . . . oculis . . . phantasma: Note the emphasis again on the sense of sight and its errors.

et non (*noveram*) CDGO Maur. Ver. : *non* S Knöll Skut. Pell.

et non noveram deum esse spiritum: Jn. 4: 24, 'spiritus est deus.' The word *spiritus* is avoided when this topic is raised again at its proper place in the narrative at 7. 1. 1–7. 2. 3: *spiritus* is by preference the word that designates the third person of the trinity, and is otherwise used in phrases like *spiritus hominis*, without again occurring in this predicative sense of God. A notion of a corporeal God is blamed on *curiositas* at *Io. ep. tr.* 7. 10.

et si infinita sit . . .: See on 1. 3. 3.

secundum quod essemus: Cf. 1. 2. 2, 'me qui non essem nisi esses in me'.

essemus OS Knöll Skut. Ver. : *essemus similes deo* CDG Maur.

imaginem: See on 3. 2. 2.

prorsus ignorabam: Much commoner verb for defects of knowledge in *conf.* than 'non noveram': See particularly on 1. 20. 31, 'fugiebam dolorem, abiectionem, ignorantiam', where *ignorantia* [2] is what the child flees from in search of *veritas*, only to find *error*.

3. 7. 13

At times, A. could claim that the OT *patres* did no wrong: *b. coniug.* 25. 33, 'quae cum ita sint, haeretici quidem sive manichaei sive quicumque alii patribus veteris testamenti de pluribus calumniantur uxoribus, hoc esse argumentum deputantes quo eorum convincant incontinentiam, satis superque responsum est: si tamen capiunt non esse peccatum, quod neque contra naturam committitur, quia non lasciviendi sed gignendi causa illis feminis utebantur; neque contra morem, quia illis temporibus ea factitabantur; neque contra praeceptum, quia nulla lege prohibebantur. illos vero qui inlicite feminis usi sunt vel arguit in scripturis illis divina sententia vel nobis lectio iudicandos atque vitandos, non approbandos imitandosve proponit.' At other times, he insisted only on correct interpretation: *doctr. chr.* 3. 12. 18, 'quae autem quasi flagitiosa imperitis videntur, sive tantum dicta sive etiam facta sunt vel ex dei persona vel ex hominum quorum nobis sanctitas commendatur, tota figurata sunt, quorum ad caritatis pastum enucleanda secreta sunt'; he reinforces that at *doctr. chr.*

3. 18. 26–22. 32. The same argument occurs at *s.* 51. 15. 25 and 51. 18. 28 (see above on 3. 5. 9).

Et non noveram: See on 3. 7. 12.

rectissima O Maur. Ver. Pell. : *iustissima* G : *lectissima* CDS Knöll Skut. Vega. Pellegrino rightly takes G's reading in support of *rectissima*.

eos . . . iudicari iniquos: 'noveram' still governs the indirect discourse.

iudicantibus ex humano die: 1 Cor. 4: 3, 'mihi autem minimum est ut a vobis diiudicer aut ab humano die: sed neque meipsum iudico.'

tamquam si quis . . .: The conventional rhetorical employment of the three similes from everyday life—which embrace the whole of social life seen in conventional Roman terms, from the camp to the forum to a private household—has a surprising comic turn to it.

sic sunt: Answering the similes that began with 'tamquam'.

homine . . . die . . . aedibus: Reminders of the three similes above; see 'corpore . . . die . . . domo' below.

numquid iustitia . . . mutabilis?: *Aen.* 4. 569–70 (Mercury instructing Aeneas to end his affair with Dido), 'varium et mutabile semper femina.' Women in *conf.* are: pregnant (1. 6. 9), patronized as sources of women's lore about infants (1. 6. 10), a figure of heresy (the *aenigma Salomonis*: 3. 6. 11), piously joyful (Monnica: 3. 11. 20), sex objects (6. 11. 20, 8. 1. 2), tiresome wives (6. 14. 24), spurned and freshly resolved on continence (his concubine: 6. 15. 25), giving birth (7. 6. 8), worried about household finances (from a gospel parable: 8. 3. 6, 10. 18. 27), maddened (*rabies*, of Justina: 9. 7. 16), staunch in the face of death (Monnica: 9. 11. 28), and subjected to Adam by God in Eden (13. 32. 47, 13. 34. 49). The present passage presents itself before that backdrop: the line from the *Aeneid*, profoundly unjust to Dido, has here its own comic effect—surely justice shares none of the defects that are, the phrasing inevitably reminds us, characteristic of *women*?

vita . . . brevis est: Cf. Wisd. 15: 9, 'brevis illis vita est.'

corpore . . . die . . . domo: Picks up once again the similes of 'tamquam si quis . . . .'

illis: sc. *temporibus*.

hic CDGO¹S Knöll Skut. Ver. : *his* O² Maur. G–M find the latter reading plausible, but if there is a broken antithesis here, the remedy would surely be *illic*. What strikes the eye is 'serviunt' (usually translated weakly: *BA*, 'ils acceptent'; Ryan, 'give approval to the latter'): the use is difficult to parallel, but the overtone of moral judgement attracts attention. We are slaves to custom, and sniff at the customs of other times and places.

3. 7. 14

The Manichean challenge to the OT is treated here so as to suggest that the solution lies in a correct reading of scripture (as at 3. 7. 12, dealing with their critique of an anthropomorphic God). One interim answer is given, but what is hinted at is a comprehension of the allegorical meaning of OT scripture. This prepares us to see that when A. discovers the distinction between literal and allegorical senses in scripture (at Milan: 5. 14. 24 and 6. 4. 6), then he can escape the dilemma the Manichees set for him.

Haec: The contents only of 3. 7. 13 are meant.
oculos meos . . . non videbam: Cf. 'caecus' below. The paradox of the three temptations is that those who yield to them achieve the opposite of what they seek; see 1. 20. 31.
cantabam: This example, designed to confute the Manichees' criticism, comes directly from A.'s personal experience, not from the conventionalisms that were incorporated in the similes—employed to the same purpose—at 3. 7. 13.
praecipit OS Knöll Skut. Ver. : *praecepit* CDG Maur.
utentes: *doctr. chr.* 1. 4. 4, 'uti autem [est], quod in usum venerit ad id quod amas obtinendum referre, si tamen amandum est.'
futura praenuntiantes: cf. Col. 2: 17, 'umbra futurorum' (cited in *c. Faust.* 6. 2 (quoted in notes on 3. 9. 17, 'ad futura praenuntianda')).

3. 8. 15

Cf. the discussion of the temptations/vices in 2. 6. 13, holding a comparable place in the structure of that book, as meditation on his own lapse situates itself in meditation on the nature of sin. Here too sin is divided in three possibilities ('contra naturam', 'contra mores', and 'contra morem aut pactum quorumlibet'); it would be fanciful, but possible, to see in sins against *natura* the concupiscence of the flesh; in sins against the *mores hominum*, *ambitio saeculi*; and in sins against the commandment of God, *concupiscentia oculorum*. To descry such patterns is not to impose them on the author's will, but to see the consistency of his vision.

diligere deum . . . te ipsum: Mt. 22: 37–40, 'diliges dominum deum tuum ex toto corde et ex tota anima et ex tota mente. . . . (39) diliges proximum tamquam te ipsum. (40) in his duobus praeceptis tota lex pendet et prophetae.' (Parallel NT versions at Mk. 12: 33, Lk. 10: 27, all conflations by Jesus of Deut. 6: 5 (love of God) and Lev. 19: 18 (love of neighbour).) The two great commandments occur here by way of example, reminding the reader of other issues, as he passes on quickly. That Jesus presents them as a summary of OT doctrine sharpens their usefulness against the Manichees (cf. *mor.* 1. 28. 57, 'atque in his duobus convenit mihi cum manichaeis, id est ut deum et proximum diligamus: sed hoc veteri testamento negant contineri'). Elsewhere in *conf.* they are instanced only in the post-narrative Bks. 10–13: 10. 37. 61, 12. 25. 35, 13. 17. 21, 13. 24. 36; at 13. 19. 24, *dilectio proximi* stands in sequence in place of the ninth and tenth commandments of the decalogue—against coveting the neighbour's wife and property, but cf. also *en. Ps.* 32. en. 2. s. 1. 6 quoted below on 3. 8. 16.

On the force of the prepositional phrases, see *mor.* 1. 8. 13, '"diliges", inquit, "dominum deum tuum." dic mihi etiam quaeso te, qui sit diligendi modus: vereor enim ne plus minusve quam oportet inflammer desiderio et amore domini mei. "ex toto", inquit, "corde tuo." non est satis. "ex tota anima tua." ne id quidem satis est. "ex tota mente tua." quid vis amplius? vellem fortasse, si viderem quid posset esse amplius.'

(*corde*) **et** CDS Knöll Skut. : om. GO Ver.

(*anima*) **et** CDGS Knöll Ver. : om. O Skut.

A.'s citations elsewhere omit the *et* in both cases slightly more often than not, but no conclusion may be drawn from that (Milne, 60).

flagitia: See on 3. 8. 16.

contra naturam: Rom. 1: 26, 'nam et feminae eorum immutaverunt naturalem usum in eum qui est contra naturam'; cf. 'eodem reatu'. For the sin of the Sodomites: *b. coniug.* 11. 12, 'cum ille naturalis usus, quando prolabitur ultra pacta nuptialia, id est ultra propagandi necessitatem, venialis sit in uxore, in meretrice damnabilis; iste qui est *contra naturam*, execrabiliter fit in meretrice, sed execrabilius in uxore. tantum valet ordinatio creatoris et ordo creaturae ut in rebus ad utendum concessis, etiam cum modus exceditur, longe sit tolerabilius quam in ea quae concessa non sunt vel unus vel rarus excessus. . . . cum vero vir membro mulieris non ad hoc concesso uti voluerit, turpior est uxor si in se, quam si in alia permiserit.' This argument would repudiate *a fortiori* sexual contact between members of the same sex, but only later did A. specify that this is the sin of the Sodomites: *civ.* 16. 30, 'tota illa regio impiae civitatis [Sodomae] in cinerem versa est, ubi stupra in masculos in tantam consuetudinem convaluerant, quantam leges solent aliorum factorum praebere licentiam'; cf. *civ.* 6. 8. See also *qu. vet. t.* cited in following note. (Julian seems to agree with the specification: as quoted at *c. Iul. imp.* 6. 23.)

contra naturam . . . contra mores hominum . . . aliquid contra morem aut pactum quorumlibet iubet: *c. Faust.* 22. 47, 'alia enim sunt peccata contra naturam, alia contra mores, alia contra praecepta.' *qu. vet. t.* pr., 'generalem iustitiam non violat quis, nisi libidine transgressus fuerit aut *placitum societatis humanae*, sicut est furtum, rapina, adulterium, incestum et huiusmodi; aut *naturam*, sicut est contumelia, caedes, homicidium, concubitus masculorum vel pecorum; aut *modum in concessis*, sicut est servum amplius verberare quam oportet vel filium, edere vel bibere amplius quam oportet, cum ipsa coniuge concumbere amplius quam oportet, et similia.'

libidinis perversitate: Approximately = *voluntate per libidinem perversa*.

deo regnatori CDGO[1] Skut. Ver. Pell. : *deus regnator* SO[2] Knöll.

suae (*ad ea*) CDGO Skut. Ver. : *sive* S : *cui* coni. Knöll, in order to retain 'deus regnator'. The construction seemed to demand a subject for 'iusserit', and to deprecate a double dative with 'serviendum est', hence the correction in the MSS. With Skut., take the double dative with 'serviendum est'; to follow Knöll requires, as G–M remark, reading an anacoluthon that assumes that the first clause began, 'si rex potest', and supplying a *potest* and a comma after 'creaturae suae' in this line.

maior potestas minori ad oboediendum praeponitur: Rom. 13: 1, 'omnis anima potestatibus sublimioribus subdita sit. non est enim potestas nisi a deo; quae autem sunt, a deo ordinatae sunt.'

3. 8. 16

facinoribus: Originating in *ira*, while *flagitia* arise from *libido* (4. 15. 24). The scheme is expanded to include *errores* at 4. 15. 25 (see notes there). *doctr. chr.* 3. 10. 16, 'quod autem agit indomita cupiditas ad corrumpendum animum et corpus suum, flagitium vocatur; quod autem agit ut alteri noceat, facinus dicitur.' *Facinora* are worse than *flagitia*: *exp. prop. Rom.* 6 (6) (on Rom. 1: 28–9), 'quisque perniciosam dulcedinem flagitiorum sequens, dum impedientes personas removere conatur, pergit in facinus.' See also *c. acad.* 3. 16. 36, 'flagitiis ac facinoribus', and *s.* 9. 10. 15, quoted below.

The two terms are consciously linked often in CL (see *TLL* 6. 82): Cic. *Catil.* 1. 7. 18, 'quae tecum, Catilina, sic agit et quodam modo tacita loquitur: "nullum iam aliquot annis facinus exstitit nisi per te, nullum flagitium sine te"'; Cic. *Scaur.* 13, 'omni facinore et flagitio deformatos'; cf. Cic. *Sull.* 5. 16; Sall. *Cat.* 14. 1. 2, Tac. *ann.* 3. 50.

Both *facinus* and *flagitium* are limited in their appearances in *conf.*; *flagitium* only in Bks. 1, 2, 3, and 4; *facinus* only in Bks. 2, 3, and 4. They are thus confined to the books devoted to A.'s moral downfall. *Peccatum* is far more common, occurring in every book. Elsewhere in *conf.* the most important application of *facinus* is to the pear-theft (2. 6. 12); of *flagitium*, to his moral lapses (e.g. 3. 1. 1, 'sartago flagitiosorum amorum'). Elsewhere in A., the pair is frequent.

spectatores: See on 3. 2. 2.

inrisores: See on 1. 6. 7.

principandi [1] **et spectandi** [2] **et sentiendi** [3] **libidine:** Cf. 1 Jn. 2: 16 (see on 1. 10. 16). For the senses as the domain of *concupiscentia carnis*, see 10. 30. 41 ff.

aut una aut duabus earum aut simul omnibus: A. does not treat the three temptations as a set of watertight compartments, but acknowledges that they interact in particular situations in different combinations. In the sequence of his own fall through these books, there is a broad pattern of succession from *concupiscentia carnis* (esp. in Bk. 2) to *concupiscentia oculorum* (esp. in Bk. 3) to *ambitio saeculi* (esp. in Bk. 4), but Bk. 2 is the only one of the three in which the predominant temptation reigns virtually alone. Here in Bk. 3 we have already seen how *curiositas* and *libido* both had a role in A.'s reaction to the *spectacula*.

tria et septem: This phrase confirms that he thinks of the earlier phrase ('principandi et spectandi et sentiendi libidine') as a group of three elements: 'the three against the three + seven.' 7 + 3 = 10 is rock-bottom certainty even in his Academic days: 6. 4. 6.

psalterium decem chordarum: Ps. 32: 2, 'confitemini domino in cithara, in psalterio decem chordarum psallite ei'; *en. Ps.* 32. en. 2. s. 1. 6, 'praecepta enim legis decem sunt; in decem praeceptis legis habes psalterium. perfecta res est. habes ibi dilectionem dei in tribus, et dilectionem proximi in septem. . . . haec tria [mandata] pertinent ad dilectionem dei, cuius cogita unitatem [1], veritatem [2] et voluptatem [3].' Cf. Ps. 143: 9, 'in psalterio decem chordarum psallam tibi' (*en. Ps.* 143. 16, 'in lege decem praeceptorum'). A more elaborate trinitarian exposition occurs at *s.* 9. 5. 6 'de decem chordis', 'decalogus enim legis decem praecepta habet: quae decem praecepta sic sunt distributa, ut tria pertineant ad deum, septem pertineant ad homines. tria ad deum, quae iam dixi: [1] unus est deus noster, ei nullam similitudinem debemus facere, et non fornicari post deum, qui unus est. [2] quia deus et Christus filius dei unum est cum patre, et ideo non debet a nobis accipi in vanum, ut putemus eum factum, id est, creaturam aliquam, per quem facta sunt omnia. [3] quia ipse unus deus, pater est et filius et spiritus sanctus, in spiritu sancto, hoc est, in dono dei, requies nobis sempiterna promittitur.'

altissime et dulcissime: Take the echoes closely with preceding line: cf. Ps. 91: 2, 'bonum est confiteri domino et psallere nomini tuo, altissime . . . (4) in decachordo psalterio, cum cantico in cithara.'

quod in se homines perpetrant: *s.* 9. 10. 15, 'cum vero imaginem dei, quod es tu, corrumpis in te per fornicationes et per diffluentias libidinis . . . non attendis per libidines et inlicitas fornicationes cuius imaginem violasti? . . . omnia enim peccata hominum aut ad corruptelam pertinent flagitiorum, aut ad facinora nocendi. quia deo nocere non potes in facinoribus, in flagitiis eum offendis, in corruptelis eum offendis, in te illi facis iniuriam.'

mentitur iniquitas sibi: Ps. 26: 12, 'et mentita est iniquitas sibi'; cf. 1. 5. 6.

sive corrumpendo ac pervertendo naturam suam [1], **. . . vel immoderate utendo concessis rebus** [2], **vel in non concessa flagrando in eum usum qui est contra naturam** [3]: The trinitarian pattern reappears; that the first clause marks sin against the first person of the trinity is made explicit by 'quam tu fecisti et ordinasti'; 'immoderate use of licit things' is a generic description of *curiositas*; while 'in eum usum qui est contra naturam' is a particular form of *concupiscentia carnis* (see on 3. 8. 15, esp. for echo of Rom. 1: 26).

aut . . . aut: Identifying sins against the two great commandments, i.e. against God or against neighbour.

adversus stimulum calcitrantes: Cf. Act. 26: 14, 'Saul, Saul, quid me persequeris? durum est tibi contra stimulum calcitrare.' The second line is also interpolated by some MSS at Act. 9: 5. *en. Ps.* 85. 9, 'aedificetur in hoc cor vestrum, fratres, cor christianum, cor fidele; ne incipiatis tristes facti, veluti fraudati desideriis vestris, ire in indignationem contra deum; etenim non expedit adversus stimulum calcitrare. recurrite ad scripturas.'

But the expression is proverbial: Diomedes (Keil, 1. 462. 27), 'parhoemia est vulgaris proverbii usurpatio rebus temporibusque adcommodata, cum aliud significatur quam dicitur, ut: adversum stimulum calces, quo significatur contra pessimos vel potentiores audere stultum esse.' So Ter. *Phorm.* 77 and Amm. Marcell. 18. 5. 1.

fons vitae: Ps. 35: 10, 'quoniam apud te fons vitae'; *en. Ps.* 35. 15, quis est fons vitae, nisi Christus? venit ad te in carne, ut inroraret fauces tuas sitientes; satiabit sperantem, qui inroravit sitientem.' Cf. Jer. 2: 13, 'duo enim mala fecit populus meus: me dereliquerunt fontem aquae vivae ut foderent sibi cisternas, cisternas dissipatas quae continere non valent aquas.'

The metaphorical use of *fons* in good senses is abundant in *conf.*: 4. 4. 7, 6. 1. 1, and 6. 16. 26 have 'fons misericordiarum'; 'fons tuus' at 9. 3. 6, 9. 10. 23, 12. 10. 10; 'fons vitae' at 9. 10. 23, 13. 4. 5, 13. 16. 19, 13. 21. 30, 13. 21. 31; also 12. 30. 41, 'fontem veritatis', 13. 13. 14, 'quemadmodum cervi ad fontes aquarum' (Ps. 41: 1), 13. 17. 20, 'animas sitientes tibi . . . occulto et dulci fonte inrigas'.

unus . . . unum falsum: G–M: 'The phraseology is coloured by the neo-Platonic doctrine according to which τὸ ἕν ("the One" (unum)) is man's true life-centre and sin is essentially a turning away from that true centre to become absorbed in the things of sense, which then constitute a false life centre or substitute for "the One".'

privata superbia: G–M: 'from self-regarding pride'; *BA*: 'par orgueil particulariste'. Cf. here 'privatis conciliationibus'; the adjective has been dissociated by A.'s time from the verb *privare*, and is not to be taken as the participle of that verb. Cf. *en. Ps.* 114. 3, 'dies meos dico, quos ipse mihi feci privata audacia, qua deserui eum.'

diligitur in parte unum falsum: 'There is loved, in one of the parts of the universe, a false One.'

reditur in te: See on 1. 18. 28.

purgas: The word has a variety of literal uses in A.'s writings (purifying gold, threshing grain, etc.), and is common among both Christians and Platonic philosophers for renewal of various kinds, esp. sacramental/theurgic. Cf. *en. Ps.* 113. s. 1. 5 ('interrogate ergo corda vestra: si ea circumcidit fides, si purgavit confessio') with Mart. Cap. 5. 458, 'nam purgatio est, cum *confesso facinore* animum voluntatemque purgamus'. *en. Ps.* 136. 9, 'a sordibus malarum cupiditatum purgantur', could have been written by many writers of this period, Christian or not. In *conf.* note 4. 1. 1, 'purgari nos ab istis sordibus', 6. 2. 2, 'plenum purgatioribus votis pectus', 6. 4. 6, 'purgatior acies mentis meae', 11. 29. 39, 'purgatus et liquidus igne amoris tui'.

propitius es peccatis: Cf. Ps. 77: 39, 'et propitius erit peccatis eorum, et non perdet eos'; Ps. 78: 9, 'et propitius esto peccatis nostris propter nomen tuum'; Lk. 18: 13, 'et publicanus a longe stans . . . dicens, "deus propitius esto mihi peccatori".'

gemitus compeditorum: Ps. 101: 21, 'dominus de caelo in terram prospexit ut audiret gemitum compeditorum, ut solvat filios mortificatorum'; *en. Ps.* 101. s. 2. 3, 'solvitur enim unusquisque a vinculis cupiditatum malarum, vel a nodis peccatorum suorum.'

adversus (*te*) GS Knöll Skut. Pell. : *versum* CDO Ver.

cornua falsae libertatis: Ps. 74: 5, 'nolite exaltare cornu vestrum'; *en. Ps.* 74. 7, 'si fecistis iniquitatem per cupiditatem, nolite eam defendere per elationem; confitemini, si fecistis. qui enim non confitetur, et iniquus est, exaltat cornu. . . . exaltabitur in vobis cornu Christi, si non exaltetur cornu vestrum.' For *libertas*, see on 2. 6. 14. (Knauer, 162 n. 2 and 182 n. 1, comments on the artistry of the sentence: climax through seven-fold *et* (on *et* see on 1. 1. 1), careful antitheses, employment of Psalm citations.)

amplius amando proprium nostrum: *lib. arb.* 2. 19. 53, 'voluntas autem aversa ab incommutabili et communi bono et conversa ad proprium bonum . . . peccat.'

3. 9. 17

The digression on *flagitia* and *facinora* (3. 8. 15–16) now reaches its goal: application to the case of the conduct of the OT patriarchs.

peccata proficientium: *s.* 302. 1. 1, 'quaedam enim plerumque parva et ludicra concedit pater parvulis filiis. . . . benigna et paterna indulgentia haec impertit, haec donat, quae non vult permanere in filiis suis iam grandiusculis, iam proficientibus.' The verb is common in A. for the 'progress' of the moral life, e.g. *Io. ev. tr.* 41. 12.

et sunt quaedam: Things that are apparently *peccata* but do not offend against either of the two great commands ('nec te . . . nec sociale consortium').

conciliantur: 'are procured'.

aliqua in usum vitae, congrua et tempori: G–M: '. . . "congrua" goes both with "in usum" and with the dative "tempori"'; Pellegrino agrees, wrongly, for 'in usum vitae' is governed by 'conciliantur', and 'congrua' goes with 'tempori' alone.

congrua et S Knöll Skut. Vega : *congrua* G Pell. : *congruae* CD : *congruentem* O[1] (**-nt-** in ras. O[2]) : *congrue* coni. Ver. For once, S is right. The hyperbaton of the *et* was unusual enough and the meaning easy enough to confuse, and so it fell out, leaving the *e* in CD and driving O to conjecture.

libidine habendi: This would be a *flagitium*. For an OT example (though both cases are meant to be generic), cf. *c. Faust.* 22. 71, 'de expoliatis Aegyptiis . . .: carnalis itaque adhuc ille populus erat et rerum terrenarum cupiditate occupatus . . .: et forte secundum suas voluntates et cogitationes hebraei magis permissi sunt facere ista quam iussi; sed eis deus permissionem suam per famulum suum Moysen innotescere voluit.'

libidine nocendi: A case of *facinus*; *c. Faust.* 22. 79, 'quid ergo crudele Moyses aut mandavit aut fecit . . . ? nam eum nulla crudelitate sed magna dilectione fecisse quod fecit, quis non in verbis eius agnoscat?'

viderentur CDO[1]S Knöll Skut. Ver. : *videntur* GO[2].

articulus: *cat. rud.* 3. 6, 'et per quinque temporum articulos praenuntiari venturus prophetarique non destitit', of the five ages of history before the coming of Christ. G–M unsure of applicability here; but note 'occulti' and 'occultes' below. Cf. also *civ.* 10. 14, 'sicut autem unius hominis, ita humani generis . . . recta eruditio per quosdam articulos temporum, tamquam aetatum *profecit* accessibus, ut a temporalibus ad aeterna capienda et a visibilibus ad invisibilia surgeretur.'

imperas: *civ.* 1. 26, 'cum autem deus iubet seque iubere sine ullis ambagibus intimat, quis oboedientiam in crimen vocet? quis obsequium pietatis accuset? sed non ideo sine scelere facit, quisquis deo filium immolare decreverit, quia hoc Abraham etiam laudabiliter fecit.' Vega thinks here of the marriage of Hosea (Hosea 3: 1 ff.), but A. seems not to have discussed that passage.

ad futura praenuntianda: Cf. last words of 3. 7. 14, 'futura praenuntiantes'. *c. Faust.* 6. 2, 'respondeo istos [manichaeos] omnino nescire quid intersit inter praecepta vitae agendae et praecepta vitae significandae. exempli gratia, "non concupisces" praeceptum est agendae vitae; "circumcides omnem masculum octavo die" praeceptum est significandae vitae. ex hac quippe imperitia manichaei et omnes quibus displicent litterae veteris testamenti, quidquid deus mandavit priori populo ad celebrandam umbram futurorum non intellegentes et ea modo non observari animadvertentes ex more praesentis temporis illa reprehendunt, quae utique illi *tempori congruebant*, quo ista quae nunc manifestata sunt, ventura significarentur.'

3. 10. 18

At 3. 7. 12, A. recounted how the Manichees extorted his assent; now he concludes his treatment of his conversion by sketching some of the doctrines and practices to which his association with the Manichees led him.

inriderer: See on 1. 6. 7.

nugas: A belittling word in the main, associated with childhood (6. 4. 5, 'infantiles nugas') but virtually always used to criticize some form of adult behaviour (1. 9. 15, 'maiorum nugae negotia vocantur', 4. 1. 1, 'spectaculorum nugas', 8. 11. 26, 'nugae nugarum et vanitates vanitantium')—but see on 4. 8. 13 for a rare exception. For the doctrine, cf. *en. Ps.* 140. 12, on the *crux luminis*: 'qui autem, inquiunt [manichaei], agricola est, multum laedit crucem luminis. quaeris quam crucem luminis? membra, inquiunt, illa dei quae capta sunt in illo proelio mixta sunt universo mundo, et sunt in arboribus, in herbis, in pomis, in fructibus. dei membra vexat, qui terram sulco discindit; dei membra vexat, qui herbam de terra vellit; dei membra vexat, qui pomum carpit de arbore. haec ne faciat in agro falsa homicidia, facit in faenore vera homicidia. panem mendicanti non porrigit. . . . quare? ne vitam quae est in pane, quam dicunt membrum dei, substantiam divinam, mendicus ille accipiat, et liget eam in carne. . . . sed nos, inquiunt, quia fide Manichaei inluminati sumus, orationibus et psalmis nostris, qui electi sumus, purgamus inde vitam quae est in illo pane, et mittimus illam ad thesauros caelorum. tales sunt electi, ut non sint salvandi a deo, sed salvatores dei. et ipse est Christus, dicunt, crucifixus in toto mundo. . . . ergo *buccella* ne detur mendico, et ploret in buccella membrum dei, mendicus fame moriturus est! falsa misericordia in buccellam, facit verum homicidium in hominem.' Cf. see also *c. Faust.* 20. 2 on 'Iesus patibilis', quoted in notes on 3. 6. 10, 'nominis tui'; and see L. Koenen, *Illinois Classical Studies*, 3 (1978), 176–87.

misericordiam praestandam: Cf. Rom. 9: 15, 'quid ergo dicemus? numquid iniquitas apud deum? absit: Moysi enim dicit, "miserebor cuius misereor et misericordiam praestabo cuius miserebor" [Ex. 33: 19]'.

si quis enim esuriens peteret: *mor.* 2. 15. 36, 'hinc est quod mendicanti homini, qui manichaeus non sit, panem vel aliquid frugum, vel aquam ipsam quae omnibus vilis est, dari prohibetis; ne membrum dei, quod his rebus admixtum est, suis peccatis sordidatum a reditu impediat'; sim. at *mor.* 2. 16. 53, *en. Ps.* 140. 12 (quoted above), etc.

manichaeus: Only here does he identify the sect for the first time *expressis verbis*; it could be inferred by the well-informed reader from the doctrines described as early as 3. 6. 10, but there must have been early readers for whom it was not obvious.

3. 11. 19

The first substantive appearance of Monnica in the text; the first hopeful *event*; the only segment of Bk. 3 introduced with the characteristically confessional mixture of psalm citations (Knauer, 133); a calculated conclusion to Bk. 3, expressing in a few sketchy lines both the depths to which he had fallen and a first hint of the medium by which he would ascend.

Monnica: Patricius was baptized and died while A. was in Carthage. The earlier references to M. are mainly indirect, offering a conventional picture of a mother frightened by her son's illness, eager to advance his career, and made anxious by his sexual maturity. There is in this some veiled criticism of his parents' treatment of him (see on 2. 2. 4, 2. 3. 6, 2. 3. 8). Her first appearance here after her husband's death casts her irreversibly in the role that will be hers from now on. Are we to assume a flowering of religiosity in her after P.'s death? Or a new relationship between son and mother (he now an adult of 21, with a profession and a shocking new religion)? A. does not say: he simply presents, then highlights this episode by setting it in a structurally prominent position. As some of the beginnings of books of *conf.* look back and recapitulate, so some of the endings look forward hopefully (5. 14. 25, remaining a catechumen and looking for something to turn up; 7. 21. 27, taking up Paul; 8. 12. 30, opening a new life).

Courcelle, *Les Confessions*, 129, on visions in the African church: 'Quel que soit le personnage en question, dont l'identité n'est pas toujours précisée, il se présente d'ordinaire sous des traits souriants, enjoués, *hilares*, signe d'une félicité spirituelle qu'il communique au visionnaire; il a le maintien d'un jeune homme, désigné le plus souvent comme *iuvenis*.' Courcelle shows parallels for *hilares* (*passio Perp. et Fel.* 12. 7, *pass. Montani* 13, *Hermae pastor* sim. 9. 11. 4–5), *iuvenis* (*pass. Montani* 8, *pass. Mariani* 7, *vita Cypriani* 12, *pass. Isaacis et Maximiani*), and *arrisit* (*pass. Mariani* 6).

The best study is Dulaey 158–65, especially on the style (at p. 160: note the Ciceronian combinations, where the second element is the more specific: 'hilarem atque arridentem', 'maerens et maerore confecta', 'maestitiae cotidianarumque lacrimarum', 'iussisse atque admonuisse', 'attenderet et videret'). See also O'Daly, 114–27.

misisti manum tuum: Ps. 143: 7, 'emitte manum tuam ex alto et exime me, et erue me de aquis multis'; *en. Ps.* 143. 14, 'ipse salvator corporis, manus dei'; *en. Ps.* 108. 29, 'intellegamus itaque manum dei esse Christum; unde alibi dicitur, "et *brachium* domini cui revelatum est"?' Cf. 4. 12. 19, 9. 4. 9. Knauer, 53 n. 3, suggests the relevance of passages where God dwells 'in excelsis' et sim.: 1. 18. 29, 8. 3. 4, 13. 13. 14.

caligine: See on 2. 3. 8.

eruisti animam meam: Ps. 85: 13, 'confitebor tibi, domine deus meus, in toto corde meo, et glorificabo nomen tuum in aeternum; quoniam misericordia tua magna est super me, et eruisti animam meam ex inferno inferiore.' The second half of Ps. 143: 7 (just cited) provides the verbal echo that suggests the present line. *en. Ps.* 85. 17–18 finds *inferno inferiore* problematical, implying two layers of hell; hence that expression fails to find a place here.

fleret: The verb is common in Bks. 1–10, does not occur in 11–13. See on 3. 2. 4, 'lacrimae'.

fidelis tua: See on 2. 3. 6.

ex fide et spiritu: Cf. Gal. 5: 5, 'nos enim spiritu ex fide spem iustitiae expectamus.'

lacrimas: *persev.* 20. 53, 'et in eisdem etiam libris quod de mea conversione narravi, deo me convertente ad eam fidem quam miserrima et furiosissima loquacitate vastabam, nonne ita narratum esse meministis, ut ostenderem me fidelibus et cotidianis matris meae lacrimis, ne perirem, fuisse concessum?'

rigarent: The verb is commonly used of tears in CL: *Aen.* 9. 251, 'vultum lacrimis atque ora rigabat'; not biblical.

somnium: 9. 10. 25 (Ostia), 'sileant somnia et imaginariae revelationes.' Cf. *div. qu. Simp.* 2. 3. 2, 'item quisquis videt somnium, non dicit, "vidi imaginem Augustini aut Simpliciani", sed, "vidi Augustinum aut Simplicianum", cum eo tempore quo tale aliquid vidit nos ignoraremus, usque adeo manifestum est non ipsos homines sed imagines eorum videri.' See on 6. 13. 23 for Courcelle's argument that A. took particular interest in dreams and their meaning at Milan. A. accepts that ordinary dreams are not visions of reality but products of the imagination (*cura mort.* 11. 13).

(*vivere*) **mecum** GOS Knöll Skut. Ver. : *me secum* CD **me** Maur.

cederet GO¹S Knöll Skut. Ver. : *crederet* CDO² : *concederet* Maur.

habere OS edd. : *haberet* CD : *haberē* G.

(*habere*) **mecum** CDGOS Knöll Skut. Ver. : *secum* Maur.

The text is clear enough: Monnica, now widowed, would normally live in the home of her eldest son (as she did in Italy: *cura mort.* 13. 16, 'pia mater . . . quae terra marique secuta est [see on 6. 1. 1] ut mecum viveret'). When it came to that in Carthage (see below), she at first resisted the idea of living under the same roof and/or of sharing the same table; but she quickly enough gave up that idea, under the influence of her vision and her conversation with the bishop. This is not the customary interpretation here, and the reason is textual. The Maurists here read, with little manuscript support, *ut vivere me concederet et habere secum eandem mensam in domo*. That reading implied that A. was banished from M.'s house, and most scholars have held this view (Brown, 53: 'She shut him out of her house'; Mandouze, 87: 'refusant un temps de recevoir son fils sous son toit'); the correct reading of the MSS, however, makes it clear that it was the other way around—if any such thing happened at all, which 'nolle *coeperat*' makes unlikely.

A contributing difficulty is that of plotting A.'s movements through these pages. 3. 1. 1 takes him to Carthage to study; at 4. 1. 2, he is teaching; at 4. 7. 7, he reports that he began his teaching 'in municipio quo natus sum,' i.e. Thagaste; and at 4. 7. 12, grief-stricken by his friend's death, he flees. The common assumption seems to have been that A. returned to Thagaste and found his mother at first unwilling to take him in. This view is taken as reinforced by *c. acad.* 2. 2. 3, thanking Romanianus for sharing his home with A.: 'in nostro ipso municipio ... communicatione domus' (but the same passage shows that A. had been received into Romanianus' house before as well as after the death of Patricius, and in neither case does the language require us to assume A. actually resided with R.); and see on 4. 4. 9, 'paterna domus', for a sign that A. did live at home when in Thagaste in 375/6.

But nothing in Bk. 3 suggests any return to Thagaste or any teaching. It is the book of his wastrel student years; hence this passage suggests a time when M. came to Carthage after Patricius' death. The evidence on M.'s role is ambiguous; from 3. 4. 7, 'maternis mercedibus', it would appear that she 'paid the bills' after Patricius' death, but there is no evidence that she ever acted as head of a household (the picture of domestic life at Cassiciacum suggests the opposite). A.'s entourage seems to have depended on the kindness of friends for domicile in Rome, Milan, Cassiciacum, and Ostia. The present passage describes, then, a time when M. thought of insisting on living apart from A., and may even have done so for a brief time, but where and how she might have lived then—on whose kindness she would have depended—we know not.

blasphemias: A Grecism retained in Latin versions of scripture. Elsewhere in *conf.* only of his Manichean doctrines: 4. 16. 31 contrasts *blasphemias* to *confessio* and *invocatio*; elsewhere, A. uses the verbal form as the opposite of *laudare* (*en. Ps.* 149. 2, 'caritas laudat dominum, discordia blasphemat dominum'). He implies a definition thus: *c. mend.* 19. 39, 'ideo autem peius est blasphemare quam peierare, quoniam peierando falsae rei adhibetur testis deus, blasphemando autem de ipso falsa dicuntur deo.'

regula: The wooden rule is problematic, for we are too familiar with wooden rulers to see instinctively the potential strangeness of the object. For interpretation, see S. Poque, *Riv. stor. e lett. rel.* 30 (1984), 480–8: on the basis of Vitruvius 8. 5. 3 ('chorobates autem est regula longa circiter pedem viginti'), she believes that the instrument was a levelling device used in building aqueducts. The image would then place A. and M. a few feet off the ground, on the 'straight and narrow'. The vision is recalled at the end of the garden scene in 8. 12. 30, where the phrase is 'regula fidei', a much commoner phrase in A. and one roughly equivalent to 'baptismal creed': see Poque, 488 n. 36, quoting *symb. cat.* 1. 1, 'accipite regulam fidei, quod symbolum dicitur'. On *that* reading, the slight elevation is reminiscent of 8. 2. 5, 'de loco eminentiore', whence Victorinus declaimed the baptismal creed (the builders' instrument seems to have had a channel of water running along its length to assist in levelling). Further texts on the *regula fidei* in A. in the otherwise superseded article of L. C. Ferrari, *Aug. Stud.* 6 (1975), 193–205; see also Ferrari, *Aug. Stud.* 7 (1976), 47–58, and *Aug. Stud.* 10 (1979), 3–17 (where he offers the useful reminder that we never hear of any of Augustine's own dreams).

arridentem: The only place in *conf.* where laughter has an unmistakably positive character; see note on the beginning of this paragraph.

maerore confecta: Lam. 1: 13, 'de excelso misit ignem in ossibus meis et erudivit me, expandit rete pedibus meis, convertit me retrorsum, posuit me desolatam tota die maerore confectam.' See on 'attenderet et videret' below.

docendi . . . non discendi gratia: *mag.* 1. 1, 'loquendo nos docere velle manifestum est; discere autem quomodo? . . . etiam tunc nihil aliud quam docere nos velle intellego.' There, the quibble is that all speech is of one kind; here, a more delicate observation is made; but the phraseology remains suggestively parallel.

ut adsolet: G–M: 'Cf. e.g. Jn. 21: 3, Acts 1: 11', by which they imply that the figure is Christ. The phrase certainly seems to demand some such explanation.

quo secura esset: Referring not to a gesture of the youth, but to A.'s own interpretation afterwards (Dulaey, 161).

attenderet et videret: Lam. 1: 12, 'attendite et videte si est dolor sicut dolor meus'; Is. 63: 15, 'attende de caelo et vide.'

aures tuae: Cf. Ps. 9: 38, 'praeparationem cordis eorum exaudivit auris tua'; *en. Ps.* 9. 33, 'haec est cordis praeparatio . . . de qua dicit apostolus, "si autem quod non videmus speramus, per patientiam exspectamus".'

3. 11. 20

visum: Regular for dream-vision: see on 6. 1. 1.

inquit: This is the first verbatim quotation of another person's specific spoken words in *conf.* (there are numerous earlier 'generic quotations'—the sort of thing someone might say, or the sort of thing everyone at the time was saying).

confiteor: The first 'confessional' meditation in Bk. 3, which has been otherwise devoted entirely to the description of A.'s fallen ways.

responso tuo: The dream begets two *responsa*: here and in 3. 12. 21 (the bishop).

tam vicina interpretationis falsitate: Bad reading is always close to good, a danger for all commentators.

novem ferme anni: i.e. 373–82; see on 5. 6. 10 for passages in other works citing the same number (cf. 4. 1. 1, 'per idem tempus annorum novem') and for discussion of the chronological implications.

in illo limo profundi: See on 2. 2. 2, 'de limosa concupiscentia', and cf.

7. 18. 24, 'de limo nostro', of the earth from which Adam was created. In the last line here, it is again *caligo* in which A. is caught. Ps. 68: 3, 'infixus sum in limo profundi'; *en. Ps.* 68. s. 1. 4, 'de limo enim factus est homo. sed isti, cadendo a iustitia, limus profundi facti sunt; quibus persequentibus et ad iniquitatem trahere cupientibus quisquis non consenserit, de limo suo aurum facit. merebitur enim limus ipsius converti in habitudinem caelestem.'

quales amas: 1 Tim. 5: 3–16 catalogues the virtues of Christian widows; see on 5. 9. 17, 'viduae castae ac sobriae'.

intrabant in conspectum tuum: Ps. 87: 3, 'intret in conspectu tuo oratio mea; inclina aurem tuam ad precem meam'; *en. Ps.* 87. 2, 'ingressus orationis in conspectu dei, acceptatio eius est.' The prepositional phrase recurs constantly, evoking the silent presence of God in which this work unfolds: 5. 3. 3, 6. 2. 2, 7. 8. 12, 9. 2. 2, 9. 7. 16, 9. 12. 33, 10. 2. 2, 10. 4. 5, 10. 5. 7, 10. 35. 57, 11. 2. 4, 12. 11. 11 (2 ×), 12. 11. 13.

3. 12. 21

This first bishop in the text is an African ex-Manichee who had probably been brought up a Christian ('nutritum in ecclesia': he had a Christian mother, but it was the mother who was 'seducta' by the Manichees and gave up the son to them) and who had written Manichean books (something A. never did: *de pulchro et apto* was philosophy, not theology, like the dialogues of Cassiciacum; A.'s strengths were in disputation, which he practised when and how he could (cf. 'nonnullis quaestiunculis . . . exagitassem')). He is also the first of A.'s Doppelgänger as well; like that of Victorinus, his story inserted in *conf.* mirrors and implicitly comments upon that of A.

G. Papini, *S. Agostino* (Florence, 1930), 73 and 388, thinks this was Antigonus, bp. of Madauros at the Council of Carthage of 349; Vega quotes Papini but is agnostic; Decret seems nowhere to have broached the question of his identity. Mandouze, *Pros. chr.* s.v. Antigonus, does not discuss this possibility, and is even hesitant about identifying Antigonus' see with Madauros. Frend, *Donatist Church*, 236: 'Apart from Augustine himself and Alypius, successive bishops of Constantine, Profuturus and Fortunatus, were ex-Manichees. This fact did not escape the notice of the Donatists.'

propero: This haste is a purely literary affectation; cf. on 9. 4. 7, 'quando mihi sufficiat tempus commemorandi . . . praesertim ad alia maiora properanti?'; and on 9. 8. 17, 'multa praetereo quia multum festino' (where a Ciceronian parallel is telling). Not to be taken, as Courcelle does (*Recherches*, 23–4), as a sign of rushed or thoughtless composition, or of defects of structure in the existing work. (Cf. *civ.* 3. 21, 'multa enim praetereo suscepti operis modum cogitans.')

errores: See on 1. 20. 31 for the appropriateness of this word to the faults of *curiositas*.

faciebat: The *BA* translation makes Monnica the subject of this verb; Ryan and Pusey clearly and rightly apply the practice to the unnamed bishop.

exagitassem: Alfaric, 48, sees this memory as the forerunner of A.'s many later public debates and disputations: 'Son existence entière n'est qu'une longue discussion, où ses partenaires se renouvellent constamment sans parvenir à l'épuiser, et où lui-même modifie bien parfois ses convictions mais sans rien perdre jamais de son ardeur première. Il ne se contente pas d'avoir à moitié raison. Il prétend détenir la vérité intégrale.' In Bk. 8 it is a *controversia* within himself that leads to his conversion (8. 11. 27, 'ista controversia in corde meo'). For one example of his pro-Manichee debating, see 4. 15. 26, 'et dicebam parvulis fidelibus tuis . . .'.

tantum CDGS Knöll Skut. Ver. : *et tantum* O Maur.

ipse legendo reperiet: A prophecy fulfilled in Bk. 7 (7. 9. 13, neo-Platonic texts leading to scripture), Bk. 8 (8. 12. 29–30, the garden scene), Bk. 9 (9. 4. 8 ff., reading Psalm 4).

scriptitasse: Brown, 53, interprets merely as 'had religiously copied out the authoritative books of Mani'.

substomachans taedio: A. still portrays Monnica in a less than completely flattering light. (The verb is another apparent Augustinian neologism (Hrdlicka, 16), but if so its next appearance in the literature is in a surprising place: Julian, quoted at *c. Iul. imp.* 6. 16, 'et substomacharis, si senis Manis soboles nuncupere?' Julian was capable of turning A.'s own words back at him (see on 9. 8. 18, 'meribibula'), but this is more likely a case of a recent word in circulation attested by chance in A. (see on 10. 6. 7 for another example).)

ita vivas: Knöll (followed by Skut.and Ver.) weakens the punctuation after this phrase to a comma.

filius istarum lacrimarum: Mandouze, 86 n. 5, notes the strength of 'istarum', and recommends stronger translation than is customary—than e.g. Ryan's 'son of such tears': the sense is something like 'son of tears like these of yours'.

ac si de caelo sonuisset: The oracular quality of this moment ending the book parallels the garden scene at the end of Bk. 8; cf. *ep.* 54. 2. 3, 'hoc cum matri renuntiassem, libenter amplexa est. ego vero de hac sententia [on Saturday fasting] etiam atque etiam cogitans ita semper habui, tamquam eam *caelesti oraculo* acceperim.'

BOOK FOUR

4. 1. 1–3. 6	*Teaching (at Carthage)*
4. 4. 7–12. 19	*The death of a friend (at Thagaste)*
4. 13. 20–16. 21	*Intellectual development (at Carthage)*

4. 1. 1–3. 6 portray A. at work as teacher, the prey of the *variae cupiditates* (4. 1. 1, sc. *carnis et oculorum*) to which he had yielded in the course of the narratives of Bks. 2 and 3. As G–M remark, 'artem rhetoricam' at 4. 2. 2 means that these paragraphs deal with his career at Carthage; the rhetorical contests, with awards by the proconsul, such as those described in 4. 3. 5—and implied by 4. 1. 1, 'agonem coronarum'—were surely held there. We must therefore conclude that this book is made up of reminiscences of Carthage (376/83) framing the Thagaste episode (375/6) in the mid-section of the book (see on 3. 11. 19, 'vivere mecum').

4. 4. 7–12. 19 recount the death of his friend and present an extended meditation on its meaning. The rustication to Thagaste may have been half-involuntary: *cod. theod.* 14. 9. 1 of 12 March 370 (dealing, to be sure, only with Constantinople and Rome[1]) decreed that students could remain at Rome only 'usque ad vicesimum aetatis suae annum'. It may also have been correspondingly brief (our only other information on that period is that Alypius first studied with A. then: 6. 7. 11), thereby partly accounting for the limited place this phase of A.'s career plays in his narrative. (R. A. Kaster, *Guardians of Language* (Berkeley, 1988), 129, emphasizes the infrequency with which *grammatici* ascended the ladder to become rhetoricians, as A. did on moving from Thagaste to Carthage. A.'s employment as a *grammaticus* was thus perhaps a temporary expedient.)

4. 13. 20–16. 31 depict his intellectual life at Carthage, describing the circumstances surrounding the writing of the *de pulchro et apto* (4. 13. 20–15. 27), then recalling an earlier, undated intellectual feat, his reading of the *categoriae* of Aristotle (4. 16. 28–31). The book is thus loosely bracketed by two acts of successful interpretation of difficult texts: at 4. 3. 5, when Vindicianus is presented as a student of astrology, and at 4. 16. 28, when A. reads Aristotle on the categories.

The structure of Bk. 4 offers L. Verheijen in *Collectanea Augustiniana* (New York, 1990), 175–201 (the same article in French in V.'s *Nouvelle approche de la Règle de saint Augustin* (Louvain, 1988), 2. 92–114), opportunity for stimulating speculations about A.'s habitual patterns of constructing literary works in tripartite ways.

[1] Exceptions have been noted only for Athens: A. Müller, *Philologus*, 69 (1910), 299.

4. 1. 1

The construction of the paragraph is unusually artful. The first sentence is remarkably ornate, while the ring composition from 'inrideant et confitear' to 'sed inrideant . . . autem confiteamur' sets off a passage reminiscent of 1. 1. 1.

annorum novem: In all, from 372/3 to 381/2. For the first date, see on 3. 4. 7; for the chronological problems, see on 5. 6. 10.

seducebamur et seducebamus: The antitheses here are repeated in deliberate co-ordination through 'per quos liberaremur,' juxtaposing A.'s public life as teacher of traditional rhetoric and his private activities on behalf of Manicheism. Note 'palam . . . occulte', 'hic . . . illic', 'hac . . . illac'. Cf. *Io. ev. tr.* 8. 11, 'verumtamen seducti [mathematici] seducunt.'

falsi atque fallentes: Cf. 7. 2. 3, 'deceptos deceptores' (to the analogous texts there add Macrob. *somn. Scip.* 1. 6. 64, 'Hippocrates . . . tam fallere quam falli nescit'). *quant. an.* 33. 76, 'vanitas vanitantium. vanitas enim est fallacia, vanitantes autem *vel falsi vel fallentes* vel utrique intelleguntur'; *c. acad.* 2. 5. 12, 'homini enim homo *falsus* docendus, *fallax* cavendus debet videri, quorum prius magistrum bonum, posterius discipulum cautum desiderat'; *c. litt. Pet.* 3. 17. 20, 'vel falsus vel fallens'; sim. (usu. *fallax* instead of the participle) at *duab. an.* 1. 1, *trin.* 3. 7. 12 ('ad fallendos fallaces'), *civ.* 9. 18, 11. 13, *retr.* 1. 7. 1 ('manichaeorum iactantiam de *falsa et fallaci* continentia vel abstinentia').

doctrinas . . . liberales: See Excursus on 4. 16. 30; the book is bracketed by the 'liberal arts'. The form of expression, which recurs there, is somewhat pejorative, as Pellegrino saw (reviewing Courcelle, *Les Confessions*, in *Studi Medievali* ser. 3, 4 (1963), 651): sim. at *s. Guelf.* 24. 4, *s.* 133. 4, and *ep.* 101. 1 ('litteris illis, quas variarum servi libidinum liberales vocant'), and cf. 4. 3. 4, 'quos mathematicos vocant'.

occulte: Imperial legislation, tempered by neglect, commended prudence; compare the case of Faustus (*c. Faust.* 5. 8): accused by Christians, brought before the proconsul, condemned; at the request of his accusers, his punishment was mitigated to relegation to an island, from which not long after he was released by a general amnesty of the emperor. (See on 5. 3. 3.)

superstitiosi: See on 3. 6. 10, 'incidi'.

vani: *vanus/vanitas* occurs 8 × in Bk. 1, particularly of A.'s educational hopes and ambitions; recurs in the same context twice in Bk. 3, and now occurs repeatedly in Bk. 4: note especially that the address to his soul at 4. 11. 16 begins thematically, 'noli esse vana, anima mea'. It is then a constant refrain until 8. 11. 26, just before the garden scene: 'retinebant nugae nugarum et vanitates vanitantium'. Thereafter it appears mainly in the reading of Psalm 4 (9. 4. 9, where Ps. 4: 3 reads 'ut quid diligitis vanitatem') and in the examination of conscience under the headings both of *curiositas* (10. 35. 54–7) and *ambitio saeculi* (10. 38. 63, 'de ipso vanae gloriae contemptu'). Thereafter it is rare, but note that at 11. 11. 13 ('cor eorum volitat et adhuc vanum est'), the reference is to Manichean opponents. See L. Chevallier and H. Rondet, *REAug* 3 (1957), 221–34, esp. 222.

hac Z (**h** sub linea) Maur. Knöll Skut. Ver. Pell. : *ac* CDGOS.

gloriae ... inanitatem [1] **... spectaculorum nugas** [2] **... intemperantiam libidinum** [3]: His public life alone offers a comprehensive display of the vices of 1 Jn. 2: 16.

usque ad theatricos plausus: See below, 4. 2. 3, 'theatrici carminis certamen'.

agonem coronarum: 4. 3. 5, 'coronam illam agonisticam'.

faenearum: G–M, 'perhaps "quickly fading"; ... cf. Sueton. *Nero* xii "orationis carminisque latini coronam".' An image of frailty, as at 9. 7. 16 (where see notes), 'quantum patet aura in domo faenea' (cf. Is. 40: 6, 'omnis caro faenum'); sim. at 13. 15. 18.

nugas: See on 3. 10. 18.

sordibus: See on 3. 1. 1, 'sordibus concupiscentiae'.

electi et sancti: *en. Ps.* 140. 10, 'qui sunt isti electi? quibus si dixeris: peccasti; statim illam defensionem impiam et peiorem ceteris magisque sacrilegam proferunt: "non ego peccavi, sed gens tenebrarum." quae est ista gens tenebrarum? quae bellum gessit cum deo.'

angelos: See on 3. 10. 18, 'anhelaret de illa angelos'.

inrideant: Again below here; see on 1. 6. 7.

elisi: 11. 31. 41, 'tu enim erigis elisos'.

dedecora: See e.g. 2. 4. 9, 'defectum meum ipsum amavi, ... non dedecore aliquid sed dedecus appetens [sc. in concupiscentia carnis].'

laude: Ps. 105: 47, 'salvos nos fac, domine deus noster ..., ut confiteamur nomini sancto tuo et gloriemur in laude tua.' Knauer, 80 n. 2, sees in *laus dei* a theme uniting the first half of Bk. 4.

circuire ... immolare: Ps. 26: 6, 'circuivi et immolavi in tabernaculo eius hostiam iubilationis'; *en. Ps.* 26. en. 2. 12, 'quid ergo immolamus? abundantissimum et inenarrabile gaudium, nullis verbis, voce ineffabili. haec est hostia iubilationis.' At *sol.* 2. 20. 34 and *quant. an.* 4. 6, 7. 12, *circuitus* is used of the roundabout ways of argument that leads to vision of higher things, hence with almost a positive sense; here, however, cf. 8. 2. 3, 'narravi ei circuitus erroris mei.' On imperatives at the beginning or end of sections of the text, see Knauer, 64 with n. 1 and (esp. on this passage) 72.

circuire O²S Knöll Skut. : *circumire* CDO¹ Maur. Ver.

praesenti memoria: = *contuitus* [2] (see on 12. 15. 18, 'expectatio ... contuitus ... memoria'), and see Excursus on 10. 8. 12, 'Memory in Augustine'.

praeteritos circuitus: = *memoria* [1] (or rather its contents: the past).

immolare tibi hostiam iubilationis: = *expectatio* [3] (the future life of ceaseless praise).

quid . . . sum: The question is asked twice: once of what A. is like without God, once of what he is like when he is with God.

dux in praeceps: 2. 3. 7, 'et praeceps ibam tanta caecitate'. For a better *dux*, see 2. 7. 15 and 7. 10. 16 ('te duce').

sugens lac tuum: Cf. Deut. 33: 19, 'populos ad montem vocabunt; ibi immolabunt victimas iustitiae qui inundationem maris quasi lac sugent et thesauros absconditos harenarum'; Is. 60: 16, 'et suges lac gentium et mamilla regum lactaberis.' These references echo the vocabulary here, but are not particularly apt; better, 1 Cor. 3: 2–3, 'lac vobis potum dedi, non escam, nondum enim poteratis. sed ne nunc quidem potestis, (3) adhuc enim estis carnales.' Recall the suckling child at 1. 6. 7, and cf. 7. 18. 24, 'ut infantiae nostrae lactesceret sapientia tua,' and 13. 18. 23, 'tamquam parvulus in Christo lactisque potator'. R. J. O'Connell, *Thought*, 48 (1983), 188–206, emphasizes the maternal qualities imputed to God by this image, and refers to numerous passages in Isaiah, but his treatment of evidence is entirely unsatisfactory. The image gives rise to the observation of R. Brändle and W. Neidhart, *Theol. Zschr.* 40 (1984), 166 n. 67: 'Augustin findet in seinem Suchen nach Gott nur selten Begriffe der Väterlichkeit. Viel häufiger sind Attribute oder Bezeichnungen für Gott, die in die Sphäre des Weiblichen gehören.' But A. was far less insistent on drawing gender lines than are moderns: for example, at 13. 22. 32, A. applies the same image to Paul: 'ad hoc enim dispensator ille tuus generans per evangelium filios, ne semper parvulos haberet, quos lacte nutriret et tamquam nutrix foveret.' (And see on 9. 10. 23, where Monnica has a masculine role to play.)

cibo qui non corrumpitur: Cf. Jn. 6: 27, 'operamini non cibum qui perit, sed qui permanet in vitam aeternam.' Cf. 3. 1. 1, 'fames mihi erat intus ab interiore cibo.'

et quis homo est: G–M: 'and what kind of a man is any man when he is merely a man? i.e. who is truly a man when he is only a man?'

fortes et potentes: Cf. 1 Cor. 4: 10, 'nos stulti propter Christum, vos autem prudentes in Christo; nos infirmi, vos autem fortes, vos nobiles, nos autem ignobiles.'

infirmi et inopes: Cf. Ps. 73: 21, 'egenus et inops laudabunt nomen tuum'; other echoes at 10. 38. 63 and 11. 2. 3.

4. 2. 2

artem rhetoricam: Possidius *v. Aug.* 1. 2, 'nam et grammaticam prius in sua civitate et rhetoricam in Africae capite Carthagine postea docuit.' On the implications for dating, see on 4. 1. 1.

loquacitatem: See on 1. 4. 4; *loquacitas* is apt against himself as a Manichee and merchant of verbal facility.

cupiditate: If in 4. 1. 1 the *cupiditatibus* denote the temptations of the flesh and of the eyes, here only *ambitio saeculi* is meant.

tu scis: See on 1. 5. 6, 3. 3. 6, etc.; Knauer, 76–7.

caput innocentis: Cic. *off.* 2. 14. 51 (cf. Testard, 1. 44 ff., 2. 25), 'atque etiam hoc praeceptum officii diligenter tenendum est, ne quem umquam *innocentem* iudicio *capitis* arcessas; id enim sine scelere fieri nullo pacto potest. nam quid est tam inhumanum quam eloquentiam a natura ad salutem hominum et ad conservationem datam ad bonorum pestem perniciemque convertere?'

de longinquo: On the remoteness of God, see on 1. 18. 28, esp. echoing here again Lk. 15: 13, 'in regionem longinquam'.

fidem meam: Here and just below, the only occurrences of *fides* in *conf.* without some Christian overtone; here, to show the lingering traces in his life of something absent in every essential way.

diligentibus vanitatem et quaerentibus mendacium: Ps. 4: 3, 'filii hominum, usquequo graves corde? utquid diligitis vanitatem et quaeritis mendacium?'; see on 9. 4. 9. A. twice wrote on lying (*mend.* of *c.*394/95 and *c. mend.* of *c.*420), taking an absolutist position (note *c. mend.* 20. 40: the catechumen prisoner of infidel guards, about to be put to death, is not to lie in order to gain a chance at baptism), which has discomfited later moralists (recently, S. Bok, *Lying* (New York, 1978)); the harsh judgement of his own professorial past is intimately linked with this attitude.

unam habebam: For the legalities and sentimentalities of A.'s 'marriage', see on 6. 15. 25.

sane: Virtually *quidem* (restrictive), 'to be sure': Hensellek, *Anzeiger Akad. Wien* 120 (1983), 102–3; 11 × in all in *conf.*

proles: Adeodatus was 'annorum ... ferme quindecim' at the time of his baptism in the spring of 387 (9. 6. 14) and 'in annis sedecim' at the dramatic date of *mag.* not long after; on this calculation, he was born 371/72, when A. was perhaps 17. The words 'in illis annis' thus reflect another of the chronological inconsequences of *conf.*, inasmuch as they suggest a liaison that must have begun in 371 or perhaps even 370, thus apparently probably in the first year of studies at Carthage but conceivably during the year of indolence recorded at 2. 3. 5–6 (the philoprogenitive optimism of Patricius did not have so long to wait). Adeodatus' mother was dismissed from Milan and returned to Africa in 385/6 (see on 6. 15. 25), and thus shared A.'s outward peregrinations through his entire *adulescentia*.

contra votum nascitur: A. accused the Manichees of resisting marriage and children, hypocritically: 'filios autem inviti suscipiunt' (*c. Faust.* 15. 7; almost the same words at *c. Faust.* 20. 23). Brown, 39, offers a delicate suggestion: 'Next, his son, Adeodatus, was born: this event, unwelcome at the time, may well have had the "sobering" effect which Augustine would later recommend to young husbands [*b. coniug.* 3. 3].' But A. had something specific in mind in that passage: 'deinde quia reprimitur et quodam modo verecundius aestuat concupiscentia carnis, quam temperat parentalis affectus.'

4. 2. 3

A.'s reluctance to sponsor sacrifice is not a sign of personal delicacy, or anything better ('non ex tua castitate repudiavi'): *mor.* 2. 17. 54, 'ab animalium *nece* . . . vos temperatis', reminds us that it was against Manichean doctrine to kill any living thing (see Alfaric, 220–1, on this passage). As Knauer, 72–3, observes, this kind of sacrifice, at one time an option A. considered for propitiating his god, exemplifies the context in which we are to take e.g. 5. 1. 1, 'accipe sacrificium confessionum mearum'.

certamen . . . ut vincerem: The temptation is that of *ambitio saeculi* [1]: *vera rel.* 53. 103, 'quare qui fines ipsos desiderant, prius curiositate [2] carent. . . . deinde accipiunt actionis facilitatem [1] pervicacia posita scientes maiorem esse facilioremque victoriam non resistere animositati cuiusquam.'

mandasse mihi: G–M and Hrdlicka, 78, render 'mandare' = 'to send word' and cite Eutrop. 5. 5, '[Mithridates] senatui mandavit bellum se ei . . . inlaturum'; but the usage is classical (Tac. *ann.* 14. 38, 'mandabat in urbem, nullum proeliorum finem exspectarent'). Souter more usefully suggests merely 'send (a messenger, etc.)' and adds ' (Tert. on)': on that interpretation, here render 'sent (sc. to ask) . . .'.

haruspicem: Strictly the *haruspices* were interpreters of prodigies (as at *civ.* 2. 24, 3. 11), but A. knew them as sources of magic power as well (*en. Ps.* 73. 18, 134. 20, *doctr. chr.* 2. 20. 30, *s.* 9. 11. 17 (linked to criticism of *spectacula*)).

respondisse: The claim is corroborated at 10. 35. 56, 'nec anima mea umquam responsa quaesivit umbrarum'.

ex tua castitate: God's, not man's, as Madec rightly emphasizes at *REAug* 7 (1961), 246, where he castigates the French translators for their unwillingness to predicate chastity of God; the same is true in English: cf. even Ryan, 'it was not out of a chaste devotion to you'; better 'not by means of the chastity that comes from you'. *Io. ep. tr.* 4. 9, 'ergo castificat nos sicut et ipse castus est; sed ille castus aeternitate, nos casti fide.' The metaphor is counterposed to *fornicari abs te* (see on 1. 13. 21; the phrase recurs here and at 5. 12. 22); but perhaps it is stronger than metaphor: *beata v.* 3. 18, 'quomodo, inquit, castus potest esse qui ab inlicito tantum concubitu sese abstinens ceteris peccatis non desinit inquinari? ille est vere castus, qui deum attendit et ad ipsum solum se tenet' (words of Adeodatus, but strongly approved by A. ad loc.). 4. 3. 4 uses *salubritas* to denote roughly the same purity.

deus cordis mei: Ps. 72: 26–7, 'defecit cor meum et caro mea, deus cordis mei, et pars mea deus meus. (27) ecce qui longe se faciunt a te, peribunt. perdidisti omnem qui fornicatur abs te.' See also on 6. 1. 1, 'deum cordis mei'; cf. 9. 13. 35.

non . . . noveram . . . non noveram: Recalls the repeated use of this phrase at 3. 7. 12–13, and reiterates one of the ignorances confessed there: 3. 7. 12, 'non noveram deum esse spiritum'; Bk. 4 is haunted by the false god: 4. 7. 12, 'vanum phantasma et error meus erat deus meus.'

fulgores corporeos: *c. Faust.* 22. 8, 'non distinguunt inter lucem quod est ipse deus et lucem quam fecit deus.'

figmentis: 'products of artifice': poetic (1. 13. 22, 1. 17. 27), Manichean doctrine (3. 6. 10, 4. 15. 27, 4. 16. 29), his own misconceptions of catholic teaching (6. 3. 4, 10. 34. 53), and, once only in a strong metaphor, man as God's creation (13. 23. 33, 'quoniam tuum sumus figmentum creati in operibus bonis'). *trin.* 4. pr. 1, 'ego certe sentio quam multa figmenta pariat cor humanum. et quid est cor meum, nisi cor humanum? sed hoc oro *deum cordis mei*, ut nihil ex eis figmentis pro solido vero eructuem in has litteras.' Frequent in *civ.* in negative sense, esp. 'figmenta poetarum' (e.g. *civ.* 2. 8).

fornicatur abs te: Cf. Ps. 72: 26, quoted above; see on 1. 13. 21; *en. Ps.* 72. 33, 'huic fornicationi contrarius est amor castus. quis est amor castus? amat iam anima sponsum suum.'

fidit in falsis et pascit ventos: Prov. 10: 4, 'qui fidit in falsis, hic pascit ventos'; Hosea 12: 1, 'Ephraim pascit ventum et sequitur aestum; tota die mendacium et vastitatem multiplicat.' The Prov. text is not in the Hebrew, and is cited only in the app. crit. of the Stuttgart Vulgate as an addition to Prov. 10: 4, with thin support (though it is in the Clementine Vulgate, hence its familiarity to modern readers and the verse number). It occurs in LXX as Prov. 9: 12a where it immediately precedes the *aenigma Salomonis* (see on 3. 6. 11) and is catalogued by La Bonnardière, *Biblia Augustiniana: Proverbes*, 212 as Prov. 9: 12b (but the passage from *bapt.* 6. 12. 18 should be struck out: it arises from an Erasmian interpolation that the Maurists failed to eject). Used in various contexts, the verse consistently suggests to A. that those who employ falsehood become themselves food for demons: cf. *c. Faust.* 20. 22, 'illi quippe superbi et impii spiritus non nidore ac fumo, sicut nonnulli vani opinantur, sed hominum pascuntur erroribus'; sim. at *c. Sec.* 26, *c. Cresc.* 3. 9. 9, and *ep.* 105. 2. 6. (Identification by A. Vaccari, *Scritti di erudizione e di filologia*, 2 (Rome, 1958), 12–13.)

daemonibus, quibus . . . sacrificabam: See on 1. 17. 27, 'non enim uno modo sacrificatur transgressoribus angelis.'

superstitione: See on 3. 6. 10.

4. 3. 4

Astrology recurs at 7. 6. 8, esp. with the mention there of Vindicianus (mentioned but not named at 4. 3. 5). This paragraph continues the motif of *curiositas* from Bk. 3; cf. 10. 35. 56, 'nec curo nosse transitus siderum', from the discussion there of *curiositas*; and *ord.* 2. 15. 42, 'quae similiter definiendo ac secernendo in ordinem nectens astrologiam genuit, magnum religiosis argumentum *tormentumque curiosis*.'

Astrology recurs pastorally throughout Augustine's career. The same vocabulary, and the same scriptural quotations, appear over and over. The place of astrology in African life is vividly depicted at *cons. ev.* 1. 23. 36: if the Romans were consistent, 'et mathematicos vel genethliacos maxime delerent, qui Saturnum, quem sapientium effectorem isti dicerent, maleficum deum inter alia sidera constituerent.' At Carthage, the fear of him was so profound that the *vicus Saturni* was often referred to euphemistically as the *vicus Senis*. The presence of soothsayers was also a persistent pastoral concern. Consider e.g. *Io. ev. tr.* 7. 7, 'non quando nobis dolet caput, curramus ad praecantatores, ad sortilegos et remedia vanitatis. fratres mei, non vos plangam? cotidie invenio ista; et quid faciam? nondum persuadeo christianis in Christo spem esse ponendam?' Cf. *Io. ev. tr.* 8. 8 f. and 10. 5.

How early in life A. read Cicero's *div.* is uncertain; before *civ.* the only probable sighting is at *conf.* 10. 16. 25 (see notes there for the uncertainty). It is not impossible, of course, that A. knew this work; to accept that proposition we must also accept that A. could be immune to the arguments even of figures he respected highly, and this from early in life. See on 8. 7. 17, 'petieram a te castitatem', for the history of his resistance to those (including Cicero) who preached continence. Many of his moments of 'conversion' took the form of re-reading, and now accepting, books and authors he had long known.

planos: used in this sense (= πλάνος) by Cic., Hor., Petron., Plin. iun., and Aul. Gell.; *OLD:* 'one who practises deceit or imposture, esp. as a means of making a living'. Amerbach's Basle edition of 1506 wrongly conjectured *planetarios* (surviving as a ghost word in Lewis and Short).

mathematicos: Cf. *c. acad.* 1. 7. 21, for the learned derision of Flaccianus (procos. Afr. 393?, a pupil of A.? (*c. acad.* 1. 6. 18)) directed against the Carthaginian soothsayer Albicerius; Courcelle, *Recherches*, 76 n. 3, suggests Flaccianus probably was behind the events described here in *conf.* (Mandouze, *Pros. chr.* s.v. Flaccianus, tends to agree); cf. 4. 3. 5, 'sine verborum cultu', for a hint that Albicerius is the *divinator* meant here (for at *c. acad.* 1. 7. 21, F.'s derision arises from his obvious cultural superiority to the less educated abilities of Albicerius); Flaccianus was at some other date the source for A.'s acquisition of the Christian Sibylline texts (*civ.* 18. 23). Possidius *indiculum* (ed. Wilmart, *MA* 2. 185, line 49) notes a letter to Flaccianus, which was apparently extant *c.*825: Courcelle, *Recherches*, edn. 2, 262.

Two of A.'s most detailed discussions of the subject of astrology precede *conf.* by a relatively short period, and another probably followed not long after:

div. qu. 45. 2, 'adversus eos autem qui nunc appellantur mathematici, volentes actus nostros corporibus caelestibus subdere, et nos vendere stellis, ipsumque pretium quo vendimur a nobis accipere'; he instances the case of twins with different fates, though sharing the same *constellatio* (see on 7. 6. 8), to confute their pretensions.

doctr. chr. 2. 21. 32, 'neque illi ab hoc genere perniciosae superstitionis segregandi sunt qui genethliaci propter natalium dierum considerationes, nunc autem vulgo mathematici vocantur. nam et ipsi, quamvis veram stellarum positionem cum quisque nascitur consectentur et aliquando etiam pervestigent tamen quod inde conantur vel actiones nostras vel actionum eventa praedicere, nimis errant et vendunt imperitis hominibus miserabilem servitutem. nam quisque liber ad huiusmodi mathematicum cum ingressus fuerit, dat pecuniam ut servus inde exeat aut *Martis* aut *Veneris*, vel potius omnium siderum; quibus illi qui primi erraverunt erroremque posteris propinaverunt, vel bestiarum propter similitudinem vel hominum ad ipsos homines honorandos imposuerunt vocabula.' The example of twins is used in refutation at *doctr. chr.* 2. 22. 33: Esau and Jacob are named (as at *div. qu. Simp.* 1. 2. 3, where the astrological interpretation is trumped by Pauline grace); for conventional antecedents of the argument, cf. Amb. *exam.* 4. 4. 14 and Persius 6. 18–19. Cf. *doctr. chr.* 2. 23. 35, 'hinc enim fit ut, occulto quodam iudicio divino cupidi malarum rerum homines tradantur inludendi et decipiendi pro meritis voluntatum suarum, inludentibus eos atque decipientibus praevaricatoribus angelis. . . . quibus inlusionibus et deceptionibus evenit, ut istis superstitiosis et perniciosis divinationum generibus multa praeterita et futura dicantur nec aliter accidant quam dicuntur multaque observantibus secundum observationes suas eveniant, quibus implicati *curiosiores* fiant et sese magis magisque inserant multiplicibus laqueis perniciosissimi erroris. hoc genus *fornicationis animae* salubriter divina scriptura non tacuit.' (A. quotes Deut. 13: 1–3, 1 Kgs. 28: 14–20, Sirach 46: 23.)

Gn. litt. 2. 17. 35, 'de fatis autem qualeslibet eorum argutias et quasi de mathesi documentorum experimenta, quae illi *apotelesmata* vocant, omnino a nostrae fidei sanitate respuamus; talibus enim disputationibus etiam orandi causas nobis auferre conantur et impia perversitate in malis factis, quae rectissime reprehenduntur, ingerunt accusandum potius deum auctorem siderum quam hominem scelerum.' Jacob and Esau appear at 2. 17. 36.

The other most substantive discussion is at at *civ.* 5. 1; cf. *epp.* 246. 2, 55. 4. 6 ff., *Gal. exp.* 34. For a particularly vivid document giving life to the polemics, see *en. Ps.* 61. 23 quoted below. The attack is eventually modified to take in the Pelagians: *c. ep. pel.* 2. 6. 12, 'fatum quippe qui adfirmant, de siderum positione ad tempus quo concipitur quisque vel nascitur, quas constellationes vocant, non solum actus et eventa verum etiam ipsas nostras voluntates pendere contendunt; dei vero gratia non solum omnia sidera et omnes caelos, verum etiam omnes angelos supergreditur.'

plane: Here, 'utterly, absolutely, quite' (*OLD*). Translators have strayed: Ryan: 'openly to consult'; but Valentinian II had made consultation of *mathematici* a capital offence (*cod. theod.* 9. 16. 8). Carena's translation assumes a connection in sense to 'planos': 'Perciò quegli altri vagabondi, che chiamano matematici, non desistevo che vagamente dal consultarli.'

quod quasi: 'for the reason that it was as if'—with subjunctive in the protasis of a contrary-to-fact condition. In 'quod tamen' below, *quod* is merely a connecting relative.

bonum est enim confiteri tibi: Cf. Ps. 91: 2, 'bonum est confiteri domino'; *en. Ps.* 91. 3, 'deinde multi non accusant satanam, sed accusant fatum. fatum meum me duxit, dicit. . . . quaeris ab illo quid sit fatum, et dicit: stellae malae. quaeris ab illo quis fecit stellas, quis ordinavit stellas; non habet quid tibi respondeat, nisi deus. restat ergo ut . . . deum accuset.' This verse and the following quotation of Ps. 40: 5 recur in similar contexts because astrological doctrine implies that human responsibility for sin is non-existent. For A., the opposite is the case, and Ps. 40: 5, confessionally, is the thematic statement of this throughout Augustine's life.

'miserere . . . tibi': Ps. 40: 5, 'ego dixi: "domine, miserere mei, sana animam meam, quoniam peccavi tibi"'; *en. Ps.* 40. 6, 'in factis meis, in peccatis meis non accuso fortunam, non dico: "hoc mihi fecit fatum"; non dico: "adulterum me fecit Venus, et latronem me fecit Mars, et avarum me fecit Saturnus".'

Ps. 40: 5 appears 3 × in *conf.*, in each case introduced by *confiteri* (cf. 4. 12. 19 and 5. 10. 18). The latter passage is the only one in *conf.* where A. connects the appeal of Manicheism to the freedom from guilt for sin it could offer; and it is certainly significant (Knauer, 166) that astrology could offer a similar freedom, hence probable that the appeal of astrology both arose for him at this time alongside his Manicheism and is reported here in detail for its analogous function. He does once link the Manichees to astrological speculation: *cont.* 5. 14, 'et alii quidem qui sua consueverunt excusare peccata, fato se ad peccandum queruntur impelli, tamquam hoc decreverint sidera et caelum prius talia decernendo peccaverit. . . . haec manichaeorum est immundissima insania.' For the verse elsewhere, see *lib. arb.* 3. 2. 5 (implicit link to *curiositas*), *s.* 20. 2 (391? *c.*395?).

licentiam peccandi: Cf. Sirach 15: 21, in a discussion of purgatory and penance at *ench.* 19. 70, '"nemini" enim "dedit *laxamentum peccandi*", quamvis miserando deleat iam facta peccata si non satisfactio congrua negligatur.'

'ecce sanus . . . contingat': Jn. 5: 14, 'ecce iam sanus factus es: noli peccare, ne quid tibi deterius contingat'—to the lame man made to walk on the sabbath.

salubritatem: Picks up the medical metaphor implied by 'cura animam meam' and the John quotation, and continued and enlarged (with ironic twists) by the treatment of the medical proconsul in the next paragraph—'medicinae artis peritissimus'.

Venus . . . Mars: *en. Ps.* 140. 9, 'totum hoc doctum et magnum, defensio peccati est. eris adulter, quia sic habes Venerem; eris homicida, quia sic habes Martem. Mars ergo homicida, non tu; et Venus adultera, non tu'; cf. *s. Mai* 17. 1, *civ.* 5. 1 (Mars), *en. Ps.* 31. en. 2. 16 (Mars and Venus), *en. Ps.* 128. 9 (Saturn, Mars, and Venus), *en. Ps.* 61. 23 (where even a converted *mathematicus* is represented using the standard examples of Mars and Venus: quoted below). For the conventionality of the forms of expression, cf. Plot. 2. 3. 6. 1–4, *Ἄρεα δὲ τόνδε ἢ Ἀφροδίτην θεμένους μοιχείας ποιεῖν, εἰ ὡδὶ εἶεν, ὥσπερ ἐκ τῆς τῶν ἀνθρώπων ἀκολασίας αὐτοὺς ἐμπιπλάντας ὧν πρὸς ἀλλήλους δέονται, πῶς οὐ πολλὴν ἀλογίαν ἔχει;* Immediate influence is unlikely, for Plotinus is rather subtler than A., who has to find a way for Mars to stand for violence, rather than for collaboration in adultery.

caro et sanguis: Mt. 16: 17, 'beatus es, Simon Bariona, quia caro et sanguis non revelavit tibi, sed pater meus, qui in caelis est'; 1 Cor. 15: 50, 'quia caro et sanguis regnum dei possidere non possunt'. *Caro et sanguis* is not so common in scripture as its equivalent in English. One could read the Matthew text to say that the revelation that comes from *caro et sanguis* would be owed to a purely human means of divination or prognostication.

superba putredo: Cf. 2. 6. 14, 'o putredo, o monstrum vitae et mortis profunditas!' The exclamation there arises from a diagnosis of the sin of pride. Cf. also 2. 1. 1 ('computrui') and 4. 11. 16 ('reflorescent putria tua').

suavitas: Also of God at 2. 6. 13, 9. 1. 1, 10. 17. 26. Cf. Ps. 144: 7, 'memoriam abundantiae *suavitatis* tuae eructabunt, et *iustitia* tua exsultabunt.'

qui reddes unicuique: Ps. 61: 13, 'quia tu reddes unicuique secundum opera eius'; Mt. 16: 27, 'venturus est in gloria patris sui cum angelis suis, et tunc reddet unicuique secundum opus eius'; Rom. 2: 6 (see Knauer, 61 n. 1), 'in die irae et revelationis iusti iudicii dei, qui reddet unicuique secundum opera eius'; Job 34: 11 (VL), 'quia reddit homini opus suum, et iuxta viam suam unusquisque reperiet.'

Just here in Ps. 61 A.'s homily on the text records a dramatic moment: *en. Ps.* 61. 23, in an appendix 'post tractatum de psalmo, cum mathematicus in populo monstraretur': 'diu mathematicus fuit; seductus seducens, deceptus decipiens, inlexit, fefellit, multa mendacia locutus est contra deum. . . . iste dicebat quia adulterium non faciebat voluntas propria, sed Venus; et homicidium non faciebat voluntas propria, sed Mars; et iustum non faciebat deus, sed Iovis: . . . portat secum codices incendendos, per quos fuerat incendendus ut, illis in ignem missis, ipse in refrigerium transeat.' Despite the spontaneousness implied by the unusual stage-direction in the text there, it is likely that this last paragraph of *en. Ps.* 61 was integral to the sermon, with the converted *mathematicus* lurking to be brought out as a show-piece. Clearly there was much suspicion of the *bona fides* of this conversion, but A. vouched for him.

cor contritum: Ps. 50: 18–19, 'quoniam si voluisses sacrificium, dedissem utique; holocaustis non delectaberis. (19) sacrificium deo spiritus contribulatus; cor contritum et humilatum deus non spernit.' This is the *sacrificium* (cf. above) that A. offers as an alternative to what all the unbelievers—himself when young, the astrologers who lured him, and all the others—had to offer.

humilatum: For the spelling adopted in text and commentary, see D. De Bruyne, *MA* 2. 558–61.

4. 3. 5

eo tempore: If Vindicianus is the proconsul, this event dates to 379/82, and hence belongs unequivocally to the Carthage years of A.'s life (see on 4. 1. 1), not long before Faustus' appearance recounted in Bk. 5 (note 'rhetoricam' below, which is what he taught at Carthage). The episode here is carefully matched against the comparable discussion in 7. 6. 8; in both there is the intersection of *curiositas* and *ambitio saeculi*.

vir sagax: 7. 6. 8, 'acutus senex', where Vindicianus is named; also *ep.* 138. 1. 3, 'magnus ille nostrorum temporum medicus Vindicianus . . . acerrimus'. On Vindicianus, cf. *PLRE* 1. 967: *comes archiatrorum c.*379, serving Valentinian II; translated some Hippocrates; not otherwise attested as proconsul. See Jones, *LRE* 387, on appointment to administrative office as a reward for serving as physician to the court. O'Daly, 81 n. 5, suggests that V. may have been A.'s source for a fair amount of sophisticated medical information, notably his view of the mechanics of sense-perception; in this O'Daly follows Agaësse-Solignac (*BA* 48. 711–13) tracing echoes of V.'s *gynaecia* at *nat. et or. an.* 4. 2. 2 ff. An anecdote of Vindicianus from 411 shows him again the bastion of science against superstition for A. (but still thought to possess some 'inlicita potentia'): *ep.* 138. 1. 3.

peritissimus: On *peritia*, see on 1. 14. 23; often associated by A. in this neutral way with medicine (cf. e.g. *civ.* 22. 30 and elsewhere).

proconsul GO²S Knöll Skut. : *proconsule* CDO¹ : *pro consule* Ver. Pell. At 8. 4. 9, the editors read *proconsule*, where the MSS divide in essentially the same way except that S adds the final *-e*. On the most generous reading of the texts, A. used the connected form far more often than the etymological, and editorial practice in our editions probably masks the realities. In the editions contained in the Würzburg data base, there are eleven occurrences of the separated form, in five different works of A.; the connected form is cited 97 × from 18 different works, including four of the five that also display the separated form.

qui resistis . . . gratiam: Jas. 4: 6, 1 Pet. 5: 5 < Prov. 3: 34 (LXX): 'deus superbis resistit, humilibus autem dat gratiam.' See on 1. 1. 1.

tamen etiam: The particles suggest Vindicianus was not Christian.

iucundi: Cf. 5. 6. 10, 'hominem gratum et iucundum verbis', but also 8. 5. 10, 'deus, sola certa iucunditas'.

conloquio: Of private and intimate conversation: 4. 8. 13 ('conloqui et conridere et vicissim benivole obsequi'), 6. 7. 11, 9. 9. 19, 9. 10. 23, 12. 16. 23.

genethliacorum: *doctr. chr.* 2. 21. 32 (quoted above at length on 4. 3. 4), 'genethliaci propter natalium dierum considerationes, nunc autem vulgo mathematici vocantur.'

monuit: Of benevolent advice from one in (often parental or pseudo-parental) authority, and perhaps especially advice that is not particularly welcome: 1. 9. 14 ('recte mihi vivere puero id proponebatur, obtemperare monentibus'), 2. 3. 7 ('monitus muliebres'), 2. 3. 8, 6. 7. 11, 9. 5. 13, 10. 30. 41. A. makes no direct response, only asks the (perhaps teasing, perhaps hostile) question given below.

professionem . . . deferre: G–M allege this 'the technical phrase for making a return to the censor of one's occupation'. They seem to be inferring an idiom from the context of a single passage. *TLL* 5. 1. 314–18 s.v. *deferre* does not corroborate. *Deferre* is often used with *(in) censum,* but not with *professio*.

Hippocraten: Courcelle, *LLW* 195, quotes several texts to show that A. had access to Hippocrates in Greek in later years. Already in the late 390s, he knew of the problem of ps.-Hippocratean writings (*c. Faust.* 33. 6).

gravis: *gravis/graviter* 30 × in *conf.* 1–9, only 2 × in 10–13. Cf. 9. 6. 14, 'ingenio praeveniebat multos graves et doctos viros'.

vim sortis: See on 8. 12. 30 for A.'s own crucial experiment with the *sortes*, and note esp. *ep.* 55. 20. 37 quoted there.

instinctu: The word is infrequent in A. and almost always used in just this indefinite sense in a similar context: cf. 7. 6. 10, 'tu enim, domine, . . . occulto instinctu agis, ut dum quisque consulit hoc audiat quod eum oportet audire.' For parallels, see *civ.* 2. 26, 5. 7, 8. 26, and esp. 21. 8 (differentiating *monstra*, *ostenta*, *portenta*, and *prodigia*): 'sed viderint eorum coniectores, quo modo ex eis sive fallantur sive *instinctu spirituum,* quibus cura est tali poena dignos animos hominum noxiae *curiositatis* retibus implicare, etiam vera praedicant sive multa dicendo aliquando in aliquid veritatis incurrant.' For the classical origins of the usage, see Cic. *div.* 1. 6. 12, 'est enim vis et natura quaedam quae tum observatis longo tempore significationibus tum aliquo instinctu inflatuque divino futura praenuntiat' (similarly at *div.* 1. 18. 34, Cic. makes *instinctus* an explanation of accurate prognostication that comes without *ars*—see next note).

non arte sed sorte: Cf. 4. 4. 6, 'forte vel sorte, non arte,' and see on 7. 6. 9, 'non arte dici sed sorte'; 7. 6. 10, 'non ergo arte sed sorte vera diceret'—the latter texts from the revived discussion of astrology and divination at 7. 6. 8 ff.; *div.*

qu. 45. 2, 'quod si non arte de codicibus exit saepe versus futura praenuntians, quid mirum si ex animo loquentis, *non arte, sed sorte* exit aliqua predictio futurorum?' The distinction goes back to Greek wordplay on *τέχνη* and *τύχη* (Löfstedt, *Symb. Osl.* 56 (1981) 106, citing E. R. Dodds on Plato's *Gorgias* (Oxford, 1959), 192: *Gorgias* 448 C, *ἐμπειρία μὲν γὰρ ποιεῖ τὸν αἰῶνα ἡμῶν πορεύεσθαι κατὰ τέχνην, ἀπειρία δὲ κατὰ τύχην*). *BA* 13. 416 n. 1, quoting passages from Plot. 4. 4. 32–43: 'Ces vues de Vindicianus laissent percer une mentalité néo-platonicienne; en effet, il n'attribue pas au seul hasard les prédictions heureuses, mais reconnaît une certaine consonance de l'univers aux divers états de l'âme.'

4. 3. 6

deliniasti: 'Sketched', suggesting an imprecise figure; cf. *civ.* 17. 8, 'scriptura sancta etiam rebus gestis prophetans quodam modo in eo figuram deliniat futurorum'.

Nebridius: One of A.'s two important companions (the other is Alypius, not mentioned until 6. 7. 11) at Milan. A. always speaks warmly of N., though there was undoubtedly always some distance between them in their attitudes, right up to N.'s premature death. For fuller notice, see on 6. 7. 11. N.'s family home was near Carthage, reducing further any chance that this episode occurred at Thagaste.

castus GS Knöll Skut. Pell. Vega (wrongly reporting *O* in support of the reading) : *cautus* CDO Ver. Cf. 9. 3. 6, 'castitate perfecta atque continentia'—of N. G–M think *cautus* better here, since *castitas* is better predicated of him after conversion. *BA* defends *castus* by arguing that it is not *continentia* that is meant here, but rather just that he was untainted by this superstition: cf. *civ.* 8. 18, 'homo castus et ab artium magicarum sceleribus alienus', and see above on 4. 2. 3, 'ex tua castitate' (with Madec's note cited there). But the *BA* translation weakly offers only: 'jeune homme de grande vertu et de grande religion'. Juergens–Schaub (in Skutella (1969), 389) cite 1. 13. 21, 2. 5. 14, 4. 2. 3, 5. 12. 22, passages where, as they say, 'omnis a deo aversio *fornicatio* appellatur' (see on 1. 13. 21, 'fornicabar abs te'). *Cautus* is infrequent in A. and hardly offers unqualified praise—it has an overtone of scrupulosity when in 7. 6. 8 the astrologers are said to count the minutes of a horoscope 'cautissima observatione'. (A similar substitution of *caut-* for *cast-* appears at *c. acad.* 2. 2. 5—see Introduction.)

nullum certum quale quaerebam documentum: The first sign of this hankering for certitude (the emphasis at 3. 6. 10 is rather different), which is presented as reaching its most acute stage at the end of Bk. 5 and thenceforward as defining one of the main conditions of A.'s philosophical and religious inquiries; cf. 5. 14. 25.

4. 4. 7

The death of A.'s friend, centrepiece of Bk. 4, performs several functions.

(1) It marks a stage on the trajectory of temptation, a falling into sin whose punishment is such that we react by sinning further:

(*a*) since Bk. 2, we have seen the disordering of the power of friendship in his life (*concupiscentia carnis*);

(*b*) since Bk. 3, we have seen that his compassion and pity are out of joint (*concupiscentia oculorum*);

(*c*) here now he is punished for his sins in an apt way, and the result (4. 7. 12) is that he pursues his career to Carthage (*ambitio saeculi*).

(2) It displays him living in a world falsified with respect to both of the two great commandments (see 3. 8. 15):

(*a*) belief that God is only a phantasma (4. 7. 12, 'vanum phantasma et error meus erat deus meus': cf. 3. 6. 10);

(*b*) the inability to love his neighbour (4. 7. 12, 'o dementiam nescientem diligere homines humaniter!'), though 'amare amabam' (3. 1. 1) had characterized his ambition since his first coming to Carthage.

(3) It displays A. in the grip of violent and genuine grief, condign punishment for one who had studied long in the school of sham pity (3. 2. 2).

Two deaths in *conf.* affect A. deeply, his friend's here and that of Monnica: for comparison of the two, see on 9. 11. 28. Many other persons who figure in this narrative die in its pages (especially in Bk. 9: see on 9. 1. 1), but no other receives comparable attention. The only other death in A.'s life to show a comparable effect in his texts is that of Marcellinus, executed in 413 in circumstances that left A. devastated (*ep.* 151).

L'innominato: Courcelle, *Recherches*, 41, 'Augustin a sans doute jugé que le nom de ce jeune homme de Thagaste, mort à la fleur de l'âge, était sans intérêt pour les lecteurs et la postérité.' Would A. neglect a chance to pay edifying tribute to a deceased member of Alypius' flock (with so exemplary an attitude to baptism)? But Courcelle examines this as one of the many unnamed figures in the text; consider instead the pattern of the named ones. Leaving aside persons A. never met, there remain in *conf.* only Adeodatus, Alypius, Ambrose, Elpidius, Evodius, Faustus, Firminus, Monnica, Nebridius, Patricius, Ponticianus, Romanianus, Simplicianus, Symmachus, Verecundus, Vindicianus, 16 in all. The list is sharply pared to include only individuals who were, witting or unwitting (e.g. Faustus and Symmachus), agents of conversion; the sole exception is Patricius, not named until Bk. 9, there only once incidentally in the narrative of Monnica's life (9. 9. 19) and once formally in a request that his readers remember Monnica (herself named only there in all A.'s writings) and Patricius at the altar. There are few names in these books of downfall (only A., in such a mood and for such a reason, could write the paean to friendship that appears at 4. 8. 13 and mention not a single friend by name!), more later as

A.'s life takes a better turn (Vindicianus has been described at 4. 3. 5, but is not named until 7. 6. 8; Bk. 6 is notable for the way the coterie of friends comes together—Augustine, Alypius, Nebridius; in Bk. 8, Simplicianus and Ponticianus appear). Finally, both the friend here and the mother of Adeodatus are, to put it bluntly, people A. treated shabbily—A. even refuses himself the consolation of thinking that he treated this one as a true friend ('sed nondum erat sic amicus'): it is not hard to believe that he felt he had no right to draw even their names into his story. (See on this W. Steidle, *Romanitas-Christianitas* (Festschrift J. Straub: Berlin, 1982), 446. The pattern of biblical names cited in *conf.* is interesting in its own right: see on 7. 21. 27.)

The friend is never mentioned elsewhere. But could A. have written this passage (on Mt. 5: 29, 'si oculus tuus dexter scandalizat te ...') without thinking of him? *s. dom. m.* 1. 13. 38, 'cogit quaerere diligentius quid dixerit oculum. in qua quaestione nihil mihi occurrit congruentius quam dilectissimum amicum. nam hoc est utique quod membrum recte possumus appellare, quod vehementer diligimus, et hunc consiliarium, quia oculus est tamquam demonstrans iter, et in rebus divinis, quia dexter est. . . . in rebus autem divinis consiliarius scandalizans est, si in aliquam perniciosam haeresim nomine religionis atque doctrinae conatur inducere.' On that assumption, note that the younger A. becomes the scandalous *consiliarius*, to be tossed aside—which is just what the dying friend did when A. sought to lure him back to heresy, 'nomine religionis atque doctrinae'.

illis annis ... coeperam: This memory dates from the year teaching at Thagaste (375/6), aged 21 (as was the friend: 'coaevum'). On the chronological displacement, see on 4. 1. 1.

municipio: See on 2. 3. 5.

in scholam ieramus pariterque luseramus: Cf. the description of childhood activities at 1. 9. 15.

vera amicitia: A.'s notion of friendship had remained consistent since Cassiciacum, but the old textual underpinnings were carefully pulled out and new ones installed: *c. acad.* 3. 6. 13, 'mecum enim familiarissimus amicus meus non solum de probabilitate humanae vitae verum etiam de ipsa religione concordat, quod est veri amici manifestissimum indicium, si quidem amicitia rectissime atque sanctissime definita est [cf. Cic. *amic.* 6. 20], "rerum humanarum et divinarum cum benevolentia et caritate consensio".' The revised version takes on wider and fuller expression at *trin.* 3. 4. 9, 'illic enim dei voluntas, qui facit angelos suos spiritus et ministros suos ignem ardentem, [Ps. 103: 4] in spiritibus summa pace atque *amicitia* copulatis, et in unam voluntatem quodam *spiritali caritatis igne* conflatis, tamquam in excelsa et sancta et secreta sede praesidens. . . . inde se, quibusdam ordinatissimis creaturae motibus, primo spiritalibus, deinde corporalibus, *per cuncta diffundit* [Rom. 5: 5], et utitur omnibus ad incommutabile arbitrium sententiae suae, sive incorporeis sive corporeis rebus.' That A. never wrote in Christian terms a treatise *de amicitia* (as one could well imagine Ambrose, for example, writing), despite the powerful role he gives to friendship both in his description of his life and in the metaphors he uses (see esp. on 2. 2. 2, 'luminosus limes *amicitiae*'), is a measure of a loss that A. never knew how to redress, perhaps because it affected him too intimately for him to articulate and thus to control. (The same may be said of *s.* 385. 2. 3, a discussion of friendship that is, however, not in its present form A. but Caesarius, though depending (see Verbraken) on a lost sermon of A.'s; quoted below on 4. 8. 13, 'alia erant'.)

The Ciceronian definition recurs in A. at *ep.* 258. 1, in a letter to another friend of A.'s youth, Marcianus, who had been similarly estranged from Christianity when he and A. were young, but who had by the time of the letter (Mandouze, *Pros. chr.* s.v. Marcianus 2, puts it close to 395) changed his sentiments, though not so far as to accept baptism (*ep.* 258. 5). There the harsh judgement on the inauthenticity of the first friendship is the same: *ep.* 258. 1–3, 'antiquissimo amico, quem tamen non habebam, quam diu in Christo non tenebam. . . . (3) nolo autem suscenseas nec tibi videatur absurdum quod illo tempore cum in vana mundi huius aestuarem, quamvis me multum amare videreris, nondum eras amicus meus, quando nec ipse mihi amicus eram sed potius inimicus. diligebam quippe iniquitatem.'

haerentes OS Knöll Skut. Ver. : *cohaerentes* G : *inhaerentes* CD Maur. For *haereo* in a good sense, but not of man clinging to God, see 6. 10. 16 (2 ×) and 9. 4. 8.

tibi CDO Ver. Pell. : *sibi* GS Knöll Skut. Vega.

caritate diffusa: Rom. 5: 5, 'spes autem non confundit, quia caritas dei diffusa est in cordibus nostris per spiritum sanctum qui datus est nobis.' The quotation is repeated to bracket Bk. 13, at 13. 7. 8 (see notes there) and 13. 31. 46. Note the linking of *caritas*, the *sanctus spiritus*, and the spirit of friendship: it was particularly friendship that A. thought perverted by the fault of *concupiscentia carnis* (2. 2. 2). A.-M. La Bonnardière, *Aug. Mag.* 2. 657–65, reports the verse at least 200 × in A., from early (*mor.*) to latest (*c. Iul. imp.*), a constant intimation of the presence of the Spirit in the church for A.

cocta GO[1]S Knöll Skut. Ver. : *coacta* CDO[2].

a fide vera: The Christian milieu of A.'s upbringing is underscored: the friend is another so brought up.

superstitiosas: See on 3. 6. 10.

mater: Monnica's only appearance in Bk. 4, a reminder of the ending of Bk. 3 (3. 11. 19–12. 21).

errabat: For *error* as particularly a form of deviation in the order of knowledge (and hence corresponding to *concupiscentia oculorum*), recall 1. 20. 31 ('dolores, confusiones, errores'), and cf. 4. 7. 12 ('vanum phantasma et error meus erat deus meus'); *error*/*errare* occur by far more often in Bk. 4 than in any other (13 × in 4, 33 × in all others together), right from the outset (4. 1. 1, 'praeteritos circuitus erroris mei').

fugitivorum: At 4. 7. 12, A. flees again ('et tamen fugi de patria'), but such flight is in vain (4. 9. 14, 'quo it aut quo fugit nisi a te placido ad te iratum?'). Cf. 2. 6. 14, 3. 3. 5 ('fugitivam libertatem'), 6. 11. 20. For 'tu imminens dorso fugitivorum tuorum', cf. 4. 16. 30, 'dorsum enim habebam ad *lumen*'.

deus ultionum: Ps. 93: 1, 'deus ultionum dominus, deus ultionum fidenter egit' (A.'s text, against the usual *libere egit*).

ultionum . . . misericordiarum: In one act, divine punishment for Augustine (duly merited (in his eyes) by the deeds he has narrated) and divine mercy for the friend (snatched away just after baptism, before the Manichean Augustine had a chance to work on him again).

miris modis: The same combination at 5. 7. 13, 7. 21. 27, 10. 40. 65, 13. 11. 12; 'miris et occultis modis' at 5. 6. 10 and 6. 12. 22.

4. 4. 8

Quis laudes tuas: Cf. Ps. 105: 2, 'quis loquetur potentias domini, auditas faciet omnes laudes eius?'

laudes tuas: Only here in *conf.* in this sense constrained by the relative clause, 'deeds worthy of praise'; elsewhere 'acts or praise directed to you': 1. 17. 27 ('laudes tuae, domine, . . . per scripturas tuas'), 5. 1. 1, 6. 7. 12, 9. 7. 16, 10. 6. 8.

Carena relies on De Marchi, 312, and translates: 'Chi puó da solo enumerare i tuoi vanti, che in sé solo ha conosciuto,' effectively reading a comma after *unus* and none after *uno* (Skut. and other editors print a comma after *uno* only). *Unus* and *uno* are brought together for euphonic effect, but the first belongs to the main clause, the second to the relative clause.

investigabilis abyssus: Cf. Rom. 11: 33, 'o altitudo divitiarum sapientiae, et scientiae dei: quam incomprehensibilia sunt iudicia eius et *investigabiles* viae eius!' Sirach 42: 18, '*abyssum* et cor hominum *investigavit*.'

The Romans passage is frequently quoted in texts dealing with grace and free will as a way of ending inconclusive analysis with avowal of the paradox of grace and freedom (e.g. *div. qu. Simp.* 1. 2. 22, *pecc. mer.* 1. 21. 29, *spir. et litt.* 34. 60, 36. 66, *c. ep. pel.* 4. 6. 16, *gr. et lib. arb.* 22. 44, *corrept.* 8. 17–19, *praed. sanct.* 2. 4, 8. 16; *persev.* 12. 30, *ss.* 26. 12. 13, 27. 6. 6–7. 7, and 294. 7. 7). Here at a moment of obscure turning in his own life, the dynamics of punishment and grace are beyond reach.

investigabilis: See on 2. 9. 17.

abyssus iudiciorum tuorum: Cf. Ps. 35: 7, 'iustitia tua sicut montes dei, iudicia tua sicut abyssus multa.' Other abysses of divine judgement at 7. 6. 10 and 13. 2. 3. *en. Ps.* 41. 13, 'abyssus enim est profunditas quaedam impenetrabilis, incomprehensibilis; et maxime solet dici in aquarum multitudine. . . . denique quodam loco dictum est: "iudicia tua abyssus multa"; hoc volente scriptura commendare, quia iudicia dei non comprehenduntur.'

baptizatus est: *breviarium Hipponense* 32, 'ut aegrotantes, si pro se respondere non possunt, cum voluntatis eorum testimonium sui periculo proprio dixerint, baptizentur'; = *conc. Carth.* III (397) can. 45. Cf. *adult. coniug.* 1. 26. 33, 'catechumenis ergo in huius vitae ultimo constitutis, seu morbo seu casu aliquo si compressi sint, ut, quamvis adhuc vivant, petere sibi tamen baptismum vel ad interrogata respondere non possint, prosit eis quod eorum fide christiana iam nota voluntas est, ut eo modo baptizentur quo modo baptizantur infantes, quorum voluntas adhuc nulla patuit. [He allows, however, that some may be disinclined to baptize the unconscious for fear that they 'contrarium gerant voluntatis arbitrium'] . . . si voluntas eius incerta est, multo satius est nolenti dare quam volenti negare.' G–M suggest this was written 'doubtless with the present instance in mind'; but by 419/20 the bishop A. had seen many more sick-bed baptisms.

recreatus est et salvus factus: Similar expressions at 1. 11. 17 (A.'s childhood illness and brush with baptism), 4. 8. 13 (A.'s grief after this friend's death), 5. 10. 18 (A.'s illness on coming to Rome).

temptavi . . . inridere: *haer.* 46. 17, 'baptismum in aqua nihil cuiquam perhibent [manichaei] salutis afferre: nec quemquam eorum quos decipiunt baptizandum putant.'

at ille: This veneration for baptism is a token of the power that ritual held over the late antique imagination. What could be smiled over by a non-initiate was a different matter altogether for the initiate, however initiation was attained.

libertate: See on 2. 6. 14 for the dangers of *libertas* to A.; here closer to 'impertinence' than 'freedom' or 'liberty.'

dementiae: 4. 7. 12, 'o dementiam nescientem diligere homines humaniter!'

febribus: See on 5. 9. 16, 'ingravescentibus febribus'.

defungitur: Regular in *conf.* for death (3. 4. 7, 6. 2. 2, 8. 2. 3, 9. 8. 17, 9. 11. 27), and elsewhere in A. (e.g. *civ.* 5. 18, *c. Faust.* 14. 1), but the euphemism is not attested earlier than Pliny the elder, and in CL only in perfect tenses.

4. 4. 9

The depiction of grief is almost literally God-less: only in a vain attempt to enjoin hope does God appear here.

contenebratum est cor meum: Lam. 5: 17, 'propterea maestum factum est cor nostrum, ideo contenebrati sunt oculi nostri.' In the few other passages where *contenebrare* occurs, it is set against the illumination that comes from God, e.g. *en. Ps.* 103. s. 4. 2, 'erigunt ipsam mentem suam, non ad se sed ad artificem suum, . . . et unde recedentes contenebrantur, et quo revertentes inluminantur'; sim. at *en. Ps.* 57. 22, 65. 13, 75. 8.

mors: Bk. 4, largely because of this extended meditation on his friend's passing, is more marked by death (counting *mors/morior* and related nouns and adjectives) than any other in *conf.*; Bk. 9 comes closest (and see introd. n. to Bk. 9 for the transformation of death there). Bks. 10–13 show a marked decline (and a corresponding upswing in *vita/vivo*, etc.).

patria supplicium: A first hint that at the end of this episode A. will leave Thagaste for Carthage (4. 7. 12).

paterna domus: Here a sign that at Thagaste A. took up residence in his father's house, though Patricius was 3–4 years dead (see on 3. 4. 7); no mention of Monnica, or of her having any say whether A. would reside there (see on 3. 11. 19). The phraseology is further redolent of the story of the prodigal (see on 1. 18. 28).

cruciatum: Elsewhere of school punishments (1. 14. 23), of Monnica's grief when A. abandons her in Carthage (5. 8. 15), and of A.'s stifled grief for Monnica (9. 12. 31).

expetebant GS Knöll Skut. Vega Pell. : *expectebant* CDO Ver. *Expetere* is 'to look for', *expectare* only 'to wait for, to have the opinion that he would turn up'; the former is much better with *oculi*; cf. 4. 7. 12, 'minus enim eum *quaerebant* oculi mei'.

oderam omnia: Things A. 'hates' in *conf.* (see on 3. 1. 1): his schoolwork in general (1. 12. 19), Greek in particular (1. 13. 20, 1. 14. 23), 'securitatem et viam sine muscipulis' (3. 1. 1), students who did not pay their fees (5. 12. 22—'quamvis non perfecto odio'! See notes there), himself and his own iniquity (8. 7. 16–17); then nothing at all through Bks. 9 to 13 except 12. 14. 17, 'odi hostes eius [sc. scripturae sacrae] vehementer.'

nec mihi iam dicere poterant [sc. *omnia*]**:** This virtual personification of 'all things' is not unparalleled: cf. 9. 10. 25 ('quoniam si quis audiat, *dicunt haec omnia*') and 10. 6. 8 ('et caelum et terra et *omnia* quae in eis sunt, ecce undique mihi *dicunt*').

quaestio: Cf. 2. 10. 18, 'factus sum mihi regio egestatis,' and 10. 33. 50, 'me, in cuius oculis mihi quaestio factus sum, et ipse est languor meus'. Note (Knauer, 150 n. 2) at 4. 11. 16 the echo of Ps. 102: 3, 'sanabuntur omnes languores tui'; the passage from 10. 33. 50 provides its own commentary on the present text.

interrogabam: This *interrogatio* anticipates the formal apostrophe of 4. 11. 16; or rather parallels it, for the present passage represents his thoughts of 375/6, while 4. 11. 16 presents a line of thought contemporary to *conf.* The Psalm text is anachronistic. A. does not say that in 375/76 he saw in the Psalm a parallel to his situation; rather, he uses the Psalm's words to embody in his text an anxiety that he then felt and to comment upon indirectly. The Psalm text's presence here implies, without making explicit, an extra-textual answer to a question posed within, not merely the text, but within the past life narrated in the text. If A. had known at the time to apply Ps. 41 ('et si dicebam "spera in deum"': i.e. if the word of God had intruded on the event as it here intrudes upon the narrative), his soul would have known how to respond to the *interrogatio*—with hope and confession, the instruments with which he wrote this book twenty years later.

quare tristis esset: Ps. 41: 6, 'quare tristis es, anima mea, et quare conturbas me? spera in deum, quoniam confitebor illi'; *en. Ps.* 41. 10, 'ait [homo] sibi constituto inter has tristitias, et comparans haec illis ad quae videnda ingressus est, et post quae visa egressus est: "quare", inquit, "tristis es, anima mea, et quare conturbas me?" . . . et quasi responderet illi anima eius in silentio: "quare conturbo te, nisi quia nondum sum ibi, ubi est dulce illud quo sic rapta sum quasi per transitum? numquid iam bibo de fonte illo, nihil metuens? iam nullum scandalum pertimesco? iam de cupiditatibus omnibus tamquam edomitis victisque secura sum?" . . . sed "spera in deum" respondet conturbanti se animae suae, et quasi rationem reddenti perturbationis suae, propter mala quibus abundat hic mundus.' (A. presents Ps. 41 (*en. Ps.* 41. 1) as the chant of 'catechumeni [qui] ad gratiam sancti lavacri festinant', here expressing the desire that brings them to the church: echoing that Psalm here captures both the attitudes that should have motivated A.) Ps. 41 plays an important role at 13. 12. 13–14. 15, where see the psalm text and additional notes. Ps. 42: 5, 'utquid tristis es, anima mea, et utquid conturbas me? spera in dominum, quoniam confitebor illi.' See the similar exposition at *en. Ps.* 42. 5–6.

phantasma: See on 3. 6. 10. *BA* ad loc.: 'Le "dieu matérialisé", auquel croit alors Augustin, lui paraît manquer de réalité en regard de l'ami qui vient de mourir: un dieu irréel ne console pas d'une perte réelle.'

fletus erat dulcis: On tears in *conf.*, see on 3. 2. 4 and see next paragraph here. With the other echoes here of Ps. 41, it is apt to think of Ps. 41: 4, 'fuerunt mihi lacrimae meae panes die ac nocte'; cf. then *en. Ps.* 41. 6, 'fuerunt mihi, inquit, lacrimae meae, non amaritudo [cf. 4. 5. 10, 'suavis fructus de amaritudine vitae'], sed panis. suaves erant mihi ipsae lacrimae'; sim. at *en. Ps.* 127. 10, 'dulces sunt et ipsae lacrimae . . . quia factae sunt tibi et ipsae panis die ac nocte.'

in deliciis animi mei: Ps. 138: 11, 'fortasse tenebrae conculcabunt me et nox inluminatio in deliciis meis'; *en. Ps.* 138. 14, 'quia in nocte me desperaveram posse transire tantum mare, et tantam viam superare, et venire ad extremum perseverando usque in finem'.

4. 5. 10

A *quaestio de fletu dulci*:

(1) Is it sweet because we are derelict here? But then it would not be sweet, for there would be no remnant of hope.

(2) Because we hope God will hear? True for prayer (see the fig tree in 8. 12. 29), but not true here, where he had no hope of recovering his friend.

(3) Is it in fact bitter, but pleasant only by comparison with our revulsion from the thing we had possessed before with pleasure?

The first two alternatives are presented with refutations; the third, left unrefuted, expresses by implication his position about the tears that came with his friend's death. He no longer loved the friend in the inappropriate way in which he had loved him in life, and did not know how to react except in this preference for bitter tears.

qui veritas es: Jn. 14: 6, 'ego sum via, et veritas, et vita'; see on 1. 5. 6, and cf. (in identical words) 3. 6. 10, 5. 3. 5, 10. 23. 33.

et admovere aurem cordis mei ori tuo: The heart has ears to hear (more often than eyes): 1. 5. 5, 4. 11. 16, 4. 15. 27. What is here possibility becomes act at 13. 6. 7, 'tibi admoveo cor meum, ne me vana doceat'.

ubique: See on 1. 3. 3, 'totus ubique'.

longe: See on 1. 18. 28.

tu in te manes: Wisd. 7: 27, 'in se ipsa manens innovat omnia.' Other echoes at 7. 9. 13, 7. 11. 17, 9. 10. 24 ('et quid simile verbo tuo, domino nostro, in se permanenti sine vetustate atque innovanti omnia'). This is parody of Christian doctrine, a Manichean alternative—that the Logos would remain serene and undisturbed, apart from human affairs, and men would be left to their *miseria*. A.'s distaste for such an argument predates his conversion: cf. 6. 5. 7.

speramus: The present passage marks the entry of Christian hope into the text. Cf. 4. 4. 9, 'spera in deum' (a scriptural voice he did not hear at the time); the present passage (hope become a question and a possibility); 4. 6. 11, 'spes mea' (the narrator's confidence); 4. 16. 31, 'in velamento alarum tuarum speremus' (itself the object of prayer). In later books, *spes* becomes a recurring motif: in Bk. 6 a possibility (6. 11. 18, 'magna spes oborta est') and a contrast to secular hopes (6. 11. 19, 'relicta spe saeculi'); in Bk. 10 a way of going beyond the present remnants of sin (10. 30. 42, 'sperans perfecturum te in me'); in Bks. 11–13, a recurrent asseveration (from Rom. 8: 24, 'spe enim salvi facti sumus', echoed at 11. 9. 11, 13. 13. 14, 13. 14. 15).

istuc $C^1D^2GO^2S$ Ver. Knauer, 154 n. 1 : *istud* C^2O^1 Knöll Skut. : *istis* D^2.

miser . . . eram: The thought is continued in 4. 6. 11.

tunc O^1SV Skut. Ver. : *dum* CD AHF : *cum* O^2 : *tunc, dum* G EM Knöll Maur. : *dum tunc* BP : *nunc* Z.

4. 6. 11

Reflection on confession in the midst of confession again leads to the tension between *quaerere* and *confiteri*: both play a part in *conf.*, the former subordinated to the latter.

amicitia rerum mortalium: 1. 13. 21, 'amicitia enim mundi huius fornicatio est abs te.'

requiescebam: First occurrence of *requiesco/requies* since 1. 1. 1, 'inquietum est cor nostrum'. What he regrets is not that life was bitter to him (it always is: see citations below), but that he found repose in it: only in God is true repose (6. 16. 26, 'et tu solus requies'), but the place of repose is the place to which our love leads us (13. 9. 10, 'requies nostra locus noster', and see notes there on love's 'gravity'). The discussion of A.'s grief here is repeatedly marked by restlessness (4. 7. 12, 'nec requies erat nec consilium', 4. 10. 15, 'et requiescere amat in eis, quae amat', 4. 12. 18, 'requiescite in eo et quieti eritis . . .; non est requies ubi quaeritis eam').

in amaritudine: Cf. Job 3: 20 (VL), 'utquid enim datur eis qui in amaritudine sunt lux?' (*adn. Iob* on 3: 20: '"eis qui in amaritudine sunt lux": peccatorum honor'); Is. 38: 15, 'recogitabo omnes annos meos in amaritudine animae meae.' The bitterness of alienation from God also at 4. 9. 14, 6. 10. 17, 7. 3. 5. Cf. *en. Ps.* 85. 6, 'solus enim tu es iucunditas: amaritudine plenus est mundus.'

ita miser eram: The focus of this paragraph, to specify the degree of wretchedness (→ 'sic eram omnino, memini').

nescio an vellem: Takes away any possibility of reading his love for his friend as disinterested *caritas*.

de Oreste et Pylade: Cic. *amic.* 7. 24, 'qui clamores tota cavea nuper in hospitis et amici mei M. Pacuvii nova fabula! cum ignorante rege uter Orestes esset, Pylades Orestem se esse diceret, ut pro illo necaretur, Orestes autem, ita ut erat, Orestem se esse perseveraret. stantes plaudebant in re ficta; quid arbitramur in vera facturos esse?' The same episode at Cic. *fin.* 5. 22. 63, and, deriving from Cic. *fin.*, at Amb. *off.* 1. 41. 216, in both cases emphasizing the theatrical response. It is no coincidence that the *exemplum* is one met in so theatrical a context; whether A. ever saw a representation of the story (or, improbably, knew the text of Pacuvius), it came to him in authoritative texts as just the sort of tear-jerker (cf. Nietzsche's 'augenverdreherisch' quoted below) that enthralled him at Carthage (thus representing the triumph of sentimentality over any authentic love: see on 3. 2. 2), and the *exemplum* is just the sort of mythic context in which A. would have attempted to understand his own grief at the time.

vel simul CDGO Ver. Pell. : *simul* S Knöll Skut. Vega Pell. : '*vel simul* . . . risponde bene al senso: giacché non è possibile vivere insieme, *almeno* morire insieme.'

qua D[1]GO[1] Ver. : *quia* CD[2]O[2]S Maur. Knöll Skut. Vega Pell. Cf. Ovid, *trist.* 4. 4. 75–6 (on Orestes/Pylades): 'nec tamen hunc sua mors, nec mors sua terruit illum: | alter ob alterius funera maestus erat.'

taedium vivendi . . . moriendi metus: Ovid, *met.* 10. 482, 'mortisque metus et taedia vitae'; cf. J. Kern and O. Hammerstein, 'I'm tired of livin' but I'm feared of dyin' ('Ole Man River'). Cf. below 'et ideo mihi horrori . . .', explaining his feeling.

ecce cor meum: 6. 6. 9, 'vide cor meum, domine, qui voluisti ut hoc recordarer et confiterer tibi'; see on 1. 5. 5, 'ecce'.

spes mea: Ps. 70: 5, 'domine, spes mea a iuventute mea'; *en. Ps.* 70. s. 1. 7, 'ante enim non in te sperabam; quamvis tu fueris protector meus, qui me salvum perduxisti ad tempus quo in te discerem sperare.' (Taken literally in the Augustinian scheme of the ages of man (see on 1. 8. 13), this text would remind A. that it was in early *iuventus* (6. 1. 1, citing Ps. 70: 5, and 7. 1. 1) that he acquired this hope.)

mundas . . . immunditia: 1. 5. 6, 'ab occultis meis munda me, domine.'

evellens de laqueo pedes meos: Ps. 24: 15, 'oculi mei semper ad dominum, quoniam ipse evellet de laqueo pedes meos'; *en. Ps.* 24. 15, 'nec timeam pericula terrena, dum terram non intueor; quoniam ille quem intueor, evellet de laqueo pedes meos.' Knauer, 178: 'Der Vers wird also in zwei parallelgeordnete Kola zerlegt, die durch das hinzugefügte zweite Verbum ("dirigo") einander angeglichen werden. Da das Zitat an einen Gottesnamen angeschlossen wird, der bereits durch einen Relativsatz näher bestimmt ist, bilden die Partizipien einen ruhig auslaufenden Schluss der Periode.'

quidam dixit: Horace *carm.* 1. 3. 5–8:

navis, quae tibi creditum
debes Vergilium, finibus Atticis
reddas incolumem precor,
et serves animae dimidium meae.

suae CDGOS Knöll Ver. Pell. : *meae* Maur. Skut. The idea had become conventional: Hier. *ep.* 3. 3, 'partem animae meae' (of a dead friend: see fuller quotation on 4. 10. 15). For other echoes of the motif in classical and patristic literature, see Pellegrino, *Les Confessions*, 121 n. 5, and cf. Ennodius, *ep.* 3. 2. 2, 'qui maior animae fuit portio'. Both *exempla* of friendship given are pre-Christian.

unam fuisse animam: 9. 12. 30, 'vita, quae una facta erat'. Ovid, *trist.* 4. 4. 72 (on Orestes/Pylades), 'qui duo corporibus, mentibus unus erant'; *ord.* 2. 18. 48, 'amici quid aliud quam unum esse conantur? et quanto magis unum, tanto magis amici sunt'; Cassian, *conl.* 1. 1 (written at least twenty years after *conf.*), speaking of himself and Germanus, '. . . ut cuncti ad significandam sodalitatis ac propositi nostri parilitatem pronuntiarent unam mentem atque animam duobus inesse corporibus'.

et ideo forte: *retr.* 2. 6. 2 (the only qualification he offers there to *conf.* 1–9), 'in quarto libro, cum de amici morte animi mei miseriam confiterer, dicens quod anima nostra una quodammodo facta fuerat ex duabus, "et ideo", inquam, "forte mori metuebam, ne totus ille moreretur quem multum amaveram." quae mihi quasi declamatio levis quam gravis confessio videtur, quamvis utcumque temperata sit haec ineptia in eo quod additum est "forte".' Writing thirty years later, giving no sign of feeling the spell of this episode, A. found that the *conf.* text distanced itself too little from the atmosphere of classical sentimentalism about friendship (note that the Horace tag is quoted here not as something he thought at the time, but as a general principle true at the time he wrote). A. was thus at least as dismayed as Nietzsche with the rhetorical posturing of this passage. But a closer inspection of the passage shows that the attitude in 'ne totus ille . . .' was not that of A. of 397, but of A. of 376, aged 21, a sentimental age. A. judged himself to have exercised elsewhere a retroactive censorship on what he said of his youth, either identifying attitudes as inappropriate or censuring them—and he judged himself as having done so inadequately here.

Nietzsche's criticism is in a letter to Overbeck, 31 March 1885 (writing from Nice, between *Thus Spake Zarathustra* and *Beyond Good and Evil*[2]): 'Ich las jetzt zur Erholung[3] die Confessionen des h⟨eiligen⟩ Augustin, mit großem Bedauern daß Du nicht bei mir warst. Oh dieser alte Rhetor! Wie falsch und augenverdreherisch! Wie habe ich gelacht! (zb. über den "Diebstahl" seiner Jugend, im Grunde eine Studenten-Geschichte). Welche psychologische Falschheit! (zb. als er vom Tode seines besten Freundes redet, mit dem er eine Seele gewesen sei, "er habe sich entschlossen, weiter zu leben, damit auf diese Weise sein Freund nicht ganz sterbe." So etwas ist ekelhaft verlogen.) Philosophischer Wert gleich Null. Verpöbelter Platonismus, das will sagen, eine Denkweise, welche für die höchste seelische Aristokratie erfunden wurde, zurecht gemacht für Sklaven-Naturen. Übrigens sieht man, bei diesem Buche, dem Christenthum in den Bauch: ich stehe dabei mit der Neugierde eines radikalen Arztes und Physiologen.' (*Nietzsche Briefwechsel: Kritische Gesamtausgabe* (Berlin/New York, 1982), 3. 3. 34).

[2] That latter work reflects his reading of *conf.* in its sections 50, 200, and 247—the last of which containing a reference to *conf.* 6. 3. 3 unnoticed in the editions of Nietzsche, but already remarked by E. Norden, *Die antike Kunstprosa* (5th edn., Stuttgart, 1958), Nachträge 1. 3.

[3] Perhaps German intellectuals should not take to the *Confessions* for light reading: B. Brecht, *Tagebücher 1920–1922* (Frankfurt, 1975), 212: 'Ich lese aus Mangel an Schundromanen die Bekenntnisse des Augustinus . . . Sehr komisch ist bei ihm die so typische Haltung aller Gelehrten ihren Entdeckungen gegenüber: eine eifersüchtige, geizige, ja schadenfrohe Haltung. Er behandelt seine Religion wie sein Steckenpferd' (adduced by R. Herzog, in K. Stierle *et al.*, *Das Gespräch* (Munich, 1984), 237 n. 98.)

4. 7. 12

On this paragraph, see also C. Carena, *Riv. stor. e lett. rel.* 3 (1967), 65–70, emphasizing the link to the argument of the third book of Cicero's *Tusculans* (see on 6. 16. 26 for A.'s reading of Cicero reflected in *conf.*). The parallels Carena draws to Cicero and others are mainly ones of topic and treatment, or else conventional turns of phrase, but he well captures the atmosphere of the passage. It is undeniably apt that A.'s treatment of his state of mind at this time should reflect the analysis he *should* have given it at the time, from his classical readings.

Alfaric, 50–1, uses this passage to inveigh against A. the insincere friend: hostile to his father, mistreating his mother, casting off his concubine, tearless at his mother's death-bed, locking himself up in Thagaste and refusing to visit a friend's death-bed in Carthage (Nebridius: *ep.* 10. 1–2), confining his *caritas* to God and living himself shut away like a monk. There is certainly an element of truth to that: *caritas* never took A. out into the towns and crossroads in the manner of Francis of Assisi.

concisam GOS Knöll Skut. Ver. : *conscissam* CD Maur. The loss of his mistress evokes similar language: 6. 15. 25, 'cor, ubi adhaerebat, concisum'.

non in amoenis nemoribus . . .: When A. speaks of the senses, he regularly invokes them in the deliberate rhetorical order here (or its reverse): sight, sound, smell, taste, touch (see further on 10. 6. 8; cf. his treatment of the temptations of the flesh at 10. 30. 41–10. 34. 53). That the senses are the arena of carnal love is suggested in 4. 10. 15, 'non in eis figatur glutine amore *per sensus corporis*'. We have already seen A. chronicle his early life as a sequence of surrenders to *concupiscentia carnis* and *concupiscentia oculorum*. Here, on the death of his friend, he faced a crisis, in which he sought repose. After tears (4. 6. 11), he now sought rest ('adquiescebat') elsewhere: (1) 'non in amoenis nemoribus' (sight), (2) 'non in ludis atque cantibus' (sound), (3) 'nec in suave olentibus locis' (smell), (4) 'nec in conviviis apparatis' (taste), (5) 'neque in voluptate cubilis et lecti' (touch). That pattern gives sense to the anticlimax in the final element: 'non denique in libris atque carminibus'. First, *carmina* must mean something other than 'song' (as in 'wine, women, and song'), which is already represented by *cantus*. Sense is found if books and poems are taken as the instruments of *curiositas*—that pursuit of wisdom whose derangement A. has been tracking since 3. 4. 7. Confirmation comes from *civ.* 10. 9 (on theurgy), 'fiebant autem simplici fide atque fiducia pietatis, non incantationibus et *carminibus nefariae curiositatis* arte compositis, quam vel magian vel detestabiliore nomine goetian vel honorabiliore theurgian vocant.' A. thus says here that his first recourse in crisis was to give way to the temptations to which he had already submitted, and he presents those vices schematically and in the order in which they have already been presented in *conf.* But there is no rest for A. in these familiar faults, so he will be driven further (see on 'veni Carthaginem' below).

Carena, art. cit. 69, rightly notes that the structure of the sentence anticipates that of the verse of Paul (Rom. 13: 13) that A. will encounter in the garden scene at 8. 12. 29: but see there for the reflection of the three temptations in the Pauline verse. For the phrases that make up the sentence, Carena supplies the following classical parallels:

amoenis nemoribus: *Aen.* 6. 638–9, 'amoena virecta fortunatorum nemorum'.
ludis atque cantibus: Cic. *Tusc.* 3. 20. 46, 'et corporum complexum et ludos atque cantus'; Hor. *carm. saec.* 22, 'ut cantus referatque ludos'; Val. Flacc. 5. 443, 'ludos . . . et cantus'.
suave olentibus locis: Hor. *serm.* 1. 4. 76, 'suave locus voci resonat'; Catullus 61. 7, 'suave olentis amaraci'; Priap. 3. 13, 'suave olentia mala'.
conviviis apparatis: Cic. *off.* 3. 14. 58, 'apparatum convivium'; Livy 24. 16. 17, 'apparata convivia'.

neque CDGO Maur. Ver. Pell. : *nec* S Knöll Skut. Vega.

aliquantula requies: See on 4. 6. 11, 'requiescebam'.

sarcina: Worldly ways are a burden in *conf.* (6. 6. 9, 8. 5. 12, 8. 7. 18, 10. 40. 65), and later, the same noun was often attached to his churchly office (e.g. *ep.* 31. 4, 69. 1 ('episcopatus sarcinam')). An implicit alternative underlies his use of the term: 9. 1. 1, 'quo subderem cervicem leni iugo tuo et umeros levi sarcinae tuae' (< Mt. 11: 30, where 'et sarcina mea levis est' is A.'s preferred text (Milne, 36)): also echoed, emphasizing *iugum* and without the word *sarcina* at 8. 4. 9, 10. 36. 58, 13. 15. 17. Cf. *en. Ps.* 59. 8, 'alia sarcina premit et aggravat te; Christi autem sarcina sublevat te; alia sarcina pondus habet; Christi sarcina pennas habet.' On the word in A., see M. Jourjon, *Rech. sc. rel.* 43 (1955), 258–62; S. Poque, *Le langage symbolique* (Paris, 1984), 1. 64–6.

levanda: Ps. 24: 1, 'ad te domine levavi animam meam'; *en. Ps.* 24. 2, '"levavi" . . . desiderio spiritali, quae carnalibus desideriis conculcabatur in terra.' Cf. *en. Ps.* 85. 7, 'quocumque ergo se converterit, in terrenis rebus amaritudinem invenit; unde dulcescat non habet, nisi levet se ad deum.' Note here not 'levavi' but 'levanda erat . . . sed nec volebam nec valebam': as often, in retrospect the solution to a problem is implied by the scriptural echo, a solution to which A. was blind at the time.

phantasma: Cf. 4. 4. 9, 'phantasma', and see on 3. 6. 10.

ibi ponere: i.e. *apud phantasma quod mihi erat deus*.

per inane labebatur: Carena, art. cit. 70, detects Lucretian flavour in the phrase.

ego mihi remanseram infelix locus: Cf. 2. 10. 18, 'factus sum mihi regio egestatis'; and 13. 9. 10, 'requies nostra *locus* noster'.

fugeret: Flight from self and flight from God are equally dangerous, reminiscent of the prodigal's career (Lk. 15: 13, 'profectus est in regionem longinquam': echoed again at 4. 16. 30; see on 1. 18. 28; in the most literal sense, A. is fleeing his father's house (4. 4. 9) and land, to which he will in the most literal sense return in 388). A. in flight (all pre-conversion): 2. 6. 14 ('servus fugiens dominum'), 3. 3. 5 ('fugitivam libertatem'), 4. 4. 7 ('tu imminens dorso fugitivorum tuorum'), 4. 9. 14 ('quo it aut quo fugit'), 5. 2. 2 ('eant et fugiant a te inquieti iniqui . . . quo enim fugerunt, cum fugerent a facie tua?), 6. 11. 20 ('et ab ea [sc. beata vita] fugiens quaerebam eam'), 8. 7. 16 ('et videbam et horrebam [cf. here *horrebant*], et quo a me fugerem non erat').

quo a me ipso fugerem: Hor. *carm*. 2. 16. 18–20:

> quid terras alio calentis
> sole mutamus? patriae quis exsul
> se quoque fugit?

(Cf. Hor. *ep.* 1. 11. 27, 'caelum non animum mutant qui trans mare currunt.')

Cf. Lucr. 3. 1053–75, on those whom the fear of death inspires to flight, esp. 1068–72:

> hoc se quisque modo fugit, at quem scilicet, ut fit,
> effugere haud potis est . . .
> . . . morbi quia causam non tenet aeger;
> quam bene si videat, iam rebus quisque relictis
> naturam primum studeat cognoscere rerum.

Lucretius' solution, like A.'s, came as culmination, after considering the delights of the flesh. L.'s lines depict the frightened man racing distractedly to his villa, seeking release in sleep, then rushing back to the city again, all without clear purpose: against that, knowledge of the *natura rerum* is the key to happiness. A. probably knew the passage, since it contains both the anticipatory echo of Virgil's famous homage (quoted at *civ.* 7. 9) to L. from *geo.* 2. 490–2 and the virtual title of L.'s poem. (Reference not in Hagendahl: suggested by Verheijen in his edn. ad loc.) Seneca (cited by G–M) had already turned poetry into philosophic prose: Sen. *ep.* 28. 1 echoes Horace: 'animam debes mutare non caelum. . . . quaeris quare te fuga ista non adiuvet? tecum fugis'; and Sen. *tranq.* 2. 14–15 (quotes Lucr. 3. 1068).

veni Carthaginem: Cf. 1. 13. 22, 'Aenean aliquando Carthaginem venisse'; 3. 1. 1, 'veni Carthaginem'. The secrecy of his later departure from Carthage for Rome is dramatically emphasized (5. 8. 15), but the earlier flight from Thagaste was similarly surreptitious (*c. acad.* 2. 2. 3, quoted below).

At 5. 8. 14, he is, moreover, carefully circumstantial in discussing his motives, while denying that ambition had anything to do with it; here, the reasons given within the limits of the present text are inadequate to explain the direction he took. A radical break with the immediate past was a logical way to escape his grief, but nothing in *conf.* says why or with what hope it was to Carthage that he went. A text of *c. acad.*, on the other hand, reveals the power of ambition: *c. acad.* 2. 2. 3, 'tu [sc. Romanianus] Carthaginem *inlustri-oris*[4] *professionis gratia* remeantem, cum tibi et meorum nulli consilium meum spemque aperuissem, quamvis aliquantum illo tibi insito—quia ibi iam docebam—patriae amore cunctatus es, tamen ubi evincere *adulescentis cupiditatem ad ea quae videbantur meliora tendentis* nequivisti, ex dehortatore in adiutorem mira benevolentiae moderatione conversus es.'

The pattern then, though scarcely obtrusive, is this: *concupiscentia carnis* prevails in A.'s life and satisfies him for a time; *concupiscentia oculorum* takes its turn, not effacing the earlier vices but adding to them; when a further crisis reveals that both have lost their power to provide relief and release, the third temptation of 1 Jn. 2: 16, *ambitio saeculi*, inspires a despairing and panicky gesture. (For the three temptations as defence mechanisms in the face of death, cf. *s. Guelf.* 31. 4, quoting 1 Jn. 2: 16 and adding, 'contemnunt plerumque homines mortem per concupiscentiam carnis, contemnunt mortem per concupiscentiam oculorum, contemnunt mortem per ambitionem saeculi: sed omnia ista de saeculo sunt.') Henceforth, until the reverse process of liberation begins in Bk. 6, A.'s life will be an inextricable *mélange* of the effects of all three temptations; thus in a few lines (4. 8. 13), he will speak of 'pristina genera delectationum, quibus cedebat dolor meus ille'. This mixture dilutes schematic clarity of pattern, and rhetorical presentation is sacrificed to narrative verisimilitude. It is easy to see and say that Bk. 2 (or Bk. 8) is 'about' *concupiscentia carnis*; but Bk. 4 (or Bk. 6) is only partly 'about' *ambitio saeculi*.

4. 8. 13

de die in diem: Ps. 60: 9, 'sic psallam . . . ut reddam vota mea de die in diem'; sim. at Ps. 95: 2, Sirach 5: 8 ('et ne differas de diem in diem'), Is. 58: 2, 2 Cor. 4: 16, 2 Pet. 2: 8; in *conf.* close echoes of Sirach at 6. 11. 20, 7. 8. 12, 8. 7. 18, once in each of those hopeful but reluctant books. Is the expression classical? *OLD* indicates not, offering only *diem ex (de) die*. Hrdlicka, 170, declares the phrase ecclesiastical: Tert. *adv. Marc.* 5. 11, plus scriptural texts cited above. Thus time passes *de die in diem*, that is, in a way that scripture repeatedly shows A. how to sanctify; but he does nothing. The implication is that there was some good to be got out of all this (with the evocation of friendship's delights, this is one of the brightest moments in all these early books), and the echo, however faint, hints at the greater good that was there for the taking.

harenam: Ov. *trist.* 5. 6. 43, 'his qui contentus non est, in litus harenas, in segetes spicas . . . fundat'; Ov. *her.* 5. 115, 'quid harenae semina mandas' (Otto, *Sprichwörter*, s.v. *harena* 2, p. 159).

[4] *inlustrios* here in *CCSL* is a typographical error.

diligendo moriturum acsi non moriturum: Echoing 4. 6. 11 ('quem quasi non moriturum dilexeram).

acsi: = *quasi* (*TLL* 2. 1083).

maxime quippe me reparabant: Statement of this book's variation on the theme of friendship: here are the false friends of this time, the Manichees ('cum quibus amabam quod (i.e. the *phantasma* of 4. 7. 12) pro te amabam'), and here the *curiositas* ('pruriens in auribus') that characterizes their way of life.

ingens fabula: G–M: 'sc. the Manichaean system'. Not earlier in *conf.* of Manichees, but frequent from here on: 5. 3. 3 ('illis manichaeorum longis fabulis'), 5. 7. 12, 5. 9. 17, 5. 10. 19, 6. 5. 7. 2 Tim. 4: 3–4, 'erit enim tempus, cum sanam doctrinam non sustinebunt sed ad sua desideria coacervabunt sibi magistros prurientes auribus, (4) et a veritate quidem auditum avertent, ad fabulas autem convertentur.' See on 1. 10. 16 (note there 'curiositate').

confricatione: See on 2. 8. 16.

alia erant: To counter the falsehood of these friends, he limns the traces of authentic friendship that were present (compare 1. 20. 31, closing that book with a comparable reversal of field). This is A.'s gentlest gesture to his old friends among the Manichees.

To read this passage repeatedly in different moods arouses different responses: it can be argued that the passage is suspect, too generous-minded for *conf.* (note 'nugari', for example; see on 'nugas' at 3. 10. 18 for how consistently belittling the word and its congeners are elsewhere). More consistently Augustinian in spirit, and similar in content, is *s.* 385. 2. 3 (Caesarius depending on A. (Verbraken): see above on 4. 4. 7), 'est quaedam *amicitia adhuc carnalis* per consuetudinem cohabitandi, conloquendi, simul conversandi; ut contristetur homo, quando deseritur ab amico cum quo solet conloqui et habere coniunctiones. conveniunt duo homines, ambulant secum triduo, et iam nolunt a se recedere. et ista quaedam amicitiae dulcedo est: honesta quidem, sed adhuc discutiamus illam, quia gradus amoris huius quaerimus. . . . est ergo ista amicitia consuetudinis, non rationis. habent illam et pecora. . . . est alia superior amicitia, non consuetudinis, sed rationis, qua diligimus hominem propter fidem et mutuam benevolentiam in ista vita mortali.'

redamantium: G–M: 'the word "redamare" was first introduced by Cicero, *amic.* 14. 49, "quid enim tam absurdum quam delectari multis inanibus rebus . . . animante virtute praedito, eo qui vel amare vel, ut ita dicam, redamare possit, non admodum delectari?" but, like many of the words coined by him, it first reappears in the late writers like Aug., Macrob. *sat.* 6. 6. 8, Symmachus *ep.* 3. 2, etc.' Cf. also Apuleius, *Plat.* 2. 13 and Amb. *in Luc.* 5. 75, 'commune est omnibus, etiam peccatoribus, redamare'.

conflare CDGO Skut. Ver. : *flagrare* S Knöll. G–M: '*conflare*, "fuse together". Note that this, being active, could lack a subject, and goes less well with "fomitibus".' But all these infinitives from *conloqui* down are not strictly verbal but nominal: they are the *alia* which *animum capere*; and the notion, joining souls together, is perfectly apt; cf. *en. Ps.* 78. 2, 'in animam unam et cor unum caritatis igne conflati'; *en. Ps.* 95. 15, 'sed misericordia dei undique conlegit fracturas, et conflavit igne caritatis, et fecit unum quod fractum erat.' Cf. Cic. *amic.* 25. 92, 'nam cum amicitiae vis sit in eo ut unus quasi animus fiat ex pluribus' (see also *de off.* 1. 17. 56 and Cic. often elsewhere: Otto, *Sprichwörter*, s.v. *animus* 1, pp. 25–6).

4. 9. 14

conscientia: The conscience, restless and voiced: 8. 7. 18 ('et increparet in me conscientia mea, "ubi est lingua mea?"'), 10. 3. 4, 10. 30. 41; before God: 5. 6. 11 ('domine deus meus, arbiter conscientiae meae': 10. 2. 2, 'cuius oculis nuda est abyssus humanae conscientiae'). Conscience takes a voice in Latin perhaps as a reflection of Rom. 9: 1 ('veritatem dico in Christo, non mentior, testimonium mihi perhibente conscientia mea [συμμαρτυρούσης μοι τῆς συνειδήσεώς μου] in spiritu sancto': cf. *civ.* 1. 19, 'habent quippe intus gloriam castitatis, testimonium conscientiae'). For the conscience in writing, see 1. 18. 29.

si non amaverit: Cf. 4. 7. 12, 'nescientem homines diligere humaniter'.

nihil quaerens: This friendship explicitly excludes *concupiscentia carnis:* cf. 2. 2. 2 and 3. 1. 1.

amaritudinem: See on 4. 6. 11.

mors viventium: 4. 4. 9, 'et quidquid aspiciebam mors erat.'

beatus qui amat te . . .: The two great commandments (see 3. 8. 15), applied. Tob. 13: 18, 'beati omnes qui diligunt te, et qui gaudent super pace tua'; Mt. 5: 43–4, 'audistis quia dictum est, diliges proximum tuum, et odio habebis inimicum tuum. (44) ego autem dico vobis, diligite inimicos vestros, benefacite his qui oderunt vos.' *s. dom. m.* 1. 21. 69, 'perfectio autem misericordiae, qua plurimum animae laboranti consulitur, ultra dilectionem inimici porrigi non potest. et ideo sic clauditur, "estote ergo vos perfecti, sicut pater vester, qui in caelis est, perfectus est".'

et quis est iste: For the same mannerism, cf. 4. 3. 4, 'et quis est hic nisi deus noster'.

qui fecit caelum et terram: See on 1. 2. 2, 'deus, qui fecit caelum et terram'; cf. Ps. 145: 5–7, 'spes illius in dominum deum ipsius, (6) qui fecit caelum et terram, mare et omnia quae in eis sunt, (7) qui custodit veritatem in aeternum' (see 'et lex tua veritas . . .' below).

et implet ea: Jer. 23: 24, 'numquid non caelum et terram ego impleo?' See on 1. 2. 2.

amittit . . . dimittit: The same word-play at 9. 1. 1.

quia (*dimittit*) DS Knöll Skut. Ver. Vega Pell. : *qui* CG Maur. : *qui** O. The appeal of *qui* is great, but it is the *facilior lectio*.

quo it aut quo fugit: Ps. 138: 7, 'quo ibo a spiritu tuo? et quo a facie tua fugiam?'; *en. Ps.* 138. 10, 'locum quaerit quo fugiat ab ira dei. quis est locus recepturus fugitivum dei? . . . quis fallit deum? quem non videt deus? . . . vertit se hac atque illac, quasi quaerens locum fugae suae.' See also on 4. 7. 12, 'fugi de patria'.

placido . . . iratum: *en. Ps.* 94. 2, 'quo iturus est, quo fugiturus est ab illo irato, nisi ad ipsum placatum? . . . non ergo loco quisque longe est a deo, sed dissimilitudine'; sim. at *en. Ps.* 30. en. 2. s. 1. 8, 138. 12, 146. 20.

et lex tua veritas et veritas tu: Ps. 118: 142, 'iustitia tua iustitia in aeternum et lex tua veritas'; *en. Ps.* 118. s. 28. 5, 'quomodo enim non veritas lex, per quam cognitio peccati, et quae testimonium perhibet iustitiae dei? sic enim dicit apostolus: ". . . testificata per legem et prophetas" [Rom. 3: 21].' The Law, i.e. scripture, is certainly meant in A.'s reading of the psalm, and the overtone is clear in the parallel passage at 2. 5. 10 ('domine deus noster, et veritas tua et lex tua': see notes there). That is also a passage about friendship and sin (the pear-theft).

4. 10. 15

This paragraph stands at the centre of Bk. 4. It transcends the narrative of his lost friend, offering prayer followed by the apostrophe to his own soul of paragraphs 16–19. For the extension and complexity of the development, this passage to the end of paragraph 19 has nothing like itself anywhere else in *conf.* If the flight to Carthage (4. 7. 12) marked the final fall into the last grave temptation, then we have reached a nadir. The last section of this book (4. 13. 20–16. 31) exemplifies the workings of *ambitio saeculi* in his life and shows (along with Bk. 5) the hopelessness of his first attempts to ascend again.

Created things pass away; mortal words pass away; the senses are inadequate to grasp things and words and hold them in place; only the Word that does not pass away can speak with authority—and its authority overpowers the transience that is otherwise baffling. The dizzying awareness of transience is the fate of man at his nadir; significantly, it also becomes the centre of the problematic of the *de pulchro et apto* described in 4. 13. 20 ff. (see Testard quoted on 'in pulchris' below).

Deus virtutum: Ps. 79: 8, 'domine, deus virtutum, converte nos et ostende faciem tuam, et salvi erimus.'[5] The 'turnings' of man provide word-play at greater length at the conclusion of this book (4. 16. 31) in a sequence of related words: 'perversitas . . . aversi . . . perversi . . . revertamur . . . ut non

[5] This is the scriptural verse that the student Licentius was heard singing in an indecorous place at *ord.* 1. 8. 22.

evertamur' (Knauer, 72). Cf. also 4. 4. 7 (introducing the episode of his friend's death), 'qui convertis nos ad te miris modis'.

virtutum: Knauer, 47–8, takes 'virtues' (not 'hosts', 'miracles') as the likeliest rendering, arguing from *en. Ps.* 67. 15, 'cum itaque dixisset: dominus dabit verbum evangelizantibus virtute multa.... rex est virtutum dilecti.' The phrase 'dominus virtutum' often escapes commentary in A.'s Psalm sermons, but there are several clear passages in which A. easily accepts the equation *virtutes* = *angeli*: in *conf.*, 7. 13. 19, 'in excelsis omnes angeli tui, omnes virtutes tuae'. See *en. Ps.* 45. 11, '"dominus virtutum nobiscum"; ... non quicumque homo, non potestas quaelibet, non denique angelus, non aliqua creatura, sive terrena sive caelestis, sed "dominus virtutum nobiscum".... qui misit angelos, venit post angelos, venit ut ei servirent angeli, venit ut homines faceret aequales angelis'; cf. *en. Ps.* 23. 10, 102. 28. We should be cautious, however, in enforcing any such reading; the phrase *deus virtutum* uses the Latin word for 'acts of strength' as a periphrasis for 'those through whom God performs acts of strength,' and in the process becomes a familiar expression that may not do more than underline the power of God, and even that perhaps only faintly. When, for example, in devout English one speaks of 'Almighty God', the word 'almighty' has far less force than it would in other contexts, because the epithet has become conventional and familiar.

ad dolores: sc. *patiendos*. The passage is elliptical and may be expanded thus: 'When the soul of man is fixed (= *requiescat*) somewhere other than in God, it reposes in pain that must be suffered, even if it reposes in things that are beautiful—for those things are outside God, outside the soul.' For *figere* as a sign of stability and hence of the repose that A. thinks he wrongly sought at this time in things that were not God, cf. just in Bk. 4: 4. 3. 5, 'adsiduus et fixus inhaerebam', 4. 11. 16, 'ibi fige mansionem tuam', and 4. 12. 18, 'et in illo fixae stabiliuntur'.

in pulchris: The phrase provides the logical link to A.'s philosophical inquiry into beauty, reported at 4. 13. 20 ff., beginning 'et amabam pulchra inferiora'. Cf. Testard 1. 51 n. 1: 'Il est frappant de voir en effet comment cette partie annonce par les questions envisagées et le vocabulaire, les pages qui traiteront directement de l'ouvrage perdu.... Finalement tout le problème du *de pulchro et apto* est là: Augustin voulut remonter du monde à Dieu en vertu d'une méthode analogique, mais il lui manqua de mettre entre le monde et Dieu le rapport de créature à Créateur, et la distance de matière à esprit, qui sauvegardent à la fois l'immanence et la transcendance divines.'

quae tamen nulla essent, nisi essent abs te: Cf. 1. 2. 2, 'sine te non esset quidquid est.'

oriuntur et occidunt: Cf. Sall. *Iug.* 2. 3 (Hagendahl, 240), 'postremo corporis et fortunae bonorum ut initium sic finis est, omniaque orta occidunt et aucta senescunt.' That passage is quoted verbatim at *epp.* 143. 6 (to Marcellinus),

166. 5. 14 (to Jerome), and closely echoed at *civ.* 13. 16 and 13. 17. The idea is conventional.

modus: See on 1. 7. 12 (*modus/ordo/species*), and cf. 'ipse est modus eius' below. We might say 'a condition of their natures'.

partes . . . universum: *universum* is not here 'the universe' (for which see 1. 20. 31, 3. 8. 16, 3. 8. 15, and elsewhere) but 'a whole' (cf. 3. 8. 15, 'turpis enim omnis pars universo suo non congruens', 4. 13. 20, 'sicut pars corporis ad universum suum' (the parallel discussion in the treatment of the *de pulchro et apto*)); *civ.* 19. 16, 'omnis pars ad universi cuius pars est integritatem refertur'.

ecce sic . . . succedat aliud: These two sentences have a double function: formally, they provide a parallel example of something that is evanescent and time-bound, constituting a whole only through a succession of parts; but because the parallel is drawn from human speech, the example has the second effect of saying that ordinary words are no more fixed or reliable than material things. The paragraph will end with a single voice that speaks reliably, though it speaks only to establish transience (for that paradox, cf. 1. 4. 4, 'stabilis et incomprehensibilis'). For similar analysis of sound/speech, 11. 27. 34 ff. (and see notes there).

peragitur: Used elsewhere of the passage of time (1. 6. 10, 11. 23. 30) and of the slipping away of words here and at 11. 27. 36 (but there words illustrate the passage of time itself).

sermo noster: To what does this refer? this book? human language generally? The latter is probably 'intended', but even so this book exemplifies the idea, for a book is dynamic rather than static, meaningful only when read consecutively, one word slipping away to make room for another. Beyond this book, beyond *sermo noster*, is the appealing ideal of a lasting and permanent language, where words have neither beginning nor end (cf. 'in verbo enim tuo . . .' below). But in the mean time, the praise that is called for ('laudet te ex illis anima mea') is thus contingent and imperfect, relying on defective instruments.

laudet te: Ps. 145: 2, 'lauda, anima mea, dominum'; cf. the similar near-echo of Ps. 145: 6 in 4. 9. 14, 'deus, qui fecit caelum et terram', and cf. 5. 1. 1.

ex illis: sc. *pulchris*—the parenthesis on human language does not break the main line of thought.

deus, creator omnium: Amb. *hymn.* 1. 2. 1: see on 9. 12. 32, where the hymn is quoted *in extenso*; the phrase occurs at 2. 6. 12, 6. 4. 5, 9. 6. 14, 10. 34. 52, 11. 27. 35 (where it is the object of metrical analysis in the later extended discussion of the transience of time-bound human language).

in eis CDGO Maur. Isnenghi Ver. : *eis* S Knöll Skut. Vega Pell.

figatur CDGO[1] Isnenghi Ver. : *infigatur* O[2] (del.) S Maur. Knöll Skut. Vega Pell. Isnenghi adduces parallels for the decomposition of verbs: 4. 13. 20, 'esset in eis', 10. 34. 52, 'haereo in . . . insidiis'.

glutine D²OS Maur. Knöll Skut. Ver. : *glutino* CD¹G.

amore OS Knöll Skut. Ver. : *amoris* CDG Maur.

Infigere is not used elsewhere in *conf.*; it is relatively infrequent in A. generally, occurring mainly in scriptural quotations and discussions thereof. Note esp. Ps. 68: 3, 'infixus sum in limo profundi, et non est substantia.' The verse is echoed to describe A.'s Manichean affiliation at 3. 11. 20 ('in illo limo profundi ac tenebris falsitatis . . . volutatus sum'), while at 7. 18. 24 *limus* is the clay out of which man is made and which Christ assumes at the incarnation. At *en. Ps.* 68. s. 1. 4–5, the verse is taken as spoken by Christ, and thus incarnationally, but there is an implied interpretation of *infixus in limo* that is apposite here: *en. Ps.* 68. s. 1. 5, 'deus fecit hominem; substantiam fecit; atque utinam maneret in eo quod deus fecit! si maneret homo in eo quod deus fecit, non in illo infixus esset quem deus genuit. porro autem quia per iniquitatem homo lapsus est a substantia in qua factus est (iniquitas quippe ipsa non est substantia; non enim iniquitas est natura quam formavit deus, sed iniquitas est perversitas quam fecit homo); venit filius dei ad limum profundi, et infixus est; et non erat substantia in qua infixus est, quia in iniquitate illorum infixus est.' That verse is thus faintly present here, by contraries: what seem to be *pulchra* are, for the one who would repose (become mired) in them, *limus*.

glutine: *trin.* 10. 5. 7, 'cum ergo aliud sit non se nosse, aliud non se cogitare, tanta vis est amoris ut ea quae cum amore diu cogitaverit eisque curae glutine inhaeserit attrahat secum etiam cum ad se cogitandam quodam modo redit'; so *trin.* 10. 8. 11, 'cohaeserunt enim mirabiliter glutine amoris.' (The manuscripts regularly vary between *glutine* and *glutino* for abl. in all A.'s works, but *civ.* 15. 27 attests *gluten* as nom. sg.) For the collocation with a noun of 'loving', see the parallel at Hier. *ep.* 3. 3, 'postquam glutine caritatis haerentem impia distraxit avulsio' (cf. 4. 4. 7, 'eam tu agglutinas inter haerentes tibi'). That passage of Jerome refers to the death of a friend in terms similar to those of 4. 4. 7: 'ex duobus oculis unum perdidi, Innocentium enim, partem animae meae, repentinus febrium ardor abstraxit.'

amore: Skut. compares 10. 10. 24, 'veritate pabulo': there, although there are MSS for *veritatis*, the correct reading is clearly the ablative.

pestilentiosis: The word itself not attested before Ulpian; for such plaguey unhealthfulness elsewhere in *conf.*, see 1. 14. 23, 2. 3. 8, 5. 10. 20.

ipsa esse vult: The 'will to exist' is part of the pattern of the trinity reflected in man: 13. 11. 12, 'volo esse et scire'. That desire is natural, therefore, but the further desire to repose 'in eis' (sc. *pulchris*) is not, and is the source of trouble.

requiescere: See on 4. 6. 11, 'requiescebam'.

sensu carnis: The world of unreliable things is described in unreliable language, having been made the object of knowledge through unreliable senses. That inability of the senses to find a pattern in the beauty of material things becomes the—inadequate—pretext for the ascent beyond the senses attempted by the *de pulchro et apto*.

modus eius: See above on 'modus'.

in verbo . . . tuo: Christ: cf. 4. 11. 16 ('"numquid ego aliquo discedo?" ait verbum dei'), 4. 12. 19 (the end of this section of Bk. 4), and 4. 15. 25, in discussing *de pulchro et apto*. We reach here the only firm ground in this paragraph, itself explaining and defining the baffling transience that has led the soul to its present predicament.

hinc et huc usque: The creating Word sets the *modus* for the created things; cf. above and see on 1. 7. 12. For the *ipsissima verba* of God in quotation marks (here and in next paragraph), see on 6. 16. 26.

4. 11. 16

The only previous parallel for this apostrophe is the much briefer 'cave immunditiam, anima mea' (3. 2. 3); contrast *sol.*, where 'Augustine' is the junior partner in dialogue with *Ratio*. For the sense, cf. 4. 4. 9, 'factus eram ipse mihi magna quaestio et interrogabam animam meam, quare tristis esset et quare conturbaret me valde, et nihil noverat respondere mihi'; sim. at 10. 6. 10 and 11. 15. 19. Note that at 10. 30. 42 ('ut anima mea sequatur me ad te'), the *anima* is not identical with the 'self': hence the possibility of apostrophe.

The line between past and present is blurred. A. speaks as if in the present, using words only available to him in the present (cf. 4. 13. 20, 'haec tunc non noveram'), but the address is apt to the condition in which he found himself. A. 'fell' at specific times between his sixteenth and twenty-second year (370–6): but that fall took place in response to a pattern of temptations that continued to exist and plague him (as the examination of conscience in Bk. 10 makes clear). The apostrophe here does not address the *anima* as it was twenty years earlier; rather A. turns from contemplating his fall, as he completes its description, to address the soul by way of admonition against the future. That what he says here in paragraphs 15–19 is what he should have said at the time adds irony: paragraphs 20–27 will show what actually happened.

Cf. the similar, though with much change over nearly thirty years, apostrophe at *trin.* 15. 27. 50 (one paragraph from the end of that work), where he appeals to his converted soul to convert again, to know that the goal is not yet reached: 'o tu, anima mea, ubi te esse sentis, ubi iaces aut ubi stas donec ab eo qui propitius factus est omnibus iniquitatibus tuis sanentur omnes languores tuos? [Ps. 102: 3: see below] agnoscis te certe in illo esse stabulo quo Samaritanus ille perduxit eum quem reperit multis a latronibus inflictis vulneribus semivivum.'

tumultu: instrumental ablative: 'do not be deafened by the uproar.'

verbum ipsum clamat: 4. 12. 19, 'clamans ut *redeamus* ad eum'; cf. esp.

10. 27. 38, 'vocasti et clamasti et rupisti surditatem meam'; sim. at 7. 10. 16, 9. 4. 9, 13. 29. 44, hence, introducing direct discourse attributed to God.

ibi: i.e. *apud verbum dei* (see below, 'ibi . . . ibi').

est locus: See esp. on 13. 9. 10, 'requies nostra locus noster'; at 7. 7. 11 and 10. 40. 65, on the verge of mystic experience in both cases, he has trouble finding a *locus*, finding it only in God. Contrast 4. 7. 12, 'ego mihi remanseram infelix locus.'

discedunt . . . succedunt: 4. 10. 15, 'oriuntur et occidunt . . . et omnia intereunt.'

partibus . . . universitas: See on 4. 10. 15 above ('universum'), and below, 4. 11. 17 ('totum cuius hae partes sunt') and 4. 13. 20 ('sicut pars corporis ad universum suum').

ibi fige mansionem tuam: Cf. Jn. 14: 23, 'si quis diligit me, sermonem meum servabit, et pater meus diliget eum, et ad eum veniemus, et mansiones apud eum faciemus.' *s.* 119. 3. 3, 'quis comprehendat verbum manens? omnia verba nostra sonant et transeunt [cf. 4. 10. 15]. quis comprehendat verbum manens, nisi qui in ipso manet? . . . noli sequi flumen carnis. caro quippe ista fluvius est: non enim manet. . . . vis manere? verbum autem domini manet in aeternum [Is. 40: 8: see on 'semper stantem . . .' below].'

fige: See on 4. 10. 15.

saltem: 'at least, if only'; you turn to truth, *faute de mieux*, out of weariness with frauds and lies.

fallaciis: 4. 1. 1, 'falsi atque fallentes in variis cupiditatibus'.

veritati commenda: The parallel with 'ibi commenda quidquid inde habes' makes it explicit that *veritas* = *verbum dei*; see on 1. 5. 6.

et [7 ×]**:** See on 1. 1. 1, 'et laudare te vult homo'; for the anaphora, note that a similar construction marks the passage from Mt. echoed in 'et sanabuntur . . .' (see below).

reflorescent: *en. Ps.* 27. 7, '"et refloruit caro mea", id est, et resurrexit caro mea.'

putria C¹DO¹S Maur. Knöll Skut. Ver. : *putrida* C²GO². See on 4. 3. 4, 'superba putredo', and cf. 2. 1. 1, 'computrui'.

sanabuntur omnes languores tui: Mt. 4: 23, 'et circumibat Iesus totam Galilaeam, docens in synagogis eorum, et praedicans evangelium regni, et sanans omnem languorem, et omnem infirmitatem in populo'; cf. Ps. 102: 3–4 (cited repeatedly in Bk. 10; see Knauer, 146–50): 'qui propitius fit omnibus iniquitatibus tuis, qui sanat omnes languores tuos, (4) qui redimet de interitu vitam tuam'; but note the apostrophe at Ps. 102: 2, 'benedic anima mea dominum et noli oblivisci omnes retributiones eius.' Knauer, 150 n. 1, compares 10. 33. 50, 'in cuius oculis mihi quaestio factus sum et ipse est languor meus', with the present passage in tandem with 4. 4. 9, 'factus eram ipse mihi magna quaestio'.

et renovabuntur: 7. 9. 14, 'participatione manentis in se sapientiae [2] renovantur, ut sapientes sint,' 11. 9. 11 (*sanatio languorum* with *renovatio*), 13. 22. 32 (*renovatio* according to the image of God in man: echoed variously at 13. 23. 33, 13. 26. 40, and 13. 34. 49).

sempter stantem ac permanentem deum: Cf. Ps. 101: 13, 'tu autem domine in aeternum permanes'; 1 Pet. 1: 23–5, 'per verbum dei vivi et permanentis, (24) quia omnis caro ut faenum et omnis gloria eius tamquam flos faeni: exaruit faenum, et flos decidit. (25) verbum autem domini manet in aeternum. hoc est autem verbum quod evangelizatum est in vos'; Heb. 1: 11, 'tu autem [domine] permanebis'; 1 Jn. 2: 17, 'qui autem facit voluntatem dei manet in aeternum.'

4. 11. 17

This paragraph reprises 4. 10. 15 (see notes below), which was addressed to God. Prayer is itself a way of learning from God: cf. *s. dom. m.* 2. 3. 14, 'fit ergo in oratione conversio cordis ad eum [cf. *conversam* here] . . . et in ipsa conversione purgatio interioris oculi.' A. receives from God the words to say back to God: from that, A. learns the words (in many ways identical) to say to his own soul. That these are now A's own words *may* explain the relatively unscriptural language of this paragraph: scripture did its job in 4. 10. 15, now it is A.'s turn, using his own words—except at the very end of this paragraph.

conversam OS Knöll Vega Ver. : *conversa* CDG Maur. Skut. Pell.

in parte . . . totum: *BA* ad loc. sees in this dialectical opposition of part and whole a reminiscence of Porphyry, *sent.* 40 (translated at length at *BA* 13. 680–1); the resemblance is doctrinal and atmospheric more than verbal, but the passage of Porphyry is one with numerous points of contact with A.'s thought.

sensus carnis . . . modum: 4. 10. 15, 'sic est modus eorum . . . tardus est enim sensus carnis, quoniam sensus carnis est: ipse est modus eius', and see on 1. 7. 12.

syllabas: 4. 10. 15, 'ecce sic peragitur et sermo noster'; cf. the meditation on the transience of speech and sound at 11. 27. 34 ff.

et non sunt omnia simul ea quibus constat: 'And all the things that one thing comprises do not exist at once (*simul*).' 4. 10. 15, 'quae non sunt omnes simul'. (The parentheses here are Verheijen's.)

(*quibus*) **constat** CDGO[2] Maur. Knöll Skut. Ver. : *constant* O[1]S.

sed longe his melior: 4. 10. 15 ends with a comparable appeal to divine authority, 'in verbo enim tuo . . . ibi audiunt'. Cf. Ps. 99: 3, 'scitote quoniam dominus ipse est deus. ipse fecit nos, et non ipsi nos' (see on 9. 10. 25 and cf. 4. 11. 16, 'mansionem . . . permanebunt . . . permanentem'): owning the authority of God is to see him as creator, and vice versa. *en. Ps.* 99. 15, 'non debemus superbire: totum bonum quod habemus, ab *artifice* nostro habemus.' Cf. first words of 4. 12. 18.

non discedit: 4. 10. 15, 'decedat . . . succedat'.

4. 12. 18

The represented discourse (and the quotation marks) should be carried (with Pellegrino) from 'hunc amemus' through to 4. 12. 19, '. . . ascendendo contra deum'. G–M: 'The summary of the preacher's message continues to the end of sec. 19.' They are mistaken only in saying 'the preacher's'. Pusey had already punctuated thus: Skutella and Verheijen limit the quotation marks to 'hunc amemus . . . non est longe' in this paragraph; Vega limits them to 'hunc amemus'.) Note that the second-person discourse switches here to the plural and remains plural wherever it occurs until 'dic eis' recurs at the end of 19; 'cui confitetur anima mea' in 19 remains, though this slightly stretches the frame (see below on 'redite, praevaricatores'), within the represented discourse. That this is represented discourse allows for some further repetitions from shortly before: we have first the prayer to God (15), then the discourse to the soul (16–19) containing within it the discourse of the soul to other souls (18–19).

This whole passage thus exemplifies obedience to the two great commandments (see on 4. 9. 14).

Si placent corpora: *en. Ps.* 148. 15, 'non ergo ita tibi placeat quod fecit, ut recedas ab eo qui fecit. . . . si pulchra sunt quae fecit, quanto pulchrior est qui fecit?' Sim. at *s.* 21. 2, *en. Ps.* 32. en. 2. s. 1. 6.

deum ex illis lauda: Cf. Ps. 145: 2, 'lauda, anima mea, dominum' (see on 4. 10. 15).

artificem CGO[1] Knöll Skut. Ver. : *artifice* DO[2]S. See on 4. 11. 17.

fixae: See on 4. 10. 15.

rape ad eum tecum quas potes: The second-person singular continues the address (from the beginning of 4. 11. 16) to the *anima*.

rape: This exhortation precedes the represented discourse that runs through paragraph 19, so at the end of 19 there immediately follows 'et sic eos rape tecum ad deum'. The vivid evocative force of the verb is frequently employed: 1. 17. 27, 2. 2. 2, 3. 4. 8, 4. 15. 27 ('rapiebar foras'), 5. 8. 15, 6. 16. 26, 7. 17. 23 ('sed rapiebar ad te decore tuo'), 8. 4. 9 ('age, domine, fac excita et revoca nos, accende et rape, flagra, dulcesce: amemus, curramus'), 8. 8. 19, 9. 10. 25 ('et haec una [visio] rapiat et absorbeat et recondat in interiora gaudia spectatorem suum'), 10. 35. 56, 11. 11. 13. For the mystical flavour, see du Roy, 76.

Hunc amemus CDOS Knöll Skut. Ver. : *hunc amemus hunc amemus* G Maur. Pell.

The love of God is generally the bond of society and specifically the guarantee of meaning in speech: cf. 10. 3. 3–4 and see on 4. 14. 21 below. Cf. Ps. 99: 3 (see on 4. 11. 17), 'ipse fecit nos, et non nos' (see on 9. 10. 24–5); Act. 17: 27–8, 'quaerere deum, . . . quamvis non longe sit ab unoquoque nostrum; (28) in ipso enim vivimus et movemur et sumus.' See on 1. 18. 28, 'longe'; cf. 9. 11. 28 (Monnica), 'nihil longe a deo'. *en. Ps.* 137. 2, 'non enim est ille longe ab unoquoque nostrum positus, in quo vivimus, movemur et sumus. longe te a deo non facit, nisi iniquitas sola.'

ex illo in illo sunt: Rom. 11: 36, 'quoniam ex ipso et per ipsum et in ipso omnia ipsi gloria in saecula. amen.' See on 1. 2. 2, 'ex quo omnia'.

ecce ubi est: J. le Clercq (d. 1736) suggested (reprinted at *PL* 47. 206) that 'est' should be followed by a question mark; Verheijen uses a colon. Le Clercq emended the following 'ubi' to 'ibi'.

intimus cordi est, sed cor erravit ab eo: 10. 27. 38, 'et ecce intus eras et ego foris et ibi te quaerebam et in ista formosa quae fecisti deformis inruebam.' (The antecedent of *intimus* is *ipse*.)

cor erravit: Ps. 118: 176, 'erravi sicut ovis perdita'; *en. Ps.* 118. s. 32. 7 applies this text to the parable of the good shepherd; but see on 'vias difficiles' below.

redite, praevaricatores, ad cor: Is. 46: 8, 'redite, praevaricatores, ad cor'; *en. Ps.* 57. 1, 'sed quia homines appetentes ea quae foris sunt etiam a seipsis exsules facti sunt, data est etiam conscripta lex; non quia in cordibus scripta non erat, sed quia tu fugitivus eras cordis tui, ab illo qui ubique est comprehenderis, et ad teipsum intro revocaris. propterea scripta lex quid clamat eis qui deseruerunt legem scriptam in cordibus suis? "redite, praevaricatores, ad cor".' *s. Caill.* 2. 11. 5 (on the prodigal, when the son returns home), 'iam ergo iste obtriverat cor in regione egestatis: *redierat enim ad cor,* ut obtereret cor.' The same citation treated similarly at *en. Ps.* 57. 3, 70. s. 1. 14, 76. 15, and *Io. ev. tr.* 18. 10; as regular admonition, 'redi ad cor' at *ss.* 13. 3. 3, 16. 7. 7. Cf. 4. 11. 16, 'verbum ipsum clamat ut redeas.' Note the sequence: 'Augustine' says to his 'soul' that the Word says that the soul should return to the place of imperturbable quiet (i.e. to its own heart that has found the Word); now the 'soul' says on its own to other souls what it has heard from the Word by way of 'Augustine': 'redite . . . ad cor'. (On *redire*, see on 1. 18. 28.) Better to say that 'Augustine' addresses his 'soul' in 16–17, and the distinction between speaker and hearer is slight but just sustainable; when his 'soul' turns to address other souls that distinction becomes nugatory and it is as though 'Augustine' were speaking to them. This stretches, but does not break, the frame surrounding this paragraph.

inhaerete: See on 7. 11. 17, 'mihi autem *inhaerere* deo bonum est,' echoing Wisd. 7: 27.

requiescete . . . quieti: See on 4. 6. 11, 'requiescebam', and recall 1. 1. 1, 'inquietum est cor nostrum donec requiescat in te' (cf. above, 'redite . . . ad *cor*'); 4. 11. 16, 'locus quietis imperturbabilis'. Cf. here 'non est requies ubi quaeratis eam'.

quo itis: Cf. Ps. 138: 7, 'quo ibo a spiritu tuo et quo a facie tua fugiam?' Cf. 4. 9. 14, 'quo it aut quo fugit'; it is not that A. here again evokes the scriptural text, so much as that he repeats for the other souls the doctrine evolved earlier.

deserto illo: 4. 11. 16, 'et ibi est locus quietis imperturbabilis ubi non deseritur amor, si ipse non deserat' (and cf. yet again 'redite . . . ad cor' here).

quo vobis: G–M: 'lit. "to what end is it for you?"', i.e. why?

vias difficiles et laboriosas: Wisd. 5:7, 'repleti sumus iniquitatis via et perditionis et ambulavimus vias difficiles; viam autem domini ignoravimus.' The 'vias difficiles' are little different from 'vias cum muscipulis' (see on 3. 1. 1). Cf. Is. 9: 2, cited below, and see 5. 2. 2, 'post suas vias difficiles'.

beatam vitam: The 'blessed life', important from A.'s earliest writings (e.g. *beata v.* itself, written at Cassiciacum), appears in three discrete passages in *conf.:* here, at 6. 10. 17–11. 20 (in and surrounding the artful passage of represented indecision at 6. 11. 18–19), and at 10. 20. 29–23. 34, at the culmination of the search for God in memory; see on 10. 20. 29 for fuller discussion. For the background, see Holte, *Béatitude et Sagesse* (Paris, 1962), and W. Beierwaltes, *Regio beatitudinis: Augustine's Concept of Happiness* (Villanova, 1981: also in German in the *Sitzungsber. Akad. Heidelberg*, (1981), 6).

in regione mortis: Is. 9: 2 (Vg.; quoted from LXX at Mt. 4: 16), 'populus qui ambulabat in tenebris vidit lucem magnam; habitantibus in regione umbrae mortis lux orta est eis'; *en. Ps.* 87. 6, 'ex tenebris et morte impietatis educti sunt.'

4. 12. 19

The represented discourse continues from 4. 12. 18.

The turn to Christ was anticipated at the end of 4. 10. 15, just before the address to the *anima* began. That Christ is central to this passage is a sign of its importance to the structure of *conf.* as a whole. At the nadir of descent, this rhetorically complex passage builds to a climax that presents the full drama of incarnation and redemption in Christ; every major stage of the ascent to follow is measured by its 'Christ-content': see on 7. 9. 13 ff. (esp. 7. 18. 24), 8. 12. 30, 9. 10. 25, and 10. 43. 70. On A.'s view of incarnation as redemption, see generally van Bavel.

The general structure of a baptismal creed lies behind this paragraph, as it also does e. g. *doctr. chr.* 1 (see H.-J. Sieben, *REAug* 21 (1975), 72–90). The text of the Apostle's Creed used in baptism at Hippo in A.'s time was not written down, though it may be reconstructed from allusions and interpretations (see *s.* 214. 1); hence a diffidence in following that familiar text too closely, though the general shape may be descried.

descendit: Jn. 6: 33, 'panis enim dei est, qui descendit de caelo et dat vitam mundo'; for *descendere* as a verb of the incarnation, see *BA* 72. 284 n. 54 with refs.

occidit eam: Cf. 2 Tim. 1: 10, 'qui destruxit quidem mortem'.

clamans: Christ steals Jupiter's thunder (see on 1. 16. 25); see also on 4. 11. 16, 'verbum ipsum clamat ut redeas'; note here 'clamans ut redeamus' repeated below.

unde processit: Ps. 18: 6–7, 'in sole posuit tabernaculum suum et ipse tamquam sponsus procedens de thalamo suo exultavit ut gigans ad currendam viam. (7) a summo caelo egressio eius et occursus eius usque ad summum eius, nec est qui se abscondat a calore eius' [*en. Ps.*, in the places quoted next, has 'exultavit sicut', but Ps. Rom. has 'ut' as here]. Consistently taken as an allegory of the incarnation: *en. Ps.* 18. en. 1. 6, 'et ipse procedens de utero virginali, ubi deus naturae humanae *tamquam sponsus sponsae copulatus est* [cf. 'nupsit' here]'; *en. Ps.* 18. en. 2. 6, 'hoc est, "ille tamquam sponsus", cum verbum caro factum est, in utero virginali thalamum invenit; atque inde naturae coniunctus humanae, tamquam de castissimo procedens cubili, humilis misericordia infra omnes, fortis maiestate super omnes. hoc est enim, "gigans exsultavit ad currendam viam", natus est, crevit, docuit, passus est, resurrexit, ascendit; cucurrit viam, non haesit in via'; cf. *c. litt. Pet.* 2. 32. 74, *c. s. arrian.* 8. 6 ('geminae gigans substantiae': the wording echoes Amb. *hymn.* 6. 17–20 (Walpole): 'procedat e thalamo suo, | pudoris aula regia, | geminae gigans substantiae; | alacris occurrat viam').

gigans D¹GO¹S Ps. Rom. Knöll Skut. Ver. : *gigas* CD²O² Maur.

tardavit: Ps. 39: 18, 'deus meus, ne tardaveris!'; Heb. 10: 38, 'qui venturus est veniet et non tardabit' (echoing Hab. 2: 3, 'quia veniens veniet et non tardabit').

descensu, ascensu: Eph. 4: 8–9, 'ascendens in altum captivam duxit captivitatem, dedit dona hominibus. (9) quod autem *ascendit* quid est, nisi quia et *descendit* primum in inferiores partes terrae; qui *descendit* ipse est et qui *ascendit* super omnes caelos.' Sim. at *ss.* 78. 6, 91. 6. 7, 261. 1. 1.

et discessit ab oculis: Lk. 24: 51, 'et factum est cum benediceret eis, recessit ab eis et ferebatur in caelum' (text from *cons. ev.* 3. 25. 83; there is support for *discessit* in the African VL tradition: Milne, 119). Act. 1: 9, 'et cum haec dixisset, videntibus illis elevatus est et nubes suscepit eum ab oculis eorum.'

redeamus ad cor: See on 4. 12. 18, quoting Is. 46: 8, 'redite, praevaricatores, ad cor'. The juxtaposition of 'redeamus ad eum' and this line identifies the heart as the place where Christ is found.

ecce hic est: Mt. 24: 23 (with parallel at Mk. 13: 21), 'tunc si quis vobis dixerit, "ecce hic Christus" aut "illuc", nolite credere.' The gospel texts and the present expression represent the radical uncertainty of post-ascension Christianity: the belief that Christ is present matches exactly the sense of absence.

mundus per eum factus est: Jn. 1: 10–11, 'in mundo erat, et mundus per ipsum factus est, et mundus eum non cognovit, (11) in propria venit et sui eum non receperunt.'

venit in hunc mundum peccatores salvos facere: 1 Tim. 1: 15, 'fidelis sermo et omni acceptione dignus, quia Christus Iesus venit in mundum peccatores salvos facere.'

confitetur: The only present, indicative, active, non-interrogative occurrence of this verb in Bk. 4: the final assertion of this lengthy and central represented discourse, followed by questions and exhortations.

et sanat eam: Ps. 40: 5, 'ego dixi "domine miserere mei, sana animam meam quia peccavi tibi"'; *en. Ps.* 40. 6, 'flagellandum iudicas omnem filium quem recepturus es, qui nec unico pepercisti. ille quidem sine peccato flagellatus est; ego autem dico, miserere mei, sana animam meam, quoniam peccavi tibi.' Cf. 4. 3. 4 ('cura animam meam, quoniam peccavi tibi') and 5. 10. 18 ('ut sanares animam meam, quoniam peccabat tibi'). For 'healing' in Bk. 4, cf. also 4. 11. 16, 'sanabuntur omnes languores tui.'

filii CGO² Maur. Pell. : *fili* DO¹S Knöll Skut. Vega Ver. For the spelling, see on 1. 16. 26. Ps. 4: 3, 'filii hominum usquequo graves corde? ut quid diligitis vanitatem et quaeritis mendacium?' See on 9. 4. 9.

posuistis in caelo os vestrum: Ps. 72: 8–9, 'cogitaverunt et locuti sunt nequitiam, iniquitatem in excelso locuti sunt: (9) posuerunt in caelum os suum et lingua eorum transiit super terram.' (The verse is applied to the prodigal by Amb., *de interpell. Iob et David* 4. 5. 12.) *en. Ps.* 8. 13, 'vide nunc etiam volucres caeli, *superbos,* de quibus dicitur: "posuerunt in caelum os suum".' Sim. at *en. Ps.* 72. 15, 103. s. 3. 5. This is the *ascensio contra deum* of which A. speaks here.

ut ascendatis, et ascendatis ad deum O (+ BPZAHM) Maur. Knöll Isnenghi Ver. : *ut ascendatis ad deum* CDS (+ F) Skut. Vega Pell. : *ut ascendatis et ascenditis ad deum* E : *ut ascendatis ad eum et ad deum ascendatis* G.

Isnenghi: 'um emporzusteigen, *und zwar zu Gott* emporsteigen zu können'.

ascendatis: Cf. 5. 3. 5, and see on 9. 10. 24, 'et adhuc ascendebamus', and cf. the parallel passages from the 'ascent' in Bk. 10: 10. 7. 11, 10. 8. 12, 10. 17. 26, and cf. also 13. 9. 10 (quoted below on 'ardens igne caritatis').

cecidistis: Cf. *s.* 261. 1. 1, cited above.

(*contra*) **deum** GOS Knöll Skut. Ver. : *eum* CD Maur.

Dic eis . . . rape: See on 4. 12. 18, 'rape'.

in convalle plorationis: Ps. 83: 6–7, 'beatus vir cuius est auxilium abs te domine; ascensus in corde eius disposuit, (7) in convalle plorationis, in locum quem disposuisti eis'; *en. Ps.* 83. 10–11, 'quanto ergo plus amaveris, tanto plus ascendes. . . . disponit ascensus in corde. ubi disponit ascensus? in corde, in convalle plorationis. ecce habetis torcular convallem plorationis; ipsae lacrimae piae contribulatorum, mustum sunt amantium. . . . (11) quare autem in convalle plorationis? et ex qua convalle plorationis ad illum locum gaudii veniemus? ... afflixit nos lege, pressit nos lege, ostendit nobis torcular; vidimus pressuram, carnis nostrae tribulationem cognovimus, ingemuimus rebellante peccato adversus mentem nostram, clamavimus: "miser ego homo"; sub lege gemuimus; ... adveniet gratia post legem; ... hic enim per gratiam multae virtutes dantur: ... multae virtutes, sed hic necessariae; et ab his virtutibus imus in virtutem. quam virtutem? Christum, dei virtutem, et dei sapientiam.' Cf. 9. 2. 2.

ardens igne caritatis: That heat and love are associated is conventional (6. 7. 12, 'et ad me ardentius diligendum'), but that divine love (*caritas*) burns with its own flame is equally important to A.: see 13. 9. 10, for the heat of divine love connected with the ascent to God; see also on 2. 1. 1, 'exarsi', and for similar fires, see 3. 4. 8, 5. 3. 4, 9. 10. 24, 10. 29. 40, 11. 29. 39. 'Ardor' carries over to the act of confession at 12. 18. 27 ('mihi ardenter confitenti'); the fire is pentecostal, and enflames those who receive it at 13. 19. 25.

4. 13. 20

On beauty in A.'s thought, see K. Svoboda, *L'Esthétique de saint Augustin et ses sources* (Brno, 1933), and R. J. O'Connell, *Art and the Christian Intelligence in Saint Augustine* (Cambridge, Mass., 1979); there is also a fragmentary study by J. Tscholl, *Augustiniana* 14 (1964), 72–104 (the 'Fortsetzung' there promised has apparently not appeared); see below for studies of the *de pulchro et apto*. The notes that follow will attempt only to place that work on an intellectual trajectory, doing justice to the biographical facts represented here and to the way in which A. chose to narrate them.

The formal studies of A.'s aesthetic do not sufficiently emphasize the link, everywhere important when the subject arises in *conf.*, to A.'s trinitarian speculations; see on 1. 7. 12 for the fundamental triad, *modus/species/ordo*. In Bk. 4, it should be borne in mind that *pulchritudo* corresponds to *species* (*util. cred.* 16. 34, '*species* [2] rerum omnium, quam profecto ex aliquo verissimae pulchritudinis fonte manare credendum est', and see here 'decus et species': beauty is in the realm of the eyes, where the temptation is again *curiositas* [2], and hence is associated with the second person of the trinity. That link is already foreshadowed by the central place of Christ in this book (see on 4. 10. 15, 4. 11. 16, and 4. 12. 19); if we accept the link *pulchritudo* = *Christus*, then those passages imply that the beauty that is Christ, present in the beauty of created things, cries out with invitation to those who see them; but for various reasons amply set forth in this book, A. at the time was unable to heed that call, and confined his love to material, earthly beauty. Hence he found no way leading upward from the depths.

Of particular interest are the patterns of idea and expression that link the composition of the *de pulchro et apto* to later episodes in *conf.* This early philosophical essay is presented as a first attempt at an intellectual ascent to God, whose later stages are more dramatically displayed at 7. 10. 16, 7. 17. 23, 9. 10. 24–5, and 10. 6. 8–27. 38. This selective presentation by A. at the time of writing *conf.* does not guarantee that these particular episodes were linked in any conscious way as he lived them; their relationship is at least as much, if not exclusively, a pattern imposed in reflective memory years later. Writing the *de pulchro et apto*, after all, is the only intellectual task of A.'s for which we have such discussion from the time he began teaching at Thagaste in 375 until his move to Italy in 383, and even here the presentation is sketchy and impressionistic in the extreme.

Beauty in *conf.* embraces the earthly (e.g. 2. 2. 3, 'novissimarum rerum fugaces pulchritudines', 12. 2. 2), the divine (e.g. 3. 6. 10, 'mi pater summe bone, pulchritudo pulchrorum omnium', 10. 27. 38, 'sero te amavi, pulchritudo tam antiqua et tam nova'), and the earthly as a sign of the divine (esp. citing Rom. 1: 20, as at 7. 17. 23, and cf. 13. 20. 28, 13. 28. 43).

The place of beauty in A.'s works is affected by his reading of Plot. 1. 6, an event that necessarily masks from us the exact state of A.'s thought on the subject at the time of *de pulchro et apto*. At the same time, the appeal of Plotinus' treatise must itself have been enhanced by A.'s prior interest. The importance of the subject and its link to the highest concerns are clear at *c. acad.* 2. 3. 7, 'quid est enim philosophia? amor sapientiae. quid philocalia?[6] amor pulchritudinis. quaere de graecis. quid ergo sapientia? nonne ipsa vera est pulchritudo? germanae igitur istae prorsus, et eodem parente procreatae: sed illa visco libidinis detracta caelo suo, et inclusa cavea populari viciniam tamen nominis tenuit.' The Platonic overtone—particularly the cave—is noticeable; and this passage evoked this later qualification: *retr.* 1. 1. 3, 'in secundo autem libro prorsus inepta est et insulsa illa quasi fabula de philocalia et philosophia, quod sint germanae et eodem parente procreatae. aut enim philocalia quae dicitur nonnisi in nugis est, et ob hoc philosophiae nulla ratione germana; aut si propterea est hoc nomen honorandum, quia latine interpretatum amorem significat pulchritudinis, et est vera ac summa sapientiae pulchritudo, eadem ipsa est in rebus incorporalibus atque summis philocalia quae philosophia, neque ullo modo sunt quasi sorores duae.'

Only when the creature achieves its own measure of beauty is it possible to see God (*ord.* 2. 19. 51; cf. *imm. an.* 8. 13, 16. 25—a text so Platonic (or Porphyrian) that some think it little more than notes on A.'s reading). So the summary of the stages of the ascent of the mind to God at *quant. an.* 35. 79 rephrases itself entirely in terms of beauty: 'ascendentibus igitur sursum versus primus actus docendi causa dicatur *animatio*, secundus *sensus*, tertius *ars*, quartus *virtus*, quintus *tranquillitas*, sextus *ingressio*, septimus *contemplatio*. possunt et hoc modo appellari: de corpore, per corpus, circa corpus, ad

[6] *Philocalia* perhaps first in Latin here.

seipsam, in seipsa, ad deum, apud deum. possunt et sic: pulchre de alio, pulchre per aliud, pulchre circa aliud, pulchre ad pulchrum, pulchre in pulchro, pulchre ad pulchritudinem, pulchre apud pulchritudinem.'

The most practical conclusion drawn from this doctrine in A.'s early works is the aesthetic theory of the nature of evil (e.g. *lib. arb.* 1. 16. 35, 3. 25. 77). Beauty appears in speculative contexts after *conf.* (e.g. *trin.* 6. 10. 12, 'in illa enim trinitate summa origo est rerum omnium [1] et perfectissima pulchritudo [2] et beatissima delectatio [3]'), but beauty gradually loses its hold on A.'s ways of thinking. For the apogee of its role in *conf.*, see on 10. 27. 38.

non noveram: Cf. 3. 7. 12–13, where 'non noveram' is the formula that defined his Manicheism by introducing the answers he later found acceptable to questions that he answered otherwise at the time.

amabam pulchra inferiora: The theme was anticipated at 4. 10. 15, 'tametsi figitur in pulchris extra te et extra se'.

in profundum: Cf. Is. 31: 6, 'convertimini, sicut in profundum recesseratis, filii Israhel'; Prov. 18: 3, 'impius cum in profundum venerit peccatorum contemnit'; Is. 7: 11, 'pete tibi signum a domino deo tuo, in profundum inferni sive in excelsum supra.'

amicis meis: The first and last sections of Bk. 4 each open thus with invocations of the 'friends' with whom he lived and argued (4. 1. 1, 'cum amicis meis per me ac mecum deceptis'; 4. 8. 13, 'aliorum amicorum solacia, cum quibus amabam quod pro te amabam'; they appear again at 5. 8. 14, urging A. to go to Rome to advance his career); the middle section is devoted to the one friend he lost.

num amamus aliquid nisi pulchrum?: The thought is trite in A.: *mus.* 6. 13. 38, 'dic, oro te, num possumus amare nisi pulchra'; *en. Ps.* 79. 14, 'quare autem amas ista nisi quia pulchrae sunt?' Here, at the nadir of his fall, A. thought himself a lover of beauty; at the zenith of his ascent, his lament is only that love of beauty came to him too late: 10. 27. 38, 'sero te amavi, pulchritudo tam antiqua et tam nova'. We are left to infer that his first love of beauty was love of the wrong thing (see on 1. 1. 1, 'aliud enim pro alio potest invocare nesciens').

quid est quod nos allicit . . . amamus: Sim. at *ss.* 34. 2. 4, 243. 8. 7, and *retr.* 1. 11. 2.

aliud esse . . .: See on 4. 10. 15, 'partes . . . universum'.

apte adcommodaretur: Testard, 1. 49–67, attempting to prove that the background of the *de pulchro et apto* was more Ciceronian than Manichean or Platonic, offers a raft of Ciceronian texts in which *aptum* and *adcommodatum* occur together, e.g. *off.* 1. 40. 142, 'illa est eutaxia in qua intellegitur ordinis conservatio. . . . *nam et ordinem sic definiunt, compositionem rerum aptis et adcommodatis locis*'; cf. *fin.* 4. 17. 46, 5. 6. 17, 5. 8. 23, 5. 9. 24, *de or.* 3. 55. 210, *part. or.* 15. 54, *nat. deor.* 2. 55. 139.

The only passage in the rest of A.'s surviving *œuvre* that significantly parallels this employment of *aptus* shows how the philosophical heritage of youth returned, fully Christianized, but still useful in addressing secular intellectuals: *ep.* 138. 1. 5 (to Marcellinus), 'haec quaestio quam late pateat, profecto videt, quisquis pulchri aptique distantiam sparsam quodam modo in universitate rerum valet neque neglegit intueri. pulchrum enim per se ipsum consideratur atque laudatur, cui turpe ac deforme contrarium est. aptum vero, cui ex adverso est ineptum, quasi religatum pendet aliunde nec ex semet ipso sed ex eo cui conectitur, iudicatur.' Only incidental is *Io. ev. tr.* 26. 13 (of a healthy body), 'sit pulchrum, sit aptum, sit sanum'. See, however, on 2. 5. 10, 'in contactu carnis' and 'congruentia' and texts cited there, for an analogous notion.

A.'s doctrine in this lost work could be paraphrased so: 'Things exist, they are beautiful in themselves, and they are parts of a beautiful whole by their seemly arrangement together.' *That* paraphrase is a neat embodiment of A.'s familiar triadic pattern: existence [1], beautiful appearance [*species*, 2], and providential arrangement [*ordo*, 3]. *That* equation in turn functions as a reminder that one of the three dialogues recorded from Cassiciacum is *de ordine*, an extended discussion of providential arrangement in the world of created things.

apte: 3. 7. 13, 'quod non *apte* conveniat'. Cf. *trin.* 4. 2. 4, where A. expressed Gk. ἁρμονία as *coaptatio* (and uses the word also at *civ.* 22.24).

scaturrivit: 'welled up' (G–M) spontaneously: G–M canvass the possibility of an ironic interpretation, but think 'the solemn phrase *ex intimo corde meo* precludes the idea that A. is smiling at his own youthful enthusiasm; he is impressed, rather, with the semi-independent force of the inner mind—a neo-Platonic thought.' The matter is perhaps not so simple. For 'ex intimo corde meo', compare 4. 12. 18, 'intimus cordi est [veritas], sed cor erravit ab eo'; on that reading, A.'s 'inmost heart' at this period is void of truth, and what comes bubbling forth is not necessarily judged positively.

libros 'de pulchro et apto': Other works of A. cited in *conf.*: 9. 4. 7, 'libri disputati cum praesentibus (i.e. *c. acad.*, *beata v.*, *ord.*) et cum ipso me solo coram te (*sol.*) . . . cum absente Nebridio . . . epistulae'; 9. 6. 14, 'de magistro'. On the *de pulchro et apto*, the best concise study is the note by A. Solignac at *BA* 13. 670–3; cf. Mandouze, 461–3, and D. A. Cress, *Aug. Stud.* 7 (1976), 153–63; see also Alfaric, 224, Pincherle, *Rassegna di filosofia* 2 (1953), 19, and T. Katô, *REAug* 12 (1966), 229–40, who suggest Manichean influences (*non liquet*).

Solignac displays possible parallels in Platonic texts (*symp.* 211 D, *Phaedrus* 249 D, 264 C, esp. *Hipp. mai.* 290 A-296 E, where P. studies explicitly the relation of τὸ καλόν and τὸ πρέπον; cf. also Cic. *off.* 1. 28. 97 ff. (Svoboda, *L'Esthétique de S. A.* 10–16) and *orator* 21. 70 ff., where he renders τὸ πρέπον as *decorum*. Solignac (*BA* 13. 671): 'il y avait néanmoins dans ces réflexions l'ébauche de la thèse du *frui* et de l'*uti*, c'est-à-dire de la distinction du bien absolu dont on peut et doit jouir, et du bien relatif, dont on ne doit qu'user.'

Svoboda, 10–16, suggests neo-Pythagorean links; Solignac, *RA* 1 (1958), 129–37, agrees; see also Solignac quoted on the monad/dyad at 4. 15. 24. Finally, Solignac offers what is only speculation, but stimulating and attractive none the less, a scheme for the arrangement of the work in three books: Bk. 1, body; Bk. 2, soul (4. 15. 24, 'et converti me ad animi naturam'); Bk. 3, God (4. 15. 26, 'sed ego conabar ad te et repellebar abs te'). He is followed by Mayer, *Zeichen*, 1. 97, who adds: 'Der augustinische "ascensus" ist in dieser Einteilung nicht zu verkennen' (see further on 4. 15. 24). Svoboda, 16, surmises plausibly that the work was in dialogue form (perhaps suggested above by 'et dicebam amicis meis').

Of the many Ciceronian texts cited by Testard (see above on 'apte'), a few go beyond 'aptitude' to *pulchritudo* and do usefully parallel the higher ambitions that motivated A.'s treatise (and see *off.* 1. 4. 14 and *fin.* 2. 14. 47, quoted on 1. 7. 12): *nat. deor.* 2. 22. 58, *fam.* 9. 24. 3, *Tusc.* 4. 13. 31, *off.* 1. 28. 98 ('ut enim pulchritudo corporis apta compositione membrorum movet oculos et delectat hoc ipso, quod inter se omnes partes cum quodam lepore consentiunt, sic hoc decorum quod elucet in vita movet approbationem eorum quibuscum vivitur, ordine et constantia et moderatione dictorum omnium atque factorum'). On a similarly lower level are the conjunctions of *ornate* and *apte* as adverbs appropriate to the orator's tasks of *elocutio* and *actio* throughout *de or.* 3, e.g. 3. 36. 144, 'duas tibi reliquas feceras, quem ad modum primum ornate, deinde etiam apte diceremus'; such texts suggest an influence for the lines drawn to create the categories, but not for the substance of the doctrine that A. would have expounded.

nam excidit mihi: Courcelle, *Recherches*, 35 n. 3, 'cet oubli surprenant est peut-être affecté'; but the verb is well-chosen: Cic. *de or.* 1. 21. 94, (Antonius) 'scripsi etiam illud quodam in libello, qui me imprudento et invito excidit et pervenit in manus hominum'; sim. of books at *de or.* 1. 47. 206 and *Att.* 3. 12. 2 ('quo modo exciderit nescio').

nescio quo modo: Cf. A. *ep.* 159. 1, 'nescio quo modo aberravit', of a letter of Evodius that A. cannot find in his files.

4. 14. 21

Writing the *de pulchro et apto* was an exercise in *curiositas*; dedicating it to Hierius (next paragraph) was an exercise in *ambitio saeculi*.

Quid . . . me movit: The answer to the rhetorical question is undoubtedly *ambitio saeculi:* cf. 'sed magis quia . . .'.

Hierium: On Hierius, cf. *PLRE* 1. 431 s.v. Hierius 5. Never mentioned again by A., he is otherwise known from two *subscriptiones* in MSS of Ps.-

Quintilian. Marrou, 163, comments on the remarkable combination in H. of rhetoric with philosophy, of Greek with Latin; the former A. found compatible with his own zeal, the latter incompatible with his tastes. The most useful external *comparandi* for H., however, are Ammianus and Claudian. The late fourth century was an age when Greeks might make a career in the west *in Latin* (unlike Plotinus and Porphyry).

Within *conf.*, the comparison is more subtle: Hierius is a counter-image to Ambrose (himself first sought for his rhetoric, valued for his philosophy: 5. 13. 23), to Marius Victorinus (8. 2. 3, 'quondam rhetor urbis Romae', and of course a noted philosopher and theologian), and to A. himself. But Hierius is the man who was not there: A. never (apparently) met him or had any response to the dedication of the book. H. appears in this text mainly as an external referent for the ambitiousness of A.

et quaedam verba eius audieram: These words should caution any reader who tries too confidently to say what A.'s 'sources' were for his ideas at any period before his adherence to catholic Christianity. What Hierius' philosophy would have amounted to is a matter that defies even wild surmise. How many other contemporary authorities there were, at home in Carthage and *per litteras* from abroad, of whom A. had 'heard some of their words' at this time we can only guess.

Syro, docto: L. Herrmann, *Latomus*, 13 (1954), 37–9, would emend to *miro doctore*.

dictor CDGS Maur. Knöll Skut. Ver.: *doctor* O.

studium sapientiae: i.e. *philosophia* (*vera rel.* 5. 8, 'non aliam esse philosophiam, id est sapientiae studium, et aliam religionem').

mihi placebat: Cf. 'quia placebat aliis'; as in the pear-tree episode, camaraderie is influential.

absit!: Why so forceful? This is the clue to the seriousness of the issue. Is love something that one human manufactures in another? No. At 10. 3. 3 ff., A. says that his own confession (where the principal business is the *praise* of God) is heard in other hearts and believed only when the antecedent act of divine *caritas* prepares those hearts (10. 3. 4); there camaraderie is replaced by a more authentic friendship.

laudatur . . .: The subject, indirectly, is confession. 'You may praise a *man* and he will be loved in his absence. Does this love pass from the mouth of the one who praises into the heart of the one who hears? *Absit*! But one lover is enkindled by another. For this reason is the man who is praised loved, when he is thought to have been proclaimed by a sincere heart, that is, when one who loves him praises him.'

non fallaci corde: Ironic when applied to the love A. felt for Hierius, for *fallacia* had power in A.'s life at the time (cf. esp. 4. 11. 16, 'saltem fatigata falliciis', 4. 1. 1, 'falsi atque fallentes', and 4. 3. 5, 'hanc . . . fallaciam' (i.e. astrology)).

praedicari: echoes the *praedicante*, *praedicatus*, and *praedicatoris* of 1. 1. 1. *Praedicatio* is an instrument, as here and at 1. 1. 1, but the instrument is not the thing itself; the derangement in this case comes from *ambitio saeculi*.

4. 14. 22

Cf. 4. 9. 14 on good and bad friendship; lost friendship is replaced in some sense by hero-worship. A. did 'love' Hierius for being what he would like to be himself; not so charioteers and actors. Loving men for *what* they are is the derangement; his love of Hierius is of a sort with his love of actors.

Sic enim tunc amabam homines: Cf. 4. 7. 12, 'o dementiam nescientem diligere homines humaniter!'

fallitur: Cf. 4. 14. 21, 'non fallaci corde'.

sed tamen: Courcelle, *Les Confessions*, 111 n. 3, characterizes these lines as 'développement diatribique [cynico-stoicienne]', comparable to the exempla instanced at 3. 7. 13 (the man who tries to wear his helmet on his foot, etc.) and 8. 3. 7 (the emperor rejoicing at his victory). G–M cite Plato, *rep.* 10. 605 E: *ἦ καλῶς οὖν, ἦν δ' ἐγω, οὗτος ὁ ἔπαινος ἔχει, τὸ ὁρῶντα τοιοῦτον ἄνδρα, οἷον ἑαυτόν τις μὴ ἀξιοῖ εἶναι ἀλλ' αἰσχύνοιτο ἄν, μὴ βδελύττεσθαι ἀλλὰ χαίρειν τε καὶ ἐπαινεῖν;*

A. claims that he sought praise ('laudari vellem') out of *ambitio saeculi*, and further, that in his ambition lay a desire to be himself the object of exactly such praise and emulation as he now bestowed on Hierius; the lesser praise (bestowed by the *curiosi* on characters they would not themselves want to emulate) was not what he desired. (The connection to *curiositas* is evident from the choice of *histriones* for the example: admiration for actors marked of A.'s own *curiositas* at 3. 2. 2–4.) Put another way, he sought not only praise but envy.

cur non: sc. 'amabatur Hierius'.

pondera . . . amorum: See on 13. 9. 10, 'pondus meum amor meus'.

quid est . . . homo?: The sentence is complex and gnomic. In the disorderly loves of the fallen world there are times when we love in one way what at the same time we hate in another way. This paradox is not resolved, but leads rather to the exclamation below, asserting the intelligibility of man and at the same time the inaccessibility of that intelligibility to man.

grande profundum est ipse homo: *en. Ps.* 41. 13, 'si profunditas est abyssus, putamus non cor hominis abyssus est? quid enim est profundius hac abysso? loqui homines possunt, videri possunt per operationem membrorum, audiri in sermone; sed cuius cogitatio penetratur, cuius cor inspicitur? quid intus gerat, quid intus possit, quid intus agat, quid intus disponat, quid intus velit, quid intus nolit, quis comprehendet? . . . quanta profunditas infirmitatis latebat in Petro, quando quid in se ageretur intus nesciebat, et se moriturum cum domino vel pro domino temere promittebat! quanta abyssus erat! ... ergo omnis homo licet sanctus, licet iustus, licet in multis proficiens, abyssus est.' The idea is central to Bk. 10; cf. 10. 2. 2 (*abyssus* is used of mortals only there, 'abyssus humanae conscientiae'), and see on 10. 8. 15. Contrast the extremist Pascal (*Pensées*, 418–20), 'Il est dangereux de trop faire voir à l'homme combien il est égal aux bêtes, sans lui montrer sa grandeur. Il est encore dangereux de lui trop faire voir sa grandeur sans sa bassesse. Il est encore plus dangereux de lui laisser ignorer l'un et l'autre. . . . (420) S'il se vante, je l'abaisse; s'il s'abaisse, je le vante; et le contredis toujours, jusqu'à ce qu'il comprenne qu'il est *un monstre incompréhensible*.'

cuius etiam capillos: Mt. 10: 30, 'vestri autem et capilli capitis omnes numerati sunt'; *en. Ps.* 36. s. 3. 13, 'capilli capitis vestri numerati sunt. ergo est securitas, sed si intus sit deus.'

4. 14. 23

The treatment of the dedication to Hierius and the treatment of the contents of the work itself are roughly equal in length.

typho: See on 7. 9. 13 (and cf. 4. 14. 23).

circumferebar: Eph. 4: 14, 'ut iam non simus parvuli fluctuantes et circumferamur omni vento doctrinae in *fallacia* hominum, in astutia ad circumventionem erroris'; *en. Ps.* 47. 6, 'fundamentum nostrum in Sion sit; ibi stabiliri debemus, non perflari omni vento doctrinae.' Courcelle, *Les Confessions*, 426 n. 3, cites this and connects it to the unique scriptural foundation for the ethereal quality of demons at Eph. 2: 2, 'principem potestatis aeris huius', and to the expression *ventos pascere* (Hosea 12: 1, echoed at 4. 2. 2); he also cites *c. acad.* 1. 7. 20, 'ab huius aeris animalibus quibusdam vilissimis, quos daemonas vocant'; but at *conf.* 5. 5. 9, *this* verse recurs when A. is expressing benign toleration of less-educated Christians who err venially in doctrine: 'sed etiam talis infirmitas in fidei cunabulis a caritate matre sustinetur, donec assurgat novus homo in virum perfectum et *circumferri* non possit *omni vento* doctrinae.' Nothing there suggests that the verse speaks to A. of demons.

gubernabar: Of God's power over A. at 6. 5. 8, 6. 7. 12, 11. 2. 2.

accenderer: For love and fire, see on 4. 12. 19.

in eo GO^{1}S Knöll Skut. Ver. : *in eum* CDO^2 Maur.

soliditati veritatis: Cf. below 'cor vanum et inane soliditatis tuae'; the two phrases are linked by 3. 6. 10 (of the Manichees), 'cor inane veri'. Cf. *trin.* 2. pr., 'solidissima veritas'. *Soliditas* is contrasted to *ventositas* (compare here the next lines) at another passage where *veritas* is not the issue: *trin.* 8. 7. 11, 'potentior est enim et tutior solidissima humilitas quam ventosissima celsitudo.'

opinantium: *util. cred.* 11. 25, 'unum [genus hominum] est opinantium, id est eorum qui se arbitrantur scire quod nesciunt.'

obnubilatur: Elsewhere at 2. 2. 2, 3. 1. 1, 7. 1. 1.

ob os contemplationis meae: G–M: 'An elaborate variation, in the manner of the late rhetoric, on the phrase "ob oculos mentis".' The phrase *ante os* offers the closest classical parallel: Cic. *rep.* 3. 15, 'ut esset posteris ante os documentum . . . sceleris sempiternum.' No biblical echo seems relevant: in scripture, *os* is almost exclusively used of the mouth, esp. as organ of speech. Solignac (*BA* 13. 12) wishes to take 'animo versabam, ob os contemplationis meae' as indication 'selon beaucoup de vraisemblance' that this work was already written in the manner of *sol.* and *imm. an.*: 'Augustin s'est livré de bonne heure à ses entretiens avec lui-même.' He sees *conf.* as a further step in this line of stylistic development. The present text does not make clear to our satisfaction whether A. means to describe the content of the work or the method that led to its composition.

nullo conlaudatore: Amb. *exam.* 1. 9. 34, 'bonorum operum proprium est ut externo commendatore non egeant, sed gratiam suam, cum videntur, ipsa testentur.' Ryan's translation presses the text to indicate that the work found no receptive audience: 'Although there was no one to join me in praising *the book*, I myself admired *it*'; others translate noncommittally. The phrase evokes a sense of intellectual isolation, of a time when A. had outstripped his colleagues and contemporaries and was pursuing lines of thought that he could not share with others; if that is so, his eventual departure for Rome may have had a more praiseworthy motive than any he allows when that stage is reached in the narrative (5. 8. 14), namely the desire for community of intellectual discourse (A.'s disappointment with Faustus could have reinforced such a motive).

It is difficult to escape the impression (fortified by 'tamen' just above) that A. means to say something kind about his juvenile effort. It was *curiositas* that led him to the topic and its treatment; it was *ambitio saeculi* that made him dedicate the book to Hierius; he knew nothing of truth, for his heart was empty; *and yet*, all alone, he found himself in contemplation of the beautiful and the fitting. That he was alone in his admiration of the beautiful and the fitting is appropriate for this stage of his life when his friendships are all disordered and wrongful; only in Bk. 6 do his friends acquire names and begin to join him in the contemplation of authentic values; and he only truly reaches his goal when he has those friends with him, in a growing crowd (by the end of Bk. 9, at Monnica's death, loneliness is dispersed and a comparative throng shares his grief in greater and lesser degrees). See on 4. 15. 24 for a fuller treatment of the way the treatment of the *de pulchro et apto* anticipates later movements of *conf.*

mirabar: Knauer, 60, notes that 'mirabilia' in the next sentence (4. 15. 24) hints a connection; A. was looking in the right direction, but 'falsa opinio' prevented him from seeing the truth, and so he looked away just at the last moment.

4. 15. 24

This paragraph begins to outline and analyse the lost treatise. This episode is part of a larger narrative whose further stages throw light back on this passage. See on 4. 13. 20 for bibliography and a summary of modern opinions regarding those contents; this note will confine itself to setting this analysis in the context provided by other passages in *conf.*

Courcelle's *Recherches* already focused (157–67) on what C. called 'vaines tentatives d'extases plotiniennes', and debate since has further refined our sense of what A. sought and felt at that Milan period of his life (see further on 7. 10. 16 and 7. 17. 23). But those episodes themselves form as well the next main stages on the path opened here, and the pattern is incomplete without attention not only to the 'vision of Ostia' (9. 10. 24–5), but also to the structure and contents of the first half of Bk. 10 (10. 1. 1–27. 38) and, to a lesser extent, to the structure and contents of Bks. 11–13.

The most important sentence of this summary is the last: 'non enim noveram. . . .' We have already seen (see on 3. 7. 12–13, and cf. the opening of 4. 13. 20) that 'non noveram' is a phrase A. uses to specify the defects in his understanding at a prior stage in his life. Here the criticism is twofold: that he did not know that evil was not itself a substance, and that he thought that the human mind itself was the *summum bonum*. The problems thus posed are all-encompassing: the nature of God, the nature of created being (for if evil is a substance, then created being can be evil).

Both these errors date to his turn to Manicheism (3. 7. 12) and recur later. The identification of the Christian God with the *summum bonum* is the subject of 7. 1. 1 ff. (summarized at 7. 4. 6, 'qui *summum et optimum bonum* es'), but there the problem of evil is expressly revived as unresolved (7. 3. 4–5); the discovery of the *platonicorum libri* intervenes and leads to the first mystic ascent of 7. 10. 16; that ascent (see notes there) is incomplete and unsuccessful. There then follows a development (7. 12. 18–16. 22) of the neo-Platonic doctrine of evil as the *privatio boni*; this leads to the second ascent (7. 17. 23). In the Introduction and in the commentary to Bk. 7 it is argued that this second ascent is a complete and successful ascent in Plotinian terms as A. understood them at the time.

Thus the intellectual effort that produced the *de pulchro et apto* is presented as a first, halting and failed attempt at ascending in the mind towards the divine source of beauty and order. Two specific gaps in his knowledge are adduced to explain the inadequacy of this ascent; when those subjects are revived, the resolution of one issue precedes a renewed attempt at the ascent (and, failure that it is, it is at least a less dismal failure than what is portrayed here); the resolution of the second issue is then presented, and immediately followed by a further (by this count, *third*) attempt at the ascent, markedly more successful than either of those that preceeded.

The link between these issues and other crucial concerns of *conf.* is clearer if we recall again the trinitarian connection suggested above (on 4. 13. 20). The *de pulchro* [2] *et apto* [3] can be construed (and here it must be emphasized that the *intentions* of A. at the time he wrote the work are beyond recapture: we only see the work as he presents it later, when such trinitarian patterns are uppermost in his mind) as a legitimate attempt to write about the manifestation in the world of the second and third persons of the trinity. What is lacking is an accurate knowledge of the first person, hence of the *modus* of things as they are. To escape his predicament A. needs (*a*) to discover the true *summum bonum*, (*b*) to understand the 'being' of things—i.e. that aspect of created things that reflects the workings of the first person of the trinity—by discovering that *esse* is good and evil is insubstantial, and (*c*) to do that he needs to espouse the virtue of *humilitas* (the opposite of *ambitio saeculi*—that is made explicit at 4. 15. 26 with the echo in the words 'superbis resistis').

in arte tua: Parallel expressions strongly suggest that this phrase marks the role of the second person of the trinity in giving to created things their particular beauty: *vera rel.* 31. 57, 'nam haec est illa incommutabilis veritas quae lex omnium artium recte dicitur et ars omnipotentis artificis'; *lib. arb.* 3. 15. 42, 'ars ipsa per quam facta sunt omnia, hoc est summa et incommutabilis sapientia dei'; sim. at *div. qu.* 68, *c. Faust.* 21. 5, *trin.* 6. 10. 11–12, and *civ.* 11. 7, 'cognitio quippe creaturae in se ipsa decoloratior est, ut ita dicam, quam cum in dei sapientia cognoscitur, velut in arte qua facta est.' The present passage could thus be paraphrased, 'I did not know that the point on which the whole subject depended was the role played by the second person of the trinity.' (For a more conservative view, see J. Stiglmayr, *Zschr. Ask. u. Myst.* 6 (1931), 159–61.)

qui facis mirabilia solus: Ps. 71: 18, 'benedictus dominus, deus Israhel, qui facit mirabilia solus'; *en. Ps.* 71. 20, 'quoniam quicumque faciunt, ipse in eis operatur qui facit mirabilia solus.' Ps. 135: 4, 'qui facit mirabilia magna solus'; *en. Ps.* 135. 5 takes the rest of that psalm as the gloss of verse 4: Ps. 135: 5–10, 'qui fecit caelos in intellectu . . . (6) qui firmavit terram super aquas . . . (7) qui fecit luminaria magna solus . . . (8) solem in potestatem diei . . . (9) lunam et stellas in potestatem noctis . . . (10) qui percussit Aegyptum cum primogenitis eorum quoniam in aeternum misericordia eius.'

ibat: Passing through bodily forms implies an action that takes the mind outside itself: another sign of the disorder in this investigation, confirmed by 10. 27. 38, 'et ecce intus eras et ego foris et ibi quaerebam, et in ista formosa quae fecisti deformis inruebam.'

animus CDGO[1]S Skut. Ver. : *animus meus* O[2] Knöll Maur. Cf. the similar case below.

formas [2].

aptum . . . adcommodatum: See on 4. 13. 20.

definiebam et distinguebam: 10. 14. 22, 'et quidquid de his disputare potuero *dividendo* singula per species sui cuiusque generis et *definiendo*'; the verbs imply (and slightly belittle) the technical activities of dialectic, and are common in that sense in the polemics with Julian (e.g. *c. Iul. imp.* 5. 50, 'ubi sunt definitiones tuae, quas tanta loquacitate distinguis?').

ad animi naturam: Here is the sticking point in this 'ascent' (as G–M saw ad loc.).

ipsa vis veri: Authentic experience (of the truth [2]—made explicit at 4. 15. 25) was there, if only he were capable of it. The imagery echoes that of the Platonic cave, with an ironic twist upon A.'s *concupiscentia oculorum*: his vision fails him, for as he struggles to see he fails to see that which he most seeks.

incorporea: *BA* 13. 451 n. 1 suggests the ἀσώματον of Porphyry, *sent.* 19, 35, 37; cf. Courcelle, *Les Confessions*, 52 n. 1, putting this text beside 7. 10. 16, 'reverberasti infirmitatem', and 7. 17. 23 'repercussa infirmitate', from the two ecstasies of Bk. 7: in the present passage A. hardly looks in the right direction at all before turning his eyes away voluntarily to corporeal things.

animum O[1]S Skut. Ver. : *animum meum* CDGO[2] Maur. Knöll.

inque illa unitatem: The doctrine presented here is the exact reverse of A.'s mature trinitarian teaching. The difference is that between saying that 'Oneness is God' and 'God is Oneness.' That A. made his way from the first to the second of those positions at least in part through the mediation of Plotinus should not be surprising. The first formulation entails a harsh reading of 'Duality', a sign to him later of the deficiencies of this analysis.

deus meus, ex quo sunt omnia: God the creator; see on 1. 2. 2 for this conflation of 1 Cor. 8: 6 and Rom. 11: 36. The scriptural echo says concisely and authoritatively, without becoming didactic and prosy, what the mature A. believes true doctrine is *and how it is known* (i.e. from scripture), in contradiction to the error under review.

'monadem' . . . 'dyadem': The distinction had roots in Pythagoreanism (E. R. Dodds, *CQ* 22 (1928), 129–42, at 135 f.; *BA* 13.672–3; Diels, *Doxographi graeci*, 302 A–B, 280 A–B, and Porphyry, *vita Pyth.* 38), and in Platonism (J. Dillon, *The Middle Platonists* (Ithaca, 1977), 3). Three African texts of the fifth century show the power of these ideas. The earliest in date is by A.'s own student at Carthage between 376 and 383 (cf. *cura mort.* 11. 13), Favonius Eulogius, probably writing at Carthage between 390 and 410 (see R.-E. van Weddingen, in the prolegomena to his edn. of Favonius (Collection Latomus 27: Brussels, 1957), 5–8). His exposition of the monad and the dyad leads off his (very selective) commentary (Favonius, *disp.* 5–6); he was followed and criticized by Macrobius (as van Weddingen saw, and as A.

Cameron's redating of Macrobius (*JRS* 56 (1966), 24–38) confirms): see Macrob. *somn. Scip.* 1. 6. 7–9. Slightly later in date was Martianus Capella (A. Cameron, *CP* 81 (1986), 320–8): see Mart. Cap. 7. 731–32.

A. was well-informed in this area (Solignac, *RA* 1 (1958), 129–37); his enthusiasm resembles that Cicero expressed, notably at *Tusc.* 4. 1. 2–2. 4. For his praise of Pythagoras, from *ord.* to *civ.*, see on 6. 14. 24. A few months after *ord.*, A. wrote to Nebridius (*ep.* 3. 2) of the monad, now linked to 'intelligible number' (for *numerus* [2], see on 5. 4. 7), but the dyad does not appear. How far A. may have adapted the philosophical notion to his Manicheism is suggested by the speculations of Brown, 52: 'When Augustine wrote a treatise of aesthetics at the age of 26, he will reflect, in an acceptable, classical form, this exotic and potent [Manichean] myth. Here again, the Good is a "monad", "like a mind without the tension of male and female"; while it is evil that is active, "a divided thing", "reasonless", "anger and lust".' Just as the mature, orthodox A. presents us Christianity in dress that is often Ciceronian and always marked by a native Latin *gravitas*, so also the present passage makes it likely that his arguments on behalf of Manicheism would have had a comparably reassuring classical sobriety. To have heard A. the Manichee preach his creed would not have been as exotic as a modern catalogue of their doctrines might lead us to expect.

sexu OS Knöll Skut. Ver. : *sensu* CDG. G–M: 'This is doubtless the right reading . . . cf. Macrob. *somn. Scip.* 1. 6. 7, "unum autem, quod monas, id est unitas, dicitur, et mas idem et femina est."'; see Solignac, at *BA* 13. 672–3, quoting Nicomachus of Gerasa, while Katô, *REAug* 12 (1966), 235–8, situates the notion of sexless mind in the Manichean tradition.

nec ullam substantiam . . . incommutabile bonum: See *vera rel.* 30. 56–31. 57, partially quoted on 'in arte tua' above.

From here to 4. 16. 29, 'mutability' and 'immutability' become important for the first time in *conf.*; they then disappear until 7. 1. 1 (see on that passage for parallels), when the immutability of God reappears as the one sure thing that A. had held on to through all his wanderings.

4. 15. 25

facinora . . . flagitia: See on 3. 8. 16. Here the pattern is also trinitarian: 'facinora' (*ambitio saeculi*—'se iactat insolenter et turbide', as regards the *animus*), 'flagitia' (*concupiscentia carnis*—'carnales . . . voluptates', as regards the *anima*), and 'errores et falsae opiniones' (*curiositas*—affecting the *mens rationalis*). The addition to the pattern here is *curiositas*, whose *errores* are what hold him back. The *de pulchro et apto* thus comes to embody all of A.'s failings: his hankering after the beauty of this world, his longing to be thought great in the eyes of men, and his eagerness to have all the answers.

rationalis mens: The existing literature on A.'s terminology involving 'mind' and 'soul' and similar concepts does not detect a systematic pattern of their employment in A. (see G. J. P. O'Daly, *Aug.-Lex.* 1. 315–16, summarizing his *Augustine's Philosophy of Mind*, 7–8). The triad *animus/anima/mens rationalis* does not occur in the tables of the standard works on A.'s triadic trinitarian theology (e.g. du Roy). *BA* 16. 581–3 has a general note à propos of *trin.* 8. 6. 9, on his 'terminologie . . . assez flottante': *Anima* is 'le principe vital', sentient life that is, i.e. common to beasts and animals; it is invisible, but can be known by the senses. *Animus* is the human soul as opposed to the animal, a spiritual substance, source of intellectual and rational knowledge, but it is also the locus of memory and imagination. *Mens* is much the vaguest in his usage, creeping in and out of *trin.* according to convenience (O'Daly, *Aug.-Lex.* 316: 'The mind . . . is a "pars animi", viz. its best part. . . . "Animus" can, however, also be reserved for "mind".'). *Spiritus* is 'plus difficile à définir'; it lends itself to trichotomies such as *spiritus/anima/corpus* (cf. Pauline πνεῦμα, ψυχή, σῶμα at 1 Thess. 5: 23), and scriptural data confuse the issue (e.g. *Gn. litt.* 12. 7. 18, 'dicitur spiritus [from Eccles. 3: 21 just quoted there] et ipsa mens rationalis, ubi est quidam tamquam oculus animae'). *Gn. litt.* 12. 24. 51 rouses a whole school of controversies: probably representing a Porphyrian tripartition of the soul, implying a hierarchy *mens/spiritus/anima*, where *animus* could substitute for *spiritus*. That passage has acquired a separate literature (see *BA* 16. 581–3 and 49. 559).

There is no other discussion by A. of *mens rationalis* as such (4. 15. 24, metaphorically at 12. 15. 20). The phrase is not uncommon and is as ill-defined as *mens* alone: earliest *lib. arb.* 2. 6. 13, most interesting *s.* 145. 2 (poss. contemporary with *conf.*), 'anima humana, hoc est, mens rationalis ad imaginem dei facta'; *anima* and *mens rationalis* equated also at *Io. ev. tr.* 23. 5–6, with the qualification that the *mens rationalis* is a way of describing the specifically human *anima* (implicitly at *c. Max.* 2. 25, *gr. et lib. arb.* 13. 25, *en. Ps.* 118. s. 18. 4, though those passages may only specify the *mens rationalis* itself as the distinguishing feature of man); at *persev.* 24. 67, it is a part of the *anima*: 'sive animam sive in ipsa anima mentem rationalem', as at *en. Ps.* 145. 5; cf. *Gn. litt.* 8. 25. 47).

Investigation into sources does not take precedence over careful attention to the patterns A. himself uses, nor should patterns from one place in his work be forced on to another passage. Here the trinitarian analogy applies in the first instance to the categories of sin (as so often in *conf.*) and the other terms are employed in the first instance because of their applicability to each category of sin, without a priori reference to a notion of trinitarian appropriateness for them here.

alio lumine: Anticipated by 4. 15. 24, 'et inruebat in oculos ipsa vis veri', now made explicit.

particeps veritatis [2]: Sim. at 7. 9. 14 (in the wake of reading the *platonicorum libri* and of 'reading' Jn. 1) and 7. 19. 25 (of the errors in his Christology at the time when he had read the *platonicorum libri* and undertook the 'ascents' of Bk. 7).

non est ipsa natura veritatis: i.e. he wrongly identified 'Mind' and 'Truth'; cf. 4. 15. 24.

quoniam tu inluminabis: There begins here an important concatenation of scripture at an important moment, giving weight and substance to the critique of his youthful book. (Jn. 1 texts recur at 7. 9. 13 f.—though there again they are texts that A. did *not* read at the time.)

Ps. 17: 29, 'quoniam tu inluminabis lucernam meam, domine, deus meus, inluminabis tenebras meas' (the verse also at 6. 1. 1, 7. 1. 2); *en. Ps.* 17. 29, 'nos enim peccatis nostris tenebrae sumus.' In the present context, the 'literal sense' (if we may call it thus) of the citation is that the *lucerna* is the mind and that God, who gives it light, is not identical with mind—that refutes the false position reported at 4. 15. 24 and implied here. But the 'allegorical sense' is that here in his life that light lay in the future; that assertion about the future is repeated on the way to Bks. 6 and 7, but not again thereafter in the narrative—a sign of how we are to interpret the progress reported in Bk. 7. The beginning of Bk. 11 is in many ways a new beginning, and there in a stock-taking (11. 2. 2) A. speaks of the 'primordia inluminationis tuae et reliquias tenebrarum mearum'. Thereafter he twice looks forward again to enlightenment in these terms: 11. 25. 32 and 13. 8. 9.

domine. deus meus: The punctuation reflects A.'s reading of the underlying Psalm verse (cf. *en. Ps.* 17. 29).

de plenitudine tua: Jn. 1: 16–17, 'et de plenitudine eius nos omnes accepimus, et gratiam pro gratia. (17) lex enim per Moysen data est, gratia et veritas per Iesum Christum facta est.'

lumen verum: Jn. 1: 9, 'erat lumen verum quod inluminat omnem hominem venientem in hunc mundum'; *Io. ev. tr.* 2. 7, 'intellegat caritas vestra: veniebat enim ad mentes informes, ad corda saucia, ad aciem animae lippientis. . . . quia et qui saucios habent oculos idonei sunt videre parietem inluminatum et inlustratum a sole; vel montem vel arborem, aut aliquid huiusmodi idonei sunt videre; et in alio inlustrato demonstratus illis ortus ille, cui videndo adhuc minus idoneam aciem gerunt. sic ergo illi omnes, ad quos Christus venerat, minus idonei erant eum videre.'

quia in te . . . obumbratio: Jas. 1: 17, 'omne datum optimum et omne donum perfectum desursum est, descendens a patre luminum, apud quem non est transmutatio nec vicissitudinis obumbratio.' It is *a patre luminum* that provides the link to this phrase (as at 3. 6. 10, where see notes): God is both illuminated and illuminating. At *civ.* 11. 21, the echo evokes eternity; at *trin.* 1. 1. 2, it suggests eternity's condition, immutability; while at *Gn. litt.* 5. 18. 36 (closest in date here), it signifies the unvarying serenity of divine knowledge (as against the morning and evening knowledge of creatures): 'ipsi autem deo non audeo dicere alio modo innotuisse, cum ea fecisset, quam illo quo ea noverat ut faceret, apud quem non est commutatio nec momenti obumbratio.'

4. 15. 26

conabar . . . repellebar: Courcelle, *Les Confessions*, 45 n. 2, rejects in advance the interpretation advanced here: 'Mais il ne s'agit pas, cette fois [7. 3. 5], de contemplation proprement dite, non plus que IV, 15, 26.' The apt parallel is to the other passages Courcelle cites there, 7. 10. 16, 'et inde admonitus redire ad memetipsum', and 7. 20. 26, 'repulsus sensi', of the two 'ascents'; but see on 7. 21. 27, 'conari', for the confirmation of the mystical quality of the verb here. Cf. also 5. 10. 20, 'conaretur . . . repercutiebatur', and 10. 42. 67. The vocabulary of all the failed contemplative ascents is that of attempt and grasping; Ostia (9. 10. 24–5), on the other hand, is a gift.

repellebar: Cf. Ps. 42: 2, 'quia tu es deus meus, fortitudo mea; utquid me repulisti, et utquid tristis incedo, dum affligit me inimicus?' Knauer, 76 n. 1, is unsure of the appositeness of this citation; the word itself is common in Psalm language: 43. 10, 43. 23, 59. 3, 59. 12, 73. 1, 88. 39, 93. 14, 107. 12, 118. 10.

quoniam superbis resistis: See on 1. 1. 1 for this thematic and familiar echo of Jas. 4: 6 etc. (cf. 'resistebas' below); here A. leaves it to the reader to supply the conclusion, 'humilibus autem das gratiam': once A. learned humility, the gift would come. A form of pride—*ambitio saeculi*—is the particular sin of Bk. 4 (see on 4. 7. 12); the same verse cited already at 4. 3. 5.

adsererem . . . me id esse naturaliter quod tu es: A. alludes to his attempt to identify *mens rationalis* with the *summum bonum* (4. 15. 14), already criticized at 4. 15. 25.

cum enim ego essem . . .: The implied criticism is deliberately presented as one that he should have seen and accepted at the time. By insisting that the *mens rationalis* was itself divine, he had implied that God shared his own mind's imperfection and perfectibility, hence its mutability.

ventosae cervici: G–M: the 'bold combination of metaphors . . . is probably deliberate.' Brown, 205: 'He had to return to a life which resembled only too closely the life of a minor public figure, such as he had rejected with great abruptness, in 386. His career in Milan had been a *ventosa professio*—"a puffed-up occupation" [*c. acad.* 1. 1. 3]; but it was easy for the life of an African bishop, also, to be a "puffed-up existence", *ventosa tempora*'—*ep.* 23. 6, 'non enim cogito in ecclesiasticis honoribus tempora ventosa transigere, sed cogito me principi pastorum omnium rationem de commissis ovibus redditurum.'

imaginabar formas corporeas: See on 4. 15. 24 for the place of phrases such as this in identifying the 'vision' A. was seeking and finding at this time.

nondum revertebar ad te: Cf. again 10. 27. 38, 'et ecce intus eras et ego foris et . . . in ista formosa quae fecisti deformis inruebam.' Cf. Ps. 77: 38–9, 'ipse autem est misericors, et propitius fiet peccatis eorum, et non disperdet eos, et abundabit ut avertat iram suam et non accendet omnem iram suam, (39) et recordatus est quia caro sunt, spiritus vadens et non revertens': see on 1. 13. 20. Cf. also Prov. 2: 19, quoted at *en. Ps.* 77. 24, of the *via iniquitatis*: 'omnes qui ambulant in ea, non revertentur.' This is the pull that replaces *vadens* with *ambulans* in *en. Ps.* and here. Note also how this text mimics the anaphoric *et*-structure of the psalm.

nondum CDGO Maur. Ver. : *non* S (following the Psalter text) Knöll Skut. Vega Pell.

a mea vanitate fingebantur ex corpore: sc. the *phantasmata splendida* (3. 6. 10) of his Manichean beliefs.

parvulis: Mt. 11: 25, 'quia abscondisti haec a sapientibus et prudentibus, et revelasti ea parvulis'; 1 Cor. 3: 1, 'et ego fratres, non potui vobis loqui quasi spiritalibus, sed quasi carnalibus: tamquam parvulis in Christo' (for that verse, see on 13. 13. 14). Those echoes add a layer of self-criticism to the passage: 'your *parvuli*, baptized citizens of the City of God, probably already knew better, and there I was, learned and clever, babbling nonsense.' See Van Bavel, *Augustiniana*, 7 (1957), 256–7.

fidelibus tuis: For the special sense, 'baptized Christians', see on 2. 3. 6.

civibus meis: i.e. 'in civitate dei': Vega finds that interpretation strained, and thinks that the earthly city, and specifically Thagaste, is in mind.

garrulus: See on 1. 4. 4 for *loquax* of the Manichees; and see on 7. 20. 26 for the verb form, *garrire*.

contendebam: G–M: 'The construction is "contendebam magis quam confitebar"', i.e. *contendebam* is the appropriate verb for 'assert, allege, maintain' (*OLD* 6, but note other uses 'to contend in words', 'to speak seriously or passionately'), hence it governs the first, false assertion; 'confitebar' governs the second, true assertion: 'contendebam substantiam errare magis quam confitebar [substantiam] deviasse.'

coactam errare: G–M: 'The reference is to the captivity of the light elements according to the Manichaean system.'

et poena errare confitebar: G–M: '"and that my error was penal" (lit., "that I erred by reason of punishment").' Cf. *lib. arb.* 3. 18. 52, 'illa est enim peccati poena iustissima, ut amittat quisque quod bene uti noluit. . . id est autem ut . . . qui recte facere cum posset noluit amittat posse cum velit.'

4. 15. 27

eram aetate: To take him literally this falls in the period 13 November 380 and 13 November 382, when he had been back in Carthage teaching four or five years, two or three years before he left for Rome—and nearing the end of the 'nine years' he spent as an adherent of the Manichees (see on 5. 6. 10—that period ended *c.*382/3).

corporalia figmenta: See 4. 2. 3, 'figmentis'.

dulcis [3] **veritas** [2]: See on 1. 20. 31 for the triple sequence *dulcedo*, *honor*, *fiducia*, where *dulcedo* looks to the world of sensible beauty to which concupiscence of the flesh is a disordered response.

melodiam tuam: The metaphor has shifted from sight to hearing; cf. *ep.* 138. 1. 5, 'universi saeculi pulchritudo, cuius particulae sunt quae suis quibusque temporibus *apta* sunt, velut magnum carmen cuiusdam ineffabilis modulatoris excurrat, atque inde transeant in aeternam contemplationem speciei qui deum rite colunt, etiam cum tempus est fidei.'

gaudio gaudere propter vocem sponsi: Jn. 3: 29, 'qui habet sponsam, sponsus est; amicus autem sponsi, qui stat et audit eum, gaudio gaudet propter vocem sponsi'; *Io. ev. tr.* 13. 12, 'bene te humilas, merito non cadis, merito stas, merito audis eum, et gaudio gaudes propter vocem sponsi'; *Io. ev. tr.* 14. 2, 'qui enim vult gaudere de se tristis erit; qui autem de deo vult gaudere semper gaudebit, quia deus sempiternus est.' See also 11. 8. 10.

erroris [2]: See on 4. 15. 25, 'errores'.

rapiebar: See on 4. 12. 18, 'rape'.

pondere [3] **superbiae** [1] **meae:** See on 13. 9. 10, 'pondus meum amor meus'. All three temptations afflict him here.

in ima decidebam: See on 2. 3. 6, 'ima'.

non enim dabas auditui meo gaudium et laetitiam: Ps. 50: 10, 'auditui meo dabis exsultationem et laetitiam, et exsultabunt ossa humilata' (Knauer, 134, remarks the change from future to imperfect, balancing the change from positive to negative; hence the implication is a hopeful one of 'not yet'); *en. Ps.* 50. 13, 'gaudebo audiendo te, non loquendo contra te. . . . feliciores sunt qui audiunt quam qui loquuntur. . . . inde . . . Iohannes ille Baptista, . . . ait, "qui habet sponsam, sponsus est; amicus autem sponsi stat et audit eum." stantem se fecit et audientem, non cadentem et loquentem. . . . audistis auditum; ubi est exsultatio et laetitia? continuo sequitur: "stat et audit eum, et gaudio gaudet propter vocem sponsi. auditui meo dabis exsultationem et laetitiam, et exsultabunt ossa humilata".'

humilata: 4. 15. 26–7 are marked by humility (the implicit cure for A.'s illness), both *expressis verbis* and by implication from the scriptural texts and situations echoed.

4. 16. 28

proderat: Recurs at first line of each of the last paragraphs of this book and twice more (4. 16. 29, 4. 16. 30 (2 ×), 4. 16. 31 (2 ×)). The question is left without explicit answer, save for the exasperation of 'nimia perversitas!' at 4. 16. 31. Eccles. 2: 15, 'et dixi in corde meo, si unus et stulti et meus occasus erit, quid mihi prodest quod maiorem sapientiae dedi operam?'

annos natus ferme viginti: A. turned 20 in November 374. Bk. 3 dealt with the excitement of his nineteenth year (372/3) and left him in Carthage (no later than 375). The central section of Bk. 4 on the death of his friend comes from the year teaching back at Thagaste (375/6), while the treatment of the *de pulchro et apto* is carefully dated (4. 15. 27) to 380/2 (age *c.*26/7). Now we revert in this flashback to a period not long after his reading of the *Hortensius*. The intellectual excitement and the intellectual achievement of that biographical moment bracket these two books of spiritual decline. There are few discussions of this chronological rearrangement (Mandouze, 111 n. 6, for instance, merely suggests that this emphasizes that A.'s philosophical training continued even while he was a Manichee neophyte—this hardly needs emphasizing *here*); see on 4. 16. 29 for one suggestion. The narrative here may be partly corroborated by *sol.* 2. 12. 22, where *Ratio* speaks of the existence of 'aliquid in subiecto . . . ut in sole lux, ut in igne calor . . .' and *Augustinus* responds 'ista quidem vetustissima nobis sunt et ab ineunte adulescentia studiosissime percepta et cognita.'

aristotelica: Years later, Julian of Eclanum attacked A. with relish as the 'Aristoteles Poenorum' (*c. Iul. imp.* 3. 199); Courcelle, *Recherches*, 247 n. 1, assumes Julian has this passage in mind (note that J. refers to Monnica's 'drinking problem' at *c. Iul. imp.* 1. 68, and refers to *conf.* 5 at *c. Iul. imp.* 1. 25). See also *c. Iul. imp.* 2. 51, 'quomodo ergo te ipsi saltem pelagiani intellecturi sunt, nisi prius ad scholas dialecticorum, ubicumque terrarum potuerint inveniri, propter haec discenda mittantur? an forte categorias Aristotelis, antequam tuos libros legant, eis exponens ipse lecturus es?' Other mentions of Arist. *cat.* at *c. Iul.* 1. 4. 12, 2. 10. 34, 2. 10. 37, 3. 2. 7, 5. 14. 51, and 6. 20. 64; and other attacks on A.'s dialectic at *c. Iul. imp.* 1. 72 and 5. 11 (Jul. calls A. 'philosophaster Poenorum'); A. responds in kind at *c. Iul. imp.* 2. 26. Others saw too much dialectic in A.: Maximus of Madaura (*ep.* 16. 3) and Volusianus (*ep.* 135. 1). Aristotle is the proverbially difficult philosopher at *util. cred.* 6. 13 ('libros Aristotelis reconditos et obscuros').

quaedam: See on 3. 4. 7, 'cuiusdam Ciceronis'.

decem categorias: The first work of the Aristotelian *organon*; its authenticity is not assured. The appeal of this work lay in its claim to comprehend all predication in its ten categories (see below on 'loquentes de substantiis'). A. seems also to have known Aristotle's discussion of 'future contingents' in his *peri herm.* 9 (*c. Faust.* 26. 4–5: see H. Chadwick, *Augustine* (Oxford, 1986), 8).

rhetor Carthaginiensis: We know none of A.'s academic teachers (see on 1. 9. 14, 'in scholam'). Despite this hint of some philosophical guidance from a nameless professor at Carthage (we know nothing of the 'alii qui docti habebantur'), A. was essentially self-taught. (Mandouze, 466: 'Ni la culture universitaire, ni plus tard la culture ecclésiastique ne devaient réussir à tuer en lui ce penchant pour les tranquilles audaces de l'autodidacte.')

typho: At 7. 9. 13 (where see notes), this word again characterizes a man putting in A.'s hands a work of Greek philosophy; at 3. 3. 6, it characterizes

A.'s own attitude at the moment when he is about to discover the *Hortensius*. In Bk. 4 the adjective also appears of A. himself at 4. 14. 23.

magistris . . . depingentibus: Teachers draw in the dust at *civ.* 11. 29 and already at Amb. *exam.* 6. 4. 23. What A. reads as a sign of intuition on his part is at the same time a fragment of the history of literacy. Where other students could approach the text only through the medium of oral discussion, A. had mastered the skills of purely textual manipulation.

loquentes de substantiis: There are various ways of listing the categories. The list here: 'de substantiis' (= *substantia*), 'qualis sit' (= *qualitas*), 'et statura' (= *quantitas*), 'cognatio' (= *relatio*), 'ubi', 'quando', 'aut stet an sedeat' (= *positio*), 'aut calciatus vel armatus sit' (= *habitus*), 'aut aliquid faciat aut patiatur aliquid' (= *actio/passio*).

How A. encountered this famous text is a puzzle. In the Aristotelian order (*cat.* 4. 1^b25–2^a4, *top.* 1. 9. 103^b20–23), the categories are: *substantia* (οὐσία), *quantitas* (ποσότης), *qualitas* (ποιότης), *relatio* (πρός τι), *ubi* (τόπος), *quando* (χρόνος), *positio* (κεῖσθαι), *habitus* (ἕξις), *actio* (πράττειν), *passio* (παθεῖν). Cf. *trin.* 5. 7. 8, for *substantia, quantitas, qualitas, relatio, situs, habitus, quando, ubi, facere, pati.*[7] The Aristotelian order is reproduced in ps.-Aug. *categoriae decem* (144. 17–19 Minio-Paluello) and in Cassiodorus (*inst.* 2. 10—with some idiosyncratic variations). But there is another tradition, in Ps.-Archytas, Mart. Cap. 4. 363 f., Boethius (*in Porph. isag.* 1 and 2. 1. 4), and Calcidius (*in Tim.* 305 (307. 3–6 Waszink)), according to which quality is the first accident of substance, as here; see I. P. Sheldon-Williams, ed. *Johannes Scottus Periphyseon Liber I* (Dublin, 1968), 232 n. 98. On the evidence of the order, therefore, A. is in the tradition from Porphyry and Calcidius (and leading to Boethius): fully neo-Platonic (and for that matter Porphyrian more than Plotinian: C. Evangeliou, *Aristotle's Categories and Porphyry* (Leiden, 1988)). For a contrary view, see Minio-Paluello's edn. of the *cat. dec.* in *Aristoteles Latinus* 1. 1–5 (Bruges–Paris, 1961), 129–75, with preface there, pp. lxxvii–xcvi; and Minio-Paluello, *CQ* 39 (1945), 63–74. Against Minio-Paluello (and Marrou, 34, followed by *BA* 13. 88) and with P. Hadot, *Marius Victorinus* (Paris, 1971), 187–8, we should doubt Cassiodorus' claim that Victorinus made a translation. Another order, close to the one given here, is in Quintilian (3. 6. 23): *essentia, qualitas, quantitas, ad aliquid, ubi, quando, facere, pati, habere, keisthai* (*positio*). Minio-P., art. cit. 65, says that Varro must have been the first translator, and that no other is known until 350, hence Quint. must/may depend on him or the Greek text or some doxographic source. That Varro may stand in this tradition requires us to pause long enough to remember the other revivals of his name and lore in A.'s lifetime: A. himself quotes him extensively (see Hagendahl for details), and Martianus Capella imitates him in more ways than one; see below, Excursus on 4. 16. 30.

[7] For evidence that A. used the work wittingly in his *imm. an.*, see du Roy, 178 n. 2.

an (*sedeat*) GS Knöll Skut. Vega Pell.: *aut* CDO Ver. The *aut* must be reserved for connecting different categories; note that in the next phrase *vel* is used for connecting two examples of the same category. (There, only G substitutes *aut* for *vel*.)

4. 16. 29

proderat: 4. 16. 28, 4. 16. 30 (2 ×), 4. 16. 31 (2 ×).

simplicem atque incommutabilem: See on 2. 6. 13.

illis . . . comprehensum: Here the pertinence of his youthful reading of Aristotle becomes clear: Aristotle had provided the theoretical instruments, themselves defective by virtue of his ignorance of the truth about God, that led years later to the failed attempt to come nearer to God through writing *de pulchro et apto*.

falsitas: 4. 1. 1, 'falsi atque fallentes'.

ut terra spinas et tribulos pareret: Gn. 3: 17–19, 'maledicta terra in operibus tuis; in tristitia edes illam omnes dies vitae tuae; (18) spinas et tribulos edet tibi et edes foenum agri. (19) in sudore faciei tui edes panem tuum, donec convertaris in terram ex qua sumptus es, quia terra es et in terram ibis.' (Text from *Gn. litt.* 11. 38. 51; but at verse 18, *pariet* is attested from *Gn. litt.* 3. 18. 28 in place of *edet*; in verse 19, VL reads *sudore* regularly, but sometimes (*en. Ps.* 7. 16, 40. 6, 84. 7, *s.* 45. 4) A. reads *labore*.) The subject is the punishment for original sin. The echo here sets the misery of the moment (mild intellectual confusion) in a wider context. Pricks and thorns also at 2. 2. 3 (post-paradise), 2. 3. 6 (libido). Cf. *Gn. c. man.* 2. 20. 30, 'spinae ac tribuli sunt punctiones tortuosarum quaestionum aut cogitationes de provisione huius vitae: quae plerumque, nisi exstirpentur et de agro dei proiiciantur, suffocant verbum, ne fructificet in homine'; sim. at *en. Ps.* 102. 17.

4. 16. 30

Knauer, 86, notes a change here: paragraphs 28–9 deal with the categories, 30–1 with the *artes liberales*. As far as substance goes, that is not a great change, for the *artes* and philosophy are intimately linked: see Excursus below. The narrative effect is more interesting: the reading of Aristotle was specifically dated (see on 4. 16. 28), and cast the reader's attention back to an earlier period in A.'s life; the reading of the *libri artium* here described is not dated but leads back to the time of the *de pulchro et apto*. Given A.'s view of the *disciplinae*, this is a shorthand way of saying that his philosophical training had continued in anticipation of just the 'ascent' that the *de pulchro et apto* attempted. (Cf. 2. 3. 8 for Monnica's confidence *c.*370 that book-learning would draw A. to the church.)

proderat: 4. 16. 28, 4. 16. 29, 4. 16. 31 (2 ×), and below here.

artium quas liberales vocant: See on 4. 1. 1, 'doctrinas . . . liberales'.

nequissimus . . . servus: The *artes* are a specific against concupiscence of various sorts: see Excursus below.

per me ipsum legi et intellexi: 4. 16. 28, 'legi eas solus et intellexi'.

gaudebam: (1) This joy antedates the frustrated longing for joy of 4. 15. 27, 'cupiens et audire te et gaudio gaudere propter vocem sponsi, et non poteram'; (2) this joy is conditioned by the truth and certainty that were present in the *libri artium*, though A. did not know the source of that (as of all) truth at the time.

dorsum . . . ad lumen: G–M adduce Plot. 5. 5. 7. 16–19, *οὕτω τοίνυν καὶ ἡ τοῦ νοῦ ὄψις· ὁρᾷ μὲν καὶ αὕτη δι' ἄλλου φωτὸς τὰ πεφωτισμένα ἐκείνῃ τῇ πρώτῃ φύσει, καὶ ἐν ἐκείνοις ὄντος ὁρᾷ· νεύουσα μέντοι πρὸς τὴν τῶν καταλαμπομένων φύσιν ἧττον αὐτὸ ὁρᾷ.* The Platonic cave provides a no less apt parallel (see Courcelle, *Les Confessions*, 51 n. 4: cf. Plato, *rep.* 7. 514 A–B). But cf. Jeremiah 2: 27, 'verterunt ad me tergum et non faciem'; for this text as a signpost of A.'s 'paganism', see on 2. 3. 6; and note that here ('sed non inde sacrificabam tibi'), A. specifies that his cleverness did not lead him to sacrifice *to God*—we should remember that the Jeremiah passage is taken as a sign of sacrifice directed elsewhere.

et (*ad ea*) GS Knöll Skut. Vega Pell. : om. CDO Ver.

quidquid de arte loquendi . . . de numeris: Suggesting grammar, rhetoric, geometry, music, and arithmetic. (But dialectic may perhaps (see Marrou, 189) be meant by 'disserendi' and grammar and rhetoric subsumed in 'loquendi'; then the only one of the seven missing is astronomy: but that subject has been on the table already at 4. 3. 4–6—another element binding Bk. 4 together.) Of A.'s own project for writing *libri disciplinarum*, *mus.* survives intact, and parts at least of treatises on grammar, rhetoric, and dialectic are possibly preserved. A.'s other lists (see Marrou, 189) from *ord.* 2. 12. 35, 2. 16. 44, 2. 18. 47 (gram., dial., rhet., mus., geom., astro., phil.—omitting arithmetic), *ord.* 2. 4. 13, 2. 5. 14 (gram., dial., rhet., arith., mus., geom., astro.), *quant. an.* 33. 72 (gram., dial., rhet., arith., mus., geom., astro.), and *retr.* 1. 6 (gram., mus., dial., rhet., geom., arith., phil.—omitting astronomy) all manage to add up to seven; the present passage is perhaps meant to be suggestive rather than comprehensive. (There is another partial listing by implication at 10. 9. 16, comprising gram., dial., and rhet.)

dispiciendi acumen: *c. acad.* 1. 7. 20, 'a quibus [daemonibus] nos superari acumine ac subtilitate sensuum posse concedo, ratione autem nego' (cf. Hensellek, *Anzeiger Akad. Wien* 114 (1977), 146–7); cf. *ep.* 9. 3, 'sensus acerrimus'; *SLA* s.v. *acumen*.

sed non inde sacrificabam tibi: Ps. 53: 8, 'voluntarie sacrificabo tibi et confitebor nomini tuo, domine, quoniam bonum est.'

fortitudinem . . . custodiebam: Ps. 58: 10, 'fortitudinem meam ad te custodiam, quia deus susceptor meus es'; *en. Ps.* 58. s. 1. 18, 'ego autem fortitudinem meam ad te custodiam; quia si recedo, cado; si accedo, fortior fio. . . . non habet ex se *lumen*, non habet ex se vires: totum autem quod pulchrum est in anima, virtus et sapientia est; sed nec sapit sibi. . . . est quaedam origo fonsque virtutis, est quaedam radix sapientiae, est quaedam, ut ita dicam, si et hoc dicendum est, *regio* incommutabilis veritatis; ab hac anima recedens tenebratur, accedens inluminatur. . . . ergo fortitudinem meam ad te custodiam: non a te recedam, non de me praesumam.'

profectus sum . . . cupiditates: Lk. 15: 12–13, 30 (on the prodigal), 'pater, da mihi portionem substantiae quae mihi contigit. et divisit illis substantiam. (13) et non post multos dies, congregatis omnibus, adolescentior filius peregre profectus est in regionem longinquam, et ibi dissipavit substantiam suam vivendo luxuriose. . . . (30) postquam filius tuus hic, qui devoravit substantiam suam cum meretricibus . . .' For 'vivendo luxuriose' cf. here 'meretrices cupiditates', but also above 'nequissimus malarum cupiditatum servus'. See on 1. 18. 28, 'in longinqua regione'; cf. 13. 2. 2, 'in longinquam dissimilitudinem tuam'; *civ.* 9. 17 (immediately after quoting Plotinus 1. 6. 8), 'si ergo deo quanto similior, tanto fit quisque propinquior, nulla est ab illo alia longinquitas quam eius dissimilitudo'—so finally cf. 7. 10. 16, 'regio dissimilitudinis'.

regionem: Best translation 'land' (in Latin NT = Gk. χώρα): of an area larger than the *civitas* of Hippo, smaller than the *provincia* of Africa (*s. Guelf.* 1. 8). The word with various attributive genitives in A. has attracted attention (esp. 7. 10. 16, 'regio dissimilitudinis'), but its frequency in various broadly congruent contexts in the years before *conf.* (later: appears much less often) should be underlined (even leaving aside an abundant variety of half-literal passages like *s. Guelf.* 9. 1, 'regio nostra, terra; regio angelorum caelum, venit ergo dominus noster ad istam regionem ex alia regione: ad regionem mortis, de regione vitae: ad regionem laboris, de regione felicitatis'). The background is philosophic (see on 'regio dissimilitudinis' at 7. 10. 16), reinforced by scripture, esp. Lk. 15: 13 (echoed clearly as here and at 1. 18. 28 at: *qu. ev.* 2. 33, 2. 51; *adn. Iob* on 9: 25, *en. Ps.* 24. 5, 47. 3, 118. s. 5. 2, 118. s. 8. 2, 123. 9, 131. 12, 138. 5, 138. 9, *s. Caill.* 2. 11); A. Adam, *Zschr. für Kirchengesch.* 69 (1958), 10–12, claimed without success a Manichee influence.

For this use at Cassiciacum and shortly after: *c. acad.* 2. 9. 22, 'quasi in regionem suae originis rediens triumphaturum de libidinibus . . . securior rediturus in caelum' (*retr.* 1. 1. 3, 'iturus autem quam rediturus dixissem securius; . . . sine controversia ergo quaedam originalis regio beatitudinis animi deus ipse est'), *beata v.* 1. 1, 'in beatae vitae regionem' (as at *beata v.* 1. 3 and *s. Lam.* 4), *quant. an.* 1. 2, 13. 22, *lib. arb.* 2. 11. 30, *mag.* 8. 21. It is to be found irregularly in later years, and in increasingly Christianized expressions: *c. Faust.* 21. 10 ('regionem dei'), *util. ieiun.* 5. 7 ('namque et pagani ieiunant aliquando, nec regionem quo tendimus norunt'), *Io. ev. tr.* 22. 3

('quasi de regione infidelitatis ad regionem fidei'), *trin.* 10. 5. 7 ('in regionem incorporeae naturae'), *Gn. litt.* 12. 26. 54 ('regionem intellectualium vel intellegibilium' (see on 9. 10. 24–5 for this text on mystical vision)), *Gn. litt.* 12. 34. 65 ('omnis etiam spiritalis quasi regio ubi animae bene est merito paradisus dici potest'), *en. Ps.* 38. 22 ('beatam illam regionem, beatam patriam, beatam domum, ubi participes sunt sancti vitae sempiternae atque incommutabilis veritatis'), *s.* 19. 4 ('de regione felicitatis Christus venit': so *civ.* 20. 21), *c. Iul. imp.* 3. 161 ('in regione beatitudinis illius'). Elsewhere in *conf.:* 1. 18. 28 ('regionem longinquam'), 2. 10. 18 ('regio egestatis'), 4. 12. 18 ('regione mortis'), 7. 7. 11 ('media regio salutis meae'), 7. 10. 16 ('regio dissimilitudinis'), 9. 10. 24 ('regionem ubertatis indeficientis').

non enim: G–M: 'The force of the *enim* is somewhat obscure; probably it justifies "bona": "(and how good a thing it was I did not know), for I was not aware that these works were understood with difficulty even by the diligent and clever until I endeavoured to explain them to pupils of ability, and found that the best of them could only follow my explanation somewhat slowly".' The parentheses around the last sentence but one are new, then, as is the reduced punctuation after 'sacrificabam tibi', leaving only one interposed sentence. Then 'bona res' corresponds to 'donum tuum' before the parenthesis, and poses the question afresh (as at the beginning of paragraphs 28, 29, 30, and 31). The force of the 'enim' is thus that of 'you see', explaining the grounds for putting the question again by reviving the problem posed before the parenthesis. The *BA* translation starts a new paragraph with 'non enim'; other translators do not especially clarify the matter.

ille excellentissimus: Knauer, 46 n. 1, thinks this was the friend whose death was reported at 4. 4. 7. This would bind up the book well (that episode is dated to 375/6, whereas at the beginning of 4. 15. 28 we are put back to 374/5; at 4. 4. 7 he tells us he introduced his friend to his Manichaean ways: his school-friend would not improbably benefit from A.'s instruction in other ways), but cannot be proved on the faintness of this echo. Even if Knauer's suspicion is right, it is hard to see that we are meant by A. to entertain it.

EXCURSUS: THE *LIBERALES DISCIPLINAE*

Modern treatments of late antique education fail to convey adequately a sense of its mystical dimension, the way in which the training of the mind according to the *enkyklios paideia* was meant to bring about enlightenment. The best treatment is still, despite weaknesses, Marrou, 187–327; a typical statement, from an author not at all disinclined to see in A. a doctrine of philosophical ascent, is R. J. O'Connell, *St. Augustine's Early Theory of Man* (Cambridge, Mass., 1968), 191: 'The accent at Cassiciacum is, in the main, alarmingly intellectualistic, even more so than in certain passages of the *Enneads*. Part of the reason may be the influence of Varro . . ., an influence that threatens at times to flatten out Plotinian intellectualism into a kind of numerical rationalism.' Few seem to have considered that what seems to us flat numerical rationalism can be exalted by a Plotinian metaphysic into a high mystical practice. An important rectification of this neglect has been achieved by I. Hadot, *Arts libéraux et philosophie dans la pensée antique* (Paris, 1984), who demonstrates that the *disciplinae* in the form in which we have them are not the cadaver of an ancient system of 'general education',[8] but a sophisticated philosophical construction animated by neo-Platonic doctrines, with a particular role for the *de regressu animae* of Porphyry.[9] A text of Cicero well illustrates the tradition that we have overlooked:

nec enim omnibus iidem illi sapientes arbitrati sunt eundem cursum in caelum patere: nam vitiis et sceleribus contaminatos deprimi in tenebras atque in caeno iacere docuerunt, castos autem [animos], puros, integros, incorruptos, bonis etiam studiis atque artibus expolitos leni quodam et facili lapsu ad deos, id est ad naturam sui similem pervolare. (Cic. *cons.* fr. 12M = Lact. *inst.* 3. 19. 6)

A revealing text from Cic. *fin.* shows the Epicurean Torquatus attacking the *disciplinae* as ineffective and useless: if, as is suggested below on 6. 16. 26, A. was (re-)reading Cic. *fin.* at Milan in 386, the attack hints, by what it criticizes, at the implicit purpose of his own enterprise a few months later:

qui quod tibi parum videtur eruditus, ea causa est, quod nullam eruditionem esse duxi nisi quae beatae vitae disciplinam iuvaret. (72) an ille tempus aut in poetis evolvendis [i.e. employing grammar, rhetoric, and dialectic], ut ego et Triarius te hortatore facimus, consumeret, in quibus nulla solida utilitas omnisque puerilis est delectatio, aut se, ut Plato, in musicis, geometria, numeris, astris [i.e. music, geometry, arithmetic, astronomy] contereret, quae et a falsis initiis profecta vera esse non possunt et, si essent vera, nihil afferrent quo iucundius, id est quo melius viveremus, eas ergo artes persequeretur, vivendi artem tantam tamque et operosam et perinde fructuosam relinqueret? non ergo Epicurus ineruditus, sed ii indocti, qui, quae pueros non didicisse turpe est, ea putant usque ad senectutem esse discenda. (Cic. *fin.* 1. 21. 71–2)

Whether Hadot is correct in dissociating the *disciplinae* entirely from the patronage of Varro is doubtful; not only does the title of his work (*disciplinarum libri*: Cassiod. *inst.* 2. 3. 2) recur in A.'s own project (*retr.* 1. 6 cited below), not only does Licentius' poem to Augustine begin with a decidedly 'mystical' allusion to Varro's work (*ep.* 26, in which L.'s poem begins: 'arcanum Varronis iter scrutando profundi | mens hebet adversamque fugit conterrita lucem': on that poem, see A. Solignac, *RA* 1 (1958), 121–3, and D. Romano, *Letteratura e*

[8] French scholars, esp. Marrou, make the *disciplinae* sound suspiciously like the studies for the *baccalauréat*; others evoke a good public school, a small liberal arts college, or a rigorous Gymnasium, according to taste.

[9] D. J. O'Meara, *Pythagoras Revived* (Oxford, 1989) opens doors beyond Hadot into the history of late antique Platonism, especially that 'flat numerical rationalism', through which we will learn more and more in this vein. Of interest as well is the essay of U. Pizzani, *Atti 1986* 1. 331–61 (note Cic. *rep.* 3. 2. 3, quoted in parallel with *ord.* at 343).

storia nell'età tardoromana (Palermo, 1979), 107–26), not only does A. speak of Varro in *civ.* as if he were a practitioner of whatever A. understood to be 'liberal arts' (*civ.* 6. 2, 'ut in omni eruditione, quam nos saecularem, illi autem liberalem vocant, studiosum rerum tantum iste [Varro] doceat, quantum studiosum verborum Cicero delectat'), and not only does Claudianus Mamertus in the fifth century invoke a 'mystical' Varro (*de statu animae* 2. 9), but we have the literary filiation of the Menippean satire from Varro to two late antique enthusiasts of the *disciplinae*, Martianus and Boethius. The latter's *consolatio* is dedicated to demonstrating the ascent of the mind to God, and it is the culmination of the career of the writer who had written more widely and profoundly on the *disciplinae* than any other.

Our evidence for a high view of the *disciplinae* in A. comes from a limited period: 386 to 391, between his decision to accept baptism and his ordination to the priesthood. But the ideas were not new or newly espoused in 386, and see on 4. 13. 20 ff. for the way the writing of the *de pulchro et apto* is congruent with the principles concretely attested from the later period. While direct evidence fades from view after *vera rel.* and ordination, there is no explicit contrary exposition until *doctr. chr.*, following A.'s further ordination as bishop. Most important, there is no reason to think that these ideas about the liberal disciplines were in any essential way incompatible with the practice and belief of Christianity in A.'s time. Though he himself moved beyond them, that is at least in part because he believed that those ideas about the *disciplinae* were assimilated into and made less urgently necessary by the *disciplina ecclesiae*.

No systematic treatment of the *disciplinae* survives from A.'s hand. Our most direct testimony comes in this sketchy paragraph written forty years after the fact:

per idem tempus quo Mediolani fui baptismum percepturus, etiam disciplinarum libros conatus sum scribere, interrogans eos qui mecum erant atque ab huiusmodi studiis non abhorrebant, *per corporalia cupiens ad incorporalia quibusdam quasi passibus certis vel pervenire vel ducere*. sed earum solum de grammatica librum absolvere potui, quem postea de armario nostro perdidi, et de musica sex volumina, quantum attinet ad eam partem quae rhythmus vocatur. sed eosdem sex libros iam baptizatus iamque ex Italia regressus in Africam scripsi, inchoaveram quippe tantummodo istam apud Mediolanium disciplinam. de aliis vero quinque disciplinis illic similiter inchoatis—de dialectica, de rhetorica, de geometrica, de arithmetica, *de philosophia*—sola principia remanserunt, quae tamen etiam ipsa perdidimus; sed haberi ab aliquibus existimo. (retr. 1. 6)

But progress in the *disciplinae* is everywhere a subject at Cassiciacum in *ord.*, the most positive and prescriptive of the three dialogues (where *c. acad.* decries a false way and *beata v.* a true goal, *ord.* offers concrete guidance). From the outset of that treatise, the pursuit of the *disciplinae* is tied to the most elevated of goals:

cuius erroris maxima causa est, quod homo sibi ipse est incognitus. qui tamen ut se noscat, magna opus habet consuetudine recedendi a sensibus et *animum in se ipsum conligendi* atque in se ipso retinendi. *quod hi tantum adsequuntur, qui plagas quasdam opinionum*, quas vitae cotidianae cursus infligit, aut solitudine [cf. A.'s own *sol.*, begun shortly after the dramatic date of *ord.*; and see on 10. 43. 70, 'solitudinem'] aut *liberalibus medicant disciplinis.*[10] (*ord.* 1. 1. 3)

The role of the *disciplinae* is part of an explicit exhortation to young Licentius, who showed great promise—whether he ever lived up to it or not:

nam eruditio disciplinarum liberalium modesta sane ac succincta et alacriores et perseverantiores et comptiores exhibet amatores amplectendae veritati, ut et ardentius appetant et constantius insequantur et inhaereant postremo dulcius, quae vocatur, Licenti, beata vita. (*ord.* 1. 8. 24)

A. then advises him to moralize the story of Pyramus and Thisbe with a view to 'foedae libidinis ... exsecratio',[11] and 'in laudem puri et sinceri amoris, quo animae dotatae disciplinis et virtute formosae copulantur intellectui per philosophiam et non solum mortem fugiunt, verum etiam vita beatissima perfruuntur.' There the allegorization of a mythological love story approaches the aims and technique of Martianus Capella. On another day at Cassiciacum, A. shakes his head at the zeal Licentius shows for his verse-making, and seeks again to call him back: *c. acad.* 3. 4. 7, 'ad istarum disciplinarum, quibus excoluntur animi, circulum [a semi-calque on ἐγκύκλιος παιδεία] revocare vos cupio.'

Programmatic statements continue (e.g. *ord.* 2. 5. 14, 'talis enim eruditio ... talem philosophiae militem nutrit vel etiam ducem ut ad summum illum modum [see on 1. 7. 12], ultra quod requirere aliquid nec possit nec debeat nec cupiat, qua vult evolet atque perveniat multosque perducat'), with the first hint of compromise with a less intellectual Christianity: *ord.* 2. 5. 15, 'ego autem si quid meos monere possum, quantum mihi apparet quantumque sentio, censeo illos disciplinis omnibus erudiendos. ... si autem aut pigriores sunt aut aliis negotiis praeoccupati aut iam duri ad discendum, *fidei sibi praesidia parent*, quo illos vinculo ad sese trahat atque ab his horrendis et involutissimis malis liberet ille qui neminem sibi per mysteria bene credentem perire permittit.' Religion so far appears only as the philosopher's safety-net, but its role will grow.

The mystical remoteness of the goal and the ardour of its technique are emphasized (*ord.* 2. 7. 24, 'a multitudinis vel suspicione remotissima disciplina'),[12] while syncretism with Christianity is pressed further (*ord.* 2. 8. 25, 'haec autem disciplina ipsa dei lex est'). Ambivalence continues:

[10] This passage attracted, forty years later, a reproachful word in *retr.* 1. 3. 2, 'displicet mihi ... quod multum tribui *liberalibus disciplinis*, quas multi sancti multum nesciunt, quidam etiam qui sciunt eas sancti non sunt.'

[11] Thus in the text here, 'nequissimus malarum cupiditatum servus' describes the condition the *artes* are expected to treat; cf. *ord.* 2. 16. 44.

[12] The text of *ord.* from this point to 2. 18. 48 has been studied by Hadot, 101–36, detailing the Platonic resonances of this central text. The greatest weakness of H.'s treatment of A. is that she does not go beyond those explicit and programmatic pages to document the presence of the doctrine elsewhere in A.'s early thought.

ad discendum item necessario dupliciter ducimur, auctoritate atque ratione. tempore auctoritas, re autem ratio prior est. aliud est enim quod in agendo anteponitur, aliud quod pluris in appetendo aestimatur. . . . ad quam cognitionem *in hac vita* pervenire *pauci*, ultra quam vero etiam *post hanc vitam* nemo progredi potest. qui autem sola auctoritate contenti bonis tantum moribus rectisque votis constanter operam dederint aut contemnentes aut non valentes *disciplinis liberalibus atque optimis* erudiri, beatos eos quidem, cum inter homines vivunt, nescio quo modo appellem, tamen inconcusse credo, mox ut hoc corpus reliquerint, eos quo bene magis minusve vixerunt, eo facilius aut difficilius liberari.[13] (*ord.* 2. 9. 26)

That passage introduces A.'s most formal exposition of the invention of the *disciplinae* by reason, and the logic of their relations (*ord.* 2. 12. 35–15. 43: detailed study at Hadot, 109–24). The section concludes with a fuller definition, marked by a triad of *materia/species/motus* (not in du Roy) and ending (*ord.* 2. 16. 44), 'cum enim *artes illae omnes liberales partim ad usum vitae partim ad cognitionem rerum contemplationemque discantur*, usum earum assequi difficillimum est nisi ei qui ab ipsa pueritia ingeniosissimus instantissime atque constantissime operam dederit.'

But then, in a way that would have ramifications right up to the writing of *conf.*, he fundamentally undercuts everything he has just said by a long address, *ord.* 2. 17. 45–20. 52, to Monnica. She is the great exception, the one who knows the essential things by direct experience that has no need for the *disciplinae*. See on 9. 10. 24 for further discussion of this role played by M. repeatedly in A.'s surviving works.

We saw already how the *soliloquia* could be conceived as a complement in method to the *disciplinae* (*ord.* 1. 1. 3, quoted above). The praise of the *disciplinae* is strong through the second book of that work as well, with emphasis on the specific individual disciplines that train the mind (e.g. *sol.* 2. 11. 19, on grammar). A definition, transporting us to another level entirely from the ordinary modern sense of grammar, is then offered: *sol.* 2. 12. 22, '[Ratio] ita est aliquid in subiecto, ut ab eo nequeat separari, ut in hoc ligno forma et species quam videmus, ut in sole lux, ut in igne calor, *ut in animo disciplina*, et si qua sunt alia similia.' In other words, *disciplina* is to the *animus* as *forma* or *species* is to *materia*, or in other words again, as *Logos* is to the Father. This assimilation of related ideas is essential to the eventual neglect of the idealization of the *disciplinae* into A.'s Christianity. Given the identification of *veritas* with Christ familiar from Jn. 14: 6, it is no less important when Ratio goes on (*sol.* 2. 13. 24), 'est autem disciplina veritas et semper, ut in initio libri huius ratio persuasit, veritas manet. semper igitur animus manet'. On that view *disciplina* is itself the sign of life in the soul. Thus the goal to which they aspire is correspondingly lofty: *sol.* 2. 20. 35, '[Ratio] tales sunt qui bene disciplinis liberalibus eruditi,

[13] Hadot, 103, shows that this paragraph resembles *civ.* 10. 29, a recognized fragment of Porphyry, *de regressu animae*; cf. *ord.* 2. 11. 31, 'nam ut *progressus* animae usque ad mortalia lapsus est, ita *regressus* esse in rationem debet.'

siquidem illas sine dubio in se oblivione obrutas eruunt discendo et quodam modo refodiunt; non tamen contenti sunt nec se tenent, donec totam faciem veritatis, cuius quidam in illis artibus splendor iam subrutilat, latissime atque plenissime intueantur.'[14]

Quant. an. is of great importance in documenting A.'s views on the mystical ascent of the mind to God at this early period. The *disciplinae* are not without their role: *quant. an.* 15. 25, 'nam et exercet animum hoc genus disciplinarum ad subtiliora cernenda ne luce illorum repercussus et eam sustinere non valens in easdem tenebras quas fugere cupiebat libenter refugiat, et adfert argumenta, nisi fallor, certissima, quibus quod fuerit inventum atque confectum impudentem habeat dubitationem, quantum homini talia vestigare permissum est.' *Quant. an.* 33. 70–6 presents a fully worked-out pattern of ascent in seven stages (*gradus*): (1) vivification of the body; (2) life of the soul through the senses; (3) life of the mind; (4) detachment from the material world; (5) maintaining the self in detachment; (6) straining to achieve the final vision; (7) repose in that vision.

The third stage is pertinent here, offering a long list of examples of the contents of the *disciplinae*: *quant. an.* 33. 72, 'tot artes opificum, agrorum cultus, exstructiones urbium, variorum aedificiorum ac moliminum multimoda miracula, inventiones tot signorum in litteris, in verbis, in gestu, in cuiuscemodi sono, in picturis atque figmentis, tot gentium linguas, tot instituta, tot nova, tot instaurata, tantum librorum numerum et cuiuscemodi monumentorum ad custodiendam memoriam ... modulandi peritiam, dimetiendi subtilitatem, numerandi disciplinam, praeteritorum ac futurorum ex praesentibus coniecturam.' A.'s summary (at *quant. an.* 35. 79) has already been quoted on 4. 13. 20.

These ideas give context to the project conceived at Milan, when A. returned from Cassiciacum to take baptismal instruction, of writing *libri disciplinarum*. This is not a worn-out pedant displaying his expertise, but a serious undertaking by a mature man who sees in this prospect his own serious and original contribution to philosophical literature. The magnitude of the task he set for himself must be emphasized: to have acquired sufficient expertise to write *in extenso* all the *disciplinae* with the originality and thoughtfulness that characterize *mus.* would have been a heroic achievement not otherwise exampled in what we know of the ancient authors: Varro and Martianus Capella are the only *comparandi*. Further, he sees the ascent of the mind to God

[14] This passage attracted comment at *retr.* 1. 4. 4, 'item quodam loco dixi quod "disciplinis liberalibus ... refodiunt." sed hoc quoque improbo. credibilius est enim propterea vera respondere de quibusdam disciplinis etiam imperitos earum, quando bene interrogantur, quia praesens est eis, quantum id capere possunt, lumen rationis aeternae, ubi haec immutabilia vera conspiciunt, non quia ea noverant aliquando et obliti sunt, quod Platoni vel talibus visum est. contra quorum opinionem, quantum pro suscepto onere dabatur occasio, in libro duodecimo de trinitate disserui [*trin.* 12. 15. 24].'

that these 'disciplinae' make possible as closely related to the other union with God that his impending baptism will forge. Cult and culture are drawn intimately together.

The attempt to identify surviving fragments of A.'s project has received much attention. See Marrou, 570–9, and most recently V. A. Law, *RA* 19 (1984), 155–83, in favour of the *ars breviata* as representing A.'s *gram.* B. D. Jackson's edn. (Boston, 1975) of *dial.* makes a case on stylistic grounds there. For recent, measured, discussion, see U. Pizzani, *Atti 1986* 1. 348–54; his distinction at 354 between works dealing with the trivium and the quadrivium (of which *mus.* is the only surviving example), with the former confined to jejune scholasticism while the latter expand more broadly into contemplative speculation, is the best argument to 'save the phenomena', if the surviving fragments are to be taken as authentic and indicative. Doubts should remain; note that *mus.* as we have it is in dialogue form, a link with the literary projects of Cassiciacum that should not be minimized. From 'interrogans eos' at *retr.* 1. 6 it is possible to infer that the works were all to be in dialogue form; cf. *retr.* 1. 4. 1, describing *sol.*: 'scripsi etiam duo volumina . . . me interrogans mihique respondens.' The sixth book of *mus.* is the place where the ordinariness of schoolbooks gives way to something different—a difference that has perplexed many modern readers into a patronizing disdain for A.'s willingness to lump odd philosophical speculation into a treatise whose main use ought to be—if the advantage of modern readers is the primary consideration—conveying accurate information about ancient notions of music.[15] The sixth book begins (*mus.* 6. 1. 1) by saying that the preceding five books:

non ob aliud suscipiendum putavimus nisi ut adolescentes vel cuiuslibet aetatis homines quos bono ingenio donavit deus non praepropere, sed quibusdam gradibus a sensibus carnis atque a carnalibus litteris, quibus eos non haerere difficile est, duce ratione avellerentur, atque uni deo et domino rerum omnium, qui humanis mentibus nulla natura interposita *praesidet* [see on 10. 6. 10] incommutabilis veritatis amore adhaerescerent. . . . nam turba cetera de scholis linguarum tumultuantium, et ad plaudentium strepitum vulgari levitate laetantium, si forte inruerit in has litteras, aut contemnet omnes, aut illos quinque libros sufficere sibi arbitrabitur: istum vero *in quo fructus illorum est*, vel abiiciet quasi non necessarium, vel differet quasi post necessarium.[16] reliquos vero qui ad ista intellegenda eruditi non sunt, si sacramentis christianae puritatis inbuti, in unum et verum deum summa caritate nitentes, cuncta puerilia transvolaverunt, fraterne admoneo ne ad ista descendant, et cum hic laborare coeperint, de tarditate sua conquerantur, ignorantes itinera difficilia et molesta pedibus suis, volando se posse etiam ignorata transire. si autem ii legunt qui et infirmis aut inexercitatis gressibus hac ambulare non possunt et nullas pietatis alas habent, quibus ista neglecta praetervolent, non se inserant inconvenienti negotio; sed praeceptis

[15] It also drove Marrou, 580–3, to an unnecessary hypothesis about two editions of the work. Ancient works that scandalize the expectations of even great scholars often get rewritten this way.

[16] This prophecy has surely proved accurate.

saluberrimae religionis et nido fidei christianae pennas nutriant, quibus subvecti laborem ac pulverem huius itineris evadant, magis ipsius patriae quam viarum flexuosarum amore flagrantes. his enim haec scripta sunt, qui litteris saecularibus dediti, magnis implicantur erroribus, et bona ingenia in nugis conterunt, nescientes quid ibi delectet. quod si animadverterent, viderent qua effugerent illa retia et quisnam esset beatissimae securitatis locus.

The development of that book follows lines that should by now be familiar (notably the warning against sexual impurity at *mus.* 6. 11. 33), and it ends (*mus.* 6. 17. 59) much as it begins.

It has been suggested that one other surviving work may be related to the project of the *libri disciplinarum*: *mag.*, viewed in this way as an introductory treatise expounding for beginners the underlying philosophy, of course heavily Christianized—the main argument of the work is that Christ is the only true *magister*. On that reading, the following passage is of interest, not only for what it says about the *disciplinae* but for its connections to the vocabulary of the mystical ascent familiar from *conf.* 7 and the discussion of *interior veritas* from 10. 6. 10 ('intus cum veritate conferunt'):

num hoc magistri profitentur, ut cogitata eorum ac non ipsae *disciplinae* quas loquendo se tradere putant percipiantur atque teneantur? nam quis tam stulte curiosus est, qui filium suum mittat in scholam ut quid magister cogitet discat? at istas omnes disciplinas quas se docere profitentur ipsiusque virtutis atque sapientiae [cf. 1 Cor. 1: 24] cum verbis explicaverint, tum illi qui discipuli vocantur, utrum vera dicta sint, apud semetipsos considerant, *interiorem* scilicet illam *veritatem* pro viribus intuentes. . . . (46) sed de tota utilitate verborum, quae si bene consideretur non parva est, alias, si deus siverit, requiremus.[17] nunc enim ne plus eis quam oportet tribueremus, admonui te, ut iam non crederemus tantum sed etiam intellegere inciperemus quam vere scriptum sit auctoritate divina, ne nobis quemquam magistrum dicamus in terris, quod unus omnium magister in caelis sit [Mt. 23: 8–10: cf. *conf.* 11. 8. 10]. (*mag.* 14. 45–6)

Equally telling is *mag.* 8. 21, 'dabis igitur veniam, si praeludo tecum non ludendi gratia, sed exercendi vires et mentis aciem, quibus regionis illius, ubi beata vita est, calorem ac lucem non modo sustinere, verum et amare possimus.'

There are other texts that evince the preoccupation with the *disciplinae*, esp. two passages from letters to Nebridius that must be dated to the period before A.'s priestly ordination in 391. The first has a notably Christianized notion of *disciplina* set in a trinitarian context, without losing the connection to the educational programme: *ep.* 11. 4, '*species* quae proprie filio tribuitur, ea pertinet etiam ad *disciplinam* et ad artem quandam, si bene hoc vocabulo in his rebus utimur, et ad intellegentiam qua ipse animus rerum cogitatione *formatur*. itaque quoniam per illam susceptionem hominis id actum est, ut quaedam nobis disciplina vivendi et exemplum praecepti sub quarundam sententiarum

[17] Where? In the *libri disciplinarum*? As it turned out, this task is most nearly performed, in a different spirit, by *doctr. chr.*

maiestate ac perspicuitate insinuaretur, non sine ratione hoc totum filio tribuitur. ... ergo disciplina hominibus erat necessaria, qua *inbuerentur* [sacramental: see on 8. 2. 4, 'imbutus'] et qua ad modum *formarentur* [philosophical, but also linked to sacramental initiation: see on 1. 11. 18, 'effigiem'].' The mode of instruction is then found in the incarnation; in *ep.* 12, A. said the same thing more concisely: '*disciplina* ipsa et *forma* dei, per quam facta sunt omnia quae facta sunt, filius nuncupatur. quidquid autem per susceptum illum hominem gestum est, ad eruditionem informationemque nostram gestum est.'

The last work containing this association of ideas concerning the *disciplinae* in a form recognizably continuing that of Cassiciacum is *vera rel.*, written at Thagaste before A. was ordained at Hippo in 391. Here the integration of the idea into an ecclesiastical setting is almost complete, especially with reference to sacramental practice. *vera rel.* 17. 33, 'iam vero ipse totius doctrinae modus, partim apertissimus, partim similitudinibus in dictis, in factis, in *sacramentis* ad omnem animae instructionem exercitationemque adcommodatus quid aliud quam rationalis *disciplinae* regulam implevit? nam et *mysteriorum* expositio ad ea dirigitur quae apertissime dicta sunt, et si ea tantum essent quae facillime intelleguntur, nec studiose quaereretur nec suaviter inveniretur veritas, neque si essent in scripturis et in sacramentis non essent signacula veritatis, satis cum cognitione actio conveniret.' Later in the 390s A. still linked the *disciplinae* to the second element in his trinitarian triads: cf. *div. qu.* 38, 'natura [1] disciplina [2], usus [3]' (see du Roy, 299 n. 2) and *doctr. chr.* 2. 40. 60 (on 'Egyptian gold'), 'sed etiam liberales disciplinas usui veritatis aptiores et quaedam morum praecepta utilissima continent'—with the *disciplinae* culminating in *philosophia*.

In after years, A. was more reserved, as the passages already cited from *retr.* have indicated. Writing in 408/9 to Memor, the father of the Julian who would become the bane of A.'s old age, A. introduced his friend and eventual biographer Possidius as (*ep.* 101. 1) 'non litteris illis quas variarum servi libidinum liberales vocant, sed dominico pane nutritus': now the *disciplinae* have been linked with *concupiscentia*, where before (e.g. *ord.* 2. 16. 44) they were a way of overcoming it. Memor sought from A. a course in the Christian liberal arts for his son, but A. was no longer so enthusiastic, and he wrote in the body of the letter (*ep.* 101. 2), 'quid enim aliud dicendum est eis, qui cum sint iniqui et impii, *liberaliter* sibi videntur eruditi, nisi quod *in litteris vere liberalibus* legimus: "si vos filius liberaverit, tunc vere liberi eritis?" [Jn. 8: 36] . . . non ergo illae innumerabiles et impiae fabulae quibus vanorum plena sunt carmina poetarum ullo modo nostrae consonant libertati [gram.], non oratorum inflata et expolita mendacia [rhet.], non denique ipsorum philosophorum garrulae argutiae [dial.], qui vel deum prorsus non cognoverunt . . . "et servierunt creaturae potius quam creatori" [Rom. 1: 21–5: see on 7. 9. 14]. absit omnino ut istorum vanitates et insaniae mendaces [*insaniae mendaces* at 6. 11. 18, 8. 2. 4, 9. 2. 2], ventosae nugae ac superbus error recte *liberales litterae* nominentur.' Truly *liberales disciplinae* may still be envisioned, but A. seems to have lost any expectation of finding them among the practitioners of his old profession. He is willing to pass on a copy of *mus.*, but purports to have a somewhat disdainful view of its contents (*ep.* 101. 3): 'initio nostri otii cum a curis maioribus magisque necessariis vacabat animus, volui per ista quae a nobis desiderasti scripta *proludere*, quando conscripsi de solo rhythmo sex libros.' But even at this date, he singles out the function of the sixth book (*ep.* 101. 4): 'sextum sane librum quem emendatum repperi, *ubi est omnis fructus ceterorum*' (echoing *mus.* 6. 1. 1, quoted above).

In summary: Everything A. wrote from his conversion in 386 down to but not including *vera rel.* on the threshold of ordination can be interpreted either as anti-Manichean or pro-*disciplina*. *Vera rel.*, the first work since Cassiciacum to carry a dedication—and dedicated to the same man as *c. acad.* (Romanianus)—marks a genuinely new start for A. But in the period 386–91, A. left a body of work that makes unambiguously clear that his expectations of the *liberales disciplinae* were high ones, at least philosophical and better to be described as mystical, expectations that he demonstrably shared with a near-contemporary with whom he has little else in common: where A. writes of the *disciplinae* in connection with his own baptism, Martianus can now be seen linking them to the theurgical cult practices of neo-Platonism (see D. Shanzer, *A . . . Commentary on Martianus . . . Book* 1 (Berkeley, 1986)).

What should be borne in mind is the triad of fifth-century writers: Augustine, Macrobius, Martianus Capella. All perhaps Africans (on Macrobius, see Cameron, *JRS* 56 (1966), 24–38), none having anything overtly to say about the others, but all in some way in dialogue. If A.'s *civ.* is a response at a distance to the ideas of Cicero's *rep.* (expressly targeted in *civ.* 2 and reprised in *civ.* 19, and cf. *civ.* 22. 28 for a tripartite dialogue of Aug. and C. with Plato's *rep.*!), it is at least a coincidence that we have Macrobius, writing perhaps *c.*430, giving us a sustained imitation of Cicero's *rep.*, animated by an ethos directly opposed to the Christian view of A. Given that relationship, it is then even more striking that we should have in A. and Martianus (Martianus' work dates, if we follow Shanzer, 29–44, but do not accept all S.'s most tendentious arguments, to the mid-fifth century, 430/80: see A. Cameron, *CP* 81 (1986), 320–8) two eager readers and imitators of Varro. That A. imitates sober, learned Varro and Martianus imitates satiric Varro should not confuse us: Apuleius had already shown that learning, satire, and mysticism may coexist in a single work (see now J. Winkler, *Auctor & Actor* (Berkeley, 1985)), and the first two qualities, at least, demonstrably coexisted in Varro.

4. 16. 31

proderat: 4. 16. 28, 4. 16. 29, 4. 16. 30 (2 ×), and below here. The rhetorical question finally offers a concise statement of the Manichean doctrine he entertained at the time, emphasizing the wrongness in his view of God. The only answer is the exclamation 'nimia perversitas!'

corpus esses lucidum et immensum: Cf. *Io. ev. tr.* 102. 4, 'homo animalis sic audit quaecumque audit de dei natura, ut aliud quam corpus cogitare non possit, quamlibet amplissimum vel immensum, quamlibet lucidum ac speciosum, corpus tamen.'

perversitas: Cf. in this paragraph 'aversi ... perversi ... revertamur ... evertamur' within a space of ten words below.

confiteri ... tuas: Cf. on 1. 15. 24, 'in confitendo tibi miserationes tuas' (echoing Ps. 106: 8, etc.); and compare 5. 1. 1, 'et confiteatur tibi miserationes tuas'.

latrare adversum te: Jdt. 11: 15, 'ego adducam te per mediam Hierusalem et habebis omnem populum Israhel sicut oves quibus non est pastor, et non latrabit vel unus canis contra te.' Cf. 9. 4. 11, 'ex quibus [manichaeis] fueram, pestis, latrator amarus et caecus'; Manichees are also *caeci latratores* at *mor.* 1. 19. 36.

nodosissimi libri enodati: He seeks the same cleverness of Faustus: 5. 6. 10, 'si qua forte maiora quaererem enodatissime expedirentur'.

in doctrina pietatis errarem: The same measure employed at 5. 5. 8 ('doctrinam religionis') and 5. 5. 9 ('doctrinae pietatis formam').

parvulis: See on 4. 15. 26, 'parvulis'.

ut in nido ... nutrirent: Ps. 83: 4, 'etenim passer invenit sibi domum et turtur nidum sibi ubi ponat pullos suos'; cf. Job 39: 26 (VL), 'numquid in sapientia tua plumescit accipiter, expansis pennis immobilis, respiciens ad austrum'; *adn. Iob* on 39: 26, 'sicut in sapientia dei, quae est Christus, novus homo paulatim innovatur, conversationem habiturus in caelestibus.' (*Plumesco* only here in Vg.) Knauer, 86–8, suggests Mt. 23: 37, 'quoties volui congregare filios tuos [Hierusalem], quemadmodum gallina congregat pullos suos sub alas.' *s.* 51. 5. 6, 'ego autem miser, cum me ad volandum idoneum putarem, reliqui nidum, et prius cecidi quam volarem. sed dominus misericors me, a transeuntibus ne conculcarer et morerer, levavit, et in nido reposuit.'?

in velamento alarum tuarum: Ps. 60: 5, 'cooperiar *in velamento alarum tuarum*'; Ps. 16: 8, 'in tegmine alarum tuarum *protege* me'; Ps. 35: 8, 'filii autem hominum in protectione alarum tuarum *sperabunt*'; Ps. 56: 2, 'in umbra alarum tuarum sperabo'; Ps. 62: 8, 'et in velamento alarum tuarum exsultabo'; Ps. 90: 4, 'et sub alis eius sperabis'. Knauer, 87: 'Diese Psalmenverse sind also völlig miteinander verschmolzen'; he cites (in lieu of numerous passages in *en. Ps.* in the same vein) *ep.* 148. 4. 13, 'nam de membris dei, quae assidue scriptura commemorat, ne quisquam secundum carnis huius formam et figuram nos esse crederet similes deo, propterea eadem scriptura et alas habere deum dixit, quas nos utique non habemus. sicut ergo, alas cum audimus, protectionem intellegimus ... et si quid aliud eadem scriptura tale commemorat, puto spiritaliter intellegendum.' The nest is the Church, in whose protection the little one finds himself (cf. *en. Ps.* 83. 7). Cf. the treatment of the same image at 10. 4. 6 ('id [verbum tuum facere] ago sub alis tuis . . .').

usque ad canos tu portabis: Is. 46: 4, 'usque ad senectam ego ipse et usque ad canos ego portabo, ego feci et ego feram et ego portabo et salvabo [dicit deus].' Sirach 6: 18, 'fili, a iuventute tua excipe doctrinam et usque ad canos invenies sapientiam.' The faintest of reminders of the search for *sapientia* under way, and fruitless, since 3. 4. 7 (*Hortensius*).

tu portabis CDGO Ver. : *tu portabis tu portabis* S Knöll Skut. Vega Pell.

infirmitas: Returns as leitmotif of Bk. 10 (10. 3. 4, 10. 5. 6, 10. 35. 57); Mt. 4: 23, 'Iesus sanans omnem languorem et omnem infirmitatem'.

bonum nostrum: Cf. 4. 15. 24, 'non enim noveram . . .'.

revertamur: See on 4. 15. 26, 'spiritus ambulans non revertebar ad te.'

tu ipse es: Cf. Ps. 101: 28 (see on 7. 21. 27), 'tu vero idem ipse es, et anni tui non deficient'; other echoes at 7. 20. 26, 7. 21. 27, 8. 3. 6, 10. 4. 6.

timemus OS Knöll Skut. Ver. : *timebimus* CDG.

redeamus: See on 1. 18. 28.

non ruit domus nostra: Cf. 11. 31. 41 (end of book), 'humiles corde sunt domus tua . . . et non cadunt quorum celsitudo tu es.' So here 'aeternitas tua' looks ahead to Bk. 11 as a whole. Is this 'house' the eternal home of the pre-existent soul? So R. J. O'Connell, *Atti 1986* 2. 48, identifying it further with the *caelum caeli* of 12. 2. 2 (for O'Connell's thesis, see on 1. 6. 7).

BOOK FIVE

Bk. 5 is tripartite after a retrospective and prospective prologue; each section corresponds to one of the three great cities in which A. lived. The period covered is from *c.* 382 to late 384. It begins with A. at a nadir, and ends with his outward situation if anything worse; but the first glimpse of Ambrose is a hint that better things lie ahead.

5. 1. 1–2. 2	*Prologue*
5. 3. 3–7. 13	*Faustus (at Carthage)*
5. 3. 3–5. 9	*Anticipation*
5. 6. 10–7. 13	*Realization*
5. 8. 14–11. 21	*Rome*
5. 8. 14–10. 18	*Leaving Carthage, illness at Rome*
5. 10. 19–11. 21	*Appeal of the Academics*
5. 12. 22–14. 25	*Milan*

The renewed emphasis on *confessio* is one sign among several that a hinge in the work's structure has been reached. Where the first four books have carried us through some twenty-eight years, the next five will cover the five years 383–7; the narrative passages become correspondingly more detailed and circumstantial (so G. Misch, followed by Pellegrino, *Les Confessions*, 127).

5. 1. 1

This paragraph renews the prayer for the gift of licit speech (1. 5. 5, 'miserere ut loquar'), the gift of confession.

Accipe . . . similis tibi: Knauer, 153: 'Die fünf Bücher 5–9 sind also durch Psalmzitate in ihren Prooemien innig miteinander verbunden und sie werden dadurch deutlich gegen den ersten Teil der Konfessionen (1.–4. Buch) und die vier folgenden Bücher (10–13) abgegrenzt.' This is the most formal preface since Bk. 1; note especially the imperatives (Knauer, 73, finds no further imperatives in Bks. 5–7) and the first appearance in the text of *confessiones* referring to the present work. (Elsewhere only at 5. 10. 20 before the conversion events at Milan; after that: 9. 8. 17, 9. 13. 37, 10. 3. 3, 10. 4. 6, 10. 34. 51, and 3 × in Bk. 12.)

accipe sacrificium: Mal. 1: 10, '"non est voluntas mea in vobis", dicit dominus omnipotens, "nec accipiam sacrificium de manibus vestris"'; v. sim. at 8. 1. 1 and 9. 1. 1, both picking up the quotation of Ps. 34: 10. See J. Ratzinger, *REAug* 3 (1957), 389–91, esp. 391: 'dieser Opferbegriff . . . die ganze Rückkehrbewegung der Menschheit zu Gott in sich begreift und keine Trennung von Kult und Leben zuläßt.'

sacrificium: *civ.* 10. 5–6, 'sacrificium ergo visibile invisibilis sacrificii sacramentum, id est sacrum signum est. . . . (6) proinde verum sacrificium est omne opus quo agitur ut sancta societate inhaereamus deo, relatum scilicet ad illum finem boni quo veraciter beati esse possimus. . . . anima ipsa, cum se refert ad deum, ut igne amoris eius accensa formam concupiscentiae saecularis amittat eique tamquam incommutabili formae subdita reformetur, hinc ei placens, quod ex eius pulchritudine acceperit, fit sacrificium.'

de manu linguae meae: Cf. perhaps Prov. 18: 21, 'mors et vita in manibus linguae' (Knauer, 151, says without evidence that *manu* is VL, but the half-dozen citations at La Bonnardière, *Biblia Augustiniana: Proverbes*, 218–19, all show the plural). Cf. 11. 2. 2 ('lingua calami'), 11. 11. 13 ('manus oris mei'). *en. Ps.* 72. 30, 'numquid lingua habet manus? sed quid est: "in manibus linguae"? in potestate linguae. quid est: in potestate linguae? "ex ore tuo iustificaberis, et ex ore tuo condemnaberis" [Mt. 12: 37]'; sim. at *en. Ps.* 120. 11.

excitasti: 1. 1. 1, 'tu excitas ut laudare te delectat'; here 'confiteatur' parallels the 'laudare' of 1. 1. 1; see also 8. 4. 9, 12. 12. 15. *Confessio* has the power to excite in turn those who read this text: 10. 3. 4, 'confessiones praeteritorum malorum meorum . . . cum leguntur et audiuntur, excitant cor', as at 11. 1. 1 (and echoed at *retr.* 2. 6. 1).

confiteatur nomini tuo: Ps. 53: 8, 'voluntarie *sacrificabo* tibi et confitebor nomini tuo, domine, quoniam bonum est.'

sana omnia ossa mea: Ps. 6: 3, 'miserere mei, domine, . . . sana me, domine, quoniam turbata sunt ossa mea'; *en. Ps.* 6. 4, 'id est firmamentum animae meae, vel fortitudo; hoc enim ossa significant.'

et dicant: Better as *ut dicant*, suggests Löfstedt, *Symb. Osl.* 56 (1981), 106: 'natürlich nicht notwendig, sie empfiehlt sich aber wegen der Konzinnität.' But for A.'s preference for *et* in *conf.*, see on 1. 1. 1.

domine, quis similis tibi?: Ps. 34: 10, 'omnia ossa mea dicent: "domine, quis similis tibi?"'; *en. Ps.* 34. s. 1. 13–14, 'quis digne de his verbis aliquid dicat? ego puto tantum pronuntianda esse, non exponenda. quid quaeris illud aut illud? . . . (14) o corpus Christi sancta ecclesia, omnia ossa tua dicant: "domine, quis similis tibi?"'

duritia hominum: Rom. 2: 5, 'secundum autem duritiam tuam et impaenitens cor, thesaurizas tibi iram in die irae.'

et non est qui . . . a calore tuo: Ps. 18: 7, 'a summo caelo egressio eius, et occursus eius usque ad summum eius, et non est qui se abscondat a calore suo.' Cf. 9. 4. 8 and 9. 7. 15, 'nos adhuc frigidi a calore *spiritus* tui'. *en. Ps.* 18. en. 2. 7, 'postea vero quam excurrit inde et recurrendo remeavit, misit spiritum suum. visae sunt illis super quos venit linguae divisae velut ignis. sicut ignis venit spiritus sanctus, faenum carnis consumpturus, aurum cocturus et purgaturus; sicut ignis venit, et ideo sequitur, "et non est qui se abscondat a calore eius".' Sim. at *c. litt. Pet.* 2. 32. 74.

te laudet anima mea: Ps. 145: 2, 'lauda, anima mea, dominum'; cf. 4. 10. 15 ('laudet te ex illis anima mea, deus, creator omnium').

confiteatur tibi miserationes tuas: Ps. 106: 8, 'confiteantur domino misericordiae eius, et mirabilia eius filiis hominum.' (For God's mercy as the subject, cf. 4. 16. 30 and 5. 10. 20.)

spiritus omnis per os conversum ad te: Ps. 150: 7 (last verse of last psalm), 'omnis spiritus laudet dominum'; Tob. 3: 14, 'ad te domine faciem meam converto.' (Cf. just above, 4. 16. 30, 'dorsum enim habebam ad lumen', etc.)

exsurgat: 3. 4. 7 (reading the *Hortensius* and echoing the prodigal), 'et surgere coeperam ut ad te redirem.'

innitens ... mirabiliter: Introducing an idea (that God can be known through the senses and the faculty of reason) definitively expressed at Rom. 1: 20, itself of great importance in Bk. 7 (see on 7. 9. 14) and to be cited soon (5. 3. 5). The avoidance of scriptural language here is striking. *vera rel.* 24. 45, 'nam in quem locum quisque ceciderit, ibi debet incumbere ut surgat. ergo ipsis carnalibus formis quibus detinemur *nitendum* est ad eas cognoscendas quas caro non nuntiat.'

transiens ad te: See on 1. 16. 25; the verb recurs dealing with the mystic ascent: 9. 10. 25 (Ostia), 10. 7. 11, 10. 17. 26.

qui fecisti haec mirabiliter: Cf. Ps. 71: 18, 'qui facit mirabilia solus'; Ps. 135: 4, 'qui facit mirabilia magna solus'; v. sim. at 4. 15. 24.

refectio: Cf. Ps. 22: 2, 'super aquam refectionis educavit me'; *en. Ps.* 22. 2, 'super aquam baptismi, quo reficiuntur qui integritatem viresque amiserant, educavit me.' Verbs of re-fashioning begin to occur in Bk. 5, a sign that A.'s condition, outward appearances notwithstanding, is improving: 5. 1. 1 (and notes), 'refectio', 5. 7. 13, 'reficientem', 5. 3. 4, 'recreans', 5. 10. 18, 'recreasti'.

vera fortitudo: Cf. 'lassitudine', but also 'ossa mea' above and *en. Ps.* quoted there.

5. 2. 2

inquieti iniqui OS Knöll Skut. Ver. : *inquieti et iniqui* CDG Maur. This is the closest A. comes to identifying 'inquietude' with wickedness; at 1. 1. 1 ('inquietum est cor nostrum'), restlessness is an ill whose cause is only implicit ('testimonium peccati sui').

et ecce pulchra sunt: *civ.* 11. 23, 'sicut pictura cum colore nigro loco suo posito, ita universitas rerum, si quis possit intueri, etiam cum peccatoribus pulchra est, quamvis per se ipsos consideratos sua deformitas turpet.'

quid nocuerunt tibi: *civ.* 12. 3, 'idcirco vitium, quo resistunt deo qui eius appellantur inimici, non est deo sed ipsis malum. . . . nulla quippe mala deo noxia, sed mutabilibus corruptibilibusque naturis.'

quo enim fugerent: Ps. 138: 7–8, 'quo ibo a spiritu tuo? et quo a facie tua fugiam? (8) si ascendero in caelum, tu ibi es; si descendero in infernum, ades.' See on 2. 6. 14, 'servus fugiens dominum suum'; other close echoes at 4. 7. 12, 4. 9. 14. *en. Ps.* 138. 10, 'ecce invenis *in longinquo fugitivum* [cf. Lk. 15 on the prodigal] non latere oculos eius a quo fugit. . . . quo ergo ibit fugitivus iste a facie dei? vertit se hac atque illac, quasi quaerens locum fugae suae.' Cf. Ps. 67: 2, 'exsurgat deus et dispergantur *inimici* eius, et *fugiant* qui oderunt eum a *facie* eius'; *en. Ps.* 67. 2, 'fuga quippe animi, est timor. . . . facies quippe eius appellata est praesentia eius per ecclesiam eius.'

excaecati: Rom. 11: 7–11, 'ceteri vero *excaecati* sunt, (8) sicut scriptum est; dedit illis deus spiritum compunctionis, oculos ut non videant, aures ut non audiant, usque in hodiernum diem. . . . (11) dico ergo, numquid sic *offenderunt* ut caderent? absit, sed illorum delicto salus gentibus ut illos aemulentur.'

quia non deseris aliquid: Cf. Wisd. 11: 25, 'diligis enim omnia quae sunt, et nihil odisti horum quae fecisti, nec enim odium habens aliquid constituisses.' Taken in support of the aesthetic argument at *c. Adim.* 17. 3, *div. qu. Simp.* 1. 2. 8, 1. 2. 18, *Io. ev. tr.* 110. 6. See on 1. 10. 16, 'ordinator et creator rerum omnium naturalium, peccatorum autem tantum ordinator'.

ubique: See on 1. 3. 3, 'ubique totus'.

longe: See on 1. 18. 28. Ps. 72: 27, 'ecce qui longe se faciunt a te, peribunt. perdidisti omnem qui *fornicatur* [see on 1. 13. 21] abs te'; *en. Ps.* 72. 33, 'illi vero longe recesserunt, quia non solum terrena desideraverunt, sed ea a daemonibus et a diabolo petierunt. . . . et quid est, longe a deo fieri? "perdidisti omnem qui fornicatur abs te".'

proicientium se in te: Cf. 8. 11. 27 ('continentia' speaks), 'proice te in eum, . . . proice te securus, excipiet te et sanabit te.'

vias suas difficiles: Cf. Wisd. 5: 7, 'lassati sumus in via iniquitatis et perditionis et ambulavimus vias difficiles, *viam autem domini* ignoravimus.' Cf. 4. 12. 18, 'adhuc ambulare vias difficiles'; see on 3. 1. 1, 'viam sine muscipulis'.

terges lacrimas eorum: Is. 25: 8, 'et auferet dominus deus lacrimam ab omni facie'; cf. Apoc. 7: 17 (= Apoc. 21: 4), 'et absterget deus omnem lacrimam ex oculis eorum.' *en. Ps.* 127. 10, 'sed veniet deus tuus, de quo dicitur: "ubi est?" et absterget lacrimas, et ipse pro pane lacrimarum succedet, et te in aeternum saginabit; quia erit nobiscum verbum dei, quo pascuntur angeli.' On tears in *conf.*, see on 3. 2. 4.

caro et sanguis: See on 4. 3. 4.

reficis: See on 5. 1. 1, 'refectio et vera fortitudo'.

ubi ergo eram: Cf. 2. 2. 4, 'ubi eram? et quam longe exulabam a deliciis domus tuae anno illo sexto decimo anno aetatis carnis meae'; there, desolation; here, with an answer that offers reassurance.

ego . . . inveniebam: Self-knowledge only in knowledge of God; *sol.* 2. 1. 1, 'noverim me, noverim te'; 10. 27. 38, 'et ecce intus eras et ego foris et ibi te quaerebam et in ista formosa quae fecisti deformis inruebam.'

5. 3. 3

Proloquar: literally (and perhaps here mainly) 'reveal, expound, expose'; but there are numerous passages in A. that use the verb with an overtone of accusation or denunciation, and that sense cannot be ruled out here; see on 1. 5. 6. Mandouze, *Lectio III–V*, 43, offers the improbable translation 'Je vais *commencer par* parler de cette vingt-neuvième année de mon âge.' (Emphasis his: he aims to defend A. against charge of confusion, because Bk. 5 goes beyond the chronological terminus of 13 November 383, A.'s twenty-ninth birthday.)

in conspectu dei mei: The phrase is 'confessional' in its evocation of the silent presence of God before which A. speaks; see on 1. 16. 26.

annum illum undetricensimum: 13 November 382–13 November 383. The dating by his age connects this passage to those before and below that mark his progress through the Manichees: at 3. 4. 7, his age was given as eighteen at the time he fell in with them; at 4. 1. 1, his nine years with them ran from age eighteen to age twenty-seven; and at 5. 6. 10, the nine years will be recalled as he reaches the end of his rope on account of Faustus: see there for a fuller treatment of the chronological difficulties.

episcopus: *haer.* 46. 16, 'nam ex electis suis habent duodecim quos appellant magistros, et tertium decimum principem ipsorum; episcopos autem septuaginta duos, qui ordinantur a magistris, et presbyteros, qui ordinantur ab episcopis indefiniti. habent etiam episcopi diaconos.' Though A. had been among the Manichees for nine years, he had apparently not met any of the highest ranking members of the sect, neither the *magistri* nor the *episcopi*; that may reflect the concentration of the sect's leaders in Italy (see below on Faustus' career), and may further reflect their discretion in not showing themselves indiscriminately even to their own followers at a time of potential persecution.

Faustus nomine: The form of introduction calls attention to the name itself, 'Lucky', again marked below ('nominatus'); and repeated at 5. 7. 13, 'ita ille nominatus apparuit'. A. uses the word only once in its ordinary acceptation (and that in a 'pagan' sense: *ench.* 21. 79, 'secundum vanas doctrinas hominum fausta vel infausta . . . tempora'), and it is rare among Christian writers, apparently out of distaste for its religious overtones, though it does survive in other later Latin (e.g. Amm. 15. 8. 21, 'imperatorem clementem appellam et faustum'). See on 4. 4. 7 for the way A. chooses which of the names of his past to include in *conf.* Though presented as 'laqueus diaboli', F. marked the beginning of the end of bad times for A.: 5. 7. 13, 'meum [laqueum] quo captus eram relaxare iam coeperat.'

A.'s thumbnail description of his old associate: *c. Faust.* 1. 1, 'Faustus quidam fuit gente Afer, civitate Milevitanus, eloquio suavis, ingenio callidus, secta manichaeus ac per hoc nefando errore perversus.' See Mandouze, *Pros. chr.* s.v. Faustus 2; Faustus on Faustus, quoted at *c. Faust.* 5. 1: 'ego patrem dimisi et matrem, uxorem, filios, et cetera quae evangelium iubet; . . . ego aurum argentumque reieci et aes in zonis habere destiti, cotidiano contentus cibo nec de crastino curans nec unde venter impleatur aut corpus operiatur sollicitudinem gerens.' A. rejects this self-description at *c. Faust.* 5. 5, making him sound positively dissolute compared to schismatic ascetic Manichees (the Mattarii) and even in comparison to his own father, whom A. reproaches F. for having abandoned. F. came from Milev,[1] of family origins probably more humble than A.'s. He was older than A., though perhaps only by a few years. He was married and had children when he renounced all to convert to Manicheism. There is no sign that he had been Christian or raised among Christians, but we have no reliable information about his pre-Manichean career (and especially none about his education). That he had become one of their bishops by this date (382) is a sign of his success in the sect; he had spent at least part of his time with the sect at Rome (A. would use Manichean connections there too: 5. 10. 18). Around 385/6, he was denounced as a Manichee and relegated to exile on a desolate Mediterranean island (A. suggests that the mildness of the punishment was due to intervention by the orthodox: *c. Faust.* 5. 8). He was amnestied not long after, spending perhaps only a year on the island, but while there found time to write the work that A. would refute in detail *c.*400 in his own *c. Faust.*, by which time F. was dead.

laqueus diaboli: Cf. 1 Tim. 3: 7, 'ut non in opprobrium incidat et laqueum diaboli'; 2 Tim. 2: 26, 'et resipiscant a diaboli laqueis.' See on 3. 6. 10, and cf. 5. 7. 13. Knauer, 139: 'gliedernd heben diese Zitate Anfang und Ende des Abschnittes hervor.'

vasculo sermonis: Words are *vasa* at 1. 16. 26 and 5. 6. 10.

praelocuta: 'said beforehand'; cf. e.g. *Gn. litt.* 4. 14. 25, 'dicam sane quod sentio, haec duo indubitata praeloquens, . . .'.

disciplinis liberalibus eruditus: See Excursus on 4. 16. 30 for the high esteem that A. would have accorded such a reputation.

et quoniam: Though A. has suggested that Faustus is about to be introduced, he does not properly appear until 5. 6. 10, where there are numerous verbal echoes of the introduction to the fifth book up to this point; instead we now have a detailed discussion of A.'s mental state in the period immediately leading up to the appearance of F. There is a change of atmosphere from the last episode narrated in chronological order in Bk. 4, that of writing the *de*

[1] In Numidia, a hundred miles or so west of Thagaste, not far from Constantine (mod. Cirta); Decret, *Aspects* 197–8 thinks it plausible, but impossible to prove, that F. was introduced to Manicheism in his native city.

pulchro et apto: enthusiasm and loquacity have given way to doubts. A year or two have elapsed (at 4. 15. 27, he says he was 26 or 27 then, now he is 28).

multa philosophorum legeram: See Solignac at *BA* 13. 92–3, resuming his landmark article on A.'s use of the doxographic tradition (*RA* 1 (1958), 113–48): 'il est permis de croire que c'est à de telles lectures qu'il est fait allusion dans les deux textes des Confessions'—i.e. the present text and 5. 14. 25, 'plerosque sensisse philosophos magis magisque considerans atque comparans iudicabam.' The value of Solignac's study is not that it demonstrates a source, but that it reminds us of the penumbra of lost sources and influences, many of them humble, that shaped A.'s mind. Marius Victorinus is assigned a similar background: 8. 2. 3, 'omnium liberalium doctrinarum peritissimus quique philosophorum tam multa legerat'. The multiplicity of philosophical doctrine often underscored (cf. 6. 5. 7, 'pugnacitas calumniosarum quaestionum per tam multa quae legeram inter se confligentium philosophorum') is in willed contrast to the unity and clarity of ecclesiastical doctrine; in a welter of answers, whose abundance confuses, Christianity (as received and as constructed by A.) provides an embracing, and thus explaining, and thus reassuring, unity.

ut possent aestimare saeculum: Wisd. 13: 8–9, 'iterum autem nec his debet ignosci: (9) si enim tantum potuerunt valere ut possent aestimare saeculum, quomodo eius dominum non facilius invenerunt?' At *exp. prop. Rom.* 3, this text appears embedded in a discussion of Rom. 1: 18–23, which it resembles: for the later text, cf. here 5. 3. 5. At *doctr. chr.* 2. 21. 32–22. 33, the Wisd. text is quoted against *curiositas* and astrology; cf. also *doctr. chr.* 2. 29. 46 and *Gn. litt.* 5. 16. 34 (after quoting Wisd., he adds 'qui fundavit terram *propinquat mentibus nostris*'). See *ep.* 55. 4. 6, quoted at length for its report on Manichean astronomical fantasies at 5. 3. 6 below.

magnus es, domine: Ps. 144: 3, 95: 4, etc.; see on 1. 1. 1. Cf. Ps. 137: 6, 'quoniam magna est gloria domini, quoniam excelsus dominus, et humilia respicit et excelsa a longe cognoscit'; Lk. 18: 14, 'quoniam qui se exaltat, humilabitur, et qui se humilat, exaltabitur'. *s.* 351. 1, 'non ergo mirum si publicanus magis curatus abscessit, quem non puduit ostendere quod dolebat. in rebus quippe visibilibus, ut excelsa quisque contingat, in excelsum erigitur: deus autem cum sit omnium excellentissimus, non elatione, sed humilitate contingitur. unde propheta dicit: "*prope* est dominus his qui obtriverunt cor" [Ps. 33: 19: see on 'obtritis corde' below]. et iterum: "excelsus dominus, et humilia respicit, et excelsa a longe cognoscit." excelsa ipsa posuit pro superbis. . . . quisquis itaque paenitentiae recusat humilitatem, deo *propinquare* non cogitat.' Sim. at *en. Ps.* 39. 20, 137. 11, etc.; see Knauer, 52–3, and see further passages cited immediately below on Ps. 33: 19; and on *longe*, see on 1. 18. 28; 9. 11. 28 (Monnica speaks), 'nihil longe est a deo.'

obtritis corde: Ps. 33: 19, 'iuxta est dominus his qui obtriverunt cor, et humiles spiritu salvos faciet'; *en. Ps.* 33. s. 2. 23, 'magna mysteria, fratres. deus super omnia est; erigis te, et non illum tangis; humilas te, et ipse ad te descendit.' Sim. at *en. Ps.* 31. en. 2. 11, 31. en. 2. 18, 74. 2 (quoting Ps. 137: 6), *s.* 21. 2; cf. *s. Caill.* 2. 11. 5 (on the return of the prodigal): 'ideo dicitur in alio psalmo, "dixi, *proloquar* adversus me delictum meum domino" [Ps. 31: 5].... quam proxima est dei misericordia confitenti! non enim longe est deus a contritis corde; sic enim habes scriptum, "prope est dominus eis qui obtriverunt cor." iam ergo iste obtriverat cor in *regione egestatis* [2. 11. 18]; redierat enim ad cor [4. 12. 18], ut obtereret cor.'

sidereas plagas: *trin.* 4. pr. 1, 'laudabiliorque est animus cui nota est vel infirmitas sua quam qui, ea non respecta, vias siderum scrutatur ...' (adduced by M. Jackson, *Atti 1986*, 1. 417).

5. 3. 4

The return to astronomy here recalls the bout with astrology recounted at 4. 3. 4–5, and anticipates the final return of the subject at 7. 6. 8–10. A.'s belief that the truth about the stars is intimately tied to the truth about higher and deeper things is, among many other things, a reflection of his long-standing fascination with Cicero. Cic.'s *nat. deor.*, for example, is impregnated with astronomical speculation and lore.

The word *et* occurs 24 times in this paragraph (though at 'et quales se ipsi fecerant' and 'et trucidant exaltationes suas' the sense would be improved if we read *nec* in place of *et*); see on 1. 1. 1.

defectus luminarium: Cf. *c. Fel.* 1. 10, 'non legitur in evangelio dominum dixisse, "mitto vobis paracletum qui vos doceat de cursu solis et lunae." christianos enim facere volebat, non mathematicos.' There were three solar eclipses visible in Africa in A.'s early life (359 (probably fairly pronounced in effect), 378, and 381 (the latter two partial eclipses)) that could have been subject for debate (381 is chronologically convenient): L. C. Ferrari, *REAug* 19 (1973), 263–76. The influence eclipses could hold on the popular mind is seen in A.'s contemporary Maximus of Turin, whose Christian flock deprecated a lunar eclipse with clamorous prayer, seeking to defend the moon against the hostile influences of unspecified *magi* (Max. Taur. *ss.* 30–1 (Mutzenbecher)). Ferrari is less convincing, at *Augustiniana* 27 (1977), 139–50, reprised in his *The Conversions of Saint Augustine* (Villanova, 1984), 40, when he suggests that Halley's comet's appearance in March–April 374 might have influenced A.'s conversion to Manicheism; *inter alia*, we have learned since Ferrari wrote that not *every* appearance of Halley's comet is memorably vivid. A. noticed that the *Hortensius* spoke of eclipses as well (*civ.* 3. 15), and believed that eclipse-prediction had begun with Thales of Miletus (*civ.* 8. 2; as far as Greece goes, this is more or less true: see the successful prediction by Thales in 585 BC reported at Herod. 1. 74, a feat that some (see B. Frischer, *The Sculpted Word* (Berkeley, 1981), 17 n. 26) think sealed T.'s reputation as σοφός). The Manichean explanation and the achievements of the astronomers received fuller comment at *ep.* 55. 4. 6–7, 'et qui subtilius ista scrutati sunt, incrementa et decrementa lunaria ex conversione globi eius coniecerunt, non quod aliquid substantiae vel recedat ei cum augetur, vel decedat cum minuitur, quod *delira* [cf. 5. 3. 6, 'delirans'] imperitia Manichaei opinantes repleri eam dixerunt, sicut repletur navis, ex fugitiva dei parte, quam commixtam principibus tenebrarum et eorum sordibus inquinatam corde atque ore sacrilego . . . loqui non dubitant. hinc ergo impleri lunam dicunt, cum eadem pars dei magnis laboribus ab inquinamento purgata de toto mundo atque ab omnibus cloacis fugiens redditur deo lugenti dum redeat; repleri vero per mensem dimidium et alio dimidio in solem refundi velut in aliam navem. . . . (7) illi autem qui haec certis numeris indagarunt, ita ut *defectus solis et lunae* non solum cur fierent, sed etiam *quando futuri essent, longe ante praedicerent* et eos determinatis intervallis temporum canonica supputatione praefigerent litterisque mandarent, quas modo qui legunt atque intellegunt, nihilo minus eos praedicunt.' See also *Gn. litt.* 1. 19. 39, *trin.* 3. 2. 7, 3. 9. 19, *s.* 68. 1. 2, *s. Mai* 126. 4–5 (using Rom. 1: 18–20 to specify the failure of the scientists to see beyond creation to the creator: see below on 5. 3. 5), *ep.* 199. 10. 34, and *gr. et pecc. or.* 2. 23. 27.

mirantur haec homines: Cf. 10. 8. 15 ('et eunt homines mirari alta montium'); the implicit reproach is clear in both cases.

non . . . religiose quaerunt: See on 5. 3. 5, 'non pie quaerunt'.

tu fecisti eos: Cf. Ps. 99: 3, 'scitote quoniam dominus ipse est deus; ipse fecit nos, et non nos'; see on 9. 10. 25.

volatilia . . . pisces maris . . . pecora campi: Ps. 8: 8–9, 'omnia subiecisti sub pedibus eius, oves et boves universas, insuper et pecora campi, (9) volucres caeli, et pisces maris, qui perambulant semitas maris.' The Psalm text evokes from A. one of his fullest expositions of the three temptations of 1 Jn. 2: 16 (quoted on 10. 30. 41): *volucres* [1], *pisces* [2], *pecora* [3]. A similar reading, without the allusion to 1 Jn. 2: 16, at *Gn. c. man.* 1. 20. 31: 'cum enim non reguntur isti motus, erumpunt et pergunt in foedissimas consuetudines, et per diversas perniciosasque delectationes nos rapiunt, et faciunt similes omni generi bestiarum.'

ignis edax consumas: cf. Deut. 4: 24, 'quia dominus deus tuus ignis consumens est, deus aemulator'; Deut. 9: 3, 'ignis devorans atque consumens, qui conterat eos et deleat atque disperdat ante faciem tuam velociter'; Heb. 12: 29, 'etenim deus noster ignis consumens est'; and cf. *Aen.* 2. 758 (of Troy as Aeneas returns to the battle): 'ilicet ignis edax summa ad fastigia vento | volvitur; exsuperant flammae.'

recreans: Cf. 5. 1. 1, 'refectio'.

5. 3. 5

non noverunt: Repeated twice below; see on 3. 7. 12, 'non noveram'—the phrase a sign of unknown high doctrine, and at the same time of later 'confessional' interpretation.

viam: note '*hanc* viam' twice below; see on 3. 1. 1 and 7. 7. 11: Jn. 14: 6, 'ego sum via et veritas et vita.'

per quod fecisti ea: Jn. 1: 1–3, 'in principio erat verbum. . . . (3) omnia per ipsum facta sunt, et sine ipso factum est nihil.'

numerant (4 ×) **. . . numerus . . . numeratus:** For 'number' as the correlate to the second person of the trinity in the triad *mensura/numerus/pondus*, see on 5. 4. 7; recall here 5. 3. 3, 'curiosa peritia numerent'.

et sapientiae tuae non est numerus: Ps. 146: 5, 'magnus dominus noster [echoed at 5. 3. 3] et magna virtus eius et sapientiae eius non est numerus'; see on 1. 1. 1. The echoes through the first paragraphs of Bk. 5 back to the opening of Bk. 1 suggest that there is a new beginning here; cf. also the imperatives in 5. 1. 1, the naming of Faustus, and the echo of Mt. 7: 7 in 'non pie quaerunt' below.

factus est nobis sapientia: 1 Cor. 1: 30, 'in Christo Iesu, qui factus est sapientia nobis a deo, et iustitia et sanctificatio et redemptio'; 1 Cor. 1: 24, 'Christum dei virtutem et dei sapientiam'.

et solvit tributum Caesari: Mt. 17: 24–7 (Vg. modified against *en. Ps.* 137. 16, where the episode is retold), 'et cum venissent Capharnaum, accesserunt qui tributum accipiebant ad Petrum et dixerunt, "magister vester non solvit tributum?" (25) ait, "etiam." et cum intrasset domum, praevenit eum Iesus dicens, "quid tibi videtur, Simon? reges terrae a quibus exigunt tributum? a filiis suis an ab alienis?" (26) cum autem ille dixisset, "ab alienis", dixit illi Iesus, "ergo liberi sunt filii. (27) sed tamen ne scandalizemus eos, vade, mitte hamum in mare, et qui primus piscis ascenderit, aperi os eius invenies staterem. illum sumens, da illis pro me et te".' Cf. *en. Ps.* 118. s. 31. 1, 'numquid eorum rex milites suos prohibuit impendere et exhibere quae debentur *regibus terrae*? nonne de hoc sibi calumniam molientibus Iudaeis ait: "reddite *Caesari* quae Caesaris sunt, et deo quae dei sunt?" nonne *tributum* de ore piscis etiam ipse per*solvit*?' The editors (Knöll, Skut., Ver., Pell.) cite instead Mk. 12: 17 and/or Mt. 22: 21 ('Render unto Caesar') wrongly.

hanc viam: Cf. *doctr. chr.* 1. 11. 11, 'cum ergo ipsa [sapientia] sit patria, viam se quoque nobis fecit ad patriam.'

excelsos: 5. 3. 3, 'magnus es, domine, et humilia respicis, *excelsa* autem a longe congnoscis' (cf. Ps. 137: 6, 'quoniam excelsus dominus, et humilia respicit').

et ecce ruerunt in terram: Is. 14: 12–13, 'quomodo cecidisti de caelo, Lucifer, qui mane oriebaris? corruisti in terram . . . (13) qui dicebas in corde tuo: "in caelum conscendam, super astra dei exaltabo solium meum".'

et obscuratum est insipiens cor eorum: Rom. 1: 21, 'et obscuratum est insipiens cor eorum'; already echoing Is. 44: 20, 'insipiens cor'; Prov. 12: 23, 'et cor insipientium provocat stultitiam'; Ps. 13: 1 (= Ps. 52: 1), 'dixit insipiens in corde suo, "non est deus".' The rest of this paragraph quotes and supplements the text of Rom. 1: 21–5, the classic Pauline text on the failings of secular wisdom in the domain of theology. For A.'s text and the passage's role in *conf.*, see on 7. 9. 14.

veritatem: Jn. 14: 6, 'ego sum via et veritas et vita' (and cf. just below, 'qui veritas es').

non pie quaerunt: Mt. 7: 7, 'quaerite et invenietis' (see on 1. 1. 1). At 1. 1. 1, the implication was that *credere* must precede *quaerere*; here that assumption is given an ethical twist by the adverb. The phrase here (with slight variants) is common in A.: explicitly related to *curiositas* (in a way that illumines A.'s way of thinking of the Manichees as 'curious') at *div. qu.* 68. 1, 'cum videatur apostolus corripuisse *curiosos,* dicendo "o homo, tu quis es, qui respondeas deo?" [Rom. 9: 20] de hoc ipso illi quaestionem movent, et in ea sententia non desinunt esse *curiosi,* qua obiurgata est ipsa *curiositas:* et *impii* quidem cum contumelia ... et obiurgasse *quaerentes,* quia non poterat quod quaerebatur exponere.... quidam autem bona et *pia* mente scripturas legentes *quaerunt* quid hic possit vel malidicentibus vel calumniantibus responderi' (hence the contrast between *curiositas* and *pietas* at *s.* 265. 3. 4, 'curiositas abscedat, pietas succedat'). 'Impious search' is contrasted to 'confessio' at *lib. arb.* 3. 2. 5, 'quia *non pie quaerunt* velocioresque sunt ad excusationem quam *ad confessionem peccatorum suorum*'. A heuristic list of other passages (note how many are early and anti-Manichean): *quant. an.* 14. 24, *mor.* 1. 10. 17, *util. cred.* 3. 9, *Gn. c. man.* 1. 5. 9, 2. 2. 3 (with Mt. 7: 7), *c. ep. fund.* 23. 25 (quoting Mt. 7: 7), *s. dom. m.* 1. 4. 11, *cat. rud.* 4. 8, *c. Faust.* 3. 2 (with Mt. 7: 7 emphasized), 3. 5, 3. 4, 11. 6, 12. 25, 12. 30, 16. 22, *ep.* 44. 5. 12, *trin.* 1. 3. 5 ('"*quaerite* faciem eius semper": et hoc placitum *pium* atque tutum coram domino deo nostro cum omnibus inierim'), *trin.* 3. pro. 1, *c. adv. leg.* 1. 13. 17 (from a late work, but one with many affinities to A.'s anti-Manichean polemics), *s.* 261. 2. 2; cf. *c. Adim.* 3. 4, 'sed sanctae scripturae non temerarios et superbos accusatores, sed diligentes et *pios* lectores desiderant' (implying *non pius* = *superbus*). See finally *s.* 51. 5. 6, important for its reflections on the scriptural difficulties the Manichees posed for the young Augustine: 'loquor vobis, aliquando deceptus, cum primo puer ad divinas scripturas ante vellem afferre acumen discutiendi, quam *pietatem quaerendi*' (and the next clauses echo Mt. 7: 7. closely). See R. D. Di Lorenzo, *Aug. Stud.* 14 (1983), 117–28.

perversissima caecitate: 5. 2. 2, 'excaecati'.

5. 3. 6

ex ipsa creatura: 5. 3. 5, 'multa vera de creatura dicunt.'

occurrebat mihi ratio: sc. *in philosophis legendis*; the first mention of *ratio* in this sense in *conf.* Manicheism had commended itself to A. by asking tough questions of a doctrine otherwise given to him by faith: Courcelle, *Recherches*, 65: 'Ainsi, le motif fondamental pour lequel Augustin a embrassé le manichéisme, est son appétit rationaliste'—an observation both true and important, not least because it explains how it was to the challenge of rational argument that Manicheism was vulnerable for A. For A. as bishop, what the Manichees had offered was false *ratio*: he only uses the word here to record when that *ratio* led him to truth—*vera dicta*.

Manichaei: *Manichaeus* is capitalized in this text where it stands for the founder of the sect, but not where it is adjective or substantive for his followers.

delirans: Notably elsewhere in *conf.* of 'curious' behaviour: 3. 6. 10 (Manichees), 7. 6. 8 and 7. 6. 10 (*mathematici*); at 1. 17. 27, 'in quibus . . . deliramentis' evokes the frivolities on which A. wasted his talents in his schooldays.

solistitiorum C^{1}D^{1}OS Knöll Skut. Ver. : *solstitiorum* C^{2}D^{2}G Maur.

defectuum C^{2}GOS Knöll Skut. Ver. : *defectum* C^{1}D. See on 5. 3. 4.

in libris saecularis sapientiae: Evidently books he had encountered in pursuing the *disciplinae liberales*.

non occurrebat . . . erat: sc. *quod credere iuberer* as subject of both verbs.

5. 4. 7

deus veritatis: Ps. 30: 6, 'redemisti me, domine, deus veritatis'; *en. Ps.* 30. en. 2. s. 1. 11, 'faciens quod promisisti, non fallens in pollicitatione tua, deus veritatis'.

beatus autem: See on 1. 6. 10, 'quid ad me'; *ench.* 5. 16, 'quando nobis Maronis ille versus placet: "felix qui potuit rerum cognoscere causas" [*geo.* 2. 490], non nobis videatur ad felicitatem consequendam pertinere si sciamus causas magnarum in mundo corporalium motionum quae abditissimis naturae sinibus occuluntur: "unde tremor terris, qua vi maria alta tumescant obicibus ruptis, rursusque in se ipsa residant" [*geo.* 2. 479–80], et cetera huiusmodi. sed bonarum et malarum rerum causas nosse debemus. . . . nam si causae corporalium motionum noscendae nobis essent, nullas magis nosse quam nostrae valetudinis deberemus. cum vero eis ignoratis medicos quaerimus, quis non videat quod de secretis caeli et terrae nos latet quanta sit patientia nesciendum?' Cf. *geo.* 2. 493, 'fortunatus et ille deos qui novit agrestis.' On this echo, see M. Jackson, *Atti 1986*, 1. 413–17.

cognoscens te: Rom. 1: 21, 'cognoscentes deum non sicut deum glorificaverunt aut gratias egerunt sed evanuerunt in cogitationibus suis.' The echo links back to 5. 3. 5, forward to 7. 9. 14, and outward to scripture; 'sicut te' for *sicut deum* (which appear in a few minor MSS and Maur.) makes the words both scriptural and A.'s own.

sicut enim: Answered by 'sic fidelis homo' below.

arborem: The example seems random and arbitrary, but Ferrari (*REAug* 16 (1970), 235), sets this in the context of A.'s frequent references to trees in *conf.*

metitur . . . numerat . . . diligit: The verbs anticipate the trinitarian pattern of *mensura*, *numerus*, and *pondus* below.

fidelis homo: Prov. 17: 6 (VL), 'fidelis hominis totus mundus divitiarum est' (not in Hebrew or Vg.). Cited and followed (as here) by 2 Cor. 6: 10, at *c. Adim.* 18. 1, *epp.* 153. 6. 26, 157. 32, *civ.* 20. 7, and *en. Ps.* 143. 18 and 48. s. 1. 3, 'numquid ipse Paulus non habebat divitias? habebat plane. quas divitias? de quibus dicit alio loco scriptura: "fideli homini mundus totus divitiarum est." audi et ipsum confitentem, quasi nihil habentes et omnia possidentes. qui vult ergo esse dives, non haereat parti, et totum possidebit: illi *inhaereat* qui totum creavit.' (Verse identified by A. Vaccari, *Scritti di erudizione e di filologia*, 2 (Rome, 1958), 13.) On *fidelis*, see on 2. 3. 6.

quasi nihil habens omnia possidet: 2 Cor. 6: 10, 'quasi nihil habentes et omnia possidentes'. Not an excuse for wealth, but an exposition of true wealth: *en. Ps.* 103. s. 3. 16, 'prorsus totum mundum dimisit Petrus, et totum mundum Petrus accepit, quasi nihil habentes, et omnia possidentes.'

cui serviunt omnia: Ps. 118: 91, 'omnia serviunt tibi.'

septentrionum gyros: 10. 8. 15, 'gyros siderum'; cf. *en. Ps.* 139. 13, 'in gyrum eant erroris, ubi iter est sine fine. . . . qui in gyrum it, nusquam finit.'

qui omnia in mensura et numero et pondere disposuisti: Wisd. 11: 21, 'omnia in mensura et numero et pondere disposuisti.' The principal *conf.* text to be set beside this one is 13. 9. 10, 'pondus meum amor meus', where see further notes. The trinitarian implications of this passage, and the indirect implications for *conf.* as a whole, are considerable. See La Bonnardière, *Biblia Augustiniana: Sagesse*, 90–8, and W. Beierwaltes, *REAug* 15 (1969), 51–61; earlier, see also W. J. Roche, *New Scholasticism* 15 (1941), 350–76. Du Roy, 279–80, 421–4 offers sustained discussion of these passages in a trinitarian context (421: 'C'est avant tout le texte scripturaire de *Sap.* XI, 21 qui lui donne occasion d'exposer la structure trinitaire du créé'), noting especially the treatment of *modus/species/ordo* in *nat. b.* (quoted on 1. 7. 12); cf. also *BA* 49. 636: 'Mesure, nombre et poids, mode d'être, forme et ordre, définissent ainsi la *structure métaphysique* de l'être créé; ce n'est pas une structure statique, mais bien *relationnelle*.' The Platonic antecedents are worth tracing (e.g. God as μέτριον of all things at *Laws* 716 C), but do not seem determinative. Here the traceable influences are scriptural, whatever preconditioning influences of a purely secular, scientific variety there may have been (Solignac, *RA* 1 (1958), 137, suggests Nicomachus of Gerasa; cf. also on the monad/dyad at 4. 15. 24). La Bonnardière, *Biblia Augustiniana: Sagesse*, 295–6, lists the verse 31 × in A.; only the most notable and instructive are cited here:

(1) Adumbration of this triad occurs in A.'s earliest work from Cassiciacum: *ord.* 2. 15. 42, 'etiam ibi per constantissimas temporum vices, per astrorum ratos definitosque cursus, per intervallorum spatia moderata intellexit nihil aliud quam illam *dimensionem* [1] *numerosque* [2] dominari, quae similiter definiendo ac secernendo in *ordinem* [3] nectens astrologiam genuit, magnum religiosis argumentum tormentumque curiosis' (du Roy and La Bonnardière omit this passage and begin with *Gn. c. man.* 1. 16. 26, quoted below). For the substitution of *ordo* for *pondus*, see *mus.* 6. 11. 29: 'delectatio quippe quasi pondus est animae. delectatio ergo ordinat animam'; from much later, *civ.* 5. 11 equates the two triads explicitly. The same conflation with the other triad at *lib. arb.* 2. 20. 54, 'omnem quippe rem, ubi *mensuram* et *numerum* et *ordinem* videris, deo artifici tribuere ne cuncteris.' (Perhaps, as du Roy, 326–8, argues, this comes from the second phase of composition of *lib. arb.*, hence is later in date than the next quotation.)

(2) The pertinence of the conflation is demonstrated by the first clear citation of the verse: *Gn. c. man.* 1. 16. 26, 'in omnibus tamen cum mensuras et numeros et ordinem vides, artificem quaere. nec alium invenies, nisi ubi *summa mensura*, *summus numerus* et *summus ordo* est, *id est deum* de quo verissime dictum est, quod "omnia in mensura et numero et pondere disposuisti".' A further echo at *Gn. c. man.* 1. 21. 32, and the two triads dance together at *c. Faust.* 21. 6, where he traces the vestiges of this triad at all levels, from God to human beings to irrational animals, to all *corpora*; du Roy, 422–3, shows how it is at this period (close to that of of *conf.*) that A. begins strongly to find the qualities of measure, number, and weight by analogy in God himself.

(3) A passage in *Gn. litt.* further links the two triads to the motif of the mind's ascent to God. *Gn. litt.* 4. 3. 7–4. 9, esp. 4. 3. 8, 'magnum est paucisque concessum, excedere omnia quae metiri possunt, ut videatur *mensura sine mensura*, excedere omnia quae numerari possunt, ut videatur *numerus sine numero*, excedere omnia quae pendi possunt, ut videatur *pondus sine pondere*.' Cf. also *Gn. litt.* 3. 16. 25, 'habent enim omnia, quamdiu sunt, mensuras, numeros, ordines suos; quae cuncta merito considerata laudantur nec sine occulta pro suo genere moderatione pulchritudinis temporalis etiam ex alio in aliud transeundo mutantur. *quod si stultos latet, sublucet proficientibus clarumque perfectis est*.'

(4) In *trin.*, a more abstract paragraph suggests some of the mechanism behind the link to mystical vision in the passage just cited: these triads are invoked to explain the mechanics of vision itself. *trin.* 11. 11. 18, 'sed quia

numerose cogitari possunt quae singillatim sunt impressa memoriae, videtur ad memoriam mensura, ad visionem vero numerus pertinere quia licet innumerabilis sit multiplicitas talium visionum, singulis tamen in memoria praescriptus est intransgressibilis modus. mensura igitur in memoria, in visionibus numerus apparet, sicut in ipsis corporibus visibilibus mensura quaedam est cui numerosissime coaptatur sensus videndi, et ex uno visibili multorum cernentium formatur aspectus, ita ut etiam unus propter duorum oculorum numerum plerumque unam rem geminata specie videat, sicut supra docuimus. . . . voluntas vero quae ista coniungit et ordinat et quadam unitate copulat, nec sentiendi aut cogitandi appetitum nisi in his rebus unde visiones formantur acquiescens conlocat, ponderi similis est.' (This passage closes *trin.* 11; so measure/number/weight is the pivotal triad to connect the exterior person to the inner. Cf. also *trin.* 3. 9. 16, 6. 10. 12; *trin.* 11. 8. 14 equates *mensura* with *modus*.)

(5) Other texts from later years: *Io. ev. tr.* 1. 13; *en. Ps.* 146. 11 (412). By this time, the application of the text, however, seems nearly restricted to the second person of the trinity (du Roy, 424), and the text and the theme gradually lapse, having done their work for A.

5. 5. 8

The matters in which Mani was contradicted by the scientists were not of the doctrine of faith. But Mani had stuck his neck out; his failure here made it possible to see the error of his whole undertaking. This is an argument perfectly intelligible in a nineteenth-century rationalist on his way out of church forever because the palaeontology he heard there was out of date. To be sure, A. writes as one who has chosen *pietas* (5. 3. 5, 'non pie quaerunt') over *scientia*, but that was not the direction in which he was moving at the moment represented here. Academic scepticism lay ahead. The pervading irony of Bk. 5 is that motion in a wrong direction was the beginning of motion in the right direction.

quaerebat: G–M: 'used in the sense of "request" with accusative of person and prolative infinitive, is unusual'.

Manichaeum nescio quem: on the patronizing 'nescio quis' see on 1. 13. 20; for Mani, see 5. 3. 6, 'cum dictis Manichaei'.

dixisti: replacing *dixit* in the original Job (see next note): at *trin.* 14. 1. 1, he introduces this quotation by saying, 'ubi legitur dei sapientiam dixisse homini'.

ecce pietas est sapientia: Job 28: 28 (VL), 'dixit autem homini, ecce pietas est sapientia.' *trin.* 12. 14. 22, '"ecce pietas est sapientia; abstinere autem a malis scientia est." in hac differentia intellegendum est ad contemplationem sapientiam, ad actionem scientiam pertinere. pietatem quippe hoc loco posuit dei cultum, quae graece dicitur *theosebeia*' (the same Gk. equivalent and lengthy discussion at *ench.* 1. 2 and *spir. et litt.* 11. 18 (with explicit link to Eucharistic liturgy)). The discussion draws together the 'wisdom' that A. had pursued since adolescence with formal religious cult; see Excursus on 4. 16. 30 for suggestions how this link was made outside Christianity at this time. Cf. 8. 1. 2, where the same verse occurs in a context where A. has transcended the *vanitas* of the Manichees but is still liable to the reproach of Rom. 1: 21.

vanitas: As a characteristic of Manichean ideas at 4. 2. 2 (see notes there) and 4. 7. 12.

devius: 2. 10. 18, 'erravi, deus meus, nimis devius ab stabilitate tua'; *devius* at least hints at the thoroughgoing notion of Christ as *via*: see on 7. 7. 11.

multum locutus: Cf. Prov. 10: 19, 'ex multiloquio non effugies peccatum'; *c. Cresc.* 1. 1. 2, 'multiloquium autem est superflua locutio, vitium scilicet loquendi amore contractum. plerumque autem loqui amant, etiam qui nesciunt quid loquantur vel quomodo loquantur'. The Proverbs text recurs in self-reproach in the last paragraph of *trin.* (15. 28. 51) and at *retr.* pr. 2.

spiritum sanctum, consolatorem et ditatorem fidelium tuorum: See on 3.6.10, 'paracleti'.

personaliter: Cf. 'divinae personae' below.

doctrinam religionis: 4. 16. 31, 'doctrina pietatis'; 5. 5. 9, 'doctrinae pietatis formam'.

tribuere sibi: Cf. 5. 3. 5, 'sibi tribuendo quae tua sunt'.

5. 5. 9

illum aut illum: = English 'so-and-so', as an indefinite expression replacing a proper name; cf. 'aliud pro alio'.

domine creator omnium: Cf. Amb. *hymn.* 1. 2. 1, often echoed—see on 9. 12. 32; if the echo is heard here, perhaps it is partly linked by the astronomical context.

situs et habitus: Two of the Aristotelian categories (4. 16. 28, 'aut ubi sit constitutus ... aut calciatus vel armatus sit'), suggesting knowledge according to the worldly philosophers' way of knowing.

doctrinae pietatis formam: See on 5. 5. 8.

etiam talis infirmitas: i.e. even pertinacious affirmation of error is bearable where faith and *caritas* prevail. Mani's real fault is not in his errors of fact, but in the claims he made for himself as purveyor of error.

a caritate matre: *caritas* is aptly mentioned here, for *caritas* is the theological virtue most characteristic of the true Holy Spirit whom Mani emulates; see on 13. 6. 7, where the same phrase, 'mater caritas', appears again Spirit-linked.

adsurgat: Of 'growing up': cf. *ord.* 1. 6. 16, 'videbam adulescentem ... in amicum quoque iam mihi surgere atque grandescere'.

novus homo: Eph. 4: 24, 'et induite novum hominem qui secundum deum creatus est'; *en. Ps.* 8. 10, 'sed novus ex vetere nascitur, quoniam spiritalis regeneratio mutatione vitae terrenae atque saecularis inchoatur.'

virum perfectum: The scriptural echo allows a glimpse of eschatological repose: Eph. 4: 13–14, 'donec occurramus omnes in unitatem fidei, in agnitionem filii dei, in virum perfectum, in mensuram aetatis plenitudinis Christi, (14) ut iam non simus parvuli fluctuantes, et circumferamur omni vento doctrinae in nequitia hominum.' *en. Ps.* 101. s. 1. 2, 'Christus et ecclesia utrumque unus, unus quidam vir perfectus in forma plenitudinis suae'; *en. Ps.* 64. 7, 'qui autem manent in compage Christi et membra eius sunt, faciunt quodammodo unum virum, de quo dicit apostolus, "donec ...".'

nondum liquido compereram: These lines lead to the arrival of Faustus, expressing the exact form of the doubts that A. was saving to present to the visiting teacher—resuming the argument about astronomical phenomena from 5. 3. 4.

5. 6. 10

Since 4. 1. 1 A. has portrayed his life as awkwardly divided between his public profession and his private cult. Here the two halves come together. Faustus arrives, and differs from the other Manichees merely by being a little better than others in the skills A. earned his living by selling. For most, this would be enough, but not for A. He was too good at his craft, and at the same time thought little enough of it; it was his disdain for the craft at which he excelled that made him a seeker of *sapientia*.

per annos ferme ipsos novem: A. says repeatedly that he stayed with the Manichees nine years; there are contradictions in his record, and doubts have been raised whether the break was as sharp at the time of his meeting with Faustus as he would have us believe. 'Nine years' explicitly also at 3. 11. 20, 4. 1. 1 ('ab undevicensimo anno [372/3] aetatis meae usque ad duodetricensimum [381/2]'; cf. 5. 3. 3, 'proloquar ... annum illum undetricensimum aetatis' [382/3]), *mor.* 1. 18. 34 ('nihil opus erit novem annis quibus me ludificastis'), *mor.* 2. 19. 68 ('novem annos totos magna cura et diligentia vos audivi'), and *util. cred.* 1. 2. The interpretation that fits all the evidence is that A.'s association with the Manichees began during 373 at Carthage, and that Faustus came some time early in the year 383; nine full years he spent as a wholehearted enthusiast before his doubts took command at the time of Faustus's visit. Mandouze, *Lectio III–V*, 53: 'La raison majeure qui a poussé Augustin à la décision de rupture avec le manichéisme n'est point d'ordre religieux. Elle est d'ordre rationnel, voire scientifique.' To cling to that argument is to ask for trouble here. For all his cerebration, Manicheism was to A. a *religious* allegiance: even when his head got in trouble over the tenets of the sect, his allegiance was of another sort and there was then no competing *religious* cause to which to attach himself. When he finally detached himself, it was by accepting the proposition that in default of an adequate *religious* experience, a *philosophical* position (extremely unsatisfactory on many counts) must be *temporarily* elected (he says (5. 14. 25) that he *never* planned to remain an Academic), while waiting for another *religious* certitude to arise. That decision was, however, accompanied by a specific shift of religious position, back to the catholic catechumenate (see on 5. 14. 25). His *religious* separation from the Manichees thus dates to his departure from Rome, with growing *intellectual* alienation going back several years earlier (Pellegrino, *Les Confessions*, 129, implicitly reads this passage as confirming that 'les doutes avaient surgi bien tôt': that goes too far) and coming to a head with the Faustus episode. Hence the truth to Courcelle's surmise (*Les Confessions*, 17–19: but see on 6. 11. 18 for strong reservations about his handling of the evidence he quotes in support of that surmise) that Manichean 'reflexes' continued to jerk his knees in Milan—they are of no account *because* he never regressed to that cult, but of course when he arrived in Milan, in the middle of his Academic period, they probably loomed larger in his mind. If he had chosen to go back that way, they would appear now as the first steps back: to us they are only stumbles in a dark passage.

audivi: The correct verb for his status as an *auditor* in the sect.

extento: The verb occurs in *conf.* mainly from Phil. 3: 13, 'in ea quae sunt ante extentus', of a mystic longing connected to the third person of the trinity through the triad *memoria/contuitus/expectatio* (note here 'expectabam': see on 12. 15. 18). See on 9. 10. 23, 'in ea quae ante sunt extenti', and cf. 12. 16. 23, 'extento in eam sursum corde', and 13. 13. 14. Here the participle modifies *desiderium*, a lesser form of the love whose supreme form is *caritas*. Translators vary: Ryan, 'intense longing'; *BA*, 'un désir toujours plus tendu'; Pusey, 'I had longed but too intensely'; Pine-Coffin, 'with the keenest expectation'. Vega takes 'extento' closely with 'per annos ferme novem ipsos', to suggest that A.'s doubts started early; this is unlikely. The lexica insist on a pedestrian meaning: *OLD* s.v. 6: 'to extend the duration of, prolong, continue'; cf. *TLL* 5. 2. 1980: 'in longum tempus continuando productus': rare but attested. Hence, 'a desire too long prolonged', creating an expectation probably impossible to fill; the same expression of a desire that could be filled because the object was divine at *Io. ep. tr.* 4. 6, 'sic deus differendo extendit desiderium, desiderando extendit animum, extendendo facit capacem' (going on to quote Phil. 3: 13–14).

hominem gratum et iucundum verbis: The accusation that an opponent is merely facile is a familiar one (from the world of classical rhetoric itself, of course). Cf. *nat. et or. an.* 2. 1. 1, describing the work of Vincentius Victor on the origin of the soul: 'quibus lectis vidi *hominem* in sermone quidem non solum usque ad sufficientiam, verum etiam usque ad redundantiam profluentem, sed in rebus de quibus loqui voluit nondum sicut oportet instructum.'

From 391, we have two passages reflecting this encounter with F.: *util. cred.* 1. 2, 'diu multumque de imperitorum erroribus latissime ac vehementissime disputabant—quod *cuivis mediocriter erudito esse facillimum sero didici*'; *util. cred.* 8. 20, 'ut enim a vobis trans mare abscessi, iam cunctabundus atque haesitans, quid mihi tenendum, quid dimittendum esset—quae mihi cunctatio in dies maior oboriebatur, ex quo illum hominem, cuius nobis adventus, ut nosti, ad explicanda omnia quae nos movebant quasi de caelo *promittebatur*, audivi eumque excepta quadam eloquentia talem quales ceteros esse cognovi.'

garrientem: Of fluent but inauthentic speech: see on 7. 20. 26, and cf. 6. 4. 5; see also on 1. 4. 4, 'loquaces', and 5. 5. 8, 'multum locutus', for that fault attached specifically to the Manichees.

pretiosorum GS Knöll Skut. Vega Pell. : *pretiosiorum* CDO Maur. Ver. The determining factor is the echo of 1. 16. 26, 'non accuso verba, quasi *vasa* electa atque *pretiosa*'. See also here the last line of this paragraph, 'verbis sicut vasis urbanis et rusticanis'.

nec ideo vera quia diserta: *nat. et or. an.* 1. 3. 3, 'habet [Vincentius Victor] enim eloquium, quo possit explicare quae sentit. unde cum illo agendum est eique optandum ut recta sentiat, ne faciat esse delectabilia quae sunt inutilia et quae *diserta* dixerit *vera* dixisse videatur.'

boni rerum existimatores: *doctr. chr.* 1. 27. 28, 'ille autem iuste et sancte vivit, qui *rerum integer aestimator* est; ipse est autem qui *ordinatam* [see on 1. 7. 12] habet dilectionem, ne aut diligat quod non est diligendum, aut non diligat quod diligendum est.'

aliud genus hominum: The description given here fits the Academic sceptics well enough, but they are not introduced as a serious alternative until 5. 10. 19, after A.'s move to Rome. The present passage may hint that he was already aware of their position before he left Carthage; but describing them without a name leaves them here only a shadowy schematic antithesis to Manichean credulity, setting up two poles between which A.'s own attitude at the time is portrayed as more moderate and sensible—not to say conventional.

docueras CDG Knöll : *docuerat* OS Skut. Ver. That A. had by now been taught this message marks for him real progress over 3. 5. 9 (see there on 'indigna'), where on looking into the truth of scripture he was put off by the absence of *comptus atque uber sermo*.

et propterea credo: The parentheses were suggested by G–M, who remark that 'docueras' requires an object of the thing taught if the sentence is to be punctuated with other editors by a full stop after 'claruerit': 'quod tu me docueris' on that reading lacks a reference. With the parentheses, 'quod tu me docueris' can anticipate what comes below, 'nec eo debere . . .', and 'abs te didiceram' marks a new beginning for the thought after the parenthesis.

nec quisquam praeter alius doctor est veri: This is the thesis of *mag.*, as summarized at *retr.* 1. 12, 'in quo disputatur et quaeritur et invenitur, magistrum non esse qui docet hominem scientiam nisi deum, secundum illud etiam quod in evangelio scriptum est: "unus est magister vester Christus" [Mt. 23: 10].' The idea is congenial for an inveterate autodidact, but there is more to it than that.

iam ergo: Cf. 5. 3. 3, '[suaviloquentiam] ego iam tametsi laudabam, discernebam tamen a veritate rerum.'

nec eo . . . nec eo . . . rursus nec ideo . . . nec ideo: Cf. 1. 18. 29, on barbarisms and solecisms.

5. 6. 11

expectaveram: Cf. 5. 6. 10, 'extento desiderio', and the connections perhaps implied there to the triad *memoria/contuitus/expectatio*.

vel etiam prae multis: This is how people in Africa would have remembered A.'s relations with Faustus: F. arrives with A. in the lead among those praising him, later in friendly literary relations with him (5. 7. 13, 'coepi cum eo . . . agere vitam'). There was no visible break with Manicheism before A. left Africa, rather the opposite. It was not accidental that A. chose, about the time of writing *conf.*, F.'s own treatise for his largest systematic refutation of Manichean doctrines—his own past contained not only Manicheism, but Manicheism as Faustus presented it, and there were suspicions that A. was not entirely cleansed of the poison. Long years later, Julian of Eclanum threw the association up to A.: *c. Iul. imp.* 1. 25, 'Faustus, quem in libris confessionis tuae praeceptorem tuum loqueris'; also at *c. Iul. imp.* 1. 69.

ferebam: G–M think this = either 'ferebam laudibus', or else (quoting Tac. *ann.* 16. 2, 'nec aliud per illos dies populus credulitate, prudentes diversa fama tulere') 'talked about.' *BA:* 'je le louais et l'exaltais.'

sed moleste habebam: Cf. 6. 3. 3 (Ambrose reading silently), where A. makes allowances for the preoccupation of the prospective teacher. That contrast assumes a connection between the two events and emphasizes that in this case, though with difficulty, A. does win the ear of Faustus (not entirely privately, but in the company of a few intimates), only to find that F. has no answers for his questions. Though in 5. 6. 10 the emphasis seemed to be on rhetoric ('decorum eloquium'), what A. says below ('et quia aderat . . .') corrects that impression: F. had no formal rhetorical training, but only a certain amount of native skill and practice (note 'inde') beyond the elementary grammatical training. See further on 6. 3. 3, esp. 'coniectabamus'.

(*expertus sum*) **prius:** G–M: 'The force of the comparative seems to be "before coming to the more difficult questions".'

expertem liberalium disciplinarum: See Excursus on 4. 16. 30: for A. at the time, this unsophistication on F.'s part would have been itself the explanation for his failure to answer A.'s questions—it may even have prevented A. from giving him a full and fair hearing, as suggested at the beginning of 5. 7. 12 below.

aliquas tullianas orationes ... volumina: The list of *auctores* need not be taken as literally accurate, but gives an impressionistic recollection of the bits of cultural baggage that F. managed to produce.

Senecae: The only prose text of Seneca that A. ever cites or echoes (Hagendahl 245–9) is the *de superstitione*, which comes to us only in fragments;[2] it is quoted several times in *civ.* 6. 10–11. 'Superstition' was a charge the Manicheans levelled against mainstream Christianity, a charge that apparently had some effect on the young A., and *superstitio* is a word that came often to A.'s pen (see on 3. 6. 10). It is at least likely that this would have been one of the few books of Seneca that F. knew, and not unlikely that it was through F. that A. came to know this one isolated Senecan text. (I. Hadot, *Aug.-Lex.* 1. 288, asserts that A. must have known Seneca well, arguing in this curious way: 'Das in *conf.* 5, 11 angeführte Beispiel des Faustus zeigt zweierlei: erstens, daß selbst ein nur wenig gebildeter Mann zur Zeit des A. gewöhnlich einige philosophische Abhandlungen Senecas gelesen hatte, und zweitens, daß A. selbst eine gute Kenntnis der Werke Senecas gehabt haben muss. Wie hätte er sonst feststellen können, daß Faustus nur eine sehr geringe Anzahl ("paucissimos ... libros") annäischer Schriften gelesen hatte?') *Paucissimi* for A. corresponds to English 'few': at one extreme, cf. *en. Ps.* 47. 9, 'unus uel duo uel paucissimi'; at *s.* 203. 1. 1, the 12 days from Christmas to Epiphany are *paucissimi*; at *Io. ev. tr.* 122. 9, 'paucissimis verbis interpositis' denotes the omission in quotation of one verse of Matthew, containing 21 words.

recolo: Wijdeveld, *REAug* 6 (1960), 315–16, would put the question mark here, and a comma after 'meae'; the *BA* translation follows him. But to end the question with 'recolo' leaves us lacking a second person of address. Courcelle, *Recherches*, 45, takes the passage, as is his custom, literally: 'il est vrai qu'il ne paraissait pas fort sûr de l'exactitude de son souvenir.'

[2] Otherwise from Seneca, A. shows two and a half lines of tragedies, the five lines of Cleanthes translated at Sen. *ep.* 107. 11, ending in the famous 'ducunt volentem fata, nolentem trahunt', and (at *ep.* 153. 5. 14) the quasi-proverbial 'omnes odit, qui malos odit'—that phrase occurs several times in Sen. *de ira* and *de benef.*, but A. cites neither, rather introducing Seneca there as coeval with the apostles and 'cuius etiam quaedam ad Paulum apostolum leguntur epistulae'—not even claiming to have read those letters himself. I. Opelt, *Atti 1986*, 1. 364–5, tries to make much of S.'s name in *gramm.* (Keil, 5. 496) and a phrase in *ep.* 3*. 2 that appears in S.'s *de ira*, but as Divjak notes ad loc., that phrase is proverbial and proves nothing.

arbiter conscientiae meae: *ep.* 23. 3, 'conscientiarum arbiter iudicare [coeperit]'.

coram te: Num. 10: 9, 'et erit recordatio vestra coram domino deo vestro'; cf. Act. 8: 21, quoted below on 5. 7. 12, 'etsi non rectum ad te'.

abdito secreto providentiae tuae: 5. 7. 13, 'in abdito providentiae tuae'.

errores: The results of *curiositas*: see on 4. 15. 25, 'errores et falsae opiniones'.

ante faciem meam: Cf. Ps. 49: 21, 'arguam te et statuam contra faciem tuam'; *en. Ps.* 49. 28, 'quid enim vis latere teipsum? in *dorso* tuo tibi es, non te vides; facio ut te videas; quod post dorsum posuisti, ante faciem ponam'; *s.* 142. 3. 3, 'quod nolebat anima videre, ponitur ei ante oculos; et quod post dorsum habere cupiebat, ad faciem illi admovetur.' Cf. 4. 16. 30, 'dorsum enim habebam ad lumen', and 2. 3. 6, 'qui ponunt ad te tergum et non faciem', both interpreted in light of Jer. 2: 27, 'verterunt ad me tergum et non faciem'; and see 8. 7. 16, 'et constituebas me ante faciem meam, ut viderem quam turpis essem.' *en. Ps.* 113. s. 1. 7, 'increpat enim dominus quosdam qui dorsum ad eum posuerunt et non faciem. . . . et sic obliviscatur ea quae retro sunt, ut in ea quae ante sunt extendatur' (Phil. 3: 13: see above on 5. 6. 10, 'extento').

5. 7. 12

earum artium: sc. *liberalium*: see on 5. 6. 11; cf. 7. 6. 9, 'falsa, non *artis imperitia* sed sortis mendacio'.

aperire: The verb of opening books, doors, and mysteries; of books, most notably at 8. 12. 29, 'arripui, aperui et legi in silentio capitulum'; of doors metaphorically at 13. 38. 53, where the work ends, 'a te petatur, in te quaeratur, ad te pulsetur: sic, sic accipietur, sic invenietur, sic aperietur'; of mysteries, 2. 10. 18, 6. 4. 6, 6. 11. 18, 9. 7. 16.

quorum quidem ignarus: A.'s judgement *at the time* was that F. was not clever enough to answer his questions. A. now sees this was the wrong grounds on which to judge, for lack of cleverness is no sin; it was F.'s Manichean error that put him wrong. His justice to F. here lies in seeing this.

libri . . . eorum: See S. N. C. Lieu, *Manichaeism*, generally, and for detailed examination of A.'s handling of their doctrines, Decret, *Aspects* and *L'Afrique*; P. Alfaric, *Les Écritures manichéennes* (Paris, 1919), is excellent but dated; for the pertinence here, see above on 5. 3. 4. What works of Mani himself A. knew is a delicate question; the most circumspect survey of A.'s knowledge of Mani scripture is that of Alfaric, *L'Évolution intellectuelle*, 217. A. certainly knew the two most celebrated and authoritative texts, the 'Fundamental Epistle' (*c. ep. fund.* 5. 6, 25. 28; cf. *c. Faust.* 13. 6, 13. 18) and the 'Treasure' (*nat. b.* 44, quoting at length from the seventh book). We do not know whether, when A. implies in *c. ep. fund.* 6. 7 that he has read *aliae epistulae*, he knows the other texts of the Manichean 'Torah'. At the same time A. often spoke with a confidence that implied an extensive knowledge (at least extensive by comparison with that of his one-time co-religionists): *mor.* 2. 12. 25, where 'non hoc sonant libri manichaei' implies that A. has a broad knowledge of Manichean texts; at *c. Sec.* 3 he threatens to overwhelm his opponent with footnotes: 'innumerabilibus locis de libris Manichaei recitabo . . .'; and at *c. Faust.* 20. 21, he catches Faustus speaking of the *umbrae defunctorum*: 'sunt ergo umbrae defunctorum? numquam hoc in vestris sermonibus audivimus, numquam in litteris legimus; immo contradicere soletis talibus opinionibus.'

mihi eum: Indirect object and accusative subject for indirect discourse not resumed until 'an certe vel par etiam'.

etiam inde: i.e. from the side of the Manichees.

sarcinam: See on 4. 7. 12.

loquaces: See on 1. 4. 4, 'quoniam loquaces muti sunt' (and cf. here 'et dicentes nihil') and cf. 3. 6. 10, 7. 2. 3.

etsi non rectum ad te: Ps. 77: 37, 'cor autem ipsorum non erat rectum cum eo' (Act. 8: 21, 'cor enim tuum non est rectum coram deo'); *en. Ps.* 77. 21, 'talis ergo erat generatio prava et amaricans, etiam cum deum quaerere videretur, diligens in ore et in lingua mentiens; in corde autem non recta cum deo, ubi ea potius diligebat propter quae dei adiutorium requirebat.' The biblical echo in a concessive clause has the effect of emphasizing the concession.

A. goes as far as he can to praise F., but does not undermine, and indeed strengthens, his judgement that F. was fundamentally in the wrong. This pattern recurs with A.'s judgements on the Platonists; his ability to praise the virtues of his opponents without losing grip on his adverse judgement of them has disconcerted many.

non usquequaque . . . placuit: So here a specific memory: it was the modesty with which F. failed to live up to his advance reputation that endeared him the more to A., even as A. was disappointed with what F. had to say. Here again, evidence that there would have been those in Africa who remembered A.'s growing friendship with the Manichean bishop.

pulchrior est enim . . . cupiebam: A variation on 1. 6. 10, 'gaudeat etiam sic . . .'; and cf. 10. 8. 15, 'et eunt homines mirari alta montium. . . .' The connection here is evidently that it was still the *pulchriora* of created nature to which A.'s curiosity lead him (see on *de pulchro et apto* at 4. 13. 20).

5. 7. 13

de ceteris eorum doctoribus: i.e. other Manichean teachers besides Faustus; A. might have chosen to go on looking for a guru, but he deemed F. the best chance.

ita: As described in 5. 6. 10–5. 7. 12.

nominatus: See on 5. 3. 3, 'Faustus nomine'.

agere vitam: Not necessarily 'live with' but certainly of a convivial friendship; see on 5. 6. 11.

legere cum eo: Another Manichean leader (a presbyter, therefore clearly not F., who was a bishop; date unknown) clung to association with A. out of mutual interest in *studia liberalia* (*mor.* 2. 19. 71).

conatus omnis meus: This Micawberish decisive indecision encompasses the last half of Bk. 5, restated at 5. 10. 18 and at 5. 14. 25, and evoked in retrospect at 6. 10. 17 and 6. 11. 18, then one last time at 8. 7. 18, where he has notably not achieved his goal, just when Ponticianus leaves and he and Alypius are about to go into the garden. At this time, A. had a clear idea of the sort of thing he expected to turn up; it turned up in the period narrated in Bks. 6 and 7, but what also turned up was the inadequacy of that expectation, and the need to look further, to find what Bk. 8 held for him: the recurrence at 8. 7. 18 emphasizes that.

The present passage may suggest that A. hoped to become one of the Manichean elect himself; *util. cred.* 1. 2 says that he remained an auditor to preserve his prospects for a worldly career; that is surely a sign that he at least considered, and perhaps even was recruited for, elect status. Did he think that his incontinence prevented him? Is that part of the reason for his resentment of the imperfections of the elect he met in Rome (see on 5. 10. 18)? Was his break with the Manichees partly caused by growing pressure to present himself for 'election'? Was the Milan conversion to continence a renunciation that he had been meditating, but had been unable to make, for years? It is not to be expected that A. would be entirely forthcoming, or even entirely conscious of his own motives, on that score. That the Manichean and Christian alternatives posed themselves in some way equally to A. is implicit at *s.* 132. 1. 1, where he speaks of Christian 'catechumeni vel *audientes*'. In both sects through these years he stood as one who *heard* (a significant passivity for one otherwise so loquacious) but did not participate.

laqueus mortis: Ps. 17: 6, 'praevenerunt me laquei mortis'; *en. Ps.* 17. 6, 'tales autem homines capiunt in perditionem quibus male persuaserunt iactatione iustitiae; cuius non re, sed nomine, adversus gentes gloriantur.' Cf. Prov. 21: 6, 'qui congregat thesauros lingua mendacii; vanus est et impingetur ad laqueos mortis'; see on 5. 3. 3, 'laqueus diaboli'.

meum: sc. *laqueum*.

nec volens nec sciens: On the trinitarian pattern, a human being consists of three faculties: *esse/nosse/velle* (see on 13. 11. 12); accordingly this passage shows A. out of control, not yet an integrated self, to whom good things happen at the level of *esse*, undetected at the level of *nosse* or *velle*.

manus . . . tuae: In plural of God, only here and 13. 17. 20, 'aridam terram manus tuae formaverunt', where the allegorical interpretation of Gn. reads 'arida terra' appositely as (13. 17. 21) 'animas sitientes tibi'. Here 'praeter manum tuam reficientem quae fecisti' evokes the context of creation (and redemption).

matris meae: Not mentioned since 4. 4. 7, upon the death of his friend.

a domino . . . diriguntur: Ps. 36: 23, 'a domino gressus hominis dirigentur, et viam eius volet'; *en. Ps.* 36. s. 2. 15–16, 'ipse homo ut velit viam domini, ab ipso domino diriguntur gressus eius. nam si dominus non dirigeret gressus hominis, tam pravi erant ut semper per prava irent, et semitas curvas sequendo redire non possent. . . . (16) iam cum sequeris *viam Christi*, non tibi saeculi prosperitates promittas.' At *c. Faust.* 1. 1, A. writes 'ut omnes qui haec legent intellegant quam nihil sit acutum ingenium et lingua expolita, nisi a domino gressus hominis dirigantur.' (Cf. Ps. 39: 3, quoted on 5. 8. 14, 'gressus', below.)

reficientem: 5. 1. 1, 'quia fecisti haec mirabiliter [cf. 'miris modis' here]: et ibi *refectio* et vera fortitudo'; cf. 5. 3. 4, 'recreans'.

5. 8. 14

The departure from Carthage and the arrival at Rome are really one episode, linked by the constant reference to Monnica's hopes and prayers, even when A. is at Rome and she in Africa. The time at Rome was important: Mandouze, 107 n. 6, 'Tout compte fait—et contrairement à ce que donnent à penser la plupart des biographes d'Augustin—l'année romaine fut, bien plus que ne devait l'être le séjour milanais, le moment des contradictions et des incertitudes les plus grandes: c'est cependant à Rome que tous les éléments de la décision ultérieure devaient être rassemblés.'

Egisti: 5. 7. 13, '*egisti* mecum miris modis . . . tu illud *egisti*, deus meus, nam a domino gressus hominis diriguntur.' We move accordingly in a circle from Ps. 36: 23, echoed at the end of 5. 7. 13, back to the same echo (plus Ps. 39: 3) at the end of this paragraph, showing God's merciful direction of A.'s 'steps' towards, at one and the same time, the *falsa felicitas* (see the end of this paragraph) he thought he was finding and the *terra viventium* (see also below) that he did find.

Romam pergere: Probably in 383, within a few months of the meeting with Faustus (see on 5. 6. 10); to date the journey to 384 would leave a narrow window of time in which A. would make the trip in the spring (after the sailing season began), languished and recovered from his illness at Rome, made his impression on Symmachus, and journeyed again to Milan—all while leaving at least some time to acquire his impressions of teaching conditions at Rome.

altissimi . . . praesentissima: 6. 3. 4, 'altissime et proxime, secretissime et praesentissime' (cf. 1. 4. 4, 'secretissime et praesentissime'); also vocative at 1. 7. 12, 3. 8. 16, and 6. 12. 22.

recessus: Nominative; on substance more likely plural (so Ryan, *BA*, e.g.), on grammar more likely singular (used of memory in singular at 10. 8. 13, in plural at 10. 40. 65); A. is indifferent.

misericordia: This section of Bk. 5 is marked by hints of divine mercy taking—at last—a hand: cf. 5. 8. 15, 'misericorditer', 5. 9. 16, 'miserebaris', 5. 9. 17, 'deus misericordiarum, . . . quoniam in saeculum misericordia tua'.

non ideo: See on 4. 7. 12 for the evidence that as late as 386 he recalled and admitted that the departure from Thagaste to Carthage in 376 (thus seven years earlier than the move to Rome) was motivated by *ambitio saeculi*; the present passage provides an argument from silence to confirm that reading—only *here* does he feel he can deny the obvious imputation. Here he is beyond the reach of the three temptations; apathy has set in. He is not redeemed, but no longer driven to sin; he suffers the punishment for his sins by going on sinning; and he waits for something to turn up. In the midst of what he now sees as foul sinfulness, it was a moral fastidiousness that made him move: a *vestigium* of divine order, surely, but surely a faint one. The lingering force of the third of the great temptations is left to a concessive clause: 'quamquam et ista ducebant animum tunc meum'. The unnamed 'friends' are akin to those mentioned in 4. 1. 1, 4. 8. 13, 4. 13. 20; 'friends' recur in another vein in Bk. 6, where from 6. 7. 11 they are a sign of common progress towards the light.[3]

quietius . . . ordinatiore: see on 13. 9. 10, 'minus ordinata inquieta sunt; ordinantur et quiescunt'.

apud Carthaginem: See on 3. 3. 6, 'eversionibus', where he confirms what he says here ('meos esse nolui'), that he had not been among their number.

verum autem tu: So far in this paragraph, Augustine's will in this matter; from here on, God's will.

spes mea et portio mea in terra viventium: Ps. 141: 6, 'tu es spes mea, portio mea in terra viventium'. Knauer, 38, reports that the verse occurs 19 times in A. (based on the Beuron files), though infrequent before *conf.* (6 ×), and comparatively rare in writings of other *patres*. This 'terra' is not that of those who die, but indicates (*s. dom. m.* 1. 2. 4) 'quandam soliditatem et stabilitatem hereditatis perpetuae'. 'Denn' (Knauer, 39) 'dort ist das künftige Leben, hier nur die Hoffnung.' Most often the verse is adduced to illustrate *terra* in some other biblical passage, so the weight is on *terra* not *spes*. Here it is invoked as A. pursues his hopes from one land to another. *en. Ps.* 5. 1, 'titulus . . . est: "pro ea quae hereditatem accipit." . . . "beati mites, quia ipsi hereditate possidebunt terram" [Mt. 5: 4]. quam terram, nisi de qua dicitur, spes mea es tu, portio mea in terra viventium.' See Knauer, 38, on the artistry and balance of these lines.

[3] At *c. litt. Pet.* 3. 25. 30, A. deals with P.'s claim that A. left Africa only in 386 when Manichees were being persecuted.

pro salute animae meae: Ps. 34: 3, 'dic animae meae, salus tua ego sum'; see on 1. 5. 5.

vitam mortuam: Courcelle, *Les Confessions*, 485 n. 1, suggests two Augustinian passages and a Hermetic source, none of which seems exactly apt: 10. 17. 26 'in homine vivente mortaliter' (not exactly the same), *civ.* 20. 8, 'succedunt nascendo morientibus', and the *Hermetica*, tract 7. 2 (ed. Nock–Festugière, 1. 81), *τὸν ζῶντα θάνατον*. More pertinent is the contrast with 'terra viventium', and see on 'terram sapiebant' below. Petilian uses the phrase in an intriguingly apt way (*c. litt. Pet.* 2. 7. 14, 'qui sic vivit ut reus sit, vita mortua cruciatur'); cf. *s. Lambot* 6, 'quo ipsa vita mortua est', but there *vita* = Christ and the context is different.

hinc insana facientes, inde vana pollicentes: Cf. 5. 10. 18, 'falsis illis atque fallentibus sanctis [manichaeorum]'; the 'homines' just above are thus his Manichean friends.

gressus: See on 5. 7. 13. Ps. 39: 3, 'et statuit supra petram pedes meos et direxit gressus meos'; *en. Ps.* 39. 3, 'opus est tamen adhuc ut ambulemus, ut ad aliquid perveniamus.' (There he goes on to quote Phil. 3: 13–14—see on 5. 6. 10, 'extento'.)

terram sapiebant: Phil. 3: 18–19, 'inimicos crucis Christi, (19) quorum finis interitus, quorum deus venter, et gloria in confusione eorum qui terrena sapiunt.' Most apt of Manichees, for whom the crucifixion was a phantasm (5. 9. 16, 'in cruce phantasmatis'), and for whom the bellies of the elect were the manufactory of God; cf. above, 'terra', and the Ps. text.

falsam felicitatem: For A.'s suspicion of all 'felicitas', see above on 5. 4. 7.

5. 8. 15

A. came to Carthage in 3. 1. 1, there to find concupiscent love; here he leaves, not with a purpose of abandoning concupiscent love, but with that unintended effect. Monnica, like Dido, becomes herself the derelict lover, victim of her own inordinate attachment, but the physical circumstances reveal a decisive change—Aeneas leaves Dido to die on her own pyre, A. leaves Monnica to pray at a tomb where the power of redemption is specially present. A.'s departure, like that of Aeneas, is guided by divine authority, but here the guidance is unheard and invisible. C. Bennett, *REAug* 34 (1988), 61: 'Any reflections on this scene should be tempered by recognizing how deliberate a literary reference this is—and how deliberate a literary re-enactment it was at the time. . . . Augustine's allusions to the *Aeneid*, here as elsewhere, reflect not just literary modelling, but also a life shaped by imitation of the epic.'

For the Dido parallel, see J. J. O'Meara, *Mélanges . . . Christine Mohrmann* (Utrecht–Antwerp, 1963), 252–61, following C. Kligerman, *Jour. Amer. Psychoan. Assoc.* 5 (1957), 469–84; but M. here also resembles the mother of Euryalus (whose own words are echoed in characterizing M. at 6. 1. 1) at *Aen.* 9. 287–9:

> hanc [sc. matrem] ego nunc ignaram huius quodcumque pericli
> inque salutatam linquo (nox et tua testis
> dextera), quod nequeam lacrimas perferre parentis.

A.'s treatment of M. here is harsh. He does not stint the pathos in depicting her anguish at seeing her son abandon her (note esp. the two terse and evocative phrases bracketing the narrative: 'et illi matri' and 'et ego Romam'), but his sympathy stops there—her anguish arose from an excess of affection. For corroborating contrast, cf. the reaction of Romanianus to the same flight: *c. acad.* 2. 2. 3, 'tu etiam, cum te absente atque ignorante navigassem, nihil suscensens quod non tecum communicassem ut solerem atque aliud quidvis quam contumaciam suspicans, mansisti inconcussus in amicitia . . .'.

In one other text, A. is hard on a mother whose natural affection impedes the religious progress of her son. In the strenuousness of his attack on worldly ties, it is impossible to avoid seeing a hostility to *some* of what M. stood for. He wrote to Laetus, who had written for advice and encouragement on setting out on the religious life. A.'s principal scriptural support is Lk. 14: 26–33, 'si quis venit ad me, et non odit patrem suum, et matrem, et uxorem, et filios, . . . non potest meus esse discipulus.' *ep.* 243. 3–10,[4] 'quo enim tibi nunc quaedam mulier mater est, hoc ipso utique non est et mihi. quapropter hoc temporale ac transitorium est, sicut transisse iam vides quod te concepit, quod gestavit utero, quod peperit, quod lacte nutrivit. quod autem soror in Christo est, et tibi est et mihi . . . (4) potes hoc facillime in ipsa tua matre cognoscere. nam unde te nunc inretitum involvit et ab instituto cursu retardatum reflectit et curvat, nisi ex quo tua propria mater est? . . . (7) sed quid dicit aut quid allegat? forte decem illos menses quibus viscera eius onerasti, et dolores parturitionis ac labores educationis? hoc, hoc interfice verbo salutari; hoc perde matris, ut in vitam aeternam invenias eam . . . *carnalis enim affectus est iste et adhuc veterem hominem sonat*. . . . (10) quid interest utrum in uxore an in matre, dum tamen *Eva* in qualibet muliere caveatur? nam ista umbra pietatis *de foliis illius arboris* venit, quibus se primum parentes nostri in illa damnabili nuditate texerunt.' (See on 2. 6. 14 for the connection between the tree of paradise and the concupiscence of the flesh.)

tu sciebas: See on 1. 5. 6.

finxi me amicum nolle deserere: A. is never alone. If there really was a friend who went with him to Italy, we do not know who it was (Alypius, Nebridius, and Romanianus only joined him later); given that A. found

[4] Date uncertain: Goldbacher leaves this with the whole group of undated letters at post-395; Mandouze, *Pros. chr.* s.v. Laetus 1, inclines to pre-395 for this letter but cannot rule out a later date.

shelter with the Manichees in Rome, his travelling companion was probably of that sect.

et evasi: Duplex in sense. The first meaning is mundane—'I escaped from her'; but the clause that follows gives it weight—'I survived the lie and its consequences.' A. was always a reluctant traveller, only twice on ship in his life (this journey and the return five years later): Perler, 57–81, catalogues A.'s often conventional but often repeated descriptions of the perils and fears of the sea; and see on 5. 9. 16 below.

et (*hoc*) GO²S Knöll Skut. : om. CDO¹ Ver.

aquis maris: Cf. 13. 7. 8, 'aquas . . . affectus sunt, amores sunt'.

sordibus: Also of the stain to be removed by baptism at 1. 11. 17.

aquam gratiae tuae: cf. 9. 2. 4, 'aqua sancta', 9. 13. 35, 'aquam salutis'.

abluto: Also of baptism at 1. 11. 17; cf. 6. 16. 26, 'ablutura'.

cotidie: First here in *conf.*; 'day by day' is a mark of the separation of God and humankind, even in the midst of confession and redemption: it marks challenge, struggle, and longing: 6. 1. 1, 6. 11. 20, 6. 13. 23, 8. 6. 13 ('et cotidie suspirabam tibi'), 10. 3. 4, 10. 31. 44 ('his temptationibus cotidie conor resistere'), 10. 31. 47, 10. 35. 57, 10. 37. 60, 13. 13. 14.

memoria beati Cypriani: There were two *memoriae* at Carthage in honour of Cyprian, one at the site of his death (the *mensa Cypriani*), one where he was buried (the *Mappalia*); A. preached at both at one time or another. Because both were outside the city walls (on the evidence of Vict. Vit. *hist. pers.* 1. 5. 16), the present passage may refer to a third shrine, within the city itself. There is no other evidence, and no other obvious pretext, for such a place to have existed. See Perler, 420–1, with references; cf. also van der Meer, 477, and G. Charles-Picard, *La Carthage de saint Augustin* (Paris, 1965), 190–5.

memoria: *cura mort.* 4. 6, 'sed non ob aliud vel "memoriae" vel "monumenta" dicuntur ea quae insignite fiunt sepulchra mortuorum, nisi quia eos qui viventium oculis morte subtracti sunt, ne oblivione etiam cordibus subtrahantur, in memoriam revocant.'

cardinem: Cf. 4. 15. 24, 'tantae rei cardinem'.

non curasti: Prayer, even Monnica's prayer, can fall short of God and remain unanswered. This incontrovertible fact is perhaps the source of the confidence with which A. rebukes his mother's carnal affection here; he thinks he sees the good reason God had to ignore her prayer at this moment.

cupiditatibus meis: A. is still the man of his concupiscences (this the most explicit statement of this theme in Bk. 5), but one whose concupiscences were (memory clearly sees) in the service of a higher purpose.

raperes: See on 4. 12. 18.

more matrum: See on 6. 10. 17 (of Nebridius), 'relicta domo et *non secutura matre* . . . Mediolanum venerat'.

reliquiarium Evae: Cf. Gn. 3: 16, 'et mulieri dixit, "multiplicans multiplicabo tristitias tuas et gemitum tuum et in tristitiis paries filios".' *reliquiarium*: Gn.

45: 7, quoted by A. *qu. hept.* 1. 148, 'misit enim me deus ante vos remanere vestrum reliquias super terram et enutrire vestrum reliquiarium magnum.'

5. 9. 16

The illness follows hard on the danger of loss on the ocean-crossing (cf. 'evasi' in 5. 8. 15): the parting with M. is fraught with danger, for it could easily have brought all her prayers to nought.

The Eternal City made little impression that we can see. Though *civ.* is a meditation on the fate of the city of Rome filled with descriptions of Roman civil religion, there are few, if any, traces of his time at Rome to be found there. His reports of observed 'paganism' go back to Carthage, not Rome.[5] One recollection surfaces at *c. Adim.* 24. 1 (on Mt. 6: 34, 'nolite cogitare de crastino'), 'nam si hoc ideo dictum est ut non servetur panis in crastinum, magis hoc implent vagi Romanorum quos passivos appellant, qui annona cotidiana satiato ventre aut donant statim quod restat aut proiciunt, quam vel domini discipuli, qui etiam cum ipso domino caeli et terrae in terra ambulantes loculos habebant.' (Always a nervous eye for the underclass, arising from his Manichean experience: see on 6. 6. 9.)

flagello: Both A. and M. were punished for their part in the parting: cf. 5. 8. 15, 'iusto dolorum flagello vapularet [mater]'; elsewhere in *conf.*, *flagella* only of lashes in the name of divine justice and mercy: 2. 2. 4, 3. 3. 5, 8. 7. 18, 8. 11. 25, 9. 4. 12—much more consistently than in scriptural texts, where only vague parallels appear, e.g. Ps. 31: 10, 'multa flagella peccatoris', and the most explicit text is hardly well-known: Jdg. 8: 27, 'sed reputantes peccatis nostris haec ipsa minora esse supplicia, flagella domini quasi servi qui corripimur ad emendationem non ad perditionem nostram evenisse credamus.'

ibam iam ad inferos: Job 7: 9 (VL), 'si enim homo descenderit ad inferos, ultra non ascendet'; Ps. 138: 8, 'si descendero in infernum, ades' (see on 1. 2. 2).

et in te et in me et in alios: Parallels the two great commandments: Love *God*, Love thy *neighbour* as *thyself* (see on 3. 8. 15).

originalis peccati: See on 1. 7. 11, 'nemo mundus'.

quo omnes in Adam morimur: 1 Cor. 15: 22, 'sicut in Adam omnes moriuntur, sic et in Christo omnes vivificabuntur'; cf. Rom. 5: 12–14, 'sicut per unum hominem peccatum in hunc mundum intravit, et per peccatum mors, et ita in omnes homines mors pertransiit, in quo omnes peccaverunt. . . . (14) sed regnavit mors ab Adam.' On the interpretation of Rom. 5: 12 and the influence of Ambrosiaster on the reading 'in quo omnes peccaverunt' in A., see TeSelle, 158; cf. 9. 13. 34, 'quae in Adam moritur'.

[5] *ep.* 29. 4 reports ecclesiastical customs at St Peter's basilica, but does not demonstrably come from A.'s own experience.

donaveras: G–M: 'pardoned', cf. *s.* 216. 5. 5, 'simul et culpae esse voluit donator et poenae'.

solverat . . . inimicitias: Eph. 2: 14–16, 'ipse enim est pax nostra . . . medium parietem maceriae solvens, inimicitias in carne sua; . . . (16) et reconciliet ambos in uno corpore deo per crucem interficiens inimicitias in semetipso'; cf. Col. 2: 13 quoted on 5. 9. 17.

cruce phantasmatis: *s.* 116. 4. 4, 'qui nec crucifixum credis quia nec natum credis'; in *c. Faust.* 14 A. addresses Manichean criticism based on the verse, 'maledictus omnis qui pendet in ligno' (Deut. 21: 23, quoted at Gal. 3: 13).

febribus: Vega suggests this was malaria; note that M.'s final illness was also marked by fever (9. 11. 27, 'decubuit febribus') and was also contracted on taking up residence in the vicinity of Rome (as noted by D. Burt at the 1987 Oxford Congress). On A.'s health, see B. Legewie, *MA* 2. 5–21.

irem . . . abirem: Cf. 5. 8. 15, 'sed quare hinc abirem et illuc irem' (of the move to Rome).

nisi in ignem atque tormenta digna factis meis: Mt. 25: 41–2, 'discedite a me maledicti in ignem aeternum qui paratus est diabolo et angelis. (42) esurivi enim, et non dedistis mihi manducare, sitivi, et non dedistis mihi potum.' The Manichees refused to feed hungry beggars (see on 6. 6. 9), hence fire is apt punishment in store for A., almost dying as a Manichee.

in veritate [2] **ordinis tui** [3].

ubique praesens: See on 1. 3. 3. R. J. O'Connell, *REAug* 9 (1963), 2 n. 13 stresses the way the phrase here and *ubique totus* at 6. 3. 4 bracket the aporia that surrounded A. on his way to Milan.

sacrilego: See on 3. 3. 5.

sicut iam . . . confessus: See 1. 11. 17.

sed in dedecus meum creveram: Contrast Lk. 2: 40 ff., 'puer autem crescebat . . . (52) et Iesus autem proficiebat sapientia, et aetate, et gratia apud deum et homines.'

consilia medicinae tuae demens inridebam: At 4. 4. 8 (note there 'inridere' and 'dementiae meae'), when the unnamed friend was baptized.

bis mori: Apoc. 2: 11, 'non laedetur a morte secunda' (cf. Apoc. 20: 6, 20: 14, 21: 8); *civ.* 13. 2, 'mors igitur animae fit, cum eam deserit deus, sicut corporis, cum id deserit anima. . . . huius modi autem totius hominis mortem . . . quam secundam mortem divinorum eloquiorum appellat auctoritas.'

parturiebat spiritu: Cf. Gal. 4: 19, 'filioli mei, quos iterum parturio donec formetur Christus in vobis'.

5. 9. 17

viscera: Col. 3: 12, 'induite vos ergo . . . viscera misericordiae'; cf. Lk. 1: 78, Phil. 2: 1.

preces . . . sine intermissione: 1 Thess. 5: 17, 'sine intermissione orantes'; *en.*

Ps. 37. 14, 'ipsum enim desiderium tuum, oratio tua est; et si continuum desiderium, continua oratio. . . . est alia interior sine intermissione oratio, quae est desiderium.' At 1 Tim. 5: 5, the pious widow 'sperat in deum et instat obsecrationibus et orationibus nocte ac die'; cf. 3. 11. 20 for M.'s constant prayer.

deus misericordiarum: 2 Cor. 1: 3, 'pater misericordiarum et deus totius consolationis'.

cor contritum et humilatum: Ps. 50: 19, 'sacrificium deo spiritus contribulatus, cor contritum et humilatum deus non spernit'; cf. Ps. 146: 3, 'qui sanat contritos corde', and Ps. 33: 19, 'prope est dominus his qui obtriverunt cor' (echoed at 5. 3. 3).

viduae . . . elemosynas: Lk. 21: 2 (cf. Mk. 12: 41), 'vidit autem et quandam viduam pauperculam mittentem aera minuta duo et dixit, "vere dico vobis quia vidua haec pauper plus quam omnes misit".'

oblationem ad altare tuum: Eucharist: *ep.* 111. 8, 'nec istae possunt vel ferre oblationem ad altare dei vel invenire ibi sacerdotem per quem offerant deo.'

bis die, mane et vespere: In A.'s own experience, daily eucharist was the rule (*ep.* 29. 11, 'acta sunt vespertina quae cotidie solent'; at *ep.* 54. 3. 4 A. defends at length the practice of daily communion; cf. *civ.* 10. 20), but the practice was not universal (*Io. ev. tr.* 26. 15; the eastern church specified for noncompliance: *s. dom. m.* 2. 7. 26), though A. himself takes the 'daily bread' of the Lord's prayer (at *s. dom. m.* 2. 7. 27) to be spiritual.

ut te audiret . . . et tu illam: Divine speech preceeds human speech: hearing God, returning what is heard in prayer.

lacrimas: 3. 12. 21, 'filius istarum lacrimarum'.

immo vero aderas [1] **et exaudiebas** [2] **et faciebas ordine** [3].

praedestinaveras: The term elsewhere in *conf.* only twice in two lines at 13. 34. 49. Here, predestination appears not as a doctrine but a fact of experience, comparable to 'originale peccatum' in 5. 9. 16, and the word is comparably rare in works before the Pelagian controversy.

visionibus: Dulaey, 113: 'Quand Augustin emploie *visio*, ce qui est rare, c'est toujours pour des rêves inspirés [as here and *nat. et or. an.* 4. 18. 26 (visions of Perpetua)], ou du moins sortant de l'ordinaire [*cura mort.* 12. 15, 17. 21, *Gn. litt.* 12. 17. 37–12. 18. 39].' At 6. 13. 23, Monnica claims to be able to distinguish between divine revelations seen in dreams and the fantasies of her own spirit; see also 3. 11. 19–12. 21. Courcelle, who thinks the biographical sketches of Alypius and Monnica in Bks. 6 and 9 were originally separate essays, says at *Recherches*, 40 n. 1: 'Il paraît probable que, dans un opuscule biographique spécial, Augustin s'était étendu plus longuement sur les visions de Monique.'

chirographa tua: Col. 2: 13, 'donans vobis omnia delicta [cf. 5. 9. 16], delens quod adversum nos erat chirografum decretis'; always taken as reference to the cross.

quoniam in saecula misericordia tua: Ps. 117: 1, 'quoniam in saeculum misericordia eius'; Ps. 137: 8, 'domine, misericordia tua in aeternum'; also echoed at 11. 1. 1.

omnia debita dimittis: Mt. 18: 32, 'tunc vocavit illum dominus suus et ait illi, "serve nequam omne debitum dimisi tibi quoniam rogasti me"'; cf. Mt. 6: 12, 'dimitte nobis debita nostra'. The echo suggests that M.'s prayers conform to the model provided by Christ.

5. 10. 18

salvum fecisti filium ancillae tuae: Ps. 85: 16, 'et salvum fac filium ancillae tuae'; cf. Ps. 115: 16 (7), 'ego servus tuus, et filius ancillae tuae' (echoes at 2. 3. 7 and 9. 1. 1).

falsis illis atque fallentibus sanctis: See on 4. 1. 1, 'falsi atque fallentes', 7. 2. 3, 'deceptos deceptores'.

auditoribus: A.'s regular term for the Manichean 'hearers', but in the Coptic texts in C. R. C. Allberry, *A Manichaean Psalm-Book* (Stuttgart, 1938), the regular word is the Gk. κατηχούμενος, which *auditor* more or less tries to translate; A. only uses the Gk. word for the orthodox Christian initiatory status in which he was brought up.

is in cuius domo: H. Chirat, in a note at Pellegrino, *Les Confessions*, 133 n. 5, thinks this must mean the same Constantius who gave shelter to numerous Manichees and later himself became catholic (*c. Faust.* 5. 5, 'Constantius, modo iam frater noster catholicus christianus, qui multos vestrum Romae in domum suam congregaverat propter implenda praecepta Manichaei'), and the possibility cannot be ruled out.

quos electos vocant: Only here does he claim to have associated directly with the Manichean elect (at 4. 1. 1 he implies that he co-operated in feeding them, but that need not imply social intercourse); if only in Rome did he first associate with them, it is a sign of how involved in the movement he still was, whatever his doubts. At *c. litt. Pet.* 3. 17. 20 he denies having been a 'presbyterum ... manichaeorum' (see on 5. 7. 13 for whether he ever had such aspirations); cf. *c. Fort.* 3, 'nostis autem me non electum vestrum sed auditorem fuisse.' On the distinction, *ep.* 236. 2, 'auditores autem qui appellantur apud eos, et carnibus vescuntur et agros colunt et, si voluerint, uxores habent, quorum nihil faciunt qui electi vocantur. sed ipsi auditores ante electos genua figunt, ut eis manus supplicibus imponatur non a solis presbyteris vel episcopis aut diaconis eorum sed a quibuslibet electis.'

adhuc enim mihi videbatur: See on 7. 3. 5, 'audiebam'.

aliam ... naturam: The attitude A. here characterizes is one he may not have espoused in so many words during his time with the Manichees; Courcelle, *Recherches*, 100–1, quotes Amb. *exam.* 1. 8. 31, 'quid alienam naturam accersis ad excusationem tuorum lapsuum? ... quid naturam accusas, o homo?' The characterization assimilates this attitude to that which he later attributes to devotees of astrology (see on 4. 3. 4), and in at least one passage the two come together (*en. Ps.* 140. 9); the link between the astrologers and the Manichees for A. included their common *concupiscentia oculorum*.

ut sanares animam meam: Ps. 40: 5, 'ego dixi, domine, miserere mei; sana animam meam, quia peccavi tibi.'

adversus me . . . diviserat: Mt. 12: 26, 'et si satanas satanam eicit, adversus se divisus est.'

omnipotens: Biblical, frequent, from Gn. to Apoc.

te . . . superari: i.e. the fragment of the divine imprisoned in A.

nondum ergo . . . electis eorum: Ps. 140: 3–4, 'pone, domine, custodiam ori meo, et ostium continentiae circa labia mea, (4) ut non declines cor meum in verba maligna ad excusandas excusationes in peccatis cum hominibus operantibus iniquitatem et non combinabis cum electis eorum'; *en. Ps.* 140. 6, 'ostium et aperitur et clauditur; ergo si ostium est, et aperiatur, et claudatur; aperiatur ad confessionem peccati, claudatur ad excusationem peccati. ita enim erit ostium continentiae, non ruinae.' It seems certain that the verses are apt in part for the occurrence of 'electis', but Mandouze, 105 n. 2, disagrees.

sed tamen iam desperans: A foreshadowing, invisible to him at the time of the events described, of the course of *conf.* On matters of doctrine, he was making progress, if only by going in reverse; but it was the moral bind that kept him tied to Manicheism. Similarly, in Bk. 7 he would solve almost all his doctrinal problems (except the Christological ones), but it would remain for Bk. 8 and the moral renovation that came with the last doctrinal correction for him to be freed.

proficere: 5. 7. 13, 'proficere in illa secta'.

decreveram: See on 5. 6. 13, cf. 5. 14. 25, ending the book. Courcelle, *Recherches*, 77: 'Mais ses doutes, croit-il, ne sont que provisoires.'

5. 10. 19

academicos: A. does not say that he gave his head to the Academics, only that he began to think they *might* be right. For their position, see chiefly *c. acad.* 3. 17. 37–19. 42: at 3. 17. 38, 'quare cum in falsas opiniones ruere turba sit pronior et consuetudine corporum omnia esse corporea facillime sed noxie credatur, instituit vir acutissimus atque humanissimus dedocere potius quos patiebatur male doctos quam docere quos dociles non arbitrabatur. inde illa omnia nata sunt quae novae academiae tribuuntur.' Another version of the doubts that brought him to the Academics is given at *util. cred.* 8. 20 (quoted in Introduction); see also his self-justification at *c. litt. Pet.* 3. 21. 24–22. 26 and the characterization of the Academics at *ench.* 7. 20. A less reverent view was that of Monnica, at Cassiciacum: when she asked who the Academics were and what they were about, A. explained briefly and she responded (*beata v.* 2. 16): '"isti homines", inquit, "caducarii sunt"—quo nomine vulgo apud nos vocantur quos comitialis morbus subvertit—et simul surrexit ut abiret.'

A suggestive text that would have spoken loudly to A. in 384: Cic. *Lucullus* 3. 8, 'nam ceteri primum ante tenentur adstricti quam quid esset optimum iudicare potuerunt; deinde infirmissimo tempore aetatis aut obsecuti amico cuidam aut una alicuius quem primum audierunt oratione capti de rebus incognitis iudicant et ad quamcumque sunt disciplinam quasi tempestate delati ad eam tamquam ad saxum adhaerescunt.'

See R. Holte, *Béatitude et Sagesse* (Paris, 1962), 42–4; J. J. O'Meara, in the introduction to his translation of *c. acad.* (*Ancient Christian Writers*, volume 12 (Westminster, Md., 1950)); J. Mourant, *RA* 4 (1966), 67–96; and M. Baltes, *Aug.-Lex.* 1. 40–5. The notion that the Academics had a secret inner doctrine for adepts obviously appealed to A. at the time, and he had (unreliable) testimony to trust: Cic. *Lucullus* frg. 21M ($<$ *c. acad.* 3. 20. 43): 'quisquis autem putat hoc sensisse academicos, ipsum Ciceronem audiat. ait enim illis "morem fuisse occultandi sententiam suam nec eam cuiquam nisi qui secum ad senectutem usque vixisset aperire consuesse".' If we accept that the Manichees had won A.'s allegiance by promising rational argument, it is easy to see how Cicero could trump that card: *Lucullus* 18. 60, '"non solemus", inquit [Arcesilas], "ostendere." quae sunt tandem ista mysteria? aut cur celatis quasi turpe aliquid sententiam vestram? "ut qui audient", inquit, "*ratione* potius quam *auctoritate* ducantur".'

comprehendi: The use of καταλάμβανειν = *comprehendere* as a technical term was a topic of Cicero's *academica*: *Lucullus* 6. 18, 'ex quo efficitur nihil posse comprehendi'.

dissimulavi: 'neglect' with infinitive, cf. 1 Kgs. 23: 13, 'dissimulavit exire', and other passages adduced by G–M ad loc. Here A. tries to hold on to the sect while distancing himself from the doctrine; by the end of Bk. 7, he will himself be attempting to stay aloof from the catholic community while acquiring its doctrine—a tendency against which Simplicianus is shown reacting at 8. 2. 4.

familiaritas: *duab. an.* 9. 11, 'sed me duo quaedam maxime, quae incautam illam aetatem facile capiunt, per admirabiles attrivere circuitus: quorum est unum familiaritas nescio quomodo repens quadam imagine bonitatis tamquam sinuosum aliquod vinculum multipliciter collo involutum.'

occultat: 4. 1. 1, 'occulte'; hence the barb in *ep. Sec.* 2, 'recessionem tuam . . . quae per timorem facta est'—implying that at Rome A., in search of a great career, found the Manichees persecuted and removed himself from them out of self-interest; A.'s protest is at *c. Sec.* 1.

domine caeli et terrae: Mt. 11: 25, 'confiteor tibi, pater, domine caeli et terrae, quia abscondisti haec a sapientibus et revelasti ea parvulis.'

creator omnium visibilium et invisibilium: Nicene Creed (apud Denzinger, *Enchiridion Symbolorum* (36th edn., Freiburg, 1976), 125, in the version of Hil. Pict. *de synodis* 84 (*PL* 10. 536A)): 'patrem omnipotentem, omnium visibilium et invisibilium factorem'.

mihi turpe videbatur: A doubt also at 3. 7. 12, 'utrum forma corporea deus finiretur', and raised again with a view to a solution at 7. 1. 1, 'figura corporis humani'.

maxima et prope sola causa: If we take this assertion seriously, then it is the reading of the *platonicorum libri* at 7. 9. 13 that seems to solve the problem; cf. the list of 'causes' that made A. a Manichee at 3. 7. 12 (and see notes there). R. J. O'Connell, *REAug* 9 (1963), 1–39, plausibly suggests the influence of Plot. 6. 4–5 in shaping A.'s solution (but du Roy, 70 n. 1, is unsure whether A. had read them before Cassiciacum). The nature of evil is A.'s next concern (5. 10. 20) and at 5. 11. 21 the reliability of scripture is adduced again as well. In introducing the appeal of the Academics, he thus makes clear that none of the specific doubts and ignorances that led him to the Manichees had yet found a lasting solution.

5. 10. 20

substantiam: The word is not properly predicated either of God (*trin.* 7. 4. 9–5. 10) or of evil (see on 3. 7. 12), but at this stage A. applies it to both.

crassam ... subtilem: The adjectives are regularly opposed (*SLA* s.v. *subtilis*).

pietas: A non-Manichean quality (5. 3. 5, 'non pie quaerunt').

angustius: *c. ep. fund.* 21. 23, 'quaero utrum et ipsa terra tenebrarum unum latus habuerit et cetera infinita, sicut terra lucis. non ita credunt; timent enim ne deo videatur aequalis. dicunt ergo illam per profundum immensam et per longum: sursum versus autem supra illam spatia infinitae inanitatis affirmant. et ne ipsa vel simplum, terra autem luminis duplum tenere videatur, *angustant* eam a duobus etiam lateribus. tamquam si unus panis ... in quadras quattuor decussatim formetur, in quibus tres sint candidae, una nigra: modo de tribus candidis tolle distinctionem, et fac illas et sursum versus et deorsum versus, et undique retro infinitas: sic ab eis esse creditur terra lucis. illam vero nigram quadram fac deorsum versus et retro infinitam, supra se autem immensam inanitatem habere: si opinantur terram tenebrarum. sed haec nimis attentis et studiose inquirentibus quasi secreta demonstrant.' (Just the sort of doctrine, in other words, for one like A., pressing for answers.)

conaretur ... repercutiebar: Echoes again (cf. 4. 15. 26, 'conabar ad te et repellebar abs te') the vocabulary of the mystic ascent; here again a failure, here again in a context that emphasizes the persistence of the dilemmas first sketched at 3. 7. 12.

catholicam: First here in *conf.*, frequent from here onwards.

miserationes: The subject of the verb: God gives the act of confessing; for the significance of the inversion, see on 7. 6. 8, and cf. 4. 16. 31 and 5. 1. 1. Ps. 106: 8, 'confiteantur domino miserationes eius', Ps. 118. 77, 'veniant mihi miserationes tuae, et vivam.'

quod mihi nescienti: The parenthesis states just the position demolished in 7. 1. 1–3. 4.

nisi eam . . . diffunderetur: M. Colish, *The Stoic Tradition* (Leiden, 1985), 2. 147, 'This concept of mind and of the divine nature . . . does indeed have a strongly Stoic filiation, reflecting the Stoic notion that everything which acts is a body. Earlier this Stoic principle had been absorbed and professed positively by Tertullian against Praxeas.'

ipsumque salvatorem: The problem of the nature of evil is closely tied to that of Christology. Unravelling the first in early Bk. 7 will lead to the second in the end of Bk. 7 and in Bk. 8. On Manichean Christology see *BA* 13. 674–6. Their Christ is a being of the spirit, not the flesh: he calls souls back to their divine nature and awakens them to repentance. The historic Jesus is the *Iesus patibilis*, conceived and begotten by earth fertilized by the Holy Spirit in the ambient air. At *c. Faust.* 20. 11 A. accused the Manichees of having three Christs: 'aliusne est quem de spiritu sancto concipiens terra patibilem gignit, omni non solum suspensus ex ligno, sed etiam iacens in herba, et alius ille quem Iudaei crucifixerunt sub Pontio Pilato, et tertius ille per solem lunamque distentus?'

imaginari: i.e. create a *phantasma*—A. relishes the irony that the Christ of the Manichees really *was* a *phantasma*.

talem itaque naturam: For Manichean denial of the Virgin birth, see *c. ep. pel.* 2. 2. 3 and Evodius *de fide c. man.* 22 (*PL* 42. 1145).

carni concerneretur: 'without becoming intermingled with the flesh'; The verb is rare but attested elsewhere (*TLL* 4. 24: from Silius, Ambrose, and Maximus of Turin); A. will have liked the alliteration: could he have imagined he saw an etymology? *c. Faust.* 22. 31, 'ut cum in ceteris actibus ad humanam pacem pertinentibus mulier viro debeat servitutem, huius unius rei, qua sexus utriusque carnali sorte discernitur et carnali commixtione concernitur, similem in se habeant potestatem vir in uxorem ut uxor in virum.' (Also contrasted to *discernere* at *en. Ps.* 49. 11.) Equally intriguing is the link through Gk. συγκίρνημι (a rare and late equivalent of συγκεράννυμι: see J. le Clercq at *PL* 47. 207) to what may be at the back of A.'s mind here: Porph. *sent.* 4, τὰ καθ' αὑτὰ ἀσώματα . . . οὐ . . . συγκίρναται τοῖς σώμασιν.

in carne natum CDGO Maur. Skut. Ver. : *incarnatum* S Knöll.

ridebunt: On mockers, see on 1. 6. 7, 'inrisor'.

confessiones meas: See on 5. 1. 1.

5. 11. 21

defendi posse: The same words at 5. 14. 24, at Milan, when scripture does begin to seem defensible.

cum aliquo . . . doctissimo: He wants a catholic Faustus (5. 6. 10, '*conlatoque* conloquio'); see on 5. 13. 23, 'studiose'.

Elpidii: Another named figure (see on 4. 4. 7), from before the time under review (383/4 at Rome), but as with all named figures, one who contributed to A.'s betterment. Courcelle, *Recherches*, 65 n. 5, adduces *mor.* 2. 8. 11, 'nam etiam de quodam dicente nullam substantiam malum esse, unus de primatibus huius haeresis, quem familiarius crebriusque audiebamus, dicebat: "vellem scorpionem in manu hominis ponere ac videre utrum non subtraheret manum; quod si faceret, non verbis sed re ipsa convinceretur aliquam substantiam malum esse, quandoquidem illud animal substantiam non negaret." et dicebat hoc *non coram illo*, sed cum ad eum nos commoti referremus quod ille dixisset: respondebat ergo, ut dixi, pueriliter pueris.' Courcelle thinks the *quidam* may be Elpidius, but doubts Alfaric's identification (Alfaric, 254 n. 3) of the Manichean doctor with Faustus: Courcelle puts the episode back to an earlier period. That text demonstrates that on two of the three main questions that made A. a Manichee (see 3. 7. 12), he had access while in Africa to catholic arguments essentially identical to those that would eventually convince him—and they did not. What is to be observed in Milan is not what he heard, but why it moved him.

scripturas . . . falsatas: Cf. e.g. *c. Faust.* 18. 7, 10. 3.

nulla proferrent: *mor.* 1. 29. 61, 'proferendus est namque tibi alius codex eadem continens . . . ubi sola desint quae hic immissa esse criminaris.'

moles illae: The 'masses' envisioned at 5. 10. 20.

5. 12. 22

ergo: For *ambitio saeculi* as relief from other troubles, see on 4. 7. 12.

eversiones: See on 3. 3. 6, cf. 5. 8. 14.

ne mercedem magistro reddant: The witness on whom A. depends here might not be entirely reliable. Elsewhere, A. implicitly accepts the idea that students who had not yet paid their fees might well abandon a teacher if they found him less than satisfactorily competent: *util. cred.* 6. 13, 'itaque in quaestiuncula magistro deficienti et quid respondeat non habenti suscensemus potius quam illum mutum vitio Maronis putamus. iam si ad defensionem suam peccatum tanti auctoris adserere voluerit, vix apud eum discipuli vel datis mercedibus remanebunt.' For comparison with conditions known to Libanius, see A. Müller, *Philologus*, 69 (1910), 314.

non perfecto odio: Ps. 138: 22, 'perfecto odio oderam illos, inimici facti sunt mihi'; *en. Ps.* 138. 28, 'oderam in eis iniquitates eorum, diligebam conditionem tuam. hoc est perfecto odio odisse, ut nec propter vitia homines oderis, nec vitia propter homines diligas. . . . ut hoc in eis oderit quod iniqui sunt, hoc diligat quod homines sunt.' A. does not say that he actually suffered this fraud, but he was warned of it. The pertinence is that his best motive for the move to Rome (5. 8. 14) was that he wanted more peaceable students. Mandouze, 108 n. 5, reproaches A.'s self-reproach: 'Le reproche qu'Augustin se fait à lui-même est ici exactement opposé à celui qu'il formulait contre soi au sujet du vol des poires. En effet, tandis qu'il regrettait la "gratuité" de l'acte de maraudage, il s'accuse ici d'avoir été "intéressé". Nouvelle preuve que l'évêque d'Hippone ne manque jamais dans les *Confessions* de présenter l'interprétation la plus défavorable pour tous ses actes de la période "païenne".' That reading is tendentious, but the parallel with the pear-tree is apt. A. has moved to the other side, and *adolescentes* and their pranks seem to him as victim insufferable, for reasons equally concupiscent and inadequate. Both sinner and victim can be a long way from God.

fornicantur abs te: Ps. 72: 27, 'perdidisti omnem qui fornicatur abs te'; see on 1. 13. 21 and cf. 4. 2. 3. *en. Ps.* 140. 18, 'profecisti, iam unum deum colis: optime; non fornicaris ab eo ad idola, ad *mathematicos,* ad sortilegos, ad aruspices, ad augures, ad maleficos; ista enim fornicatio est a domino deo.'

volatica: 7. 6. 8, 'arborum volatica folia'.

lucrum luteum: Tit. 1: 7, 'turpis lucri cupidum'; 1 Pet. 5: 2, 'neque turpis lucri gratia'.

redeunti . . . humanae: Another echo of the prodigal son: see on 1. 18. 28.

ei (*vero*): sc. *doctrinae ipsae*.

veritatem [2] **et ubertatem certi boni** [1] **et pacem castissimam** [3].

5. 13. 23

The present paragraph is echoed and conflated with other passages in Possidius, *v. Aug.* 1. 3–6, 'huius interea verbi dei praedicatoris frequentissimis in ecclesia *disputationibus* adstans *in populo intendebat suspensus* atque adfixus. (4) verum aliquando manichaeorum apud Carthaginem adulescens fuerat errore seductus, et ideo ceteris suspensior aderat, ne quid vel pro ipsa vel contra ipsam haeresim diceretur. (5) et provenit dei liberatoris clementia sui sacerdotis cor pertractantis, ut contra illum errorem incidentes legis solverentur quaestiones, atque ita edoctus sensim atque paulatim haeresis illa miseratione divina eius ex animo pulsa est: protinusque in fide catholica confirmatus, proficiendi in religione eidem amoris ardor innatus est, quo propinquantibus diebus sanctis paschae salutis aquam perciperet. (6) et factum est divina praestante opitulatione, ut per illum tantum ac talem antistitem Ambrosium et doctrinam salutarem ecclesiae catholicae et divina perciperet sacramenta.' For an earlier version of this period from A.'s hand, see *util. cred.* 8. 20, quoted in the Introduction.

We can surmise what A. was looking for at this stage from *quant. an.* 33. 70 (post-baptism, but marked by the atmosphere of the earlier search), 'o utinam doctissimum aliquem neque id tantum, sed etiam eloquentissimum et omnino sapientissimum perfectumque hominem de hoc ambo interrogare possemus!' The conjunction of learning, wisdom, *and* eloquence is noteworthy and characteristic; as here, Ambrose's greater learning marks him off from Faustus, with whom he is otherwise implicitly compared throughout, e.g. 'et delectabar suavitate sermonis, quamquam eruditioris'.

Itaque: This speaks of a time when he understood how things stood for teachers at Rome, but before he had any particular success or personal experience of the problems.

Mediolanio: Ausonius, *ordo urbium nobilium*, 7, 'Mediolanum' (Aus. was in retirement in Gaul by the time A. reached Milan):

> et Mediolani mira omnia, copia rerum
> innumerae cultaeque domus, facunda virorum
> ingenia et mores laeti; tum duplice muro
> amplificata loci species populique voluptas
> circus et inclusi moles cuneata theatri,
> templa Palatinaeque arces opulensque moneta
> et regio Herculei celebris sub honore lavacri,
> cunctaque marmoreis ornata peristyla signis
> moeniaque in valli formam circumdata limbo:
> omnia quae magnis operum velut aemula formis
> excellunt, nec iuncta premit vicinia Romae.

See R. Krautheimer, *Three Christian Capitals* (Berkeley, 1983).

praefectum urbis: Symmachus was in office summer 384–January/February 385 (*PLRE* 1, Symmachus 4); Courcelle, *Recherches*, 78–9, dates this encounter to autumn 384. That S. owned substantial property in Africa and had served there as proconsul briefly in 373 (during A.'s student days in Carthage) increased the opportunities for acquaintance between him and A.'s Manichean friends.

A.'s appointment is notoriously contemporaneous with S.'s famous *relatio* on the restoration of the Altar of Victory. For a view of that speech and its context stripped of romantic melodrama, see my comments in *Traditio* 35 (1979), 73–6; a more traditional view at length in J. Wyztes, *Der letzte Kampf des Heidentums in Rom* (Leiden, 1977). We are inclined to underline the disjunction between 'pagan' prefect and Christian bishop, but if we recall that Amb. and S. were probably related, however formally and distantly,[6] S. may perfectly well have recommended A. to Amb. with a letter not unlike

[6] *exc. fratr.* 1. 32 (Amb. addressing his dead brother), 'qui, cum a viro nobili revocareris, Symmacho, tuo parente, quod ardere bello Italia diceretur, . . . respondisti hanc ipsam tibi causam esse veniendi, ne nostro deesses periculo, ut consortem te fraterni discriminis exhiberes'; perhaps corroborated by Symm. *ep.* 1. 63, 'Saturus communis frater'.

those that survive (S. *epp.* 3. 30–37, all to Amb., even including one (3. 33) apparently asking Amb.'s help in gaining financial relief for a supporter of the supposedly pro-'pagan' usurper Eugenius), and see on a late example from that group (dating to 396/7?) P. Bruggisser, *Hermes* 115 (1987), 106–15. (Was it as prefect or merely as distinguished connoisseur that S. was consulted in this appointment? Seeck, in his edn. of S., p. 57, suggests the latter, though the text here is against that view; Jones, *LRE* 999–1000, thinks the formal appointment would have been made by the city council of Milan.) The hypothesis advanced by J. Rougé, *REA* 63 (1963), 61, and tentatively endorsed by Brown, 70, that Trygetius (from the Cassiciacum dialogues) was the son of a boyhood friend of S.'s founders on chronology: *that* Trygetius (or at least *Trygeti filius*) was already a candidate for the praetorship in 376 (see *PLRE* 1. 923), while A.'s pupil is young in 386/7; and the express description (*beata v.* 1. 6, 'Trygetius et Licentius cives et discipuli mei') puts his origins back to Africa and Thagaste.

There are curious echoes of S.'s *relatio* in A. The one at *sol.* 1. 13. 23, 'sed non ad eam [sapientiam] una via pervenitur', was retracted at *retr.* 1. 4. 3, but cf. also *vera rel.* 28. 51, 'quales patriarchae ac prophetae inveniuntur ab eis qui non pueriliter insiliunt, sed pie diligenterque pertractant divinarum et humanarum rerum tam bonum et *tam grande secretum*.' For both phrases, cf. S. *rel.* 3. 10, 'uno itinere non potest perveniri ad tam grande secretum.'

evectione publica: Jones, *LRE* 402, 830–1, on the rules (and exceptions) governing issuance of these passes; abuse was deplored when the abuser was disliked (e.g. Amm. 21. 16. 18 on Christian bishops), but was a notable sign of high favour.

ambivi: The verb of *ambitio saeculi*; cf. 6. 6. 9, 'ambiebam'.

ebrios: Drunks elsewhere (some metaphorical) in *conf.*: 1. 16. 26, 6. 2. 2, 6. 6. 10 (the drunken beggar), 6. 8. 13, 8. 3. 7, 8. 12. 29 (in Rom. 13: 13), 9. 3. 6, 10. 31. 45 (A. immune to the temptation).

utrique nesciebamus: This is the clearest sign that there was some lingering attachment to Manicheism up to the time of leaving Rome for Milan (Mandouze, 107 n. 8: see on 5. 6. 10). If he were already detached in all but outward show, he would have known perfectly well that going to Milan was a way of getting away from the Manichees (*inter alia*, he seems not to have associated with any—except Romanianus—when he arrived there). See below on 5. 14. 25 on the way A. always had a church to go to.

dictione proposita: A. had the handicap of a provincial accent: *ord.* 2. 17. 45, 'me enim ipsum . . . adhuc in multis verborum sonis Itali exagitant'.

et veni Mediolanium ad Ambrosium episcopum: Ambrose, now in his mid- to late forties, had been bishop of Milan since 374; for sketch with special ref. to A., see E. Dassmann, *Aug.-Lex.* 1. 270–85. The part Amb. played in A.'s life is difficult to characterize, and has raised controversy; see esp. on 6. 3. 3. Beyond the pages of *conf.*, most noticeable is the revived interest in Amb.'s works that characterized A.'s anti-Pelagian period, when quoting Amb. for his side was a guarantee of A.'s own orthodoxy (it was then that A. elicited the biography we have of Amb. from Paulinus of Milan—see on 6. 3. 3). In those days A. quotes many of Amb.'s works repeatedly: e.g. *c. ep. pel.* 4. 11. 29–11. 31 quotes the *exp. Esaiae*, *c. Nov.*, *apol. proph. David*, *arca Noe*, *fuga saec.*, *in Luc.*, and the *bono mort.*; Amb. is abundantly present in *c. Iul.* and *c. Iul. imp.*: *c. Iul.* 1. 3. 10, 'sed adhuc audi alium excellentem dei dispensatorem, quem veneror ut patrem: in Christo enim Iesu per evangelium ipse me genuit, et eo Christi ministro lavacrum regenerationis accepi. beatum loquor Ambrosium'; in *c. Iul.* there are citations from at least *apol. proph. David*, *exp. Ps. 48*, *in Tob.*, *Isaac. arca Noe*, *c. Nov.*, *parad.*, and the *de philosophia sive de sacramento regenerationis*. Counting occurrences of just the name (not periphrases like *episcopus Mediolanensis*) clarifies the picture: 335 × in all the works, of which 58 in *c. Iul.*, 181 in *c. Iul. imp.* alone; on the other hand, only 23 × in all works before 410, of which 11 × in *conf.*—but Amb. was already a touchstone of orthodoxy for A. in controversy at *epp.* 44. 4. 7 (397) and 93. 6. 20 (407/8). Possidius also recalled A.'s way of recounting sayings of Amb. to his disciples: Possid. *v. Aug.* 24. 17, 'Ambrosium se praesente in ecclesia tractavisse nobis aliquando rettulerat'; *v. Aug.* 27. 4, 'servandum ... quod instituto sanctae memoriae Ambrosii compererat, ut ...' (introducing several related sayings). P. Rousseau, *Augustiniana* 27 (1977), 153, compares these passages to 'the apophthegmata of a master preserved by his disciple'. For a sketch of Amb.'s influence, see J. Patout Burns in *Collectanea Augustiniana* (New York, 1990), 373–86.

cuius tunc eloquia: The 'tunc' here (taken with 8. 2. 3, 'tunc episcopi': cf. Courcelle, *Les Confessions*, 580 n. 1) gives a possible *terminus post quem* for composition of *conf.* of 4 April 397, the death of Amb.; see Introduction.

adipem frumenti: Eucharist: cf. Ps. 147: 14, 'qui posuit fines tuos pacem et adipe frumenti satians te'; *en. Ps.* 147. 21, 'qui ibi erit adeps frumenti, nisi panis ille qui descendit de caelo ad nos? in patria ipsa quomodo saturabit, qui in peregrinatione sic pavit?'; cf. also Ps. 80: 17 with *en. Ps.* 80. 22.

laetitiam: Ps. 4: 7–8 (see at 9. 4. 10), 'dedisti *laetitiam* in corde meo (8) a tempore frumenti, vini et *olei* sui multiplicati sunt'; *en. Ps.* 4. 9, 'est enim et frumentum dei, siquidem est panis vivus qui de caelo descendit. est et vinum dei, nam *inebriabuntur* inquit *ab ubertate domus tuae.* est et oleum dei, de quo dictum est: *impinguasti in oleo caput meum*.' Sim. interp. of Ps. 103: 13, 'et vinum laetificat cor hominis ut exhilaret faciem eius in oleo', at *en. Ps.* 103. s. 3. 13; see also *en. Ps.* 22. 5, 44. 19.

sobriam vini ebrietatem: Amb. *hymn.* 1. 7. 23–6 (Walpole):

> Christusque nobis sit cibus
> potusque nobis sit fides,
> laeti bibamus sobriam
> ebrietatem spiritus.

Behind this and similar expressions in Amb. lies the Philonic νηφάλια μέθη: see Hans Lewy, *Sobria ebrietas* (Giessen, 1929), esp. 146–57 for Amb. and 157–64 for A. The vocabulary (bread, oil, wine) offers, as just noted, allusion to the sacraments that Amb. also ministered to his people (for clearer allusions with the same vocabulary in Amb., see Lewy, 152–4; note esp. *sacr.* 5. 3. 17, 'et ideo praeclara ebrietas quae sobrietatem mentis operatur. haec sunt quae de sacramentis breviter percucurrimus'; sim. at *exp. Ps.* 1. 33). Already in Amb. the phrase was beginning to shift from its original Platonic meaning (what Lewy, 149, calls 'seine spezifische Bedeutung als Terminus für einen ekstatischen Vorgang') to a Christianized adaptation (Lewy, 'Bezeichnung der Grundstimmung des in Gott ruhenden Frommen'). For similar evocations of divine and sober drunkenness in A., cf. *ss.* 23. 12. 12 (wrong ref. at Lewy, 163), 34. 1. 2 ('humilitas excelsa, et ebrietas sobria'), 143. 3. 3 (not in Lewy; 'a carnalibus cupiditatibus mente sobria et spiritalibus desideriis ebria'), 225. 4. 4, *en. Ps.* 33. s. 2. 12 (not in Lewy: 'sanam insaniam et sobriam ebrietatem illius David'), 35: 14 (on Ps. 35. 9, 'inebriabuntur ab ubertate domus tuae'), and *agon.* 9. 10 ('et desideremus ipsum vitae fontem, ubi sobria ebrietate inundemur et inrigemur').[7] Lewy later (*Chaldæan Oracles and Theurgy* (Cairo, 1955), 198–9) found divine drunkenness in a fragment of the Chaldean oracles preserved by Psellus, but there are grave difficulties of priority there, and the distinctive *sobria ebrietas* is missing.

The echo of Amb.'s hymn (and note 'laetitiam' just preceding) is, as several writers observe, high praise, but perhaps the highest is that it comes in the midst of a sentence whose other compliments all come from scripture. The phrase affected Paulinus of Nola with similar force, leading to the echoes at *carm.* 24. 685–6 and 27. 104–5.

nesciens: Answered by 'nesciens' at the end of the paragraph.

ille homo dei: Deut. 33: 1 (Moses), 1 Kgs. 9: 7 (Samuel), 3 Kgs. 13: 4 (anon.), 4 Kgs. 1: 9 (Elias, ut saep. in 4 Kgs.), 2 Chron. 8: 14 (David). Knauer, 183 n. 1: 'Die Demonstrativa "quidam" und "ille" werden ohne wesentlichen Unterschied gebraucht': but here it almost has the effect of insisting on a scripture quotation.

peregrinationem: Knauer, *Hermes*, 85 (1957), 226: 'nicht nur den "äußeren" Weg bis nach Mailand, sondern auch den "innern" seiner bisherigen geistigen Entwicklung'.

satis episcopaliter: 1 Tim. 3: 2, 'oportet ergo episcopum inreprehensibilem esse, unius uxoris virum ... hospitalem, doctorem.' Amb. *Abr.* 1. 5. 32, 'apostolus principaliter eam [sc. hospitalitatem] in episcopo esse oportere ...: ut praesto sit advenientibus et occurrat obviam et itinera exploret.' A.'s view of the bishop's responsibility: *s.* 355. 1. 2, 'vidi necesse habere episcopum exhibere humanitatem assiduam quibusque venientibus sive

[7] There are similar expressions as well, e.g. *s. Lambot* 6, 'sancta ebrietas'.

transeuntibus: quod si non fecisset episcopus, inhumanus diceretur.' The adverb is only here in A.; the adjective from which it derives is far and away most commonly applied to episcopal authority to bind, loose, and judge, hence those expressions that do occur of 'episcopal' magnanimity are in a judicial context.

dilexit . . . amare: Courcelle, *Recherches*, 86 n. 3: 'On notera l'opposition entre *dilexit* (sentiments chrétiens d'Ambroise) et *amare* (sentiments purement humains d'Augustin)'; at 6. 1. 1, '[Monnica] diligebat [Ambrosium]'.

desperabam: These are the most desperate straits of *conf.*; the verb *despero* (or its present participle) is used with A. as its subject 6 × in Bk. 5 of *conf.*, and elsewhere only at 6. 1. 1 (where a different note is sounded by the first words of the book: 'spes mea a iuventute mea'), at 6. 11. 18 (in the retrospective representation of the development of his views), and in a contrafactual case much later (10. 43. 70, 'alioquin desperarem').

studiose: C. Mohrmann, *Vig. Chr.* 5 (1951), 252: not as de Labriolle's translation ('assidû') nor Courcelle's ('en amateur'); but that A. is motivated 'par un intérêt tout à fait *technique*'. Alfaric, 370, puts well A.'s response at this time: 'Augustin trouvait la doctrine d'Ambroise très admissible, mais sans se sentir obligé de l'admettre.'

disputantem: The verb itself indicates what A. was looking for: not a preacher, but a *disputator*; 5. 6. 11, 'aviditas mea . . . delectabatur quidem motu affectuque disputantis'; cf. *util. cred.* 8. 20, 'nonnullae disputationes Mediolanensis episcopi'. W. Hensellek, *Anzeiger Akad. Wien* 115 (1978), 20, finds the noun uncommon in the sense of 'preaching'.

suspendebar: Of intense attention at 4. 16. 28 (of his enthusiasm for Arist. *categ.*), 6. 1. 1 (of Monnica in the presence of Amb.), and 6. 3. 3.

intentus . . . incuriosus: (1) A. has these two attitudes in the wrong order: just where curiosity would stand him in good stead, it fails him, and his *intentio* is thus misdirected; (2) but there is a connection between *intentio* (see on 11. 18. 23) and *curiositas*: both are functions of that part of the soul that corresponds to the second person of the trinity.

suavitate sermonis: *Suavitas* is the instrument of *delectatio*, a necessity for even the humble style of oratory: *doctr. chr.* 4. 26. 56, 'quid etiam quaerit nisi credi, qui aliquid, licet submisso eloquio, discentibus narrat? et quis eum velit audire nisi auditorem nonnulla etiam suavitate detineat?' Cf. Cic. *or.* 21. 69, 'docere necessitatis est, delectare suavitatis, flectere victoriae', as misquoted at *doctr. chr.* 4. 12. 27 (for 'docere' Cic. has 'probare'). The difference between form and content in speech runs through Bk. 5: 5. 3. 3 (on *suaviloquentia*), 5. 6. 10, 5. 14. 24.

quamquam . . . quam Fausti erat: These words encourage us to look more closely at the verbal and substantive echoes that link the introduction of F. in Bk. 5 with the appearance and actions of Ambrose in Bk. 6; see on 6. 3. 3.

ceterum . . . comparatio: This judgement is not that of the A. of 384, but the

A. of 397; in 384 A. had not yet begun to compare F. and A. on substantive grounds.

ille (*autem*) CDO¹S Knöll Skut. Ver. : *iste* GO² Maur.

sed longe . . . salus: Ps. 118: 155, 'longe a peccatoribus salus, quoniam iustificationes tuas non exquisierunt'; *en. Ps.* 118. s. 30. 3, 'quis te discernit a peccatoribus, ut non a te longe sed tecum sit salus? hoc te nempe discernit, quia id quod isti non egerunt ipse fecisti, hoc est, dei iustificationes exquisisti. quid autem habes quod non accepisti?' The sub-text is that proper search (not the 'non pie quaerunt' of the Manichees (5. 3. 5) but the 'quaerentes enim invenientes eum' of 1. 1. 1) is the hope for drawing nearer to God; on 'longe', see on 1. 18. 28.

5. 14. 24

See *util. cred.* 8. 20, quoted in Introduction. *doctr. chr.* 3. 5. 9, '"littera occidit, spiritus autem vivificat." . . . ea demum est miserabilis animi servitus, signa pro rebus accipere; et supra creaturam corpoream, oculum mentis ad hauriendum aeternum lumen levare non posse.' (Note the association of the vocabulary of the mystical ascent with correct scriptural interpretation.)

It is the great achievement of Courcelle's *Recherches* (98–132) to have traced the surviving sermons of Ambrose that can reasonably be dated to the time of A.'s sojourn in Milan, and their echoes in A. For 2 Cor. 3: 6, not always *ad verbum* (with C.'s dates): *Iacob* 1. 4. 13 (early 386), *interpell. Iob et David* 1. 5. 12 (June 387: 'inveteravit littera, evertit enim [Christus] et subruit intellectum secundum litteram et statuit intelligentiam spiritalem. intellectus ergo legis ille carnalis evanuit: facta est spiritalis'), *in Luc.* 3. 28, 6. 29, 9. 1 and 9. 37 (387). Earlier: *fide* 3. 5. 37 (380), *spir. sanct.* 2. 4. 30 (381). Later: *in ps.* at least 6 × (389 and later). See L. F. Pizzolato, *La dottrina esegetica di sant'Ambrogio* (Milan, 1978), 194–201.

The other most important works that A. was exposed to were the *de Isaac vel anima* (386) and the *de bono mortis* (386), but their relevance is Platonico-philosophical rather than exegetico-theological. In some cases, the most we can do is note a congruence to A.'s situation at the time or to his later thought; e.g. the *bono mort.* spoke to the fear of death of which he speaks at 6. 16. 26, the *Isaac* spoke to the nature of spiritual existence (cf. 7. 1. 1 ff.), and the *apol. proph. David* (387?) spoke to the sinfulness of all. Even where, as in the last case, we cannot be sure that A. heard the surviving sermon (s), the presence of the doctrines in Amb.'s work is at least suggestive of what A. could have heard in sermons that were not transmitted to us.

See on 5. 13. 23, *studiose*, for A.'s attitude at first.

inanis cura: He knew it an 'inanis cura' at the time: see on 5. 13. 23.

diserte . . . vere: Of Faustus at 5. 6. 10, 'vera . . . diserta'.

gradatim: Cf. 5. 13. 23, 'sensim'.

defendi: See on 5. 13. 23; cf. 5. 11. 21, 'quae . . . defendi posse non existimabam'.

uno atque altero et saepius: G–M: '"one and another and in fact a considerable number [saepius]". The older editions put a comma at *altero*, making the clause parenthetic "— and oftenest by way of allegory —" . . . Pusey renders, "I had heard one or two places of the OT resolved, and oftentimes in a figure".' G–M clearly right: it was allegoresis that made the OT intelligible to A.

aenigmate: C. P. Mayer, *Aug.-Lex.* 1. 140–1: '"Allegoria" und ae.[nigma] verhalten sich zueinander wie Genus und Species: "omne aenigma allegoria est, non omnis allegoria aenigma est" (*trin.* 15. 9. 15).' For A.'s view of the problem, and its solution, five years later, when he was still cautious and sceptical about the ability of the church to make these things clear, with verbal echoes that link to the situation here, see *util. cred.* 2. 4, 'et quia sunt ibi quaedam quae suboffendant animos ignaros et neglegentes sui—quae maxima turba est—populariter accusari possunt; *defendi* autem populariter, propter mysteria quae his continentur, non a multis admodum *possunt*. qui vero pauci hoc facere noverunt non amant propatula et famigerula quaedam in disputatione certamina.'

aenigmate soluto: *doctr. chr.* 2. 16. 23, 'ad solvenda aenigmata scripturarum', 2. 29. 45, 'ad aenigmata scripturarum solvenda'; cf. *Gn. c. man.* 2. 2. 3.

cum ad litteram acciperem: 2 Cor. 3: 5–6, 'sed sufficientia nostra ex deo est; (6) qui et idoneos nos fecit ministros *novi testamenti*, non littera, sed spiritu: littera enim occidit, spiritus autem vivificat.' (Quoted *ad verbum* in representing Amb.'s views at 6. 4. 6, where see notes on the place of the verse in A.'s *œuvre*.)

desperationem: For despair, see on 5. 13. 23; note that the effect here is only to deepen his depression by persisting in a despair that he now finds unworthy.

credideram: 3. 7. 12, 'et utrum iusti existimandi essent'.

legem et prophetas: The OT, as at Mt. 5: 17, 7: 12, Lk. 16: 16; and cf. A.'s later work, *contra adversarium legis et prophetarum*—a defence of the OT.

catholicam viam: For *catholica*, see on 5. 10. 20; for *via* = *Christus*, see on 7. 7. 11.

illud: i.e. Manicheism.

catholica: Perhaps sc. *via* from above, but cf. 6. 4. 5, 'catholicam tuam', and often elsewhere in A., where *catholica* simply = *catholica ecclesia*.

non mihi victa . . . nondum etiam victrix: Epigram is his best measure of the progress that despair brought him. The battle is once again open, and in doubt. At the moment represented by the end of Bk. 4, he was a confident and loquacious defender of Manicheism. He may owe something here to Lucan's famous line (*b. c.* 1. 128), 'victrix causa deis placuit sed victa Catoni'.

5. 14. 25

intendi animum: Nothing more is said of this vain attempt to confound the Manichees by main force of argument; the resolutions for his problems will come to him from others, not from his own learned researches.

spiritalem substantiam: The anthropomorphic question (3. 7. 12); still in the dark on this at 6. 3. 4, and emerging at 7. 1. 1.

omnique natura quam sensus carnis attingeret: A hint of the limitations of philosophy; cf. 9. 10. 24 (Ostia), 'carnalium sensuum', where the senses of the flesh are finally transcended.

philosophos: See 5. 4. 3–5. 9.

sicut existimantur: Cf. 5. 10. 19, 'academicos . . . ut vulgo habentur'.

fluctuans: 6. 1. 1, 'illam ancipitem fluctuationem', 6. 5. 8, 'fluctuabam'.

decrevi: G–M give a sober and straightforward list of his reasons. They offered no spiritual sustenance, only fables about sun and moon (3. 10. 18, *util. cred.* 1. 3); the morality of the elect was hypocritical (*mor.* 2. 19. 68–72); they were better at refutation than in proof (7. 2. 3, *util. cred.* 1. 2); their claims about the falsification of scripture were unfounded (5. 11. 21, *util. cred.* 3. 7); their astronomy was bogus (5. 3. 6–5. 9); Faustus was a failure (5. 7. 13). Such a list is no more than a report about the state of his conscious mind at the time of his separation from the Manichees. Other factors to be kept in mind are his removal from the original circle of Manichean friends and associates he knew in Africa, followed by his apparent removal from any close contact with Manichean cultists on his removal from Rome to Milan, and the possibility, hinted at by his suggestion that the emotional satisfaction he sought in Manicheism was the relief it offered from a sense of sin and guilt, that in some visceral way beyond the reach of his conscious arguments or our analysis he simply felt unsatisfied.

sine salutari nomine Christi: Philosophy was already rejected for lacking the *nomen Christi* at 3. 4. 8, on reading the *Hortensius*. He has now abdicated the doctrines that *curiositas* inspired in him, but not the *curiositas* itself: he has instead returned to where the *Hortensius* left him, now seeking a new satiety for his *curiositas*.

curationem languoris: Mt. 9: 35, 'et curans omnem languorem et omnem infirmitatem'; *languores* most recently at 4. 11. 16 (the address to his soul, promising a healing that comes 'a veritate') and thematically through Bk. 10 (see on 10. 3. 3).

catechumenus: He had been a catechumen since boyhood (1. 11. 17); cf. again *util. cred.* 8. 20 (in Introduction: 'decreveramque tamdiu esse catechumenus in ecclesia cui traditus a parentibus eram, donec aut invenirem . . .'), and for the turn of phrase and mind, *Gn. litt.* 10. 3. 6, 'ut . . . acceptabilem de hac re sententiam nanciscamur, ut eam tenere, donec certum aliquid elucescat'; see on 5. 7. 13, with discussion of the repetition of this theme at 5. 10. 18, here, 6. 10. 17, 6. 11. 18, and 8. 7. 18; and cf. again Ps. 36: 23 (also see on 5. 7. 13).

The catholic church had some appeals to A. that he does not mention here: *c. ep. fund.* 4. 5–5. 6, 'in catholica enim ecclesia, ut omittam sincerissimam sapientiam, ad cuius cognitionem pauci spiritales in hac vita perveniunt, ut eam ex minima quidem parte, quia homines sunt, sed tamen sine dubitatione cognoscant—ceteram quippe turbam non intellegendi vivacitas, sed credendi simplicitas tutissimam facit—ut ergo hanc omittam sapientiam, quam in ecclesia esse catholica non creditis, multa sunt alia quae in eius gremio me iustissime teneant. *tenet* consensio populorum atque gentium; *tenet* auctoritas miraculis inchoata, spe nutrita, caritate aucta, vetustate firmata; *tenet* ab ipsa sede Petri apostoli, cui pascendas oves suas post resurrectionem dominus commendavit, usque ad praesentem episcopatum successio sacerdotum; *tenet* postremo ipsum catholicae nomen, quod non sine causa inter tam multas haereses sic ista ecclesia sola obtinuit. . . . (6) ego vero evangelio non crederem, nisi me catholicae ecclesiae commoveret auctoritas.' The last sentence is famous, but, like the former paragraph, seems empirical and factual rather than prescriptive; how *else* any Christian might believe in 'The Gospel' (i.e. the Gospel as transmitted in the four canonical texts, as distinct from variant apocryphal versions) is not clear.[8]

So A. began to go to the catholic church again. A. probably never 'missed church' a week in his life. He was a catholic catechumen until 18, followed the Manichees enthusiastically for years after, and still probably participated, at least outwardly, in their cult while living with them at Rome. On going to Milan he takes up orthodox Christianity again. It is typical of late antique men that they are rarely (if ever) cultless.

[8] On that text, see G. Bonner, *St. Augustine of Hippo* (London, 1963), 231–2.

BOOK SIX

A. at Milan in 385. The central section of the book takes the focus away from A. for the first time, to recount the early life of Alypius; cf. 9. 8. 17–9. 22 for sim. treatment of Monnica: the two 'digressions', each a conversion story in its own right, bracket A.'s central conversion story—in all three cases an admonitory word (see Brown, 32) is decisive in effect.

The first section recounts doctrinal progress under the tutelage of Ambrose (with a reminder of the moral distance that remained to go), while the last section describes in detail the difficult personal choices that remained. The plans for the future that preoccupy A. now show the vanity of such undertakings and the underlying failure of his ambition. In *this* regard, measured against the three temptations, the book represents undiluted, though often painful, progress.

6. 1. 1–6. 9	*A. (and Monnica) at Milan*
6. 6. 1–3. 4	*Encounters with Ambrose*
6. 4. 5–5. 8	*Doctrinal lessons*
6. 6. 9	*A beggar in the street*
6. 6. 10–10. 17	*Friends (esp. Alypius, but also Nebridius)*
6. 11. 18–16. 26	*Perplexities*
6. 11. 18–11. 19	*State of mind: interior monologue*
6. 12. 20–13. 23	*Marriage plans*
6. 14. 24	*Philosophical plans*
6. 15. 25	*Dismissal of A.'s concubine*
6. 16. 26	*Dismay*

6. 1. 1

The density of scriptural allusion here is a mark that an important stage has been reached.

We seem to have reached spring 385, after the seas opened again to allow Monnica to take ship (*BA* 13. 141 n., arguing that his change of view regarding scripture must be dated to 385, leaving winter and spring 386 for reflection). Thus Courcelle, *Recherches*, 86–7, marks these pages for the first weeks of June 385. Courcelle, *Les Confessions*, 65: 'Les sermons d'Ambroise ont pour effet provisoire de mettre Augustin en un péril plus grand, du point de vue de la foi chrétienne, qu'au temps où il était manichéen.' In other words, they plunged him headlong ('précipité trop avidement') into the Platonists and the Christology of Porphyry. C. errs here chiefly because he takes the last lines of this paragraph too literally; he misreads the calendar besides, for A. clearly imbibed the Ambrosian view of scripture before he ever went near the *platonicorum libri*.

Who came with Monnica? At least Navigius and probably Lartidianus and Rusticus. What about Nebridius? Trygetius? Licentius? If A. made a brilliant career, there would be pickings for the others. See M. G. Mara, *Lectio VI–IX*, 76; C. Lepelley, *Bull. litt. eccl.* 88 (1987), 243, speaks of 'une manière de roman balzacien avant la lettre . . . la suite logique d'une stratégie familiale élaborée de longue date', from the time when Patricius and Monnica began to promote A.'s education and prospects in the great world. In A.'s rise, they would profit all.

Spes mea a iuventute mea: Ps. 70: 5, 'quoniam tu es patientia mea, domine, domine, spes mea a iuventute mea, in te confirmatus sum *ex utero de ventre matris meae* tu es protector meus'; *en. Ps.* 70. s. 1. 7, 'ante enim non in te sperabam; quamvis tu fueris protector meus, qui me salvum perduxisti ad tempus quo in te discerem sperare.' A. was about to cross (aged thirty) one of the boundaries, which he marked carefully, between the six ages of mortal life (see on 1. 8. 13): 7. 1. 1, 'ibam in *iuventutem*'.

Knauer, 41: 'und daher scheint es, als habe Augustin durch den Teil des Psalmverses, der die ersten Worte des 6. Buches bildet, sofort angeben wollen, in welchem Zustand er sich befand: er wartet auf das "certum"'; K. observes the frequency of *spes* and its derivatives in Bk. 6 (it has a similar role in Bks. 10 and 13). Besides this paragraph, cf. 6. 6. 3, 6. 6. 10, 6. 7. 11, 6. 7. 12, 6. 11. 18, 6. 11. 19, 6. 16. 26, and *per contra* 6. 15. 25, 'desperatius dolebat'. Cf. on 5. 13. 23, on despair in Bk. 5.

et quo recesseras: Cf. Ps. 9: 22, 'utquid, domine, recessisti longe? despicis in opportunitatibus, in tribulationibus'; *en. Ps.* 9. 20, 'id est, opportune despicis, et facis tribulationes ad inflammandos animos desiderio adventus tui. his enim iucundior est *fons ille vitae*, qui multum sitierint.'

an vero: On *et* (8 × in succession here), see on 1. 1. 1. For the citation, see on 10. 17. 26: Job 35: 11 (VL), 'ubi est deus qui fecit me? qui distribuit custodias nocturnas, qui separat me a quadrupedibus terrae, et a volatilibus caeli sapientiorem me fecit'; *adn. Iob* on 35: 11, '"qui separat me a quadrupedibus terrae et a volatilibus caeli sapientiorem me fecit." sic enim quaerendus est dominus in adflictionibus vitae huius, ut non ab eo terrena bona desideremus, quia iam bestiis meliores sumus antequam illa accipiamus.' See L. Verheijen, *Augustinianum* 17 (1977), 541–4. But cf. also 5. 3. 4, 'volatilia . . . pisces . . . pecora', and the echo discussed there of Ps. 8: 8. If the three temptations are in mind here, the beasts of earth and the birds of air represent *concupiscentia carnis* and *ambitio saeculi* respectively.

a volatilibus CDO Ver. : *volatilibus* GS Knöll Skut.

et ambulabam per tenebras: Is. 50: 10, 'qui ambulavit in tenebris et non est lumen ei, speret in nomine domini'; Ps. 34: 6, 'fiat tamquam pulvis ante faciem venti, et *angelus* domini tribulans eos; fiat via eorum tenebrae et lubricum'; *en. Ps.* 34 s. 1 9, 'sunt ista duo mala magnae poenae hominum: tenebrae, ignorantia; lubricum, luxuria.'

foris: See on 1. 18. 28, and cf. esp. 10. 27. 38, 'et ecce intus eras et ego foris et ibi te quaerebam'. *vera rel.* 39. 72, 'noli foras ire, in te ipsum redi. in interiore homine habitat veritas.'

deum cordis mei: Ps. 72: 26, 'deus cordis mei, pars mea deus meus'; *civ.* 10. 25, 'non subiecit, deus cordis et carnis meae, sed deus cordis mei. per cor quippe caro mundatur. unde dicit dominus, "mundate, quae intus sunt, et quae foris sunt munda erunt" [Mt. 23: 26].' See also 4. 2. 3, 9. 13. 35.

et veneram in profundum maris: Cf. Ps. 67: 23, 'convertam in profundum maris'; *en. Ps.* 67. 31, 'sed ibi convertit eos qui in profundo huius saeculi iacent demersi pondere peccatorum.' Metaphor becomes reality with the narrative of M.'s voyage, then returns as metaphor. M. appears comforting mariners, but it is A. who is lost at sea ('et invenit me periclitantem ... graviter').

desperabam de inventione veri: A standard way of describing his predicament at the time; cf. *ep.* 1. 3 (to Hermogenianus, in the winter of 386/7, commenting on A.'s own *c. acad.*), 'non tam me delectat, ut scribis, quod academicos vicerim—scribis enim hoc amantius forte quam verius—quam quod mihi abruperim odiosissimum retinaculum quo a philosophiae ubere *desperatione veri*, quod est animi pabulum, refrenabar.'

terra marique me sequens: *Aen.* 9. 492 (the mother of Euryalus speaks: see also on 5. 8. 15), 'hoc sum terraque marique secuta?' *cura mort.* 13. 16, 'si rebus viventium interessent animae mortuorum et ipsae nos, quando eas videmus, adloquerentur in somnis, ut de aliis taceam, me ipsum pia mater nulla nocte desereret, *quae terra marique secuta est ut mecum viveret*.' C. Bennett, *REAug* 34 (1988), 64 n. 33: 'Although Euryalus' mother was an obscure character in the *Aeneid*, knowing who she was (or what her name was [she is not named in the *Aeneid* itself]) seems to have been one of those pieces of *Aeneid*-trivia that the educated prided themselves on knowing (*ord.* 2. 12. 37, 'qui si non responderint, quid vocata sit mater Euryali, accusantur inscitiae, cum ipsi eos a quibus ea rogantur vanos et ineptos nec curiosos audeant appellare?').' (See K. Schelkle, *Virgil in der Deutung Augustins* (Stuttgart, 1939), 160–1, for other examples of this grammarian's *curiositas* (already satirized by Juv. 7. 230–6).)

consolabatur ... consolari: Deponent, then passive; both uses classical, but the mixture is unusual. To the manifold roles of M. in *conf.*, add that of Paul, consoling mariners on the authority of a vision: Act. 27: 23–4, 'adstitit enim mihi hac nocte angelus dei cuius sum ego et cui deservio (24) dicens, "ne timeas Paule".' (See 6. 13. 23 for the source of M.'s confidence in her own visions.)

visum: Regularly in *conf.* for 'dream-vision': 3. 11. 20 (Monnica's vision *in* *regula*: confirmed to be a dream from 6. 13. 23), 6. 13. 23 (Monnica again), 9. 7. 16 (Ambrose's discovery of the relics of Protasius and Gervasius). Cicero attempts to regularize *visum* as a translation for *phantasia* (*acad. post.* 1. 11. 40), but his own usage of the word includes dreams, which should be *phantasmata* (e.g. *Lucullus* 15. 47, 27. 88).

non quasi: Knöll, Skut. and Ver. follow Maur. in printing a comma after 'non'. On that reading, A. is made to say that M. was *not* delighted at his news, because she expected it; with the comma deleted, M. is allowed her joy, while A. officiously explains that she really was not at all surprised because of the hope she had that made her so like the mother in the gospel. With 'nulla ergo' below, A. offers a qualification in accord with his explanation: there was nothing inordinate about her rejoicing to leave her a prey to emotion.

cum iam: The *cum* is causal; to salvage the comma after *non*, Pusey had to make the *cum* concessive: 'she was not overjoyed, as at something unexpected; although she was now assured . . .'—*that* reading of her motives makes no sense at all; *BA* ignores the *cum*: 'elle ne sursauta pas de joie, comme à une nouvelle inattendue; désormais pourtant, elle était rassurée . . .'

exilivit CDGO Maur. Ver. : *exiluit* S Knöll Skut. The same verb at 2. 3. 6, of Monnica at the thought of grandchildren, and at 9. 7. 16, of a blind man at Milan at the news of the discovery of the relics of Protasius and Gervasius; cf. 'exultatione' below.

tamquam mortuum sed resuscitandum: *cura mort.* 18. 22, 'ut corpori mortuo sed tamen resurrecturo et in aeternitate mansuro'.

offerebat GOS Knöll Skut. Ver. : *efferebat* CD Maur. (cf. Lk. 7: 12 (Vg.), 'efferebatur'). G–M: 'A. may well have intended to imply that M., in contrast with the widow of Nain, was consciously seeking Christ's aid for her son.'

ut diceres filio viduae: Lk. 7: 12–15 (Vg. modified by *adn. Iob* on 29: 13: see Milne 96), 'cum autem appropinquaret portae civitatis et ecce ferebatur mortuus filius unicus matris suae, quae erat vidua, et turba civitatis multa cum illa. (13) quam cum vidisset dominus, misericordia motus super ea dixit illi, "noli flere", (14) et accessit et tetigit loculum. hii autem qui portabant steterunt, et ait, "adulescens tibi dico surge", (15) et resedit qui erat mortuus et coepit loqui et dedit illum matri suae.' *s.* 98. 2. 2, 'de iuvene illo resuscitato gavisa est mater vidua: de hominibus in spiritu cotidie suscitatis gaudet mater *ecclesia*.' See also *en. Ps.* 97. 1, *lib. arb.* 3. 23. 67 (*c.*395, on infant baptism), 'quid enim filio viduae profuit fides sua, quam utique mortuus non habebat, cui tamen profuit matris ut resurgeret?' (Cf. *s. dom. m.* 1. 12. 35 and *s.* 98. 5. 5, on three kinds of resurrection.) C. Bennett, *REAug* 34 (1988), 64: 'Monnica . . . is transformed before our very eyes into a different character, the widow mother of Luke 7. . . . The juxtaposition of Vergilian and Biblical figures is a critique of the 'pagan' text, but not merely that. It is also a reinterpretation. Read aright, the desires and griefs of Vergil's character point to the same longings and values as Scripture, and the Scriptural figures illuminate the real significance of the Vergilian for us.'

revivesceret OS Knöll Skut. Ver. : *revivesce et* G : *revivisceret* CD Maur.

(*certa erat*) **et:** = *etiam*.

pleno fiduciae: an expression thought 'étonnamment rare' by L. G. Engels, *Graecitas et Latinitas Christianorum Primaeva: Supplementa*, III (Nijmegen 1970), 109 n. 4; he can instance in A. only *en. Ps.* 55. 15, 'confessionem veram, piam, plenam fiduciae,' but for the near-match 'plena fiducia' and the like, cf. *doctr. chr.* 1. 8. 8, *en. Ps.* 143. 7, *s.* 211. 1. 1, *c. Faust.* 12. 32, *Gal. exp.* 42.

fidelem catholicum: See on 2. 3. 6; the confidence descends from her earlier vision at 3. 11. 19.

fons misericordiarum: Same expression at 4. 4. 7, prefacing the story of his friend's death, and at the end of this book, 6. 16. 26.

accelerares: Ps. 30: 3, 'accelera ut eximas me'; *en. Ps.* 30. en. 2. s. 1. 8, 'ad hoc enim positum est verbum, ut hoc totum quod nobis diu videtur quamdiu volvitur saeculum, intellegas punctum esse. non est diu quod habet extremum.'

adiutorium: Ps. 69: 2, 'deus in adiutorium meum intende'; Ps. 37: 23, 'intende in adiutorium meum, domine salutis meae.'

et inluminares tenebras meas: Ps. 17: 29, 'quoniam tu inluminabis lucernam meam, deus meus, inluminabis tenebras meas'; *en. Ps.* 17. 29, 'nos enim peccatis nostris tenebrae sumus.' Sim. at 7. 1. 2, 11. 2. 2, 11. 25. 32, 13. 8. 9.

currere . . . suspendi: Historic infinitives, connected to 'preces et lacrimas' above, all without a finite verb.

ad fontem salientis aquae in vitam aeternam: Jn. 4: 14, 'sed aqua quam dabo ei fiet in eo fons aquae salientis in vitam aeternam'; *Io. ev. tr.* 15. 16, 'hinc eam hauriunt homines hydria cupiditatum. . . . cum pervenerit quisque ad voluptatem saeculi huius, cibus est, potus est, lavacrum est, spectaculum est, concubitus est; numquid non iterum sitiet? ergo "de hac aqua qui biberit, iterum", inquit, "sitiet": si autem acceperit a me aquam, "non sitiet in aeternum".' An exact parallel is *ep.* 25. 2, where Paulinus of Nola uses the same expression of A.: 'os enim tuum fistulam aquae vivae et venam fontis aeterni merito dixerim, quia fons in te aquae salientis in vitam aeternam Christus effectus est.' Though Pusey wants to take this as a baptismal reference, that specification is difficult to corroborate from A.: *en. Ps.* 62. 8, 'et fecit nobis consolationem in deserto, mittendo ad nos praedicatores verbi sui; et dedit nobis aquam in deserto, implens spiritu sancto praedicatores suos, ut fieret in eis fons aquae salientis in vitam aeternam'; sim. at *en. Ps.* 1. 3; four other citations (*en. Ps.* 17. 16, 67. 35, 73. 17, and 147. 26) have no baptismal overtones. It is Amb. as preacher to whom M. looks now (and his preaching will reach A. before ever his baptizing can). That interpretation goes well with 'angelum' just below.

diligebat: See on 5. 13. 23 for *diligere* as a less carnal verb for love than *amare*.
sicut angelum dei: Gal. 4: 14, 'sed sicut angelum dei excepistis me, sicut Christum Iesum.'
fluctuationem: 5. 14. 25, 'inter omnia fluctuans', 6. 5. 8, 'fluctuabam'.
criticam CDGO[1] Maur. Knöll Skut. Ver. : *creticam* O[2]S. How familiar would the word be to A.'s audience? He introduces it as a technical term; *TLL* 4. 1211 presents it as uncommon.

6. 2. 2

Nothing here says that A. or M. ever took this query to Ambrose. The plain sense of the text is that the *ostiarius* turned M. away with some variant of, 'Sorry, ma'am, bishop's orders,' and that was enough for her. No private consultation need be inferred. (The remark of Amb. reported has no connection with the question of *laetitiae* (see on 6. 6. 9, 'laetitiam') and an encounter of that sort could have been public, at a time when the bishop was receiving callers of various kinds.)

It is commoner to assume that A. went to ask Amb. about the practice on M.'s behalf. Courcelle, *Recherches*, 91 n. 3, typically reconstructs a sequence of events: M. is prohibited by the shrine-keeper from her usual practice; she is shocked; A. goes to Amb., who says his piece, which A. reports to M., who is then willing to obey. Her first instinct was negative reaction, and the request to Amb. was not in the main for an explanation but for permission to continue the practice. The only ground for this scenario lies not in *conf.* but in A.'s letters (see below).

The practice of the *laetitiae* made orthodox Christians vulnerable to Manichean criticism: see Faustus' attack quoted at *c. Faust.* 20. 4, 'defunctorum umbras vino placatis et dapibus', and A.'s response at *c. Faust.* 20. 21, 'qui autem se in memoriis martyrum inebriant, quomodo a nobis approbari possunt, cum eos, etiamsi in domibus suis id faciant, sana doctrina condemnet? sed aliud est quod docemus, aliud quod sustinemus.'

memorias: See on 5. 8. 15, 'memoria beati Cypriani'.
ostiario: Not a servant, but a functionary in minor orders known from the third century on. The only other ecclesiastical *ostiarius* of whom we hear in A. is a rather disreputable character at *ep.* 26*. 1, who had been ordained a deacon in a way A. judged fraudulent; when the circumstances came to light, A. appointed him as *ostiarius* at the shrine of saint Theogenes at Hippo, where he lasted only a short time before being expelled from office by the *presbyteri* of Hippo while A. was out of town. The appointment was meant to provide him with a modest livelihood. (The *ostiarius* is better attested after rather than before A.'s time, but the office was so modest that it is probably under-reported; see H. Leclercq, 'Portier', *DACL* 14. 1525–33.)

hoc episcopum vetuisse: Amb.'s only surviving comment: *de Helia et ieiunio*, 17. 62, 'et haec vota ad deum pervenire iudicant, sicut illi qui calices ad sepulchra martyrum deferunt atque illic in vesperam bibunt: aliter se exaudiri posse non credunt.' From A.'s description of the state of affairs at the Vatican basilica of St Peter, one surmises why the *ostiarius* was so firm in dealing with a foreigner: *ep.* 29. 10 (to Alypius), 'deinde hortatus sum [Hipponenses], ut transmarinarum ecclesiarum, in quibus partim ista recepta numquam sunt, partim iam per bonos rectores populo obtemperante correcta, imitatores esse vellemus. et quoniam de basilica beati apostoli Petri cotidianae vinulentiae proferebantur exempla, dixi primo audisse nos saepe esse prohibitum, sed quod remotus sit locus ab episcopi conversatione et in tanta civitate magna sit carnalium multitudo peregrinis praesertim, qui novi subinde veniunt tanto violentius quanto inscitius illam consuetudinem retinentibus, tam immanem pestem nondum compesci sedarique potuisse.'

A. wrote to Aurelius of Carthage *c.*393 (Goldbacher links this to the council of Hippo of 393, occasion of A.'s preaching *f. et symb.*) on the best strategies for controlling these folkways in Africa. These puritanical efforts by the young clerics were not enthusiastically received. (See van der Meer, 498–526, for survey of burial practices and ceremonies.) A.'s criticism of these customs almost a decade after M. was dissuaded from them need not be distorted to exempt her from any retrospective criticism (as van der Meer, 516–17, does). M. was to A. no plaster saint, the more so because the present passage in *conf.* has the effect of showing her lack of *concupiscentia carnis*; note 'consuetudine': see 8. 5. 10 for A.'s analysis of the role of 'habit' in inveterate concupiscence; M. has the habit but not the concupiscence, hence the habit is easily broken. See esp. *ep.* 22. 1. 3–4: 'comissationes enim et ebrietates ita concessae et licitae putantur, ut in honore etiam beatissimorum martyrum non solum per dies sollemnes, quod ipsum quis non lugendum videat, qui haec non carneis oculis inspicit, sed etiam cotidie celebrentur. . . . saltem de sanctorum corporum sepulchris, saltem de locis sacramentorum, de domibus orationum tantum dedecus arceatur. . . . (4) haec si prima Africa temptaret auferre, a ceteris terris imitatione digna esse deberet; cum vero et per Italiae maximam partem et in aliis omnibus aut prope omnibus transmarinis ecclesiis, partim quia numquam facta sunt, partim quia vel orta vel inveterata *sanctorum et vere de vita futura cogitantium episcoporum* diligentia et animadversione extincta atque deleta sunt, . . . dubitare[1] quo modo possumus tantam morum labem vel proposito tam lato exemplo emendare?' A. suggests (*ep.* 22. 1. 6) that the offerings be replaced by alms for the poor. The official ban is *breviarium Hipponense*, canon 29 (393), 'ut nulli episcopi vel clerici in ecclesia conviventur, nisi forte transeuntes hospitiorum

[1] The ellipse here marks a lacuna detected by Goldbacher.

necessitate illic reficiant; populi etiam ab huiusmodi conviviis, quantum potest fieri, prohibeantur.'

A.'s own local triumph, and the making of his reputation as a figure of authority at Hippo (it may almost have coincided with his elevation as 'coadjutor' bishop), is recounted in *ep.* 29 (to Alypius: *c.*395): *ep.* 29. 2, 'cum post profectionem tuam nobis nuntiatum esset tumultuari homines et dicere se ferre non posse ut illa sollemnitas prohiberetur, quam laetitiam nominantes *vinulentiae* nomen frustra conantur abscondere, sicut etiam te praesente iam iam nuntiabatur, opportune nobis accidit occulta ordinatione omnipotentis dei, ut quarta feria illud in evangelio capitulum consequenter tractaretur: "nolite dare sanctum canibus neque proieceritis margaritas vestras ante porcos." tractatum est ergo de canibus et de porcis ita ut et pervicaci latratu adversus dei praecepta rixantes et voluptatum carnalium sordibus dediti erubescere cogerentur, conclusumque ita, ut viderent quam esset nefarium intra ecclesiae parietes id agere nomine religionis quod in suis domibus si agere perseverarent, sancto et margaritis ecclesiasticis eos arceri oporteret.' (See van der Meer, 515–25, and Mandouze, 643–52.) But twenty years after the ban, at *civ.* 8. 27, he still attacks the custom as something alive, at least in some regions.

The instinct of grave-cult was strong and persistent, even among Christians, as witness *cura mort.* A disdain for such conduct was strongly rooted in the doctrine of Christianity, but difficult to inculcate in practice; the official church hierarchy was indisputably closer to the gospel message than was the flock that followed. Of course prayer, esp. eucharistic prayer, for the dead was accepted on all sides (see on 9. 11. 27). On the history, see J. Quasten, *HThR* 33 (1940), 253–66, who loyally defends A.'s campaign against excesses while showing sensitively the depth of genuine *pietas* at work in the custom; the classic text is an inscription from Satafis (Quasten, 257, from Diehl, *ILCV* 1570 (1. 301)).[2]

Courcelle, *Recherches*, 90: 'C'est cet état d'esprit autoritaire qu'Augustin avait détesté de tout temps dans l'Église catholique; c'est le grief manichéen qui l'avait détourné, dès sa jeunesse, de la foi de sa mère. Vers juin 385, nous apprenons donc que le premier réflexe d'Augustin devant la réponse d'Ambroise est un réflexe hostile, un réflexe manichéen: il se cabre à l'idée

[2] Rebecca West, *Black Lamb and Grey Falcon* (New York, 1943), 765, recalls a scene in an Orthodox cemetery in Bitolj (Macedonia) in the 1930s: 'I saw a peasant woman sitting on a grave under the trees with a dish of wheat and milk on her lap, the sunlight dappling the white kerchief on her head. Another peasant woman came by, who must have been from another village, for her dress was different. I think they were total strangers. They greeted each other, and the woman with the dish held it out to the new-comer and gave her a spoon, and she took some sups of it. To me it was an enchantment; for when St. Monica came to Milan over fifteen hundred years ago . . . [t]hat protocol-loving saint, Ambrose, had forbidden the practice because it was too like picknicking for his type of mind. To see these women gently munching to the glory of God was like finding that I could walk into the past as into another room.'

que cet homme impose une règle, alors qu'on lui demande une raison. Suivre la coutume n'est pas une raison.' Courcelle returned to the question at *Les Confessions*, 21, but seems to be straining. Recall that the narrative of Bk. 5 shows *ratio* beginning the process of detaching A. from Manicheism and pushing him into Academicism.

On two occasions, A. described a separate question of church practice and discipline on which he consulted Amb. at Monnica's request, that of fasting. Courcelle wondered (*Recherches*, 91) why this query is not reported in *conf.* The answer is surely that the problem of the *laetitiae* was one where Amb. commanded a higher, more demanding discipline; in the case of the fast, he was recommending laxity. M.'s willingness to obey was more remarkable in the former case.

ep. 36. 14. 32 (date controversial, probably rather late), 'indicabo tibi quid mihi de hoc requirenti responderit venerandus Ambrosius, a quo baptizatus sum, Mediolanensis episcopus. nam cum in eadem civitate mater mea mecum esset et *nobis adhuc catechumenis parum ista curantibus* illa sollicitudinem gereret, utrum secundum morem nostrae civitatis sibi esset sabbato ieiunandum, an ecclesiae Mediolanensis more prandendum, ut hac eam cunctatione liberarem, interrogavi hoc supra dictum hominem dei. at ille: "quid possum", inquit, "hinc docere amplius, quam ipse facio?" ubi ego putaveram nihil eum ista responsione praecepisse, nisi ut sabbato pranderemus; hoc quippe ipsum facere sciebam. sed ille secutus adiecit: "quando hic sum, non ieiuno sabbato; quando Romae sum, ieiuno sabbato. et ad quamcumque ecclesiam veneritis", inquit, "eius morem servate, si pati scandalum non vultis aut facere." hoc responsum rettuli ad matrem eique suffecit nec dubitavit esse oboediendum; hoc etiam *nos* secuti sumus.'

ep. 54. 2. 3, has a narrative using many of the same words (esp. those of Amb.), adding: 'hoc cum matri renuntiassem, *libenter amplexa est*. ego vero de hac sententia etiam atque etiam cogitans ita semper habui, tamquam eam *caelesti oraculo* [cf. 6. 3. 4, 'oraculo'] acceperim. sensi enim saepe dolens et gemens multas infirmorum perturbationes fieri per quorundam fratrum contentiosam obstinationem vel superstitiosam timiditatem, qui in rebus huius modi, quae neque sanctae scripturae auctoritate neque universalis ecclesiae traditione neque vitae corrigendae utilitate ad certum possunt terminum pervenire, . . . tam litigiosas excitant quaestiones, ut nisi quod ipsi faciunt, nihil rectum aestiment.'

vinulentia: M. appears here centre stage engaged in a practice that many thought bibulous; cf. 9. 8. 18 for the story of her childhood temptation to just such a fault (and see there for why this was a serious matter among women). In after years, Julian (*c. Iul. imp.* 1. 68: see on 9. 8. 18) would attend sharply to just this aspect of her personality.

dignationem: G–M: 'here . . . used (abstract for concrete) for the sip of wine which she took "for courtesy" before handing the cup to others. In *beata v.*

2. 10 "quando dignaberis" appears to mean "when you will be in the position of host".' Their reading of the context here is unexceptionable, but in the passage cited, we need understand nothing more than 's'il vous plaît'. *TLL* s.v. merely speculates '(i.e. libationem?)'; the word also occurs elsewhere, e.g. *epp.* 28. 4. 6, 151. 1, *s.* 43. 5. 6, but not in a similar sense, and cf. *s.* 58.1.2, 'sic enim debet vivere, qui invenit talem patrem ut dignus sit venire ad eius haereditatem. dicimus autem communiter, "pater noster". quanta dignatio? hoc dicit imperator, hoc dicit mendicus; hoc dicit servus, hoc dicit dominus eius. . . . intellegunt ergo se esse fratres, quando unum habent patrem.' In *conf.*, it occurs at 10. 25. 36, 'tu dedisti hanc dignationem memoriae meae, ut maneas in ea' (which closely resembles Amb. *Iacob* 1. 6. 25, 'sed vereris dubios vitae anfractus et adversarii insidias, cum habeas auxilium dei, habeas tantam eius dignationem, ut filio proprio pro te non pepercerit?').

partiretur: J. le Clercq (see *PL* 47. 207) emended to *potaretur*.

antistite: Elsewhere in *conf.* at 6. 7. 12 (Alypius), 9. 5. 13 (Amb.), 9. 7. 16 (Amb.): elsewhere in A., an infrequent substitute for 'episcopus'. The word had the advantage of being a native Latin term, and in many late antique authors (e.g. Ammianus) that virtue made this word preferable to the Grecism, but not so ordinarily for A., who also uses *antistes* for non-Christian religious leaders (*civ.* 8. 5, 10. 11). The several appearances in *conf.* ('episcopus' appears only 8 × in *conf.*) raise the possibility that the deliberately ambitious style of the work makes him—perhaps unconsciously—just slightly more fastidious in diction.

parentalia: The authentic Parentalia were celebrated on 13–21 February, with offerings to the shades of ancestors (Ov. *fast.* 2. 533–616). On Christian *refrigerium*, see *BA* 13. 676–7 n. and Quasten art. cit. above. Courcelle, *Recherches*, 87 n. 1, offers lengthy bibliography. J. W. Halporn, *JbAC* 19 (1976), 82–108, dates *s. Guelf.* 29 to this same period, in which A. attempts to deal with the explicitly 'pagan' character of some other drinking parties. For criticism of 'pagan' practices from 410/11, see *s.* 361. 6. 6.

gentilium: The word *paganus* never appears in *conf.*, and is rare elsewhere in A.; for reasons, see my note at *Class. Folia* 31 (1977), 163–9.

ut et S Maur. Knöll Skut. : *et ut* CDGO Ver.

(*et*) **sic** BPZ Maur. Knöll Skut. : *si* CDGOS Ver. Vega suggests that *sic* became *si* because the next word begins *com-*. The reading *si* follows the MS tradition impeccably, but the resulting text can scarcely be translated. It is, in view of the manuscripts, *possible* that the text printed here (as Skut. and Vega, etc.) is an attempt in one later family of MSS to clean up something impossible in the best manuscripts; but it is also *possible* that it is authentic and it is clearly better than the paradosis. On this reading, her change of heart makes two things possible: she could better dispose of the money she spent on grave-offerings, and eucharist could be celebrated at the *memoriae* (because the eucharistic fast could be observed).

in conspectu tuo: See on 1. 16. 26, 'deus meus, in cuius conspectu'.

conversationem: The word in this acceptation is classical, but owes its frequency in Christian authors to biblical usage (esp. 8 × in 1 and 2 Petr.): H. Hoppenbrouwers, *Graecitas et Latinitas Christianorum Primaeva*, 1 (Nijmegen, 1964), 47–95.

in bonis operibus: Cf. 1 Tim. 5: 9–10, 'vidua eligatur non minus sexaginta annorum, quae fuerit unius viri uxor, (10) in operibus bonis testimonium habens, si filios educavit.'

fervens spiritu: Cf. Act. 18: 25, 'et fervens spiritu loquebatur'; Rom. 12: 11–12, 'spiritu ferventes, domino servientes, (12) spe gaudentes, in tribulatione patientes, oratione instantes'.

viam vitae: Cf. Ps. 15: 11, 'notas mihi fecisti vias vitae', quoted verbatim at Act. 2: 28; Prov. 6: 23, 'quia mandatum lucerna est, et lex lux, et via vitae increpatio disciplinae'; Prov. 10: 17, 'via vitae custodienti disciplinam'; Prov. 15: 10, 'doctrina mala deserenti viam vitae, qui increpationes odit morietur'. The *via vitae* is undoubtedly partly to be explained as a periphrasis for Christ (see on 7. 7. 11), but the Proverbs use of *disciplina* is important. This story about M. dealt with discipline, submission to church authority, walls and Christians (8. 2. 4); it is that discipline that A. is beginning to face here as a guide to the *via vitae*.

6. 3. 3

This paragraph has drawn a variety of incompatible explanations, tinged with disapproval (from which A. carefully refrains: see 'quolibet tamen animo' below) and suspicions that A. exaggerates. Courcelle, *Recherches*, 92: 'l'évêque catholique de Milan, si occupé fût-il, n'eût sans doute pas refusé au rhéteur officiel de sa ville quelques heures de conversation intime, comme l'évêque manichéen Faustus, fort accaparé lui aussi, avait fini par accorder à Augustin, alors rhéteur à Carthage, l'honneur d'une audience particulière.' But not *all* readers are put off by the portrayal: J. H. Newman, *Letters*, 6 (Oxford, 1984), 246 (to H. Wilberforce (19 May 1838)): 'Rogers well suggests that St. Aug.'s account of St. Ambrose's conduct to him, (sitting still and reading a book) is a remarkable and happy specimen by way of contrast of the Catholic mode of effecting conversions.' It is also important to remember that Amb. was away from Milan at a crucial time, in the summer of 386, to attend on the fate of Priscillian at Trier (cf. H. Chadwick, *Priscillian of Avila* (Oxford, 1976), 136–7; J. R. Palanque, *Saint Ambroise et l'empire Romain* (Paris, 1933), 516–18, 579).

Newman's insight over Courcelle is in the phrase 'by way of contrast'. Amb. here is presented in a way that contradistinguishes him from Faustus in Bk. 5. A. approached F. to obtain from him private guidance and explanation of his difficulties; instead of profound wisdom, he found only a superficial eloquence. A. approached Amb. to examine his outward eloquence, but was led instead to surmise deeper wisdom. But when A. then sought to approach Amb. as he had approached Faustus, for private tutelage (or, as Courcelle, *Recherches*, 155 n. 2, put it, a 'lecture à deux': cf. 4. 8. 13, where friendship includes 'simul legere libros dulciloquos'), he found himself rebuffed and compelled to extract Amb.'s teaching from his public, episcopal proclamations (6. 3. 4). (Neither at 5. 6. 11 nor here did he want or get a completely private hearing from his teacher: here the plural verbs ('adessemus', 'discedebamus', 'coniectabamus') convey that; cf. 5. 6. 11, 'et aures eius *cum familiaribus meis* eoque tempore occupare coepi'.) The private study-circle of master and disciples was a feature of late antique life both inside and outside Christian communities; A. here marks a deliberate turn away from that tradition to a more public, less gnostic form of instruction. On the tradition, G. Fowden, *The Egyptian Hermes* (Cambridge, 1986), 190–1, with notes on the relevance to Christianity in Brown, *Body and Society*, 104–7. See below on 7. 9. 13 for reservations about the conventional view of the neo-Platonic 'circle' at Milan, which seems to have few of the features that Fowden and Brown identify in other contexts.

A.'s Manicheism may have played a part in Amb.'s reserve, and A. may himself have seen this, as another passage (clearly echoing the situation here, with perhaps a less sophisticated interpretation but offering both an excuse for the clergy and hope for the seeker) shows: *mor.* 1. 1. 1, 'nec si ea discere cupiens, in aliquos forte inciderit vel episcopos vel presbyteros vel cuiuscemodi ecclesiae catholicae antistites et ministros, qui aut passim caveant nudare mysteria aut contenti simplici fide altiora cognoscere non curarint, desperet ibi scientiam esse veritatis, ubi neque omnes a quibus quaeritur docere possunt, neque omnes qui quaerunt discere digni sunt. et diligentia igitur et pietas adhibenda est: altero fiet ut scientes inveniamus, altero ut scire mereamur.' Convenient conversions at court were not new to Amb.: Amb. *exp. Ps.* 118. 20. 48—see Brown, 82.

J. Mazzeo, *Jour. Hist. Ideas* 23 (1962), 192: 'Ambrose's "good reason" for silence was nothing else than listening to the instruction of the inner teacher.' He quotes Ignatius, *ep.* ad Eph. 15. 1–3, but could have quoted *en. Ps.* 126. 3, 'tamquam vobis ex hoc loco doctores sumus; sed sub illo uno magistro in hac schola vobiscum condiscipuli sumus' (sim. at *s.* 278. 11. 11, *s. Guelf.* 32. 4; cf. *s.* 298. 5. 5, 'superiore loco propter praeconium praesidemus, sed in una schola communem magistrum in caelis habemus'), or *s.* 52. 1. 1, 'ab illo [domino] enim exspectavit cor meum tamquam iussionem proferendi sermonis, ut hinc cum intellegerem loqui me velle quod recitari ipse voluisset.' For a larger collection of texts on that subject, see Madec, *BA* 6. 545–8.

A further juxtaposition is often overlooked. Ambrose is presented here as a man who divides his time between service to his flock, refreshing his physical powers, and nourishing his mind with reading (cf. 'reparandae menti suae nanciscebatur'). At 11. 2. 2, *bishop* Augustine proposes to divide his time similarly: 'et olim inardesco meditari in lege tua . . . et nolo in aliud horae diffluant quas invenio liberas a necessitatibus *reficiendi corporis* et intentionis *animi* et *servitutis* quam debemus *hominibus* et quam non debemus et tamen reddimus.' (A similar partition of his preoccupations at *ep.* 151. 13.) Similarly, Amb. here is portrayed facing warily the temptations to pride to be found in episcopal office; at 10. 36. 59 ff., the same temptations are the subject of A.'s meditation on the lingering threat he faced from the temptation of *ambitio saeculi*, and that theme recurs in Bk. 13 (see on 13. 25. 38) as well. The parallel is so strong that we must infer that Amb. is portrayed here as a model of episcopal conduct against which A. later measures himself. Amb. emerges as anything but aloof, but as *too* accessible and too prone to be distracted by the demands of his flock.[3]

But why would a bishop read silently? A. sees the bishop as the Christian orator *par excellence* (see *doctr. chr.*, esp. Bk. 4); at *sol.* 2. 14. 26, *Ratio* evokes 'ille *in quo ipsam eloquentiam, quam mortuam dolebamus, perfectam revixisse cognovimus*': see below for further quotation to indicate that Ambrose is meant. As far as A.'s own overt, public discussion of episcopal office is concerned, his oratorical function looms largest. (The priestly function of the bishop is one about which he has much less to say, but A.'s reticence about liturgical matters may be responsible for a distortion in our view here.) As the present passage suggests, Amb. must have been A.'s main model from life. Amb. shared A.'s view that the bishop's role centred on his teaching from the pulpit, and began his own *de officiis* (389) with a long disquisition on the merits of silence and of knowing your place and when to speak:

Amb. *off.* 1. 1. 2–2. 5, '... officium docendi, quod nobis refugientibus imposuit sacerdotii necessitudo. ... (3) non prophetarum gratiam, non virtutem evangelistarum, non pastorum circumspectionem, sed tantummodo *intentionem et diligentiam circa scripturas divinas opto adsequi* ... ut docendi studio possim discere.... (4) ego enim, raptus de tribunalibus atque administrationis infulis ad sacerdotium, docere vos coepi quod ipse non didici. itaque factum est ut prius docere inciperem quam discere, discendum igitur mihi simul et docendum, quoniam non vacavit ante discere. (5) *quid autem prae ceteris debemus discere quam tacere, ut possimus loqui*, ne prius me vox condemnet mea quam absolvat aliena? ... complures vidi loquendo peccatum incidisse, vix quemquam tacendo, ideoque tacere nosse quam loqui difficilius est.' The discussion continues in that vein as a proem to the *de officiis* to the end of

[3] The pattern A. constructs for episcopal conduct reflects an old model: Cic. *or.* 42. 143, 'alteri [oratores], cum domesticum tempus in cognoscendis componendisque causis, forense in agendis, *reliquum in se ipsis reficiendis omne consumerent*, quem habebant instituendi aut docendi locum?' So cf. Macrob. *sat.* 1. 2. 1–5, where we even have the hesitation to press a learned query on the time of a busy orator; but the atmosphere there is tellingly different from here. At 6. 11. 18, A. represents his own activities as orator and professor in a similar distribution: 'sed quando salutamus amicos maiores, quorum suffragiis opus habemus? quando praeparamus quod emant scholastici? quando reparamus nos ipsos relaxando animo ab intentione curarum?'

1. 6. 22, when Amb. turns to address his Ciceronian model. In a similar vein is Amb. *exp. Ps.* 36. 66, 'ideo meditare semper, loquere quae dei sunt, sedens in domo [Deut. 6: 7]. domum possumus accipere ecclesiam, possumum domum accipere interiorem in nobis, *ut intra nos loquamur*.'

That A. himself took the lesson of such behaviour is corroborated indirectly by his own words: *s.* 179. 1. 1–2. 2 (date uncertain but probably earlier than 410), 'verbi dei enim inanis est forinsecus praedicator, qui non est intus auditor. nec ita aversi ab humanitate et fideli consideratione sumus ut pericula nostra non intelligamus, qui verbum dei populis praedicamus. . . . nam ut noveritis, fratres, quam tutiore loco stetis quam nos. . . (2) ego qui vobis assidue loquor . . . tunc solidum gaudeo, dum audio.' Cf. also *en. Ps.* 85. 7, 'oratio tua locutio est ad deum; *quando legis, deus tibi loquitur*; quando oras, deo loqueris.'

In order, therefore, for the Christian bishop to teach authoritatively in the church, he must himself first *hear* attentively the word of God speaking in scripture. That message the bishop preaches openly in church, and it contains the fullness of Christian doctrine. There is no secret gnosis reserved for the elect. Both these ideas A. recalled as momentary stumbling blocks to him, his expectations conditioned by his Manicheism, but they both became essential elements of his own self-image as bishop. This view of the bishop's role appears stylized in the *vita Ambrosii* written under A.'s patronage by Paulinus of Milan *c.*422:[4] *v. Amb.* 17, 'is constitutus in ecclesia, tractante episcopo, vidit (ut ipse postmodum loquebatur) *angelum* ad aures episcopi tractantis loquentem, ut *verba angeli populo episcopus renuntiare videretur*.'

That Christianity has no secret gnosis remained important for A. The dialectic that he weaves throughout his career between *fides* and *intellegentia* reveals itself in one text (*Io. ev. tr.* 98) to be a sophisticated way of escaping the temptations of gnosis while retaining many of the benefits of such a system. A.'s Christianity makes a distinction not between kinds of doctrine but between kinds of believers—those who have penetrated further have themselves changed, but they have been given no essential teaching that was withheld from them before. For 1 Cor. 3: 1 interpreted in this light, see on 12. 27. 37. (The implicit issues raised here about the position of the bishop in the church return for their most important development later: see on 10. 36. 59.)

There is controversy about a parallel text in A. that would throw light on the present passage, showing that the frustration at Amb.'s aloofness lasted through the winter of 386–7: *sol.* 2. 14. 26, '[Ratio] . . . ante oculos nostros . . . ille *in quo ipsam eloquentiam, quam mortuam dolebamus, perfectam revixisse*

[4] The later date for Paulinus' *vita* is preferable in spite of É. Lamirande, *Paulin de Milan et la 'Vita Ambrosii'* (Paris/Montreal, 1983), 21–4. Though Lamirande prefers a date of 412–13, he does not positively rule out 422. He does not discuss the part A. played in the work's genesis; surely it is inescapable that A. wanted the *vita* as a buttress for his own reiterated appeals to Amb.'s authority at this period. More satisfactory is A. Paredi, *Sacris Erudiri* 14 (1963), 212–13 (good on the anti-Pelagian context of the *vita*).

cognovimus. illene nos sinet, cum scriptis suis vivendi modum docuerit, vivendi ignorare naturam? [A.] non arbitror equidem et multum inde spero, sed unum doleo, quod vel erga se vel erga sapientiam studium nostrum non ei, ut volumus, valemus aperire. nam profecto ille miseraretur sitim nostram et exundaret multo citius quam nunc. securus enim est, quod sibi iam totum de animae immortalitate persuasit, nec scit aliquos esse fortasse, qui huius ignorationis miseriam satis cognoverint, et quibus praesertim rogantibus non subvenire *crudele* sit.' This passage was applied to Amb. by the Maurists, and M. Ihm ('Studia Ambrosiana', *Jahrbücher für classische Philologie*, 17 (Supplementband, 1. Heft: Leipzig, 1890), 76) astutely suggested that the reference in *scriptis* is to the *de philosophia*; but Courcelle, *Recherches*, 206–7, rejected the identification in favour of Mallius Theodorus (on the grounds that what is required here is a philosopher rather than a theologian, and one known for his writings) and, though Madec, 252–6, rightly sees that Courcelle completely misunderstood the nature and contents of the *de philosophia*, he nevertheless follows Courcelle. There is no reason not to apply the *sol.* passage to Amb.: (1) on the bishop as model of eloquence, see above; (2) his writings abundantly provided a 'modus vivendi', specifically the life of continent pursuit of wisdom; (3) A. had been unable to share his inmost feelings with him; (4) A. had almost certainly heard his sermon of 386, *de Isaac vel de anima*, on the immortality of the soul. The 'cruelty' A. attributes then to this remote figure is a sign of the intensity of his frustration during that Cassiciacum winter, which is certainly compatible with the re-evocations of 6. 3. 4 and 6. 11. 18 ('quando quaeretur [veritas]? non vacat Ambrosio').

ad quaerendum intentus: A.'s situation is not entirely bad: he is not yet ready to pray, but at least ready to seek—see on 1. 1. 1.

et ad disserendum inquietus: A. seeks dialogue that he never gets from Amb. He must learn to seek by listening, but if there is 'Manichean reflex' here it is in his own eagerness to speak and quarrel—the *loquacitas* of Manicheism he often rebukes (1. 4. 4, 3. 6. 10, 7. 2. 3, etc.).

caelibatus: *caelebs* is infrequent in A., and its use suggests that it is the masculine equivalent of *innupta*, 'unmarried' (*c. Faust.* 29. 4, *virg.* 27. 27, *b. vid.* 5. 7, *en. Ps.* 40. 4); hence here an observation without overt moral colouration. Cf. 6. 12. 22, 'ut me affirmarem . . . caelibem vitam nullo modo posse degere'; to this point, continence is a puzzle not apparently related to A.'s own search. It does not seem that in early 385, hearing Amb. at Milan, A. imagined that his own future would be conditioned by a decision to emulate the continence of the 'pagan' and Christian sages; A. still sought an intellectual resolution to his perplexities. Bk. 8 testifies to the redirection of his quest under Amb.'s influence. Here in *conf.*, however, he tells us less than the whole story; later he will suggest (see on 8. 7. 17, 'petieram') that the search for continence had been with him ever since adolescence, but for the moment he merely says that there was in Amb.'s own conduct that which was a stumbling block for A. and a sign of things to come. Psychologically, A. clearly seeks in Amb. something of a father figure (cf. 5. 13. 23, 'paterne', and *c. Iul.* 1. 3. 10, '[Ambrosium] excellentem dei dispensatorem, quem ut patrem veneror'), but if Amb. is celibate, on some subconscious level that makes it impossible for A. to accept him as a father.

quid autem: Courcelle, *Recherches*, 219–21, takes this text as evidence that at some time after conversion A. and Amb. found the opportunity for conversations of a more personal nature; not one word in the text suggests that. The words 'nec conicere noveram nec expertus eram' suggest that A.'s later enlightenment came from conjecture and his *own* episcopal experience.

et adversus CDGO Maur. Ver. : *adversus* S Knöll Skut.

excellentiae: In *conf.* of God and things divine (1. 20. 31, 2. 6. 13, 3. 7. 14, 6. 5. 8) and of human affairs (both in a good sense [4. 16. 30, 7. 19. 25 (2 ×), 8. 6. 14) and a bad sense (here and 10. 38. 63, 1. 19. 30, 2. 6. 13).

ruminaret: *c. Faust.* 6. 7, 'immundum quippe illud animal in lege positum est eo quod non ruminet; . . . sunt autem homines qui per hoc animal significantur, immundi proprio vitio, non natura; qui cum libenter audiant verba sapientiae, postea de his omnino non cogitant. quod enim utile audieris, velut ab intestino memoriae tamquam ad *os* cogitationis recordandi dulcedine revocare quid est aliud quam spiritaliter quodam modo *ruminare*?'

foveam: The image is biblical (cf. Ps. 56: 7) and familiar in A.'s other works, but appears only three times in *conf.*, and on each occasion one of the three friends, A., Alypius, and Nebridius, is imperilled: 6. 7. 12, 'proripuit [Alypius] se ex fovea tam alta'; 9. 3. 6, 'in illam foveam perniciosissimi erroris inciderat [Nebridius]'. Parallels suggest that with this image A. has in mind specifically the Manichees, sinners through concupiscence of the eyes and hence most aptly the blind leading the blind (Mt. 15: 14, 'caecus autem si caeco ducatum praestet, ambo in foveam cadunt'), preparing traps for their brethren and falling in themselves, rejoicing in their iniquity: *en. Ps.* 56. 14, 'omnis qui parat foveam fratri suo, necesse est ut ipse incidat in eam. . . . ipsa laetitia iniqua facientis, ipsa est fovea.'

non enim quaerere ab eo poteram: The same frustration was encountered on first seeing Faustus, but then overcome: see on 5. 6. 11, 'sed moleste habebam'.

negotiosorum AH²BPZ Maur. Knöll Skut. : *negotiorum* CDGOS Ver. A crux. (1) The paradosis *negotiorum* is solid. (2) How to read it is less obvious: punctuate with a comma and assume asyndeton: 'crowds of affairs, of men who . . .'? (3) If *negotiosorum* is read, an easier sense is given, not unparalleled in A. (*c. Faust.* 22. 56, 'actuosi et negotiosi homines'; the adj. 23 × in all in A.). (4) *Catervis* is much more common with genitives of people than of things. It is hard to resist the idea that *negotiosorum* is at least the best emendation to date.

sed cum legebat: Silent reading. This passage has attracted much notice, seeming to show Amb. as the veritable inventor of silent reading (J. Balogh, *Philologus*, 82 (1926), 84–109, 202–40), but it has since been shown (B. M. W. Knox, *GRBS* 9 (1968), 421–36) that the practice was known from much earlier in antiquity (Cicero, *Tusc.* 5. 40. 116, describes the pleasure the deaf can derive from reading, without hearing, poetry). Reading aloud remained a—probably the—common practice for many centuries (sometimes a necessary one: in *ep.* 101. 3 A. says that his *de musica* is hard to understand unless you have somebody to read it out loud who can vocalize the quantities properly); being-read-to also attested at *Io. ev. tr.* 112. 1. A near contemporary recommended reading aloud softly as an aid to memorization (Mart. Cap. 5. 539, in the persona of Rhetoric). The written word was prompt-copy for a text that was still essentially oral in nature; only when read aloud did the symbols on the page become words in a useful sense. With some kinds of texts, A. was a perfectly competent consumer of a purely textual artefact (see on 4. 16. 28, the *categ.* of Aristotle); but his earlier, purely textual, encounter with scripture (3. 5. 9) had failed, and here he betrays his need for a more traditional, orally mediated reading of the text. There remains to be written a history of reading that would do justice to the variety of techniques and situations across the centuries, on the one hand, and to the way reading techniques affected the ways people thought and acted.

The significance of this passage is to be found elsewhere; see above, introd. n. (Balogh already noted that if Amb. invented the skill, A. was a quick learner; cf. 8. 12. 29, 'aperui et legi in silentio'; G. Lawless, *REAug* 26 (1980), 55, also notes that 10. 8. 13 offers the parallel of 'silent singing': 'et quiescente lingua ac silente gutture canto quantum volo'; and cf. 11. 27. 36, 'et voce atque ore cessante peragimus cogitando carmina et versus et quemque sermonem'). A better context is the history of vocal and silent prayer, on which see J. Balogh, *Archiv für Religionswissenschaft* 23 (1925), 345–8. For example, Cassian, *conl.* 9. 35, 'in abscondito oramus, quando corde tantum et intenta mente petitiones nostras soli pandemus deo, ita ut ne ipsae quidem adversae valeant potestates genus nostrae petitionis agnoscere. propter quod summo est orandum silentio, non solum ne fratres adstantes nostris susurris vel clamoribus avocemus . . . sed ut ipsos quoque inimicos nostros, qui orantibus nobis maxime insidiantur, lateat nostrae petitionis intentio.' From Amb., cf. *exp. Ps.* 118. 17. 9, 'Anna, cum oraret tacita clamabat, labia non movebat et interioris voce pia mentis excitabat Iesum,' reflecting the text of 1 Sam. 1: 13, 'Anna loquebatur in corde suo, tantumque labia illius movebantur, et vox penitus non audiebatur.'

coniectabamus: The reason given here is conjecture dating from the time of the event. It is almost correct. Amb. was moved not simply by the desire not to get bogged down explaining things (and hence waste of time), but by the desire to hear for himself, and to reserve his explanations for the pulpit (see above). Note that A. here recalls no displeasure on his part at this frustration; contrast 5. 6. 11, 'sed moleste habebam', for a contrasting recollection when faced with a similar frustration reaching Faustus; see above on *sol.* 2. 14. 26 for possible different feelings closer to the event.

obtundebatur: *OLD* (as G–M): 'to make (the voice) husky or hoarse'; Cic. *de or.* 2. 70. 282, 'cum vocem in dicendo obtundisset'.

6. 3. 4

See Courcelle, *Recherches*, 98–132, on the sermons A. could have heard at Milan 385/7 (see on 5. 14. 24). In such studies, two errors must be avoided: (1) of conflating too quickly what we learn there with the biographical narrative of *conf.*; this narrative is directed by definite rhetorical strategies, and what it contains and what it omits is an important distinction that must be borne in mind; (2) of reading only the works of Amb. that present sermons A. might actually have heard in 385/7; we have just seen, for instance, on 6. 3. 3 that Amb.'s *de officiis* of 389 offers useful illumination to *conf.*

An earlier passage should be kept in mind: 5. 10. 20, 'cum enim conaretur animus meus recurrere in catholicam fidem, repercutiebar, quia non erat catholica fides quam esse arbitrabar.' It would be wrong here to speak of what A. reports as 'conversion' of any kind. A. is not changing his mind about any doctrine of faith, but correcting his impression of Christian doctrine, finding that doctrine compatible with what he believed. This remains an aspect of his narrative as late as 7. 19. 25, when A. and Alypius sort out their Christological views. Of the three Manichean challenges to Christianity whose force A. acknowledged at 3. 7. 12, only the last of them, to the authority of the Old Testament, is in any way affected by the teachings of Amb. reported here (6. 4. 6 on 2 Cor. 3: 6).

oraculo: Other *oracula* in *conf.* are scriptural in authority: 8. 12. 29, of the text that converted Antonius; 11. 9. 11, the text of Ps. 103: 24; 12. 15. 22, Moses and his books ('oracula sancti spiritus'). This is high praise for Amb. in his role as master of divine eloquence.

pectore illius: See on 3. 4. 7, 'pectus'. Cf. *ep.* 31. 8 (addressing Paulinus of Nola), 'nam *pectus* tuum tale domini *oraculum* est'.

nisi . . . audiendum: e.g. the inquiries A. made on behalf of Monnica.

aestus: 6. 3. 3.

in populo . . . tractantem: Unlike Faustus, with Amb. the public proclamation has a considerable positive effect, public, in church, on Sunday. 2 Tim. 2: 15, 'sollicite cura teipsum probabilem exhibere deo, operarium inconfusibilem *recte tractantem verbum veritatis*.'

omni die dominico: Cf. *beata v.* 1. 4, 'saepe' (see below).

deceptores: 7. 2. 3, 'adversus illos deceptos deceptores'.

comperi: Cf. 6. 11. 18, 'nefas habent docti eius credere deum figura humani corporis terminatum.' *beata v.* 1. 4 reports a slightly more advanced view derived from Amb.'s sermons and the words of Mallius Theodorus: 'animadverti enim et *saepe* in sacerdotis nostri et aliquando in sermonibus tuis, cum de deo cogitaretur, nihil omnino corporis esse cogitandum, neque cum de anima.' In *conf.* that more accurate statement of the immateriality of God is impossible before Bk. 7; here A. presumably encounters a non-anthropomorphic view that encourages but does not satisfy him fully.

ad imaginem tuam: Gn. 1: 26, 'faciamus hominem ad imaginem nostram'; Gn. 9: 6, 'ad imaginem quippe dei factus est homo'; Sirach 17: 1, 'deus creavit de terra hominem et secundum imaginem suam fecit illum.' The doctrine of creation of the first human in God's image and likeness is a potent one for A., and its appearance in *conf.* is carefully controlled and restrained: hinted at in 3. 7. 12 (as the Christian truth contrary to the error attributed to Christianity by the Manichees), here, then twice just as the *platonicorum libri* are coming into view in Bk. 7 (7. 9. 11, 7. 9. 15), then not at all until the important and climactic passage beginning at 13. 22. 32.

Amb.'s teaching freed A. from anthropomorphism, but Amb. thought of image and likeness as a goal to be reached, not an inherent quality (hence emphasized the *ad* of Gn. 1: 26: for A.'s development on this subject, see on 13. 22. 32). See Markus, *REAug* 10 (1964), 137–40, and cf. G. B. Ladner, *The Idea of Reform* (Cambridge, Mass., 1959), 185 ff.; H. Somers, *REAug* 7 (1961), 105–25, who neatly deploys the stages in A.'s exegesis of Gn. 1: 26, from *Gn. c. man.*, to *Gn. litt. imp.*, to *Gn. litt.*, to the last stage, the late-added conclusion to *Gn. litt. imp.*, showing the increasing influence of Christian sources, esp. Greek ones, later in A.'s career. At the time of *conf.*, the principal Christian influence is his contact with anti-Arian polemic (perhaps even including Marius Victorinus' *adversus Arium*, though this is unlikely). There is value in G. A. McCool's study of Amb.'s influence on A. in this regard, *Theol. Stud.* 20 (1959), 62–81, even if the attempt to show direct and immediate dependence on Amb. is judged unsuccessful: the analogies of approach retain their interest.

catholica: Not adj. but noun; see on 5. 14. 24.

regenerasti: The verb is regularly (and in *conf.*, only: 8. 2. 4, 9. 3. 6, 9. 13. 34) associated with baptism. Cf. *pecc. mer.* 2. 27. 43, 'sacramentum autem baptismi profecto sacramentum regenerationis est.'

te determinatum scripsi : *te terminatum* Maur. Pusey G–M : *determinatum* MSS Knöll Skut. Ver. Cf. 5. 10. 19, 'figuram te habere . . . et . . . liniamentis corporalibus terminari'; 6. 4. 5, 'te creatorem . . . undique terminatum membrorum humanorum figura'; 6. 11. 18, 'credere deum figura humani corporis terminatum'. As G–M rightly saw, 'The *te* appears indispensable to the sense,' but despite the parallels in *conf. determinatum* is more frequent elsewhere (7 ×: *ep.* 118. 4. 31, *s.* 23. 5. 5, *Io. ev. tr.* 53. 2, *Gn. litt.* 6. 12. 20, *nat. b.*

41 ('nec ipsi qui hoc agebant formis suis determinati essent, nisi *modus* ibi esset'), *en. Ps.* 30. en. 2. s. 1. 7, *s. Mai* 158. 4) than *terminatum* (4 ×: *epp.* 148. 1. 1, 148. 1. 2, *Io. ev. tr.* 40. 4, *s.* 23. 6. 6); note the variation within two paragraphs of *s.* 23 (happily confirmed by an ed. crit.).

The accusation of anthropomorphism against orthodox Christianity was not wholly baseless. Apart from the language of the OT, there were sects literal-minded enough to assert anthropomorphism openly (*haer.* 50; for the difficulties, cf. 'Audianer', *RAC* 1. 910–15).

tenuiter atque in aenigmate: 1 Cor. 13: 12, 'videmus nunc per speculum in aenigmate'; *en. Ps.* 130. 14, 'adhuc ex parte, in aenigmate'.

discere S Maur. Knöll Skut. : *dicere* CDGO Ver. A. does not associate *dicere* and *quaerere* elsewhere.

altissime et proxime: 5. 8. 14, 'altissimi tui recessus et praesentissima in nos misericordia'; 1. 4. 4, 'secretissime et praesentissime'.

alia maiora et alia minora: Cf. *ep.* 118. 4. 23, 'qui enim didicerit deum non distendi aut diffundi per locos neque finitos neque infinitos, quasi in aliqua parte maior sit in aliqua minor, sed *totum ubique* esse praesentem ... nequaquam eum movebit quod de infinito aere sensit, quicumque sensit quod ipse esset deus.' On 'ubique totus', see on 1. 3. 3.

6. 4. 5

pulsans: Mt. 7: 7, 'pulsate, et aperietur vobis'; i.e. it is better to have begun the sequence invoke–seek–find–praise (1. 1. 1), than to go off in the ways of *curiositas*; the same text in the last words of *conf.* (13. 38. 53).

proponerem: G–M: = *quaestionem proponerem*. 'The reference is apparently to the attitude which A. would have taken up if he had had the opportunity of discussing his difficulties with Ambrose.' For imperfect for pluperfect, cf. 6. 11. 20, 'dares si pulsarem', and 8. 6. 13, 'posset, si vellet'. That reading is preferable to that of the translators: Pusey: 'I should have knocked'; *BA:* 'j'aurais dû frapper à la porte.' The preceding sentence (the last of 6. 3. 4) is a 'confessional' interruption that obscures the connection to the last sentence but one. There, A. described and repented his earlier hard-headedness; here, he suggests what his attitude would now have been, after the impact of Amb.'s sermons.

ita: 'in such a way'—i.e. in the way he mistakenly thought Christians took the doctrine.

rodebat: 8. 7. 18, 'rodebar intus et confundebar' (cf. 'confundebar et convertebar' here).

quid certi retinerem: The indirect question is governed by 'cura'.

certi: The yearning for certainty carries over from Bk. 5, e.g. 5. 14. 25, 'donec aliquid certi eluceret'.

animositate: See on 2. 3. 5.

garrisse: 5. 6. 10 (of Faustus), 7. 20. 26 (of himself upon reading the Platonic books), 9. 1. 1 (of prayer (!) after the garden scene), 12. 28. 38 (of incautious exegetes). On indiscretion of speech as a Manichean characteristic, see on 1. 4. 4, 'quoniam loquaces muti sunt'.

confundebar et convertebar: Ps. 6: 11, 'convertantur et confundantur valde velociter'. There the collocation is a prayer directed against others, leading to their abasement; here, for himself, it is narrative that leads to *gaudium*; the order of the two verbs is reversed. As Manichee he was acting just as the adversary of the Psalmist would act; and the fate the Psalmist prayed for came down upon him—and it was good for him.

corpus unici tui: Col. 1: 18, 'et ipse est caput corporis ecclesiae, qui est principium, primogenitus ex mortuis ut sit in omnibus ipse primatum tenens'; Col. 1: 24, 'qui nunc gaudeo in passionibus pro vobis et adimpleo ea quae desunt passionum Christi in carne mea pro corpore eius, quod est ecclesia.' *Unicus* is not in the Vulgate, but not infrequent in A. only for Christ: also at 7. 9. 14, 9. 8. 17, 10. 43. 69, 10. 43. 70, 11. 2. 4, 13. 2. 3, 13. 13. 14, 13. 34. 49 (see on 1. 7. 12 for *une* of the Father; *BA* 72. 234 n. 34 with parallels); *en. Ps.* 102. 6, 'unicus filius dei'.

nomen Christi: See on 3. 4. 8. The allusion is to A.'s initiation as catechumen (1. 11. 17).

nugas: The slur is directed against others at 3. 10. 18 ('perductus ad eas nugas ut crederem ficum plorare'), 7. 6. 8 ('flagrabant in eas nugas (mathematicorum)').

te creatorem omnium: Amb. *hymn.* 1. 2. 1 (see 9. 12. 32).

undique terminatum: See on 6. 3. 4, 'te determinatum'.

6. 4. 6

He heard from Ambrose a principle that would lay open all of Christian teaching to him, but he allowed it to work only so far as to weaken his attachment to anything else and to tear down the barriers that separated him from Christianity. He could have looked upon the truth now with 'purgatior acies mentis', and he did not. Cf. *util. cred.* 8. 20, quoted in Introduction. At 5. 14. 24, A. reported the first impression Amb.'s sermons made: a similar reliance on 2 Cor. 3: 6, with perhaps a difference. There he reports that Amb. showed, by the practical application of that verse (echoed there not as attributed to Amb., but as part of A.'s own interpretation in making sense of the episode), that various passages of scripture were not taken by catholics in the way A. had thought they must be; here, Amb. is reported as expounding directly the principle implied by 2 Cor. 3: 6.

illo oculo . . . audiebam: A.'s erroneous reading of scripture is put in visual, literal terms, while the key to his correct reading comes through the sense of hearing.

quo . . . absurda: 5. 14. 24, 'non absurde obiecta refellerent'.

popularibus: Cf. 6. 3. 4, 'in populo . . . tractantem'; more than a third of A.'s own surviving *œuvre* consists of transcripts of sermons (chiefly *sermones*, *en. Ps.* and *Io. ev. tr.*), and he calls them 'sermones populares' at *civ.* 17. 17; see on 6. 3. 3.

littera occidit, spiritus autem vivificat: 2 Cor. 3: 6, 'littera occidit, spiritus autem vivificat'; for Amb.'s sermons citing this text, see Courcelle, *Recherches*, 98 n. 4, and Solignac, *BA* 13. 141 n. 1, and see on 5. 14. 24. This text *could* have played a much more visible role in A.'s exegetical principle and practice than it did, and we must be careful how we interpret it. The *distinction* is one that A. often employed, as early as 388/90 in *Gn. c. man.* 2. 2. 3, 'sane quisquis voluerit omnia quae dicta sunt secundum litteram accipere, id est non aliter intellegere quam littera sonat, et potuerit evitare blasphemias et omnia congruentia fidei catholicae praedicare, non solum ei non est invidendum sed praecipuus multumque laudabilis intellector habendus est. si autem nullus exitus datur, ut pie et digne deo quae scripta sunt intellegantur, nisi figurate atque in aenigmatibus proposita ista credamus, habentes auctoritatem apostolicam, a quibus tam multa de libris veteris testamenti solvuntur aenigmata, modum quem intendimus teneamus, adiuvante illo qui nos petere, quaerere et pulsare adhortatur, ut omnes istas figuras rerum secundum catholicam fidem, sive quae ad historiam sive quae ad prophetiam pertinent, explicemus, non praeiudicantes meliori diligentiorique tractatui, sive per nos sive per alios dominus revelare dignatur.'

But that passage takes a strongly pro-literal view, seeing allegory as an escape from difficulties and praising the man who can expound scripture without recourse to it—hardly the view of Amb. endorsed here. The first significant appearance of the scripture citation is in the first work A. wrote as an ordained clergyman: *util. cred.* 3. 9, 'in quibus tamen legis praeceptis atque mandatis, quibus nunc christianos uti fas non est, quale vel sabbatum est vel circumcisio vel sacrificia et si quid huiusmodi est, tanta mysteria continentur ut omnis pius intellegat nihil esse perniciosius quam quidquid ibi est accipi ad litteram, id est ad verbum; nihil autem salubrius quam spiritu *revelari*. inde est: "littera occidit, spiritus autem vivificat".' So also *exp. prop. Rom.* 10 (11), *div. qu. Simp.* 1. 1. 17.

That view persists through *conf.* and fills the pages of *doctr. chr.* (e.g. 3. 5. 9), but it was not A.'s last word on the subject. In commenting on the *util. cred.* in *retr.*, A. quibbles with his own interpretation: *retr.* 1. 14. 1, 'in hoc libro dixi, "in quibus tamen legis praeceptis . . . inde est, 'littera occidit, spiritus autem vivificat'." sed aliter exposui verba ista apostoli Pauli et, quantum mihi videtur vel potius ipsis rebus apparet, multo convenientius in eo libro qui inscribitur "de spiritu et littera" [*spir. et litt.* 5. 7, 'non de figuratis locutionibus dictum, quamvis et illic congruenter accipiatur, sed potius de lege aperte quod malum est prohibente'], quamvis non sit et sensus iste respuendus.' In *spir. et litt.* (esp. 4. 6–5. 8) the passage becomes a key to interpretation not merely of texts but of conduct and history. This re-reading of the verse saves the OT from over-allegorism: *spir. et litt.* 14. 23 makes this clear, for there are for A. precepts of the Law of the OT that are not simply 'umbrae futuri' but that remain characteristics of just action (the decalogue is instanced). A similar spirit motivates a work of controverted authenticity, the *speculum quis ignorat*: most recently vindicated for A. by A. Mutzenbecher, *REAug* 30 (1984), 60–83, and A.-M. La Bonnardière, *Augustin et la Bible* (Paris, 1986), 401–9, it had been impugned by G. de Plinval (*Aug. Mag.* 1. 187–92), who saw in its commandment-centred approach a virtual Pelagianism. But the tendency is exactly that of *spir. et litt.*, the first *anti*-Pelagian work from A.'s pen. (On A.'s exegesis there is an extensive literature (for survey, La Bonnardière's *Augustin et la Bible* is excellent); on its pertinence here, cf. J. Pépin, *RA* 1 (1958), 243–86.)

remoto mystico velamento: 2 Cor. 3: 14–16, 'usque in hodiernum enim diem idipsum velamen in lectione veteris testamenti manet, non revelatum, quoniam in Christo evacuatur; (15) sed usque in hodiernum diem, cum legitur Moyses velamen positum est super cor eorum; (16) cum autem conversus fuerit ad dominum auferetur velamen.' The image of the 'veil' serves in lieu of letter/spirit in many of A.'s early works; Mayer, *Zeichen*, 2. 465, cites from the anti-Manichean period alone numerous echoes, e.g. *Gn. c. man.* 1. 22. 33 ('velamen enim aufertur quando similitudinis et allegoriae cooperimento ablato veritas nudatur ut possit videri'), 2. 26. 40, *Gal. exp.* 22, s. dom. m. 2. 20. 68, *c. ep. fund.* 11. 12, *div. qu. Simp.* 2 pr., *c. Faust.* 6. 3, 12. 23, and *adn. Iob* on 26: 8 and 38: 32.

tenebam enim cor: G–M: 'A complex word-play. There is, first, an allusion to the suspense of judgement which the Academics advocated ("refrenationem et quasi suspensionem assensionis," *c. acad.* 2. 5. 12), further "timens praecipitium" seems to hint that "adsensio" plays on "ascensio" as "suspendium" does on "suspensio." The word-play may be roughly reproduced thus: "I hung back from every assent, dreading precipitation, and died by hanging instead".'

septem et tria decem: This truth is a benchmark of certitude for A., early and late in life, even when all else is in doubt: *c. acad.* 2. 3. 9 (quoted in next note), *lib. arb.* 2. 8. 21, 'septem autem et tria decem sunt et non solum nunc sed etiam semper neque ullo modo aliquando septem et tria non fuerunt decem aut aliquando septem et tria non erunt decem'; sim. at *lib. arb.* 2. 12. 34, *ep.* 162. 2 (414/15). On the numerology, *doctr. chr.* 2. 16. 25, 'porro autem denarius numerus creatoris atque creaturae significat scientiam; nam trinitas creatoris est, septenarius autem numerus creaturam indicat propter vitam et corpus. nam in illa tria sunt, unde etiam "toto corde, tota anima, tota mente" diligendus est deus; in corpore autem manifestissima quattuor apparent, quibus constat, elementa.'

comprehendi: On this word's Academic sense, see on 5. 10. 19.

purgatior acies: *c. acad.* 2. 3. 9, 'ego enim nunc aliud nihil ago quam me ipse *purgo* a vanis perniciosisque opinionibus. . . . cavete ne quid vos nosse arbitremini, nisi quod ita didiceritis saltem, ut nostis unum duo tria quattuor simul conlecta in summam fieri decem.' Cf. Plot. 1. 6. 9. 25–32, *ἐὰν δὲ ἴῃ ἐπὶ τὴν θέαν λημῶν κακίαις καὶ οὐ κεκαθαρμένος ἢ ἀσθενής, ἀνανδρίᾳ οὐ δυνάμενος τὰ πάνυ λαμπρὰ βλέπειν, οὐδὲν βλέπει, κἂν ἄλλος δεικνύῃ παρὸν τὸ ὁραθῆναι δυνάμενον. τὸ γὰρ ὁρῶν πρὸς τὸ ὁρώμενον συγγενὲς καὶ ὅμοιν ποιησάμενον δεῖ ἐπιβάλλειν τῇ θέᾳ. οὐ γὰρ ἂν πώποτε εἶδεν ὀφθαλμὸς ἥλιον ἡλιοειδὴς μὴ γεγενημένος, οὐδὲ τὸ καλὸν ἂν ἴδοι ψυχὴ μὴ καλὴ γενομένη.* He could thus now have attempted with some success the 'ascent' of the mind (as he is reported to have done at 7. 10. 16 and 7. 17. 23), but he did no such thing. If this be taken as a Plotinian echo, its purpose is to situate this moment on the trajectory leading to and past those 'tentatives.' For the purification of vision that does come, see on 7. 8. 12, 'collyrio'.

veritatem tuam semper manentem: Cf. Ps. 116: 2, 'et veritas domini manet in aeternum'; *en. Ps.* 116, 'sive in eis quae promisit iustis, sive in eis quae minatus est impiis.'

ne falsa crederet: What he sought in Manicheism was solid reasonable argument, but he discovered it to be mere, even vulgar, faith. When disabused of that faith, his earlier faith in reason and argument supervenes and leads him towards, but not into, the Christian community.

resistens manibus tuis: Cf. Ps. 16: 8, 'a resistentibus dextrae tuae custode me'; cf. Dan. 4: 32, 'et non est qui resistat manui eius, et dicat ei, quare fecisti?'

medicamenta fidei: *vera rel.* 24. 45, 'animae medicina . . . tribuitur . . . in auctoritatem atque rationem. auctoritas fidem flagitat et rationi praeparat hominem. ratio ad intellectum cognitionemque perducit, quamquam neque auctoritatem ratio penitus deserit, cum consideratur cui credendum sit, et certe summa est ipsius iam cognitae atque perspicuae veritatis auctoritas. sed quia in temporalia devenimus et eorum amore ab aeternis impedimur, quaedam temporalis medicina, quae non scientes sed credentes ad salutem vocat, non naturae excellentia, sed ipsius temporis ordine prior est. nam in quem locum quisque ceciderit, ibi debet incumbere ut surgat.'

6. 5. 7

sive esset quid: Setting aside the question of natural theology: 'whether it was something [demonstrable] and there was, as it happened [*forte*], no one to whom it was [demonstrable/demonstrated], or whether it just was not something [demonstrable] at all.'

illic: i.e. *inter manichaeos*.

absurdissima: Though himself enamoured of paradox, A. retains the label *absurdus* for outwardly contradictory propositions that he himself chooses to disown, and he is fond of introducing such dismissals with the phrase, 'nihil absurdius quam' (in *conf.* cf. 12. 2. 2, 'non absurde', and 12. 29. 40, 'non est absurdus'). In the present context, 'fabulosissima' denotes the origin of the Manichean doctrines, 'absurdissima' their irrational quality. Contrast the disintegrating *absurditas* of scripture at 6. 5. 8.

credenda imperari: In one way A. became an Academic in 385 and never gave up that sect. He ceased to believe that philosophical argument and demonstration will ever lead to the essential truth. The most that can be expected of the rationalist Manichees is that they will offer *scientia*, mock *credulitas*, and then proffer their own implausible *credenda*.

consideranti: Agrees with 'mihi' below.

quam innumerabilia crederem: Cf. the whole of *f. invis.*, written in 400, one of A.'s post-*conf.* literary tasks, an extended footnote to this chapter, a sign of the strength he conceded in the arguments of those who felt that Christianity was all authority and no reason. (See also *util. cred.* 9. 21–14. 32.) *f. invis.* 1. 1, 'sunt qui putant christianam religionem propterea ridendam potius quam tenendam, quia in ea non res quae videatur ostenditur, sed fides rerum quae non videntur hominibus imperatur. ... admonendi sunt quam multa non solum credant, verum etiam sciant, quae talibus oculis videri non possunt. quae cum sint *innumerabilia* in ipso animo nostro ...'

de quibus parentibus ortus essem: See on 1. 6. 7, 'non enim ego memini', from the infancy narrative, on his reliance on others to tell him of his own life; *f. invis.* 2. 4, 'coguntur fateri incertos sibi esse parentes suos.'

persuasisti mihi: Faith comes *first* as faith in the authority of scripture; not in any doctrines taught by or through scripture, but just in scripture itself (this is a form of faith in the incarnate Word, incarnate in the scriptural text: see on 6. 5. 8, 'quae via'), as the necessary preliminary to learning any doctrine from the scriptural text. From here forward, everything A. does and hears and says in the rest of Bks. 6–8 are the acts of a man who accepts the authority of scripture. (Note the discretion of this paragraph: specific scriptural authority is cited not at all.)

quos ... fundasti: Cf. *culmen auctoritatis* et sim. of scripture often elsewhere (6. 5. 8, 6. 11. 19, 12. 16. 23, and 12. 26. 36). It is at first glance naïve of A. to suggest that Christian scripture holds a special and privileged position of authority that even a Platonist reader (as when such a phrase occurs at *civ.* 10. 32) might be expected to recognize. The kernel of truth is that Christianity was a profoundly and publicly textual and historical community. The scriptures were not an esoteric possession, jealously guarded, and their message was factual and historic in its form, depending on neither mystery nor myth nor mere argument and theory. The advantage was one that Christians jealously guarded and pressed, and, as here, proclaimed proudly. It is thus not mere brag and bluster, but a serious attempt to characterize and explain literary success, when he begins *civ.* 11. 1, 'civitatem dei dicimus, cuius ea scriptura testis est quae non fortuitis motibus animorum, sed plane summae dispositione providentiae super omnes omnium gentium litteras omnia sibi genera ingeniorum humanorum divina excellens auctoritate subiecit.' *Auctoritas* was, *inter alia*, an antidote to Academic scepticism, as in the sketch of his opinions at *util. cred.* 8. 20 ('saepe rursus intuens . . .': quoted in Introduction). The standard study is K. H. Lütcke, *'Auctoritas' bei Augustin* (Stuttgart, 1968).

per tam multa . . . philosophorum: 5. 3. 3, 'multa philosophorum legeram.'

ut aliquando . . . pertinere: The antecedent faith in the existence of God and divine care for human affairs is one commonly held among the philosophers and sects of the time (cf. Cic., *nat. deor.* 2. 1. 3, 'primum docent [nostri] esse deos, deinde quales sint, tum mundum ab eis administrari, postremo consulere eos rebus humanis'). A. does not regard *that* faith as a specifically Christian one, nor does he introduce it to indicate that he regards it as a sign of progress; it is the obverse of one failure he never risked. (The first lines of 6. 5. 8 further qualify this element of faith.) See Boeth. *cons.* 1. P6. 5 and 1. P6. 20 for a similar reduction to the bare minimum of belief in God and his concern for human affairs.

6. 5. 8

id: The minimalist faith just outlined at the end of 6. 5. 7.

ignorabam: An ignorance cured in Bk. 7.

quae via: i.e. Christ—this looks ahead to the end of Bk. 7 and the substance of Bk. 8; see on 7. 7. 11.

cum essemus infirmi: Rom. 5: 6, 'ut quid enim Christus, cum adhuc infirmi essemus, secundum tempus pro impiis mortuus est?' Reception of scripture is thus tied to the incarnation (see on 6. 5. 7).

veritatem: i.e. Christ.

auctoritate sanctarum litterarum: For A. 'authority' is in its origin empirical rather than innate; the respect shown to scriptural texts is itself the proof of authority.

absurditatem: The noun form is first attested in A., in whom it is abundant: cf. *TLL* s.v. and Hrdlicka, 11; see on 6. 4. 7, 'absurdissima'.

exposita audissem: A rare hint of pre-Ambrosian Christian preaching to which A. gave heed; see 5. 11. 21 for a more explicit reminiscence.

sacramentorum altitudinem: No attempt to distinguish rigorously between *sacramentum* as 'mysterious signification within a text' and as 'religious ritual, e.g. baptism' will succeed in dealing with A.'s texts. Many texts can be placed under one or the other of those headings, but there are others that defy such categorization. The bishop's role is often characterized in aptly ambiguous form, e.g. 11. 2. 2, 'et sacramentum tuum dispensare populo tuo' (cf.

10. 30. 41, 6. 7. 12). Those texts probably speak to the bishop's priestly function, but as preacher he was equally responsible for mediating *sacramenta* (cf. 6. 9. 15, '[Alypius] futurus dispensator verbi tui') to his flock. The underlying distinction to be kept in mind is that of 'things' and 'signs' (*doctr. chr.* 1): texts and rituals are both 'signs' hence 'sacraments' in that sense. Scripture is the *sacramentum par excellence* (*BA* 13. 532 n.: 'le lieu des *sacramenta*'). See on 1. 11. 17, 'sacramentis'; cf. also G–M on 4. 2. 3, citing *c. Faust.* 19. 16, 'sacramenta legis et prophetarum'.

verbis apertissimis . . . intentionem eorum: The taste for a text both outwardly accessible and inwardly mysterious is part of A.'s thought, though there is little to say just when the taste arose. Nothing in A.'s own career as student and professor of classical literature aims in such a direction; as he presents himself, he was a literal and downright young man.[5] *doctr. chr.* 2–3 is devoted to the method, which is part of the appeal of Christianity at *ep.* 137. 5. 18, 'modus autem ipse dicendi quo sancta scriptura contexitur, quam omnibus accessibilis, quamvis paucissimis penetrabilis! ea quae aperta continet, quasi amicus familiaris sine fuco ad cor loquitur indoctorum atque doctorum; ea vero quae in mysteriis occultat, nec ipsa eloquio superbo erigit . . . sed invitat omnes *humili sermone*, quos non solum manifesta pascat, sed etiam secreta exerceat veritate.' (The notion of a *sermo humilis* goes back to the classical doctrine of the three styles, and animated A.'s own rhetorical theory: *doctr. chr.* 4. 17. 34 ff.)

leves corde: Sirach 19: 4, 'qui credit cito, levis corde est.'

sinu: Cf. 'gremio', apparently the same. The church accepts all and has something to offer for all (unlike a gnostic cult), even though relatively few come through the needle's eye—but more, he hastens to add, than would ever get through were this *auctoritas* not provided.

per angusta foramina: Mt. 19: 24, 'facilius est camelum per foramen acus transire quam divitem intrare in regnum caelorum'; and cf. Mt. 7: 13 quoted below.

auctoritatis: A. seems to have forgotten that the subject of the sentence (hence even of 'emineret' here) is itself 'auctoritas'.

suspirabam . . . fluctuabam: The same collocation of verbs of Alypius at 6. 10. 17.

viam saeculi latam: Mt. 7: 13–14, 'intrate per angustam portam, quia lata porta et spatiosa via quae ducit ad perditionem et multi sunt qui intrant per eam. (14) quam angusta porta et arta via quae ducit ad vitam! et pauci sunt qui inveniunt eum.' Cf. the broad ways leading to perdition at 6. 14. 24 and 7. 6. 8; cf. also 6. 6. 9, 'inhiabam'.

[5] We should remember, however, that the wilful creation of puzzles and complex wordplay is not at all alien to the poetry of such fourth-century Latin writers as Optatianus Porphyrius, Ausonius, and even (to follow the arguments of M. Malamud, *A Poetics of Transformation* (Ithaca, 1989)) Prudentius.

6. 6. 9

Inhiabam: He sees himself firmly in the grip of *ambitio saeculi* (and probably not *concupiscentia carnis*: marriage by this time represented career advancement) beginning to go sour; Bk. 6 will mark several stages of disillusionment with such hopes. Cf. in this paragraph 'cupiditatibus', 'talibus conatibus nostris', 'cupiditatum', 'ambiebam', 'ambitionibus', 'placere inde quaerebam hominibus'.

inridebas: See on 1. 6. 7, 'inrisor meus'.

dulcescere: See on 1. 4. 4, 'dulcedo mea sancta'.

vide ... domine: Lam. 1: 9–11, 'vide domine adflictionem meum ... vide domine'.

vide cor meum: 4. 6. 11, 'ecce cor meum, deus meus, ecce intus; vide, quia memini, spes mea.'

tibi inhaereat anima mea: Ps. 72: 28, 'mihi autem ad-/inhaerere deo bonum est' (see on 7. 11. 17).

visco: Also at 6. 12. 22, and cf. 10. 36. 42; the image more explicitly: *s.* 255. 7. 7, 'visco malarum cupiditatum involutas pennas habens'; *s.* 311. 4. 4, 'quod amas in terra, impedimentum est: viscum est pennarum spiritalium, hoc est virtutum, quibus volatur ad deum'; also at *trin.* 8. 2. 3, quoted on 7. 10. 16.

relictis omnibus: Lk. 5: 11, 'et subductis ad terram navibus, relictis omnibus secuti sunt illum' (of the call of Peter, James, and John); Lk. 5: 28, 'et relictis omnibus surgens secutus est eum' (the call of Levi/Matthew).

converteretur: Ps. 21: 28, 'et convertentur ad dominum universi fines terrae'; Ps. 50: 15, 'et impii ad te convertentur'.

qui es super omnia: Rom. 9: 5, 'qui est super omnia deus benedictus in saecula'.

sine quo nulla essent omnia: Sim. at 4. 10. 15 and 10. 27. 38.

converteretur et sanaretur: Is. 6: 10, 'excaeca cor populi huius.... ne forte videat oculis suis, et auribus suis audiat, et corde suo intellegat, et convertatur, et sanem eum'—quoted (via LXX) at Mt. 13: 15.

miser: The adj. to describe A. in *conf.* at 1. 16. 26, 2. 6. 12, 2. 8. 16, 3. 2. 4, 3. 10. 18, 4. 5. 10, 4. 6. 11, 4. 15. 24, 6. 6. 9, 8. 7. 17—all in the pre-conversion books; after that only at 10. 28. 39, 'misericors es, miser sum', and 10. 40. 65, 'hic esse valeo nec volo, illic volo nec valeo, miser utrubique'—at the beginning and end of the long examination of conscience that catalogues the remnants of fallenness and unconversion A. sees in himself.

pararem ... laudes: The only occasion on which we are certain that A. delivered such a panegyric was the accession to the consulship of Bauto in January 385: *c. litt. Pet.* 3. 25. 30, 'cum ego Mediolanium ante Bautonem consulem venerim eique consuli calendis Ianuariis laudem in tanto conventu conspectuque hominum pro mea tunc rhetorica professione recitaverim'. The argument of Courcelle, *Recherches*, 80–2, that this passage must refer to the decennalia of Valentinian II, has been confounded by C. Lepelley, *Atti 1986*, 1. 109, drawing on the researches of A. Chastagnol to show that the decennalia would have taken place not in 385, as had been generally assumed, but in 384, too soon after A.'s arrival.

cogitationum . . . aestuaret: Cf. 2. 2. 3, 'exaestuarent fluctus', with 7. 7. 11, 'fluctibus cogitationis'.

pauperem mendicum: The sight of a beggar would have rubbed a moral sore spot for A. at this period no matter what other reflections it aroused. Manichean doctrine, often stingingly rebuked by A. in after years, taught that beggars should not be aided: e.g. *mor.* 2. 15. 36, quoted on 3. 11. 18, 'si quis enim esuriens peteret'.

amicis: His friends are always around, still unnamed; see on 6. 7. 11.

omnibus talibus conatibus nostris: His Manichean efforts similarly went for naught at 5. 7. 13, 'conatus omnis meus . . .'

sarcinam: 4. 7. 12, 'sarcina miseriae'.

laetitiam: The word is used *in bono* and *in malo*, but esp. (see on 6. 2. 2) of the African rites at tombs of martyrs that A. tried to abolish.

emendicatis: The word is unusual, and occurs notably in contexts of *ambitio*: for details, see on 10. 38. 63. It is ironic there, and it occurs here in a similar context, with another ironic twist. Where people usually 'beg for' the signs of power and position, all this *mendicus* has acquired is a little loose change, which suffices him.

anfractibus et circuitibus: A. pursues these *circuitus* as early as 4. 1. 1 and as late as 8. 2. 3. For the vocabulary, *mag.* 10. 31, 'quanto tandem circuitu res tantilla peracta sit, meministine quaeso? . . . [Adeodatus:] vellem quidem tantis ambagibus atque anfractibus esset ad certa perventum.'

sed placere: Ps. 52: 6, 'quoniam deus dissipavit ossa hominibus placentium'; *en. Ps.* 52. 9, 'volentes placere hominibus, timuerunt perdere locum. . . . inde dissipata sunt ossa eorum, illius [sc. Christi in cruce] ossa nemo confregit.'

baculo disciplinae tuae: Cf. Ps. 22: 4, 'virga tua et baculus tuus, ipsa me consolata sunt'; *en. Ps.* 22. 4, 'disciplina tua tamquam virga ad gregem ovium, et tamquam baculus iam ad grandiores filios et ab animali vita ad spiritalem crescentes, ipsa me non afflixerunt, magis consolata sunt; quia memor es mei.'

confringebas ossa mea: Ps. 41: 11, 'utquid contristatus incedo, dum affligit me inimicus, dum confringit ossa mea'; *en. Ps.* 41. 18, 'ossa enim fortes sunt, et aliquando ipsi fortes temptationibus cedunt.' Cf. Is. 38: 13, 'sperabam usque ad mane quasi leo; sic contrivit omnia ossa mea.'

6. 6. 10

Recedant ergo ab anima mea: Cf. Jer. 6: 8, 'erudire Hierusalem, ne forte recedat anima mea a te, ne forte ponam te desertam terram inhabitabilem.'

qui dicunt: Who? the worldly-wise? Just below, 'interest vero' admits some truth to the objection, but places it in a different perspective.

gloria: Cf. 1 Cor. 1: 31, 'qui gloriatur in domino glorietur'; 2. Cor. 10: 17, 'qui autem gloriatur in domino glorietur.'

vertebat: The opposite of 'convertebat'.

vide: Cf. 'vide cor meum', 6. 6. 9: hence the imperative here is probably addressed to God, as at 10. 33. 50 and 11. 2. 4.

felicior: Almost always of worldly happiness (as 'hilaritate' following); see on 5. 4. 7.

bene optando: G–M: 'by wishing good luck (to the passers-by)'.

typhum: i.e. *ambitionem saeculi*. On *typhus*, see on 3. 3. 6.

caris meis: See on 6. 6. 9, 'amicis'; still unnamed, still not part of anything good.

in his: G–M: = 'on such occasions as this'; so *BA*, Ryan, Vega. Pusey, 'And I often marked in them how it fared with me.'

dolebam: Cf. 6. 6. 9, 'multos *dolores*'.

6. 7. 11

On the naming of friends in *conf.*, see on 4. 4. 7. That he now names his two great friends of this period is a mark of the progress that A. records; they are almost the only friends named in Bk. 7 (the exception is Firminus, in a self-contained episode): see on 7. 6. 8.

Courcelle attempted to make this *vita Alypii* the cardinal point in the genesis of *conf.* as a whole. His attempt depends on an elaborate reconstruction of the correspondence of A. and Alypius with Paulinus of Nola, and assumes that a single impulse can be given a privileged position in 'explaining' the genesis of the work. Courcelle's argument that the *vita Alypii* had a separate life apart from *conf.* is not convincing (see Excursus following notes on this paragraph), and Paulinus' curiosity was no more than one impulse among many in pushing A. towards *conf.*

This is a story of Alypius' conversions. He was converted from the circus at Carthage, and from the gladiatorial games at Rome, turning from the life of pleasure to follow philosophy and Augustine. But notice that Al.'s 'garden scene' brings about the first conversion, and he falls again when he gets to Rome—one might take from this the almost subliminal message that garden scenes are not enough when there is no true faith, and the end of 6. 8. 13, 'sed longe postea', would confirm that. His story most resembles A.'s own narrative of Bk. 3 (the *Hortensius*, etc.).[6] In the schematic moral terms of the three

[6] Highlighting incidentally what A. is otherwise content to let go unnoticed, that his own conversion to philosophy in Bk. 3 must have brought some improvement in the tone of his conduct: only one woman, less time at the *spectacula*, a more regular life suitable to his profession.

temptations, *concupiscentia carnis* was not a serious problem for Alypius (6. 12. 21), but *ambitio saeculi* (8. 6. 13; but see on 6. 8. 13) and *curiositas* (6. 7. 12 f.) both had their power over him.

A. places himself at this stage of his life in the close company of two friends, both from his African days, both of whom lived and praised a life of sexual continence, even from African days (see on 6. 7. 12). The continence of Nebridius was mentioned in the narrative at 4. 3. 6, but nothing in the events narrated showed A. aware of continence as a moral issue. Yet the ideal of continence was part of the philosophical tradition he encountered in the *Hortensius,* and it was part of the Manichean creed. In both cases, however, it was something he could put aside, something that *as he now represents it c.397* did not become an issue for him. What his real feelings were on joining the Manichees and taking a, probably orthodox Christian, concubine at the same time, we can only guess (and probably should not do so). (Not mentioned in *conf.* is the special role of Romanianus as a model of willed, 'converted' continence (*c. acad.* 2. 2. 3): there are reasons for the omission (see on 6. 14. 24), but we must remember that the model was at hand in these days.)

For fuller biographical details, see Mandouze, *Pros. chr.* s.v. Alypius; and E. Feldmann, A. Schindler, and O. Wermelinger, *Aug.-Lex.* 1. 245–67. He was born after 354 of a better family than A.'s ('parentibus primatibus municipalibus'), and died after 427/8. This sketch is the only information for any period before his time with A. in Milan. He returned to Thagaste with A. and became bishop there before 4 March 395; at some time in the intervening years, he visited Jerome and the sites of the Holy Land. A. never travelled, Al. travelled much; and Al. has the reputation, right or wrong, chiefly arising from the management of the great conference with the Donatists of 411, of having been the legalist, the 'enforcer', to A.'s pastoral and more (at least outwardly) genial role.

See also Mandouze, *Pros. chr.* s.v. Nebridius (born at least before 373, died 388/91, probably by 389), and for fuller treatment (and convenient summary of the correspondence), J. J. Gavigan, *Cath. Hist. Rev.* 32 (1946), 47–58. He was born (it is only a guess that he was about the same age as A.) and bred at and near Carthage, where his father had property; there he met A., and from there he followed A. to Italy. Of amiable character, retired and studious, he was willing to assist Verecundus in his teaching (8. 6. 13, 9. 3. 6). The testimonial here given to Nebridius attracts attention because he otherwise seems to have played little part in the outward drama. He was absent the day of Ponticianus' visit reported in Bk. 8, he was not at Cassiciacum, and not part of the community at Thagaste. Passages of *conf.* in which he features include 4. 3. 6 and 7. 6. 8 (on his opposition to astrology), 6. 10. 17, 6. 16. 26, 8. 6. 13–14 (with A. and Alypius at Milan), 7. 2. 3 (against the Manichees early), and 9. 3. 6–9. 4. 7 (his absence from Cassiciacum). He returned to his family property, living there in perfect chastity and converting his household. A.'s correspondence with him offers valuable additional evidence for the tone and direction of the philosophical inquiries of these years. For a sentimental recollection of N.'s ways, see *ep.* 98. 8, 'recordatus sum Nebridium amicum meum, qui cum esset rerum obscurarum ad doctrinam pietatis maxime pertinentium diligentissimus et acerrimus inquisitor, valde oderat de quaestione magna responsionem brevem. et quisquis id poposcisset, aegerrime ferebat eumque, si eius persona pateretur, vultu indignabundus et voce cohibebat, indignum deputans qui talia quaereret, cum de re tanta quam multa dici possent deberentque nesciret.'

maxime ac familiarissime: These words hint that the circle was larger; at 6. 14. 24, Romanianus is mentioned, and at Cassiciacum we encounter A.'s brother and cousins. This intimate coexistence is probably not meant to embrace such public friends as Zenobius and Mallius Theodorus.

docere coepi: This dates to 375/6, the year of the death of A.'s other great friend (4. 4. 7).

et postea: A. returned to Carthage in 376 and remained until 383/4.

spectacula: 3. 2. 2, 'spectacula theatrica' (no mention of circus before the present text).

in insaniam S Maur. Knöll Skut. : *in insania* CD : *insania* GO Ver.

rhetoricam ibi professus: 4. 2. 2, 'docebam in illis annis artem rhetoricam.'

exitiabiliter: *TLL* 5. 2. 1526 reports the adv. only here, *civ.* 1. 17, 'dei misericordiam desperando exitiabiliter paenitens', *ep.* 188. 1. 3 (to which add *op. mon.* 25. 32), and in two later authors (Capreolus (AD 430/7) and Gildas (6th cent.)).

EXCURSUS: ALYPIUS, PAULINUS, AND THE GENESIS OF THE *CONFESSIONS*

Courcelle (*Recherches*, 31–2, reprised at *Les Confessions*, 559–607) saw in the *vita Alypii* of 6. 7. 11 a key to the genesis of *conf.* as a whole. This is the appropriate place to review the evidence, and to see how far we can agree. It must be borne in mind, however, that C. elsewhere assumes that *conf.* as we have the work is only a torso, and constructs imaginary unwritten sections of the work for which there is no good evidence.

The story briefly is this. In summer 395, Al. wrote to Paulinus on the occasion of his establishment at Nola. Al. indicated that when he was receiving baptismal catechesis in Milan, he heard well of Paulinus; he offered friendship, and sent along five books of A. *contra manichaeos* (Courcelle suggests: *vera rel.*, *Gn. c. man.* (in two books), *mor.* (also in two books)) as rejoinder for the 'five writings' of the Manichees (cf. *c. Fel.* 1. 14, 'quinque auctores'), seeking a copy of Eusebius *chron.* Paulinus responded in the autumn of 395, sending the Eusebius.

Then Paulinus (Paul. Nol. *ep.* 3. 4, to Alypius): 'specialiter autem hoc a te peto, ... ut pro hac historia temporum referas mihi omnem tuae sanctitatis historiam, ut "qui genus, unde sis domo" [*Aen.* 8. 114] tanto vocatus a domino, quibus exordiis "segregatus ab utero matris tuae" [Gal. 1: 15] ad matrem filiorum dei prole laetantem abiurata carnis et sanguinis stirpe transieris et in "genus regale ac sacerdotale" [1 Pet. 2: 9] sis translatus edisseras. quod enim indicasti iam de humilitatis nostrae nomine apud Mediolanium te didicisse, cum illic initiareris, fateor curiosius me velle condiscere, ut omni parte te noverim, quo magis gratuler, si a suspiciendo mihi patre nostro Ambrosio vel ad fidem invitatus es vel ad sacerdotium consecratus, ut eundem ambo habere videamur auctorem.'

A. seems to have answered for Al., with the requested narrative postponed by the premature departure of the courier, Romanianus: A. *ep.* 27. 5, 'est etiam aliud quo istum fratrem amplius diligas; nam est cognatus venerabilis et vere beati episcopi Alypii, quem toto pectore amplecteris, et merito. nam quisquis de illo viro benigne cogitat, de magna dei misericordia et de mirabilibus dei muneribus cogitat. itaque cum legisset [sc. Alypius] petitionem tuam, qua desiderare te indicasti ut historiam suam tibi scribat, et volebat facere propter benivolentiam tuam et nolebat propter verecundiam suam. quem cum viderem inter amorem pudoremque fluctuantem, onus ab illo in umeros meos transtuli; nam hoc mihi etiam per epistulam iussit. cito ergo, si dominus adiuverit, totum Alypium inseram praecordiis tuis. ... quod iam fecissem iamque illum legeres, nisi profectio fratris inprovisa repente placuisset, quem sic commendo cordi et linguae tuae, ut ita comiter ei te praebeas, quasi non nunc illum, sed mecum ante didiceris.'

Courcelle, *Recherches*, 31–2: 'La suite de la correspondance est perdue; mais la biographie d'Alypius a été remployée dans les *Confessions.* Il y a tout lieu de penser que Paulin, mis en goût par cet opuscule, a incité Augustin à narrer tout au long l'histoire de sa vie, de sa conversion et de son ordination, intimement liées à l'histoire d'Alypius. Lorsqu'Augustin mentionne les "spirituels" qui pourront sourire amicalement en apprenant les bizarres erreurs où il est tombé dans sa jeunesse [5. 10. 20], il songe sûrement à Paulin surtout.'[7] At *Les Confessions*, 559–607, 'Appendice III: La correspondance avec Paulin de Nole et la genèse des *Confessions*', Courcelle develops extensively his view of the interaction between correspondence and 'confession'. There are four surviving letters of Paulinus, eight of Augustine; H. Lietzmann had inferred an additional nine lost letters. Courcelle finds reason to surmise no fewer than 18 *more*, and prints an elaborate chart (*Les Confessions*, 601–3) to accompany his discussion. The claims C. makes for his reconstruction are considerable, e.g. 607, 'tout le développement du livre X, relatif aux tentations, est issu

[7] *Recherches*, 32 n. 1: 'On remarquera aussi qu'Augustin, dans son plan primitif, pensait narrer l'histoire de son ordination, comme Paulin l'espérait d'Alypius.' This is unlikely: Paulinus had asked about Alypius' ordination only under the impression that he had been ordained by Amb.

directement des impressions échangées entre Paulin et Augustin.' The main argument against his reconstruction is simply that A. did not address or dedicate *conf.* to Paulinus. The second strongest argument is that there is no mention anywhere in the surviving correspondance of *conf.* or of the 'biography of Alypius' after the fact, nor does A. say anything about Paulinus in *retr.* The letter to Alypius may well have helped trigger memory and rumination, but to reduce the genesis of this elaborate work to a mechanical response to a friend's curiosity (note 'curiosius' (!) in Paulinus' letter quoted above) is to misread the work as a whole completely.

On Courcelle's view, Al. cannot out of modesty write about himself; so A. agrees to write about him, to make sure that none of God's gifts are omitted; so—to follow Courcelle—A. is quickly persuaded to write about himself. Well, perhaps. But a crucial transformation takes place, turning A. the confessor away from an audience and towards God, discovering the whole rhetorical strategy of *conf.*, quite different from what Al. must have conceived in trying to write about himself, or from what A. meant to do in writing about Al., or certainly from anything Paulinus expected.

For all the ferocious attention to detail in Courcelle's treatment of this correspondence, two elements remained undervalued. First, the place in the correspondence and the genesis of *conf.* of A.'s request to Paulinus (*ep.* 31. 8, of late 396/early 397), asking him to send a copy of Amb.'s *de philosophia sive de sacramento regenerationis* ('libros ... adversus nonnullos imperitissimos et superbissimos, qui de Platonis libris dominum profecisse contendunt'); see the Introduction to this commentary for the importance of that work in A.'s formation, and in the reformation of his views of Platonic philosophy along Ambrosian lines at about the time he was writing *conf.* In Courcelle's reconstruction, Paulinus only complied with this request in late 399 (item *c* in Courcelle's reconstruction: *Les Confessions*, 568–9), but the lack of mention of that request in *epp.* 42 and 45 to Paulinus of 398 and 399 leaves open the possibility that A. had already obtained the work; and at all events, 399 is well within the conventional range of dates (397–401: see also Introduction for discussion) within which *conf.* is agreed to have been written.

Second, there has been no detailed comparative study of the style of Paulinus' letters and *conf.* Paulinus is well above average in stylishness for a Christian writer of the time (far more elegant and subtle than Amb., e.g.), and the text of his letters (see, e.g., A.'s *ep.* 30) is a tissue of scriptural language to an even greater degree than is true of *conf.* A. must have been impressed by this, and it was surely one element in the forging of the distinctive style of *conf.*

6. 7. 12

Courcelle, *Recherches*, 59, 'il ne paraît guère douteux non plus qu'Augustin n'ait "arrangé" l'épisode d'après *l'exemplum* célèbre de Polémon adolescent;

celui-ci, dans des circonstances analogues, fut converti à la philosophie par Xénocrate. Augustin connaît bien cette anecdote, qui figurait peut-être, précisément, dans l'*Hortensius*.' Nothing requires the episode to have come from the *Hortensius*, however; Solignac, *BA* 13. 543 n., thinks the source the Celsus-doxography; it may be read in Val. Max. 6. 9, a comical little tale of a sermon being taken to heart by the drunkard, who first takes off his crown, then rearranges his clothes and straightens his face. If there is anything to the link Courcelle suggests between this episode and that, it seems clear from *ep.* 144. 2 that A. means to say both that something good can happen entirely outside Christianity, and that it is from the Christian God: 'Xenocrates Polemonem, ut scribitis et *nos ex illis litteris recordamur*, de fruge temperantiae disputando non solum ebriosum verum et tunc ebrium ad mores alios repente convertit. quamquam ergo ille, sicut prudenter et veraciter intellexistis, non deo fuerit adquisitus sed tantum a dominatu luxuriae liberatus, *tamen ne idipsum quidem, quod melius in eo factum est, humano operi tribuerim sed divino*.' (Also mentioned at *c. Iul.* 1. 4. 12, and cf. *c. Iul.* 1. 7. 35.)

antistitem: Otherwise only of Amb.; see on 6. 2. 2.

sacramenti tui: G–M: 'i.e. of the Christian faith': that hardly seems to be an adequate representation, for 'antistitem' has a verbal force, and a following genitive is epexegetic of that active sense. See on 6. 5. 8 for range of meanings of 'sacrament'.

exponerem CDGOS Vega : *exponerem et* Maur. Knöll Skut. (with the support of one minor MS) : *exponerem et cum* Ver. The difficulty is not the conjunction or lack of one, but the punctuation of the whole sentence. As edited in this century (G–M, Skut., Vega, Pellegrino, Verheijen), a single sentence is punctuated to run from *quam dum* to *cogitaverim*; the only finite verb in this 'sentence' is *scis*. This means that the editors, who generally print some conjunction, either *et* or Verheijen's conjecture *et cum*, before *opportune* (Vega alone omits the conjunction but retains the punctuation), believe that the *dum*-clause provides a circumstance that usefully qualifies the statement, 'You know, God, that I was not thinking of Alypius.' The weaknesses of that reading are immediately obvious when one attempts to translate the passage.

corripe . . . te: Prov. 9: 8 (VL), 'corripe sapientem et amabit te' (quoted in that form at *ep.* 210. 2; *trin.* 14. 1. 2 follows Vg.).

carbones ardentes: Ps. 139: 11, 'decident supra eos carbones ignis in terra' (at *en. Ps.* 139. 14, the *carbones* destroy the vices and clear the ground, making way for the positive construction of divine grace: exactly Alypius' situation here); Ps. 119: 4, 'cum carbonibus desolatoriis'; *en. Ps.* 119. 5, 'erant autem in illo multa quae male fronduerant, multae carnales cogitationes, saeculares multi amores; ipsi uruntur carbonibus desolatoriis, ut fiat purus locus desolatus, in cuius loci puritate faciat deus aedificium suum; . . . accedunt carbones desolatorii, et deiciunt quod male fuerat aedificatum, et desolato loco accedit structura felicitatis perpetuae. *videte ergo quare dicti sunt carbones*: quia qui se convertunt ad dominum, de mortuis reviviscunt. carbones autem quando accenduntur, antequam accenderentur, exstincti erant. nam exstincti carbones, mortui dicuntur; *ardentes*, vivi appellantur.' (The words 'carbonum ignis ardentium' occur at Ezech. 1: 13 (cited by Ver. here), but the context is not relevant.)

mentem spei bonae: See on 1. 16. 26, 'bonae spei puer'.

taceat laudes: Cf. Ps. 106: 8, 15, 21, 31: 'confiteantur domino miserationes eius et mirabilia eius filiis hominum'.

quae ... confitentur: It is not A. that confesses, but the mercy of God working through him; cf. 1. 5. 5, 'miserere ut loquar'.

ex fovea tam alta: 6. 3. 3, 'foveam periculi mei'.

et resiluerunt: This sudden conversion on encountering a chance word of rebuke is not without its later lapse in the next paragraph, 6. 8. 13.

illa mecum superstitione involutus est: The implication is that Alypius' adherence to Manicheism was on altogether different grounds from that of A.: for Alypius no expiation of his guilt, especially not sexual guilt, but high-minded purity.

ostentationem: G–M adduce *c. Faust.* 5. 1 (Faustus speaking), 'omnia mea dimisi, patrem, matrem, uxorem, filios, aurum, argentum, manducare, bibere, delicias, voluptates.'

veram et germanam: The Manichees preached much about continence, but what continence they had was not of a good source, not authentic, and much of what they purported to have was false. For Manichean views of sexuality in general, see on 8. 1. 2; at *mor.* 2. 18. 65 and 2. 19. 68–72, A. took them to task for the inconsistencies of their position and the peccadilloes of their leaders, with lurid stories.

pretiosas animas captans: Prov. 6: 26, 'mulier autem virorum prestiosas animas captat' (as cited at *civ.* 2. 5).

6. 8. 13

This episode corresponds to the pear-theft (2. 4. 9 ff.), similarly impossible without the company and urging of friends left unnamed.

incantatam: G–M: perhaps 'the path in life which his parents had inaugurated for him,' or merely 'dinned it into him.' But the normal usage of the verb has a clear sense of 'cast a spell, bewitch'. *BA:* 'la voie terrestre dont ses parents lui avaient chanté les charmes'. The notion of bewitchment/enchantment is borne out by 'viam' postponed to the end of the phrase: in light of the weight that *via* carries in *conf.*, esp. in Bk. 7, the *terrena via* is no trivial, or value-neutral, thing, but is in a way the worst thing of all. What does emerge is that Alypius was less enslaved to *ambitio saeculi* than A.: he entered upon that path under his parents' influence and here is said only not to have abandoned it.

TLL 7. 1. 847 s.v. *incanto* instances this passage under the heading, 'laxius et non technice . . . fere i. q. decantare, declamare'. But the other passages adduced there do not much resemble this one: they all variously deprecate the object of the incantation, while it certainly does not seem that Alypius' parents opposed his earthly career. Other passages in A. offer a potential parallel to the sense suggested by G–M, but where there is an express object of the verb, it is something unmistakably divine and good: *en. Ps.* 57. 8, 'Stephanus martyr praedicabat veritatem, et tamquam tenebrosis mentibus, ut eas in lucem educeret, incantabat'; sim. at *en. Ps.* 48. s. 1. 1, *Gn. c. man.* 2. 26. 40 (but no express object). In the end, the usage is unparalleled.

gladiatorii spectaculi: Again Alypius takes delight in a form of spectacle that A. had not (to our knowledge) patronized (see on 6. 7. 12). A. attended the blood sports himself in youth (*en. Ps.* 147. 7, 'et aliquando nos quoque ibi sedimus et insanivimus'), but they had not the hold on him that the theatre did; see W. Weismann, *Aug.-Lex.* 1. 300–3.

pervium CDGOS and all other MSS with one exception (see next reading) Knöll Skut. Vega Ver. : *per viam* M[2] : *pervius* Maur. Pell. *Pervius* is rare in A., and in every case (7. 1. 2, *en. Ps.* 67. 5, *civ.* 4. 2; used twice in the same way by Julian, quoted at *c. Iul. imp.* 6. 9, 6. 33) it is adjectival and unambiguously means not 'in the way, across the path' but 'permeable, passable, open to transit'; every sense of the word listed by *OLD* shares this common, root meaning. But the translators have simply compelled the passage to say what they infer it ought to mean and so reversed the sense of this word: Ryan: 'whom he chanced to meet'; Pusey: 'he was one day by chance met'; *BA:* 'le rencontrèrent'; Pell.: 'incontrò per strada'. Reading *pervium*/*pervius* = *obviam* with the translators encounters a further insuperable difficulty. Though any form of *pervius* is rare in A., *obviam* is, if infrequent, perfectly common in similar circumstances, occurring no less than 3 × elsewhere in *conf.*: e.g. 6. 9. 15, 'fit eis obviam quidam architectus'; cf. 7. 6. 8, 8. 11. 26.

G–M offer a more complicated and on the surface appealing theory, emending to *perviam*, to mean *obiter* (instancing Charisius *inst. gram.* 2 (Keil, 1. 209), who mentions that Augustus found fault with Tiberius for using *perviam* in the sense of *obiter*). But that latter allegation from a grammarian is the sole citation in that sense in *OLD* (*TLL* unavailable here). It is risky to assume that a sense known only to have been objected to by a famous grammarian is the exact sense required here, and the mere existence of *perviam* elsewhere is doubtful: it apparently appears only in a conjecture in Plautus, *Pseudolus* 760, where the meaning offered by *OLD* for the phrase 'perviam est' is 'there is a way through.' (Other passages cited in older dictionaries do not preserve the reading in current critical editions.)

Sc. perhaps 'amphitheatrum': 'when the stadium happened to be open to them as they were coming back from the mid-day meal'. On that reading, the friends do not, on the way to the show, happen to encounter Alypius, but he is dragged by his friends inside when they decide on the spur of the moment to pass the afternoon that way.

adero itaque absens: The risks of an absent presence are evidently considerable.

quo (*ubi*) BPZF Maur. Knöll Skut. Pell. : *quod* CDGOS Vega Ver. *quo* may be an emendation, but a good one.

curiositate: The temptation afflicting someone in the presence of *spectacula* (here and at 6. 7. 11; see on 3. 2. 2); note 'concupivit'. *Curiositas*, once it takes root, is not a thing of the physical eyes. When they are blocked off, the temptation can enter through the ears (as here).

aperuit oculos CDGO Maur. Ver. Vega : *aperuit* S Knöll Skut. Pell.

de se praesumserat: Cf. Jdt. 6: 15, 'ostende quoniam non derelinquis praesumentes de te; et praesumentes de se . . . humilas.'

nesciebat: 'he was unaware what he was doing'. The verb is abundant through *conf.* 1–10 (e.g. 14 × in Bk. 4, 12 × in Bk. 6, 24 × in Bk. 10), occurs 12 × in Bk. 11, but then only 3 × in Bk. 12 and not at all in Bk. 13: an unconscious measure of the progress he makes.

docuisti CDGO Maur. Ver. Pell. : *docuisti eum* S Knöll Skut. Vega.

non sui habere sed tui fiduciam: Is. 57: 13, 'qui autem fiduciam habet mei, hereditabit terram'; Prov. 3: 5, 'habe fiduciam in domino ex toto corde tuo, et ne innitaris prudentiae tuae.'

6. 9. 14

Courcelle, *Recherches*, 45, identifies an apparent want of chronology in these paragraphs: first student days at Carthage, then student days at Rome, then Carthage again, then Rome when A. catches up with him. The only disorder, however, is the placing of the story of the gladiatorial show at Rome, and that is merely the logical conclusion of the *curiositas* story that begins the chronological sequence.

The strongest feature of the argument that the Alypius-*vita* was written beforehand is just that *this* episode obtrudes. It has no real parallel in A's narrative of his own past; it is comical; it has a moral, but a rather low-powered one, and none of the breast-beating we might expect if A. told it of himself—all the guilt and shame he would have felt for what he had not done, when he never felt guilt and shame for things he had done, etc. But it was, after all, a form of *curiositas* that got Alypius into his troubles: going to see what was afoot, then picking up the axe . . .

aeditimis: Not an especially common word, always in CL with religious connections: 'temple-guards'. The form is preferred as more archaic to *aedituus* (Varro and Gellius; cf. *civ.* 6. 7), but the latter form prevails from Cicero on and appears 3 × in Vg. in specifically religious contexts, e.g. Ezech. 44: 11.

quam . . . credulitate: 'how a man should not easily in trials be condemned by another man out of rash credulousness'.

credulitate CDG Maur. Knöll Skut. Ver. (who regards the reading as correct but not in the archetype) : *crudelitate* OS.

vico argentario: A case of precocious mass-production: *civ.* 7. 4, 'ridemus quidem, cum eos [deos] . . . tamquam opifices in vico argentario, ubi unum vasculum, ut perfectum exeat, per multos artifices transit, cum ab uno perfecto perfici posset.'

furem manifestum: The crime was indeed *furtum*, and being caught in possession of the axe (taken as evidence that Al. had done the hacking the workmen had heard) was adequate to suggest that he was *fur manifestus:* the legal texts collected by E. Schepses, *SDHI* 5 (1939), 141–51, are unambiguous.

6. 9. 15

Sed hactenus docendus fuit: 'but the lesson only had to go so far.'

cuius testis eras tu solus: Cf. Wisd. 1: 6, 'quoniam renum illius testis est deus et cordis illius scrutator est verus et linguae eius auditor'; cf. Jer. 29: 23, 'ego sum iudex et testis, dicit dominus.'

supplicium: The case was liable to criminal rather than civil proceeding (Schepses, art. cit., 151–3); the nature of the *supplicium* is not clear, but the context implies either imprisonment or corporal punishment.

architectus: A respectable official, probably with the rank *clarissimus comes* (Symm. *ep.* 5. 76, *rel.* 25. 26, etc.); *cod. theod.* 13. 4. 1 (334), 'architectis quam plurimis opus est; sed quia non sunt, sublimitas tua in provinciis Africanis ad hoc studium eos impellat, qui ad annos ferme duodeviginti nati liberales litteras degustaverint. quibus ut hoc gratum sit, tam ipsos quam eorum parentes ab his quae personis iniungi solent volumus esse immunes, ipsisque qui discent salarium competens statui.'

in domo . . . senatoris: Alypius' innocence does not get him a hearing; his social connections save him.

parvus: *BA* translates as 'jeune'; Pusey as 'young'; both infer that the slave was so young and innocent as to fail to see the harm that his testimony could do to his master.

futurus dispensator: Particularly apt: Tit. 1: 7, 'oportet enim episcopum sine crimine esse, sicut dei dispensatorem'; see on 6. 5. 8, 'sacramentorum'.

dispensator . . . examinator: On the pre-eminence of preaching and service to his flock (hearing cases) in a bishop's activities, see on 6. 3. 3, 11. 2. 2. On bishop as judge, *op. mon.* 29. 37, 'tumultuosissimas perplexitates causarum alienarum pati de negotiis saecularibus vel iudicando dirimendis vel interveniendo praecidendis'.

6. 10. 16

The escape to a legal career was strictly illegal if Alypius was really a *curialis*, as he seems to have been: *cod. theod.* 12. 1, various titles, *ad infinitum*.

adhaesit: Cf. 'inhaerebat' below.

ter iam adsederat: 8. 6. 13, 'post *adsessionem tertiam*'; 'adsederat' in the present passage speaks of the moment at which he left Rome for Milan, so the passage in Bk. 8 bespeaks a time when he had been unsuccessful in finding work there. Aul. Gell. 12. 13 on *adsessores*: 'iuris studiosi quos adhibere in consilium iudicaturi solent'; cf. 'consiliarius'. See S. Poque, *Le langage symbolique*, 122–3.

indoles: Cf. 6. 7. 11, 'propter magnam virtutis indolem'.

comiti largitionum Italicianarum: The *comes Italicianarum* was a dignitary of moderate standing, subordinate to the *comes sacrarum largitionum* (*not. dig.* 7); *cod. theod.* 6. 19. 1 (400), 'eos, qui consularitatis functi sunt dignitate [not the consulship, but the office of *consularis*, i.e. governor of certain provinces—see on 6. 11. 19], comitibus Italicianarum et Gallicianarum iure praeferimus, si quidem haud exiguus sit titulus meritorum regere et gubernare provincias.' The department of *largitiones* dealt mainly with collection of old cash-taxes (*aurum coronarium*, *conlatio lustralis*, *conlatio glebalis*) and currency generally (e.g. the mints); see Jones, *LRE* 429–30.

pretiis praetorianis: G–M cite Pusey, 'who interprets it as referring to the privilege of purchasing at a special tariff enjoyed by officials connected with the court, . . . [i.e.] "at palace prices".' *BA*, 13. 555 n., think of the court's expense fund and imagine Alypius refusing to have the official scribes do copying for his behalf and charging it to the office. These are plausible speculations, but *praetorianis* is odd unless it is a technical term—but then it is odd that it is not attested elsewhere. (But note that at 6. 11. 18, A. is worrying about where he will acquire *codices*. His friend's integrity may have been a minor nuisance.)

codices: Always of 'book-as-artefact', never 'book-as-text'.

qui in parvo . . .: Lk. 16: 10–12 (parable of the unjust steward), 'qui fidelis est in minimo et in maiori fidelis est, et qui in modico iniquus est et in maiori iniquus est. (11) si ergo in iniquo mammona fideles non fuistis, verum quis credet vobis? (12) et si in alieno fideles non fuistis, vestrum quis dabit vobis?'

veritatis = *Christi*.

verum CDO^{1}S Knöll Skut. Ver. : *vestrum* GO^2 : *quod verum est* BPZ Maur.

vestrum CDGOS Knöll Skut. Ver. : *quod vestrum est* BPZ Maur. For both these readings, see the text as quoted at *en. Ps.* 136. 2, and supplement with Milne, 108.

mecumque nutabat . . . modus: Brings us back to Alypius' introduction at the beginning of 6. 7. 11, 'congemescebamus in his qui simul amice vivebamus'.

6. 10. 17

Nebridius: see on 6. 7. 11 and 4. 3. 6.

paterno rure: Hor. *epod.* 2. 1–8:

beatus ille, qui procul negotiis
ut prisca gens mortalium,
paterna rura bubus exercet suis,
solutus omni faenore,
neque excitatur classico miles truci,
neque horret iratum mare,
forumque vitat et superba civium
potentiorum limina.

The poem was well known to A., for at *mus.* 5. 9. 18–13. 27, the first verse is quoted and put through metrical transformations; at *qu. hept.* 3. 74 there is an echo of 2. 6, 'iratum mare'; see Keseling, *Philol. Wochenschr.* 51 (1931), 1279. The epode would be attractive for the echo of Ps. 1: 1 ('beatus vir qui non abiit in consilio impiorum') with which it begins, and it aptly describes the *otium cum philosophia* to which N. returned after Milan.

But a distinctly Christian counter-model is mixed into these lines: Mt. 19: 29, 'et omnis qui *reliquit domos* vel fratres aut sorores aut *patrem* aut *matrem* aut filios aut *agros* propter nomen meum, centuplum accipiet et vitam aeternam possidebit.' That text is the tag end of Mt. 19: 21 ff., 'Go, sell all you have, and come follow me', quoted from the Anthony conversion story at 8. 12. 29, just before A. picks up the codex of Paul.

non secutura matre: Courcelle, *Recherches*, 86 n. 5: 'Il paraît probable qu'elle [Monique] a voyagé en compagnie de Nebridius, car Augustin précise (6. 10. 17) que la mère de Nebridius avait refusé au dernier moment d'accompagner celui-ci.' The hypothesis is not implausible, though C.'s misreading and faulty translation of the present text beget error. There is no 'last moment' here, only the implied contrast ('with his mother not to follow [the way some people's mothers might follow]'). See on 1. 11. 17, Excursus on mothers and fathers in *conf.*: it is impossible to avoid the suspicion that there is some contrast implied with Monnica here. N.'s conversion resembles A.'s in many ways, but he has no Monnica to pray him home.

veritatis atque sapientiae: The phrase both reports their aim and hints at the manner of its fulfilment. The friends search, both knowing and not knowing what it is they seek, not seeing the (to A. of 397, ineluctable) epithets of Christ in these words. Cf. 'inquisitor' with 9. 3. 6, '[Nebridius] inquisitor ardentissimus veritatis' and 6. 11. 19, 'ad solam inquisitionem veritatis'. For A. of 397, *beata vita* and *veritas* are not two different things (if they ever had been for him: *c. acad.* 1. 3. 7, 'placuit enim Ciceroni nostro "beatum esse, qui veritatem investigat, etiamsi ad eius inventionem non valeat pervenire." [*Hort.* frg. 101 M] . . . quam ob rem si et sapientem necessario beatum esse credendum est et veritatis sola inquisitio perfectum sapientiae munus est, quid dubitamus existimare beatam vitam etiam per se ipsa investigatione veritatis posse contingere?').

suspirabat . . . fluctuabat: Same verbs of A. already at 6. 5. 8.

ad te OS Knöll Skut. Ver. : *a te* CDG Maur.

ad te . . . opportuno: Ps. 103: 27, 'omnia a te expectant, domine, ut des illis cibum in tempore opportuno'; Ps. 144: 15–16, 'oculi omnium in te sperant et tu das escam illis in opportunitate, (16) aperis tu manum tuam et imples omne animal [Ps. Veron.: 'omnem animam'] benedictione.' The echo recurs at 6. 14. 24, when they consider retirement from the city. Knauer, 139: 'beide Male möchte eine "societas" gleichgesinnter Freunde versuchen, gemeinsam die innere Ruhe zu finden, wobei es charakteristisch ist, wie Augustin die Tempora beider Psalmverse . . . ändert in den Konkunktiv des Imperfekts bzw. das Partizip des Futurs: sind doch die Freunde in beiden Fällen von einer solchen "benedictio" Gottes weit entfernt.'

amaritudine: See on 2. 1. 1. Here no particular scriptural passage is evoked, but A. often draws upon a private 'lexicon' created by consciousness of the frequency of occurrence in scripture and the contexts there.

tenebrae: 4. 15. 25, 'deus meus, inluminabis tenebras meas' (=Ps. 17: 29; cf. 6. 1. 1 and 7. 1. 2 for same verse); 6. 1. 1, 'et ambulabam per tenebras et lubricum et quaerebam te foris a me et non inveniebam deum cordis mei.'

gementes: 6. 7. 11, 'congemescebamus'.

quamdiu haec: 8. 12. 28, 'quamdiu, quamdiu cras et cras'; that passage and this bracket the search for truth from the naming of the friends to the garden scene. Alypius and Nebridius are drawn into the dismay that A. depicts afflicting himself at 6. 6. 9–10 (on seeing the beggar).

quia . . . apprehenderemus: Here the friends are drawn into the Micawberish conclusion of Bk. 5: see on 5. 14. 25, and cf. 5. 6. 13 and 5. 10. 18.

6. 11. 18–19

Courcelle, *Les Confessions*, 18: 'il s'agit sans aucun doute non d'un résumé abstrait et de caractère historique, mais des réflexions que faisait Augustin trentenaire, lorsqu'il partageait à Milan les incertitudes d'Alypius et de Nébridius, eux-mêmes à peine sortis du manichéisme. Le monologue intérieur, qui se prolonge pendant deux pages, décrit sous forme de paragraphes antithétiques les tergiversations d'Augustin entre divers partis possibles, dont les deux extrêmes sont d'une part la conversion à une vie continente catholique, d'autre part un mariage brillant et une carrière profane.'

This elaborately perverse view of these paragraphs (C. goes on at length, adding *inter alia* the view that A. actively considered returning to Manicheism at this time!) has already been gently disowned by the *BA* annotators ad loc. A true view of this text must do justice to the ironic quality of the self-representation. These lines take into account the whole development from the reading of the *Hortensius* to this first moment of renewed hope, and in them accumulate all the false hopes and vain expectations that had accompanied A. on that trajectory. This is almost a parody of his past, often using similar words to those earlier in the narrative, as the notes below show. Take the words 'ecce Faustus veniet'—Courcelle, *Recherches*, 45 n. 2, thinks those words require us to consider whether Faustus planned to come to Milan in 386, while Mandouze, 279 n. 9, calls this 'exclamation anachronique'—but that view only partly abandons Courcelle's error. There is no anachronism at all, for A. neither said this in Milan nor represents himself as having said it there. We reach a genuine present tense only at the end of paragraph 19. (Note that the discussion of life-after-death at the beginning of 6. 11. 19 anticipates the end of Bk. 6 at 6. 16. 26, as the mention of marriage later in 6. 11. 19 anticipates 6. 12. 20–15. 25.)

The theme of this meditation—so little time, so much to know—recurs throughout *ep.* 21, when A. shortly after taking priestly orders asks Valerius for a short leave from duty to study scripture. Similar anxiety closer to the date here is *sol.* 2. 14. 26: '[Ratio] sed quid facimus? an incepta omittimus, et exspectamus ecquid nobis librorum alienorum in manus incidat, quod huic quaestioni satisfaciat? nam et multos ante nostram aetatem scriptos esse arbitror, quos non legimus, et nunc, ut nihil quod nescimus opinemur, manifestum habemus, et carmine de hac re scribi et soluta oratione et ab his viris quorum nec scripta latere nos possunt et eorum ingenia talia novimus, ut nos in eorum litteris quod volumus inventuros desperare non possimus, praesertim cum hic ante oculos nostros sit ille[8] in quo ipsam eloquentiam, quam mortuam dolebamus, perfectam revixisse cognovimus.' Most noticeable between that text and this is the disappearance of the hope that solutions would come from new books by learned friends. (On the interior monologue, see on 10. 8. 14, 'dico apud me'; see also 7. 5. 7 and 8. 7. 18 for similar representations.)

recolens: The key, ignored by Courcelle, to the retrospective aspect of this representation.

quam longum tempus: A similar retrospect on his progress and failures from the time of reading the Hortensius appears at 8. 7. 17; the age (18) is also insisted upon at 3. 4. 7, 4. 1. 1.

studio sapientiae: See on 3. 4. 7; 'sapientiae studio' recurs below to frame this meditation.

[8] Amb.? See on 6. 3. 3.

omnes vanarum cupiditatum spes inanes et insanias mendaces: Ps. 39: 5, 'beatus vir cuius est nomen domini spes eius et non respexit in vanitates et insanias mendaces'; *en. Ps.* 39. 7, 'ecce qua volebas ire, ecce turba viae latae [see on 6. 5. 8, 'viam saeculi latam']: non frustra ipsa ducit ad amphitheatrum, non frustra ipsa ducit ad mortem. . . . sed turbae strepunt, turbae festinant. . . . noli imitari, noli averti; vanitates sunt, et insaniae mendaces. sit dominus deus tuus spes tua.'

tricenariam aetatem: Reached by A. on 13 November 384; on Monnica's arrival as being probably in early 385, see on 6. 1. 1; 8. 7. 17 says '19th year plus 12', which gets us more or less to the same 384/85, though the circumstances are those of August 386.

in eodem luto haesitans: Ter. *Phormio* 780, 'in eodem luto haesitas'; a copybook formula (cf. Otto, *Sprichwörter*, 201–2): quoted by Lactantius *inst.* 2. 8. 24, 7. 2. 3, Hier. *ep.* 143. 2, *c. pel.* 1. 11, 1. 24, *in Is.* 10 pr., and Marius Mercator (*PL* 48. 158). Cf. 2. 2. 2, 'limosa concupiscentia carnis'.

fruendi: For the moral error in 'enjoying' created things, see *doctr. chr.* 1. 4. 4 f.

fugientibus et dissipantibus: Two scriptural echoes already heard elsewhere: the prodigal (1. 18. 28, 'in longinqua regione vivens prodige dissiparet', 4. 16. 30, 'in longinquam regionem ut eam dissiparem in meretrices cupiditates') and Is. 11: 12 (1. 3. 3, 'nec tu dissiparis, sed conligis nos').

cras: See on 8. 12. 28, 'cras et cras'.

ecce Faustus veniet: 5. 6. 10, 'nimis extento desiderio venturum expectabam istum Faustum'; G–M descry ironical allusion to Jn. 4: 25: 'dicit ei mulier: scio quia Messias venit (qui dicitur Christus): cum ergo venerit ille, nobis annuntiabit omnia.'

o magni viri academici: 5. 10. 19, 'prudentiores illos . . . quos academicos appellant . . . nec aliquid veri ab homine comprehendi' (see notes there for *comprehendere* in Academic vocabulary). The sarcasm recurs with different targets at *en. Ps.* 31. en. 2. 16, 'at enim *magni viri* sunt, qui defendunt peccata sua; *magni* sunt et qui numerant sidera, et qui computant stellas et tempora, et dicunt quis quando vel peccet vel bene vivat, et quando Mars faciat homicidam, et Venus adulteram; *magni*, docti viri, et electi videntur in hoc saeculo.'

comprehendi: See on 5. 10. 19 (and cf. 6. 4. 6, 'comprehendi').

immo . . . non desperemus: Mandouze, 91 n. 2, takes this as progress over the 'dubitans de omnibus' of 5. 14. 25; for *despero/desperatio* see on 5. 13. 23; after this passage, the only occurrence in any form in the narrative books is 6. 15. 25, of the dismissal of his concubine.

quae absurda videbantur: 6. 4. 6, 'quo antea videbantur absurda'.

possunt . . . intellegi: 6. 3. 4, 'non sic intellegi'.

in eo gradu: See on 5. 14. 25, 'statui ergo tamdiu esse catechumenus . . . donec aliquid certi' (cf. 6. 10. 17, 'quia non elucebat certum aliquid').

perspicua veritas: 13. 18. 23, 'perspicuae veritatis luce'.

ubi quaeretur: Mt. 7: 7, 'quaerite et invenietis' (*quaer-* occurs 6 × in these lines).

non vacat Ambrosio: 6. 3. 3, 'secludentibus me ab eius aure atque ore catervis'; 6. 3. 4, 'nulla dabatur copia sciscitandi quae cupiebam'.

codices: Courcelle, *Recherches*, 155 n. 2: 'Il paraît clair que les *codices* en question ne sont pas des *libri ecclesiastici*.' He thinks that A. here is pining for a 'lecture à deux' (see on 6. 3. 3) of the *Enneads* with Amb.: this goes too far. There is no reason to think that ecclesiastical books would have been on principle excluded at this time.

pro salute animae: Ps. 34: 3, 'dic animae meae, salus tua ego sum'; the same phrase as here at 5. 8. 14.

non docet catholica fides: 5. 10. 20, 'non erat catholica fides quam esse arbitrabar'; 5. 14. 24, 'fidem catholicam ... iam non impudenter adseri existimabam.'

nefas ... terminatum: Cf. 6. 3. 4, 'non ... humani corporis forma te determinatum'.

pulsare: Mt. 7: 7 again; cf. 1. 1. 1, 13. 38. 53.

amicos maiores: See 6. 11. 19, 'amicorum maiorum'.

sed quando ... curarum: Cf. 6. 3. 3 and 11. 2. 2 on the bishop's daily round, and see on 6. 3. 3 for the parallel already there between the bishop's and the orator's activities (esp. Cic. *or.* 42. 143 quoted there): distracted by involvement with others, by time spent on preparing things to say to the public, and by refreshing the body for its labours.

quando praeparamus: The anxiety of the teacher facing the morrow: *cura mort.* 11. 13, 'Carthaginis rhetor Eulogius ... cum rhetoricos Ciceronis libros discipulis suis traderet, recensens lectionem quam postridie fuerat traditurus, quendam locum offendit obscurum, quo non intellecto vix potuit dormire sollicitus.'

6. 11. 19

quid si mors: Envisions the Epicurean theory; cf. 6. 16. 26, ending the book, and 7. 5. 7, 'ingravidato curis mordacissimis de timore mortis et non inventa veritate'; 'et hoc quaerendum' suggests intellectual investigation of the theory.

culmen auctoritatis: See on 6. 5. 7, 'quos tanta ...'

spe saeculi: i.e. *ambitio saeculi*, the topic from here to the end of this paragraph. This exposition of the power of *ambitio saeculi* sets the stage for dismantling that *cupiditas* through the rest of the book. *Not* that A. will withdraw from all worldly pursuits, but that his *cupiditas* for them will die. The phrase 'relicta spe saeculi' recurs with variations in the participle at 8. 7. 18, 8. 12. 30, 9. 10. 26, and resonates almost thirty years later: *s.* 355. 1. 2, 'iuvenis veni ad istam civitatem. . . . spem quippe omnem saeculi reliqueram, et quod esse potui, esse nolui.'

praesidatus: Auson. *prof. Burd.* 15. 18, 'honore gesti praesidatus inclitus'. See Jones, *LRE* 527–9, on the standing of *praesides*: the office was originally equestrian and *infra dignitatem* for senators, but that distinction began to blur in the fourth century; it remained, however, the lowest-ranking of provincial governorships. There were approximately 22 provinces governed by *consulares*, three by *correctores*, and 31 by *praesides*. The 'vel' suggests that this was only the lower limit of his ambitions at the time, with good reason: see C. Lepelley, *Bull. litt. eccl.* 88 (1987), 229–46, and *Atti 1986*, 108 (quoting Symm. *ep.* 1. 20, 'iter ad capessendos magistratus saepe litteris promovetur'). A. used the same term for the office exercised by Pontius Pilate (*symb. cat.* 3. 7).

et ducenda uxor: As the context makes clear, the principal cupidity a wife would satisfy was *ambitio saeculi*: see *sol.* 1. 11. 18, quoted on 6. 14. 24 below, and cf. *beata v.* 1. 4, '*uxoris honorisque* inlecebra detinebar.'

multi magni viri: Cf. 6. 12. 21, 'exemplis eorum qui coniugati coluissent sapientiam et promeruissent deum'.

6. 11. 20

de die in diem: See on 4. 8. 13; the phrase once each in these central books (here, 7. 8. 12, 8. 7. 18). *ep.* 261. 2, 'ut de die in diem non solum quantum accedam, sed utrum omnino aliquid accedam, latere me fatear.' (See also *en. Ps.* 102. 16 [quoted at 8. 12. 28] and 144. 11.)

in sede sua: See 10. 20. 29–23. 34.

putabam: Continence first appears here as the central issue retarding 'conversion'; it is prepared in the text by the examples of Ambrose (6. 3. 3) and of Alypius (6. 7. 12: to be picked up again at 6. 12. 21, 'erat . . . castissimus'), and earlier of Nebridius (4. 3. 6). The issue was sharpened by the example and preaching of Amb. (see Introduction).

ad eandem infirmitatem sanandam: Cf. Mt. 4: 23, 'et circumibat Iesus totam Galilaiam docens in synagogis eorum et praedicans evangelium regni et *sanans* omnem languorem et omnem *infirmitatem* in populo'; cf. Ps. 102: 3, 'qui propitius fit omnibus iniquitatibus tuis, qui sanat omnes languores tuos' (see on 10. 3. 3).

propriarum virium: As Mandouze, 191 n. 3, says, this is another of A.'s misconceptions of Christian teaching; it could even be called a 'Manichean reflex', in so far as he had thought all along that continence was a heroic achievement of which he, in his weakness, was incapable.

neminem posse . . . dederis: Wisd. 8: 21, 'cum scirem quia nemo esse potest continens nisi deus det'; this is the first appearance (so A. Zumkeller, *Aug.-Lex.* 1. 39) of the claim that the virtue of *continentia* is 'donum dei' (cf.

10. 29. 40, 10. 31. 45, associated with 'da quod iubes' in both cases). The next dated citation is not until *pecc. mer.* 2. 5. 5 (412), with an important echo of *conf.*: 'cum iubet dicendo, "post concupiscentias tuas non eas" [Sirach 18: 30], nosque dicimus, "scimus quoniam nemo esse potest continens, nisi deus det", quid aliud dicimus quam da quod iubes?' (This is the second of three such 'dialogues' in that passage of *pecc. mer.* ending with 'quid aliud dicimus quam da quod iubes?') From that date on, the text is quoted often, in every anti-Pelagian context, to the last years (e.g. *persev.* 17. 43); see on 10. 29. 40, 'da quod iubes'. See La Bonnardière, *Biblia Augustiniana: Sagesse*, 103–7, on the verse in the Pelagian controversy; at 106 n. 124: 'Nous insistons sur le fait que saint Augustin ne réduit pas la signification de *Sap.* 8. 21 à la seule continence sexuelle, dans les textes où il cite le *Da quod iubes*.' For the sentiment, see also Tyconius, *liber reg.* 3 (ed. Robinson, pp. 19–20).

gemitu interno: Ps. 37: 9, 'gemitu cordis meae'; *en. Ps.* 37. 13, 'est enim gemitus occultus qui ab homine non auditur.' Ps. 37: 10–11, 'et ante te est omne desiderium meum et gemitus meus non est absconditus a te. (11) cor meum conturbatum est et deseruit me fortitudo mea'; see *en. Ps.* 37. 14, quoted on 5. 9. 17, 'preces . . . sine intermissione'.

pulsarem: Mt. 7: 7.

in te iactarem curam meam: Ps. 54: 23, 'iacta in dominum curam tuam, et ipse te enutriet'; *en. Ps.* 54. 24 chooses to emphasize *dominum*: we are not to repose hope in any other individual (e.g. Paul—A. quotes 1 Cor. 1: 13, 'numquid Paulus pro vobis crucifixus est?'). Here in *conf.*, the analogous argument would be that A. must know the *dominus* first before he can place hope in him, and hence be released; hence see on 8. 12. 29 for a Christological interpretation of the garden scene.

6. 12. 21

amore sapientiae: = *philosophia* (see 3. 4. 8 and see on 6. 10. 17).

diu desideraremus: The idea had probably been to feather their nest with government jobs, then retire to leisured philosophy—not an irrational aim (cf. Mallius Theodorus in this context: see on 7. 9. 13).

erat enim ipse: An episode from the *vita Alypii* not contained in the paragraphs devoted entirely to him; pertinent in the context of these paragraphs on the marriage/continence debate, but revealing the thoroughness of the integration of the Alypius paragraphs in the text as a whole. Surely a few remarks on his friend's noteworthy chastity would have had a part in any independent account of his life.

promeruissent deum: Heb. 13: 16, 'beneficientiae autem et communionis nolite oblivisci, talibus enim hostiis promeretur deus'; the verb and the expression *promereri deum* occur only there in Vg. The natural interpretation is sharply limited at *div. qu. Simp.* 1. 2. 3, 'Isaac . . . qui utique nullis operibus promeruerat deum'.

The antonym for *promeruere* most commonly occurring is *offendere*; this is a hint that, to accord with the rest of A.'s thought, the verb should not be translated 'deserve, earn', but something weaker: *OLD*'s weakest reading is 'to gain the favour of, win over' (cf. *BA* here: 'acquis la faveur divine', against Ryan's stronger 'had gained merit before God', with Pusey between: 'and served God acceptably'), but if A. ever looked at the Greek of Hebrews, he found there εὐαρεστεῖται, which the 1979 Vatican Nova Vulgata more lucidly renders 'oblectatur', a sense more neatly opposed to that of 'offendere.' The rare passages in which the verb is used in a different expression seems to confirm this: *civ.* 21. 27, '"quique sui memores alios fecere merendo" [*Aen.* 6. 664], id est, qui promeruerunt alios eosque sui memores promerendo fecerunt'.

suavitate . . . catenam: 3. 1. 1, 'quanto felle mihi suavitatem illam [amoris] . . . aspersisti'; for a possible echo of Persius (Courcelle, *Les Confessions*, 116 n. 4), see on 8. 11. 25, 'versans me in vinculo meo'.

tamquam: *BA* renders 'c'est à dire'.

manum G AHV BPZ EFM Maur. Knöll Skut. Vega Pell. : *manu* C[1]DOS : *manus* C[2] Ver. ('manus *conieci*' [sic]). Ver., p. xl, alleges *util. ieiun.* 9. 11, 'cogitandum nobis est quia repellit manus medici qui secatur', but the contexts are not closely parallel. *Inter alia*, *manus* in *conf.* is much more often singular than plural, esp. of the *manus dei*.

serpens: Gn. 3: 1, 3: 14.

dulces laqueos: See on 3. 6. 10, 'laquei diaboli', 5. 3. 3, '[Faustus] magnus laqueus diaboli, et multi *implicabantur* in eo', and esp. 10. 36. 59, 'instat adversarius verae beatitudinis nostrae ubique *spargens* in laqueis "euge, euge".'

6. 12. 22

Monnica's well-intentioned plan to get Augustine baptized and married has the effect of detaching him from a long-term relationship that could have lasted much longer; and of raising the 'issue' of marriage vs. continence in a way that an ex-Manichee reader of Cicero, hearer of Ambrose, and friend of Alypius and Nebridius might not unreasonably resolve in favour of continence.

miraretur: Cf. 'admiratio' below; both *miror* and *admiror*, of course, evoke the sense of sight; see therefore on 'libidine . . . curiositatis' below. See also on 13. 21. 30.

visco illius voluptatis: See on 6. 6. 9.

caelibem vitam: Cf. 6. 3. 3.

delectationes consuetudinis meae: Mandouze, 178 n. 7, thinks this is A.'s way of saying that what he had known was in many respects closer to marriage than to what Al. had known; 'honestum nomen matrimonii' underlines the dilemma of A.'s liaison at the time. It should be borne in mind, however, that *consuetudo* is a freighted word for A., and not a favourable one: cf. 'consuetudo' below and see on 8. 5. 10.

oportere CDS Knöll Skut. Ver. : *oporteret* GO.

libidine ... curiositatis: Alypius' characteristic vice was *curiositas* [2] (6. 7. 11–12), not concupiscence.

vinculo: Cf. 6. 12. 21, 'catenam'; for *vinculum* of *concupiscentia carnis*, see 3. 1. 1 ('et perveni occulte ad vinculum fruendi'), 8. 6. 13, 8. 11. 25, and cf. both 8. 1. 1 and 9. 1. 1 ('dirupisti vincula mea'); in a clearly broader but related sense, cf. 5. 9. 16, 'multa et gravia super originalis peccati vinculum'.

servitutem: 8. 1. 2 (of the burdens of his life at Milan), 8. 5. 10 (of *concupiscentia carnis*).

sponsionem: Cf. Wisd. 1: 16, 'impii ... et sponsiones posuerunt ad illam [mortem]'; not cited elsewhere in A., according to La Bonnardière, *Biblia Augustiniana: Sagesse*; cf. Is. 28: 18, 'et delebitur foedus vestrum cum morte et pactum vestrum cum inferno non stabit.' The key word is *sponsio*, leading from marriage to death through the scriptural echo.

et qui amat periculum: Sirach 3: 27, 'cor durum habebit male in novissimo qui amat periculum, in illo peribit.' The two scriptural echoes, both clear and deliberate, taken together constitute a stern rebuke to Alypius as he thought of marriage in 385/6.

neutrum enim nostrum ... tenuiter: Another misconception (comparable to A.'s misreadings of Christian doctrine): what they thought of marriage had little enough to do with real marriage, for it was another *phantasma*.

altissime: Ps. 91: 2, 'bonum est confiteri domino et psallere nomini tuo, altissime'; 1. 7. 12, 3. 8. 16, 6. 3. 4.

deserens: Heb. 13: 4–5, 'honorabile [sit] conubium in omnibus et torus immaculatus; fornicatores enim et adulteros iudicabit deus. (5) sint mores sine avaritia, contenti praesentibus. ipse enim dixit, "non te deseram, neque derelinquam".'

humum: No verbal scriptural echo, but cf. Gn. 2: 7 (creation of Adam from dust).

6. 13. 23

Note the passives through here. 'instabatur ... promittebatur ... instabatur ... petebatur ... expectabatur': control is slipping out of A.'s hands; contrast 6. 6. 9, 'inhiabam honoribus, lucris, coniugio'; so also the hankering for a sign ('ut ei per visum ostenderes').

maxime matre dante operam: C. Lepelley, *Atti 1986*, 1. 110, shrewdly takes it as a sign of the relative social standing of A.'s family that M., 'cette petite provinciale africaine', could play such a part among the first families of Milan.

baptismus: He says nothing here of how he felt about baptism. Baptism was probably part of the marriage bargain with the distinguished catholic family; but it could all seem slightly unreal and irrelevant, as far A. was concerned, since it was a matter of at least two years before the marriage and 'iam coniugatum' implies that baptism need not come first—might even be postponed further?

rogatu meo: A. already takes M.'s religion, and her aptitude for visions, seriously enough. We lack a name for the degree of faith that acknowledges validity in another's Christianity, without taking it on for oneself. See 3. 11. 19–12. 21, for her vision of A.'s future; there were others: 5. 9. 17, 'absit ut tu falleres eam in illis visionibus et responsis tuis, quae iam commemoravi et quae non commemoravi.' Mandouze, 179 n. 2, observes that the want of a vision did not impede Monnica from pushing ahead with plans.

Courcelle, *Recherches*, 103–5, speaks of an 'enquête sur les faits relatifs aux songes' at Milan, and connects that inquiry to this vision of M.'s. He cites also *cura mort.* 11. 13 (story of a father appearing to his son in a dream to tell him where an important financial document could be found, of whose very existence the son was unaware; and at about the same time, as A. learned *later*,[9] A.'s former pupil Eulogius, rhetor at Carthage, had a dream in which A. appeared to explain a passage in Cicero), *epp.* 9. 5 (to Nebridius: 'sed huic epistulae adiunge illam quam tibi nuper de imaginibus et de memoria misi'), 159. 3 (to Evodius), 169. 3. 11 (also to Evodius), and *quant. an.* 18. 31 (on deaf-mutes). The present passage makes it more likely that A. was always willing to accept the veridical quality of certain dreams, and it tells us nothing special about his interests or activities at Milan. Courcelle, *Recherches*, 2nd edn., 263–4 (and cf. 368–82), quotes *civ.* 18. 18, with a curious story from A.'s Milan days about rumours about women who could turn men into beasts (A. himself compares the story to Apuleius). For a more nuanced view of the development of A.'s ideas from 386 to 391 in the face of neo-Platonic (and Porphyrian) doctrines, see Dulaey, 73–88.

solebat G Knöll Skut. : *solet* CDOS Ver. (who cites, p. xxxix, *sol.* 1. 14. 25 as a parallel: but the cases are not comparable).

aetas . . . nubilis: Justinian, *inst.* 1. 22, says that males achieve puberty at age 14, and that females are 'viripotentes' at 12. The legal schools had divided (see *inst.*, ed. Moyle (5th edn., Oxford, 1912), 166) over whether to wait for physical evidence of maturity in each case or to set a specific age; in any case, patient abstinence of the sort envisioned here was perhaps in practice the exception rather than the rule (A. Rousselle, *Porneia*, 32–3, 73–4).

[9] This gives no support to C.'s claim that A. was interested in these things already at Milan.

6. 14. 24

The larger circle of friends (6. 5. 7, 6. 6. 9, 6. 11. 18, 6. 11. 19, and see on 6. 16. 26) recurs; see here 'amici' and 'amicitiae sinceritatem'; cf. this integration with the jarring juxtaposition in Bk. 4. (see on 4. 7. 12).

The other text to reflect this undertaking is *c. acad.* 2. 2. 4 (addressing Romanianus), 'nam cum praesens praesenti tibi exposuissem interiores motus animi mei vehementerque ac saepius assererem nullam mihi videri prosperam fortunam, nisi quae *otium philosophandi* daret, . . . sed me tanto meorum onere, quorum ex officio meo vita penderet, multisque necessitatibus vel pudoris vel ineptae meorum miseriae refrenari, tam magno es elatus gaudio, tam sancto huius vitae inflammatus ardore, ut te diceres, si tu ab illarum importunarum litium vinculis aliquo modo eximereris, omnia mea vincula etiam patrimonii tui mecum participatione rupturum.' The link between money-worries and the yearning for the otiose life is clear in both the earlier text and the present passage.

The project imagined here implies the near- ('paene') renunciation of *ambitio saeculi* (see on 6. 6. 9–7. 11, 6. 11. 19); all that is left of the old enthusiasm is weariness and regret ('suspiria et gemitus'). He is detached from *that* now at heart, but clings to it outwardly until after the garden scene (rather as he had been detached at heart from Manicheism and *curiositas* by the Academics); *concupiscentia carnis* clings to him.

Courcelle, *Recherches*, 179, arguing again that A. was not completely detached from the Manichees and that Romanianus remained attached, saw in the studious retreat envisaged here an analogue to the Manichean community at Rome led by the *auditor* Constantius (cf. *mor.* 2. 20. 74), which dissolved in embarrassing circumstances while A. was at Milan. The first difficulties with that view are these: (1) even if there was mimesis at work, this has no bearing on the question whether A. was entirely detached from the Manichees; (2) it is less clear than it might be whether A. knew anything of the Rome commune *before* he went to Milan. The tenor of the report suggests otherwise, since A. is eager to show himself a good witness, but makes no claim for knowing anything of the group except *ex post facto* ('. . . Romae autem me absente quid gestum sit. . . . et ego quidem postea Romae cum essem, omnia vera me audisse firmavi'). Similarly, Constantius is mentioned again at *c. Faust.* 5. 5, still alive as a catholic Christian; again there is no claim of eyewitness testimony by A.

Courcelle went on to see a Platonic analogue: 'Plus récemment, Augustin avait pu lire dans les "livres des Platoniciens" que Plotin et ses amis . . . méditèrent de fonder une Platonopolis.' But on the testimony of *conf.*, whatever he found in the *platonicorum libri* (perhaps the *vita Plotini*?) at the time recounted at 7. 9. 13 could not yet have influenced him at this stage. Hence, more credit should be given to the suggestion that A. knew of Pythagorean communities mentioned by Iamblichus (*v. Pyth.* 30). Varro would be the intermediate source (cf. W. Theiler, reviewing Courcelle, *Recherches*, in *Gnomon*, 25 (1953), 116, followed by Solignac, *BA* 13. 566–7 n.; for Pythagoras in A.'s thought earlier, see on 4. 13. 20). For praise of Pythagoras, cf. *ord.* 2. 20. 53, 'illa venerabilis ac prope divina . . . Pythagorae disciplina' (on which cf. esp. *retr.* 1. 3. 3, 'nec illud placet, quod Pythagorae philosopho tantum laudis dedi, ut qui hanc audit vel legit possit putare, me credidisse nullos errores in Pythagorica esse doctrina, cum sint plures idemque capitales', but from years after Cassiciacum cf. also *cons. ev.* 1. 7. 12, 'Pythagoras, quo in illa contemplativa virtute nihil tunc habuit Graecia clarius'); *c. acad.* 3. 17. 37, 'Plato, vir sapientissimus et eruditissimus temporum suorum . . . dicitur post mortem Socratis magistri sui, quem singulariter dilexerat, a Pythagoreis etiam multa didicisse'; *civ.* 8. 4 speaks similarly of Plato's visit to Magna Graecia and is otherwise verbally close to the *c. acad.* passage.

Testard, 1. 98–9, thought the exercise Ciceronian; a view less compelling than the Pythagorean model, but his best point is worth keeping: that *in principle*, this community admitted women/wives, which is difficult to imagine of a Manichean community. In practice, they were too much of an obstacle, but only in practice. By the time he wrote *Les Confessions*, 21–5, Courcelle had read Testard and acknowledged that a Ciceronian influence could have been mediated through Faustus (!), but he then conceded that 'elle évoque surtout . . . la "vie pythagoricienne", qui comportait aussi le lien amical, la mise en commun des biens et l'élection de "magistrats" pour leur gestion.' But it is not necessary to insist on written sources; Courcelle, *Recherches*, 108, is right to say that the visit of Ponticianus was important for revealing to A. the existence of Christian communities (monasteries) that did the same thing, only better (8. 6. 15).

We should not lose sight of what his friends and associates would have noticed at the time: that the retreat to Cassiciacum in late 386 bore a strong resemblance to the project spoken of here in perhaps late 385, differing mainly in its impermanence; but even that impermanence may not have been obvious to those friends and associates. Cassiciacum may have seemed to others as a nucleus and a beginning of a reorganized project, now undertaken by a smaller cadre consisting entirely of unmarried men (but not excluding women: Monnica). (Mandouze, 125 n. 6, reports the suggestion from an unpublished thesis by a student of his that Alypius, and perhaps Navigius, functioned at Cassiciacum rather as the 'magistrates' envisioned here.) Further evidence that the project was fresh in mind in some form comes from *sol.* 1. 11. 18, where *Ratio* summarizes A.'s position, deliberating thus: 'iam de uxore nihil disputo, fortasse enim non potest ut ducatur existere talis necessitas. quamquam si eius amplo patrimonio certum sit sustentari posse omnes, *quos tecum in uno loco vivere otiose cupis*, ipsa etiam concorditer id sinente [this phrase seems to counter exactly the problem of the *mulierculae* raised here], praesertim si generis nobilitate tanta polleat [this suggests that the 'fiancée' arranged for (6. 13. 23) may have been of a good family], ut honores illos quos esse posse necessarios iam dedisti, per eam facile adipisci possis, nescio utrum pertineat ad officium tuum ista contemnere.'

The ambition of the otiose life in community lingered, at least for Nebridius, even after both he and A. were back in Africa. How seriously A. took N.'s feelings is unclear, for one often feels in his letters to N. that he was saying what N. wanted to hear, perhaps out of a desire to influence his conversion, perhaps out of pity for his failing health and isolation; but cf. *ep.* 11. 1 to N. (388–91), 'cum me vehementer agitaret quaestio a te dudum cum quadam etiam familiari obiurgatione proposita, quonam pacto una vivere possemus, et de hoc solo statuissem et rescribere tibi . . .'

One feature of the narrative should be retained, without worrying how closely it follows the facts. This is the last episode before the garden scene in which A. consciously plans and controls his destiny. Already here, he is oddly passive (see on 6. 13. 23 for the verbs there), and the banishment of his concubine is described the same way (6. 15. 25, 'avulsa a latere meo tamquam impedimento coniugii'). The period narrated by Bks. 7 and 8 is one marked by helplessness in the face of divine providence, with volition and control restored by the act of divine grace in the Milan garden.

conloquentes . . . remoti a turbis: cf. 9. 10. 23 (Ostia), 'ubi remoti a turbis . . . conloquebamur'. Note both A.'s inclination to the idyllic private conversation and his practice, already seen in the paragraphs on the *de pulchro et apto*, of describing the efforts at enlightenment of his earlier life in *terms* used as well to describe the more successful efforts later on. Thus he places this project on the same intellectual terrain as the Ostia event.

decem ferme homines: Mandouze, 193 n. 1, reckons this number to include A., Nebridius, Alypius, Romanianus, Verecundus, Adeodatus, Navigius, A.'s two cousins, Licentius, and Trygetius, i.e. the future participants and near-participants at Cassiciacum, but others less intimately bound may have been at least in prospect. Note also 'praedivites', plural: in Mandouze's list, only Romanianus would qualify for that adjective; note also 'multi amici'; note also 'mulierculae': A.'s fiancée, Monnica, and Romanianus' presumed wife are the only women associated with any of the figures Mandouze names.

Romanianus: A.'s financial benefactor; for a portrait of the rich man's life, see *c. acad.* 1. 1. 2, 'edentem te munera ursorum et numquam ibi antea visa spectacula civibus nostris . . . conlocarentur statuae, influerent honores, adderentur etiam potestates . . . conviviis cotidianis mensae opimae struerentur'; for his benefactions to A., see *c. acad.* 2. 2. 3. From *c. acad.* 1. 1. 3 we know him a Manichee, and he remained one in 390 (*vera rel.*: dedicated to him and still seeking his conversion); *ep.* 27. 5 (sending a letter to Paulinus in 395 (Courcelle, *Les Confessions*, 601, dates it to 397) by hand of Rom.) implies progress but not conversion. Mandouze, *Pros. chr.* s.v., thinks him related to

Alypius and A., but the evidence is thin. But though Romanianus was, in some sense or another of an imprecise phrase, 'like a second father' to A., this is the only mention of him in *conf.* Why this reticence? The answer lies in R.'s own religious history.

A Thagaste inscription (*CIL* 8. 17226, [COR]NELIVS ROMANIANVS) led A. Gabillon, *REAug* 24 (1978), 58–70, to examine A.'s *ep.* 259 (AD 408, addressed to 'Cornelius') and to show it highly probable that the addressee of the letter is indeed this Romanianus. The story implied there is unpleasant. R. had resisted A.'s various urgent pleas to take baptism, but had later accepted the sacrament at what seemed death's door (*ep.* 259. 3, 'in extremo vitae periculo baptizatus'). On his wife's death, he wrote to A. to ask for a letter of consolation, which A. brusquely refuses him out of indignation at his dissolute way of life (*ep.* 259. 3, 'plebs mulierum excubat lateribus tuis; crescit in dies pellicum numerus') out of keeping with the recipient's youthful turn to chastity (*ep.* 259. 3, 'cum esses non dicam catechumenus sed in errore nobiscum perniciosissimo constitutus iuvenis iunioribus nobis, ab hoc te vitio temperantissima voluntate correxeras, quo non post longum tempus sordidius revolutus'). That rebuke demonstrates (1) that the recipient had been a Manichee with A.; (2) that he had had his own successful 'conversion' from sexual profligacy (attested for R. at *c. acad.* 2. 1. 2, 'quis ... tam subito umquam tantum intonuit tantumque lumine mentis emicuit, ut sub uno fremitu rationis et quodam coruscamine temperantiae uno die illa pridie saevissima penitus libido moreretur?'; more discreetly hinted at in a letter of A.'s to Paulinus of Nola carried by R. (*ep.* 31): and the recipient's acquaintance with Paulinus is mentioned at *ep.* 259. 1, while at 259. 5 A. echoes a verse by Paulinus); (3) that if the recipient was R., he was not much older than A. (his son Licentius, after all, was not much older than Adeodatus, and R.'s patronage of A. would be more intelligible if they were separated more by wealth and social status and not much by age). If all this is true, then at the time of *conf.*, A.'s relations with R. were somewhat strained. The hoped-for conversion and baptism had not been forthcoming; the irregularities of private life were perhaps noticeable. (Nothing is known of Licentius' later life, but it seems probable that he either delayed or avoided baptism as well, hence his disappearance from A.'s correspondence and from the narrative here and in Bk. 9.) More important, R. himself was an awkward reminder that not every conversion from sexual profligacy was necessarily permanent. He could not, perhaps, be excised from A.'s story of his early life entirely; but he hardly appears. (The breach between A. and R. probably began with A.'s lapse from Manicheism into Academic scepticism, for Romanianus could not follow him there: *c. acad.* 2. 3. 8, 'saepius enim suscensuisti academicis, eo quidem gravius, quo minus eruditus esses'.)

graves aestus negotiorum suorum: *c. acad.* 1. 1. 2 on R.'s financial affairs:

'non enim tibi alienis exemplis persuadendum est quam fluxa et fragilia et plena calamitatum sint omnia quae bona mortales putant.'

sed posteaquam: The objections of the *mulierculae* are the only obstacle given (and even they are unspecified); Mandouze, 193 n. 7, says that if A. writing *c.* 397 had any substantive objections to the scheme, he would not have failed to mention them. This is mainly true, but A. writing *c.* 397 probably thought that the ease with which the whole enterprise fell apart was itself the most telling criticism; and the diminutive, whatever you make of it for A.'s view of 'la femme' (Mandouze, 193 n. 7: see on 1. 6. 10, 'muliercularum'), is also meant as an ironic comment on the weakness of the plan. The echo of Matthew that follows offers its own criticism.

ad suspiria: De Marchi, 312, would delete the *ad* (noting the inconcinnity of construction with the *ad* that follows almost immediately).

latas et tritas vias saeculi: Cf. 6. 5. 8, 'viam saeculi latam', and cf. Mt. 7: 13–14, quoted there. See also *en. Ps.* 39. 7, quoted on 6. 11. 18.

multae cogitationes: Prov. 19: 21, 'multae cogitationes in corde viri, consilium autem domini manet in aeternum'; see La Bonnardière, *Biblia Augustiniana: Proverbes*, 156–8. A.'s most explicit exegesis of the text (*trin.* 15. 20. 38, 'neque enim ob aliud scriptum est, "multae cogitationes . . . manet in aeternum", nisi ut intellegamus sive credamus sicut aeternum deum, ita aeternum eius esse consilium, ac per hoc immutabile sicut ipse est') sees in it a hint of the immutability of God, a vital doctrine to which A. held through his Manichean years and which formed a basis for all his future theological developments; see on 7. 1. 1. (Cf. also Ps. 32: 11, 'consilium vero domini manet in aeternum, cogitationes cordis eius in saecula saeculorum'.)

nostra et tua: sc. *consilia* in both cases. It is not impossible that for the first case he had the philosophical 'commune' in mind and for the second some combination of the communities at Cassiciacum, Thagaste, and Hippo.

daturus . . . benedictione: Ps. 144: 15–16, 'oculi omnium in te sperant, domine, et tu das escam illis in opportuno tempore. (16) aperis tu manum tuam et imples omnem animam benedictionem.' See on 6. 10. 17, 'ut dares eis escam in tempore opportuno'.

6. 15. 25

The woman with whom A. lived from 370/1 (see on 4. 2. 2) was his wife in all but name, and his situation was not especially irregular. On the facts, see *BA* 13. 677–9, 'Augustin et la mère d'Adéodat', A. Zumkeller, in *Signum Pietatis* (Würzburg, 1989), 21–35, and P. Brown, 'Augustine and Sexuality', *Colloquy 46 of the Center for Hermeneutical Studies in Hellenistic and Modern Culture* (Berkeley, 1983), 2: 'Augustine was well aware that his late Roman readers would have thought that, as a young rhetor, he had come so close, for so long, to achieving an intimate, stable and eminently respectable relationship; and that they knew that this relationship had crumbled, not through "animal passion", but under the glacial weight of the late Roman caste system.' The *concilium Toletanum* (400), canon 17, defined the church's position: 'si quis habens uxorem fidelem, si concubinam habeat, non communicet. ceterum, qui non habet uxorem, et pro uxore concubinam habet, a communione non repellatur. tantum ut unius mulieris, aut uxoris aut concubinae, ut ei placuerit, sit coniunctione contentus.' At the same time, it should be borne in mind that Amb. may have diapproved of A.'s way of life: *Abr.*, 1. 3. 19, 'discant ergo homines coniugia non spernere nec sibi sociare inpares, ne huiusmodi suscipiant liberos quos heredes habere non possint, ut vel transfundendae hereditatis contemplatione, si nullo contuitu pudoris moventur, digno studeant matrimonio.' In his own later sermons, A. is strict against concubines for the married, or for those who take them cynically as a stopgap before marriage (e.g. *s.* 224. 3. 3).

The new liaison he contracted out of incontinence (see 'procuravi aliam' below) scarcely lasts any time at all. Mandouze, 178 n. 4, assuming that the sermons A. heard from Amb. at the outset of Bk. 6 were those of the *exameron*, concludes that the 'la terrible séparation[10] [from the first woman] est donc très probablement postérieure au 4 Avril 386.' The 'chill despair' ('frigidius sed desperatius') is with him through the spring and summer that lead to the garden scene (for the second rupture then, see on 8. 12. 30).

In a nearly contemporaneous passage that must refer to this relationship, all the criticism is reserved for the heartless male, and the woman retains her dignity: *b. coniug.* 5. 5, 'solet etiam quaeri, cum masculus et femina, nec ille maritus nec illa uxor alterius, sibimet non filiorum procreandorum sed propter incontinentiam solius concubitus causa copulantur, ea fide media ut nec ille cum altera nec illa cum altero id faciat, utrum nuptiae sint vocandae. et potest quidem fortasse non absurde hoc appellari conubium, si usque ad mortem alicuius eorum id inter eos placuerit et prolis generationem, quamvis non ea causa coniuncti sint, non tamen vitaverint,[11] ut vel nolint sibi nasci filios vel etiam opere aliquo malo agant ne nascantur. ceterum si vel utrumque vel unum horum desit, non invenio quemadmodum has nuptias appellare possimus. etenim si aliquam sibi vir ad tempus adhibuerit, donec aliam dignam *vel*

[10] 'Terrible séparation'! As is well known, the banishing of mistresses is taken seriously in France: see L. Bertrand, *Celle qui fut aimée d'Augustin* (Paris, 1935)— already sketched in a chapter of his *Autour de saint Augustin* (Paris, 1921); P. Villemain, *Confessions de Numida, l'Innommée de saint Augustin* (Paris, 1957—with 80 illustrations and a preface by H.-I. Marrou). The word 'mistress' is of course entirely wrong, as Mandouze, 178 n. 3 and 178 n. 7, notes, even as he uses 'maîtresse' in his text, finding it perhaps less distasteful than 'concubine'—the other designator he uses, one often found elsewhere, is the touching 'mère d'Adéodat'.

[11] As B. Shaw points out (*Past and Present*, 115 (1987), 45), A. lived with his concubine for a decade and a half and had only one surviving child: A.'s knowledge of contemporary contraceptive techniques thus probably derived from practical expertise (he expressly mentions the 'rhythm method' at *mor.* 2. 18. 65; Shaw does not notice that passage, but has other refs. at 45 n. 189 suggesting A. was knowledgeable).

honoribus vel facultatibus suis inveniat quam comparem ducat, ipso animo adulter est, nec cum illa quam cupit invenire, sed cum ista cum qua sic cubat ut cum ea non habeat maritale consortium. unde et ipsa hoc sciens ac volens impudice utique miscetur ei cum quo non habet foedus uxorium. *verumtamen si ei tori fidem servet et, cum ille uxorem duxerit, nubere ipsa non cogitet atque a tali prorsus opere continere se praeparet, adulteram quidem fortassis facile appellare non audeam*; non peccare tamen quis dixerit, cum eam viro cuius uxor non est misceri sciat? iam vero si ex illo concubitu, quantum ad ipsam attinet, nonnisi filios velit et, quidquid ultra causam procreandi patitur invita patiatur, multis quidem ista matronis anteponenda est.' Sim. reflections on a smaller scale at *b. coniug.* 14. 16. Similarly, it is hard to believe that this case was not in A.'s mind when he wrote *c.*413 at *f. et op.* 19. 35, 'de concubina quoque, si professa fuerit nullum se alium cognituram, etiamsi ab illo cui subdita est dimittatur, merito dubitatur utrum ad percipiendum baptismum non debeat admitti.' The text is from a discussion of divorce questions, and A. is again mustering as much generosity as he can.

The relationship should not be romanticized or thought in any way untypical of the age, for despite these sensitive remarks, A. could also write: *Gn. litt.* 9. 5. 9, 'aut si ad hoc adiutorium gignendi filios non est facta mulier viro, ad quod ergo adiutorium facta est? si quae simul operaretur terram, nondum erat labor ut adiumento indigeret, et si opus esset, melius adiutorium masculus fieret. hoc et de solacio dici potest, si solitudinis fortasse taedebat. quanto enim congruentius ad convivendum et conloquendum duo amici pariter quam vir et mulier habitarent?' Similarly, the hard-heartedness for which moderns reproach him as with one voice here needs to be compared to what he said in expounding the sermon on the mount: *s. dom. m.* 1. 15. 40–1 deals with 'hating your relatives', i.e. deploring the earthly qualities of your connections, while hoping to be joined somehow spiritually in heaven (*retr.* 1. 19. 5, to be sure, softens this); *s. dom. m.* 1. 15. 41, 'sic invenitur bonus christianus diligere in una femina creaturam dei . . . odisse autem coniunctionem copulationemque corruptibilem atque mortalem, hoc est diligere in ea quod homo est, odisse quod uxor est. ita etiam diligit inimicum, non in quantum inimicus est, sed in quantum homo est'.[12]

mea peccata multiplicabantur: Sirach 23: 2–3, 'quis superponet in cogitatu meo flagella et in corde meo doctrinam sapientiae . . . (3) et ne adincrescant ignorantiae meae et multiplicentur delicta mea, et peccata mea abundent.'

cor . . . sanguinem: Cf. 4. 7. 12, 'portabam enim concisam et cruentam animam meam impatientem portari a me, et ubi eam ponerem non inveniebam'—on losing his friend. A similar image with a prescription for the cure: *ep.* 263. 2, 'haec cum cogitantur et consuetudinis violentia

[12] On this passage, see now G. Bonner, *Homo Spiritalis* (Festschrift L. Verheijen: Würzburg, 1987), 276–94.

requiritur, *cor pungitur* et tamquam sanguis cordis fletus exoritur. sed *sursum sit cor* [liturgical!: see on 12. 16. 23] et sicci erunt oculi.'

et illa in Africam redierat: The woman offers besides another model of *continentia* in A.'s circle—so Mandouze, 180 n. 5, who adds 'sa décision ultérieure de se faire moine . . . s'apparente, si on la rapproche de celle de la mère d'Adéodat, avec ce "commun accord" par lequel on voit dans l'histoire de l'Église . . . des époux prendre chacun de leur côté le chemin du couvent.' Stripped of anachronism, the point is valid.

non amator coniugii sed libidinis servus: Cf. 4. 2. 2, 'sane experirer exemplo meo quid distaret inter coniugalis placiti modum, quod foederatum esset generandi gratia, et pactum libidinosi amoris, ubi proles etiam contra votum nascitur, quamvis iam nata cogat se diligi.'

procuravi aliam: No mention is ever made again of the second concubine. See on 8. 1. 2 for the sternness with which A. held that such relationships were an impediment to baptism.

vulnus illud meum: G–M adduce *nat. b.* 20, 'item in corpore melius est vulnus cum dolore quam putredo sine dolore, quae specialiter corruptio dicitur.'

6. 16. 26

The conclusion summarizes the state of affairs late in 385 or early in 386: disaffection with *ambitio saeculi*, no progress with *curiositas* (a subject discussed in this book only in the Alypius narrative), and a looming consciousness that the central problem is *concupiscentia carnis*.

Tibi laus, tibi gloria: The resonance is liturgical rather than scriptural, but cf. 1 Chron. 29: 11–12, 'tua est domine magnificentia, et potentia, et tibi laus: cuncta enim quae in caelo sunt et in terra, tua sunt; . . . (12) tuae divitiae, et tua est gloria.'

dextera tua: i.e. Christ; *en. Ps.* 108. 29, 'intellegamus itaque manum dei esse Christum,' and frequently in *conf.* (see on 11. 2. 4, 'Christum . . . virum dexterae tuae').

raptura me de caeno: Cf. Ps. 39: 4, 'eduxit me de lacu miseriae et de luto limi'; *en. Ps.* 39. 3, 'quis est lacus miseriae? profunditas iniquitatis, ex carnalibus concupiscentiis. . . . limus luti, concupiscentiae carnales, tenebrae iniquitatum.'

ablutura CDO^{2}G Knöll Skut. : *ablutura me* Maur. : *ablatura* O^{1}S Ver. (cf. 8. 1. 2, 'dextra tua suscepit me et inde ablatum posuisti ubi convalescerem'; but somewhat redundant after 'raptura'); the verb *abluo* elsewhere in *conf.* only at 1. 11. 17, 5. 8. 15 and 6. 13. 23, always baptismal in force.

gurgite: 2. 2. 2, 'gurgite flagitiorum', 6. 7. 11, 'gurges tamen morum Carthaginiensium'.

metus mortis: For the concern, see 6. 12. 19, 'quid si mors ipsa omnem curam cum sensu amputabit et finiet?' Of course, Amb. *bono mort.* dates from around this time as well, but for all the relevance of the sermon and its topic, A. makes no specific mention. The two powerful deaths in *conf.* (emphatically not including Patricius) are roughly equally separated by the passage on the fear of death here (from middle of Bk. 4 to end of Bk. 6 = 2½ books, the same as from here to the middle, or a bit past the middle, of Bk. 9).

The topic was urgent in A.'s circle. From Cassiciacum, we have *sol.* 2. 14. 26, 'ille autem alius [Zenobius?] novit quidem pro familiaritate ardorem nostrum; sed ita longe abest, et ita nunc constituti sumus ut vix ad eum vel epistulae mittendae facultas sit. quem credo iam otio transalpino perfecisse carmen quo mortis metus excantatus effugiat, et antiqua glacie duratus animae stupor frigusque pellatur.' In 387/8, A.'s description of the mind's ascent to God has an important place for the fear of death: *quant. an.* 33. 73, '*in ipso enim purgationis negotio subest metus mortis saepe non magnus, saepe vero vehementissimus*; non magnus tum cum robustissime creditur (nam videre hoc utrum sit verum, non nisi perpurgatae animae licet) tanta dei providentia iustitiaque gubernari omnia, ut nulli mors inique accidere possit, etiamsi eam forte iniquus intulerit. vehementer autem formidatur mors in hoc iam gradu, cum et illud eo creditur infirmius, quo sollicitius quaeritur. . . . deinde quo magis sentit anima, eo ipso quo proficit, quantum intersit inter puram et contaminatam, eo magis timet, ne deposito isto corpore minus eam deus possit quam se ipsam ferre pollutam. *nihil autem difficilius quam et metuere mortem et ab inlecebris huius mundi, sicut pericula ipsa postulant, temperare*.' The Pauline verse 'absorpta est mors in victoriam' (1 Cor. 15: 54) is often met with in the works before *conf.* (e.g. *div. qu.* 70—offering a strained reading). The topic continues (contemporary with *conf.*): *qu. ev.* 1. 47, 'cupiditati quae in curiositate est opponitur timor mortis; sicut enim in illa cognoscendarum rerum est aviditas, ita in ista metus amittendae talis notitiae' (on that reading, which goes on to provide congruent fears for the other two temptations of 1 Jn. 2: 16, the exacerbation of the fear makes this a good time to conquer *curiositas*, and that is exactly what happens in Bk. 7). In 409, A. recalls the argument clearly: *ep.* 104. 1. 3 (to Nectarius), 'mortem autem malorum omnium esse finem habent quidem vestrae litterae sed nec ipsae omnes; epicureorum est quippe ista sententia et si qui alii mortalem animam putant. at illi quos Tullius quasi consulares philosophos appellat[13] . . . non extingui animam sed emigrare censent. . . . hoc congruit et litteris sacris, quarum me cupio litteratorem.' Disdain for the fear of death passes

[13] *Hort.* frg. 102 (also at *c. Iul.* 4. 15. 76); contrast the non-Platonist philosophers who doubt the immortality of the soul, characterized at *Tusc.* 1. 23. 55 as 'plebei philosophi'. See M. Ruch, *REAug* 5 (1959), 99–102; J. Glucker, *REAug* 11 (1965), 229–34; W. Görler, *Untersuchungen zu Ciceros Philosophie* (Heidelberg, 1974), 103–4.

with age: *Io. ev. tr.* 123. 5, 'mortis naturalem timorem' (Van Bavel, 136–7), and cf. *civ.* 13. 6.

Bks. 7 and 10 also end with fear: 7. 21. 27, 10. 43. 70.

disputabam: Cannot govern the indirect discourse that follows; hence the colon introduced here to allow an implicit verb of saying.

Epicurum: For an idea how the argument would have gone, cf. *c. acad.* 3. 7. 15–9. 18, a doxographical discussion of the ancient philosophical traditions in competition that has the look and feel of a pre-digested piece of exposition, not anything that has any real intrinsic relation to a dialogue at Cassiciacum. The passage is the most serious critique of Academic philosophy. Solignac, *BA* 13. 572 n. 1: 'la *doxa* d'Épicure qu'Augustin rapporte ici ne se trouve pas dans le texte de Cicéron, ni, sauf erreur, dans les autres ouvrages de Cicéron. Ce renseignement semble venir des doxographes par l'intermédiaire du manuel de Celsus', though A. probably also knew Lucretius (*RA* 15 (1980), 161–2). Epicurus was a learned possibility, not a living one, in A.'s world; mentioned in five of A.'s sermons (one devoted to Act. 17: 18 ff., where Paul encounters Stoics and Epicureans on the Areopagus), in five letters (esp. 118 to Dioscorus on various queries), and otherwise chiefly in the philosophical treatises *c. acad.* and *util. cred.*

The text here seems deliberately to suggest that A. and Alypius were reading Cicero's *de finibus bonorum et malorum* (see Testard, 1. 99–101; this same phrase recurs in this same unattributed way at *civ.* 19. 1); this is the first distinct reference of that sort since 3. 4. 7, where the *Hortensius* was invoked. Cicero had what the poet-schoolmaster William Cory called the ability 'to express assent or dissent in graduated terms'.[14] This produces a text that could well be read by one, like A., making no attempt at historical distancing, as a lively dialogue with no forced conclusion. The presentation of opposing views is fair and comprehensive, and one can well see A. and Al. reading this book with genuinely open minds, considering the merits of the various schools—the more so because the surviving Latin literature of their day offered few better such expositions.

There is perhaps a pattern to be detected in A.'s readings as he reports them here. Cicero's first philosophical works after the death of Tullia in 45 were (see Cic. *div.* 2. 1. 1–2): *consolatio*, *Hortensius*, *academica*, *de finibus*, *Tusculanae*, *de natura deorum*. We are shown A. reading the *Hortensius* in Bk. 3, approving the Academic school in Bk. 5,[15] debating 'de finibus bonorum et malorum' here, and retiring to a country villa for debates on essential issues in Bk. 9 (the five books of the Tusculans deal with, in order, *mors*, *dolor*,

[14] This quality of C. is a theme, elegantly developed, of W. Görler, *Untersuchungen zu Ciceros Philosophie* (Heidelberg, 1974).

[15] By way of illuminating exception, recall that there is no trace of Ciceronian 'consolation' in Bk. 4 after the death of A.'s friend.

aegritudo, *animi perturbationes*, and the question whether 'sufficit ad beate vivendum virtus'); and cf. *ord.* 1. 8. 26, where they write down their discussion of the night before and then A. adds: 'nihilque a me aliud actum est illo die, ut valetudini parcerem, nisi quod ante cenam cum ipsis dimidium volumen Vergili audire cotidie solitus eram' with Cic. *Tusc.* 1. 49. 119 (also at the end of the first day's discussion in a narrative of country-house philosophizing: for *Tusc.*/Cassiciacum parallels, see on 9. 4. 7, 'libri disputati'), 'sed nunc quidem valetudini tribuamus aliquid, cras autem . . . agamus haec. . . .' Whether to press the pattern and see in Bks. 10–13 (and in *trin.* beyond) a counterpoint to the *de natura deorum* is perhaps to take this observation too literally (and note that the concerns and the doctrines of *nat. deor.* can be descried in 7. 1. 1 f.: see C. Baguette, *REAug* 16 (1970), 47–77: see on 7. 1. 1). But at least it seems clear that A. plants clues through his narrative to show him following in Ciceronian footsteps. Cicero is the preparation for the 'Platonic' doctrines that he will discover in Bk. 7. See on 8. 1. 1 for the way another authoritative text plays a similar role in that book. (Amb. *off.* dates from 388/9 and offers a model for Ciceronian imitation-with-difference, but we do not know how soon A. would have seen or heard of that work; it stands to Cicero's *de officiis* even more closely than A.'s *civ.* stands to Cicero's *de republica*.)

voluptate . . . honestatis: Cic. *Lucullus* 46. 140, 'unum igitur par quod depugnet reliquum est: voluptas cum honestate. . . . audi contra illos qui nomen *honestatis* a se ne intellegi quidem dicant . . . fontem omnium bonorum in corpore esse, hanc normam hanc regulam hanc praescriptionem esse naturae.'

en. Ps. 73. 25, on Epicurus: 'et forte qui dicit, cum mortuus fuero, postea nihil ero, et litteras didicit, et ab Epicuro didicit hoc, *nescio* quo deliro philosopho, vel potius amatore vanitatis, non sapientiae, quem ipsi etiam philosophi porcum nominaverunt; qui *voluptatem corporis summum bonum dixit*'; cf. Cic. *fin.* 1. 12. 40, 'extremum autem esse bonorum voluptatem ex hoc facillime perspici potest: constituamus aliquem magnis, multis, perpetuis fruentem et animo et corpore voluptatibus, nullo dolore nec impediente nec impendente, quem tandem hoc statu praestabiliorem aut magis expetendum possumus dicere?' Pertinently here, A. was also aware of the Epicurean doctrine that souls decayed more rapidly than bodies after death (*s.* 150. 5. 6, 'nam epicurei et de corpore et de anima hoc idem sentiunt, quod utrumque mortale est. et quod est gravius et detestabilius, prius dicunt animam post mortem dissolvi quam corpus').

pulchritudinis: See on 4. 13. 20 (*de pulchro et apto*) and 10. 27. 38.

vena: 3. 1. 1, 'venam . . . amicitiae'; 3. 2. 3, 'de illa vena amicitiae': the vein that literally pumps the lifeblood of human society, the *vena* that is soiled by *concupiscentia carnis*.

cum amicis: The place of friendship in Bk. 6 is here recapitulated (in a paragraph that names both Nebridius and Alypius again). The connection is made to both *ambitio saeculi* directly and, by way of preparing the ground for Bk. 8, to *concupiscentia carnis* indirectly. Cf. Cic. *fin.* 1. 20. 65, 'restat locus huic disputationi vel maxime necessarius de amicitia. . . . Epicurus una in domo, et ea quidem angusta, quam magnos quantaque amoris conspiratione consentientes tenuit amicorum greges! quod fit etiam nunc ab epicureis.'

in quantalibet CDGO Maur. Ver. : *quantalibet* S Knöll Skut.

gratis diligebam: Cic. *fin.* 2. 26. 83, of Epicureans again: 'hoc foedus facere si potuerunt, faciant etiam illud, ut aequitatem, modestiam, virtutes omnes per se ipsas gratis diligant.'

o tortuosas vias: 2. 3. 6, 'vias distortas', 4. 12. 18, 'vias difficiles et laboriosas', 10. 36. 59, 'perversa et distorta via'; and recall the 'broad way of the world' from 6. 5. 8 and 6. 14. 24.

vae animae audaci: Is. 3: 9, 'vae animae eorum'.

quae speravit: See on 6. 1. 1 for *spes* throughout this book.

versa . . . requies: The theme of restlessness (from 1. 1. 1) takes vivid form in a metaphor from insomnia.

in tergum: See on 2. 3. 6 and 4. 16. 30.

tu solus requies: Cf. 4. 11. 16 (the address to the soul), 'audi et tu [anima]: verbum ipsum clamat ut redeas, et ibi est locus quietis imperturbabilis, ubi non deseritur amor si ipse non deserat.'

ecce ades: Ps. 138: 8, 'si ascendero in caelum, tu ibi es; si descendero in infernum, ades'; see on 1. 2. 2.

constituis CDGO Maur. Skut. Ver. Pell. : *constitues* S Knöll.

constituis nos in via tua: Cf. Ps. 31: 8, 'statuam te in via hac qua gradieris'; Ps. 85: 11, 'deduc me, domine, in via tua, et ambulabo in veritate tua'; *en. Ps.* 85. 15, 'via tua, veritas tua, vita tua, Christus'; see on 7. 7. 11.

currite: 1 Cor. 9: 24, 'sic currite ut comprehendatis [bravium]'.

ego feram et ego perducam et ibi ego feram: Is. 46: 4, 'ego feci et ego feram et ego portabo et salvabo.' The book ends extraordinarily, with the divine voice speaking, but not in an explicit, obvious, or direct biblical quotation; neither 1 Cor. nor Is. is especially close, and what the editors print in quotation marks does not match either or both. There are five places in *conf.* (4. 10. 15, 4. 11. 16, here, 7. 10. 16, 13. 29. 44) where God is 'quoted' in direct discourse and where the quotation marks we insert contain no scriptural quotation; but in the last case, the quotation marks contain words that authorize us explicitly to look everywhere else at scriptural quotations as 'God speaking'—obviously in a different way: 13. 29. 44, 'nempe quod scriptura mea dicit, ego dico. See V. Warnach, *Aug. Mag.* 1. 438 ff., on God speaking to human ears.

BOOK SEVEN

The years 385 and 386. The disease of *curiositas* is now brought to a crisis and largely cured. The central reading of the *platonicorum libri* and the 'tentatives d'extase' that follow have been the object of much attention and controversy.

7. 1. 1–8. 12	*Problems of God and evil: A.'s state of mind*
7. 6. 8–10	*Fate and the stars (cf. 4. 3. 4—6)*
7. 9. 13–10. 15	*Reading the platonicorum libri*
7. 10. 16–21. 27	*Reactions*
7. 10. 16	*Frustrated ascent*
7. 11. 17–16. 22	*The problem of evil*
7. 17. 23	*Ascent*
7. 18. 24–21. 27	*Christological confusion: turning to Paul*

The search for 'wisdom' dating to A.'s nineteenth year has been presented mainly as an intellectual inquiry. That inquiry comes to an end with Bk. 7, and we should not hesitate to grant that the formulations achieved here are for the most part those that A. will carry at least to his bishopric. What is surprising and unexpected, however, is that intellectual success proves so unsatisfactory. Bk. 8 relocates the inquiry on moral rather than intellectual terrain and finds resolution there; the Christological uncertainties of Bk. 7 and their resolution in Bk. 8 are an important link between the two.

The book reflects a structure of A.'s thought that du Roy has patiently excavated. A. comes, through reading the *platonicorum libri*, to a knowledge of the triune Godhead, a knowledge he may be thought to have shared on most essentials with some non-Christian philosophers. Then he began an ascent to a knowledge of the incarnate Christ, a knowledge not attained by the philosophers. This approach is the reverse of what, on A.'s own terms, it should be. On his theory, it should be the mediator of God and man, Christ Jesus, who introduces us to the full triune deity; but in practice (and du Roy shows how it is always this way for A.), it is the other way around. The trinity is accessible to philosophical speculation, the incarnate redeemer is not. Whether du Roy (esp. at 450–8) is correct to suggest that this paradox is responsible for constraining A.'s trinitarian thought within limits it might otherwise have transcended is another matter.[1]

But A. was always a master of capturing in his words what many of his readers have had trouble retaining or expressing, the tension of the middle time between redemption and resurrection, between conversion and beatific vision. The middle time is the time of paradox, and many of the perplexities to which A. gives voice, and to which he does not give satisfactory monovalent

[1] For reservations on du Roy's reading, see below on 7. 9. 13.

solutions (e.g. the dilemma of grace versus freedom: by 'satisfactory' I mean a solution that would remove an issue from further debate among his heirs) are themselves reflections of this time of paradox in which A. saw himself living. Faith seeks understanding in A., with profound awareness that the search is unending; but that awareness came slowly. Where at the outset (e.g. at the time of which he writes here), satisfactory monovalent solutions were certainly what he sought, by the time at which he writes here a fruitful 'disenchantment'[2] has supervened. Scholarship that reproaches A. for not finding those solutions is often itself inadequately disenchanted in that sense.

It remains true that, as du Roy repeatedly observes, it was A.'s own experience of God, as recounted in *conf.* and as abundantly displayed in his other works, that gives structure and direction to his theology. That western Christianity since has been so heavily influenced by A. means that many others have been led to think of God in ways that reflect the personal experience of this one man.

7. 1. 1

See C. Baguette, *REAug* 16 (1970), 47–77, discussing 7. 1. 1, 7. 1. 2, and 7. 5. 7 in detail. He goes too far when he takes the description of A.'s state of mind through these paragraphs as a vivid and specific autobiographical record of a particular moment. Here A. rather puts on paper the view of God that he had carried with him unconsciously through Manicheism and scepticism and that he *became aware of* in terms like those reported here only when the *platonicorum libri* brought to light a new and different view. Thus the separation of these paragraphs from the account of the reading of the Platonists is one of narrative convenience rather than chronological sequence. To Baguette's identification of Stoic themes may be raised the general objection[3] that doctrines introduced by the Stoics did not necessarily come to A. with the manufacturer's label intact. Baguette, 57, quotes a conversation with Mandouze for the suggestion that the discussion here is still marked by Manicheism; there is truth to that, in so far as A.'s idea of God in his Manichean days was itself an eclectic mixture, guided and directed by the Manichees to be sure, but containing A.'s own contributions from his reading (and there was no such thing as pure untainted Manicheism, any more than there is ever pure untainted Christianity). Baguette's best evidence is to be found in the parallels he adduces from Cic.

[2] The word is taken from Robert Markus's 1984 Villanova Augustine Lecture: *Conversion and Disenchantment in Augustine's Spiritual Career* (Villanova, 1989).

[3] Which also applies to the more recent work of M. Colish, *The Stoic Tradition* (Leiden, 1985). Much authority remains in the discussion of G. Verbeke, *L'Évolution de la doctrine du pneuma* (Paris-Louvain, 1945), esp. 489–508 on A.; Verbeke has a healthy respect for the way A.'s 'sources' were mediated to him and for the way his receptivity to them was guided by the data of the scripture texts he had in front of him.

nat. deor. There, at least, he deals with a text A. certainly knew; whether the reflections here are signs of renewed contact with that text by A. in Milan (which would carry further the pattern of Cicero readings suggested above on 6. 16. 26) remains unclear.

adulescentia . . . iuventutem: For the ages of a man, see on 1. 8. 13. Milan and 'youth' go together in memory: *s.* 318. 1, '*nobis iuvenibus* apud Mediolanum constitutis'. Advance in age brought new capacity for understanding: *vera rel.* 24. 45, 'his ergo carnalibus vel corporalibus formis inhaerere amore pueros necesse est, adolescentes vero prope necesse est, hinc iam procedente aetate non est necesse.' If *iuventus* corresponds to the fourth day of creation, then it should be (and is here in Bk. 7) the time of the bringing of light (L. Pizzolato, *Le fondazioni dello stile delle 'Confessioni'* (Milan, 1972), 83 n. 8).

vanitate: See on 4. 1. 1, 'vani'; 7. 14. 20, 8. 1. 2, and 10. 35. 57, taken together, suggest that this emptiness is associated with the concupiscence of the eyes and *curiositas*.

cogitare: Etymology (from Varro) at 10. 11. 18, 'nam cogo et cogito sic est, ut ago et agito, facio et factito . . . cogitando quasi conligere.' See *BA* 16. 606, note by Agaësse on *trin.* 10. 5. 7: 'L'idée signifiée est donc celle de "grouper," de "rassembler". En raison de l'usage le mot en est venu à désigner l'acte par lequel l'âme rassemble les connaissances éparses et latentes dans la mémoire pour les poser en quelque sorte sous son propre regard et les amener ainsi à la conscience claire.' See below on 'corporeum aliquid', and again in 7. 1. 2 and 7. 4. 6.

substantiae: Strictly speaking, inappropriately predicated of God (see on 5. 10. 20): the use of the word here is thus a clue that A. was going about his inquiry the wrong way.

non te cogitabam, deus: Earlier moments in the struggle to think about God: 4. 7. 12, 4. 16. 29, 5. 10. 19.

in figura corporis humani: He had believed that catholic doctrine was anthropomorphic (5. 10. 19), and so the Manichees had attracted him with their anti-corporeal views (3. 7. 12), but Ambrose had set him right (6. 3. 4, 6. 4. 5, 6. 11. 18, and here), so he was half-way to a solution of that one of the three Manichean 'difficulties for Catholics' catalogued at 3. 7. 12. The first, the interpretative difficulties posed by the OT, had been set aside by Ambrose's preaching (see on 6. 3. 6), and the third, the origin of evil, will be addressed shortly (7. 3. 4) and resolved later in the book (7. 11. 17 ff.).

audire aliquid de sapientia coepi: See 3. 4. 7 (the *Hortensius*).

coepi, semper CD Maur. Knöll Skut. Ver. : *coepissem per* OS : *coepi sed semper* G. The semicolon after 'humani' and a comma (for a semicolon) after 'coepi' are new corrections; the punctuation has already been questioned, and a similar solution proposed, by A. Gabillon, *Homo Spiritalis* (Festschrift L.

Verheijen: Würzburg, 1987), 440–1. Elsewhere, A. regularly uses indicative with *ex quo* (sc. *tempore*), e.g. *civ.* 3. 3, 12. 13, *bapt.* 1. 15. 23, 3. 12. 17 (2 ×).

gaudebam me hoc repperisse: At *ep.* 147. 6. 17, A. begins an extended discussion of a text from Ambrose's *in Luc.* (which A. probably heard while in Milan) that speaks directly to the question of how one sees God, but the subtext is the anthropomorphic question: how to imagine God without limbs and eyes and hands. That letter, known separately under the title *de videndo deo*, is an important and beautiful text. He concludes: *ep.* 147. 23. 52, 'tene mecum sancti viri Ambrosii sententiam iam non eius auctoritate sed ipsa veritate firmatam. neque enim et mihi propterea placet, quia *per illius os* potissimum me dominus ab errore liberavit et per illius ministerium gratiam mihi baptismi salutaris indulsit, tamquam plantatori et rigatori meo nimium faveam, sed quia de hac re et ipse hoc dixit, quod pie cogitanti et recte intellegenti loquitur etiam ille qui incrementum dat deus.' He then repeats the crucial quotation from 147. 7. 18; see also *ep.* 148. 2. 7, 148. 2. 10.

conabar: The verb of frustrated human effort, particularly of the mind struggling to ascent: see on 7. 21. 27, 'conari'.

homo et talis homo: = Augustine, as at 1. 1. 1 and elsewhere.

solum et verum deum: Jn. 17: 3, 'ut cognoscant te solum deum verum'.

incommutabilem: The immutability of God can scarcely be called a Christian doctrine, in so far as there is little explicit Christian scripture to warrant such an assertion (Num. 23: 19, 'non est deus quasi homo ut mentiatur, nec ut filius hominis, ut mutetur', and Malach. 3: 6, 'ego enim dominus, et non mutor', are the best proof-texts, scarcely compelling or authoritative), and the other texts A. adduces in such contexts (e.g. Ps. 101: 28, 'tu autem idem ipse es, et anni tui non deficient': see on 1. 6. 10) do not themselves compel such a doctrine unless there is already a predisposition in that direction. A. himself, moreover, is saying here that the doctrine was part of him when he was least Christian himself; certainly his dismay (on which the Manichees played: 3. 7. 12) at the thought of an anthropomorphic God reflects a congruent attitude. The challenge for any Christian doctrine of immutability, of course, is that there is plenty of evidence that God is mutable—if the Old Testament be read literally. And of course the scandal of the literal Old Testament was one of the factors that drove A. into the arms of the Manichees (see on 3. 7. 12).

A. does not hesitate to credit the Platonists with seeing the link between immutability and a correct notion of God as bodiless (*civ.* 8. 6). It is central to A.: Gilson, *The Christian Philosophy of Saint Augustine* (New York, 1959), 22, calls it 'the most profound and most constant element in his metaphysical thought'; B. J. Cooke, *Modern Schoolman*, 24 (1947), 42: 'The more one studies St. Augustine and notes the place of immutability in relation to the other divine attributes, the more it seems that immutability is the basis on which the other attributes rest, the root from which they spring'—Cooke's study is important for the abundance of texts quoted and discussed and for placing immutability in the wider context of A.'s thought. In *conf.*, see 1. 4. 4, 4. 15. 26 (*de pulchro et apto*), Bk. 7 almost *passim* (here, 7. 9. 14 (reading the *platonicorum libri*), both 'ascents' (7. 10. 16, 7. 17. 23), and after (7. 19. 25)), 10. 25. 36, Bk. 12 again almost *passim*, and the trinitarian discussions of Bk. 13 (esp. 13. 11. 12, 13. 16. 19). A. is the first attested writer to use, and use often, the noun and adverb forms *incommutabilitas/incommutabiliter*.

The doctrine's antecedents are philosophical (but emphatically not Ciceronian and not Stoic: Colish, *The Stoic Tradition*, 2. 148); it has a Platonic warrant (*rep.* 2. 381 B–C is a suitable proof-text) but it would be wrong to call it 'Platonic' here, for A. claims stoutly that the doctrine was part of him before ever he read the Platonists. It is present from Cassiciacum (*sol.* 2. 20. 35, *imm. an.* 5. 7–6. 10), and throughout the early works: *mus.* 6 *passim*, *quant. an.* 34. 77, *lib. arb.* 2 *passim*, *vera rel.* 3. 3 and *passim*; cf. *s.* 7. 7 (perhaps as early as 397, on Ex. 3: 14), 'esse, nomen est incommutabilitatis. omnia enim quae mutantur desinunt esse quod erant et incipiunt esse quod non erant'; *nat. b.* 1, 19, 39 (from 399; see on 1. 7. 12 for the importance of this treatise for A.'s ontological views at the time of *conf.*). The same emphasis is not so evident in A.'s later works, but see *trin.* 15. 23. 43.

medullis: Source of sighs (3. 6. 10; cf. *en. Ps.* 85. 8, 'sincerissimis medullis castisque suspiriis ipsum dilige'), *confessio* (6. 11. 12) and other truly sincere speech (*en. Ps.* 39. 20, 62. 9, 90. s. 2. 8), sacred song (9. 3. 6), and the voice of conscience (*en. Ps.* 45. 3, 'medulla conscientiae', 101. s. 1. 10, 'medulla cordis'); *en. Ps.* 65. 20, 'nihil enim interius medullis nostris: interiora ossa sunt carne, medullae interiores sunt ipsis ossibus.'

clamabat: Lam. 2: 18, 'clamavit cor eorum ad dominum super muros filiae Sion; deduc quasi torrentem lacrymas per diem et noctem, non des requiem tibi.'

phantasmata: See on 3. 6. 10.

circumvolantem turbam immunditiae: *Aen.* 3. 233 (of the Harpies gathering a second time (here see 'conglobata rursus') to harass the Trojan refugees), 'turba sonans praedam pedibus circumvolat uncis, | polluit ore dapes'.

in ictu oculi: Echoes 1 Cor. 15: 52; see on 7. 17. 23, 'in ictu trepidantis aspectus'.

corporeum . . . aliquid: The corporeality of the gods is defended by Cicero (e.g. *nat. deor.* 1. 12. 30). For the *locus* of divine activity, cf. Cic. *nat. deor.* 1. 37. 103, which distributes deities among the four elements (note here 'terreno . . . caelesti', and again 7. 1. 2: 'caelesti' evokes 'fire' because the *sidera caeli* stand for fire: see the corresponding list at 7. 5. 7, 'terra et mare et aer et sidera'), and 2. 23. 60, on the gods of Epicurus (whose denial of their corporeality is confuted at 1. 25. 71 ff.) 'in regione caeli conlocati'. O'Daly, 66, 'a concept of God that closely resembled the Stoic one which he knew from Varro, . . . not unaffected by Manichaeism, or at least . . . it grows out of Augustine's critical preoccupation with Manichaean views about God.' A similar difficulty transcending the corporeal affects Evodius' view of the soul reported and discussed at *quant. an.* 3. 4, and cf. *mor.* 1. 10. 17. Against the Manichees: *c. ep. fund.* 43. 49, 'quae suorum phantasmatum fidem secuta et divinam substantiam per locorum spatia quamvis infinita velut informem molem disiecit atque diffudit'.

sive infusum . . . diffusum: Colish, *Stoic Tradition*, 2. 148: 'In short, the physics and theology with which Augustine wrestled in *Confessions* 7 reflect a substantial modification of the more strictly Stoic view of the deity, and of being as such, to which he reports his attachment in *Confessions* 5, a modification resulting from the decidedly unStoic notion of God as immutable.'

prorsus nihil: On the paradoxes of 'nihil', *mag.* 2. 3.

tamquam spatiosum nihil: G–M: 'a nothingness which yet possessed the quality of space'.

7. 1. 2

Knowledge of God: A.'s model for knowing is that sensory data generate an image, which thought then regards. He is not concerned here with the problem of the origin of the image in God's case, but that difficulty underlies his judgement elsewhere (see on 3. 6. 10) that his view of God under the influence of Manicheism was only a *phantasma*, object of a seeming-knowledge that is not knowledge. He passes over the paradox that he already 'knew God' in a variety of ways. He 'knew' that God exists, he had various expectations (immutability, inviolability, incorruptibility) of God, and he would not accept any teaching that did not have the *nomen Christi*. But 'knowledge-about' is not the direct cognition that he now sought.

Similar problems afflicted his opinions of the soul. One issue that preoccupied debaters was the mutability of the soul under the mutilation of the body. See both *quant. an.* 31. 62 (an anecdote from Cassiciacum) and the recollection at *nat. et or. an.* 2. 6. 10 of a youthful conversation with Vincentius Victor (later divided from A. by schism) 'de amputatis membris corporis absque animae sectione'. On his later view, he needed a clarification of vision (and clarification of vision is exactly what follows here: see on 7. 8. 12, 'collyrio'): *trin.* 7. 6. 11–12, 'non enim potest [animalis homo] cogitare nisi moles et spatia, vel minuta vel grandia, volitantibus in animo eius phantasmatis tamquam imaginibus corporum. (12) ex qua immunditia donec *purgetur* credat in patrem et filium et spiritum sanctum, unum deum, . . . omnium visibilium et invisibilium conditorem, et quidquid de illo pro humana facultate digne vereque dici potest' ('quidquid': e.g. or perhaps i.e. the remainder of the Nicene Creed). His view in 386/7 was only partially dematerialized; he was willing to affirm the immaterial nature of spiritual creature, but could not understand how to retain spatial metaphors in talking about such things. Material-sounding language made him think in material terms, and *that* problem was not yet solved: *retr.* 1. 5. 2, 'quod vero dixi, "animum propterea non posse ab aeterna ratione separari, quia non ei localiter iungitur" [*imm. an.* 6. 11, in a work generally agreed to be largely a collection of unassimilated Platonism, perhaps Porphyrianism], profecto non dixissem si iam tunc essem litteris sacris ita eruditus ut recolerem quod scriptum est: "peccata vestra separant inter vos et deum" [Is. 59: 2]. unde intellegi datur etiam earum rerum posse dici separationem, quae non locis sed incorporaliter iunctae fuerant.' (Another implied link to Ambrose is dimly visible here: one of the sermons A. almost certainly heard was that which survives as the *de Isaac vel anima*; for this kind of implication, see on 6. 16. 26, 'metus mortis'.)

incrassatus corde: Cf. Mt. 13: 15 (Jesus explaining why he speaks in parables), 'incrassatum est enim cor populi huius et auribus graviter audient et oculos suos cluserunt'; cf. Act. 28: 27 (Paul explaining why his message has difficulties), 'incrassatum est cor populi huius et auribus graviter audierunt et oculos suos compresserunt.' Both passages follow Isaiah 6: 10 (God's charge to Isaiah), which in LXX is identical with the Greek of the Matthew and Acts passages. A. Maxsein, *Augustinus* 3 (1958), 323–30, describes the heart for A. in the biblical sense, not as the seat of the affections, but of just and deep thoughts, the place where we perceive truth and adhere to it. *Incrasso* itself is Christian (from Tertullian on, 7 × in Vg) = Gk. παχύνω, 'to make fat', hence in passive 'become insensitive, stupid'.

conspicuus: 'present to sight, transparent' (7. 10. 16, 'lucem . . . vulgarem et conspicuam omni carni'). Cf. *en. Ps.* 147. 16, 'cum adhuc nemo est alteri conspicuus, nemo videt cor alterius.'

intentionem: 'attention' in a high sense (reminiscent of Simone Weil, quoted on 11. 18. 23), nearly 'consciousness'; see 12. 14. 18. A.'s critique of any rash materialism survives: even if 'consciousness' is understood as the result of the interaction of material entities, that very notion of interaction itself goes beyond a common-sense materialism.

vita vitae meae: See on 1. 4. 4, 'deus meus, vita mea'; sim. at 3. 6. 10, 10. 6. 10; cf. Jn. 14. 6.

ut haberet te terra . . . nusquam: This passage offers a gloss on 1. 3. 3, 'capiunt ergo te caelum et terra . . .'; see also below.

aeris huius: Thus expressly and carefully not the *aer ingenitus* of the Manichees: *c. Fel.* 1. 18, 'immo tres sunt: pater ingenitus, terra ingenita, et aer ingenitus'; cf. *c. Faust.* 20. 2.

terrae corpus pervium: Cf. 7. 1. 1, 'sive infusum mundo'. Cf. *Aen.* 6. 724–7:

principio caelum ac terras camposque liquentes
lucentemque globum lunae Titaniaque astra
spiritus intus alit, totamque *infusa* per artus
mens agitat molem et magno se corpore miscet.

inspiratione: If A. was reading Cicero at this time (as 6. 16. 26 suggests, and cf. Cic. *nat. deor.* quoted above on 7. 1. 1), he could have found this support at *nat. deor.* 2. 7. 19: 'haec ita fieri omnibus inter se concinentibus mundi partibus profecto non possent, nisi ea uno divino et continuato spiritu continerentur.'

administrantem CDGO Ver. : *administrante* S Knöll Skut.

est (*autem*) OS Knöll Skut. Ver. : *es* CDG Maur. not without verisimilitude.

inluminaveras tenebras meas: Ps. 17: 29, 'deus meus, inluminabis tenebras meas'; *en. Ps.* 17. 29, 'nos enim peccatis nostris tenebrae sumus' (cf. 9. 4. 10, 'fuimus aliquando tenebrae'). See 6. 1. 1, 11. 2. 2, 11. 25. 32, 13. 8. 9.

7. 2. 3

deceptos deceptores: *s.* 8. 2. 3, 'in sua vanitate decepti decipiunt'; *en. Ps.* 61. 23, 'diu mathematicus fuit; seductus seducens, deceptus decipiens'; see on 4. 1. 1, 'seducebamur et seducebamus, falsi atque fallentes'. Three scriptural passages are closer to this in the Greek than in the Vulgate: 2 Tim. 3: 13, 'mali autem homines et seductores proficient in peius, errantes, et in errorem mittentes' (Gk.: *πλανῶντες καὶ πλανώμενοι*); Judg. 11: 35 (B text), *ταραχῇ ἐτάραξάς με, καὶ σὺ ἦς ἐν τῷ ταράχῳ μου*, and 2 Pet. 3: 3, *ἐν ἐμπαιγμονῇ ἐμπαῖκται* (quoted at *civ.* 20. 18 as 'inlusione inludentes', where the Vg. has 'deceptione inlusores'). Porphyry, *v. Plot.* 16. 5–8, of Christians *ἀποκαλύψεις τε προφέροντες Ζωροάστρου . . . καὶ ἄλλων τοιούτων πολλοὺς ἐξαπάτων καὶ αὐτοὶ ἠπατημένοι;* cf. Philostr. *vita Apollonii* 7. 3, *Πλάτων . . . σφαλείς τε καὶ σφήλας*; see P. Wendland, *Rhein. Mus.* 49 (1894), 309–10, with further analogous expressions from Philo, Dio Chrysostom, and Julian. A similar phrase occurs in the 'Sentences of Sextus' (ed. H. Chadwick (Cambridge, 1959)), §393: *ψεύδεσθαι φυλάττου· ἔστιν γὰρ ἀπατᾶν καὶ ἀπατᾶσθαι* (in Rufinus' Latin: 'mentiri vita: decipere est et decipi').

quoniam: Reinforces our interpretation at 1. 4. 4, 'vae tacentibus de te, quoniam loquaces muti sunt'; the Manichees are talkative, but their talkativeness is *de facto* silence, for what comes forth is not the Word, but mere words (cf. J. Ries, *Lectio III–V*, 11).

a Nebridio proponi: Nebridius' argument against the Manichees meant much to A. at every stage in his controversy with them. For texts that use the argument in one form or another, see (in chronological order: this list supplements that of Alfaric, 249) *ord.* 2. 17. 46, *mor.* 2. 12. 25–6, 'hinc enim illud exortum est, quod etiam cum studiose vos audiremus, nos magnis premebat angustiis nec ullum exitum reperiebamus, quaerentes, quid factura erat deo gens tenebrarum, si cum ea nollet cum tanta partis suae calamitate pugnarer. . . . (26) illud vero nondum dictum erat, quod nuper apud Carthaginem audivi. cum enim quidam quem maxime illo errore cupio liberari hac quaestione in easdem compingeretur angustias, ausus est dicere, scilicet regnum habuisse quosdam fines suos, qui possent invadi a gente contraria. nam ipsum deum nullo modo potuisse violari. sed dixit quod neque auctor ille vester ullo modo dicere cogeretur'; *c. Fort.* 1, 36–7, *c. Adim.* 28. 1, *c. Fel.* 1. 19, 2. 8, *nat. b.* 43, *c. Faust.* 13. 6, 21. 14, 22. 22, *c. Sec.* 20, 'ut respondeatis: quid factura erat deo gens tenebrarum, si cum ea pugnare noluisset? quam si dixeritis aliquid fuisse nocituram, fatebimini corruptibilem et violabilem deum; si autem dixeritis quod ei nocere non possit, quaeretur a vobis, cur ergo pugnavit?'—the argument was already in circulation because there A. was responding to *ep. Sec.* 6, which claimed that there were things in Manichean doctrine that defy explanation: 'sunt quaedam res quae exponi sic non possunt ut intellegantur; excedit enim divina ratio mortalium pectora. ut puta hoc ipsum, quomodo sint duae naturae *aut quare pugnaverit qui nihil poterat pati*'; see also Evodius, *de fide c. man.* 18 (*PL* 42. 1144). Cf. esp. 1 Tim 1: 17, quoted at *c. Adim.* 28. 1, where A.'s text reads 'regi autem saeculorum, invisibili, *incorruptibili*,[4] soli deo honor et gloria in saecula saeculorum'.

Courcelle, *Recherches*, 45, objects to what he sees as a chronological leap here, but the failure to mention the argument until now is characteristic of A.'s method in *conf.* He speaks of positive developments only at the moment in the narrative when they affected his inner dispositions decisively (so the recollection of Elpidius *contra manichaeos* at 5. 11. 21).

si autem . . .: A concise summary of Manichean doctrine, displayed here to show its weakness in the face of the objection raised by Nebridius—'itaque si te . . .' below represents the conclusion to which A. would force them.

substantia: Again, the use of the word (improperly applied to God) is a sign of the derangement of the argument; see on 5. 10. 20.

hanc esse animam: Translators (BA, Ryan, Vega, Carena) regularly get bogged down in this sentence and so punctuate to start a new sentence with this phrase, supplying a verb (corresponding to an understood *dicerent*); Pusey follows the text most faithfully, but the resulting English is not readily intelligible.

tuus sermo: Not *verbum*; G–M suggest this hints that the Manichean doctrine is not on the mark, that their 'Christ' was a *phantasma*. *Sermo* is, though infrequent, not impossible of divine speech, but not used by A. of the pre-existent second person of the trinity. Cf. 5. 9. 17, 11. 2. 4, 12. 13. 16, 13. 20. 26, 13. 15. 16, 13. 15. 18, and esp. 11. 9. 11, 'audiat te intus *sermocinantem* qui potest', 12. 10. 10, 'tu mihi *sermocinare*'.

sat erat ergo: Picks up the beginning of the paragraph to reiterate that the argument of Nebridius was enough.

omni modo evomendos a pressura pectoris: A potent metaphor; cf. Apoc. 3: 16 (to the Laodiceans), 'sed quia tepidus es, et nec frigidus, nec

[4] Vg. follows Gk. MSS rejected by modern editors here and reads *immortali*.

calidus, incipiam te evomere ex ore meo.' The *a* is apparently instrumental; *pectus* (*OLD* s.v. 2 a) may be 'the breast considered as a receptacle or channel for food', quoting Celsus 3. 7. 1. a, 'vomitu pectus purgare'.

7. 3. 4

incontaminabilem et inconvertibilem et nulla ex parte mutabilem: = 'incorruptibile, inviolabile, et incommutabile' (7. 1. 1).

explicatam CDO Ver. : *explicitam* GS Maur. Knöll Skut. A. *nowhere* else uses the form *explicitus, -a, -um*.

causam mali: The third of the Manichean challenges catalogued at 3. 7. 12; see on 7. 1. 1.

quod quaerebam: i.e. *causa mali*.

quaerendo: G–M take with the object; so the translators (Pusey, Ryan, 'in their search for the cause of evil'; *BA*, 'car je les voyais, dans la recherche de l'origine du mal, remplis de malice', Carena, Vega). The difficulties are that (1) A. regularly uses the present participle in such cases (1. 10. 16, 'libera nos iam invocantes te'), and (2) he regularly uses the dat./abl. gerund with a main verb not as a pseudo-participle, as in medieval Latin, but with a genuine sense of instrumentality: 'by means of . . .' In this aporia, it is decisive that the indirect question 'unde malum' occurs 3 × else in *conf.*, always governed by some form of *quaero*: 3. 7. 12, 7. 5. 7, 7. 7. 11.

repletos malitia: Rom. 1: 29, 'repletos omni iniquitate, malitia, fornicatione, avaritia, nequitia', etc., for two more verses. This paragraph of Rom. will come to the fore in 7. 9. 14; the echo here makes it clear that he includes the Manichees among the 'pagans' against whom the Pauline text is directed. Cf. Eccles. 9: 3, 'unde et corda filiorum hominum implentur malitia et contemptu in vita sua, et post haec ad inferos deducentur.'

7. 3. 5

intendebam: See on 7. 1. 2, 'intentionem'.

audiebam: Apart from the self-contained encounter with Firminus (7. 6. 8–10), this is the only verb in Bk. 7 before the appearance of the *platonicorum libri* that describes something A. was doing at the time in which someone else could have been involved. The verb *audio* is used 6 × of A.'s attendance on Ambrose's sermons in Bk. 6 alone. On what he heard, see Courcelle, *Recherches*, 99–100, 106–20; Alfaric, 370, cites Amb. *in Luc.* 8. 36, while Courcelle, *Recherches*, 100, prefers *exam.* as source: *exam.* 1. 8. 31, 'quod enim possumus non facere si nolimus, huius electionem mali nobis potius debemus quam aliis ascribere. . . . quid alienam naturam accersis ad excusationem tuorum lapsuum?' Cf. 'naturam' to 5. 10. 18, 'adhuc enim mihi videbatur non esse nos qui peccamus, sed nescio quam aliam in nobis peccare naturam'. (Courcelle adds Amb. *Iacob* (Feb. 386) 1. 1. 1 and 1. 3. 10 and *Isaac* 7.61, behind which he descries Plot. 3. 2. 10 and 6. 8. 6.)

ut male faceremus: A. narrows the focus of his investigation to moral evil; he knew the distinction: *c. Adim.* 26, 'dupliciter enim appellatur malum: unum quod homo facit, alterum quod patitur; quod facit peccatum est; quod patitur, poena.'

rectum iudicium tuum: Ps. 118: 137, 'iustus es domine et rectum iudicium tuum.'

conatus mergebar: The phrasing re-evokes the vocabulary of the ascent (e.g. 4. 15. 26, 'conabar ad te'), and thus prepares the way for the 'tentatives d'extase' of 7. 10. 16 and 7. 17. 23.

sciebam me habere voluntatem: *duab. an.* 11. 15, 'quis dubitet tunc esse peccatum, cum et velle iniustum est et liberum nolle? et ideo definitionem illam et veram et ad intellegendum esse facillimam et non modo nunc *sed tunc quoque a me potuisse dici*: "peccatum est voluntas retinendi vel consequendi quod iustitia vetat et unde liberum est abstinere"?' *lib. arb.* 3. 17. 49, 'sed quae tandem esse poterit ante voluntatem causa voluntatis?' Cf. J. Rist, *JThS* 20 (1969), 420–47, but the literature is vast.

causam CDGO² Maur. Knöll Skut. : *causa* O¹S Ver. (understanding it as ablative 'à cause de son péché' and adducing parallels for that ablative preceding the genitive it governs: *c. acad.* 3. 7. 15, *s. dom. m.* 1. 14. 39, 1. 15. 40, *b. coniug.* 10. 11).

animadvertebam CDGO Maur. Ver. : *advertebam* S Knöll Skut.

quis fecit me?: The same question, and the same answer, is a stepping stone on the ascent of the mind to God at Ostia (9. 10. 25) and the answer alone plays a similar part at 10. 6. 9, with both passages approximating the mind's ascent to God in different ways.

insevit OS edd. : *inseruit* CDG.

plantarium: Cf. *s.* 47. 15. 28, '"et excitabo eis plantarium pacis" [Ezech. 34: 29 (VL)]. testamentum pacis, plantarium pacis. germinet quod plantat deus et extirpentur quod seminavit haereticus.' (See C. Carena, *Riv. filol. istr. class.* 92 (1964), 423–7, with further parallels for the metaphor of sowing.)

amaritudinis: See on 2. 1. 1, esp. for contrast with 'dulcissimo' in the next phrase.

quod si et . . . esset: The query posed here, if it authentically dates from 385/6, betrays an awareness of catholic doctrine; the fullest treatment of the problem is at *civ.* 12. 1–9, but even there he claims to have no 'answer' to this question, only an explanation (that seems sophistic to most modern readers) of why the question is unanswerable (see *civ.* 12. 7, quoted on 2. 6. 12).

ubi nemo tibi confitetur: Cf. Ps. 6: 6, 'in inferno autem quis confitebitur tibi?'; *en. Ps.* 6. 6, 'ut infernum dixerit, caecitatem animi, quae peccantem, id est, morientem excipit et involvit.'

7. 4. 6

The purpose and significance of this paragraph are not obvious without consideration of its place in the book as a whole. Philosophically, the paragraph shows how close A. was to a solution of the twin problems of the nature of evil and of the nature of God, and prepares those problems for the resolution the *platonicorum libri* would offer. Biographically, it shows how far, at this time, A. was capable of going by means of his own efforts (cf. 'nitebar'), in the order of 'natural theology'.

The Plotinian echo suggested at *BA* 13. 589 n. (citing Plot. 6. 8) is at best incidental (anachronistically using Plotinian ideas to characterize his position at the last moment before Plotinian influence was felt), for to present express Plotinian doctrine here would undermine the structure of the book. Some justice must be done to the difficulty A. found in attempting to sort out all his ideas of the time into a linear pattern more than a decade after the fact. This book represents not so much a period as a moment (the arrival of the *platonicorum libri*) and its consequences.

The exposition from 'nullo . . . modo' is unscriptural but dogmatic. What should not be missed is the hint of trinitarian structure: incorruptible substance [1], 'voluntas' and 'potentia' unlimited [3], and knowledge unbounded ('qui nosti omnia') [2].

confitebar: In the chronological sequence of events narrated, this is the first *confessio* A. records.

summum . . . bonum: See on 2. 6. 12, and cf. the discussion of the *de pulchro et apto* at 4. 15. 24.

praeponebam: Cf. 7. 1. 1, 'praeponebam'. G–M see a parallel to Anselm's ontological proof.

esset . . . esses: The *voluntas* would only be greater than God himself if God himself were greater than God himself—because in God, self and will are identical.

7. 5. 7

Here we enter a represented aporetic interior monologue of the time, following 6. 11. 18–19 and preceding 8. 7. 18, and resembling both. Here he depicts himself as *anima naturaliter platonica*. The 'vision' that follows anticipates in form that of Ostia (9. 10. 25), suggesting that this passage is to be seen as another step on the path to intellectual ascent to God, hence the counterpart to 7. 10. 16 and 7. 17. 23, the more successful steps taken after the reading of the *platonicorum libri*.

et constituebam: Ps. 15: 8, 'providebam dominum in conspectu meo semper.'

universam creaturam: Taken up by 7. 12. 19, 'universae creaturae tuae'.

angelos: This is the first direct reference to angels in *conf.* (earlier: references to fallen angels at 1. 17. 27, 7. 3. 5; to Manichean pseudo-angels at 3. 10. 18, 4. 1. 1; and to Monnica's veneration for Ambrose 'sicut angelum dei' (6. 1. 1)). They are infrequent henceforth, with no consistent exegetical or doctrinal reference save that at 12. 11. 12, it is they who are by implication equated with the 'caelum caeli' in the interpretation there discussed. A. is always diffident in speaking of angels (Madec, *Aug.-Lex.* 1. 314–15: 'les observations d'A. sur les anges se caractérisent en effet par leur circonspection'), and they often occur as possible intermediaries to be considered but rejected (e.g. 9. 10. 25, 'neque per vocem angeli', 10. 42. 67, 'ambiendum mihi fuit ad angelos?', and 13. 38. 53, 'et hoc intellegere quis hominum dabit homini? quis angelus angelo? quis angelus homini?').

ordinavit CDG cett. Maur. : *ordinata, ut* OS Knöll Skut. Ver. Translators take 'ut imaginatio mea' as if it were 'ut in imaginatione mea' or 'secundum imaginationem meam' (BA: 'selon mon imagination', Vega: 'según mi fantasía'); better, Ryan, 'as my imagination dictated'. Pellegrino prints 'ordinata, ut' but Carena on the facing page translates 'che la mia immaginazione distribuiva pure in vari luoghi'. Given the importance of *ordo* in A.'s trinitarian thought (see on 1. 7. 12 and cf. 'formavit atque ordinavit' below), it seems clear that A. implies here that he was taking on himself a divine function—giving order to created nature in a conglomeration of *imagines* that would be both *phantasiae* and *phantasmata*. The attentive reader who catches that signal will not be surprised that the inquiry based on this 'vision' did not succeed.

distinctam: G–M: 'adorned with'; but there is little to corroborate that translation; cf. (suggesting a connection to the creative activity of the Word) 12. 3. 3, 'informem materiam formares atque distingueres', 12. 20. 29, 'quae nunc iam *distincta atque formata*', and 12. 28. 39, 'alius iam *formatas distinctasque* naturas'. Elsewhere *distinguo/distinctio* are associated about as often with *forma* [2] and *ordo* [3]: they are conjoined with *both* in both *Gn. litt. imp.* and *Gn. c. man.*; in *Gn. litt.* (e.g. 1. 17. 35, 'a formata re ad hoc distincta est illa informitas') they seem more often associated with 'formatio'. Whether they stand for one or the other or indifferently for both, their frequency of appearance in such contexts assures them a place in the economy of creation, associated with either the second or third person of the trinity, or both. *Here*, the 'formation' of *informis materia* is clearly suggested.

ex omni parte . . .: Cf. 7. 1. 2, with notes there on links to 1. 3. 3; Plot. 4. 3. 9. 36–41, cited by G–M: *κεῖται [ὁ κόσμος] γὰρ ἐν τῇ ψυχῇ ἀνεχούσῃ αὐτὸν καὶ οὐδὲν ἄμοιρόν ἐστιν αὐτῆς, ὡς ἂν ἐν ὕδασι δίκτυον τεγγόμενον ζῴη, οὐ δυνάμενον δὲ αὐτοῦ ποιεῖσθαι ἐν ᾧ ἐστιν· ἀλλὰ τὸ μὲν δίκτυον ἐκτεινομένης ἤδη τῆς θαλάσσης συνεκτέταται, ὅσον αὐτὸ δύναται.*

spongiam: *ep.* 187. 4. 14, 'sic est deus per cuncta diffusus, ut non sit qualitas mundi sed substantia creatrix mundi sine labore regens et sine onere continens mundum non tamen per spatia locorum quasi mole diffusa.' The negative example of the image of the sponge recurs at *Gn. litt.* 8. 21. 42, 'cum anima non sit natura corporea nec locali spatio corpus inpleat, sicut aqua utrem siue *spongiam*, sed miris modis ipso incorporeo nutu commixta sit uiuificando corpori, quo et inperat corpori quadam intentione, non mole.'

implet ea: The period after *ea* is from Löfstedt, *Symb. Osl.* 56 (1981), 106 (replacing a colon).

radix: 1 Tim. 6: 10, 'radix enim omnium malorum est cupiditas.'

an omnino non est: He approaches here the solution that he will finally accept only after reading the *platonicorum libri* (7. 12. 18–13. 19).

quanto non est . . . timemus: 'The evil is greater the more its object is non-existent, and we fear it none the less.'

timemus: See on 'de timore mortis' below.

(*idcirco*) **aut:** The conjunction recurs 8 × in all to mark the alternatives that define the extent of A.'s aporia.

deus fecit . . . bona: Gn. 1: 31, 'viditque deus cuncta quae fecerat et erant valde bona.'

fecit ea [1] **. . . et formavit** [2] **atque ordinavit** [3] **eam:** See on 1. 7. 12 (and note below of himself: 'adhuc *informis*').

per infinita retro spatia temporum: The hypothesis here resembles the jibe of the Manichees about God's idleness before creation rebuffed at 11. 10. 12–13. 16.

ageret: G–M have a note reading this and 'institueret' to follow (with Pusey's translation) as jussive; preferable as punctuated here and in all recent editions (including G–M).

non enim esset omnipotens: *Gn. c. man.* 1. 6. 10, 'non enim debemus esse similes istis qui omnipotentem deum non credunt aliquid de nihilo facere potuisse, cum considerant fabros et quoslibet opifices non posse aliquid fabricare nisi habuerint unde fabricarent.'

curis mordacissimis: See on 9. 1. 1.

de timore mortis: A reminder of 6. 16. 26, 'metus mortis et futuri iudicii tui'.

fides: This *fides* denotes no more than the insistence on the *nomen Christi* that goes back to 3. 4. 8, but the assertion is entirely different from the Academic scepticism of 5. 14. 25. The palpable difference is that he has heard the sermons of Ambrose showing him that the OT can be read fruitfully and found therein a confidence lacking earlier. The qualifications here ('informis . . . fluitans') are given substance by the pseudo-creed summarizing his position at 7. 7. 11.

domini et salvatoris nostri: 2 Peter 2: 20, 'si enim refugientes coinquinationes mundi in cognitione domini nostri et salvatoris Iesu Christi, his rursus implicatis superantur.'

informis: The second stage of creation (12. 3. 3 ff.) is the addition, through the agency of the second person of the trinity, of *forma* to formless matter; thus the hint here is that what was lacking in A.'s faith was the *formative* influence of the (incarnate) Christ; at 7. 18. 24, that lack will be made explicit.

7. 6. 8

Firminus is not known otherwise. Where and when did A. know him? See *BA* 13. 595 n. posing the question, Vega ad loc. arguing for Carthage. Courcelle, *Recherches*, 77 n. 6, thinks that if Carthage were the venue, A. would have known Firminus' family (cf. his relation to Alypius' father (6. 7. 11)); Mandouze, 100 n. 5, thinks the placement in *conf.* shows that his rupture with the astrologers came only after his break with the Manichees. In favour of a continued association or one beginning at Milan is the record of a letter to him in Possidius' *indiculum* (*MA* 2. 182), mentioned immediately after those to Nebridius (*epp.* 3–14), Hermogenianus (*ep.* 1), and Zenobius (*ep.* 2).

The scientific debunking of the Manichees (5. 3. 4–5. 9) has a place here. Though the scientific and mantic aspects of star-gazing might be distinct enough for the mature Augustine to see them clearly, many blurred that line. The earlier treatment of the question ended at 4. 3. 6 with an aporia that a serious interest in astronomy would have influenced.

These links suggest another aspect of the episode recounted here. The appeal of astrology to Augustine is that of *curiositas*. Its placement at the beginning of Bk. 4, the most various of the early books, is meant to mark how far A. had fallen in his Manicheism, into which he fell through youthful *curiositas*; the narrative of his astrological views is continued to show him still floundering in the grasp of the superstition. It is then abruptly dropped.

The subject recurs here abruptly, poorly integrated with the surrounding context. To read to the end of 7. 5. 7 and then to skip to the beginning of 7. 7. 11 leaves no sense of omission (but see on 7. 7. 11). The topic of surrounding chapters (the search for the nature of God through the resolution of the question of the origins of evil) is nowhere mentioned in this episode. *Prima facie*, A. admits that the episode is chronologically misplaced ('iam etiam . . . reieceram'). In sum, in Bk. 4, astrology measures the slide into *curiositas*; in Bk. 5, it is the benchmark against which another form of *curiositas* can be judged and found wanting; and here it marks liberation from *curiositas*.

But of what character is the episode? A friend is 'procured' ('procurasti' 2 ×) for A. by God. In conversation with him, A. comes suddenly to see (7. 6. 9, 'omnis illa reluctatio mea resoluta concidit') the error of his ways. In those terms it bears comparison both to the central episode of this book, the 'procuring' (7. 9. 13, 'procurasti') of the *platonicorum libri*, and to the two preliminary incidents of Bk. 8, the visits from Simplicianus and Ponticianus, both of whom recount to A. events elsewhere in a way that leads to a change of heart. Bk. 7 is mainly an account of A.'s intellectual conversion, as Bk. 8 recounts his moral conversion; in both cases, the lesser episodes and the conversions of others parallel and illustrate A.'s own experience. Though he is already detached from Manicheism, his pre-Manichean *curiositas* hounds him through these early chapters of Bk. 7, for he is still seeking the answer to two of the main questions that drove him into the Manichean camp. So, just as Bk. 3 had two kinds of temptation of the eyes (the *spectacula* and Manicheism—and another, astrology itself, appeared in Bk. 4), so Bk. 7 has two kinds of conversions from *curiositas*: first here, in a way that foreshadows the *mechanism* that will work again in 8, and second, the reading of the *platonicorum libri*.

The privacy of A.'s struggle (cf. 7. 7. 11, 'nullus hominum') is borne out in the way Bk. 7 is constructed. Nebridius appears twice in the first part (7. 2. 3 and here: in both cases he is the one immune to *concupiscentia oculorum*, just as Alypius is immune to *concupiscentia carnis*), and Alypius appears once later in a single sentence recounting his Christological view (7. 19. 25). Otherwise no named individuals appear; the only other individual even alluded to is the man 'typho turgidum' who provides the *platonicorum libri* in 7. 9. 13. Apart from the encounter with Firminus, there is no episodic narrative in this book. A. is alone with his thoughts.

mathematicorum: See on 4. 3. 4.

deliramenta: Disordered acts of intellect (1. 17. 27), associated with errors of *curiositas*: 3. 6. 10 and 5. 3. 6 (of Manicheism), here and 7. 6. 10 (of astrology).

confiteantur . . . miserationes tuae: Ps. 106: 8: see on 1. 15. 24.

intimis visceribus animae meae: At 11. 29. 39, these *viscera* stand in apposition to *cogitationes meae*; at 5. 10. 20, with the comparable construction where *miserationes* is the confessing subject, there is a parallel prepositional phrase: 'cui confitentur *ex me* miserationes tuae'.

erroris: Cf. 7. 3. 5, 'infernum . . . erroris'; see on 1. 20. 31, where *errores* have the place in the triad as punishment for *curiositas*, so often in Bks. 3–6 (e.g. 4. 7. 12, 'sed vanum phantasma et error meus erat deus meus').

vita [2] **. . . sapientia** [2] **. . . inluminans** [2] **. . . lumine** [2]: *Curiositas* [2] is the sin against the Word.

procurasti: Cf. esp. 7. 9. 13, 'procurasti'; in the astrological narrative at 4. 3. 6, 'et hoc . . . per illum [Vindicianum] procurasti mihi'. See also 3. 3. 5, 6. 15. 25, 9. 10. 23.

crebro dicenti: 8. 12. 29 (of the 'tolle lege' chant), 'dicentis et crebro repetentis'.

segnem: Wijdeveld (*REAug* 5 (1959), 33) reports a Vienna MS with *signem* and conjectures *insignem*; *BA* describes as 'conjecture habile, mais pas assez justifiée'. Firminus is *quick* to consult astrologers, but does not know their texts at all well—in other words, enthusiastic but uninstructed.

eas litteras: The technical documents of the astrologers; 7. 6. 10, 'litterisque . . . easdem litteras'; cf. here, 'librorum'.

curiosum: With 'curiosissimum' below and 7. 6. 9, 'curiositate', explicit indicators of the place of this episode in the moral architecture of *conf.*

liberaliter: Cf. 7. 6. 9, 'liberalesque doctrinas'. F. had been educated in the *disciplinae* in a way that should have led his mind upwards to philosophical wisdom (see on 4. 16. 30); he resembles A. (just as Victorinus in 8. 2. 3–5 will appear in a conversion narrative as a Doppelgänger).

constellationes: *doctr. chr.* 2. 22. 33, 'constellationes enim quas vocant notatio est siderum, quomodo se habebant cum ille nasceretur de quo isti miseri a miserioribus consuluntur'; also at *div. qu.* 45. 2 (fuller on technicalities), *Gn. litt.* 2. 17. 36, *c. Faust.* 2. 5, *s.* 56. 8. 12, *civ.* 5. 3–5. 7, and *c. ep. pel.* 2. 6. 12. Infrequent elsewhere; at Calcidius (in Tim. 118 and 148) = Eng. 'constellation'. One ancient gloss equates to Gk. καταστερισμός, but the meaning here is different.

conicere: Cf. Cic. *div.* 1. 52. 118, 'male coniecta maleque interpretata signa falsa sunt', and Plin. *nat.* 10. 49, 'praesagivere victoriam, ita coniecta interpretatione'; see *div. qu. Simp.* 1. 2. 3, 'coniectant', quoted below on 7. 6. 10, 'de Esau et Iacob', and *civ.* 5. 9, 'in his autem mathematicorum coniecturis refutandis'.

flatabant O Maur. Skut. (citing *ep.* 55. 11. 21, 'ad ipsum autem ignem amoris nutriendum et flatandum' (his app. wrongly reads *ep.* 50)) Ver. : *flabant* G : *flagitabant* CD : *flagrabant* S Knöll.

conligerent: A technical term, repeated through 7. 6. 10; cf. Amb. *exam.* 4. 4. 14, 'magnam vim dicunt esse nativitatis eamque minutis quibusdam et certis *conligi* oportere momentis ac, nisi verius *conligatur*, summam esse distantiam. . . . hoc quemadmodum possint *conligere* respondeant.'

praegnans: For the accusative, cf. 'Lampridius', *Commodus* 1. 3, 'Faustina cum esset Commodum cum fratre praegnans' (Löfstedt, *Symb. Osl.* 56 (1981), 106).

dealbatiores: G–M: 'comparatively brilliant.' They rightly prefer this to the old suggestion (going back to a note of Jean le Clercq reprinted at *PL* 47. 208: cf. *BA* ad loc.) that 'since a newly made, or mended, Roman road was white from the lime used for binding, "dealbatiores" = "well-kept".'

7. 6. 9

talis quippe narraverat: Authority in narrative reflects the moral stature of the speaker; cf. 10. 3. 3 for A.'s reflections on the authority of his own narrative.

vera pronuntiarem: Of astrological fortune-telling, as at 4. 3. 5 and 7. 6. 10.

at si CDGO Maur. Knöll Ver. : *ac si* S Skut.

conditionem servilem: For these words and all that follow to 8. 5. 12, manuscript C is defective and so D must be relied on.

unde . . . inde: The present punctuation, linking the two words as correlative, is that of A. Gabillon, *Homo Spiritalis* (Festschrift Verheijen: Würzburg, 1987), 441–2.

conlegi OS Ver. : *conligi* DG Maur. Knöll Skut.

non arte dici sed sorte: See on 4. 3. 5; cf. 7. 6. 8, 'vim sortis', and 7. 6. 10.

7. 6. 10

ruminando: 3. 6. 11, 6. 3. 3, 10. 14. 22, 11. 2. 3.

gemini: A standard Augustinian argument against astrology: *div. qu.* 45. 2, *doctr. chr.* 2. 21. 32–2. 23. 35, *Gn. litt.* 2. 17. 36, *civ.* 5. 1–8; but the argument was conventional: Persius 6. 18–19, 'geminos, horoscope, vario producis genio.'

litterisque: Cf. 7. 6. 8, 'eas litteras' and 'easdem litteras' here.

de Esau et de Iacob: Gn. 25–7; *div. qu. Simp.* 1. 2. 3 (concerning Rom. 9: 10 ff., on God's predestination of Jacob and Esau), 'ad hoc commendandum ait, "ex uno concubitu", ut nec astrologis daret locum vel eis potius quos genethliacos appellaverunt, qui de natalibus nascentium mores et eventa coniectant. . . . et facile animadvertunt, si volunt, responsa illa quae miseris venditant *nullius artis expositione sed fortuita suspicione* proferri.'

non ergo arte sed sorte: See on 7. 6. 9.

iustissime moderator universitatis [1].

occultis meritis: He comes closest here to connecting this episode with the circumambient discussion of the causes of evil; cf. 7. 3. 5, 'rectum iudicium tuum ut pateremur'.

ex abysso iusti iudicii tui: Ps. 35: 7, 'iudicia tua abyssus multa'; cf. 4. 4. 8, 13. 2. 3, 13. 12. 13.

quid est hoc: Sirach 39: 25–6, 'a saeculo usque in saeculum respicit, et nihil est mirabile in conspectu eius. (26) non est dicere, quid est hoc, aut quid est istud? omnia enim in tempore suo quaerentur.' A characteristically *curious* question, therefore; elsewhere in *conf.* at 1. 6. 10, 8. 3. 8, 8. 8. 19, 10. 6. 9, 10. 14. 21, 10. 21. 31, 13. 24. 35.

7. 7. 11

If 7. 6. 8–6. 10 were omitted, this paragraph would pick up smoothly from the end of 7. 5. 7, but *with some repetition*. In other words, the Firminus episode is not a later interpolation, but is deliberately encapsulated in the middle of a discussion that has its own continuity; see on 7. 6. 8.

adiutor meus: Ps. 29: 11, 'dominus factus est adiutor meus', Ps. 62: 8, 'quia factus es adiutor meus', Ps. 17: 3, 'deus meus adiutor meus', Ps. 58: 18, 'adiutor meus, tibi psallam', Ps. 18: 15, 'domine adiutor meus, et redemptor meus'. Also at 7. 10. 16, 8. 6. 13, 9. 1. 1 (the latter two passages also marked by liberation from *vincula*).

fluctibus: Cf. Ps. 68: 3, 'infixus sum in limo profundi, et non est substantia; veni in altitudinem maris et tempestas demersit me'; Ps. 106: 25, 'exaltati sunt fluctus eius'.

credebam: Just on the verge of intellectual discovery, A. presents here a creed of that time, summarizing his state of mind: for a similar summary at a slightly more advanced stage, see 7. 20. 26 (and cf. on 8. 1. 2). Every item in this creed has already been discussed and where relevant its acceptance by A. documented. He already believes in God the Father, in God the Son, and in the scriptures as guaranteed by the catholic church; no mention of incarnation.

auctoritas: See on 6. 11. 19, 'tam eminens culmen auctoritatis'.

viam: Jn. 14: 6, 'ego sum via et veritas et vita.' The noun *via* occurs 8 × in Bk. 7 and is consistently to be taken as denoting incarnate Logos (following 5. 3. 5, 'viam, verbum tuum'), as scriptural echoes and context make clear. This corroborates the Christological focus of the book. Apart from the present passage: 7. 9. 13, 7. 18. 24 (2 ×), 7. 20. 26 (2 ×), 7. 21. 27 (3 ×). The expression enters A.'s writing at *c. acad.* 1. 5. 13–14, 'recta via vitae sapientia nominatur' (influenced by Cic. *Tusc.* 1. 1. 1, 'ad rectam vivendi viam'), *ord.* 2. 5. 16, 'duplex enim est via', *sol.* 1. 13. 23, 'non ad eam una via pervenitur' (qualified at *retr.* 1. 4. 3: it is clear that Christ is *a* way for A. in 386, but perhaps less clear that he understands Christian doctrine to be as exclusive as he later took it to be: du Roy, 168–71); in a similar sense at *lib. arb.* 2. 9. 26, 2. 17. 45, and *vera rel.* 1. 1. First with unmistakable reference to Jn. 14: 6 probably *mor.* 1. 13. 22, but it would be hard to argue that A. was unaware of the equation already at Milan or Cassiciacum (especially bearing in mind the reluctance to mention the *nomen Christi* at Cassiciacum (see on 9. 4. 7)). Also early are *div. qu.* 38 (from Thagaste) and *f. et symb.* 4. 6, quoting Prov. 8: 22, Jn. 1: 3 and 1: 14 (see 7. 9. 13), and Phil. 2: 6–7 (see 7. 9. 14)—as one would expect, this latter is his earliest fully ecclesiastical exposition. Among many later examples, cf. *doctr. chr.* 1. 11. 11, 'cum ergo ipsa sit patria, viam se quoque nobis fecit ad patriam'; sim. at *doctr. chr.* 1. 34. 38 (linking Prov. 8: 22 and Jn. 14: 6) and *trin.* 1. 12. 24, quoting Prov. 8: 22 and Jn. 14: 6. The verse is an apt proof-text in an anti-Platonic context, e.g. *s.* 141. 1. 1 or *civ.* 10. 32, where A. takes the phrase 'universalis via animae liberandae' from Porphyry's *de regressu animae* and reiterates it over and over, emphasizing that the true 'universal way' is Christian (citing Jn. 14: 6 as he does); this is ironic in so far as the verse is one where Platonic teaching supported A.'s reconciliation with orthodoxy (even if it did not lead him as far as the incarnation).

The literal-minded consistency of A.'s exegesis by the time of his episcopal ordination and ever after makes it probable, therefore, that *via* is to be interpreted in *conf.* as a marker of the incarnate Logos wherever it occurs in a context that does not positively contradict the interpretation; numerous passages in this commentary refer to the present note with that intention. Studies include T. van Bavel, *Augustiniana* 7 (1957), 245–81; and L. C. Ferrari, *Aug. Stud.* 7 (1976), 47–58. Van Bavel, art. cit. 267–76, emphasizes the view that *via* consists of the events of Christ's earthly life; but there is a touch of special pleading here, in that van Bavel is attempting to show that A. does not unduly neglect the 'historical Jesus'.

aures tuae: Also at 3. 11. 19, 4. 5. 10, 5. 8. 15, 6. 11. 20, 9. 12. 33, 10. 35. 57, 11. 2. 3.

magnae voces: *BA*, perhaps anachronistically: 'il ne s'agit pas d'une *prière explicite*, d'une prière au sens strict du mot, mais d'un tourment intérieur à la recherche du vrai qui, par sa sincérité même, attire la miséricorde divine.'

nullus hominum: The privacy, even loneliness, in which A. laboured (see on 7. 6. 8) would not have been evident to his 'friends'—the nameless crowd that constantly surrounds him, and from which he seems in these years so detached. (At Hippo as well, he lived in a crowd, perhaps on no greater terms of real intimacy.)

rugiebam a gemitu cordis mei: Ps. 37: 9–11, 'rugiebam a gemitu cordis mei, (10) et ante te est omne desiderium meum, et gemitus meus non est absconditus a te. (11) cor meum conturbatum est, et deseruit me fortitudo mea, et lumen oculorum meorum non est mecum'; *en. Ps.* 37. 14, 'ipsum enim desiderium tuum, oratio tua est; et si continuum desiderium, continua oratio.' Cf. 11. 19. 25, 12. 29. 27.

intus enim erat, ego autem foris: See on 10. 27. 38.

intendebam: The 'looking' here resembles that of 9. 10. 25 (Ostia), with the difference that there the things of creation shrugged off A.'s gaze and passed him along to God. Here he is incapable of seeing into, or through, or beyond them. In terms of Rom. 1: 20 ff. (see on 7. 9. 14), he does not yet see God in the visible things of creation, and to that end this attempt at 'nature mysticism' is a failure.

locum ad requiescendum: See on 4. 11. 16, 'locus quietis imperturbabilis'.

superior enim eram istis: As often, to suggest the deficiencies of his position at the time, A. sketches the true doctrine he was lacking. Cf. Gn. 1: 28, 'replete terram et subiicite eam, et dominamini piscibus maris.'

media regio salutis meae: For the soul poised between God and material creation, see du Roy, 476–8. Cf. *Gn. c. man.* 2. 15. 22 ('illa medietate per quam deo subiecti erant et corpora subiecta habebant'), *epp.* 140. 2. 3, 18. 2 ('qui Christo credit non diligit *infimum*, non superbit *in medio* atque ita *summo* inhaerere fit idoneus'), *lib. arb.* 2. 19. 50, *Io. ev. tr.* 20. 11, *trin.* 12. 11. 16. The theme is easily linked (implicitly in several of these passages, explicitly in the *trin.* text) to the notion of the 'weight' (*pondus*) of the love that drags the soul down or allows it to rise (see on 13. 9. 10).

regio: See on 2. 10. 18 and 7. 10. 16.

ad imaginem tuam: Gn. 1: 26, 'faciamus hominem ad imaginem nostram'; see on 13. 22. 32.

tibi . . . corpori: *BA* 13. 681–2 n., cites *Io. ev. tr.* 1. 4, *trin.* 10. 5. 7, which loosely parallel the development here.

in cervice crassa scuti mei: Job 15: 26 (VL), 'et cucurrit contra eum contumeliose, in crassa cervice scuti sui'; *adn. Iob* 15: 26, '"et cucurrit contra eum contumeliose": adversaria faciendo quam praecepit. "in crassa cervice scuti sui": praesumens de protectione sua.' *en. Ps.* 128. 9, 'et ibi cervicem nominavit; quia sic te erigis, et non elidis oculos ad terram, et tundis pectus, et dicis, "domine, propitius esto mihi peccatori" [Lk. 18: 13]; sed iactas te de meritis tuis, "et vis mecum", inquit deus, "iudicio contendere, intrare mecum ad iudicium"; cum debeas in reatu tuo satisfacere deo, et clamare ad illum.'

redeunti: sc. *ad deum*; see on 1. 18. 28. On the punctuation here, cf. A. Gabillon, *Homo Spiritalis* (Festschrift L. Verheijen: Würzburg, 1987), 443–4.

humilasti: Ps. 88: 11, 'tu humilasti sicut vulneratum superbum'; *en. Ps.* 88. s. 1. 11, 'vulneratum enim intellegas diabolum, non penetrata carne, quam non habet, sed percusso corde, ubi superbus est.'

tumore . . . oculos meos: See on 7. 8. 12, 'collyrio'.

7. 8. 12

This paragraph is a tissue of biblical echoes and reminiscences, a signal of the importance of the stage to be reached in the next paragraph.

Tu vero . . . manes: Ps. 101: 13, 'tu vero domine in aeternum manes'; *en. Ps.* 101. s. 1. 13, 'mei dies sicut umbra declinaverunt, et tu in aeternum manes: temporalem salvet aeternus.'

non . . . nobis: Ps. 84: 6, 'non in aeternum irascaris nobis'; cf. Ps. 102. 9, 'non in finem irascetur, neque in aeternum indignabitur'.

terram et cinerem: Job 42: 6 (VL), 'ideo despexi memetipsum et distabui et aestimavi me terram et cinerem'; see on 1. 6. 7, and cf. 10. 5. 7.

et placuit: Ps. 18: 15, 'et erunt, ut complaceant eloquia oris mei et meditatio cordis mei in conspectu tuo semper, domine adiutor meus et redemptor meus'; *en. Ps.* 18. en. 2. 16, 'superba anima in conspectu hominum vult placere; humilis anima in occulto, ubi deus videt, vult placere.' Cf. Dan. 3: 40, 'sed in anima contrita et spiritu humilitatis suscipiamur; . . . sic fiat sacrificium nostrum in conspectu tuo hodie ut placeat tibi.'

deformia mea: Cf. 10. 27. 38, 'deformis inruebam'; the opposite of *forma* [2].

et stimulis: *Aen.* 11. 336–7, 'gloria Turni | obliqua invidia stimulisque agitabat amaris': regularly cited by the editors, though the Virgilian *stimuli* must be taken *in malo*, and those here *in bono*: further from the wording but closer to the thought is Eccles. 12: 11, 'verba sapientium sicut stimuli'.

et residebat . . .: The disordered use of sight (*curiositas*) leaves him with vision clouded, hankering for a clear vision of God: that is the project of Bk. 7, and it will both succeed and fail; cf. 7. 18. 24, 'verbum enim tuum, aeterna veritas ... sanans *tumorem* et nutriens amorem' (see on 3. 5. 9).

medicinae: i.e. Christ: cf. 9. 13. 35, 'exaudi me per medicinam vulnerum nostrorum, quae pependit in ligno et sedens ad dexteram tuam te interpellat pro nobis'; *en. Ps.* 118. s. 9. 2, 'eius a Iudaeis inrisam crucem totamque humilitatis christianae medicinam, qua sola *tumor* ille sanatur quo inflati cecidimus et iacentes amplius intumuimus'.

collyrio: Apoc. 3: 18, 'et collyrio ungue oculos tuos ut videas'. Elsewhere in A., *collyrium* is proper to the *incarnate* Christ (whose *dolores* were pre-eminently *salubres*) and occurs surprisingly often: *Io. ev. tr.* 2. 16, 'quia vero verbum caro factum est et habitavit in nobis, ipsa nativitate *collyrium* fecit, unde tergerentur oculi cordis nostri, et possemus videre maiestatem eius per eius humilitatem. ideo factum est verbum caro, et habitavit in nobis. sanavit oculos nostros. ... omnia enim *collyria* et medicamenta nihil sunt nisi de terra. de pulvere caecatus es, de pulvere sanaris; ergo caro te caecaverat, caro te sanat. ... verbum caro factum est: medicus iste tibi fecit *collyrium*.' The expression is already in Cyprian, *de opere et eleemosynis* 14, 'collyrio Christi', is implicit in Ambrose, *exp. Ps.* 118. 3. 22, and cf. Paul. Nol. *carm.* 19. 34, 'collyrio ... medentis Christi', with Paulinus to A. (= A.'s *ep.* 94. 1), 'lucerna semper est pedibus meis verbum tuum et lumen semitis meis. ita quotienscumque litteras beatissimae sanctitatis tuae accipio, tenebras insipientiae meae discuti sentio et quasi *collyrio* declarationis infuso oculis mentis meae purius video ignorantiae nocte depulsa et caligine dubitationis abstersa.' Similar blindness and want of *collyrium* in A. at *Io. ev. tr.* 35. 6 ('de corpore suo collyrium fecit luminibus nostris'), *en. Ps.* 39. 21, 43. 14, 65. 5; *ss.* 126. 5. 7, 136. 4 ('de sanguine suo collyrium fecit caecis'), 317. 2. 2, *s. Lam.* 25. 1 (*PLS* 2. 828); cf. *Io. ev. tr.* 18. 11, quoted on 7. 10. 16. (The word *collyrium* itself is medical Greek, but occurs in CL (Horace and Juvenal) as well as in Apoc., and so may have come to A. through any of those channels; A. is well aware of the medical use: *qu. Mt.* 13. 3—if authentic.) On purging the eye without *collyrium*, cf. *Io. ev. tr.* 1. 19, *ss.* 88. 5. 5, 88. 15. 14, 117 (on Jn. 1: central to 7. 9. 13), which deals throughout with purity of vision (esp. *s.* 117. 1. 1, 117. 3. 5; *s.* 117 also contains the Philippians and the Matthew texts quoted in 7. 9. 14), and *sol.* 1. 6. 12 ('oculi sani mens est ab omni labe corporis pura, id est a cupiditatibus rerum mortalium iam remota atque purgata')—that passage (through *sol.* 1. 6. 13, and cf. *sol.* 1. 13. 23, 'oculi tam sani et vegeti') expands on purity and strength of vision in a remarkable amalgam of neo-Platonic and Christian vocabulary, demanding *fides*, *spes*, *caritas* as the prerequisites for vision. (On the subject of vision, but without discussion of *collyrium*, see M. Miles, *Jour. of Rel.* 63 (1983), 125–42; Mayer, *Zeichen*, 2. 258.)

de die in diem: See on 6. 11. 20.

7. 9. 13

Nel mezzo del cammin. (1) With six books before and after, this one stands in the middle. (That A. could count like this is demonstrated by *Gn. litt.* 2. 13. 26, 'septem quippe dierum medius quartus est.') By count of words or lines of text, this paragraph stands at the middle of the Bk. 7.[5] (2) As long as A.'s goal was intellectual enlightenment, the reading of the *platonicorum libri* was the decisive intellectual event that reoriented his ways of thinking as nothing before or after would do. (3) The present passage has been the focus of every debate in the present century over the meaning of A.'s intellectual autobiography. What books of the Platonists did he read? What effect did they have upon him—in 386 and later? As heirs of generations of patient investigation, we can discuss those questions with greater precision than ever, but they continue to elude decisive, and universally acclaimed, answer.

The first words of this paragraph (as du Roy, 61, observes) summarize the rest of Bk. 7 (cf. 7. 20. 26, on the difference between *praesumptio* and *confessio*). God resists the proud when he refuses illumination to those who grasp at it through pride of intellect; and he gives grace to the humble when he allows the least sinner access through the *way* that is the incarnate Christ. In attending as closely as we do to the *platonicorum libri* and even to the texts of scripture, we mistake Augustine's drift in a crucial way. What he is attempting to describe is an encounter between his haughty intellect and the humbling grace of God, an encounter in which the books on the table were instruments, not in themselves indispensable.

Central to A.'s presentation of the doctrines of the Platonists is a rhetorical device that has gone comparatively unattended. He does not quote or paraphrase the Platonic books themselves (thereby making their identification difficult), but he quotes the *ipsissima verba* of Christian scripture as though they offered a fair summary of contents of a non-Christian philosophical work; incidentally the device allows close comparison of the doctrines of Platonists and Christians. He has already employed the device, in his last account of reading a philosophical classic, the *Hortensius* (at 3. 4. 8, quoting Col. 2: 8–9; see below for further parallels). The syncretism he makes is one that seemed obvious at the time.[6]

Some attention must be given to the way this 'reading' is presented. The ancient reader, no less than the modern, must have had expectations on first reading that are sharply confounded by the extended quotation from John. To be told that the books of the Platonists contained Christian scripture is a

[5] No count is exact, but using different editions and different methods of counting and measuring, the centre of the book regularly falls somewhere within ten words of the quotation here of the first words of the gospel of John.

[6] So Nebridius to A. in early 389, *ep.* 6. 1, '[epistulae tuae] mihi Christum, illae Platonem, illae Plotinum sonabunt.'

striking thing, and risks misreading on several levels. The first subtext we risk neglecting is the bishop's assertion that these books contained a great deal that was good. They were good enough to seem to be the culmination of a long search, and if they led to the further discovery that this culmination was not enough, that does not minimize the positive effect they made. He has, as always, high praise for the Platonists mixed with specific and unyielding criticism; we find it hard to take him seriously on both counts at once, but in the face of the consistency of his statements over many years (the difference separating his most Plotinian or Porphyrian remarks at Cassiciacum and his most cautious rectifications in *retr.* forty years later is slight), he must be given the benefit of the doubt. A good way to measure what change there was is to examine the thirteenth book of *trin.*, which uses Jn. 1: 1–14 at length to show the difference between philosophical and Christian doctrines of God. The incarnational verse 1: 14 is again emphasized, but there are differences in approach. The *trin.* discussion is more subtle, both granting more credit to the philosophers and drawing the line between their achievement and the fullness of what Christianity has to offer more sharply.

The way A. presents his reading here allows another parallel to emerge (noted by Mayer, *Zeichen*, 1. 150). At 3. 4. 7, he read the *Hortensius* and turned away to scripture; here, another vivid encounter with the ancient philosophical tradition depicts him immediately making comparisons to specific texts of scripture. The difference is that in Bk. 3, the turn to scripture is frustrated and he rebounds into Manicheism; here, scripture manages to maintain its grip.

Courcelle, *Recherches*, 172–3, holds that the comparison of the *platonicorum libri* with scripture was constructed by A. shortly after reading them, after the disillusionment of his 'tentatives d'extases', and presented in *conf.* more or less as then written, 'comme s'il lui avait sauté aux yeux dès le moment où il dévorait passionnément ces livres'. There is little evidence for any of this (notably the adverb 'passionnément', which better suits his reading of the *Hortensius*), and the text does not allow us to date the exercise in defining the parallels.[7] The text we have is not of 386 but of 397 or later, and the criticism of neo-Platonism implicit and explicit here was shaped over the intervening years.

The presentation of his encounter with these books is almost devoid of circumstance and narrative accoutrement (the garden scene in Bk. 8 stands in sharp contrast in this regard), but some account of the circumstances must be made. These books not were A.'s first introduction to the teachings of the Platonists (books would not have been indispensable for that purpose among the Milanese intellectuals, and he knew much simply from Cicero—to say nothing of many other sources[8]), but he clearly recalls a specific encounter with

[7] See note below on the appearance of *verbum* = λόγος in A.

[8] This is the importance of Solignac's article on the doxographic tradition: *RA* 1 (1958), 113–48.

particular *codices*. Having heard much of the Platonists, he seeks out their books (foreshadowed in his recollection at 6. 11. 18 of a need for books: 'ubi ipsos codices quaerimus?') to learn more, as full of his usual hopes as he was on turning to scripture after reading the *Hortensius*. To his surprise on plunging into them he finds not what he was looking for but something else entirely (and that kind of surprise is familiar in this text from his encounters with both Faustus and Ambrose).

It is a misreading to say that in Bk. 7 A. becomes a neo-Platonist. What he *says* is that in the midst of all his philosophizing—Platonic, neo-Platonic, and idiosyncratic—the specific texts put in front of him brought him new light and new frustration, and thus had the effect of driving him towards scriptural authority, where, in Bk. 8, the real resolution of his difficulties would be worked out. Bk. 7 teaches, in the end, that intellectual enlightenment, contrary to all A.'s youthful expectations, is not sufficient.

There is no suggestion, anywhere in A. or in any of the modern commentators, that he ever took his 'Platonism' so far as to indulge in theurgy.[9] The only possible liturgy for him now was Christianity; the ascent of the mind was non-sectarian in that important, even crucial, sense. But in one significant way, the Platonic pattern may have influenced his expectations of Christianity in a way that also goes unattended. The function of theurgy is to bring about the presence of the God, visibly. What does Christianity have that would have appeared to a half-Christian Plotinus/Porphyry reader as the rough equivalent of theurgy? Eucharist: making the god be present. The Mass as Christian theurgy? Or as Christian counterpart to theurgy? The place in which we might have expected to find such an analogy developed is Ambrose's lost work *de sacramento regenerationis sive de philosophia*.

The most ambitious theological critique of A. in our time is O. du Roy's *L'Intelligence de la foi* . . . (Paris, 1966), and it requires a response in its own theological terms.[10] The substance of his criticism is that A. puts trinity ahead of incarnation, then imprudently grants the Platonists a full doctrine of the trinity; the Platonists know all about the Fatherland, but lack only the Way to get there (du Roy, 96–106). They know the pre-existent Logos, but not the incarnate Christ. But whether A. is depicted as coming to a full understanding of the trinity anywhere in the text of *conf.* before Bks. 11–13 is a serious question. The sequence is not trinity, then incarnation, but rather incarnation (Bk. 8), then trinity (Bks. 11–13). The Platonists acquire knowledge that is accurate in so far as it goes, but utterly insufficient. That their doctrines, described to show the alleged resemblance to Christian teachings to best advantage (Porphyry would *not* have been pleased with A.'s reports of Platonic

[9] See *ep.* 118. 5. 33, quoted below, for the importance A. believed theurgy had to Plotinus' circle.

[10] Serious reservations have already been advanced by G. Madec, *REAug* 17 (1971), 119–42, esp. 128 ff.

teaching), facilitate understanding (or even did facilitate understanding for A.) of the doctrine of the trinity is irrelevant. The failure of the Platonists to grasp the truth of the incarnation would be, on that score, a divinely-managed stroke of bad luck. For the incarnation is the one piece of knowledge that has the greatest power to lead on to full and sufficient knowledge of God—that was, for A., the purpose of the incarnation.[11] The Platonists stumble on to everything else, but in a way that is wholly inadequate to a real understanding; they are closer to the demons who 'believe and shudder' (Jas. 2: 19) than to the humblest of baptized Christians.

The *platonicorum libri* enjoy unparallelled prestige among scholars, to whom they have become a talisman for locating the secret springs of A.'s spiritual life. With the hope of restoring perspective, even at the risk of tarnishing their prestige, we may consider a few of the things that A. did not do with them. He does not identify the books he read; he does not quote them at Cassiciacum (where he quotes scriptural texts and Virgil explicitly); he does not make them the objects of explicit discussion with quotation;[12] he does not write commentary upon them (the way he comments upon scripture); they never become part of his explicit, spontaneously quoted literary life;[13] there is no sign of continuous contact with them, for the extended discussion in *civ.* 8–10 reflects a return to old studies rather than a constant occupation; there is no sign in his correspondence of his handling them, recommending them, or using them as authoritative; and he does not remain in correspondence with any of his Platonic acquaintances from Milan days—the break with those times on his return to Africa is nearly total (Simplicianus is the sole, and very ecclesiastical, exception). The dialogues he records under the influence of Platonic ideas involve without exception A. and others of his African entourage, never any of the so-called neo-Platonic circle of Milan, except as dedicatees.[14] The intellectual movement of his recorded writings, beginning at Cassiciacum and lasting until his death, is consistently and continuously away from neo-Platonism.

[11] See van Bavel, *Recherches sur la christologie*.

[12] Not even (apparently) in the *libri disciplinarum*; years later he makes partial exception, esp. in *civ.* 8–10, but before ordination he was content to write his own original compositions under their influence—an importantly different practice.

[13] By contrast, the *Hortensius*, explicitly quoted, stays with him almost all his life.

[14] Our assumption that he read these books in the company of other learned Platonizers (e.g. H. Chadwick, *Augustine* (Oxford, 1986), 16: 'he became drawn into a group of laymen of high education and social standing, who met to read Plotinus and Porphyry') is hard to document. A. was more often autodidact than not (see on 4. 16. 28, on Aristotle's *categ.*): it would be like him to wrestle with Plotinus for the most part alone. Note the plangent sense of isolation at *sol.* 2. 14. 26. For similar reservations, see G. Madec, *Bull. litt. eccl.* 88 (1987), 194–205, esp. at 198: 'Je vois mal Ambroise, l'adversaire décidé de Symmaque dans l'affaire de l'autel de la Victoire, pratiquer le dialogue paisiblement dans un cercle philosophique.'

But the debate with Platonism is assuredly a subtext in almost everything A. wrote; here it will suffice to quote a few passages of particular pertinence to this description of his first serious encounter. The first signs of reserve are already there at Cassiciacum, as the second text indicates:

os illud Platonis, quod in philosophia purgatissimum est et lucidissimum, dimotis nubibus erroris emicuit maxime in Plotino, qui platonicus philosophus ita eius similis iudicatus est, ut simul eos vixisse, tantum autem interest temporis, ut in hoc ille revixisse putandus sit. (*c. acad.* 3. 18. 41)

[Ratio] si ea quae de deo dixerunt Plato et Plotinus vera sunt, satisne tibi est ita deum scire, ut illi sciebant? [A.] non continuo, si ea quae dixerunt vera sunt, etiam scisse illos ea necesse est. nam multi copiose dicunt quae nesciunt, ut ego ipse omnia quae oravi, me dixi scire cupere, quod non cuperem si iam scirem. (*sol.* 1. 4. 9)

ita si hanc vitam illi viri [Plato et platonici] nobiscum rursus agere potuissent, viderent profecto cuius auctoritate facilius consuleretur hominibus et paucis mutatis verbis atque sententiis christiani fierent, sicut plerique recentiorum nostrorumque temporum platonici fecerunt.[15] (*vera rel.* 4. 7)

The one piece of solid information about Plotinus himself that A. relates elsewhere is *ep.* 118. 5. 33, 'tunc Plotini schola Romae floruit habuitque condiscipulos multos acutissimos et sollertissimos viros. sed aliqui eorum magicarum artium *curiositate* depravati sunt, aliqui dominum Iesum Christum ipsius veritatis atque sapientiae incommutabilis, quam conabantur attingere, cognoscentes gestare personam in eius militiam transierunt.'[16]

The most famous testimony to the *rapprochement* of Platonism and Christianity, seen from a Platonic perspective, but reported by A. to reinforce the Christian view, is at *civ.* 10. 29, 'quod initium sancti evangelii, cui nomen est secundum Iohannem, quidam platonicus, sicut a sancto sene Simpliciano, qui postea Mediolanensi ecclesiae praesedit episcopus, solebamus audire, aureis litteris conscribendum et per omnes ecclesias in locis eminentissimis proponendum esse dicebat. sed ideo viluit superbis deus ille magister, quia "verbum caro factum est et habitavit in nobis".'[17] The mention of Simplicianus (see on 8. 1. 1) dates this episode to the period under review here. A. always credits the Platonists with a knowledge of the triadic nature of God (e.g. *civ.* 8. 5, 8. 10, and (more trinitarian than merely triadic) 10. 23).

As many writers have noted, the question of 'sources' can be dangerously

[15] A similar 'if Plato were alive today' thought-experiment occurs at Cic. *fin.* 4. 22. 61, and cf. *civ.* 22. 27, 'singula quaedam dixerunt Plato atque Porphyrius, quae si inter se communicare potuissent, facti essent fortasse christiani', and *ep.* 118. 3. 21, 'paucis mutatis'.

[16] How far might the view here depend on Ambrose's *de philosophia*? It would be in line with his speculations about the date of Plato and Plato's indebtedness to Jeremiah, etc. For exegesis of *ep.* 118 in the context of A.'s Platonism, without reference to Ambrose, see esp. Mandouze, 499 n. 2.

[17] The same claim that the philosophers acknowledged the truth of Jn. 1 is at *Io. ev. tr.* 2. 4. The interest of Platonists in Jn. 1: 1 is traced suggestively by A. Dyroff, *Pisciculi* (Festschrift F. J. Dölger: Münster, 1939), 86–93, to Amelius, the student of Plotinus.

misleading, for it is not only Latin translations of Greek Platonic texts that need be in question (though obviously there were some of those), but also the sermons of Ambrose (some mediating ideas from Greek Platonic texts), and well-informed oral tradition. Much debate since Courcelle has been concerned with identifying the textual sources as closely as possible, and with dating A.'s contacts with them. Better to follow the example of Mandouze, 476–8, who summarizes concisely and lucidly the most ambitious attempts to provide a schedule for A.'s readings (and hearings) of Platonic ideas, while setting that exercise aside as essentially secondary. Here is the pattern that Mandouze summarizes; see his discussion for documentation for each suggestion:

Beginning of 386	Ambrose, *de Iacob et vita beata*
Holy Week, 386	Ambrose, *exameron*[18]
May 386	Ambrose, *de Isaac vel anima*; *de bono mortis*
June/July 386	Porphyry, *Philosophy from Oracles*[19]
June/July 386	*platonicorum libri*
July/August 386	Paul

The unanswerable question remains how much of A.'s neo-Platonism he imbibed from Ambrose, and how much from his reading, and how far those two bodies of doctrine coincided. The optimism of Courcelle (that the two coincided closely) must be measured by consideration of the objections that can be brought against it, as by Theiler at *Gnomon*, 25 (1953), 114–19. Even if Theiler's *parti pris* on behalf of Porphyry is excessive, there remain problems.

quam . . . gratiam: Prov. 3: 3–4, Jas. 4: 6, 1 Pet. 5: 5: see on 1. 1. 1.

demonstrata: Used of Christ at 10. 43. 68, 'verax autem mediator, quem secreta tua misericordia demonstrasti hominibus et misisti, ut eius exemplo etiam ipsam discerent *humilitatem*, mediator ille dei et hominum, homo Christus Iesus.'

via: See on 7. 7. 11. From this paragraph onwards, *humilis/humilitas* occur 11 × in Bk. 7 alone, signs of that quality of the incarnation that human beings are meant to imitate.

quod verbum . . . homines: The central doctrine missing in the *platonicorum libri* was the incarnation (7. 9. 14), but here is a clear statement from *c.*397

[18] A particularly controversial assignment, suggested by Courcelle, *Recherches*, 102, 133, 279–80; the best case against was made by Madec, *Saint Ambroise* (Paris, 1974), 72, and *Lectio VI–IX*, 51: 'il me semble fort improbable que les sermons dont est issu l'*Hexameron* aient été prêchés au cours d'une semaine sainte aussi agitée que celle de 386 (voir la description des événements dans J.-R. Palanque, *Saint Ambroise et l'Empire romain*, pp. 160–163, qui exploite principalement l'*Epistula* 20, d'Ambroise à sa sœur Marcellina); deuxièmement, l'influence doctrinale des sermons d'Ambroise a dû se faire sentir bien avant la semaine sainte de 386, tout au long de l'année 385.'

[19] Again controversial and part of the whole debate over the relative influence of Plotinus and Porphyry.

that God sent these books to A. in order to get him to believe in the *incarnate* Christ, in order to teach him something they did not contain. The one thing he *says* he learned is the one thing that was not there.

quendam hominem: The identity of this man has exercised much curiosity. A. is careful (see on 4. 4. 7) to name in *conf.* only those who contributed to his religious pilgrimage in positive ways—a category generous enough to include the heretic Faustus. To refrain from naming this man is then a judgement against him, corroborated by the words that describe him. Was he Mallius Theodorus? That was Courcelle's identification (*Recherches*, 153–6, following *LLW* 134–40), and it remains the most compelling. It is not, however, universally accepted. O'Meara, *The Young Augustine* (London, 1954), 152, thinks the man must be Porphyry; Solignac makes a case in favour of Celsinus (see *BA* 13. 103); Mayer, *Zeichen*, 1. 128 n. 146, doubts it can be Theodorus: 'Wahrscheinlich handelt es sich um einen nicht-christlichen Neuplatoniker, der aus seinem philosophischen Dünkel—was in den Augen des Bischofs der Confessiones nur Stolz ist —, sich dem Joch des Glaubens nicht beugte.' (The other case of someone in *conf.* in whom A. later had reason to be disappointed is Romanianus: see on 6. 14. 24).

For Theodorus' career in outline see *PLRE* 1, s.v. Theodorus 27. His known career ran from *c.*376 to 399; he may well have been contemporary with A. or only slightly older. After several minor posts in the late 370s, he held one of the major financial ministries (either *comes sacrarum largitionum* or *comes rerum privatarum*) in 380. In 382 he seems to have reached the acme of a civil career, serving as praetorian prefect for Gaul. About the time of the death of Gratian in 383, he retired from public life, to reappear only as praetorian prefect under Stilicho for the central province of Illyricum, Italy, and Africa in 397–9, and he was consul in 399 along with the eastern vizier Eutropius. He was at ease among the correspondents of Symmachus (he received Symm. *ep.* 5. 4–16, none of much interest). He was the subject in 399 of a panegyric by Claudian, from whom we learn (Claudian, *M. Theod. cos.* 124) that he lived at Milan, and hear in dignified hexameters of his learned works:

> omnia Cecropiae relegis secreta senectae
> discutiens, quid quisque novum mandaverit aevo
> quantaque diversae producant agmina sectae. (67–9)

Of the effects of his work, Claudian says (94) 'in Latium spretis Academia migrat Athenis,' and that hint of Platonic loyalties is confirmed at 149–50, 'scilicet illa tui patriam praecepta Platonis | erexere magis.' (Claudian's religious inclinations and the nature of his 'paganism' are notoriously hard to describe (A. Cameron, *Claudian* (Oxford, 1970), 189–227), so while there is emphatically nothing here to suggest any Christian affiliation at all, that is not conclusive evidence.) Theodorus' treatise on metre is printed at Keil,

6. 585; he wrote an epitaph for a sister who was a 'nun' (Courcelle, *REL* 46 (1944), 66–73).

A. dedicated to Theodorus his own *beata v.* (*beata v.* 1. 1), but regretted the decision later: *retr.* 1. 2, 'displicet autem illic quod Mallio Theodoro, ad quem librum ipsum scripsi, quamvis docto et christiano viro, plus tribui quam deberem.' A. speaks (*beata v.* 1. 4) of Ambrose and Theodorus at that epoch in the same breath, but also asserts that Theodorus was a student of Plotinus: *beata v.* 1. 4, 'lectis autem Plotini paucissimis libris, cuius te esse studiosissimum accepi'. (Many have said that the wording makes the report sound indirect and suggests that T. could not have been the one to put the books into A.'s hands himself. But at 4. 3. 6 and 7. 6. 8, after all, God is said to have 'procured' variously on A.'s behalf, in ways that any observer would have to say are indirect, and it is possible that indirect example and influence could have been at work here.) Courcelle, *Recherches*, 153–6, insists that Ambrose sent A. to Theodorus, and the text of *beata v.* is his most important support for that view; but there is no direct evidence for Ambrose acting in this way (Courcelle is influenced by his reading of 6. 3. 3, which he takes literally as a sign that Ambrose had no time for A., and so passed him on to a trusted friend), and Ambrose, the author at that time already of the *de philosophia sive de sacramento regenerationis*, was unlikely to be an unmitigated enthusiast of pure Platonism. There is already a hint of warning to Theodorus, against the temptations of worldly honours, at *beata v.* 1. 3, 'superbum studium inanissimae gloriae'. (All three Cassiciacum dialogues are addressed to their recipients (Zenobius for *ord.* and Romanianus for *c. acad.*) in terms that are both flattering and didactic/hortatory.) The loudest praise for Theodorus, moreover, comes in a context (*ord.* 1. 11. 31) that is at best ambiguous, for he appears at the end of a paragraph of conventional disparagement of wealthy dilettantes, where the unavoidable suspicion arises that he is being praised as the best of a second-rate lot.

Uncertainty remains. But note that we do not have direct evidence that Theodorus was ever a *baptized* Christian (*retr.* 1. 2 calls him 'christiano' but omits the tell-tale *fidelis*—see on 2. 3. 6). In the mid-380s it may well have looked as though baptism was in his future; he had all the right friends; A.'s attitude of 397 would be perfectly intelligible if he had in the end decided not to be baptized. In the years between, A. had come to characterize unconvertable Platonists harshly: *vera rel.* 4. 7 (the immediate continuation of the passage quoted above), 'aut si hoc non faterentur nec facerent in superbia et invidia remanentes'. There is no named individual known to us more likely than T. to be the object of A.'s remarks here, indeed no named individual is even close to this probability.

typho turgidum: Both these words were used separately of A. himself reading the *Hortensius* (3. 3. 6, 'tumebam typho', 3. 5. 9, 'turgidus fastu mihi grandis videbar'—cf. Cic. *Tusc.* 3. 9. 19, 'sic igitur inflatus et tumens animus in vitio est. sapientis autem animus semper vacat vitio, numquam turgescit, numquam tumet'), together of Stoic philosophers at *s.* 348. 2. 3, 'typho turgidi'; for *typhus*, see 3. 3. 6, 4. 14. 23, 4. 16. 28, 6. 6. 10, 9. 4. 8, 10. 36. 59, 12. 25. 34. Would A. *in 386* have spoken of his youthful self this way? Probably not. Their collocation here represents a similarly retrospective criticism of the unnamed man. *Typhus* is a non-biblical Grecism already Christianized (for its background, see Courcelle, *Opuscula Selecta* (Paris, 1984), 329–72, esp. 353–64), not appearing in A. until *ep.* 22. 6 of 392, hence after his ordination; it becomes a favourite expression, esp. in the phrase *typhus superbiae* and the like (such as *en. Ps.* 55. 6, 'hominem superbum typho arrogantissimo'; cf. *en. Ps.* 102. 22, *civ.* 11. 33). Apposite here is *trin.* 4. 10. 13, quoted on 8. 2. 3, 'sacrorumque sacrilegorum particeps'; cf. *civ.* 10. 28, 'tumore inflatus vanae scientiae'. In a much later text, *typhus* is correlated to the attitude attacked by Rom. 1: 21 ff. (see on 7. 9. 14, esp. 'cothurno'): *ep.* 187. 6. 21, 'hoc sacramentum *longe est* a cordibus sapientium superborum et *ideo* non christianorum ac per hoc nec vere sapientium, illorum etiam dico sapientium qui cognoverunt deum, quia cognoscentes deum, sicut dicit apostolus, non sicut deum glorificaverunt aut gratias egerunt. nosti autem in quo sacrificio dicatur, "gratias agamus domino deo nostro". a cuius sacrificii humilitate *longe abest typhus et cothurnus illorum*'; cf. also *ep.* 102. 6. 32. *s. Frang.* 5. 5, *Gn. litt.* 11. 31. 41, *c. Iul.* 4. 16. 83, *c. ep. Parm.* 3. 5. 27.

quosdam platonicorum libros: For A.'s exact sense of *platonici*, see *civ.* 8. 12, 'recentiores tamen philosophi nobilissimi quibus Plato sectandus placuit noluerint se dici peripateticos aut academicos, sed platonicos. ex quibus sunt valde nobilitati graeci Plotinus, Iamblichus, Porphyrius; in utraque autem lingua, id est et graeca et latina, Apuleius Afer extitit platonicus nobilis. sed hi omnes et ceteri eius modi et ipse Plato diis plurimis esse sacra facienda putaverunt.' (Note that for A. at least all Platonists perform non-Christian religious cult; cf. *civ.* 8. 1.) The books are not more closely identified in *conf.* (cf. 8. 2. 3, 'me legisse quosdam libros platonicorum': but there he does say, problematically, that they were translated by Victorinus: see notes there), but at *beata v.* 1. 4, 'Plotini . . . libris' offers unambiguous testimony.[20] This has not prevented lively debate over the amount of Porphyry A. read at the time, but A.'s own choice was to leave the texts unidentified: that decision is in itself a form of commentary on them.

The central texts to that debate are Theiler's review of Courcelle, *Recherches*, at *Gnomon*, 25 (1953), 113–22, and Courcelle, *Les Confessions*, 33–42 (and cf. O'Meara's opinion on the identity of the *homo* cited above). The issue is complicated by the role Porphyry played in the transmission of Plotinus and by our ignorance of the outward form of the Latin books A. actually

[20] The Maurists read 'Platonis' in this passage, but that has since been shown to be a scribal error.

read. See du Roy, 70, for a catalogue of treatises by Plotinus variously thought to have been read by A., either at this period or slightly later, and cf. P. Hadot, *Marius Victorinus* (Paris, 1971), 201–10. The number is probably not large (*beata v.* 1. 4, 'paucissimis libris'; *c. acad.* 2. 2. 5, 'guttas paucissimas'—but those phrases have a rhetorical purpose (du Roy, 69 n. 5)); on du Roy's view, A. read 15 of the 54 Plotinian treatises; TeSelle's position resembles du Roy and he reaches a count of 19; on *paucissimus* see on 5. 6. 11), and nothing says that A. read all of each with the same attention and comprehension. (A useful *comparandum* is the case of Macrobius, who seems to have had little of Plotinus available to him, perhaps only half a dozen treatises and Porphyry's *vita*; see J. Flamant, *Macrobe et le néo-platonisme latin* (Leiden, 1977), 571–3.) The most assertive 'Plotinizer' among recent scholars has been R. J. O'Connell: see esp. *REAug* 9 (1963), 1–39. O'Connell has consistently overplayed his discoveries and the role of Plotinus in A. generally; the authorized (and deafening) reaction, emanating from Courcelle's seminar, was G. Madec, *REAug* 16 (1970), 79–137, to which O'Connell responded at *REAug* 19 (1973), 87–100. O'Connell insists on the importance of tracing 'parallel patterns' of a kind that resist philological verification; while his method gives undeniably freer rein to the search for affiliations and influences, the (mostly justified) ferocity of Madec's response is a measure of the limitations of that method.

The explicit mention of names is not insignificant in A. (see on 4. 4. 7); what is the case here? Plotinus is mentioned by name in A. in three works from Cassiciacum: *c. acad.* 3. 18. 41, *sol.* 1. 4. 9, *beata v.* 1. 4, again in *ep.* 118. 5. 33 (all quoted above), and again 11 × in Bks. 8–10 of *civ.*—and *not otherwise*. Porphyry is the object of much discussion in *civ.*, where he is named 59 ×; but elsewhere he is mentioned by name only 13 ×, in 6 different *opera/opuscula*. The earliest show A. aware of Porphyry's attacks on Christianity at *cons. ev.* 1. 15. 23 (399/400–??)[21] and at *ep.* 82. 2. 22 (*c.*405); 3 × at *ep.* 102. 2. 8 ff. (406–12), responding to Porphyrian challenges (cited years later at *praed. Sanct.* 9. 17); and *s.* 241. 6. 6–7. 7 (404 earliest, poss. after 412), defending the doctrine of the resurrection against Porphyry's 'corpus est omne fugiendum'; hence, Porphyry was never named *until* his anti-Christian polemic had given him a separate and distinct identity for A. By contrast, *Plato* is mentioned approx. 206 × by A. (outside *civ.* 68 ×), and *platonicus/platonici* of persons occur approx. 109 × (outside *civ.* 28 ×). These figures do

[21] I have argued elsewhere (*RA* 15 (1980), 173–5) that *cons. ev.* as it stands does not date from as early as 400, and suggested stating the date merely as 400–15, advancing the suspicion that the first book, with its lengthy polemic 'contra gentiles' may have been retouched later. Pending further study, there is ambiguous partial agreement with this position at Mandouze, 304 n. 4 (dating several works including *cons. ev.* to 404–12; the unpublished thesis to which M. refers there has been inaccessible to me). For the contrary view, see TeSelle, 237–58 (his chapter 'The Shadow of Porphyry'), treating *cons. ev.* as the place where, *c.*400, A. felt the shock of discovering that Porphyry had written polemically against Christianity.

not tell the whole story, but provide a framework within which to speculate, and if kept in mind offer perspective; it is often perspective that is lost in the debates on this subject.

Some Porphyry A. undoubtedly read at Milan, but whether before or after Cassiciacum is questionable. The likeliest form in which he encountered Porphyry at that time was in his *sententiae* (ἀφορμαί), possibly attached as an explanatory preface to the translation of Plotinus A. also encountered (even there we are in a dark wood, for it is likely that we have only part of the *sententiae* (see J. Bidez, *Vie de Porphyre* (Ghent, 1913), 106 n. 1)). At any rate, Plotinus and Porphyry came to A. now as a package, as *platonici*, and as a package that offered intellectual treasures that were not felt to be incompatible with Christianity. See on the place of the *conf.* in A.'s life in the introduction to this commentary for further treatment of the relevance of his revaluation of Porphyry to the making of *conf.*

The case for Porphyrian influence was given its classic expression in Theiler's *P. u. A.*, and revived more cautiously by by J. J. O'Meara, *Porphyry's Philosophy from Oracles in Augustine* (Paris, 1959), arguing that the *Philosophy from Oracles* and the work quoted by A. as *de regressu animae* were identical (see also H. Lewy, *Chaldæan Oracles and Theurgy* (Cairo, 1955), 449), and that A. knew that work from Cassiciacum at the latest. O'Meara was rebutted fiercely (but with limited success) by P. Hadot, *REAug* 6 (1960), 204–44. O'Meara is right to identify the two titles, but that does not require us to accept his insistence that the one work was known already in Italy: how could A. ever have read that work and failed to see how anti-Christian Porphyry was? The state of our knowledge of Porphyry hampers further progress. (There are almost infinite *ambages* to these debates. G. Madec, *Saint Ambroise*, offers two important precisions: at 69–70, that Ambrose's Plotinian sermons may get much of their Plotinianism in bulk from Greek Cappadocian Christian models, hence part of the explanation why those sermons never discuss *philosophia* as such and never give any direct sign of their sources or any consciousness of the syncretism that is taking place; at 167, he reminds us further of the 'probabilité d'une influence de Porphyre sur Ambroise a été établie par P. Courcelle [Courcelle, *Recherches*, 133–6]: Ambroise aurait utilisé le *De regressu animae* dans un sermon dont Claudien Mamert a cité un fragment, et dans un sermon sur *Isaïe* cité par Augustin.') The most lucid and balanced discussion of the Porphyrian question in English is that of TeSelle, 49–54; Courcelle himself strongly increased his estimate of Porphyry's influence between *Recherches* and *Les Confessions*: see *Les Confessions*, 33–42 (and see on 7. 19. 25); Madec, *Lectio VI–IX*, 61–7, has followed the movement to see more Porphyry; see also O'Meara's lucid 'Parting from Porphyry', *Atti 1986*, 2. 357–69. A fresh comprehensive survey of the state of play would be valuable. P. F. Beatrice, *Vig. Chr.* 43 (1989), 248–81, boldly limits A.'s reading of neo-Platonists not only to Porphyry, but to a single work; he accepts O'Meara's identification of the *de regressu animae* with the *Philosophy from Oracles* and goes on to insist that the famous lost work 'contra Christianos' is one and the same title with the other two. There are strong contradictions inherent in such a view, as Beatrice sees (not least the striking implication that Marius Victorinus first translated the 'contra Christianos' into Latin and *then* converted to Christianity himself; and that A. owed his own conversion to a similar encounter with what has always been taken to be the most powerful and effective of 'pagan' anti-Christian polemics), and his treatment is too sketchy to carry conviction.

The quality of the translations that A. used is also pertinent. Plotinus is notoriously difficult, and ancient translations could be highly literal. Can we make any useful comparisons? Take Plot. 1. 4. 7, the apparent source of A.'s 'last words' quoted at Possidius, *v. Aug.* 28. 11, 'et se inter haec mala cuiusdam sapientis sententia consolabatur dicentis: "non erit magnus magnum putans quod cadunt ligna et lapides, et moriuntur mortales".' Plot. 1. 4. 7. 23–4: *οὐκ ἂν ἔτι σπουδαῖος εἴη ξύλα καὶ λίθους καὶ νὴ Δία θανάτους θνητῶν μέγα ἡγούμενος.* The dependence is clear, but the verbal infidelity is also clear.[22] The same test with similar results can be performed on the 'quotation' from Plot. 1. 6. 8 (conflated with 1. 2. 3?) at *civ.* 9. 17 (quoted below on 7. 10. 16, 'in regione dissimilitudinis').

non quidem his verbis: In Bks. 7, 8, and 9, each crucial moment is reported with discourse bracketed by similar phrases: cf. A.'s lament under the fig-tree (8. 12. 28, 'non quidem his verbis') and the vision at Ostia (9. 10. 26, 'etsi non isto modo et his verbis').

suaderi rationibus: 8. 2. 3 (of Simplicianus' delight at hearing A. had been reading *platonicorum libri*): 'in istis [libris] autem omnibus modis insinuari deum et eius verbum'.

in principio: See at end of notes on this paragraph for the text of John on which A. relies. Quotations from Jn. 1 also occur at 4. 15. 25, as part of the critique of that 'tentative d'extase'. Solignac, at *BA* 13. 683–6, presents an interesting tabulation of parallels between the scriptural passages quoted here and in the next paragraph and passages of Plotinus.

verbum: = *λόγος*. This equation is not recorded in A.'s surviving works until

[22] See Pellegrino, ed. Possidius (Alba, 1955), 226 n. 14, and cf. Courcelle, *REA* 46 (1944), 205–7. The same expression is echoed at *s. Casin.* 1. 133. 7, 'doles ergo, et ploras, quia ruerunt ligna et lapides, et quia mortui sunt morituri?'; cf. *s.* 81. 9 and *civ.* 2. 2, 'lapides et ligna'. Those three texts all speak to concerns over the sack of Rome in 410; and the 'last words' come from the time of the siege of Hippo. The 'quotation' is not to be taken as homage to Plotinus, therefore, but criticism of the sort directed against the anti-Christian polemics attacked in *civ.*: here, A. is saying, is a non-Christian sage offering good advice; how much more likely, therefore, that a Christian will have hope in time of crisis? (The earlier echoes, moreover, cast at least slight doubt on the historicity of Possidius' attribution of the words to A. on his deathbed.)

388/90: *Gn. c. man.* 1. 1. 3 (noted briefly at du Roy, 270–1; fuller study at D. W. Johnson, *RA* 8 (1972), 25–53, arguing that the equation is thus more likely the result of Christian than of Platonic influence). There is a difficulty: A. himself is the source of the anecdote at *civ.* 10. 29, quoted above, about the Platonist who said that the first words of John's gospel should be written in letters of gold and put up in prominent places for all to see, and he says that it was a story Simplicianus used to tell: 'sicut a sancto sene Simpliciano ... solebamus audire'. So he had surely heard the story, and must have known the underlying doctrine and texts, at Milan. Reserve giving way to abundance marks the appearance of the word in this sense in *conf.*; it becomes suddenly frequent in this book from here, and continues in later books to be far more frequent than in earlier.

vita est: *Io. ev. tr.* 1. 17, 'faber facit arcam. primo in arte habet arcam: ... arca in opere non est vita, arca in arte vita est; quia vivit anima artificis, ubi sunt ista omnia antequam proferantur.'

hominis anima: See *civ.* 10. 2, 'saepe multumque Plotinus asserit sensum Platonis explanans, ne illam quidem quam credunt esse universitatis animam aliunde beatam esse quam nostram, idque esse lumen quod ipsa non est, sed a quo creata est et quo intellegibiliter inluminante intellegibiliter lucet. ... nec aliunde illis supernis praeberi vitam beatam et lumen intellegentiae veritatis quam unde praebetur et nobis, consonans evangelio, ubi legitur, "fuit homo missus a deo, cui nomen erat Iohannes; ... lumen verum, quod inluminat omnem hominem venientem in hunc mundum." in qua differentia satis ostenditur *animam rationalem vel intellectualem, qualis erat in Iohanne*, sibi lumen esse non posse, sed alterius veri luminis participatione lucere.' (For the same juxtaposition of a Christian text with a Plotinian doctrine, see *civ.* 10. 14.) Cf. M. Victorinus, *adv. Arium* 1. 56, 'anima enim in deserto, hoc est in mundo, exclamat quoniam scit dominum deum et vult mundari ut domino fruatur deo; et ista dicit testimonium de deo et praemissa est in mundum ad testimonium testimonii.' This phrase is the only significant deviation from the gospel text; it implies that the allegorism is to A. sure and obvious.

deus (*est*) DGO Ver. Pell. : *deus ipse* S Knöll Skut. : *dei deus* BPZ Maur.

non ibi legi: The same phrasing in 7. 9. 14 continues the contrast. It was not only the incarnation of the second person that A. did not find in the Platonists, but it was twenty years after *conf.* before he was drawn to note the lack of mention of the third person, and then only when reminded by another writer: *qu. hept.* 2. 25, 'summi philosophi gentium, quantum in eorum litteris indagatur, sine spiritu sancto philosophati sunt, quamvis de patre et filio non tacuerint, quod etiam Didymus in libro suo [*de spiritu sancto* 2 (*PL* 33. 109)] meminit, quem scripsit de spiritu sancto.'

Text of Jn. 1: 1–14 from Io. ev. tr.

Words paraphrased here, cited not *ad verbum*, or differently inflected are italicized. The punctuation follows A.'s phrasing: note 3–4, 'quod factum est in illo vita est' and 9, 'omnem hominem venientem in mundum'.

(1) In principio erat verbum, et verbum erat apud deum, et deus erat verbum. (2) hoc erat in principio apud deum. (3) omnia per ipsum facta sunt, et sine ipso factum est nihil. quod factum est (4) in illo vita est, et vita erat lux hominum, (5) et lux in tenebris lucet, et tenebrae eam non comprehenderunt. (6) fuit homo missus a deo, cui nomen erat Iohannes; (7) hic venit in testimonium, ut testimonium perhiberet de lumine, ut omnes crederent per illum. (8) non *erat ille* lumen, sed ut testimonium perhiberet de lumine. (9) erat *lux vera quae* inluminat omnem hominem venientem in mundum.[23] (10) in mundo erat, et mundus per eum factus est, et mundus eum non cognovit. (11) in sua propria venit, et sui eum non receperunt, (12) quotquot autem receperunt eum, dedit eis potestatem filios dei fieri, his qui *credunt* in nomine eius, (13) qui non ex *sanguinibus* neque ex voluntate *carnis* neque ex voluntati viri sed ex deo *nati sunt*.[24] (14) et verbum caro factum est, et habitavit in nobis, et vidimus gloriam eius, gloriam quasi unigeniti a patre, plenum gratia et veritate.

7. 9. 14

The *in extenso* citation of these central and moving texts has its effect. He could have said in cold prose, 'I found that the Platonists taught of the existence of the Logos, but not that the Logos took flesh and died for us.' He could even have quoted these texts more briefly. He did not.

quia verbum caro factum est: 10. 43. 69, 'potuimus putare verbum tuum remotum esse a coniunctione hominis et desperare de nobis, nisi caro fieret et habitaret in nobis.'

in forma patris . . .: Phil. 2: 6–11—for text, see at end of notes on this paragraph. For a palmary study of this passage through A.'s works, see A. Verwilghen, *Christologie et spiritualité selon saint Augustin: L'Hymne aux Philippiens* (Paris, 1985). Clearly echoed at 10. 43. 69, less clearly at 12. 7. 7, 12. 15. 20, 13. 2. 2, 13. 2. 3, 13. 5. 6. Jn. 1 and Phil. 2 regularly occur together in A. (for Jn. 1: 1 and Phil. 2: 6, see Verwilghen, 163; for Jn. 1: 14 and Phil. 2: 7, see Verwilghen, 230–1): e.g. *Gn. c. man.* 2. 24. 37 (first appearance of the Phil. text), *f. et symb.* 4. 5, 4. 6, *en. Ps.* 130. 9, *s.* 292. 3. 3, *civ.* 10. 29, *ench.* 10. 35, *trin.* 1. 6. 7, 1. 6. 9, 1. 7. 14 ('et haec nobis regula per omnes sacras scripturas dissolvendae huius quaestionis *ex uno capite epistulae Pauli apostoli promitur* ubi manifestius ista distinctio commendatur'), *conl. Max.* 14–15, *c. Max.* 1. 5, 1. 19

[23] A. reads here in *conf.* 'in hunc mundum' and in next phrase 'in hoc mundo'.

[24] In *conf.*, the common variant, 'natus est'.

(also with Rom. 1: 20: see below). *s.* 92. 3. 3, 'habet patriam, habet viam. habet patriam: "in principio erat verbum"; habet patriam: "cum in forma dei esset non rapinam arbitratus est esse aequalis deo"; habet viam: "verbum caro factum est"; habet viam: "semet ipsum exinanivit formam servi accipiens." ipse est patria quo imus, ipse via qua imus.' The separate totals (H.-I. Marrou, *Saint Augustin et l'augustinisme* (Paris, 1955), 84) for occurrences of these verses in A. are: Jn. 1: 1 (250+), Jn. 1: 14 (230+), Phil. 2: 7 (216).[25] On the sense of the verse, see van Bavel, 46 n. 111, pressing *forma* to equal *natura*, with support from *c. Max.* 1. 5 and Ambrose's *ep.* 46. 7; see also B. Studer, *RA* 19 (1984), 133–54, excellent on the importance of *aequalitas* in A.'s Christology.

The Phil. text was cited by Fortunatus on behalf of the Manichees (*c. Fort.* 7); his quotation begins with the words of 2. 5, 'hoc sentite in vobis, quod et in Christo Iesu': Fortunatus takes this verse to mean that the 'incarnation' occurred 'ut similitudinem animarum nostrarum ostenderet' (cf. Verwilghen, 101–2). Augustine quotes and echoes Phil. 2: 6 abundantly, but he only cites the 'hoc sentite' clauses 7 × in all his writings, and only once (*div. qu.* 71. 3) before 412 (later citations at *c. s. arrian.* 8. 6, *Io. ev. tr.* 47. 13, *ss.* 144. 3. 4, 264. 3, *c. Max.* 1. 5 (there only because the text was put on the table by Maximianus during their public debate), *spec.* [25] ad loc.); the possibility of deliberate avoidance should be kept in mind.

similitudinem GS Knöll Skut. : *similitudine* DO Ver.

habitu: *div. qu.* 73. 2, 'neque conversus aut transmutatus in hominem, amissa incommutabili stabilitate, sed quamquam verum hominem suscipiendo, ipse susceptor, "in similitudinem hominum factus", non sibi, sed eis quibus in homine apparuit habitu inventus est ut homo; id est, habendo hominem, inventus est ut homo. . . . quo nomine oportet intellegi non mutatum esse verbum susceptione hominis, sicuti nec membra veste induta mutantur.' See Mayer, *Zeichen*, 1. 324.

eum exaltavit a mortuis: Text with the acc. pronoun at *ss.* 37. 2. 2, 47. 11. 20, *en. Ps.* 109. 7 and 109. 20; without *eum* often elsewhere in A.

unigenitus: Jn. 1: 14, 'gloriam quasi unigeniti a patre'.

coaeternus: The word is not scriptural, but ecclesiastical from Tertullian on.

de plenitudine . . .: Jn. 1: 16, 'et de plenitudine eius nos omnes accepimus'; *civ.* 10. 2, 'satis ostenditur animam rationalem vel intellectualem, qualis erat in Iohanne, sibi lumen esse non posse, sed alterius veri luminis participatione lucere. hoc et ipse Iohannes fatetur, ubi ei perhibens testimonium dicit, "nos omnes de plenitudine eius accepimus".'

participatione: See on 7. 18. 24 and cf. preceding note; *en. Ps.* 121. 5, 'ubi non

[25] Marrou's numbers are best for comparative purposes, since they were gathered using a consistent method. Verwilghen's study, pursuing traces and echoes far more intently, offers a much larger total, e.g. a total of 561 'allusions' to 2. 6–7 alone; see Verwilghen, 61–96.

rapinam arbitratus est esse aequalis deo, ibi est idipsum. ut autem efficiaris tu particeps in idipsum, factus est ipse prior particeps tui, et verbum caro factum est, ut caro participet verbum.'

manentis in se sapientiae: Wisd. 7: 27, 'et permanens in se omnia innovat.'

secundum tempus . . . mortuus est: Rom. 5: 6, 'ut quid enim Christus cum adhuc infirmi essemus secundum tempus pro impiis mortuus est?'

et filio tuo unico non pepercisti: Rom. 8: 32, 'qui etiam filio suo non pepercit sed pro nobis omnibus tradidit illum'.

abscondisti: Mt. 11: 25—for text, see below.

ut venirent: Mt. 11: 28–9—for text, see below.

diriget . . . suas: Ps. 24: 9, 'diriget mites in iudicio, docebit mansuetos vias suas.'

diriget DO Ver. : *dirigit* S Maur. Knöll Skut. : *dirigeret* G.

videns . . . nostra: Ps. 24: 18, 'vide humilitatem meam et laborem meum et remitte omnia peccata mea.'

cothurno: See *ep.* 187. 6. 21, quoted on 7. 9. 13 above, esp. 'longe abest typhus et cothurnus illorum.'

discite a me: Mt. 11: 29—for text, see below; for punctuation, see on 7. 21. 27. *c. Adim.* 2. 2 (quoting this text), 'hoc sabbatum, id est hanc requiem scriptura illa significat, quam Iudaei non intellegebant et pro temporum dispensatione carnaliter sequebantur umbram, cuius umbrae quasi corpus, id est veritas nobis danda erat.'

etsi cognoscunt deum: Rom. 1: 21–2—for text, see below; for place in A.'s work, see G. Madec, *RA* 2 (1962), 273–309. The rhetorical strategy of this citation must be seen in the history of Christian assimilation of philosophical doctrines (marked at the same time by the insistence that they were originally Christian all along): see R. Holte, *Béatitude et Sagesse* (Paris, 1962), 125–41. *spir. et litt.* 12. 19–20, 'nec immerito se apostolus ex hoc articulo convertit ad eos cum detestatione commemorandos, qui vitio illo quod superius memoravi *leves et inflati ac per se ipsos velut per inane sublati*, ubi non requiescerent, sed fracti dissilirent, in figmenta idolorum tamquam in lapides deciderunt. [quotes Rom. 1: 20–3] . . . (20) illis per creaturam cognitoribus creatoris ea ipsa cognitio nihil profuit ad salutem.' So here in Bk. 7 A. is seen coming to know God the creator [1], and that is not enough.

Rom. 1: 20 is typically 'pro-Platonic' in A. (and often quoted without either the conclusion to that verse ('ut sint inexcusabiles'—apparently never cited by A. before *conf.* 10. 6. 8, where see notes) or the following verses), 1: 21 ff., 'anti-philosophical' (Madec, art. cit.), hence here he limns caution. Both halves of the quotation are applied to the Platonists explicitly at *civ.* 8. 10. See 10. 6. 9–10 for an exemplification of the procedure that A. takes Rom. 1: 20 to prescribe. As Madec showed, this proof-text was useful against Manichees (*c. Faust.* 14. 10, 20. 19, 21. 6), Pelagians (*gr. et lib. arb.* 2. 2, *c. Iul.* 4. 3. 17, 5. 4. 18, *c. Iul. imp.* 3. 106, 4. 32, *nat. et gr.* 22. 24, *epp.* 25. 2, 140. 37. 83–

4), Donatists (*un. eccl.* 19. 49), and Arians (*c. Max.* 1. 19, 2. 7). It played a notable part in his speculative theology (*trin.* 2. 5. 25, 4. 16. 21, 4. 17. 23, 6. 10. 12, 13. 19. 24, 15. 2. 3, 15. 6. 10, 15. 20. 39), and in apologetic contexts as well (*cons. ev.* 1. 34. 52 and *s.* 52. 5. 15); in *s.* 241 (405–10), it is the master text for a discussion of the doctrine of resurrection 'contra Gentiles'. For its value in the healing of *curiositas*, see on 13. 21. 31; and for a text nearly contemporary with *conf.* using Rom. 1: 20 in a purely positive sense of the Christian approach to God, *doctr. chr.* 1. 4. 4, 'si redire in patriam volumus, ubi beati esse possimus, utendum est hoc mundo, non fruendum, ut invisibilia dei per ea, quae facta sunt, intellecta conspiciantur, hoc est, ut de corporalibus temporalibusque rebus aeterna et spiritalia capiamus.'

When A. adapted this verse to his use is not certain. Courcelle (*Recherches*, 177) and Madec (art. cit. 283), who accept the report here as a factual statement of A.'s thoughts in Milan, hence think that he had already learned to use it (perhaps in the reading of Paul that follows the *platonicorum libri*? or from hearing Amb. *exam.* 1. 4. 16?), but nothing in this text requires us to make that assumption. Courcelle sees an echo at *quant. an.* 34. 77, but it is a curious one: 'ideoque divine ac singulariter in ecclesia catholica traditur nullam creaturam colendam esse animae (libentius enim loquor his verbis quibus mihi haec insinuata sunt), sed ipsum tantummodo rerum quae sunt omnium creatorem, ex quo omnia, per quem omnia, in quo omnia. . . .' The parenthesis seems to put quotation marks around some of the preceding words (probably 'nullam . . . animae'), but could just as easily apply to what follows (esp. 'ex quo . . .': see on 1. 2. 2). If that passage is not accepted as an echo,[26] the earliest is *vera rel.* 10. 19 (the anti-philosophical use of 1: 21) and 52. 101 (the pro-philosophical use of 1: 20). But that work from the end of the Thagaste period already marks the development of A.'s Platonism past the views he held while in Italy. After *vera rel.* the verse occurs frequently in A.'s writings. This text and Jn. 1: 1–14 occur together frequently in A. in what O'Meara marks as anti-Porphyrian passages (*Porphyry's Philosophy from Oracles in Augustine* (Paris, 1959), 102, 152–4, 160–2; see Madec, art. cit. 276–82).

facti sunt DO Ver. : *fiunt* GS Knöll Skut.

Text cited in 7. 9. 14

Mt. 11: 25:[27] 'confiteor tibi, pater, domine caeli et terrae, quia abscondisti haec a sapientibus et prudentibus, et revelasti ea parvulis.'

Mt. 11: 28:[28] 'venite ad me, omnes qui laboratis, et ego vos reficiam. tollite iugum meum super vos et discite a me quoniam mitis sum et humilis corde, et invenietis requiem animabus vestris.'

[26] It is not marked as such by the most recent editor, W. Hörmann in *CSEL* 89.

[27] From *ss.* 67–8.

[28] *ss.* 69–70.

Rom. 1: 21–5:[29] (20) invisibilia enim dei a creatura mundi per ea quae facta sunt intellecta conspiciuntur, et sempiterna eius virtus et divinitas,[30] ut sint inexcusabiles,[31] (21) quia cognoscentes[32] deum, non ut deum glorificaverunt aut gratias egerunt; sed evanuerunt in cogitationibus suis et obscuratum est insipiens cor eorum, (22) dicentes enim se esse sapientes stulti facti sunt.[33] (23) et immutaverunt gloriam incorruptibilis dei in similitudinem imaginis corruptibilis hominis,[34] et volucrum et quadrupedum et serpentium.[35] (24) propter hoc tradidit illos deus in concupiscentias cordis eorum, in immunditiam.[36] (25) qui transmutaverunt veritatem dei in mendacium, et coluerunt et servierunt creaturae, potius quam creatori, qui est benedictus in saecula.[37]

Philippians 2: 6–11:[38] (6) cum in forma dei esset, non rapinam arbitratus est esse aequalis deo, (7) semet ipsum exinanivit formam servi accipiens, in similitudine hominum factus et habitu inventus ut homo. (8) humilavit se, factus oboediens usque ad mortem crucis. (9) propter quod et deus eum exaltavit et donavit ei nomen quod est super omne nomen, (10) ut in nomine Iesu omne genu flectatur caelestium, terrestrium, et infernorum, (11) et omnis lingua confiteatur quia dominus Iesus Christus in gloria est dei patris.

7. 9. 15

immutatam gloriam . . .: Rom. 1: 23 (see text above, 7. 9. 14); cf. Jn. 1: 14, 'et vidimus gloriam eius'. G–M conventionally remark that this charge might not be as true of Plotinus as of his followers, adducing Porph. *de abstinentia* 4. 9, 'in which he speaks approvingly of the Egyptian animal worship'; cf. also Porph. *v. Plot.* 2. 40–2, καίπερ ἐν τοῖς Πλάτωνος καὶ Σωκράτους παραδεδομένοις γενεθλίοις θύων τε καὶ ἑστιῶν τοὺς ἑταίρους, and *civ.* 8. 12 quoted above on 7. 9. 13. But the position of Plotinus can be taken both ways; for contrasting views, cf. E. R. Dodds, *The Greeks and the Irrational* (Berkeley, 1951), 285–91 (stoutly denying that Plotinus engaged in such conduct: for a similar position with less special pleading, see A. H. Armstrong, *Cambridge History of Later Greek and Early Medieval Philosophy* (Cambridge, 1967), 203–10) with G. Fowden, *The Egyptian Hermes* (Cambridge, 1986), 129–31.

Aegyptium cibum: Gn. 25: 33–4, 'ait Iacob, "iura ergo mihi." iuravit Esau et vendidit primogenita, (34) et sic accepto pane et lentis edulio comedit et bibit et abiit parvipendens quod primogenita vendidisset.' Ps. 46: 5, 'elegit

[29] References after the last word to which they testify.

[30] *vera rel.* 52. 101.

[31] *c. Faust.* 20. 19.

[32] *spir. et litt.* 12. 19.

[33] *en. Ps.* 57. 18.

[34] *en. Ps.* 105. 19

[35] *spir. et litt.* 12. 19.

[36] *en. Ps.* 57. 18.

[37] *en. Ps.* 113. s. 2. 4.

[38] See Verwilghen, op. cit. 69–70, for authorities and for variants in A.'s text over time.

nobis hereditatem suam, speciem Iacob quam dilexit'; *en. Ps.* 46. 6, 'prior natus est Esau, et posterior natus est Iacob; sed qui posterius natus est, praelatus est primo nato, qui per gulam perdidit primogenita sua. ... ille plus amavit *quod carnaliter concupierat* quam quod spiritaliter prius nascendo meruerat; et deposuit primogenita sua, ut manducaret lenticulam. lenticulam autem invenimus *cibum* esse *Aegyptiorum*, nam ibi abundat in Aegypto. ... ergo desiderando *cibum Aegyptium*, perdidit primatum.' For the OT associations of 'Egypt', cf. *en. Ps.* 113. s. 1. 3, 'Aegyptus autem, quoniam interpretatur afflictio, vel affligens, vel comprimens, saepe in imagine ponitur huius saeculi.' A link between several scriptural moments is provided by Ps. 105: 20 (which is echoed by Paul at Rom. 1: 23): 'et fecerunt vitulum in Horeb et adoraverunt sculptile et mutaverunt gloriam suam in similitudine vituli comedentis fenum.' Ex. 32: 1–6 recounts the worship of the golden calf, and Act. 7: 39–41, evokes the story again: Act. 7: 39, 'conversi sunt corde in Aegyptum.'

See on this passage Knauer, 114–15, with rectification of earlier mistaken scholarly views. On Egypt as the home of idolatry, see also *s.* 197. 1, *un. bapt.* 4. 5 ('simulacra Aegyptiorum, ubi et instituta esse multiplicior multoque ignominiosior idolatria perhibetur'), and *civ.* 8. 26, quoting ps.-Apul. *Asclepius* as saying, 'terra Aegypti, sanctissima sedes delubrorum atque templorum' (though it is not certain that A. would have known that text as early as *conf.*); cf. also *civ.* 18. 3 (Isis), 18. 5 (Serapis), and 18. 40 (magic and astrology in Egypt generally). O'Meara, *Porphyry's Philosophy from Oracles in Augustine* (Paris, 1959), 163–5, makes the case for linking Egypt with Porphyry. The one Egyptian 'philosopher' A. consciously mentions is Hermes Trismegistus, but that only later: *civ.* 8. 23.

opprobrium: Ps. 118: 22, 'aufer a me opprobrium et contemptum, quoniam testimonia tua exquisivi'; Rom. 11: 12, 'quod si delictum illorum divitiae sunt mundi et deminutio eorum divitiae gentium, quanto magis plenitudo eorum.'

ut maior serviret minori: Gn. 25: 33, 'et maior serviet minori'; Rom. 9: 12, 'non ex operibus sed ex vocante dictum est ei quia "maior serviet minori".' The Gn. text underlies Rom. 9, which in turn is the key to extensive discussion of grace vs. merit at *div. qu. Simp.* 1. 2. 3 ff., showing that merit (here, seniority) does not prevail.

et vocasti gentes: Ps. 78: 1, 'deus, venerunt gentes in hereditatem tuam.'

ego ad te veneram ex gentibus: Courcelle, *Les Confessions*, 61 n. 3, 'ne peut être prise au sens propre, étant donné qu'Augustin n'a jamais été païen et que le contexte, relatif aux "dépouilles d'Égypte", est tout d'exégèse allégorique.' But A. would not take the expression so: (1) he did not come to Christianity out of Jewish tradition; (2) he thinks himself to have been a 'pagan' (see on 2. 3. 6 for his exegesis of Jer. 2: 27); (3) see the similar phrase emphasized in quotation below from *doctr. chr.* This is the most autobiographical sentence in the paragraph, for as a gentile A. was paradoxically less vulnerable to the seductions of idolatry than the Jews had been.

aurum: The Egyptian gold is Platonism; that equation is the clearest mark of the distance and reserve with which A., writing *c.*397, means to treat the encounter with the *platonicorum libri*. Ex. 3: 22, 'et spoliabitis Aegyptum' (cf. Ex. 11: 2, 'ut postulet vir ab amico suo et mulier a vicina sua vasa argentea et aurea'). A.'s authoritative interpretation, not long before *conf.*, is *doctr. chr.* 2. 40. 60, where the Platonists are named: 'philosophi autem qui vocantur, si qua forte vera et fidei nostrae adcommodata dixerunt, *maxime platonici*, non solum formidanda non sunt, sed ab eis etiam tamquam ab iniustis possessoribus in usum nostrum vindicanda. sicut enim Aegyptii non tantum idola habebant et onera gravia, quae populus Israhel detestaretur et fugeret, sed etiam vasa atque ornamenta de auro et argento et vestem, quae ille populus exiens de Aegypto sibi potius tamquam ad usum meliorem clanculo vindicavit, non auctoritate propria, sed praecepto dei ipsis Aegyptiis nescienter commodantibus ea quibus non bene utebantur, sic doctrinae omnes gentilium non solum simulata et superstitiosa figmenta gravesque sarcinas supervacanei laboris habent, quae *unusquisque nostrum duce Christo de societate gentilium exiens* debet abominari atque vitare, sed etiam liberales disciplinas usui veritatis aptiores et quaedam morum praecepta utilissima continent deque ipso uno deo colendo nonnulla vera inveniuntur apud eos, quod eorum tamquam aurum et argentum, quod non ipsi instituerunt, sed de quibusdam quasi metallis divinae providentiae, quae ubique infusa est, eruerunt et, quo perverse atque iniuriose ad obsequia daemonum abutuntur, cum ab eorum misera societate sese animo separat, debet ab eis auferre christianus ad usum iustum praedicandi evangelii.' See G. Folliet's note at *BA* 11. 582–4, citing *c. Faust.* 22. 91, which itself refers back to the *doctr. chr.* discussion; he identifies parallels in Origen, quoted by the Cappadocians, but see B. Altaner, *Kleine Patristische Schriften*, 197–200, arguing prudently that A. did not know the more famous Origenian model and tracing here the influence of the Latin Irenaeus (Ir. *haer.* 4. 30). For the history of 'Egyptian gold' in Christian thought, see Holte, *Béatitude et Sagesse*, 111–24; Holte may be right (122) that the interpretation was a *topos* by this time independent of any specific textual source.

et dixit Atheniensibus: Act. 17: 27–8, 'quaerere deum si forte attractent eum aut inveniant, quamvis non longe sit ab unoquoque nostrum; (28) in ipso enim vivimus et movemur et sumus, sicut et quidam vestrum poetarum dixerunt: "ipsius enim et genus sumus".' The quotation is in the words here marked and comes from Aratus' *phaenomena*. A. took the attribution to apply to the preceding words, while omitting the actual quotation (and not only here: *exp. prop. Rom.* 3 (juxtaposed with Rom. 1: 21–3), *c. litt. Pet.* 2. 30. 69, *un. bapt.* 4. 6 (with Rom. 1: 18, 21), *c. Gaud.* 2. 10. 11 (with Rom. 1: 18), *s. Mai* 126. 6 (with Rom. 1: 20–1), *c. adv. leg.* 2. 4. 13, *civ.* 8. 10 (with Rom. 1: 19 ff.)—

listed by Madec, *REAug* 16 (1970), 91–2). Here it gives him Pauline warrant for using non-Christian texts; he was not alone—cf. Athanasius *incarn.* 42. 4. F. Chatillon, *RMAL* 1 (1945), 287–305, thought the misreading deliberate, but see Courcelle, *Recherches*, 131–2, tracing Ambrosian influence, and G. Folliet, *REAug* 11 (1965), 293–5: Amb. *bono mort.* 12. 55, 'quod ideo frequentavit, ut scires hinc bonum illud philosophos transtulisse, quod summum adserunt; pande ergo illud vere bonum tutum, illud divinum, "in quo vivimus et sumus et movemur".' The link to Rom. 1: 18 ff. is not in Ambrose.

quidam secundum eos: The plain meaning for A. must have been 'some of their authorities' and not 'some among them', but it is far from clear whether he had any idea (inevitably wrong) whom Paul had in mind; the debate over this phrase (between G. Madec, *REAug* 16 (1970), 79–137, and O'Connell, *REAug* 19 (1973), 91–6) hangs on an ingenious, and mistaken, attempt of O'Connell's to put a name to *quidam*. But if A. did not have so literal-minded a notion, then he may simply never have lingered over the difficulty of the phrase.

et utique inde erant illi libri: There are three possible reasons why A. would claim the *platonicorum libri* came from Egypt.

(1) The exegesis of the Egyptian gold (see above).

(2) A. believed, following Ambrose, that Plato himself had derived his teaching from contact with Jewish scripture during a sojourn in Egypt. A. thought he recalled—as late as 397 and so perhaps even while writing *conf.*—that Ambrose had said that Plato met Jeremiah personally and was instructed by him. The notion of influence goes back to Hellenistic Judaism, and was sometimes admitted by non-Christians (A. Festugière, *La Révélation d'Hermès Trismégiste* (Paris, 1944), 1. 19–25).[39] But *retr.* 2. 4. 2, 'et in eo quod dixi de temporum historia sanctum Ambrosium solvisse quaestionem, tamquam coaetanei fuerint Plato et Hieremias, me fefellit memoria.[40] nam quid ille de hac re episcopus dixerit, in libro eius legitur quem de sacramentis sive de philosophia scripsit.' The work in question is the *de philosophia sive de sacramento regenerationis*, which A. finally acquired from Paulinus in late 390s (*ep.* 31. 8—Courcelle, *Les Confessions*, 601, dates that letter to early 397). The same argument is present in other works of Ambrose, but often only in fleeting mention (e.g. *Noe* 8. 24), and apparently the discussion in the lost work was the most extensive.

[39] A. would certainly have read this passage of Cicero as an admission of dependence: *fin.* 5. 29. 87, 'nisi enim id faceret, cur Plato Aegyptum peragravit ut a *sacerdotibus barbaris* numeros et caelestia acciperet?' Cicero also exemplifies a similar claim of unattested influence, when he suggests that the earliest Romans owed some of their wisdom to the teachings of Pythagoras, of whose activities in southern Italy they surely must have heard: *Tusc.* 4. 1. 2–4. 2. 4.

[40] This throws light on the question of just how accurately A. remembered what he had heard from Ambrose.

The error corrected by the *retr.* passage appears at *doctr. chr.* 2. 28. 43, 'quantam noster Ambrosius quaestionem solvit calumniantibus Platonis lectoribus et dilectoribus, qui dicere ausi sunt omnes domini nostri Iesu Christi sententias, quas mirari et praedicare coguntur, de Platonis libris eum didicisse, quoniam longe ante humanum adventum domini Platonem fuisse, negari non potest! nonne memoratus episcopus, considerata historia gentium, cum reperisset Platonem Hieremiae temporibus profectum fuisse in Aegyptum, ubi propheta ille tunc erat, probabilius esse ostendit quod Plato potius nostris litteris per Hieremiam fuerit imbutus, ut illa posset docere vel scribere quae iure laudantur?' A. had corrected his view by the time of *civ.* 8. 11, and goes on at 8. 12 to say that wherever Plato may have derived these doctrines the analysis and judgement of Rom. 1: 20 ff. are apposite.

(3) The most literal-minded explanation is that Plotinus himself was born in Egypt and studied at Alexandria with Ammonius Saccas (e.g. Mayer, *Zeichen* 1. 131, du Roy 68). This weakest argument is regarded as sufficient by the *doctiores*. Courcelle, *Les Confessions*, 87–8, develops it with equal literal-mindedness by noting that 8. 2. 3 mentions Isiac worship as a subject of Victorinus' attentions and then assuming that A. has this connection in mind here: 'Voilà qui éclaire, à mon avis, le passage fameux du livre VII, où apparaît la phrase, "et utique *inde* erant illi libri".' On this view A. passes a general condemnation on their authors: Plotinus, Porphyry, Iamblichus, and Victorinus as 'pagans' have put their talents to the service of Egyptian idols. Instead of opposing zoolatry, each one has been an initiate, or an interpreter, of Isiac mysteries. In a similar vein, see E. Gilson, *Med. Stud.* 8 (1946), 43–52, P. Henry, *Plotin et l'Occident* (Louvain, 1934), 98 (noting an analogy to Plot. 6. 9. 9. 9–11). Madec, *RA* 2 (1962), 279, follows O'Meara, *Porphyry's Philosophy from Oracles* (Paris, 1959), 151–76, in seeing Porphyry as a source infected with the Isiac virus; du Roy, 68, follows Courcelle; Mandouze, 502 n. 3, is more prudent: 'La seule chose sûre, c'est que . . . ce "détournement" (*inde*) du platonisme dans le sens du christianisme est considéré, par le biais d'une meilleure utilisation des richesses de celui-là, comme une restitution à celui-ci.' TeSelle, 242–51, offers a yet more nuanced version of the literal reading, seeing not a reaction to Porphyry but the influence of Ambrosiaster.

qui transmutaverunt: Rom. 1: 25 (for text see on 7. 9. 14).

7. 10. 16

This paragraph contains the first of two (by some counts: three) accounts of what Courcelle defined as 'vaines tentatives d'extases plotiniennes' (*Recherches*, 157–67, with sustained discussion, and cf. *Les Confessions*, 17–88, and *Recherches*, 2nd edn., 2, 405–40). All discussions of A.'s Platonic indebtednesses since

Courcelle have returned to these paragraphs, in which we get our fullest—if not clearest or most concrete—view of the impact the *platonicorum libri* had on him.

Courcelle's views are most concisely presented at *Recherches*, 160–4, where he presents three passages—7. 10. 16 (a), 7. 17. 23 (b), and 7. 20. 26 (c)—in parallel columns, marking seven stages, which he identifies:

(1) reading the *platonicorum libri* (ac),
(2) searching for truth and finding it above himself (abc),
(3) ascent by degrees (b),
(4) Rom. 1: 20 (bc),
(5) sense of certainty but incapacity (abc),
(6) 'faiblesse' (abc),
(7) Rom. 1: 20 (a).

This schema highlights the parallels among these paragraphs, but imposes a structure that should not be taken as originating with A.

The view taken in this commentary differs from Courcelle's in several regards (most importantly, seeing only two 'tentatives' (7. 10. 16 and 7. 17. 23) and seeing significant differences between them), and differs variously from most later studies, but can be summarized in a few words. This paragraph (7. 10. 16) presents A. seeking the ecstasy that Plotinus taught comes from the ascent of mind to union with highest being; this attempt ends in failure. The following paragraphs (7. 11. 17–16. 22) represent further intellectual ruminations at the time, which make possible a renewed attempt at union, which is described in 7. 17. 23.[41] This second attempt is successful—*on Plotinian terms as A. understood them*. But that success leaves A. disappointed and frustrated, and the remaining paragraphs of Bk. 7 (7. 18. 24–21. 27) diagnose the remaining problem and limn the treatment that is developed in Bk. 8. These events finally put A. beyond the reach of all the main Manichean objections to Christianity that had perplexed him for years (see on 3. 7. 12 and on 4. 15. 24 (where the theme of the ascent was adumbrated)); the intellectual demolition of those objections makes possible the neo-Platonic ascent reported here. As often, A. creates problems for himself with modern readers by being inadequately narrow-minded. He gives in these pages high praise, both implicit and explicit, to Plotinian doctrines and practices, but at the same time he reserves for neo-Platonism criticism that is fundamental, irrefutable (for him), and disqualifying.

We owe two important general observations about this episode to Courcelle—*Recherches*, 159: 'le récit des Confessions nous présente, non une scène déterminée, comme à Ostie, avec la date, le cadre, les circonstances

[41] See Theiler, *P.u.A.* 62, reprised in his review of Courcelle, *Recherches*, at *Gnomon*, 25 (1953), 120, holding that the first 'tentative' here in 7. 10. 16 was a success in so far as it brought an answer (reported in the paragraphs to follow) to the question 'unde malum', while the second went further. Cf. also TeSelle, 35.

précises, mais plutot une série d'élévations accompagnées des réflexions d'Augustin évêque', and *Recherches*, 160: 'Une large part des nombreuses citations des Écritures paraît porter la marque de l'évêque d'Hippone, plutôt que celle du rhéteur milanais. . . . L'absence de toute circonstance particulière rend plus probable qu'il s'agit, non d'une scène unique, mais d'une série de tentatives analogues.' The clear message of the text is that, whatever the details, the reading of the *platonicorum libri* and the experiences to which that reading led were felt by A. at the time both as a great and liberating revolution in his life and at the same time as less than completely satisfactory. The theological *diagnosis* of his dissatisfaction (failure to appreciate the doctrine of the incarnation: see on 7. 18. 24) may be in part a piece of retrospective analysis; the *prescription* (to espouse a life of continence and take baptism) is entirely Milanese and anything but anachronistic (see Bk. 8 generally, and see on 8. 1. 2 particularly).

The echoes of Plotinian texts in these paragraphs have attracted much attention. Courcelle (*Recherches*, 167) does not insist on more than Plot. 1. 6, 1. 8, and 5. 1 along with Porphyry, *de regressu animae*, but debate continues. The argument that these pages contain 'tentatives d'extase' implies that there was a success that A. did not find. But was the Ascent ever meant to be a permanent thing in Plotinus? Or was it inevitably partial, frustrated, and limited in duration? Plot. 1. 6. 9 seems to say a lasting achievement is possible, but 6. 9. 10. 1–3 recognizes that in practice it does not happen often: *πῶς οὖν οὐ μένει ἐκεῖ; ἢ ὅτι μήπω ἐξελήλυθεν ὅλος. ἔσται δὲ ὅτε καὶ τὸ συνεχὲς ἔσται τῆς θέας οὐκέτι ἐνοχλουμένῳ οὐδεμίαν ἐνόχλησιν τοῦ σώματος.* There permanence of vision seems eschatological rather than imminent. 6. 9. 11. 45–51 ends the whole collection there: *ἐκπίπτων δὲ τῆς θέας πάλιν ἐγείρας ἀρετὴν τὴν ἐν αὐτῷ καὶ κατανοήσας ἑαυτὸν ταύταις κεκοσμημένον πάλιν κουφισθήσεται δι' ἀρετῆς ἐπὶ νοῦν ἰὼν καὶ σοφίαν καὶ διὰ σοφίας ἐπ' αὐτό. καὶ οὗτος θεῶν καὶ ἀνθρώπων θείων καὶ εὐδαιμόνων βίος, ἀπαλλαγὴ τῶν ἄλλων τῶν τῇδε, βίος ἀνήδονος τῶν τῇδε, φυγὴ μόνου πρὸς μόνον.*

Of no less importance, however, for understanding the exact purport of these paragraphs and their place in A.'s thought are the parallel texts in other works of A. himself. P. Hadot, *Marius Victorinus* (Paris, 1971), 208, adduces several passages later in *conf.* (12. 11. 11–14, 12. 15. 18–22, 10. 6. 9–25. 36 (see below on 10. 1. 1)) and suggests *vera rel.* 3. 3 (beginning with a rejection of those 'qui eorum [platonicorum] libros pervicaciter diligunt', and ending at 3. 4 with a reiteration of Jn. 1: 1 for the benefit of the Platonists and the citation of 1 Jn. 2: 16, which is important to the rest of *vera rel.*) and *civ.* 8. 6 (which describes Platonic vision in the third person, that is, as something the Platonists themselves have known, with no personal link to A.: 'cuncta corpora transcenderunt quaerentes deum . . .') as well; he holds that though Plotinian phrases occur, the structure of these passages is not Plotinian. But there are numerous other pertinent texts, arranged here by subject:

Experience at Milan: beata v. 1. 4, *c. acad.* 2. 2. 5, both quoted in Introduction. Both these texts show A. connecting, in vivid and recent memory, the reading of the *platonicorum libri* with his disaffection from the lures of *ambitio saeculi*; but in the rhetorical pattern of *conf.*, that task has been accomplished in Bk. 6, and here the *platonicorum libri* work against the power of *concupiscentia oculorum*; the first passage also shows, incidentally, that in November 386 A. already described the books of Plotinus in juxtaposition to ecclesiastical authority.

Preparation for vision: sol. 2. 20. 34, 'exercitationem tuam, ut illud videre sis idoneus, operamur'; *quant. an.* 15. 25 and 33. 75 (quoted below; *quant. an.* 33. 70–6 has an extended discussion of an ascent in seven stages, but this does not exactly parallel what appears in these paragraphs). These texts indirectly reflect on the background both to this paragraph and to 7. 17. 23. Each makes sense only when seen against the development that leads to it (and 7. 11. 17–16. 22 are thus of great importance).

Defects of vision (and see below): *beata v.* 4. 35, 'adhuc vel minus sanis vel repente apertis oculis audacter converti et totum intueri trepidamus ... quamdiu quaerimus, nondum ipso fonte atque, ut illo verbo utar, plenitudine saturati nondum ... sapientes ac beati sumus.'

Later judgement of vision: On the opening page of *trin.* (which may date to shortly after *conf.*), A. takes a reserved view of the philosophical ascent of the mind. There the ascent is a possibility without authentic Christianity, but not destined for success: *trin.* 1. 1. 1, 'est item aliud hominum genus, eorum qui universam quidem creaturam, quae profecto mutabilis est, *nituntur transcendere* ut *ad incommutabilem substantiam* quae deus est *erigant intentionem;* sed *mortalitatis onere praegravati*, cum et videri volunt scire quod nesciunt et quod volunt scire non possunt, *praesumptiones opinionum suarum audacius affirmando* [cf. 7. 20. 26, 'inter praesumptionem et confessionem'] intercludunt sibimet intellegentiae vias, magis eligentes sententiam suam non corrigere perversam quam mutare defensam.' *trin.* 1. 1. 2 continues: 'ut ergo ab huiusmodi falsitatibus humanus animus purgaretur, sancta scriptura parvulis congruens nullius generis rerum verba vitavit, ex quibus quasi *gradatim* ad divina atque sublimia noster intellectus velut nutritus assurgeret. ... perniciosius et inanius evanescunt [cf. Rom. 1: 21] qui tertio illo genere erroris a veritate secluduntur, hoc suspicando de deo, quod neque in ipso neque in ulla creatura inveniri potest.'

Finally, what is at stake is not strictly an 'ascent' but an 'assumption' ('adsumpsisti'). The notion of divine initiative is certainly more Christian than Platonic (note here 'admonitus', and cf. 1. 1. 1, on the anteriority of *praedicatio* to human faith, prayer, invocation, and knowledge). The first sentence of this paragraph, for example, has three different phrases for the divine initiative. Simone Weil claims somewhere that there is no case in the Gospels of human initiative but always the divine call. This is not quite true, but it captures a distinctive atmosphere alien to that of Plotinus.

admonitus: Mayer, *Zeichen*, 1. 241, for whom *admonitio* is not merely important, but central: 'So widersprüchlich das klingt, *die Zeichen haben keine semantische Funktion*. . . . Worin besteht die utilitas verborum (signorum)? Die Antwort Augustins—sie ist im ganzen Dialog [*mag.*] verstreut—lautet: Im *admonere*. Die signa sind Warner und Mahner; ihre gnoseologische Funktion besteht in der commemoratio sensibilis des homo interior.' *mag.* 11. 38, 'intus ipsi menti praesidentem consulimus veritatem, verbis fortasse ut consulamus admoniti', *lib. arb.* 2. 14. 38, 'foris admonet intus docet'; sim. at *mor.* 2. 17. 55, *s.* 264. 4 (in a discussion with parallels to this whole section of *conf.* 7). 'Admonitio' in A.'s earliest works is especially the function of the Spirit (*beata v.* 4. 35, *sol.* 1. 1. 2 (quoted in next note)), but then the task largely shifts to the second person of the trinity (du Roy, 161–5; see also G. Madec, 'Admonitio', *Aug.-Lex.* 1. 95–9). Whether to attach the present passage to either divine person is questionable (but cf. 3. 4. 8, 'admonitio *spiritus* tui'). 'Admonitio' 16 × otherwise in *conf.*

redire: See on 1. 18. 28, and add here *sol.* 1. 1. 2, 'deus . . . pater pignoris quo admonemur redire ad te', *c. acad.* 2. 2. 5, 'prorsus totus in me cursim redibam', and *sol.* 2. 6. 9, quoted below; *ss.* 96. 2. 2, 330. 3, *s. Caill.* 2. 11. 4 ff., *qu. ev.* 2. 33, *civ.* 11. 28 (the last three dealing with the prodigal); on neo-Platonic parallels (cf. Plotinus 1. 6. 9. 7, *ἄναγε ἐπὶ σαυτὸν καὶ ἴδε*) and for a neo-Platonic and Stoic reading of A.'s exegesis, see Courcelle, *Connais-toi toi-même* (Paris, 1974), 128–63.

intima: As the site of such vision: *ep.* 10. 3 (more optimistic than the present passage), 'cur aliquando evenit ista securitas? cur tanto evenit crebrius, quanto quisque in mentis penetralibus adorat deum?' See on 7. 16. 22, 'proicientis intima sua'.

duce te: As Courcelle, *Recherches*, 128, admits, though this passage has echoes of Plotinus 1. 6 (Courcelle: 'une page qui suit de près' Plot. 1. 6. 9), this phrase is the opposite of Plot.'s (1. 6. 9. 24) *μηκέτι τοῦ δεικνύντος δεηθεὶς ἀτενίσας ἴδε* (and indeed, as Courcelle, *Recherches*, 116, noted, when Ambrose was following Plot. 1. 6 in his own *de Isaac*, he omitted any advice to such self-reliance). Cf. *sol.* 2. 6. 9, 'deus, pater noster, qui ut oremus hortaris, qui et hoc quod rogaris praestas, siquidem, cum te rogamus, melius vivimus melioresque sumus, exaudi me palpitantem in his tenebris et mihi dexteram porrige. praetende mihi lumen tuum, revoca me ab erroribus; *te duce* in me redeam et in te'; *mag.* 8. 21, 'et tamen, si dicam vitam esse quandam beatam eandemque sempiternam, quo nos *deo duce* id est ipsa veritate gradibus quibusdam infirmo gressui nostro accomodatis perduci cupiam, vereor, ne ridiculus videar, qui non rerum ipsarum quae significantur sed signorum consideratione tantam viam ingredi coeperim'; *conf.* 2. 7. 15, 'te duce'. Amb. *ep.* 37. 29 (to Simplicianus), 'solus igitur sapiens, qui *duce deo* usus est' (quoted from Philo's 'quod omnis probus liber sit'); other Philonic passages in the same sense quoted by Courcelle, *Les Confessions*, 51 n. 5.

adiutor meus: See on 7. 7. 11.

vidi: The *collyrium* (7. 8. 12) has evidently worked in part, but the vision acquired is tenuous: see below. See on 1. 6. 9, but notice that this particular form occurs 2 × in Bk. 1 (1. 6. 9, 1. 7. 10), then not again until Bk. 7, where it is frequent (7. 11. 17, 7. 12. 18, 7. 14. 20, 7. 15. 21, 7. 15. 21) in these mystical chapters, and then not again until Bk. 10's mystical chapters (10. 3. 4, 10. 8. 15, 10. 8. 15, 10. 10. 17, 10. 12. 19, 10. 16. 25, 10. 21. 30, 10. 41. 66, 'vidi enim splendorem tuum corde saucio') that parallel the Ostia 'vision'; after that, it occurs only once (11. 24. 31), in a negative protasis.

For the purposes of the present passage, of 7. 17. 23, of 9. 10. 23–5, and of 10. 1. 1–27. 38, it is useful to consider A.'s theory (most fully propounded at *Gn. litt.* 12. 7. 16 ff.: on this see M. Korger, *RA* 2 (1962), 33–57, and the note at *BA* 49. 575–85) of three kinds of vision: corporeal, spiritual, and intellectual. *Gn. litt.* 12 is not surely datable before 412–15, but this theory was already present at *c. Adim.* 28. 2 (393–6), 'nam multa genera visionis in scripturis sanctis inveniuntur: unum secundum oculos corporis; sicut vidit Abraham tres viros sub ilice Mambre. . . . alterum secundum spiritum quo imaginamur ea quae per corpus sentimus; nam et ipsa pars nostra cum divinitus *adsumitur*, multa ei revelantur non per oculos corporis aut aures aliumve sensum carnalem, sed tamen his similia: sicut vidit Petrus discum illum submitti e caelo. . . . tertium autem genus visionis est secundum mentis intuitum, quo *intellecta conspicitur* veritas atque sapientia: sine quo genere illa duo quae prius exposui vel infructuosa sunt vel etiam in errorem mittunt. . . . ex hoc tertio genere est illa visio quam commemoravi dicente apostolo, invisibilia enim dei a constitutione mundi per ea quae facta sunt intellecta conspiciuntur.' In principle, the third type of vision is infallible (*Gn. litt.* 12. 25. 52, 'in illis intellectualibus visis non fallitur'). The result can be dramatically expressed: see *Gn. litt.* 12. 26. 54, quoted in introd. n. to Bk. 10. On this reading, it makes sense to say that the failed ascent of Bk. 4 (see on 4. 13. 20) is a corporeal vision of the first type; the ascents of Bk. 7 culminating in 7. 17. 23 are spiritual visions of the second type; and the ascent of Ostia at 9. 10. 23–5 is an intellectual vision of the third type, as are by implication the later ascents of Bks. 10 and 11–13, and by intention those of later works of A.'s that exemplify the method of the 'ascent'.

oculo animae meae: See on 12. 20. 29, 'interiori oculo'.

lucem incommutabilem: Neo-Platonic sources aside, what are we invited, by the text of 7. 9. 13–15, to think this light must be? 7. 9. 13 (< Jn. 1: 9), 'sed verbum deus est lumen verum, quod inluminat omnem hominem venientem in hunc mundum' (cf. *Io. ev. tr.* 13. 5, and—quoted below—*trin.* 8. 2. 3 and *s.* 52. 6. 16). Cf. *sol.* 1. 1. 3, 'deus intellegibilis lux, in quo et a quo et per quem intellegibiliter lucent, quae intellegibiliter lucent omnia.'

oleum super aquam: See *en. Ps.* 29. en. 2. 10, quoted on 13. 9. 10.

ipsa fecit me: Cf. Ps. 99: 3, 'scitote quoniam dominus ipse est deus; ipse fecit nos, et non nos', and see on 9. 10. 25 and 10. 6. 9, where this verse recurs in later versions of the ascent.

qui novit veritatem [2], **novit eam, et qui novit eam, novit aeternitatem** [1]: The trinitarian interpretation is confirmed by the subtext, Jn. 14: 7, 'si cognovistis me, et patrem meum utique cognovistis'. The verse immediately preceding, Jn. 14: 6, is the familiar 'ego sum via et veritas et vita', so the equivalences are easy ones for A. These scriptural echoes at the centre of the 'vision' arouse suspicion: is A. Christianizing what happened *ex post facto*? He is at least explaining it in Christian terms.

caritas [3]: Cf. 'contremui *amore*'. Courcelle, *Recherches*, 129 n. 2, finds parallels in both Plotinus (1. 6. 7) and Ambrose (*Isaac* 8. 78, 'summum illud bonum, cuius nobis accenditur caritas et desiderium'), but once again the question of sources must be joined by the question of context. For A. in *conf.*, *caritas* is a distinct signpost for the third person of the trinity, and thus completes a pattern from above, which is renewed in what follows. See on 13. 7. 8.

o aeterna veritas [2] **et vera caritas** [3] **et cara aeternitas** [1]: A common triad: *cat. rud.* 26. 50, 'hoc autem ita breviter discit, ut quidquid audierit ex libris canonicis quod ad dilectionem aeternitatis [1] et veritatis [2] et sanctitatis [3] et ad dilectionem proximi referre non possit, figurate dictum vel gestum esse credat; atque ita conetur intellegere, ut ad illam geminam referat dilectionem'; *Gn. litt.* 8. 25. 47, 'aeternitate, veritate, caritate creatoris' (at *Gn. litt.* 8. 24. 45, 8. 26. 48, the triad is *aeternitas*, *veritas*, *voluntas*); again at *trin.* 4. proem, 4. 21. 30, *ep.* 169. 1. 4, 'ad eandem perveniant aeternitatem [1] veritatem [2] caritatem [3], id est ad stabilem [1] certam [2] plenamque [3] felicitatem, ubi manentibus [1] videntibus [2] amantibus [3] sint cuncta perspicua'; *civ.* 11. 28, 'quoniam igitur homines sumus ad nostri creatoris imaginem creati, cuius est vera aeternitas, aeterna veritas, aeterna et vera caritas, estque ipse aeterna et vera et cara trinitas neque confusa neque separata'; *civ.* 12. 1, *Io. ev. tr.* 97. 1 ('de dei aeternitate, veritate, sanctitate'); *s.* 71. 12. 18 (quoted on 11. 1. 1: *aeternitas/ veritas/ bonitas*); adumbrated at *mor.* 1. 30. 62, *aeternitas/ veritas/ pax*. This phrase is the focus of an important essay on the nature of mystical knowledge in A., exemplified by texts from throughout *conf.*: P. Blanchard, *RA* 2 (1962), 311–30; it has also received attention for its rhetorical *gradatio* in a delightful notice by E. Dutoit, *Augustinus* 13 (1968), 153–66.

The received punctuation in editions from the Maurists to Verheijen places an exclamation after 'cara aeternitas': the comma printed here connects with the following phrases. Knauer, 186 n. 1, would also put a comma after 'nocte', while du Roy, 75 n. 3, would put a period after 'tu es deus meus' (his disjunction of that line from the following is undermined by 10. 34. 53, quoted below). Cf. Ps. 42: 2, 'tu es deus meus et fortitudo mea.'

tibi suspiro die ac nocte: This line has the sound of a biblical echo, but it is not; du Roy, 76 n. 1, cites Ps. 41:4 (a good Psalm for mystical ascents), 'fuerunt mihi lacrimae meae panis die ac nocte' (*en. Ps.* 41. 6, 'sive totum tempus accipias diem et noctem; sive diem intellegas pro huius saeculi prosperitate, noctem vero pro huius saeculi adversitate' (sim. on 'die ac nocte' at *en. Ps.* 1. 2 on Ps. 1: 2)), but could as easily have cited Ps. 1: 2, 'et in lege eius meditabitur die ac nocte', and Jer. 9: 1, 'et plorabo die ac nocte.' *en. Ps.* 127. 10, 'redis ergo, et plangis ad deum; quia illi *suspiras,* antequam videas, et desiderio ipsius gemis; et quia in ipsius desiderio ploras, dulces sunt et ipsae lacrimae, et pro cibo tibi erunt, quia factae sunt tibi et ipsae panis die ac nocte.' See on 3. 6. 10, 'suspirabant tibi', and cf. 10. 34. 53, 'ab illa pulchritudine veniunt quae supra animas est, cui *suspirat* anima mea die ac nocte', 6. 5. 8, 6. 10. 17, 13. 13. 14.

tu adsumpsisti me: Cf. Ps. 26: 10, 'pater meus et mater mea dereliquerunt me, dominus autem adsumpsit me'; cf. *div. qu. Simp.* 2. 1. 1, 'spiritus hominis divino spiritu adsumptus', and see *Io. ev. tr.* 18. 11, quoted below on 'reverberasti'. See *BA* 13. 616–17; cf. *s.* 7. 7, 'qui enim hoc quod est et vere est digne intellexerit, et qualitercumque lumine veracissimae essentiae, vel strictim sicut coruscatione adflatus fuerit, longe se videt infra, longe remotissimum, longe *dissimillimum* sicut ille ait: "ego dixi in ecstasi mea." [Ps. 30: 23]. *adsumpta enim mente vidit* nescio quid quod plus ad illum erat.' *cons. ev.* 4. 10. 20, 'quisquis autem arbitratur homini vitam istam mortalem adhuc agenti posse contingere, ut dimoto atque discusso omni nubilo phantasiarum corporalium atque carnalium serenissima incommutabilis veritatis luce potiatur et mente penitus a consuetudine vitae huius alienata illi constanter et indeclinabiliter haereat, nec quid quaerat nec quis quaerat intellegit'; on the opacity of the self to the self, see on 10. 8. 15. Du Roy, 76 n. 2, wrestles with translation of the indirect discourse, criticizes *BA* ('que je n'étais pas encore être à le voir') and settles for 'pour me faire voir qu'*existe* ce qu'il y a à contempler, mais je n'étais pas encore capable.' See O'Connell, *St. Augustine's Confessions* (Cambridge, Mass., 1969), 1–3. Cf. Plot. 5. 1. 2. 12, ψυχὴ . . . ἀξία τοῦ σκοπεῖν. See also the close, but not identical, expression at 7. 17. 23 ('sed nondum me esse qui cohaererem'), with notes there.

qui viderem: For the form of expression, cf. *en. Ps.* 58. s. 2. 11, 'ut essem qui te invocarem, quid egi?'

reverberasti: The image is visual (for the underlying theory of vision, see on 10. 6. 9); cf. 9. 11. 27, '[Monnica] reverberans eum oculis': there in a purely human setting, the one who looks has the effect on the one looked at—the *radii* from her eyes defeat those from the other's eyes.

The word evokes a remarkable range of parallel texts (see du Roy, 77), including Ambrosian antecedents: e.g. Amb. *bono mort.* 11. 49, 'si solis radios oculi nostri ferre non possunt et si quis diutius e regione solis intenderit, caecari solere perhibetur; si creatura creaturam sine fraude atque offensione sui non potest intueri, quomodo potest sine periculo sui vibrantem cernere vultum creatoris aeterni, corporis huius opertus exuviis?'; see also Amb. *Abr.* 2. 4. 16, *in Luc.* 7. 17, *exam.* 4. 1. 1. Testard, 166–7, refers to Cic. *Tusc.* 1. 30. 73, 'sic mentis acies seipsa intuens non numquam hebescit, ob eamque causam contemplandi diligentiam amittimus' (that book of *Tusc.* is apposite in subject to that of Amb.'s *bono mort.*, and it would not be surprising if Amb.'s treatise had sent A. and some of his friends to compare Cicero's discussion).

The texts that follow all represent the difficulty of approach, the likelihood of being repelled, and the need for moral purification if the approach is to be (more or less) successful. As such, they apply in some ways to both the 'tentatives d'extase' of this book (here and 7. 17. 23), and do not address the distinction between the attempt with no success (as here) and the attempt with fleeting success (7. 17. 23). The verbal parallels among these texts are numerous; only those are highlighted that reflect the present text more or less directly. The possibility of ascent was one A. retained; see du Roy, 78 n. 1, quoting *en. Ps.* 134. 6 (see below on 7. 17. 23).

The difficulty of ascent and the impossibility of persevering in the vision to which it led were scarcely a surprise to A. at the time, and scarcely in themselves signs that the 'tentative' had failed. Both the Platonic tradition and A.'s later comments insist that relapse into the material world is inevitable. More pertinently, that relapse is an essential feature of the narrative of what is by all agreed to be the most satisfactory mystical ascent of *conf.*, that of Ostia at 9. 10. 25.

sol. 1. 9. 16–1. 10. 17, 'quaenam ergo *talium oculorum impudentia* est, velle illum solem videre? (17) . . . nonne vides hos corporis oculos etiam sanos luce solis istius saepe *repercuti* et averti atque ad illa sua obscura confugere?'; *sol.* 1. 13. 23; *quant. an.* 33. 75, 'serenum atque rectum aspectum in id quod videndum est dirigere. quod qui prius volunt facere quam mundati et sanati fuerint, ita illa luce *reverberantur* veritatis, ut . . . cum quadam libidine et voluptate miserabili in suas tenebras . . . refugiant' (sim. at *quant. an.* 15. 25); *mor.* 1. 7. 11, '[ratio,] ubi ad divinum perventum est, avertit sese; intueri non potest, palpitat, aestuat, inhiat amore, *reverberatur luce veritatis*'; *lib. arb.* 2. 16. 42, '*reverberat*'; so later *vera rel.* 20. 39, *en. Ps.* 5. 6, *c. Faust.* 15. 8, '*reverberati*'; *s.* 293. 4, 'illi continuo repercussi, . . . ad sui cordis tenebras confugerunt'; *en. Ps.* 145. 5, '*coruscatione* quadam *perstringitur*'; *cons. ev.* 4. 10. 20, '[anima] si quando adiuta excedit . . . hanc carnalem caliginem . . . tamquam *rapida coruscatione perstringitur* et in suam infirmitatem redit vivente desiderio quo rursus erigatur, nec *sufficiente munditia* qua figatur'; *Io. ev. tr.* 18. 11, 'si utcumque erexistis cor vestrum ad *videndum verbum,* et ipsius luce *reverberati* ad solita recidistis, rogate medicum ut adhibeat *collyria* mordacia, praecepta iustitiae. . . . *scis certe esse quod videas, sed idoneum non te esse qui videas*'; *en. Ps.* 41. 10, *trin.* 8. 2. 3, 'ecce in ipso primo *ictu* quo velut *coruscatione* *perstringeris*, cum dicitur, "veritas", mane si potes: sed non potes. relaberis in ista solita atque terrena. quo tandum pondere [see on 13. 9. 10], quaeso, relaberis, nisi sordium contractarum cupiditatis visco et peregrinationis erroribus'; *trin.* 12. 14. 23, 'sed veluti acie ipsa *reverberata* repellitur et fit rei non transitoriae transitoria cogitatio'; *trin.* 15. 6. 10, 'sed quia lux illa ineffabilis nostrum *reverberabat* obtutum [i.e. in the earlier books of *trin.*], . . . in creatura . . . immorati sumus a nono usque ad quartum decimum librum' (thus the fifteenth book is the renewed attempt at direct vision—also a failure); *ep.* 119. 2 (Consentius of Minorca (who had read *conf.*: *ep.* 12*. 1) writing to A.), 'nisi tu tantae rei dux ac magister adfueris, velut lippientibus oculis prospicere in eam tanti luminis *repercussa fulgore* cogitatio nostra formidat'; *s.* 52. 6. 16.

radians: Of 'radiating light' at *Io. ev. tr.* 2. 7, 'plerumque fit ut in aliquo corpore radiato cognoscatur ortus esse sol, quem oculis videre non possumus'; elsewhere more often of 'directing the gaze' (cf. *Io. ep. tr.* 6. 10, 'dirige radium dextrum sine altero, si potes. simul coeunt, simul diriguntur'). See on 10. 6. 9.

contremui amore atque horrore: 9. 4. 9, 'inhorrui timendo. . . . audivi et contremui' (of reading Ps. 4 at Cassiciacum). Such a shudder is not necessarily a pious emotion: Jas. 2: 19, 'et daemones credunt et contremiscunt'; cf. Jer. 23: 9. O. Perler, *Unterwegs zur Einheit* (Festschrift H. Stirnimann: Friburg, Switzerland, 1980), 241–52, attempts to root this phrase in Plotinian and Christian texts; the results are not decisive for *this* phrase, but are of wider interest.

in regione dissimilitudinis: Platonic and Christian themes converge and nearly merge. See Lk. 15: 13, 'et non post multos dies congregatis omnibus adolescentior filius peregre profectus est in regionem longinquam et ibi dissipavit substantiam suam vivendo luxuriose', and cf. *BA* 13. 689–93. The formula comes from Plotinus (1. 8. 13. 16–17), ἐν τῷ τῆς ἀνομοιότητος τόπῳ, where it undoubtedly derives from Plato, *politicus* 273 D 6–E 1.[42] A. quotes Plotinus expressly at *civ.* 9. 17, 'ubi est illud Plotini,[43] ubi ait, "fugiendum est igitur ad carissimam patriam, et ibi pater, et ibi omnia. quae igitur, inquit, classis aut fuga? *similem deo fieri.*" si ergo deo quanto similior, tanto fit quisque propinquior: nulla est ab illo alia longinquitas quam eius dissimilitudo.' Similar dissimilitudes at *ss.* 171. 3. 3, 369. 2 (concluding that what is needed is a mediator, specifically the 'mediator dei et hominum, homo Christus Iesus' (1 Tim. 2: 5: see on 7. 18. 24)). As *BA* 13. 691 points out, A.'s perspective was different from that of Plotinus; for P., dissimilitude

[42] Many editions of Plato print πόντον in lieu of τόπον there, citing Proclus and Simplicius against all extant Plato manuscripts, making the phrase a 'sea of disresemblance'—in the form in which A. has it, then, the phrase is authentically Plotinian just in so far as it is defectively Platonic.

[43] A conflation of 1. 6. 8 + 1. 2. 3.

emphasizes the downward movement, but for A. it is felt on the reverse, on the ascent. There are links to the Christian doctrine of human creation in the 'image and likeness' of God as well: for Ambrose's teaching, see on 6. 3. 4, and for A.'s elaboration, see on 13. 22. 32.

For *regio* in *conf.*, see on 4. 16. 30. There are no immediate biblical parallels beyond the prodigal passage in Luke, but Mayer, *Zeichen*, 1. 147 n. 124, offers a range of analogous texts, e.g. Ps. 136: 4, 'terra aliena', Ps. 87: 13, 'terra oblivionis', et sim. Just as with 'reverberatur', the range of echoes in A.'s writings is considerable, but does not speak to the distinction between 7. 10. 16 and 7. 17. 23.

sol. 2. 7. 13, 'nonne similitudinem veritatis matrem et dissimilitudinem falsitatis esse fatendum est?'; developed notably in *vera rel.*, e.g. *vera rel.* 36. 66, 'ea forma [2] est omnium quae sunt, quae summa similitudo principii et veritas [2] est, quia sine ulla *dissimilitudine* est' (see du Roy, 335–8, and cf. *vera rel.* 55. 113); *div. qu.* 23, 'similitudo patris filius dicitur'; from much later, *Io. ev. tr.* 23. 8, 'recessisti quidem a pristina similitudine, sed adhuc iaces in magna dissimilitudine'; *trin.* 7. 6. 12, 'non enim locorum intervallis sed similitudine acceditur ad deum, et dissimilitudine receditur ab eo' (sim. at *s.* 171. 3. 3, *en. Ps.* 34. s. 2. 6, *en. Ps.* 94. 2 ('ergo si *dissimilitudine* recedimus a deo, similitudine accedimus ad deum'—if *similitudine* is read as = the second person, then that phrase takes on an important double meaning), 99. 5, 146. 14). With a link to the second person of the trinity, we can see again the appositeness of *lib. arb.* 2. 11. 30 (quoted above on 'reverberasti'), and cf. *s.* 7. 7 (quoted above on 'tu adsumpsisti me').

This expression has evoked a vast literature: parallels are noted, *inter alia*, at Athanasius (*de incarnatione verbi* 43. 4: verbally identical to Plotinus), Eusebius (quoting Plato as above), Proclus (*in primum Alcibiadem* 110e), Simplicius, William of St Thierry, Bernard of Clairvaux; further esp. in Courcelle, *Les Confessions*, 624–40, *Recherches*, 2nd edn., 405–40, and *REAug* 16 (1970), 271–81. Other studies include A. E. Taylor, *ADHLMA* 9 (1934), 305–6; F. Chatillon, *Mélanges E. Podechard* (Lyons, 1945), 85–102; E. Gilson, *Med. Stud.* 9 (1947), 108–130; G. Dumeige, 'Dissemblance', *Dictionnaire de spiritualité*, 3. 1330–46 (esp. 1333–6); J. C. Didier, *Mél. sc. rel.* 8 (1951), 205–10; Takoshi Katô, *Studies in Medieval Thought* 2 (1959), 43–53; Mayer, *Zeichen*, 1. 350–2.

vocem tuam de excelso: Jer. 31: 15, 'vox in excelso audita est lamentationis.' The garden scene (8. 12. 29) also centres on a voice heard, surrounded by some mystery, and the 'vision of Ostia' is indeed an 'audition' (9. 10. 25); cf. here 'et clamasti . . .', where the divine voice is heard again. See du Roy, 81 ('Dans cette rencontre mystique, l'audition prend le relai de la vision'), who rightly takes 'cibus sum grandium . . .' in a eucharistic sense; for similar food-imagery with eucharistic overtones, see on 7. 18. 24 and 10. 43. 70. For similar direct discourse attributed to God, see on 6. 16. 26.

cibus: Metaphorically of God and Truth at 3. 1. 1, 4. 1. 1, 10. 6. 8, 13. 18. 23; for *cibus* eucharistically, cf. Jn. 6: 27, 6: 56.

cresce et manducabis me: *Io. ev. tr.* 18. 7, 'si enim adhuc parvus es, dicunt tibi angeli: cresce; nos panem manducamus, tu lacte nutrire, lacte fidei, ut pervenias ad cibum speciei.'

erudisti hominem: Ps. 38: 12, 'pro iniquitate erudisti hominem et tabescere fecisti sicut araneam animam meam'; *en. Ps.* 38. 18, 'quid tabidius aranea? . . . pone supra leviter digitum, ruina est. . . . talem fecisti animam meam, inquit, erudiendo me pro iniquitate. . . . ipsa est ergo gratia beneficii dei prima, redigere nos ad confessionem infirmitatis.' Cf. 13. 12. 13, 'erudisti hominem'.

et dixi: These remarks recall an aching discouragement. In 7. 1. 1 ff., he recalls having seen his way, as he thought, past purely material notions of God; the *platonicorum libri* propel him further in that direction, but now his frustration raises the question anew, however fleetingly. The response given is put carefully in Christian terms.

clamasti: 10. 27. 38, 'vocasti et clamasti et rupisti surditatem meam.'

de longinquo: 13. 1. 1, 'ut audirem de longinquo'. See on 'in regione dissimilitudinis' here, and see on 7. 21. 27.

ego sum qui sum: Ex. 3: 14, 'ego sum qui sum'. The echo says concisely that A. has found God the Father (the Old Testament God, eternal creator; in A.'s triadic terms, *esse* (13. 11. 12)). What he has sought in Bk. 7 (see on 7. 1. 1 for his problems in imagining an incorporeal God), he has found, but in a puzzling state of alienation that leaves room for Bk. 8 and Christ. Note as well that the words are specific, earthly words, not *ineffabilia verba* (2 Cor. 12: 4, as A. quotes that text): see on 9. 10. 25 for the contrast with Ostia.

For A. on Ex. 3: 14 see *s.* 7. 7 (quoted above on 'tu adsumpsisti me'); cf., among many other texts, *vera rel.* 49. 97, *f. et symb.* 4. 6, *en. Ps.* 1. 6, *en. Ps.* 9. 11, *doctr. chr.* 1. 32. 35 ('ille enim summe ac primitus est, qui omnino incommutabilis est et qui plenissime dicere potuit, "ego sum qui sum"'), *nat. b.* 19, *conf.* 13. 31. 46. Another forty passages are listed by E. Zum Brunn, drawing on Beuron archives, in *Dieu et l'Être: Exégèses d'Exode 3, 14 et de Coran 20, 11–24* (Paris, 1978), 164. Few antedate *conf.*, but the doctrine is fully consistent from the outset, even before ordination. At *civ.* 8. 11, A. takes this coincidence of Platonic and Christian doctrine as an indication that Plato might have read Jewish scripture (whatever the chronological problems about Jeremiah, Egypt, etc., which in the same chapter he has finally got straightened out: see on 7. 9. 15 above). See also Madec, *Lectio VI–IX*, 63.

The verse is inextricably linked with many other topics, especially that of divine immutability; see F. Körner, *Das Sein und der Mensch: Die existentielle Seinsentdeckung des jungen Augustin* (Freiburg, 1959), A. di Giovanni, *REAug* 20 (1974), 285–312; E. Zum Brunn, in *Dieu et l'Être*, 141–64, expanded by the same author's *Le Dilemme de l'Être et du Néant chez saint Augustin* (Paris, 1978: Eng. trans. as *St. Augustine: Being and Nothingness* (New York, 1988)).

quae per ea . . . conspicitur: The pro-Platonic (on A.'s reading) Rom. 1: 20 says that what he learned here was good and true, and yet everything depended on whether he went on to react in the way condemned by Rom. 1: 21 ff., or whether he put it to good use.

7. 11. 17

The ontology of Ex. 3: 14 is central to the adaptation of neo-Platonism here. If we take 7. 10. 16 as one vision, without clarity because of the *infirmitas aspectus*, then this paragraph begins the readjustment of ontology that will lead to the revelation (7. 12. 18, 'manifestatum est mihi quoniam bona sunt') of the goodness of created things and hence make possible the more successful ascent of 7. 17. 23. A right understanding of scripture and a new conception of God, culminating in the discovery of the *platonicorum libri*, made possible the attempted ascent; now a right understanding of the goodness of created things (hence an answer to the last of the Manichean questions (3. 7. 12), 'unde malum') will remove the last barrier.

In the absence of circumstantial narrative, it is safer to take this whole development from 7. 10. 16 to the end of the book as a condensed account of a single sequence of thought and inspiration (cf. 'vidi' here, which suggests as much with the verbal echo of 7. 10. 16, and cf. 7. 12. 18, 'vidi et manifestatum est mihi'), with both positive and negative aspects.

id enim vere est: *en. Ps.* 134. 4, 'est enim est, sicut bonorum bonum bonum est'; *s.* 7. 7, 'esse, nomen est incommutabilitatis'; *mor.* 2. 1. 1, 'id enim est quod esse verissime dicitur. subest enim huic verbo manentis in se atque incommutabiliter sese habentis natuare significatio.' The same sentiment often in the texts cited apposite to Ex. 3: 14 above.

mihi autem . . . bonum est: Ps. 72: 28, 'mihi autem adhaerere deo bonum est' (Ps.-text varies between *in-* and *ad-haerere* (but Veronensis has clearly *adiungi*); Vg. *adhaerere*). A favourite verse, echoed in *conf.* at 12. 9. 9, 12. 11. 12, 12. 11. 13, 12. 15. 19, 12. 15. 21, 12. 15. 22, 12. 19. 28, 13. 2. 3. This text marks an important difference between the visions of 7. 10. 16 and 7. 17. 23: the former reads, 'ut viderem esse quod viderem, et nondum me esse qui viderem'; the latter, stronger, reads, 'neque ullo modo dubitabam esse cui *cohaererem*, sed nondum me esse qui *cohaererem*.' The verse is expounded without benefit of scriptural support at *div. qu.* 54, 'haec autem est ipsa veritas; cui quia intellegendo anima rationalis iungitur, et hoc bonum est animae, recte accipitur id esse quod dictum est, mihi autem adhaerere deo bonum est.' At *civ.* 10. 18, 10. 25, and 12. 1, the verse measures the distance between Christian and Platonist views of the *beata vita*, esp. *civ.* 10. 25, '"quia ecce", inquit, "qui longe se faciunt a te, peribunt; perdidisti omnem, qui fornicatur abs te" [Ps. 72: 27], hoc est, qui multorum deorum vult esse prostibulum. unde sequitur illud propter quod et cetera de eodem psalmo dicenda visa sunt: "mihi autem adhaerere deo bonum est", non longe ire, non per plurima fornicari. adhaerere autem deo tunc perfectum erit, cum totum quod liberandum est fuerit liberatum.' In an anti-Manichean context: *c. ep. fund.* 38. 44, 'quamvis sit malum corruptio et quamvis non sit a conditore naturarum, sed ex eo sit quod de nihilo factae sunt, tamen etiam ipsa illo regente et gubernante omnia quae fecit sic ordinata est ut non noceat nisi naturis infimis ad supplicium damnatorum et exercitationem *admonitionemque redeuntium* [cf. 7. 10. 16], ut inhaereant deo incorruptibili maneantque incorrupti, quod unum est bonum nostrum, sicut per prophetam dicitur, "mihi autem inhaerere deo bonum est".' Finally, *en. Ps.* 5. 14 brackets this verse with Mt. 25: 21, 'intra in gaudium domini tui', itself central to the ascent at Ostia (9. 10. 25).

in se . . . omnia: Wisd. 7: 27, 'in seipsa manens innovat omnia'; clearly echoed here, adumbrated at 1. 4. 4, 4. 5. 10, 7. 9. 14, and returning at another moment of mystical vision at 9. 10. 24 ('in se permanenti sine vetustate atque innovanti omnia'). *lib. arb.* 2. 12. 34, 'cum illa [veritas] in se manens nec proficiat cum plus a nobis videtur nec deficiat cum minus, sed integra et incorrupta et conversus laetificet lumine et aversos puniat caecitate. quid quod etiam de ipsis mentibus nostris secundum illam iudicamus, cum de illa nullo modo iudicare possimus? . . . tantum autem mens debet intellegere quantum propius admoveri atque *inhaerere* potuerit incommutabili veritati.'[44] *en. Ps.* 109. 12 links the verse to Jn. 1: 1 ff., to Phil. 2: 6 ff., and to mystical ascent: 'veritas enim incommutabilis quod est verbum dei, deus apud deum, per quem facta sunt omnia, in se manens innovat omnia. hanc ut *videamus*, magna et perfecta cordis munditia necessaria est, quae fit per fidem. demonstrata enim forma servi, dilata est ad demonstrandum formam dei. . . . ducitur enim ad quandam *visionem* ineffabilem, quam non merebuntur impii.'

et dominus meus es: Ps. 15: 2, 'dixi domino, deus meus es tu, quoniam bonorum meorum non eges'; at *ep.* 138. 6 and 138. 7, as at several other places in *en. Ps.*, this verse is taken as a proof-text that God does not have need of sacrificial offerings; *civ.* 10. 5 goes farther: '"dixi domino, dominus meus es tu, quoniam bonorum meorum non eges" . . . ne ipsa quidem iustitia hominis deus egere credendus est.' For the association with Wisd. 7: 27, cf. Ambrose, *Isaac* 8. 78, 'ipsum autem manens in semet ipso dat aliis, nihil autem in se ex aliis suscipit. de quo propheta ait, "dixi domino meo, deus meus es tu, quoniam bonorum meorum non indiges".'

[44] The text of Wisd. is similar to Plot. 6. 5. 10. 1, *μένει οὖν ἐν ἑαυτῷ σωφρονοῦν*, a passage with numerous parallels to the part of *lib. arb.* just cited; see O'Connell, *REAug* 9 (1963), 22.

7. 12. 18

These paragraphs between the two ascents of 7. 10. 16 and 7. 17. 23 have been little attended to. There are no citations from 7. 12. 18–16. 22 in the 'index augustinien' of the first edition of Courcelle's *Recherches*, only two citations in his *Les Confessions* (setting aside 10 citations in the chapters devoted to the *Nachleben*); the best treatment is du Roy, 82–8.

manifestatum: Cf. 'et manifestatum est' near the end of the paragraph; in *conf.* always of truth and usually of divine indication: 3. 4. 8 (the substance of Col. 2: 8–9 found in the *Hortensius*), 4. 15. 26 (in the attempted 'ascent' of the *de pulchro et apto*), 5. 12. 22 (non-divine source), 10. 32. 48, 13. 18. 23, 13. 24. 37, 13. 34. 49. Here the development that follows the first 'manifestatum' is purely Plotinian, with no scriptural overtones; having completed that exposition, he turns to the crucial verb 'vidi' and uses 'manifestatum' again. What follows is the same doctrine clearly Christianized ('tu fecisti'), with a clear echo of an important scriptural text (Gn. 1: 31).

bona sunt quae corrumpuntur: Plot. 3. 2. 5. 25–32 (*civ.* 10. 14 attests that A. knew this treatise), *ὅλως δὲ τὸ κακὸν ἔλλειψιν ἀγαθοῦ θετέον· ἀνάγκη δὲ ἔλλειψιν εἶναι ἐνταῦθα ἀγαθοῦ, ὅτι ἐν ἄλλῳ. τὸ οὖν ἄλλο, ἐν ᾧ ἐστι τὸ ἀγαθον, ἕτερον ἀγαθοῦ ὂν ποιεῖ τὴν ἔλλειψιν· τοῦτο γὰρ οὐκ ἀγαθὸν ἦν. διὸ οὔτε ἀπολέσθαι τὰ κακά, ὅτι τε ἄλλα ἄλλων ἐλάττω πρὸς ἀγαθοῦ φύσιν ἕτερά τε τἄλλα τοῦ ἀγαθοῦ τὴν αἰτίαν τῆς ὑποστάσεως ἐκεῖθεν λαβόντα, τοιαῦτα δὴ γενόμενα τῷ πόρρω.* See also Plot. 1. 8.

G–M: 'But by making "privatio" equivalent to "corruptio" A. gives a special turn to the theory.' Cf. *mor.* 2. 5. 7, 'sed corruptio non est in seipsa, sed in aliqua substantia quae corrumpit: non enim substantia est ipsa corruptio. ... quod enim corrumpitur, integritate et sinceritate privatur. quod ergo non habet ullam sinceritatem qua privetur, corrumpi non potest: quod autem habet, profecto bonum est participatione sinceritatis. item quod corrumpitur, profecto pervertitur: quod autem pervertitur, ordine privatur: ordo autem bonum est.' A. goes further than Plato and Plotinus in denying the existence of evil at all (*sol.* 1. 2. 2, 'deus qui ... ostendis malum nihil esse'); closer parallels to A.'s view in later neo-Platonists, notably Proclus and Boethius (see J. Gruber, *Kommentar zu Boethius De Consolatione Philosophiae* (Berlin, 1978), 312, to which cf. H. Chadwick, *Boethius* (Oxford, 1981), 239). H. Chadwick, *Augustine* (Oxford, 1986), 20, instancing Plotinus 1. 8. 14. 49–50, *ὕλη τοίνυν καὶ ἀσθενείας ψυχῇ αἰτία καὶ κακίας αἰτία*, concludes: 'It is fair to deduce that even Plotinus failed to achieve a clear and consistent position [on the nature of evil]. After his conversion Augustine sought to correct Plotinus' mistakes.'

On the the search for an answer to the question 'unde malum', see on 3. 7. 12. For this Platonic doctrine positively contrasted to the Manichean teaching, see *civ.* 14. 5. The *argument* that evil was nothing was not new to A.

at this time, for he had heard it when among the Manichees: *mor.* 2. 8. 11, quoted on 5. 11. 21. The idea appears as well in Cic. *Tusc.* 3. 8. 18, 'nequitia ... ab eo quod nequicquam est in tali homine ex quo idem "nihili" dicitur' (which seems to be echoed at *beata v.* 2. 8; whether A. knew that passage earlier, or not, there are other signs that a renewed reading of Cicero was part of the Milan period—see on 6. 16. 26). Thirty-five years after Milan, A. recalled the doctrine as he heard it from the mouth of Ambrose: *c. Iul.* 1. 9. 44, quoting 'ille doctor meus' from the *de Isaac* 7.60: 'quid ergo est malitia, nisi boni indigentia. ... non enim sunt mala, nisi quae privantur bonis: per mala tamen factum est ut bona eminerent. ergo indigentia boni radix malitiae est.'

nocet ... noceret: 'Harm', for all the Plotinian *comparanda* here, also evokes the argument of Nebridius, made at Carthage and reported in 7. 2. 3, where divine immutability is made evidence that true good cannot be harmed. The verb, fairly common in Bks. 1–7, disappears from *conf.* after this paragraph except for 12. 11. 11 (making the same point) and 13. 21. 31 (the inability of the serpent to do harm).

privantur bono: Care must be taken to avoid both anachronism and excessive scepticism. The passage from *mor.* 2. 5. 7 quoted above is apparently the earliest appearance of what becomes A.'s characteristic vocabulary for the doctrine; but few would suggest that the thing itself is impossible without the *ipsissima verba* that he would later use so familiarly. The only other substantial text earlier than *conf.* is *lib. arb.* 3. 13. 36, and *nat. b.* 3 and 23 are more or less contemporary; see also *c. Iul. imp.* 3. 206, *civ.* 14. 11, and *c. adv. leg.* 1. 5. 7.

omnia bona tu fecisti: Gn. 1: 31, 'et vidit deus omnia quae fecit et ecce bona valde'; cf. Sirach 39: 21, 'opera domini universa bona valde'. *Gn. c. man.* 1. 21. 32, 'sane non est neglegenter praetereundum quod dictum est, "et vidit deus ... bona valde." ... si enim singula opera dei cum considerantur a prudentibus, inveniuntur habere laudabiles *mensuras* [1] *et numeros* [2] et *ordines* [3: see on 5. 4. 7] in suo quaeque genere constituta; quanto magis omnia simul, id est ipsa universitas quae istis singulis in unum conlectis impletur? omnis enim pulchritudo quae partibus constat multo est laudabilior in toto quam in parte.' See further on 13. 28. 43, 'nam singula ... bona et valde'.

7. 13. 19

By contrast to 7. 12. 18, and reverberating with the scriptural chime rung in the last line there, the context is now made as scriptural as possible. We are shown how A. now thinks he should have acted at the time; the pretext for the extensive citation of Ps. 148 (text given following on this paragraph) is surely in Ps. 148: 14 (not quoted by A. here), '*confessio* eius in terra et caelo'.

ordinem: See on 1. 7. 12, and cf. *mor.* 2. 5. 7, quoted on 7. 12. 18, and *en. Ps.* 148. 9, 'omnia ergo ista videtis qualia sunt, mutabilia, turbata, terribilia, corruptibilia; tamen habent locum suum, habent *ordinem* suum, implent et ipsa universi pulchritudinem pro modo suo, et ideo laudant dominum.'

conveniunt: Evokes the 'orderliness' underlying beauty in creation: see on 2. 5. 10.

non essent ista: Fuller remarks in the same vein are found at *lib. arb.* 3. 9. 24.

sed iam etiam . . . deberem: The turn from Plotinus is crucial: not only the scriptural echoes, but the theme of praise ('confessio') itself. See Ps. 148: 7–12 (for text see below). The reversal of order of citation from Ps. 148 (putting 7–12 before 1–5) gives the effect of movement from earth to heaven, rather than the reverse (Knauer, 99).

laudandum te ostendunt: This deictic function of created things is seen again at 9. 10. 25 (Ostia) and at 10. 6. 9.

ignis, grando, nix, glacies, spiritus tempestatis: Conceived as parts of 'terra': see 'quam terram dicimus' here.

quae faciunt verbum tuum: *en. Ps.* 148. 10–11 makes clear that A. takes the pronoun's antecedent as everything from 'ignis' to 'tempestatis', and that the function of the phrase is to make clear that these natural forces are governed by divine power and not by chance.

laudent (*nomen tuum*) S Knöll Skut. : *laudant* DGO Maur. Ver.

de caelis te laudent: Ps. 148: 1–5. Ps. 148: 6, not cited in this paragraph, is echoed at 12. 15. 19.

desiderabam . . . pendebam: The imperfects make clear that this describes not an episode but a state of mind.

superiora . . . inferiora: *en. Ps.* 148. 10, 'quare hic addidit: "quae faciunt verbum eius"? multi stulti non valentes contemplari et discernere creaturam locis suis et *ordine* suo, sub nutu et iussu dei agentem motus suos, visum est illis quia *superiora* omnia deus gubernat, *inferiora* vero despicit, abicit, deserit, ut nec curet ista, nec gubernet, nec regat.'

sed meliora omnia: The aesthetic argument to explain the presence of evil in a good world may be intellectually satisfying, but creates serious problems of theodicy. Its classic formulation in A. is *civ.* 11. 22, '[aliquibus haereticis] plurima offendunt, sicut ignis aut frigus aut fera bestia aut quid eius modi; nec attendunt quam vel in suis locis naturisque vigeant pulchroque ordine disponantur, quantumque universitati rerum pro suis portionibus decoris tamquam in communem rem publicam conferant vel nobis ipsis, si eis congruenter atque scienter utamur, commoditatis attribuant.' In fairness, A. remained sensitive to the difficulties throughout his life, and to the polemical disadvantage it forced upon him. Here (7. 16. 22), he insists 'et iustitia tua displicet iniquis', and thirty years later, at *c. Iul. imp.* 2. 22, he doggedly persists in the theodicy to which his philosophy led him: 'catholica illa [expositio] est, quae ostendit iustum deum in tot ac tantis poenis et cruciatibus parvulorum.'

Text of Ps. 148 from en. Ps. *148*

(1) Alleluia. laudate dominum de caelis; laudate eum in excelsis. (2) laudate eum omnes angeli eius; laudate eum omnes virtutes eius. (3) laudate eum sol et luna; laudate eum omnes stellae et lumen. (4) laudate eum caeli caelorum; et aquae quae super caelos sunt, (5) laudent nomen domini. quoniam ipse dixit, et facta sunt; ipse mandavit et creata sunt. (6) statuit ea in saeculum et in saeculum saeculi; praeceptum posuit et non praeteribit. (7) laudate dominum de terra, dracones et omnes abyssi, (8) ignis, grando, nix, glacies, spiritus tempestatis, quae faciunt verbum eius, (9) montes et omnes colles, ligna fructifera et omnes cedri, (10) bestiae et omnia pecora, reptilia et volatilia pennata. (11) reges terrae et omnes populi, principes et omnes iudices terrae, (12) iuvenes et virgines, seniores cum iunioribus, laudent nomen domini. (13) quia exaltatum est nomen eius solius. (14) confessio eius in terra et caelo; et exaltabit cornu populi sui. hymnus omnibus sanctis suis; filiis Israhel, populo appropinquanti sibi! alleluia.

7. 14. 20

This paragraph contains elements of a review of the development of A.'s views: 'et quia non . . .', his Manichean phase; 'inde rediens . . .', following his break with Manicheism, the view of God as an all-pervasive corporeal being (7. 1. 1–5. 7); 'sed posteaquam . . .', the clarity of vision (note 'vidi' and see on 7. 10. 16) that came in the wake of the *platonicorum libri*. For a study of Bk. 7 centring on this passage, see E. Feldmann, *Augustiniana* (= Festschrift Van Bavel), 41 (1991), 881–904.

Non est sanitas: Ps. 37: 4, 'non est sanitas in carne mea a vultu irae tuae'; cf. 'iudicio saniore' in the last line, and here, 'consopita est insania mea'; *sanitas* of a sort has been achieved (cf. the medical metaphor in 'collyrio' (7. 8. 12)).

nolebat esse tuum: These lines show how close Manichean and neo-Platonic views are to each other. Both are explanations of the presence of evil in the world designed in the first instance to protect the innocence of God. Their proximity is one explanation for the appeal of the latter to A., after years of being steeped in the former.

aliena: cf. Ps. 18: 14, 'ab occultis meis munda me, domine, et ab alienis parce servo tuo'; *en. Ps.* 18. en. 2. 13 (paraphrasing the verse), 'tolle mihi ex corde malam cogitationem, repelle a me malum suasorem; . . . diabolus suo delicto cecidit, Adam alieno deiecit.'

per infinita spatia: His pre-neo-Platonic view (7. 1. 1–2), abandoned at

7. 10. 16 ('numquid nihil est veritas, quoniam neque per finita neque per infinita locorum spatia diffusa est?').

conlocaverat in corde suo: Cf. Ezech. 14: 7, 'quia homo . . . si alienatus fuerit a me, et posuerit *idola sua in corde suo,* et scandalum iniquitatis suae statuerit contra faciem suam, et venerit ad prophetam ut interroget per eum me, ego dominus respondebo ei.'

facta erat: sc. 'anima mea' (from above). 2 Cor. 6: 16, 'qui autem consensus templo dei cum idolis? vos enim estis templum dei vivi.'

fovisti caput nescientis: R. J. O'Connell has made much of these words: *REAug* 9 (1963), 30–7, and *Thought* 48 (1983), 188–206.

clausisti oculos meos: Ps. 118: 37, 'averte oculos meos, ne videant vanitatem'; *en. Ps.* 118. s. 12. 1, 'numquid quamdiu sumus in hoc mundo, possumus non videre vanitatem? . . . ergo in spe, qua speramus nos adhaesuros contemplandae veritati, subiecti sumus interim vanitati.' See on 7. 1. 1 for the link between *vanitas* and *curiositas*, and the way the destruction of both runs through Bk. 7.

cessavi de me paululum: Pressed literally (as *BA* trans.: 'j'ai perdu conscience un instant'), this phrase may imply some mystic swoon; but a metaphorical reading is preferable in view of the non-narrative nature of the context, and in view of the answering phrase, 'et evigilavi in te'.

7. 15. 21

respexi: Cf. 7. 11. 17, 'inspexi'.

alia: i.e. 'quae non sunt deus' (cf. end of 7. 14. 20).

vidi . . . vidi: See on 7. 10. 16.

non quasi in loco: This is the denial of the 'sponge' metaphor (7. 5. 7) that arose from a corporeal view of God.

omnitenens: 11. 13. 15, 'deum omnipotentem et omnicreantem et omnitenentem'; cf. *Io. ev. tr.* 106. 5, 'sicut pater aeternus omnipotens, ita filius coaeternus omnipotens; et si omnipotens, utique omnitenens. id enim potius verbum e verbo interpretamur, si proprie volumus dicere, quod a graecis dicitur *pantokrator*: quod nostri non sic interpretarentur ut "omnipotens" dicerent, cum sit *pantokrator* omnitenens, nisi tantundem valere sentirent.' (Latin biblical translation preferred *omnipotens* for παντοκράτωρ, as at Apoc. 1: 8.)

manu veritate: BA: 'dans la main par la vérité', i.e. 'per Christum'. But of course if Christ is the *manus dei* (see on 11. 2. 4 and cf. *en. Ps.* 143. 14, 'ipse salvator corporis, manus dei'; Knauer, 121 n. 4), 'veritate' may be in apposition. De Marchi, 312–13, needlessly emends to *veritatem*.

et omnia vera sunt: *sol.* 2. 5. 8, 'verum mihi videtur esse id quod est'; at *sol.* 2. 6. 10, the false is 'quod non ita est ut videtur'.

et vidi quia: The doctrine denied by these lines is Manichean: see on

11. 10. 12, and Bk. 11 generally for the eternity of God (and his power over the flow of time) as an essential counterpart for the Christian A. of his immutability. Here he seems to be suggesting (what is similarly implicit at 7. 10. 16) that his new view of God dating from this period includes the notion of his eternity.

7. 16. 22

Apart from the doctrinal substance of the view here of matter and evil, the form of the argument marks a characteristic feature of all A.'s later thought. The fault is no longer thought to be in the world that he contemplates but in the eye with which he contemplates it. This view was implicit in his interpretation of earlier stages of his development (esp. 7. 8. 12, 'collyrio', 7. 10. 16, 'et nondum me esse qui viderem'—cf. 7. 17. 23, 'sed nondum me esse'), but this is the moment at which he first shows this view as affecting his philosophical and doctrinal positions. His inability to see the complete logical coherence of an argument is no longer in itself a reason not to accept the argument. This is a conversion in some ways—he does not after all change the essentials of his view of the world (snakes and worms are still repugnant to him); but he changes his view about his own view of the world, and that is not a small thing. (Cf. Bk. 6, where he repeatedly states that he did not change his own view, merely discovered that the catholic church's view, to his surprise, coincided with his own.)

apta . . . apti: See on 4. 13. 20 (on the *de pulchro et apto*).

detortae in infima voluntatis perversitatem: 2. 3. 6, 'perversae atque inclinatae in ima voluntatis suae'.

voluntatis perversitatem: He now accepts in mind what he had 'heard'—from Ambrose's sermons (7. 3. 5, 'audiebam')—that free will is the source of evil.

proicientis intima sua: Sirach 10: 9–10, 'quid superbit terra et cinis? (10) nihil est iniquius quam amare pecuniam; hic enim et animam suam venalem habet, quoniam in vita sua proiecit intima sua.' *mus.* 6. 13. 40, 'quare superbia intumescere, hoc illi est in extima progredi, et ut ita dicam, inanescere, quod est minus minusque esse. progredi autem in extima, quid est aliud quam intima proicere; id est, longe a se facere deum, non locorum spatio, sed mentis affectu?' *en. Ps.* 1. 4, 'hanc terram ad interiorem hominem pertinere, et inde superbia hominem proici, maxime intellegi potest in eo quod scriptum est, "quid superbit terra et cinis? quoniam in vita sua proiecit intima sua." unde enim "proiectus est" non absurde se dicitur proiecisse.' Sim. at *Gn. c. man.* 2. 5. 6. Is there a Porphyrian citation here? Unlikely, but there has been discussion: see du Roy, 84 n. 3.

intima . . . foras: For *intus/foris*, see on 10. 27. 38.

7. 17. 23

Here A. accomplishes his Plotinian ascent. Years later, he expressed the thought (common among the Platonists themselves) that a more enduring vision might be possible. *en. Ps.* 134. 6, 'tendebatis enim vos fortassis videre bonum omnium bonorum, bonum a quo sunt omnia bona, bonum sine quo nihil est bonum, et bonum quod sine ceteris bonum est; tendebatis vos ut videretis, et forte in extendenda acie mentis vestrae deficiebatis. hoc enim ex me conicio; sic patior. sed etsi est aliquis, sicut fieri potest—et valde potest—acie mentis fortior me, et *contuitum* cordis sui *diu figit* in eo quod est; laudet ille ut potest et quomodo nos non possumus, laudet.' This text led Courcelle, *Les Confessions*, 47 n. 2, to conclude: 'Le *diu figit* s'oppose directement à conf. 7. 17. 23, *sed aciem figere non evalui*.' True so far as it goes, but Courcelle wrongly inferred that the later text implied that the ascent of this paragraph had been somehow unsuccessful. Another late and optimistic (for others) text is *ep.* 147. 13. 31; for other texts on the difficulty and impermanence of such vision, see on 7. 10. 16, 'reverberasti'. For A.'s fullest description of the Plotinian ascent, see *civ.* 9. 16, 'deum quidem summum omnium creatorem, quem nos verum deum dicimus, sic a Platone praedicari asseverat, quod ipse sit solus qui non possit penuria sermonis humani quavis oratione vel modice comprehendi; vix autem sapientibus viris, cum se vigore animi quantum licuit a corpore removerunt, intellectum huius dei, id quoque interdum velut in altissimis tenebris *rapidissimo coruscamine lumen candidum intermicare*. si ergo supra omnia vere summus deus intellegibili et ineffabili quadam praesentia, etsi interdum, etsi tamquam rapidissimo coruscamine lumen candidum intermicans, adest tamen sapientium mentibus, cum se quantum licuit a corpore removerunt, nec ab eis contaminari potest.'

Already at Cassiciacum, A. had a Christianized explanation for the defects of this sort of vision: *sol.* 1. 6. 13, '[Ratio] aspectus animae ratio est. . . . aspectus rectus atque perfectus, id est quem visio sequitur, virtus vocatur; est enim virtus vel recta vel perfecta ratio. sed et ipse aspectus quamvis iam sanos oculos convertere in lucem non potest, nisi tria illa permaneant: fides, qua credatur ita se rem habere ad quam convertendus aspectus est, ut visa faciat beatum; spes qua cum bene aspexerit, se visurum esse praesumat; caritas, qua videre perfruique desideret.'

Mystical adepts agree that their experience is impossible to describe adequately; one would think that to describe someone else's mystical experience would be *a fortiori* more difficult, but few ever act that way. We speak of these things with a confidence that glosses over serious epistemological difficulties, and those difficulties are doubled here. If it is difficult for us to know what Plotinus' (or A.'s) experience was like, how was it possible for A. to know what Plotinus was talking about? We are reduced to comparing texts, but texts that are always written and read with reference to an ineffable reality presumably lying beyond them. But even that is rendered more difficult here, since whatever A. was reading had been filtered through the expectations of a third party (a translator) before it came to A. This text permits the conclusion that it was written by someone who thought he was reporting an experience that was congruent to that reported in Plotinus' texts.[45]

A minority view sees this paragraph only as a further description of the 'ascent' of 7. 10. 16: du Roy, 85: 'Ce qui pouvait sembler un nouveau récit de "tentative d'extase", n'est donc qu'un rappel de sa situation, tout naturellement amené par le contexte. C'est ainsi que compose Augustin.'

phantasma: See esp. 4. 7. 12, and see on 3. 6. 10.

frui: *doctr. chr.* 1. 4. 4 ff. (and see *BA* 11. 558–61).

rapiebar: See on 4. 12. 18. Again, the initiative is from outside and above.

decore tuo: Cf. 10. 27. 38, 'pulchritudo tam antiqua et tam nova'; 12. 15. 21, 'dilexi decorem tuum.'

pondere meo . . . pondus: See on 13. 9. 10, 'pondus meum amor meus', and on 8. 5. 10, 'consuetudo'.

memoria tui: For *memoria* as the *locus dei* in the mind's ascent, see 10. 16. 26 (and see note at beginning of Bk. 10).

neque ullo modo . . . cohaererem: See on 7. 10. 16, 'ut viderem esse quod viderem, et nondum me esse qui viderem'. The phrasing here is certainly stronger: what has come between makes a difference, specifically 7. 11. 17, 'mihi autem inhaerere deo bonum est.'

corpus quod corrumpitur: Wisd. 9: 15, 'corpus enim quod corrumpitur aggravat animam et deprimit terrena inhabitatio sensum multa cogitantem.' The verse is cited 119 × in A. (La Bonnardière, *Biblia Augustiniana: Sagesse*, 206–21, 289–92), from the whole range of his working life. Juxtaposed with 1 Cor. 15: 52 ff.: *c. ep. Parm.* 2. 14. 32, *en. Ps.* 84. 10 ('cohibens ergo aliquantum ab strepitu mundi, et avertens se ad se, et a se in illum cuius vocem audiebat interius; quasi obturans aurem contra tumultuantem vitae huius inquietudinem, et contra animam corpore quod corrumpitur aggravatam, et sensum terrena inhabitatione deprimentem multa cogitantem, ait: "audiam quid loquetur in me dominus deus; . . . quoniam loquetur pacem in plebem suam"'), 4. 9, 78.15, 83. 9, 127. 16, *ss.* 131. 7. 7, 277. 4. 4, 299. 9, 351. 3. 3, *ep.* 143. 6, *perf. iust.* 6. 14, *ench.* 23. 91; never with Rom. 1: 20 ff. The text is not specially favoured by other church fathers. A. cites it with explicit reference

[45] For a credible attempt to do justice to the problems of such a discussion, see Sorabji, *Time*, 157–73. E. Hendrikx's protest against imagining A. as a mystic in modern terms, *Augustins Verhältnis zur Mystik* (Würzburg, 1936), is now somewhat dated but retains considerable value in this regard; his cautions have the value of suggesting one reason why A., fully believing that he had experienced the full Plotinian ascent, would find it disappointing—because by temperament, he was not the sort to achieve, or find settled satisfaction in, such moments. Even with the Ostia description in 9. 10. 23–5, it is clear that A. neither expects nor aspires to such a moment again, nor does he recommend it to others as normative.

to the source only three times in all, and only four times in connection with the rest of the passage in which it finds itself. La Bonnardière, 210: 'La citation de Wisd. 9: 15 est très souvent accompagnée de la mention du terme *pondus* [*en. Ps.* 75. 4, 83. 9, 102. 6, *pecc. mer.* 1. 38. 69, *Io. ev. tr.* 69. 2] ou de l'un de ses synonymes, *sarcina, onus, gravitas, retinacula, pavimentum*.' A. is aware of, and deprecates, a possible Platonizing misreading: *civ.* 14. 3, 'aliter se habet fides nostra. nam corruptio corporis, quae adgravat animam, non peccati primi est causa sed poena; nec caro corruptibilis animam peccatricem, sed anima peccatrix fecit esse corruptibilem carnem.'

quod invisibilia tua: Rom. 1: 20 (and see below), the 'pro-Platonic' verse of that passage; see on 7. 9. 14.

inveneram . . . commutabilem: *lib. arb.* 2. 12. 33–4, 'quapropter nullo modo negaveris esse incommutabilem veritatem . . . (34) hanc ergo veritatem . . . in qua tam multa conspicimus excellentiorem putas esse quam mens nostra est . . . ? . . . mentes enim nostrae aliquando eam minus aliquando eam plus vident et ex hoc fatentur se esse mutabiles.'

gradatim . . . aspectus: For the ascent through contemplation of creation (made possible now by an accurate appreciation of the nature of evil, and hence of created being: thus this narrative is most sharply distinguished from that of 7. 10. 16), cf. 9. 10. 25, 10. 6. 8–10, and 10. 17. 26. See on 2. 6. 12 for the extensive discussion from *quant. an.* 33. 70–6 of the ascent by degrees; here it will suffice to observe that the ascent there parallels but does not duplicate that described here. The following list schematizes the present text and gives in brackets references and quotations to *quant. an.* where those are close and pertinent:

(1) [*anima vivificans*] *corpus* (*quant. an.* 33. 70)
(2) *anima per corpus sentiens* (*quant. an.* 33. 71)
(3) *interior vis animae* (combined with the next stage here: cf. *quant. an.* 33. 72)
(4) *ratiocinans potentia animae*
(5) *intellegentia animae*
(6) *ipsum incommutabile* (*quant. an.* 33. 75)
(7) *id quod est* (*quant. an.* 33. 76, 'in ipsa visione atque contemplatione veritatis, qui septimus atque ultimus animae gradus est')

Most comparable here is Plot. 5. 1. 11.

gradatim: Cf. 9. 10. 24, 'perambulavimus gradatim cuncta corporalia.'

ad sentientem Knöll Skut. Ver. (with the support of one minor manuscript) : *adsentientem* G Maur. : *assentientem* DO : *sensientem* S.

interiorem vim: On *sensus interior*, see *lib. arb.* 2. 3. 7–4. 10, and cf. O'Daly, 102–5, arguing for neo-Platonic influence. Cf. 1. 20. 31, 'interiore sensu', see further on 10. 6. 8, and for the structure underlying the sequence 'interiorem vim . . . ratiocinantem . . . iudicandum', see on 10. 6. 10.

quae se . . . mutabilem: *lib. arb.* 2. 6. 14, 'et ipsa ratio, cum modo ad verum pervenire nititur modo non nititur et aliquando pervenit aliquando non pervenit, mutabilis profecto esse convincitur.'

abduxit cogitationem a consuetudine: Cf. Cic. *Tusc.* 1. 16. 38, quoted at *ep.* 137. 2. 5, 'miratur hoc[46] mens humana et, quia non capit, fortasse nec credit. se ipsam primitus scrutata miretur, se ipsam paululum, si potest, a corpore attollat et ab eis rebus quas solet sentire per corpus, et videat ipsa quid sit quae utitur corpore. sed forte non potest; "magni" quippe "ingenii est", ut ait quidam, "sevocare mentem a sensibus et cogitationem a consuetudine abducere".'

aspergeretur DGO Ver. : *aspargeretur* S Knöll Skut. See on 9. 10. 23, 'aspersi', with other texts in which the verb describes the accession of fleeting mystical knowledge; cf. esp. *c. acad.* 2. 2. 6, 'tunc vero quantulocumque iam lumine asperso tanta se mihi philosophiae facies aperuit, ut non . . . eam demonstrare potuissem.'

incommutabile praeferendum: Sim. at 7. 4. 6.

id quod est: Cf. the corresponding element in 7. 10. 16, 'et clamasti de longinquo: immo vero "ego sum qui sum" [Ex. 3: 14]'. The difference is that here the factual indicative says that A. not only heard this claim from afar, but arrived at Being itself. See also *sol.* 2. 5. 8, quoted on 7. 15. 21, and *Io. ev. tr.* 2. 2, 'ut attingat quomodo potest *id quod est*'. True being is the decisive argument against the Manichees, as at *sol.* 1. 1. 2, 'deus universitatis conditor . . . qui paucis ad *id quod vere est* refugientibus ostendis malum nihil esse.'

in ictu trepidantis aspectus: The scriptural echo is unmarked by Courcelle, who cites instead Plot. 6. 7. 31. 8, εἶδε δὲ οἷον πληγεῖσα—which is no parallel at all. What happened at Milan was not the beatific vision; this was not corruptibility putting on incorruption in any real sense: but as against what is reported at 7. 10. 16, this is a clear statement of the plenitude of the experience. 1 Cor. 15: 50–4, 'hoc autem dico, fratres: quia caro et sanguis regnum dei possidere non possunt neque corruptio incorruptelam possidebit. (51) ecce mysterium vobis dico: omnes quidem resurgemus, sed non omnes immutabimur [this verse also at the climax of Ostia: 9. 10. 25]. (52) in momento, *in ictu oculi*, in novissima tuba: canet enim tuba, et mortui resurgent incorrupti: et nos immutabimur. (53) oportet enim corruptibile hoc induere incorruptionem: et mortale hoc induere immortalitatem. (54) cum autem mortale hoc induerit immortalitatem tunc fiet sermo qui scriptus est, "absorpta est mors in victoria" [Is. 25. 8]'. For the resurrection

[46] 'hoc' represents the doctrine of God that A. has acquired in the course of Bk. 7: *ep.* 137. 2. 4, 'non sic deus dicitur implere mundum, velut aqua, velut aer, velut ipsa lux, ut minore sui parte minorem mundi impleat partem, et maiore maiorem [cf. 1. 2. 2, 7. 5. 7]. novit *ubique totus* [6. 3. 4] esse, et *nullo contineri loco* [7. 15. 21]; novit *venire* non recedendo ubi erat, novit *abire* [7. 15. 21] non deserendo quo venerat.'

as divinization that avoids ascent by stages, see *f. et symb.* 9. 24, 'quod ergo per hos gradus fieri posse concedit, ut terra in corpus aethereum convertatur, cur non accedente dei voluntate, qua corpus humanum supera aquas potuit ambulare, celerrime id fieri posse, quemadmodum dictum est: in ictu oculi, sine ullis talibus gradibus credit, sicut fumus plerumque in flammam mira celeritate convertitur?'

A.'s literal sense of the essential phrase is precise: *s.* 277. 11. 11, 'non invenis aliquid quod huic compares celeritati, quantum attinet ad corporis celeritatem. merito huic celeritati apostolus Paulus facilitatem resurrectionis comparavit, dicens, *in ictu oculi*. ictus oculi est, non in palpebris claudendis et aperiendis: nam hoc tardius agitur quam videtur. tardius palpebram levas quam dirigis radium. citius radius tuus pervenit ad caelum quam palpebra levata ad supercilium. videtis quid sit ictus oculi: videtis quam facilitatem apostolus resurrectioni corporum dederit.' (The same reading at *s.* 362. 18. 20.)

trepidantis: *beata v.* 4. 35, as quoted at 7. 10. 16.

invisibilia tua: Rom. 1: 20 (see on 7. 9. 14). The indicative is again decisive, corroborated by 'certissimus' above, and the repetition of 'conspexi' with the phrase at 7. 20. 26 further confirmation. This echo confirms what 'ictu' suggests, by declaring that the vision has been achieved, and imposing upon it a clear Christian interpretation. A. has had the Plotinian vision of God that Rom. 1: 20 suggests is possible: it is good, and true, and valid in so far as such things are possible in this life. The inability to fix his vision there is, we know from other texts, the result of his mortality and weakness and sin, but he does not insist on that here (there is no echo of Rom. 1: 21 ff., the 'anti-Platonic' portion of that passage). He means to convey thanksgiving for what he was granted, and indirect (high) praise of Plotinus. The next paragraph turns to the search for strength, and contains within its opening sentence the (Christological) key to the search. It is not that he discovered that the Plotinian method did not work (that is Courcelle's position in essence); he discovered that it did work, and that it was not enough.

aciem figere non evalui: Madec, *RA* 2 (1962), 279, puts this phrase alongside Plot. 1. 6. 9. 24, ἀτενίσας ἴδε, which he translates 'fixe ton regard et vois' and takes it as evidence of the failure of the 'tentative' here. There seems to be a misconstrual through the French translation: ἀτενίσας suggests the effort required to direct the gaze in the right direction, not to any attempt to hold the gaze there indefinitely. *trin.* 15. 27. 50 (at the end of A.'s last contemplative essay, written perhaps a quarter century after *conf.*, addressing his soul), 'quae igitur causa est cur *acie fixa* ipsam videre non possis nisi utique infirmitas, et quis eam tibi fecit nisi utique iniquitas? quis ergo sanat omnes languores tuos, nisi qui propitius fit omnibus iniquitatibus tuis?'

repercussa: Cf. 7. 10. 16, 'reverberasti', with texts.

amantem memoriam: Cf., but more importantly contrast, 7. 10. 16, 'contremui amore et horrore.' Cf. at the beginning of this paragraph, 'iam te amabam . . . sed mecum erat memoria tui.'

7. 18. 24

Something more is required: see *civ.* 19. 23 (quoted more fully below on 7. 19. 25), 'photinianus haereticus, qui tantummodo hominem non etiam deum noverit Christum, et *ideo per eum salvus esse non possit*'. He had a rude knowledge early on (1. 11. 17, and cf. 5. 10. 20), but neither understood it in a way that facilitated acceptance nor accepted it. We have been prepared for this since 7. 9. 14, 'non ibi legi', but how much of the Christ-centredness of this and succeeding paragraphs is authentic report of his thoughts of the time and how much is pious revision a decade and more later cannot be conclusively defined. Here readers will differ, but there is nothing to suggest that he came to this position *expressis verbis* at any time before writing *conf.* What happened in 386 was thus something like this: first reading the *platonicorum libri*, then a period of enthusiasm at his discoveries, then disillusionment that his discoveries had not brought him to the goal he had sought, then anxious deliberation (culminating in the garden scene) of a moral and sacramental nature over the question whether he should present himself for baptism. In retrospect, and as a theological interpretation, he sees that what was lacking to him after reading the *platonicorum libri* was the incarnate Christ; he recalls that it was Paul he read, and Simplicianus and others with whom he spoke, and he interprets the whole movement from there until his baptism as the addition of Christ (now accurately perceived) to his beliefs and to his life. In favour of this reading, the incarnation had been part of his thoughts at Cassiciacum. When A. actually acquired a correct doctrine is open to discussion (see on 8. 12. 30), but incarnation is already central to the very Platonic conclusion to *c. acad.* (J. J. O'Meara, *Dominican Stud.* 3 (1950), 339, goes so far as to say that the 'whole book [*c. acad.*] centres around this acceptance of the Incarnation'), reporting views only a few months later than the time under review here: *c. acad.* 3. 19. 42, 'non enim est ista huius mundi philosophia, quam sacra nostra meritissime detestantur [cf. 1 Cor. 1: 20, 3: 19, Col. 2: 8], sed alterius intellegibilis, cui animas multiformibus erroris tenebris caecatas et altissimis a corpore sordibus oblitas numquam ista ratio subtilissima revocaret, nisi summus deus populari quadam clementia *divini intellectus auctoritatem usque ad ipsum corpus humanum declinaret atque submitteret*, cuius non solum praeceptis sed etiam factis excitatae animae redire in semet ipsas et resipiscere patriam etiam sine disputationum concertatione potuissent.' He is also orthodox on Christology at *ord.* 1. 10. 29 and 2. 9. 27 (see van Bavel, 6). For the ascent to God hinging on incarnation see already 4. 12. 18–19; *doctr. chr.* 1. 10. 10–11, 'non enim ad eum, qui ubique praesens est, locis movemur, sed bono studio bonisque moribus. (11) quod non possemus, nisi ipsa sapientia tantae etiam nostrae infirmitati congruere dignaretur et vivendi nobis praeberet exemplum non aliter quam in homine'; sim. at length at *Io. ev. tr.* 2. 2; for incarnation as the antidote for Platonic (specifically Porphyrian) views, see *civ.* 10. 28–9, 'hunc [sc. paternum intellectum] autem Christum esse non credis; contemnis enim eum propter corpus ex femina acceptum et propter crucis opprobrium. . . . (29) sed incarnationem incommutabilis filii dei qua salvamur, ut ad illa quae credimus vel ex quantulacumque parte intellegimus venire possimus, non vultis agnoscere. itaque videtis utcumque, etsi de longinquo [see on 7. 21. 27], etsi acie caligante, patriam in qua manendum est, sed viam qua eundum est non tenetis.' The incarnation is also the instrument to a successful ascent at *spir. et litt.* 24. 41.

One early text holding that a correct doctrine of the incarnation is linked to sacramental initiation is *ep.* 11. 2 (388/91, to Nebridius): 'accipe igitur quid mihi videatur de susceptione hominis mystica quam propter salutem nostram factam esse religio, *qua imbuti sumus*, credendum cognoscendumque commendat. quam quaestionem non facillimam omnium elegi, cui potissimum responderem . . . ex quo videtur esse consequens, ut hominem trinitas tota susceperit; . . . cur ergo in *mysteriis et sacris nostris* hominis susceptio filio tributa celebratur? haec est plenissima quaestio ita difficilis et de re tam magna ut nec sententia hic satis expedita nec eius probatio satis secura esse possit.'

The presentation of this difficulty helps us measure the intensity of A.'s contact with Ambrose and his ideas. The perplexities A. describes here could perfectly well have been clarified (at least in so far as discovering the orthodox doctrine would have offered clarification) had A. and Al. read Ambrose's *de fide* of 378–80 (where, for instance, Photinus is expressly anathematized at, e.g. *de fide* 1. 1. 6, 1. 8. 57, 2. 13. 117, etc.); but Amb. spoke at other times of Photinus, if not always in great detail, e.g. *exam.* 3. 7. 32—itself probably too late in date to be relevant here, but a good example of the kind of incidental mention A. could have encountered at almost any time; cf. also *de paradiso* (377) 12. 58, using a triad of heretics (Arius, Sabellius, and Photinus) that often recurs in A.[47]

Theological discussion of A.'s Christology continues, with occasional incidental benefits for the study of particular texts; van Bavel is most useful, but cf. W. Geerlings, *Christus Exemplum* (Mainz, 1978)—for Christ as *exemplum*, cf. 7. 19. 25 (describing his pre-conversion position) with 10. 43. 68 (view at the time of *conf.*), and of course B. Studer, *RA* 19 (1984), 133–54. Though Courcelle, *Ricerche di stor. rel.* 1 (1954), 63–71, deals expressly with the next paragraph, these two paragraphs are inseparable, and the survey he gives at 64 ff. of scholarly neglect of the passage is equally valid here.

viam: Cf. 'via' below, and see on 7. 7. 11. The unacceptability of this concept to non-Christians: *civ.* 10. 32, 'dicit Porphyrius in primo iuxta finem "de regressu animae" libro nondum receptum in unam quandam sectam quae

[47] Photinus and Arius appear as twin bogey-men ('immanes feritate lupi') in Prudentius, *psychomachia* 794–5.

universalem contineat viam animae liberandae, vel a philosophia verissima aliqua vel ab Indorum moribus ac disciplina, aut inductione Chaldaeorum aut alia qualilbet via'; Maximus of Madauros in A. *ep.* 16. 1, 'nam deus omnibus religionibus commune nomen est. ita fit ut, dum eius quasi quaedam membra carptim variis supplicationibus prosequimur, totum colere profecto videamur'; Longinianus in A. *ep.* 234. 2, 'via est in deum melior qua vir ... deorum comitatu vallatus ... aut a deo aut in deum intentione animi mentisque ire festinat'; and see on 5. 13. 23, for the echo of Symm.'s *rel.* 10 at *sol.* 1. 13. 23, criticized at *retr.* 1. 4. 3.

fruendum: See on 7. 17. 23, 'non stabam frui deo meo'.

mediatorem dei et hominum: 1 Tim. 2: 5, 'unus enim deus, unus et mediator dei et hominum, homo Christus Iesus.' Incarnation is the means by which the Logos performs this mediatorship: *civ.* 9. 15, 'quaerendus est medius, qui non solum homo, verum etiam deus sit ...; quem neque non fieri mortalem oportebat, neque permanere mortalem'; cf. *civ.* 10. 20, 10. 22. *civ.* 9. 17 (quoted above on 7. 10. 16) specifically prescribes this *mediator* for one kept apart from God by *longinquitas* and *dissimilitudo*. In so far as A. clings to this doctrine he is really capitulating to some extent to the expectations of the non-Christians, and this is one reason why a doctrine of atonement lacks a strong presence in A. The attention he gives to the notion of 'mediator' is generally taken as a sign of reaction to the ideas of Porphyry (J. J. O'Meara, *Porphyry's Philosophy from Oracles in Augustine* (Paris, 1959), 160–2; approved by Madec, *RA* 2 (1963), 279; dissenting, du Roy, 97 n. 1). The 1 Tim. text itself is never cited before *c.*394/5 (*Gal. exp.* 24, 63; *div. qu.* 80. 3), but is later frequent in A. (du Roy, 89 n. 3). See on 10. 43. 68 (in a paragraph that echoes Jn. 1 and Phil. 2). See on 10. 42. 67 f.

When does A. 'embrace the mediator'? Within the narrative of *conf.*, in the garden scene: see on 8. 12. 29–30.[48]

qui est super omnia: Rom. 9: 5, 'ex quibus Christus secundum carnem, qui est super omnia deus benedictus in saecula'.

ego sum via et veritas et vita: Jn. 14: 6: the only *ad verbum* citation of this thematically central verse in all of *conf.*, reinforcing the importance of this passage.

cibum: See on 7. 10. 16, 'cibus sum grandium'. Eucharist is 'the food that I was incapable of taking'—as clear an indication as one could want that A. sees his problem at this moment in his life as arising from his inability to participate in Christian liturgical worship.

[48] The word *mediator* itself is a relative novelty in Latin, mainly ecclesiastical (first in Cyprian) and arising out of scriptural passages such as this; it translates μεσίτης, which begins as a legal term for a legal intermediary or arbitrator and appears in a religious context in Philo and NT. In philosophical Gk., the relevant word is μεσότης (e.g. Sallustius, *de diis et mundo* 16 (ed. Nock, p. 28. 31)); hence A.'s lengthy discussion of *mediatores* in *civ.* Of surviving texts, A. offers the most abundant evidence for *mediator*.

cui capiendo: *Io. ev. tr.* 1. 12, 'ille cibus est, accipe lac ut nutriaris, ut sis validus ad capiendum cibum.'

miscentem carni: See on 5. 10. 20, 'concerneretur carni'.

quoniam verbum caro factum est: Jn. 1: 14: what was missing in the *platonicorum libri* at 7. 9. 13 is now the chief desideratum.

lactesceret: See T. van Bavel, *Augustiniana* 7 (1957), 255–6. There is a decided progression and recursion in the movement of *conf.* from the infant seeking milk from his nurses (milk provided them by God, to be sure: 1. 6. 7) to the mature man now finding milk again in a new infancy, speechless in the presence of the Word. *en. Ps.* 130. 11–12, 'dominus ergo noster Iesus Christus panis, se fecit nobis lac, incarnatus et apparens mortalis, ut in eo finiretur mors et non aberraremus a verbo, credentes in carnem quod factum est verbum. hinc crescamus, ipso lacte nutriamur; antequam validi simus ad capiendum verbum, non recedamus a fide lactis nostri. . . . (12) sic enim nos deus vult nutriri lacte, ut non ibi remaneamus; sed crescendo per lac ad solidum cibum perveniamus.' *en. Ps.* 109. 12, 'hoc est enim illud lac parvulorum, quod temperavit, panem traiciens per carnem. nam panis ille angelorum . . . factus est homo. ita nobis verbum incarnatum factum est receptibile.' Note the eucharistic language in both passages; see also *en. Ps.* 8. 5, 33. s. 1. 6, 117. 22, 119. 2, *s.* 117. 10. 16, and see also on 4. 1. 1, 'sugens lac tuum aut fruens te, cibo qui non corrumpitur'.

per quam creasti omnia: Echoes the doctrine of Jn. 1: 3, but for the word 'creasti', cf. also Col. 1: 16, 'omnia per ipsum [Christum] et in ipso creata sunt.'

humilis: A defining characteristic of the incarnation: 1. 11. 17, 'per humilitatem domini dei nostri', 7. 9. 13, 'via humilitatis', 7. 9. 14, 'humilavit se factus oboediens usque ad mortem', 7. 20. 26, 'a fundamento *humilitatis*, quod est Christus Iesus', 8. 2. 3, 'humilitatem Christi', 8. 2. 4, 'de sacramentis humilitatis verbi tui' (hence 9. 6. 14, 'iam induto humilitate sacramentis tuis congrua', of readiness for baptism), 10. 43. 68, 'ut eius exemplo etiam ipsam discerent humilitatem'.

verbum enim tuum: Here A. begins to describe the ascent of 7. 17. 23, ('[anima] erexit se ad intellegentiam suam') in terms that assimilate it to Christian doctrine.

aedificavit . . . domum: Prov. 9: 1, 'sapientia aedificavit sibi domum'; *civ.* 17. 20, 'hic certe agnoscimus dei sapientiam, hoc est verbum patri coaeternum, in utero virginali domum sibi aedificasse corpus humanum.' Cf. *Gal. exp.* 13, *s.* 225. 2. 2, *civ.* 17. 4, *conl. Max.* 8, *c. Max.* 2. 17. 2.

de limo nostro: Cf. Gn. 2: 7, 'formavit igitur dominus deus hominem de limo terrae.' That form of the text itself is attested by five passages in A. (according to VL): see especially Cypr. *de hab. virg.* 23 (quoted at *doctr. chr.* 4. 21. 47), 'portavimus imaginem eius, qui de limo est', and also *civ.* 13. 24, 'ut ex hoc limus intellegendus videretur, umore scilicet terraque concretus', *Gn. litt.*

3. 22. 34 and 6. 11. 19, *Gn. c. man.* 2. 7. 9, *en. Ps.* 89. 3; and in ordinary discussion of the creation of mankind when not directly quoting the text, *limus* and *pulvis* seem almost to alternate (cf. *Gn. litt.* 6. 2. 3, 'limo', 6. 4. 6, 'pulvere', 6. 5. 7, 'limo', 6. 5. 8, 'limo,' etc., and cf. *Gn. litt.* 7. 5. 8, 'de limo terrae vel pulvere').

tumorem: See on 7. 8. 12, 'residebat tumor meus'.

progrederentur: W. Hensellek, *Anzeiger Akad. Wien* 115 (1978), 25, shows that this is almost a technical term, the opposite of the 'redire' of 7. 10. 16: *ord.* 2. 11. 30, 'quia in istorum sensuum negotia progresso redire in semet ipsum cuique difficile est'; *util. cred.* 1. 1, 'qui nimis in haec corporalia progressi atque lapsi'. Also at 13. 17. 20, 13. 21. 31.

infirmam divinitatem: Cf. 1 Cor. 1: 25, 'et quod infirmum est dei fortius est quam homines' (as quoted at *en. Ps.* 126. 5). The daring terminology underlines the importance of kenosis (from Phil. 2: see on 7. 9. 14) in A.'s Christology; van Bavel, *Augustiniana* 7 (1957), 252.

participatione tunicae pelliciae nostrae: B. Aland, *Pietas* (Festschrift B. Kötting: Münster, 1980), 96: 'Dieser Satz ist für die philosophische Vernunft ungeheuerlich und kraß widersinnig. Denn Teilhabe, μέθεξις, ist Gründung und Verursachung des jeweils Unteren durch das Obere.' This is the only case in *conf.* where participation moves downward; see on 7. 9. 14, 'participatione'.

tunicae pelliciae nostrae: Gn. 3: 21, 'et fecit dominus deus Adae et mulieri eius tunicas pelliceas'; cf. 13. 15. 16, 'quemadmodum pellibus indueris homines cum peccato mortales fierent'. The 'tunicae pelliceae' represent mortality (itself the 'evidence of sin'—see 1. 1. 1). The view is consistent from 388–90 (*Gn. c. man.* 2. 21. 32, 'ad pelliceas tunicas, id est ad huius vitae mortalitatem') to 429/30 (*c. Iul. imp.* 4. 37); cf. *en. Ps.* 103. s. 1. 8, *s.* 362. 11. 11, and *trin.* 12. 11. 16. B. Altaner, *Historisches Jahrbuch* 70 (1951) 26–7 (= his *Kleine Patristische Schriften* (Berlin, 1967), 237) traced this reading to Origen's sixth homily on Leviticus; he was followed by J. Pépin, *Aug. Mag.* 1. 293–306, esp. 301–5; but P. F. Beatrice, in a wide-ranging article on the interpretation of Gn. 3: 21 in U. Bianchi (ed.), *La Tradizione dell'Enkrateia* (Rome, 1985), 461–3, thinks that Ambrose is the immediate influence.

7. 19. 25

For the issues and positions at stake here, see *agon.* 14. 16–32. 34, contemporary with *conf.*, which consists of a series of paragraphs, each introduced 'nec eos audiamus' outlawing various major heresies (but never naming them). For our purposes the relevant texts are: *agon.* 17. 19 (implicitly anti-Platonic); *agon.* 18. 20 (anti-Manichean, rejecting docetism); *agon.* 19. 21 (anti-Apollinarian); *agon.* 20. 22 (probably anti-Photinian), 'nec eos audiamus, qui sic dicunt ab illa aeterna sapientia susceptum esse hominem, qui de virgine natus est, quomodo et alii homines ab ea sapientes fiunt qui perfecte sapientes sunt. nesciunt enim proprium illius hominis *sacramentum* et putant hoc solum eum plus habuisse inter ceteros beatissimos, quod de virgine natus est. quod ipsum si bene considerent, fortassis credant ideo illum hoc praeter ceteros meruisse, quod aliquid proprium praeter ceteros habet etiam ista susceptio. aliud est enim sapientem tantum fieri per sapientiam dei et aliud est ipsam *personam* sustinere sapientiae dei.'

aliud putabam: Here is the most explicit statement we get of A.'s 'Christian Platonism', measured by the defects of his Christology. The view A. recalls to disown includes wholehearted acceptance of the literal truth of gospel narratives about Jesus' deeds. Only in the interpretation of those narratives did he differ from orthodox views (see on 'persona veritatis' below). He does not say here how far he was conscious at the time of the gap between himself and orthodoxy, and once again he describes his difficulty as lying mainly in his misunderstanding of the orthodox position, not in his inability to accept that position ('ego autem aliquanto . . .').

A. does not mean by all this merely a lack of precision in the formulation of doctrinal statements. To know the authentic Christ is to 'embrace' him, to be renovated by him, to reach the goal that A. had been pursuing unknowingly all his life. This will happen in Bk. 8. Right doctrine is a necessary condition for right moral conduct and acquisition of salvation. (See *civ.* 19. 23, quoted below on 'a Photini falsitate'.) Here it is impossible to read *conf.* dispassionately; to attempt to do so runs the risk of making a doctrinal quibble of the whole matter, when there is no doubt that A. was trying to show, in the language of doctrinal quibbles—that is to say, the only language he had—the traces of a deep and moving experience.

quantum de excellentis sapientiae viro: The sum of his objection to the doctrine of incarnation is expressed at 5. 10. 20. *BA* 13. 693, following Courcelle, *Ricerche di stor. rel.* 1 (1954), 67 ff., sees in Christ the man of excellent wisdom an echo of Porphyry's *Philosophy from Oracles*. Cf. *cons. ev.* 1. 7. 11, 'cur ipse dominus nihil scripserit . . . hoc enim dicunt illi vel maxime pagani, qui dominum ipsum Iesum Christum culpare aut blasphemare non audent eique tribuunt excellentissimam sapientiam, sed tamen tamquam homini . . . honorandum enim tamquam sapientissimum virum putant, colendum autem tamquam deum negant' (sim. at *Io. ev. tr.* 100. 3, attributing the view to 'pagani', and *agon.* 20. 22, quoted above); cf. *retr.* 2. 16, on *cons. ev.*, 'quorum primus liber adversus illos conscriptus est qui tamquam maxime sapientem Christum vel honorant vel honorare se fingunt'. He does not here cite the further position mentioned in *retr.*, namely that it was error by the writers of the gospels that led them to attribute divinity to Jesus, but that theory is certainly compatible with A.'s 'Photinianism' of the time.

Two cultured contemplators of Christianity come into view in A.'s letters, taking analogous positions. Longinianus (date of these letters uncertain, but just perhaps roughly contemporary with *conf.*: see Mandouze, *Pros. chr.*, and *PLRE* 2, both s. v.) writes in his own voice: *ep.* 233, 'quaero etiam quid de Christo sentias; quod enim eum non parvi pendas, adverti, sed utrum ea sola via quae ab illo demonstrata est ad vitam beatam perveniri posse existimes et aliqua ex causa non eam neglegas ire, sed differas, an et aliam vel alias ad tam opimam et prae omnibus appetendam possessionem vias esse arbitreris et aliquam earum iam te ingredi credas, nosse cupio non, ut opinor, impudenter.' At *ep.* 234. 1–3, he goes on in the same vein, using the adjective *paganus* of himself, which is unusual. Is Longinianus perhaps already half- or more than half-converted? If A. had designed and constructed an adversary from a store-bought straw-man kit, he could not have done better. He certainly looks very like the A. that Alfaric thought he had found in the period 386–91 (though L. and A. both allude to non-Christian ritual, perhaps theurgy, an extreme of non-Christianity A. never reached), the sort of man who could easily never quite convert but could well (see evidence in Mandouze, *Pros. chr.*, and *PLRE*) go on happily building baptisteries for the church. A.'s response to L. leaps to the importance of sacramental initiation: *ep.* 235. 2, 'itaque hic nodus est disputationis nostrae, quo soluto consequentia videbimus: vivatne homo bene ut sacris purgetur, an sacris purgetur ut bene vivat, an ipse quantuscumque bene vivendi in homine modus nondum sit idoneus ad beatam vitam, quae ex deo capitur, nisi accedant adiumenta sacrorum, an bene vivendi quaedam velut portio sit etiam sacra percipere, ut scilicet non aliud sit bene vivere aliud sacrate vivere, sed bene vivendi terminis etiam sacrata vita claudatur.'

Of much less interest are the objections of Volusianus, by V. himself in *ep.* 135. 2 (how God could be contained in a woman's body, or in a little baby, and how God could spend all that time away from heaven), and indirectly reported by Marcellinus in *ep.* 136. 1 (Jesus' wonder-working was not so special, considering the deeds of such as Apollonius of Tyana and Apuleius). V.'s and M.'s role in evoking *civ.* from A. has given these letters, however, a greater celebrity than those to and from Longinianus.

natus ex virgine: Jn. 1: 14 (see on 7. 9. 13). This seems to rule out (so *BA* 13. 693–8) the possibility that A.'s view here closely follows that of Porphyry, who rejected virgin birth (*civ.* 10. 29). Courcelle, *Ricerche di stor. rel.* 1 (1954), 64 n. 1, instances *ep.* 147. 7. 19, 'non quidem amisit occasionem Ambrosius ut hinc quosdam haereticos redargueret, id est photinianos, qui principium filio dei ex utero virginis tribuunt nec volunt credere quod et antea fuerit', to show that the phrase is here a rejection of the Photinians. But Courcelle insisted that this passage shows that the *Philosophy from Oracles* of Porphyry was one of the *platonicorum libri* that A. had read, though the argument seems common enough to require no single source.

exemplum: *BA* 13. 695: the phrase may apply either to virgin birth or to the phrases to follow (this seems preferable, as Courcelle saw, loc cit.). See e.g. *c. acad.* 3. 19. 42, quoted above on 7. 18. 24, related to the development of A.'s ideas in 386/7 on reason and *auctoritas*. Ideally, reason and philosophy should suffice, but in practice incarnation and revelation are necessary.

sacramenti: See on 1. 11. 17, and cf. *agon.* 20. 22 (quoted above) and *s.* 214. 6 (widely but not unanimously held to be early (391)) showing an orthodox Christology; texts cited there include Jn. 1: 1, 14, Phil. 2: 6–7.

manducavit . . . sermocinatus est: All these examples could be documented from the gospels (see below), but they are examples rather than specific references.

manducavit: Mt. 11: 19, 'venit filius hominis *manducans* et *bibens*.'

dormivit, ambulavit: Mt. 9: 24, 'et ecce motus magnus factus est in mari ita ut naviculus operiretur fluctibus, ipse vero *dormiebat*'; Jn. 6: 19, 'cum remigassent ergo stadia quasi viginti quinque aut triginta vident Iesum *ambulantem* super mare.'

exhilaratus: The word is not in the Vg.

contristatus: Mk. 3: 5, 'et circumspiciens eos cum ira *contristatus* super caecitate cordis eorum'.

sermocinatus: The word is not in the NT, but the idea is of course ubiquitous (and *sermo* is common).

quam ego iam noveram: See 7. 9. 13.

etenim nunc movere: The distribution of functions in this sentence suggests that A. is employing three categories: action, passion, and (as a separate function!) signification.

quae si falsa: The authority of scripture itself becomes an argument in favour of the truth of its contents.

totum hominem: *BA* 13. 695–6: 'l'homme tout entier: chair, âme, esprit.' *s.* 237. 4. 4, 'totum redemit qui totum creavit, *totum suscepit,* totum liberavit verbum. ibi mens hominis et intellectus, ibi anima vivificans carnem, ibi caro vera et integra; peccatum solum non ibi'; *f. et symb.* 4. 8, 'ut totum hominem suscipere dignaretur in utero virginis'; *civ.* 10. 27, 'propterea quippe totum hominem sine peccato ille suscepit, ut totum quod constat homo a peccatorum peste sanaret.'

persona veritatis: Where *veritas* must = *verbum*. The word *persona* here means 'role, personage' (*BA* 13. 697), first used thus at *Gal. exp.* 27, 'omnes fiunt filii; non natura, sicut unicus filius, qui etiam sapientia dei est; neque praepotentia et singularitate susceptionis ad habendam naturaliter et agendam *personam* sapientiae, sicut ipse mediator unum cum ipsa suscipiente sapientia sine interpositione alicuius mediatoris effectus'; the word in its fuller theological sense appears only later (list of passages at *BA* 13. 697; see also van Bavel, 13–14).

Alypius autem: A. now describes the misconceptions he and Alypius had concerning catholic doctrine; Alypius ascribed to the catholics views that were actually Apollinarian and objected to that view, while A. thought the catholic view identical with Photinianism, and found that view acceptable. A.'s own error (and by implication Alypius') has already been stated at the beginning of the paragraph; Alypius' misreading is diametrically opposite to A.'s; Nebridius erred by way of docetism (9. 3. 6: he thus remained closest to Manicheism). There has been confusion on the interpretation of this passage (see Courcelle and O'Connell cited below), but the gist of the paragraph is that A. and Alypius were themselves confused, both in their own thinking and in their understanding of what the catholic church taught. The rectification of those misunderstandings is attested by *div. qu.* 80, 'adversus apollinaristas' (quoted below), though such an essay does not so much refute the heretics as make the catholic position distinct.

Du Roy, 92 n. 1, identifies a tendency A. had to trichotomize Christological error: the three sects that define for him the possibilities of error are Manicheism, Photinianism, and Apollinarism—all three schematically together at *agon.* 17. 19–22. 24,[49] *Io. ev. tr.* 47. 9 ('instruximus etiam, quantum meministis et meminisse debetis, contra haereticos photinianos, qui solum hominem Christum sine deo esse dixerunt; contra manichaeos, qui solum sine homine deum; ex hac occasione de anima instruamus vos et contra apollinaristas, qui dicunt dominum nostrum Iesum Christum non habuisse animam humanam, id est animam rationalem, animam intellegentem, animam, inquam, in qua distamus a pecore, quod homines sumus'), and *persev.* 24. 67. But the pattern was not exclusively his; they make convenient whipping boys for Ambrose as well, and Ambrosiaster, while making no mention of Apollinarism, attacks the Photinians eight times in his Pauline commentaries, several times in conjunction with the Manichees.

postea . . . aliquanto posterius: Mandouze, 509 n. 4, thinks the adverbs are meant to show that the conversions of A. and Alypius were simultaneous, and that 'au moment de cette scène fameuse, Augustin est débarrassé de son photinianisme tout autant qu'Alypius l'est de son apollinarisme.' Though the garden scene is indeed the place to look for Augustine's achievement of an orthodox Christology, there are three problems with Mandouze's assertion: (1) the sequence of the phrases and the comparative form of the second make it more likely that Alypius came to his understanding of catholic doctrine first; (2) Alypius was never an Apollinarist (see next note); and (3) it is not so certain as we have always assumed that the garden scene had exactly the same place in the lives of A. and Alypius (see on 9. 4. 7).

[49] That work is contemporary with *conf.* and the anti-Platonic, anti-Manichean, anti-Apollinarian, and anti-Photinian passages there have many resemblances to the developments in *conf.*

Du Roy, 91 n. 4, conservatively suggests A.'s rectification came no later than Cassiciacum, from the evidence of the dialogues there; for R. J. O'Connell's attempt to postpone the date of A.'s rectification well past Cassiciacum, see Madec's refutation, *REAug* 16 (1970), 106–36.

apollinaristarum: Apollinaris of Laodicea (condemned by synods at Rome in 374 and 380, and by the Council of Constantinople in 381) believed that there was no *mens* (νοῦς) in the visible, earthly Christ: *div. qu.* 80, 'cum quidam haeretici, qui apollinaristae ex Apollinari quodam auctore suo dicti esse perhibentur, assererent dominum nostrum Iesum Christum, in quantum homo fieri dignatus est, non habuisse humanam mentem, . . . delectati sunt quidem ea perversitate, qua ille hominem in deo minuebat.' Cf. *haer.* 55, *Io. ev. tr.* 23. 6, 47. 9, *civ.* 14. 2, *en. Ps.* 29. en. 2. 2–3, 85. 4, *qu. hept.* 3. 93, *c. s. arrian.* 5. 5 ('hoc volunt intellegi, quod humanam carnem sine humana anima Christus adsumpserit. quae propria haeresis apollinaristarum est: sed etiam istos, id est, arrianos, in eorum disputationibus, non solum trinitatis diversas esse naturas, sed etiam hoc sentire deprehendimus, quod animam non habeat Christus humanam'—did Alypius' error arise from hearing of catholic Christianity from Arian or Arian-influenced sources in Milan? The link to Arianism is also made at *haer.* 55), *agon.* 19. 21, *s. Den.* 5. 7, *div. qu.* 80. 1, *persev.* 24. 67, *nat. et or. an.* 1. 18. 31, and *ep.* 140. 4. 12. At *c. Iul.* 5. 15. 55 and *c. Iul. imp.* 4. 47, Julian accuses A. of Apollinarism (another sign of his having read *conf.*? see on 4. 16. 28 for other traces); A. responds through *c. Iul. imp.* 4. 47–50. The likeliest vehicle for a clarification of Alypius' views was Amb. *incarn.* (written 381), 2. 11, 6. 49–52, clear against the Apollinarists (but making no mention of Photinians; the same treatise, *incarn.* 2. 8 and 4. 23, attacks the Manichees on Christological grounds as well); Ambrose attacked Apollinarians elsewhere, e.g. *exp. Ps.* 39. 26. For the clarification that Alypius was not himself Apollinarian, but objected to what he thought was Apollinarian in orthodoxy, see R. J. O'Connell, *REAug* 13 (1967), 209–10; Courcelle, *Les Confessions*, 64 n. 2, has a standard, but erroneous view: 'il le [dogme catholique de l'Incarnation] méconnaît et n'en pénètre pas le mystère. Au contraire, Alypius, . . . apollinariste sans le savoir, qui s'imagine penser selon la foi catholique' (Courcelle was followed by Mandouze: see following note). Apollinaris is a figure of greater importance in the history of Christology than Photinus; the classic study was H. Lietzmann, *Apollinaris von Laodicea und seine Schule* (Tübingen, 1904); a good recent summary statement is A. Grillmeier, *Christ in Christian Tradition* (London, 1965), 329–40; the most recent full-length study is E. Mühlenberg, *Apollinaris von Laodicea* (Göttingen, 1969), on which see C. Kannengiesser, *Rech. sc. rel.* 54 (1971), 27–36; active research continues, e.g. R. Hübner, *Die Schrift des Apollinarius von Laodicea gegen Photin (Pseudo-Athanasius, contra Sabellianos) und Basilius von Caesarea* (Berlin, 1989). A brief summary of A.'s mature reactions is B. Studer, 'Apollinaristae', *Aug.-Lex.* 1. 392–5.

verbum caro factum est: Jn. 1: 14.

a Photini falsitate: A. does not say that he was aware of Photinus' doctrines, or in communication with any followers of Photinus; rather his own attempts at understanding Christology had led him to views that he later learned were Photinian. The influence of Platonism (and Porphyry in particular) on the formation of those views is debatable and debated in the wake of Courcelle, *Riv. stor. rel.* 1 (1954) 63–71.

Photinus of Sirmium (deposed as bishop there and exiled in 351, and condemned at various councils after: hence also a recent controversy) doubted the pre-existent Christ and held that the Christ first existed in the virgin's womb (*ep.* 147. 7. 19; cf. *haer.* 45); *Io. ev. tr.* 26. 5, 'Photinus dixit: homo solum est Christus, non est et deus. qui sic credit, non pater eum traxit.' Other references at *en. Ps.* 67. 39, 124. 5, *ss.* 37. 12. 17, 71. 3. 5, 92. 3. 3, 182. 7. 7, 183. 1. 1, 183. 5. 8, 244. 4, 246. 4, 252. 4. 4, *util. ieiun.* 9. 11, *s. Den.* 19. 12, *s. Caill.* 2. 5. 2, *s. Mai* 95. 5, *Io. ev. tr.* 45. 5, 47. 9, 96. 3, 100. 3, *epp.* 120. 3. 15, 147. 7. 19, 185. 11. 48, *persev.* 24. 67, *agon.* 20. 22 (quoted above), *vera rel.* 5. 9. Note esp. *civ.* 19. 23 (after quoting Porphyrian oracles about Christ, e.g. Hecate saying 'viri pietate praestantissimi est illa anima; hanc colunt aliena a se veritate', and 'piissimum igitur virum . . . eum dixit et eius animam, sicut et aliorum piorum, post obitum inmortalitate dignatam et hanc colere christianos ignorantes'), 'ita laudant Christum, ut quisquis in eum talem crediderit qualis ab eis praedicatur, christianus verus non sit, sed photinianus haereticus, qui tantummodo hominem, non etiam deum noverit Christum, et *ideo per eum salvus esse non possit*.' O'Meara, *Porphyry's Philosophy from Oracles in Augustine* (Paris, 1959), 158–9, assumes that this passage indicates a Porphyrian conscience in A. at the time of his conversion, but at most it reports on A.'s views at the time of writing *conf.*, when he could well have recognized ruefully that he had then achieved a position that was no different from what he had later learned to be Porphyry's. In the present passage A. seems to describe the position as one he had reached in good conscience, thinking it adequately Christian. That would seem to rule out direct contact with unmitigated Porphyrianism. Ambrose opposed Photinians by name in his *in Luc.* 1. 13, 5. 4, and 8. 13, preached during or shortly after A.'s time in Milan; and attacked them forcefully elsewhere (*de fide* 1. 1. 6 (and elsewhere in that work: see above on 7. 18. 24), *de paradiso* 12. 58, *de spir. sanct.* 1. 16. 160 (and elsewhere in that work), *apol. David alt.* 4. 28, *epp.* 10. 12, 48. 5, and *ob. Theod.* 49, 'reges adorant, et photiniani divinitatem eius negant! clavum crucis eius diademate suo praeferunt imperatores, et arriani potestatem eius imminuunt!'

A thought experiment is conducted at *bapt.* 4. 16. 23: 'constituamus ergo duos aliquos isto modo: unum eorum verbi gratia id sentire de Christo quod Photinus opinatus est et in eius haeresi baptizari extra ecclesiae catholicae communionem, *alium vero hoc idem sentire sed in catholica baptizari, existimantem ipsam esse catholicam fidem*. istum nondum haereticum dico, nisi manifestata sibi doctrina catholicae fidei resistere maluerit et illud quod tenebat elegerit.' The second hypothetical believer in that case matches what A. himself would have been if he had accepted baptism at Easter 386 while entertaining the views he ascribes to himself here; the possibility that A. gave serious consideration to that step at that time cannot be ruled out (see on 6. 13. 23, where Monnica looks forward to A.'s marriage confidently as a time when he will be baptized). The thought that he might have been a candidate himself would have made him an attentive listener if Amb.'s *exameron* were preached during the week preceding Easter that year (that week was a regular holiday from public business: *cod. theod.* 2. 8. 19, but for the year cf. on 7. 9. 13), during the preparation of baptismal candidates. The thought experiment goes on to imagine a third candidate, knowingly Photinian but accepting baptism for worldly gain (that good marriage, that promising career—perhaps a governorship (6. 11. 19)), and A.'s scrupulous conscience may have been hounding him. The later A. knew that various forces brought candidates before him: *s.* 132. 1. 1, 'ecce pascha est, da nomen ad baptismum. si non te excitat festivitas, ducat ipsa *curiositas*, ut scias quid dictum sit, "qui manducat carnem meum, et bibit sanguinem meum, in me manet, et ego in illo" [Jn. 6: 56].' A similar thought experiment occurs at *c. ep. Parm.* 2. 16. 35: 'quid? si ergo apud eos quispiam baptizatus qui verbi gratia putaverat Christum ex eo esse coepisse, ex quo secundum carnem de virgine Maria natus est, postea veritatis sermone commotus, cum comperisset ipsum esse de quo Iohannes ait, "in principio erat verbum et verbum erat apud deum et deus erat verbum", de quo consequenter ait, "verbum caro factum est et habitavit in nobis", fateretur eis errorem suum pristinum atque in eo se fuisse cum baptizaretur sollicitus indicaret, iuberent eum denuo baptizari? numquam utique facerent, sed hominis imperitam simplicitatem, carnali opinione prius pravam, postea veritatis ratione correctam.'

sana doctrina: 1 Tim. 1: 10, 'et si quid aliud sanae doctrinae adversatur'; 2 Tim. 4: 3–4 (echoed at 1. 10. 16), 'erit enim tempus cum sanam doctrinam non sustinebunt, sed ad sua desideria coacervabunt sibi magistros, prurientes auribus, (4) et a veritate quidem auditum avertent, ad fabulas autem convertentur'; Tit. 1: 9, '[episcopus] ut potens sit exhortari in doctrina sana'; Tit. 2: 1, 'tu autem loquere quae decent sanam doctrinam.'

oportuit enim et haereses esse: 1 Cor. 11: 19, 'nam oportet et haereses esse, ut probati manifesti fiant inter vobis.' 'Heresy' is an extraordinary preoccupation when closely examined. A.'s *haer.*, written late in life, lists eighty-eight schools of error, most of them obviously small and localized. Truth is so fragile, so hemmed in by misreadings, that great precautions need be taken. The pluralism that these sects attest is the normal form of religious life in antiquity: it is only the textual, Christian view that holds that these divagations must be reduced to order. The incarnation remains a touchstone of orthodoxy through A.'s career: cf. *s.* 183. 9. 13 (416/19), 'et si discutiamus omnes haereses, invenimus eas negare Christum in carne venisse.'

infirmos: cf. Rom. 14: 1, 'infirmum vero in fide recipite' (the verse Alypius takes for himself at 8. 12. 30).

7. 20. 26

'Tunc' is the defining word, and the mention of the *platonicorum libri* the governing fact. We have here certainly not a third 'tentative d'extase' nor even exactly a continuation or renewal of what has been discussed since 7. 10. 16, but rather a recapitulation of what has been narrated in this book. The verbal echoes bind together chapters 9–20 in a single episode without narrative: 'So in those days, when I had read the books of the Platonists, and been admonished by them to seek bodiless truth, and found therein what Rom. 1: 20 tells me I should have expected to find, I was still too weak to enjoy you: "garriebam plane quasi peritus", to be sure, but had I not gone on to seek Christ, I would have perished. *Caritas* I needed, but found it not there, nor would I, such is the distinction between *praesumptio* (at whose dangerous apex I now stood: so close is terrible failure in this search to complete success) and *confessio* (which is what I learned to do in those days, and what I do now in this book).' Courcelle's reading (*Les Confessions*, 65) comes close to the truth, but blurs the line between what A. thought and intended at the time and how he later interpreted his thoughts and deeds. He *thought* he was on the straight path to orthodox Christian truth; it is only in retrospect that he saw that he was not. He did not think he had become a Plotinian sage: he thought he had become a Christian *sine parietibus* (8. 2. 4). This is confirmed by *c. acad.* 2. 2. 5, 'putabamus tamen satis nos agere' (there of the time even before reading the *platonicorum libri*).

admonitus: See on 7. 10. 16.

incorpoream: A reminder of his undoubted success in the period under review, gaining the ability to imagine God without a material body.

invisibilia tua: Rom. 1: 20 (see on 7. 9. 14, and note how close to 7. 17. 23 (esp. 'conspexi'), where see notes).

repulsus sensi: A. saw real dangers for himself at this stage: *lib. arb.* 2. 16. 43 (partly quoted on 7. 10. 16), 'vae qui derelinquunt *te ducem* . . . o suavissima lux purgatae mentis sapientia! . . . vae qui se avertunt a lumine tuo et obscuritati suae dulciter inhaerent! tamquam enim *dorsum ad te ponentes* [Jer. 2: 27, echoed at 2. 3. 6, where the notes show that the phrase is shorthand for 'practise paganism' for A.] in carnali opera velut in umbra sua defiguntur et tamen etiam ibi quod eos delectat adhuc habent de circumfulgentia lucis tuae. sed umbra dum amatur languidiorem facit oculum animi et invalidiorem ad perferendum conspectum tuum. propterea magis magisque homo tenebratur, dum sectatur libentius quidquid infirmiorem tolerabilius excipit. ex quo incipit non posse videre quod summe est.'

certus esse te: Here, as in 7. 7. 11 (therefore, before the *platonicorum libri*), A. provides a pseudo-creed to sketch the state of his mind. At 7. 7. 11, he believed God existed and was immutable and cared for humankind and that the way to salvation was in Christ and the scriptures; here the emphasis is on what he has learned in Bk. 7 of the nature of God. Note 7. 7. 11, 'et in Christo ... viam te posuisse salutis'; here the Christology is stronger; cf. 7. 18. 24, 'viam comparandi roboris'.

semper idem ipse esses: Cf. Ps. 101: 28, 'tu vero idem ipse es' (*en. Ps.* 101. s. 2. 12 cites Ex. 3: 14), and Heb. 1: 12, quoting the Psalm; see on 1. 6. 10 and cf. 4. 16. 31, 8. 3. 6, 10. 4. 6. *mor.* 2. 6. 8, 'esse enim ad manendum refertur. itaque quod summe et maxime esse dicitur, permanendo in se dicitur.'

alter DS Knöll Skut. Ver. : *aliter* GO Maur.

ex te esse omnia: Rom. 11: 36, 'quoniam ex ipso, et per ipsum, et in ipso sunt omnia: ipsi gloria in saeculum.'

certus quidem in istis eram: This certitude (anticipated at 7. 17. 23) responds to and disposes of the hesitations expressed at 5. 14. 25 (his Academic phase).

garriebam: A verb for indiscreet speech (= non-confession: 6. 4. 5, 'puerili errore et animositate tam multa incerta quasi certa garrisse'), but not always in a bad sense (9. 1. 1, 'et garriebam tibi'). A distinction is suggested in response to Ps. 76: 7, 'et meditatus sum nocte cum corde meo, garriebam et scrutabar spiritum meum'; *en. Ps.* 76. 9, 'ecce est illud garrire. observa iterum, ne deficiat spiritus tuus. non, inquit: non sic garriebam quasi foris; alio modo nunc. quomodo nunc? "garriebam et scrutabar spiritum meum." ... qui garriebat foris, ecce coepit intus garrire securus, ubi solus in silentio cogitat annos aeternos.' Cf. *s.* 242. 4. 6 (of the philosophers), 'illi garriant, nos credamus'; *ep.* 101. 2, 'ipsorum philosophorum *garrulae argutiae*' (quoted with Rom. 1: 21 ff.). For anti-Platonic contexts, cf. *cons. ev.* 1. 30. 46, 'isti fragiles contradictiunculas garrientes', *trin.* 15. 12. 21, 'multa illi philosophi [academici] garrierunt contra sensus', *civ.* 22. 25, 'sicut paucissimi garriunt', 22. 28, 'contra nos isti garriunt', 18. 40, 'vanissima praesumptione garriunt quidam.' The word is not, however, a sign of anti-Porphyrian *Tendenz* (as O'Meara, *Porphyry's Philosophy from Oracles in Augustine* (Paris, 1959), 160, thinks); Porphyry is merely a good example of the kind of person against whom A. uses the word; another target is Faustus (*c. Faust* 21. 10). Hensellek (*Anzeiger Akad. Wien* 114 (1977), 153, and 120 (1983), 79), emphasizes cases where the verb (and adj. *garrulus*) are less clearly pejorative; but a conspectus of all occurrences of the verb in A. shows a heavy preponderance of pejorative adverbs and objects (adverbial, 'tortuosissime', 'impie', 'contumaciter', 'sine ratione', 'inaniter', 'vanissima praesumptione', 'inprudenter'; objects, 'ludicrum aliquid', 'inania', 'fabulosa et falsa', 'vana', 'pudendam fabulam').

peritus: See on 1. 14. 23. We must make the effort of imagination to give this line its life. At every stage in his life Augustine had been a busy propagator of his views of the moment; he does not change now. The man who once made converts for Manicheism now goes about buttonholing friend and acquaintance with his great news, *his version of Christianity*. The remarkable thing about the change that comes over him after the garden scene is that he finally shuts up, chatters only to God, and heads for the hills. The exchange here of one form of *multiloquium* for another is change, at least.

salvatore nostro: Tit. 1: 4, 'gratia et pax a deo patre, et Christo Iesu salvatore nostro.'

periturus: If this expression represents anything A. felt at Milan in 386, it is hardly neo-Platonic.

plenus poena mea: This suggests a revival of the fear of death (see on 6. 16. 26): cf. 2. 2. 2, 'obsurdueram stridore catenae mortalitatis meae, poena superbiae animae meae' (and here 7. 18. 24, 'pelliciae tunicae'); 7. 7. 11, with its ps.-creed, also ends with a summary of his predicament/punishment—there, blindness.

et non flebam: He learns to weep properly in the garden at Milan (8. 12. 28, citing Ps. 50: 19), and he weeps again at the end of Bk. 9 (see on 9. 12. 29); here he does not yet shed these salutary tears; cf. 7. 21. 27, 'lacrimas confessionis', also citing Ps. 50: 19, and see on 3. 2. 4, esp. for the tears of Cassiciacum.

inflabar scientia: Cf. 1 Cor. 8: 1, 'scientia inflat, caritas vero aedificat'; *en. Ps.* 142. 5, 'illam scientiam comitem caritatis, magistram *humilitatis*'; 1 Cor. 13: 4, 'caritas . . . non inflatur'.

fundamento humilitatis: Cf. 1 Cor. 3: 11, 'fundamentum enim aliud nemo potest ponere praeter id quod positum est, quod est Christus Iesus.' *en. Ps.* 118. s. 25. 3, 'nec receptum deserit fundamentum; idque omnibus carnalibus suis, quibus capitur atque succumbit, delectationibus anteponit, . . . ubi si non anteponitur Christus, non illi est fundamentum.' On *humilitas* and the incarnate Christ, see on 7. 18. 24.

illi libri: The plain sense here is that he now realizes, and wishes us to think he realized at the time, that the *platonicorum libri* brought him just the vain and puffed up knowledge that Rom. 1: 21 ff. attacks.

quid interesset: The distinction is already present in Amb. *de fide* 1. 1. 4, 'in altero enim religiosa confessio est, in altero incauta praesumptio'; cf. *en. Ps.* 31. en. 2. 13, '"quoniam tacui, inveteraverunt ossa mea a clamando." si a clamando, quomodo tacuit? tacuit quiddam, non tacuit quiddam; tacuit unde proficeret, non tacuit unde deficeret; tacuit confessionem, clamavit praesumptionem. "tacui" enim dixit, non sum confessus.'

confessionem: Linked to the incarnation at *Io. ev. tr.* 14. 5, 'antequam veniret dominus Iesus, homines gloriabantur de se; venit ille homo ut minueretur hominis gloria et augeretur gloria dei. ... si sic venit ille ut dimitteret peccata, deus largiatur, homo confiteatur. etenim confessio hominis, humilitas hominis; miseratio dei, altitudo dei [Rom. 11: 33].'

inter videntes: *s.* 160. 4, 'multi enim viderunt quo, nec viderunt qua: amaverunt celsitudinis patriam, sed ignoraverunt humilitatis viam.'

viam ducentem ad beatificam patriam: *civ.* 11. 2, 'per hoc enim mediator per quod et homo; per hoc et via. quoniam si inter eum qui tendit et illud quo tendit via media est, spes est perveniendi; si autem desit, aut ignoretur qua eundum sit, quid prodest nosse quo eundum sit?' See on 7. 7. 11. The image of the *patria* has links to the counter-image of the *regio dissimilitudinis* under Plotinian and Ambrosian influences (see Plot. 1. 6. 8 and Amb. *de Isaac* 8. 78 quoted on 7. 10. 16); cf. *c. acad.* 3. 19. 42, quoted on 7. 18. 24, *beata v.* 1. 2, 'longeque a sua patria peregrinari audent, ... dulcissimae patriae'. The link to incarnational theology is clear: cf. *doctr. chr.* 1. 11. 11, 'cum ergo ipsa sit patria, viam se quoque nobis fecit ad patriam'; increasingly Christianized with time, it appears thus at *Io. ep. tr.* 7. 1: 'mundus iste omnibus fidelibus quaerentibus patriam sic est quomodo fuit eremus populo Israhel.' Cf. below, 7. 21. 27, 'videre patriam pacis'; the word is used twice of Jerusalem (10. 35. 56, 12. 16. 23) as well. The image makes its appeal to A. in Milan at a time when he is conscious of the distance that separates him from his own *patria* (he uses the word of Africa that way at 4. 4. 9, 4. 7. 12, 6. 10. 17, 9. 11. 27, 9. 11. 28); a similar sensitivity to, and rhetorical exploitation of, a similar dislocation underlies the rhetorical strategy of *civ.* (as I argued at *Aug. Stud.* 10 (1979), 75–9). For a sense of the word in A.'s lifetime, and a touching document of late Roman patriotism, see *ep.* 90 (Nectarius of Calama writing to A. *c.*408), 'quanta sit caritas patriae, quoniam nosti, praetereo. sola est enim quae parentum iure vincat affectum. ... sed quoniam crescit in dies singulos dilectus et gratia civitatis, quantumque aetas fini proxima est, tantum incolumem ac florentem relinquere patriam cupit.'

beatificam: P. Henry, *La Vision d'Ostie* (Paris, 1938), 115, argues that since the word does not much occur before Augustine, it must therefore be from Plotinus. *TLL* 2. 1794 offers before A. only Apuleius, *Plat.* 1. 5, '(deus) *beatus et beatificus*, optimus, nihil indigens' (cf. *conf.* 1. 4. 4, 'optime ... non egens'), but also reports (without further details) citations of the verb *beatifico* from the 'Itala' at Is. 3: 12, Gn. 30: 13 (from Beuron VL: Hier. *qu. Heb. in Gn.* 46. 8, 46. 11), Lk. 1: 48, Cant. 6: 8, Mal. 3: 15. The adjective occurs 19 × in A. (earliest *lib. arb.* 2. 15. 39, '[veritas] summum bonum et beatificum esse concedo', cf. *doctr. chr.* 1. 29. 30, 'beatificum bonum'—13 × modifying *bonum*); see 2. 5. 11, 'beatificis', and cf. *civ.* 9. 15 (from his discussion of Apuleius, *deo Soc.*), 'ad illud unum beatificum ⟨bonum⟩ perveniri ...; beatus et beatificus deus factus particeps humanitatis nostrae.'

non tantum cernendam sed et habitandam: The expression is directed most clearly against *curiositas*; Christ and the cross may be in the back of his mind (see *trin.* 4. 15. 20, quoted below on 7. 21. 27, 'de silvestri cacumine'), and perhaps besides a swipe (an unfair swipe?) at Plotinus (or Porphyry? cf. *civ.* 10. 32).

incidissem: Of falling on books at 2. 3. 5 (of *this* book) and 8. 2. 3; of falling among wrong-thinkers at 3. 6. 10 and 8. 1. 2; also in *beata v.* 1. 4 and *util. cred.* 1. 2 of his association with the Manichees. The consistent sense is of undirected human activity, in every case but one leading to a bad goal; the one exception (2. 3. 5) is explained by authorial humility—he does not wish to suggest that it is divine grace that leads others to *his* book.

7. 21. 27

Little attention has been paid to one claim Paul had on A.'s attention here. If A. had been seeking authentic vision of the divine following Plotinus' directions, and if he was unsatisfied with the results for whatever reason, Paul is the Christian author with pre-eminent authority to speak of visions, on account of his own vision of heaven—see 2 Cor. 5 and 12. The two chapters are linked by the theme of mystic vision at *en. Ps.* 103. s. 3. 7, and cf. *en. Ps.* 30. en. 2. s. 1. 2, 'de hoc mentis excessu, id est ecstasi, Paulus cum loqueretur, seipsum insinuans, ait: "sive enim mente excessimus deo, sive temperantes sumus vobis, caritas enim Christi compellit nos" [2 Cor. 5: 13]. . . . tantus autem fuit ille mentis excessus, ut diceret: "sive in corpore, sive extra corpus nescio, deus scit" [2 Cor. 12: 2].' On Paul in this role, see also *en. Ps.* 37. 12, 67. 36, 77. 17, etc.; emphasis is placed on Paul's hearing 'ineffabilia verba' (2 Cor. 12: 4: see on 9. 10. 25). Such vision, and nothing neo-Platonic, is what A. seems to have in mind at *quant. an.* 33. 76, 'iamvero in ipsa visione atque contemplatione veritatis, qui septimus atque ultimus animae gradus est neque iam gradus, sed quaedam mansio, quo illis gradibus pervenitur, quae sint gaudia, quae perfructio summi et veri boni cuius serenitatis atque aeternitatis adflatus, quid ego dicam? dixerunt haec quantum dicenda esse iudicaverunt magnae quaedam et incomparabiles animae, quas etiam vidisse ac videre ista credimus.' (Ambrose had called attention to Paul's visions in his *de Isaac vel anima* 4. 11 and 4. 14, preached while A. was at Milan, in a discussion of visionary philosophical ascents.)

arripui: The same verb of taking of the codex of Paul in the garden at 8. 12. 29, 'arripui, aperui, et legi'; cf. *c. acad.* 2. 2. 5, 'numquam cessavimus inhiantes in philosophiam. . . . et quoniam nondum aderat ea flamma quae summa nos *arreptura* erat, illam qua lenta aestuabamus arbitrabamur esse vel maximam. . . . itaque titubans properans haesitans *arripio apostolum Paulum*. neque enim vere, inquam, isti tanta potuissent vixissentque ita ut eos vixisse manifestum est, si eorum litterae atque rationes huic tanto bono adversarentur. perlegi totum intentissime atque *castissime*.' Note the echo here of Cic. *Tusc.* 1. 30. 73, 'itaque dubitans, circumspectans, haesitans, multa adversa reverens': Testard, reviewing Courcelle, *Les Confessions*, at *REAug* 10 (1964), 26–9.

stilum spiritus tui: Du Roy, 93 n. 2, quotes U. Riedinger, *Münchener Theol. Zschr.* 2 (1951), 431–4, in support of reading 'stilum spiritus tui' in apposition with 'Paulum'. In that case 'prae ceteris' denotes Paul's rank among the apostles (neither Riedinger nor du Roy can offer a satisfactory parallel for this expression in A.—see below). It is true that A. undertook no systematic study of all scripture at this time, but we do not need to follow Riedinger and du Roy in confining his reading exclusively to Paul, and we can believe that while taking 'prae ceteris' in its natural sense, 'and especially Paul'. The difficulty of reconciling the two testaments (mentioned a few words further on) is a problem that more naturally arises in looking at the Gospels as well as Paul, and in looking at the OT along with the NT.

stilum: Similarly metonymic at 10. 1. 1, more literal at 12. 6. 6 (for parallels, see du Roy, 93 n. 2), but in any case of the instrument of writing, and not generally of, as conventional translation here, the whole body of scripture. As du Roy observes, from scripture the Cassiciacum dialogues cite only the Gospels and Paul. (*Stilus* = *scriptura* at *en. Ps.* 67. 38.)

et prae ceteris: Hensellek, *Sitzungsber. Akad. Wien*, 376 (1981), 60–1; cf. 9. 3. 5, 9. 5. 13, and also *vera rel.* 16. 31, 'discipulo quem prae ceteris diligebat'. (That rank assigned to John, as is conventional, would make the Riedinger/du Roy reading noted above doubly difficult.)

apostolum Paulum: That he was reading Paul at this time is confirmed from *c. acad.* 2. 2. 5 (quoted above) and cf. *beata v.* 1. 4, 'conlataque cum eis, quantum potui, etiam illorum auctoritate qui divina mysteria tradiderunt'. The role of Paul in his conversion is recalled by *c. Sec.* 2, 'ego enim fateor, timore manichaeos deserui, sed timore illorum verborum quae per apostolum Paulum prolata sunt ... [quoting 1 Tim. 4: 1–4]. quibus verbis etsi alios quoque fortasse haereticos, tamen maxime manichaeos breviter aperteque descripsit.'

Courcelle, *Recherches*, 199 n. 1, attempts to construct a schedule for A.'s reading of Paul. He infers from 8. 12. 29, 'et ignorabam quid sequeretur', that A. had not yet finished reading Romans; but at 8. 5. 11 he sees the echo of Gal. 5: 17 and concludes that A. must have read Galatians a few days before the garden scene; then thinks that 10. 43. 70, echoing 2 Cor. 5: 15, indicates that he read 2 Cor. after the garden scene, though he used 2 Cor. to organize the presentation of Pauline ideas here. Courcelle concludes that 'il s'agit en réalité d'une seule et même lecture des *Épitres* de Paul'. But the workings of this book depart from that sort of literal-minded narrative. The first sentence of the present paragraph is a summary, omitting much but in some sense complete, of all the events that will be narrated in impressionistic detail in Bk. 8. There is thus no anachronism in using Paul here, and at the other end, there is no reason to think that A., performing the *sortes Paulinae,* would have known offhand the verse to follow the verse he came upon, though he may have read (or heard) Romans before. (See introd. n. to Bk. 8 on how Romans texts appear in that book.)

Paul is mentioned by name; for the pattern of names of living persons in *conf.*, see on 4. 4. 7. A. is scarcely less sparing of biblical figures. From the OT he names Abraham, Adam, David, Elias, Esau, Eve, Iacob, Isaac, Isaiah, Moses, and Noah, but from the NT apart from Christ he names *only* John the Baptist (10. 31. 46, bracketed with Noah and Elias as models of abstinence from food, and side by side with Esau and David as less praiseworthy types) and Paul, here and at 8. 4. 9, 13. 25. 38, 13. 26. 40.

On Paul in A. see recently P. Fredriksen, *Augustine on Romans* (Chico, Cal., 1982), and P. Gorday, *Principles of Patristic Exegesis: Romans 9—11 in Origen, John Chrysostom, and Augustine* (New York and Toronto, 1983). The late fourth century was a great age for reading and debating Paul in the Latin church, and a fuller study of those movements would be most welcome; for now, see A. Souter, *The Earliest Latin Commentaries on the Epistles of St. Paul* (Oxford, 1927).

adversari sibi: For an example, G–M cite the treatment of Rom. 1: 3 and 2 Cor. 5: 16 at *c. Faust.* 11. 4, 'restat ergo ut nos demonstremus quam possit utrumque verum esse et quam sibi haec duo non sint contraria.' Both the internal inconsistencies of Paul and the difficulties in harmonizing the testaments (see next note) furnished strong arguments for the Manichees.

legis et prophetarum: i.e. the OT (as often in NT, e.g. Mt. 5: 17, 7: 12, Lk. 16: 16); *mor.* 1. 16. 26–18. 34 demonstrates against the Manichees how the concord of the testaments may be understood, setting texts side by side.

eloquiorum castorum: Cf. Ps. 11: 7, 'eloquia domini, eloquia casta'; *en. Ps.* 11. 7, 'casta dicit, sine corruptione simulationis.' See on 1 Cor. 11: 19 at the end of 7. 19. 25; cf. also 13. 15. 17, 'non novi, domine, non novi alia tam casta eloquia, quae sic mihi persuaderent *confessionem*.'

exultare cum tremore didici: Ps. 2: 11, 'servite domino in timore, reges qui iudicatis terram, et exultate ei cum tremore' (cf. 10. 30. 42, 'exultans cum tremore'); *en. Ps.* 2. 9, 'ne idipsum pergeret in effusionem temeritatis, additum est "cum tremore", ut ad cautionem valeret circumspectamque sanctificationis custodiam. . . . et hoc vobis expedit ut . . . domino omnium cum timore serviatis, et exsultetis in beatitudine certissima et sincerissima, cauti et circumspicientes ne ab ea per superbiam decidatis.'

quidquid . . . legeram: The Pauline and other scriptural echoes in the first half of this paragraph depict A. now in a situation whose biblical parallels are all to the OT, to the fallen human condition, believing but awaiting the incarnation and redemption. In this way again, Bk. 8 is prepared to be the place for the 'coming of Christ' in *conf.* See on 'miser homo' below.

cum commendatione gratiae tuae: 'recommended by divine grace'; 'grace' again below ('gratia tua'); so far in A.'s life as narrated, there has been no place for this grace. On turning to Paul, it becomes an issue (as one might have surmised at 7. 9. 13, in the quotation 'resistas superbis, humilibus autem des gratiam').

quid enim habet: 1 Cor. 4: 7, 'quid autem habes quod non accepisti? si autem accepisti, quid gloriaris quasi non acceperis?' Thematic in A.'s late works, as in in *praed. sanct.* 5. 10, with a hardened position.

qui es semper idem: Cf. Ps. 101: 28 (= Heb. 1: 12; see on 1. 6. 10 and 7. 20. 26).

longinquo: For the distance, see on 1. 18. 28, and cf. 4. 16. 30 and 7. 10. 16. The 'viam' is again Christ (7. 7. 11), and the confidence of the journey toward an invisible goal is a metaphor for faith. Cf. *civ.* 10. 29, quoted on 7. 18. 24. For the persistence of the image, and its adaptation to incarnational theology, cf. *s.* 141. 1. 1 (a sermon devoted to Jn. 14: 6, much in the spirit of *conf.* 7): 'veritatem . . . viderunt quidem [huius saeculi philosophi], sed de longinquo; viderunt, sed in errore positi: et idcirco ad eam tam magnam et ineffabilem et beatificam possessionem qua via perveniretur non invenerunt'; *Io. ev. tr.* 2. 3, 'ut pervenirent ad id quod videbant de longe, a cruce Christi non recesserunt et humilitatem Christi non contempserunt.'

condelectetur: Rom. 7: 22–3, 'condelector enim legi dei secundum interiorem hominem; (23) video autem legem aliam in membris meis, repugnantem legi mentis meae et captivantem me sub lege peccati quae est in membris meis.' For extended discussion, see *div. qu. Simp.* 1. 1. 14, 'hoc autem totum ideo dicitur ut demonstretur homini captivato non esse *praesumendum* [7. 20. 26, 'inter praesumptionem et confessionem'] de viribus suis. . . . hoc enim restat in ista mortali vita *libero arbitrio* [7. 3. 5], non ut impleat homo iustitiam cum voluerit, sed ut se supplici pietate [see on 'vultum pietatis' below] convertat ad eum cuius dono eam possit implere.' The same verses are cited again at 8. 5. 11. There is an unspoken link, made explicit elsewhere, to the defects of the fallen image of God in man: *en. Ps.* 140. 15, 'adhuc bellum adversus me gero, nondum sum totus instauratus ad imaginem fabricatoris mei; coepi resculpi, et ex ea parte qua reformor displicet mihi quod deforme est.' At this period, A. takes Rom. 7 to speak generically of the Old Man, not of Paul personally, and he only changes his view much later; see M.-F. Berrouard on Rom. 7 in A.'s work at *RA* 16 (1981), 101–96, with much of interest to report on the hints of gradual change (or better, preparation for change) in A.'s works (chiefly seen in ways the text is taken to apply to already-converted Christians), but none of that seems applicable to the echoes in *conf.*; cf. *retr.* 1. 23. 1. Berrouard applies *this* passage to 'l'homme en général', and the two important echoes of Rom. 7 in Bk. 8 (8. 5. 12 and 8. 10. 22) to A. personally. That seems to create a distinction where none is necessary, for A.'s own situation is obviously in mind here. (One must realize, however, that A. early on was without any belief in baptism as an instantaneous transformation to moral perfection: *mor.* 1. 35. 80, 'et illo sacrosancto lavacro inchoatur innovatio novi hominis, ut proficiendo perficiatur, in aliis citius, in aliis tardius.')

Several scholars have remarked that there is nothing in Plotinus to match Romans 7, where the 'alien that besets us here' is our own fallen will. Plotinus is beset from outside only, fortunate fellow; purification is the stripping away and elimination of that inside us that has got there from outside. Division against himself is division of soul against body, not a lingering division within the soul itself. Plot. 1. 2. 5. 5–9, *ἀπὸ μὲν δὴ σώματος ἴσως μὲν καὶ τοῖς οἷον τόποις συνάγουσαν πρὸς ἑαυτήν, πάντως μὴν ἀπαθῶς ἔχουσαν καὶ τὰς ἀναγκαίας τῶν ἡδονῶν αἰσθήσεις μόνον ποιουμένην καὶ ἰατρεύσεις καὶ ἀπαλλαγὰς πόνων, ἵνα μὴ ἐνοχλοῖτο*. This is the description of the Likeness to which purification will lead; A. here sounds like a man who has read this passage and despairs of achieving the calm it depicts. To him, Paul in Romans has the more compelling description of the state of his soul, and offers help from outside that Plotinus did not have.

iustus es: Dan. 3: 27, 'quia iustus es in omnibus quae fecisti nobis, et universa opera tua vera, et viae tuae rectae'; sim. at Tob. 3: 2, Ps. 118: 137, and Jer. 12: 1 (in the last passage forming the basis for complaint against apparent injustice).

inique fecimus: Dan. 3: 29, 'peccavimus enim, et inique egimus recedentes a te et deliquimus in omnibus'; 3 Kings 8: 47, 'et conversi deprecati te fuerint in *captivitate* tua, dicentes, "peccavimus, inique egimus, impie gessimus".'

gravata est . . . manus tua: Ps. 31: 4, 'die ac nocte gravata est super me manus tua'; *manus dei* = *Christus* (Knauer, 121 n. 4: see on 7. 15. 21). *en. Ps.* 31. en. 2. 14, 'quid est, "gravata est super me manus tua"? magna res, fratres. respicite illam sententiam rectam inter duos, pharisaeum et publicanum. quid dictum est de pharisaeo? quoniam humilatur. quid dictum est de publicano? quoniam exaltatur. . . . ergo ut exaltantem se humilet deus, gravat super illum manum. noluit humilari confessione iniquitatis suae, humilatus est pondere manus dei.'

antiquo peccatori: *s.* 294. 15. 15, 'non primus peccavit Adam; si primum peccatorem quaeris, diabolum vide'; *en. Ps.* 39. 1, 'serpens ergo iste adulter antiquus'.

praeposito mortis: Cf. Heb. 2: 14–15, 'ut per mortem destrueret eum *qui habebat mortis imperium*, id est, diabolum, (15) et liberaret eos qui *timore mortis* [cf. 6. 16. 26!] per totam vitam obnoxii erant servitui.'

in veritate tua non stetit: Jn. 8: 44, 'ille homicida erat ex initio et in veritate non stetit, quia non est veritas in eo'—a more impressive text if we mentally substitute 'Christus' for 'veritas'. The text was problematical for A. because it seemed to imply that the devil was *created* evil (as the Manichees believed);

at *civ.* 11. 13, he is careful both to refute this possibility and, more remarkably, to insist that there is a way to take 'ab initio' as referring to the moment of creation of the devil and still avoid falling in with the Manichees.

miser homo: Rom. 7: 24, 'miser ego homo: quis me liberabit de corpore mortis huius? gratia dei per Iesum Christum dominum nostrum.' Here is the programme for Bk. 8.

in principio viarum tuarum: Prov. 8: 22, 'dominus creavit me in principio viarum suarum'; *f. et symb.* 4. 6 (also quoting Jn. 1: 14 and Phil. 2: 6), 'viarum enim eius principium caput est ecclesiae, quod est *Christus hominem indutus*, per quem vivendi exemplum nobis daretur, hoc est via certa qua perveniremus ad deum. non enim redire potuimus nisi humilitate, qui superbia lapsi sumus.' *doctr. chr.* 1. 34. 38 (between citations of Jn. 1: 14 and 14: 6), 'ille quippe qui non solum pervenientibus possessionem, sed etiam *viam se voluit praebere* venientibus ad principium viarum, voluit *carnem adsumere*. unde est illud, dominus creavit me in principio viarum suarum, ut inde inciperent, qui vellent venire.'

princeps huius mundi: Jn. 14: 30, 'venit enim princeps mundi huius, et in me non habet quidquam.'

non invenit quidquam morte dignum: Cf. 9. 13. 36, 'quaerens quid obiciat, et nihil inveniens in illo'; both allude to Pilate's inability to find anything against Jesus (Lk. 23: 4, 14, 22, Jn. 19: 4, 6). Few passages from the passion narratives find a reflection in *conf.*

evacuatum est chirographum: Col. 2: 14–15, 'delens quod adversus nos erat chirographum decreti quod erat contrarium nobis et ipsum tulit de medio affigens illud cruci (15) et expolians principatus.' 9. 13. 36 puts this verse in a eucharistic context: 'cui [sc. altari tuo] nullius diei praetermissione servierat [Monnica], unde sciret dispensari victimam sanctam qua deletum est chirographum quod erat contrarium nobis.' *lib. arb.* 3. 10. 31, 'iustissime itaque dimittere cogitur [diabolus] credentes in eum quem iniustissime occidit, ut et quod temporaliter moriuntur debitum exsolvant et quod semper vivunt in illo vivant qui pro eis quod non debebat exsolvit'. (See also *trin.* 4. 13. 17, 13. 14. 18, *ss.* 134. 5. 6, 263. 1.) This vision of atonement and redemption concludes the sketch of fall and rise presented here in biblical terms. For 'evacuatum', cf. 10. 43. 68, 'evacuaret mortem'.

hoc illae litterae non habent: Returning to the subject of 7. 9. 13–14, the defects of the *platonicorum libri*.

vultum pietatis: The failures of the philosophers in regard to *pietas* are of course the target of Rom. 1: 21 ff.; cf. *ep.* 155. 4. 17, 'pietas igitur, id est verus veri dei cultus ad omnia prodest'; *s.* 91. 3. 3, 'primo amare deum gratis: haec est enim pietas.'

lacrimas confessionis: See on 7. 20. 26, 'non flebam', and see next note.

sacrificium tuum: Ps. 50: 19, 'sacrificium deo spiritus contribulatus, cor contritum et humilatum deus non spernit'; also at 4. 3. 4, 5. 9. 17, and cf. 8. 12. 28, 'flumina oculorum meorum, *acceptabile sacrificium tuum*'.

sponsam civitatem: Apoc. 21: 2, 'et ego Iohannes vidit sanctam civitatem Hierusalem novam descendentem de caelo a deo, paratam sicut sponsam ornatam viro suo.'

arram spiritus sancti: 2 Cor. 5: 5, 'qui autem efficit nos in hoc ipsum, deus, qui dedit nobis pignus spiritus'; 2 Cor. 1: 22, 'qui et signavit nos, et dedit pignus spiritus in cordibus nostris.' *s.* 23. 8. 8–9. 9 distinguishes 'pignus' from 'arra': the former approximates 'security deposit', the latter 'down payment', and A. finds the latter more suitable for the way the Spirit is given; sim. at *ss.* 156. 15. 16, 378. Also of a groom's gift to his prospective bride, allegorically at *Io. ev. tr.* 13. 13, 13. 16, *en. Ps.* 84. 2, 122. 5, 127. 8, 148. 8; interchangeable in practice with *pignus* of the spirit at *Io. ev. tr.* 32. 5. Cf. *Io. ep. tr.* 9. 11, 'securi manebimus in eo, modo per fidem, tunc per speciem, cuius tantas arras habemus donum spiritus sancti.'

poculum pretii nostri: The phrase sounds scriptural, but no parallel presents itself. Eucharistic? Bk. 7 has made clear that the neo-Platonists know much (but not all) about Father and Son: the Spirit has been left aside. See *qu. hept.* 2. 25, quoted on 7. 9. 13, 'non ibi legi'.

nonne deo subdita: Ps. 61: 2–3, 'nonne deo subiecta erit anima mea? ab ipso enim salutare meum, (3) etenim ipse est deus meus et salutaris meus, susceptor meus, non movebor amplius'; *en. Ps.* 61. 2, 'nonne deo subicietur anima mea? audierat enim, "qui se exaltat humilabitur; et qui se humilat, exaltabitur." ... scio sub cuius alarum tegmine sperem: non movebor amplius.'

venite ad me: Mt. 11: 28–9, 'venite ad me omnes qui laboratis et onerati estis et ego reficiam vos; (29) tollite iugum meum super vos, et discite a me quia mitis sum et humilis corde, et invenietis requiem animabus vestris.' At 7. 9. 14 the same verse (along with Mt. 11: 25: see next note) explains why some things have been concealed from *sapientes* and revealed to the *humiles*. Humility is the sign and the lesson of the incarnation (for representative texts, see du Roy, 95 n. 2); above all in scripture, Phil. 2: 6–11 (see on 7. 9. 14). The context here justifies the lack of a comma after 'discere' (hence the gospel text would be taken 'learn from me that I am meek . . .'), though there is also reason to think from citations elsewhere that A. sometimes took the conjunction as causal ('learn from me, for I am meek. . . .'), e.g. *Gal. exp.* 15, 'et ideo dominus non ait: tollite iugum meum et discite a me, quoniam quatriduana de sepulcris cadavera exsuscito atque omnia daemonia de corporibus hominum morbosque depello et cetera huiusmodi, sed tollite, inquit, iugum meum et discite a me quia mitis sum et humilis corde.' The message of these verses is taken as essentially identical with the incarnation in a Christmas sermon, *s.* 184. 1. 1.

abscondisti: Mt. 11: 25, 'quia abscondisti haec a sapientibus et prudentibus et revelasti ea parvulis.' These texts close the circle opened by the reading of the *platonicorum libri*, displaying the strength and weakness of the doctrines found there.

enim: Answers the various negatives from 'non habent illae paginae': the reason why these things are not in the *platonicorum libri* is . . .

de silvestri cacumine: The context suggests Deut. 32: 49, the command to Moses to go up on Mount Nebo and see, but not reach, the promised land: no verbal echo presents itself. Cf. *c. acad.* 2. 2. 5, 'prorsus totus in me cursim redibam. respexi tamen, confiteor, quasi de itinere [= 'de silvestri cacumine'[50]) in illam religionem quae pueris nobis insita est et medullitus implicata; verum autem ipsa ad se nescientem rapiebat. itaque titubans properans haesitans arripio apostolum Paulum.' The same image emerges in a strongly Christianized judgement on the possibility of 'tentatives d'extase': *trin.* 4. 15. 20, 'sunt autem quidam qui se putant ad contemplandum deum et inhaerendum deo virtute propria posse purgari, quos ipsa superbia maxime maculat. . . . hinc enim sibi purgationem isti virtute propria pollicentur, quia nonnulli eorum potuerunt aciem mentis ultra omnem creaturam transmittere et lucem incommutabilis veritatis quantulacumque ex parte contingere: quod christianos multos, ex fide interim sola viventes, nondum potuisse derident. sed quid prodest superbienti, et ob hoc erubescenti lignum conscendere, *de longinquo prospicere patriam transmarinam*? aut quid obest humili de tanto intervallo non eam videre, in illo ligno ad eam venienti quo dedignatur ille portari?' Courcelle, *Recherches*, 225 n. 2, and Mandouze, 705 n. 1, both insist that this passage should not be retroactively applied to A. in Milan; Mandouze gives the argument thus: 'S'il n'y avait pas le *duce te* [*sol.* 2. 6. 9], on pourrait croire que l'auteur du *De Trinitate* pense quelque peu au visionnaire de Milan lorsqu'il écrit dans le même passage . . . "hinc enim purgationem . . . contingere." Mais il y a le *duce te*, et cela change tout.' The argument is marked by special pleading, the desire to rescue Milan from later criticism. But it is more natural to treat this passage as a generic criticism of the sort of thing he had himself tried at Milan, and later found inadequate on Christian terms. What we have here is retrospective criticism and distinction of types of visions, not any longer the undiluted frustration of 386. But note how he continues in *trin.* a few paragraphs further along: *trin.* 4. 17. 23, 'non sicut deum glorificaverunt aut gratias egerunt . . . [Rom. 1: 21 ff.] et cum idonei non essent in aeternitatem spiritalis incommutabilisque naturae aciem mentis tam constanter infigere, ut in ipsa sapientia creatoris atque rectoris universitatis viderent volumina saeculorum . . . atque ut ibi viderent conversiones . . . ; cum ergo ad haec ibi videnda nullo modo essent idonei, ne ad illud quidem digni habiti sunt, ut eis ista per sanctos angelos nuntiarentur.'

[50] See Courcelle, *Recherches*, 168–9, but lack of attention to the parallel has left misplaced doubts, recorded and discussed by TeSelle, 35 n. 31.

L. F. Pizzolato, *REAug* 17 (1971), 55–7, suggests the relevance of Lucretius 2. 7–10:

> sed nil dulcius est, bene quam munita tenere
> edita doctrina sapientum templa serena,
> despicere unde queas alios passimque videre
> errare atque viam palantis quaerere vitae.

The similarity is undeniable, and A. knew Lucretius (see Hagendahl, supplemented by my discussion at *RA* 15 (1980), 161–2).

et aliud est . . . et aliud: The first lines describe the plight of Plotinus, the last the remedy offered by Christ.

conari: Courcelle, *Les Confessions*, 104–5, takes this of mystical attempts, and adduces 10. 42. 67, 'multi conantes ad te redire neque per se ipsos valentes, sicut audio, temptaverunt haec et *inciderunt* in desiderium *curiosarum visionum* et digni habiti sunt inlusionibus.' Used in a similar sense at 3. 11. 20, 4. 7. 12, 4. 15. 26 ('sed ego conabar ad te et repellebar abs te': a strong clue to the importance of the mystical ascent in the discussion of the *de pulchro et apto*), 4. 16. 29, 5. 10. 20, 7. 1. 1 (2 ×), 7. 3. 5.

leone et dracone: Cf. Ps. 90: 13, 'et conculcabis leonem et draconem'; *en. Ps.* 90. s. 2. 9, 'leo aperte saevit; draco occulte insidiatur: utramque vim et potestatem habet diabolus.' See on 9. 13. 36.

viam . . . munitam: *civ.* 11. 2, 'sola est autem adversus omnes errores via munitissima, ut idem ipse sit deus et homo; quo itur, deus; qua itur, homo.' A. Gabillon, *Homo Spiritalis* (Festschrift L. Verheijen: Würzburg, 1987), 445, also notes Cic. *Mil.* 7. 17, 'proinde quasi Appius ille Caecus viam munierit, non qua populus uteretur, sed ubi impune sui posteri latrocinarentur!' The force of 'munitam' is to suggest that the road is well constructed. See on 7. 7. 11 and cf. esp. 3. 1. 1, 'via sine muscipulis': *that* way he has now to find at last.

minimum apostolorum tuorum: Cf. 1 Cor. 15: 9, 'ego sum minimus apostolorum.'

consideraveram opera tua et expaveram: Hab. 3: 2, 'consideravi opera tua, et expavi': text from *en. Ps.* 118. s. 27. 1: 'si vero tamquam *sub unius contemplationis aspectu velut audeamus cuncta contueri*, nonne fit in nobis quod ait propheta, "consideravi opera tua, et expavi"? et tamen iste non est ipsa rerum admiratione perterritus, sed eam potius dixit esse causam cur ea debuerit scrutari, quia mira sunt. cum enim dixisset, "mirabilia testimonia tua" [Ps. 118: 129], secutus adiunxit, "propter hoc scrutata est ea anima mea".' The tense here suggests that the time for such contemplation was past, that Paul and Christ would take him further. (Cf. 10. 40. 65, 'et consideravi et expavi et nihil eorum discernere potui sine te et nihil eorum te esse inveni.')

expaveram: Bks. 6 and 7 both end with expressions of fear, but by the quality of the fear in each case the difference may be measured. There, fear of death and the void; here, holy fear in the presence of God. There, no path ahead; here, a path requiring the courage to take it. Cf. also the end of Bk. 10 (10. 43. 70), 'conterritus': there the renewed uncertainty of the present at a fork in the road.